Test Item File

College Algebra

Seventh Edition

Sullivan

Chicago State University

Upper Saddle River, NJ 07458

Senior Acquisitions Editor: Eric Frank
Project Manager: Dawn Murrin
Executive Managing Editor: Vince O'Brien
Production Editor: Jeffrey Rydell
Supplement Cover Manager: Paul Gourhan
Supplement Cover Designer: Joanne Alexandris
Manufacturing Buyer: Ilene Kahn

Pearson Prentice Hall
Pearson Education, Inc.
Upper Saddle River, NJ 07458

Printed in the United States of America

10 9 8 7 6 5 4 3 2 1

ISBN 0-13-143098-X

Pearson Education Ltd., *London*
Pearson Education Australia Pty. Ltd., *Sydney*
Pearson Education Singapore, Pte. Ltd.
Pearson Education North Asia Ltd., *Hong Kong*
Pearson Education Canada, Inc., *Toronto*
Pearson Educación de Mexico, S.A. de C.V.
Pearson Education—Japan, *Tokyo*
Pearson Education Malaysia, Pte. Ltd.

CONTENTS

Ch. 0 (Chapter R) Review

0.1 Real Numbers

1 Classify Numbers

Tell which set or sets the number belongs to: natural, whole, integer, rational, irrational, real.

1) 71

A) real
B) real, rational, whole
C) real, rational, integer, whole, natural
D) real, rational, integer

2) −42

A) real, rational, integer
B) real
C) real, irrational
D) real, whole

3) $\sqrt{20}$

A) real, integer
B) real, irrational
C) real, whole
D) real, rational

4) 0

A) real, whole
B) real, rational
C) real, integer
D) real, rational, integer, whole

5) π

A) real, irrational
B) rational
C) irrational
D) real, integer

6) 9.93

A) real
B) real, rational
C) rational
D) real, rational, natural

7) 0.106106. . . (the block 106 repeats)

A) real, rational
B) real, irrational
C) real
D) real, rational, natural

8) −3.90

A) rational
B) real
C) real, rational, natural
D) real, rational

9) −0.4242. . . (the block 42 repeats)

A) real
B) real, rational
C) real, irrational
D) real, rational, natural

10) $\frac{3}{19}$

A) real
B) real, whole
C) real, rational
D) real, irrational

11) $-\sqrt{3}$

A) real, rational
B) real, irrational
C) real, whole
D) real, integer

Solve the problem.

12) List all the rational numbers in $\{-3.12, \pi, 3.4545\ldots, 0, \sqrt{5}, \frac{6}{7}, 4.562777, 15, \sqrt{9}, -10\}$.

A) $-10, 0, 15, \sqrt{9}$

B) $\pi, \sqrt{5}, \sqrt{9}$

C) $-3.12, 0, \frac{6}{7}, 3.4545\ldots, 4.562777, 15$

D) $-10, -3.12, 0, \frac{6}{7}, \sqrt{9}, 3.4545\ldots, 4.562777, 15$

13) What is the best classification for −0.714286?

A) integer, rational number, real number

B) irrational number, real number

C) whole number, integer, real number

D) rational number, real number

14) The number 0.6 belongs to which of these sets?
natural numbers, whole numbers, integer, rational numbers, irrational numbers, and real numbers.
Name all that apply.

2 Evaluate Numerical Expressions

Evaluate the expression.

1) $9 + 6 - 5$

A) 8 B) −2 C) 10 D) 20

2) $3 + \frac{2}{9}$

A) $\frac{29}{9}$ B) $\frac{27}{9}$ C) $\frac{27}{2}$ D) $\frac{29}{2}$

3) $4 \cdot [4(7-2) - 6]$

A) −4 B) −40 C) 56 D) −44

4) $-2 - (-5 + 2 \cdot 5 + 2)$

A) −5 B) −19 C) −9 D) 5

5) $\frac{24}{28} \cdot \frac{7}{8}$

A) $\frac{1}{2}$ B) $\frac{3}{4}$ C) $\frac{6}{7}$ D) $\frac{4}{3}$

6) $\frac{3}{5} + \frac{1}{10}$

A) $\frac{7}{15}$ B) $\frac{2}{5}$ C) $\frac{7}{10}$ D) $\frac{4}{15}$

7) $\frac{7}{9} - \frac{2}{6}$

A) $\frac{5}{9}$ B) $\frac{5}{54}$ C) $\frac{8}{3}$ D) $\frac{4}{9}$

8) $\dfrac{-\frac{9}{10}}{-\frac{1}{16}}$

A) $\frac{9}{160}$ B) $\frac{29}{2}$ C) $\frac{72}{5}$ D) $\frac{1}{20}$

9) $\dfrac{\frac{2}{9}}{\frac{9}{11}}$

A) $\frac{2}{11}$ B) $\frac{22}{81}$ C) $\frac{11}{2}$ D) $\frac{81}{22}$

10) $\dfrac{-\frac{6}{11}}{\frac{9}{4}}$

A) $-\frac{8}{33}$ B) $\frac{27}{22}$ C) $\frac{8}{33}$ D) $-\frac{27}{22}$

11) $\dfrac{7(4) + 2}{1 - 8(7)}$

A) $\frac{6}{11}$ B) $-\frac{42}{55}$ C) $-\frac{30}{49}$ D) $-\frac{6}{11}$

12) $\dfrac{(-6) \cdot (4 - 7) + (-6) \cdot 8}{(-6) \cdot (5 - 2)}$

A) 1 B) 2 C) $\frac{5}{3}$ D) $\frac{46}{7}$

13) $-\frac{4}{5} \div \frac{8}{11} \cdot -\frac{7}{12}$

A) $\frac{35}{66}$ B) $\frac{56}{165}$ C) $\frac{77}{120}$ D) $\frac{66}{35}$

14) $\frac{1}{3} - 4(\frac{3}{4} + 5)$

A) $-\frac{68}{3}$ B) $\frac{68}{3}$ C) $\frac{23}{3}$ D) $-\frac{23}{3}$

15) $\dfrac{\frac{2}{3} - \frac{3}{7}}{\frac{13}{14} + \frac{1}{2}}$

3 Work with Properties of Real Numbers

Name the property illustrated by the statement.

1) $(2 + 1) + 7 = (1 + 2) + 7$

A) inverse property
B) associative property
C) distributive property
D) commutative property

2) $9(x + 8) = 9x + 9 \cdot 8$

A) associative property
B) commutative property
C) distributive property
D) identity property

3) $(3 \cdot 1) \cdot 4 = 3 \cdot (1 \cdot 4)$

A) distributive property
B) identity property
C) associative property
D) commutative property

4) $7 + (-7) = 0$

A) commutative property
B) inverse property
C) associative property
D) identity property

5) $3 \cdot 1 = 3$

A) commutative property
B) identity property
C) inverse property
D) distributive property

6) $9 + 0 = 9$

A) commutative property
B) associative property
C) distributive property
D) identity property

7) $5 + 4 = 4 + 5$

A) associative property
B) identity property
C) distributive property
D) commutative property

8) $\frac{1}{8} \cdot 8 = 1$

A) identity property
B) distributive property
C) associative property
D) inverse property

9) $8 \cdot 2 = 2 \cdot 8$

A) distributive property
B) identity property
C) associative property
D) commutative property

10) $6 + (5 + 4) = (6 + 5) + 4$

A) identity property
B) distributive property
C) associative property
D) commutative property

Use the distributive property to remove the parentheses. Then simplify, if necessary.

11) $-(2x + 5y)$

A) $-2x + 5y$
B) $-2x - 5y$
C) $2x - 5y$
D) $2x + 5y$

12) $-(-2x + 8y - 3z)$

A) $2x - 8y - 3z$ B) $-2x + 8y + 3z$ C) $2x - 8y + 3z$ D) $-2x + 8y - 3z$

13) $5(k + x)$

A) $5k + x$ B) $5k + 5x$ C) $5kx$ D) $5k - 5x$

14) $10(6x + 2)$

A) $16x + 12$ B) $80x$ C) $60x + 20$ D) $60x + 2$

15) $-10(7x + 9)$

A) $-160x$ B) $-70x + 9$ C) $-3x - 1$ D) $-70x - 90$

16) $\frac{1}{3}(9x - 6)$

A) x B) $27x - 18$ C) $3x - 2$ D) $3x - 6$

17) $7(2z + 4) - 3$

A) $9z + 8$ B) $14z + 25$ C) $14z + 7$ D) $14z + 1$

18) $-4(3y + 5) - 5$

A) $-12y$ B) $-12y - 25$ C) $-12y - 40$ D) $12y + 15$

19) $(x + 6)(x - 10)$

A) $x^2 - 4x - 60$ B) $x^2 - 60x - 4$ C) $x^2 - 4x - 4$ D) $x^2 - 5x - 60$

20) $(x + 9)(x - 1)$

A) $x^2 + 8x + 8$ B) $x^2 + 7x - 9$ C) $x^2 + 8x - 9$ D) $x^2 - 9x + 8$

21) $(x + 1)(x - 1)$

A) $x^2 - 2$ B) $x^2 - 2x - 1$ C) $x^2 - 1$ D) $x^2 + 2x - 1$

22) $(x + \frac{5}{7})(x - \frac{2}{5})$

A) $x^2 + \frac{11}{35}x - \frac{2}{7}$ B) $x^2 + \frac{1}{5}x - \frac{2}{7}$ C) $x^2 - \frac{11}{35}x - \frac{2}{7}$ D) $x^2 - \frac{1}{5}x - \frac{2}{7}$

4 Additional Applications

Approximate as indicated.

1) Approximate 1.0303 rounded to three decimal places.

A) 1.029 B) 1.031 C) 1.030 D) 1.028

2) Approximate $\frac{23}{9}$ (a) rounded and (b) truncated to three decimal places.

3) Use a calculator to approximate the expression. Round the answer to two decimal places.

$(57.5 + 19.1)/56.2 + 17.9$

A) 208.54 B) 0.58 C) 75.74 D) 19.26

4) Use a calculator to approximate the expression. Round the answer to two decimal places.

$$\frac{27.7 + \pi}{3.5 + 8.3}$$

Write the statement using symbols.

5) The sum of four times x and 5 decreased by 7 is 8.

A) $4(x + 5) - 7 = 8$ B) $4x + 5 - 7 = 8$ C) $4(x + 5 - 7) = 8$ D) $7 - (4x + 5) = 8$

6) Three times the difference of x and 8 is −10.

A) $3x - 8 = -10$ B) $3 + x - 8 = -10$ C) $3(x - 8) = -10$ D) $3 - x - 8 = -10$

7) The quotient of x and the sum of 5 and x.

A) $x(5 + x)$ B) $\frac{x + 5}{x}$ C) $\frac{x}{5 + x}$ D) $x + 5 + x$

8) The sum of 24 and 12 is 36.

A) $24 + 12 = 36$ B) $24 \cdot 12 = 288$ C) $\frac{24}{12} = 2$ D) $24 - 12 = 12$

9) The product of 25 and 5 equals 125.

A) $\frac{25}{5} = 5$ B) $25 + 5 = 30$ C) $25 \cdot 5 = 125$ D) $25 - 5 = 20$

10) The difference 12 less 4 equals 8.

A) $12 \cdot 4 = 48$ B) $12 + 4 = 16$ C) $12 - 4 = 8$ D) $\frac{12}{4} = 3$

11) The quotient 15 divided by 3 is 5.

A) $15 \cdot 3 = 45$ B) $15 - 3 = 12$ C) $\frac{15}{3} = 5$ D) $15 + 3 = 18$

0.2 Algebra Review

1 Graph Inequalities

Graph the number x on the real number line.

1) $x \geq -6$

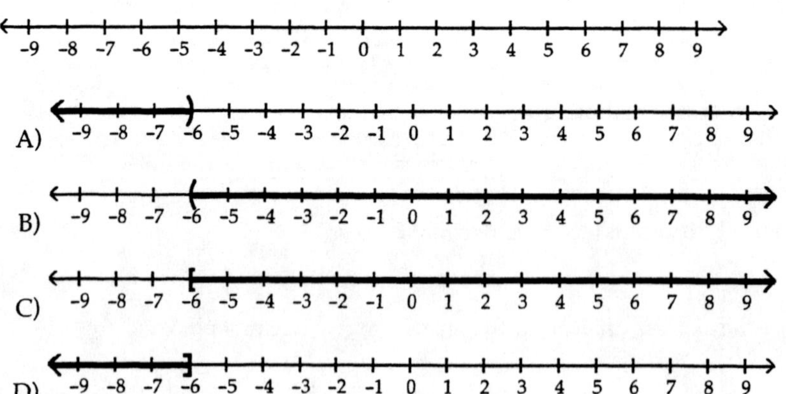

2) $x < -6$

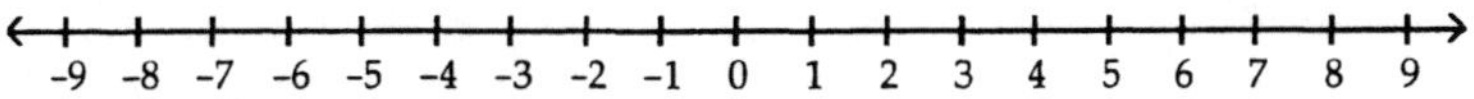

-9 -8 -7 -6 -5 -4 -3 -2 -1 0 1 2 3 4 5 6 7 8 9

A) 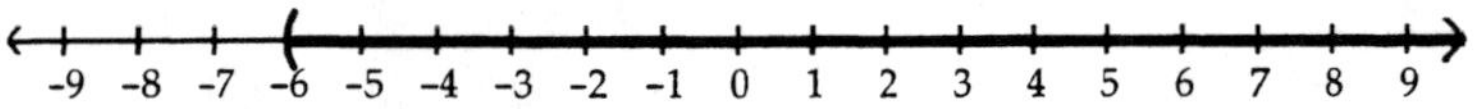 -9 -8 -7 -6 -5 -4 -3 -2 -1 0 1 2 3 4 5 6 7 8 9

B) 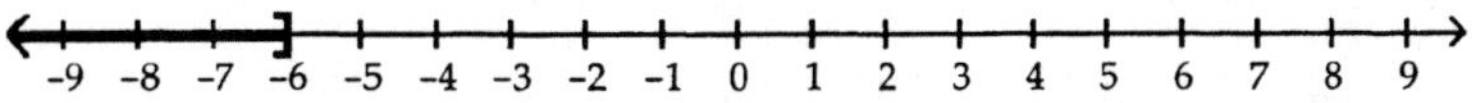 -9 -8 -7 -6 -5 -4 -3 -2 -1 0 1 2 3 4 5 6 7 8 9

C) 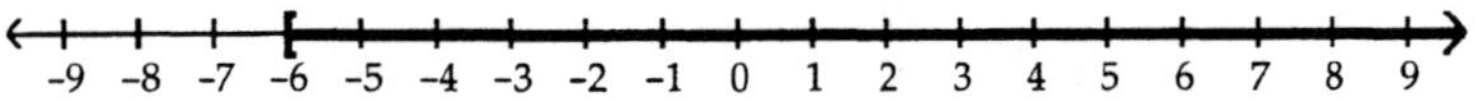 -9 -8 -7 -6 -5 -4 -3 -2 -1 0 1 2 3 4 5 6 7 8 9

D) 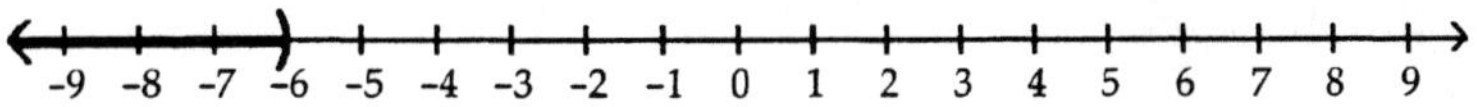 -9 -8 -7 -6 -5 -4 -3 -2 -1 0 1 2 3 4 5 6 7 8 9

3) $x > 7$

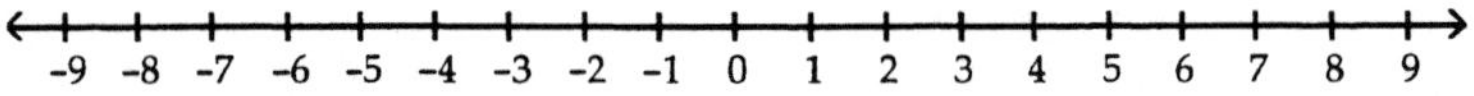

-9 -8 -7 -6 -5 -4 -3 -2 -1 0 1 2 3 4 5 6 7 8 9

A) 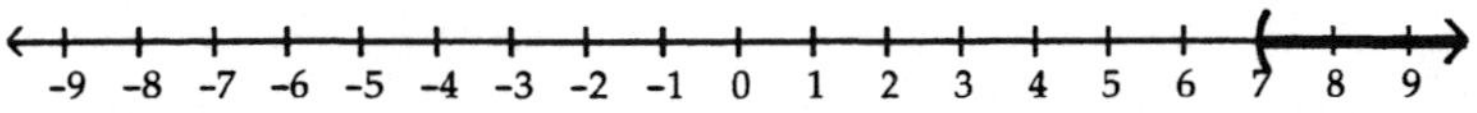 -9 -8 -7 -6 -5 -4 -3 -2 -1 0 1 2 3 4 5 6 7 8 9

B) 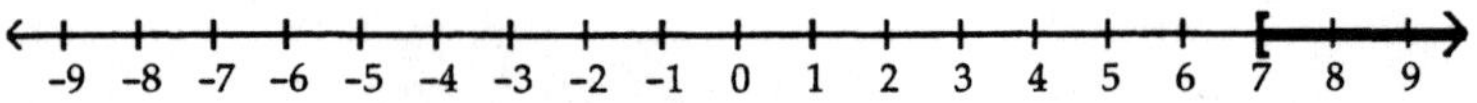 -9 -8 -7 -6 -5 -4 -3 -2 -1 0 1 2 3 4 5 6 7 8 9

C) 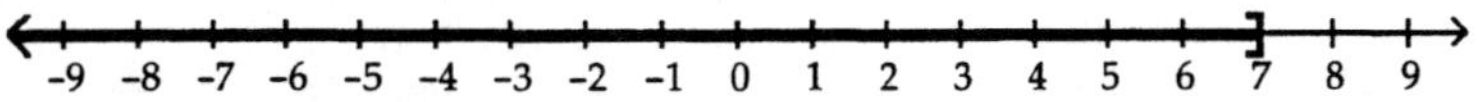 -9 -8 -7 -6 -5 -4 -3 -2 -1 0 1 2 3 4 5 6 7 8 9

D) 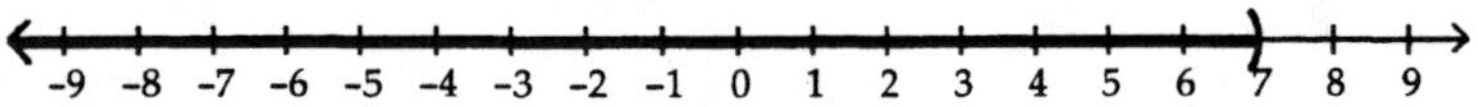 -9 -8 -7 -6 -5 -4 -3 -2 -1 0 1 2 3 4 5 6 7 8 9

4) $x \leq -1$

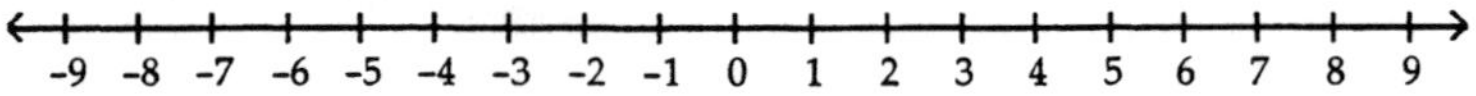

-9 -8 -7 -6 -5 -4 -3 -2 -1 0 1 2 3 4 5 6 7 8 9

A) 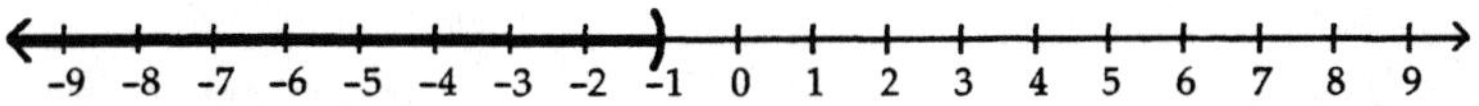 -9 -8 -7 -6 -5 -4 -3 -2 -1 0 1 2 3 4 5 6 7 8 9

B) 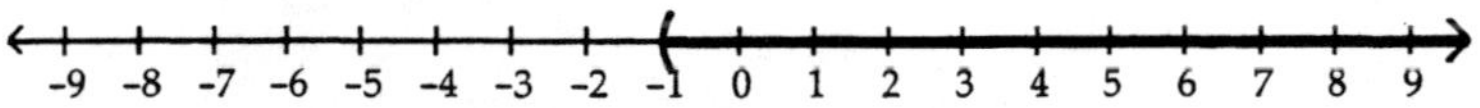 -9 -8 -7 -6 -5 -4 -3 -2 -1 0 1 2 3 4 5 6 7 8 9

C) 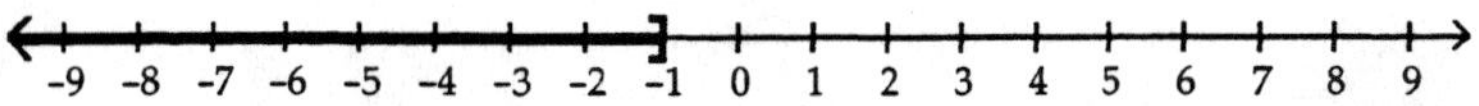 -9 -8 -7 -6 -5 -4 -3 -2 -1 0 1 2 3 4 5 6 7 8 9

D) 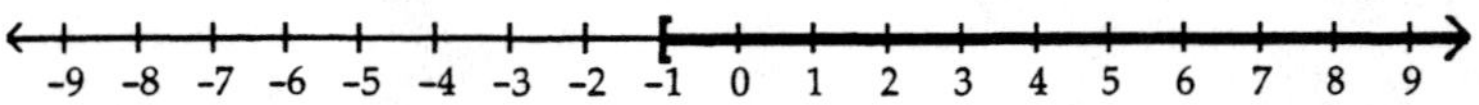 -9 -8 -7 -6 -5 -4 -3 -2 -1 0 1 2 3 4 5 6 7 8 9

2 Find Distance on the Real Number Line

Use the given real number line to compute the distance.

1) Find d(A, B)

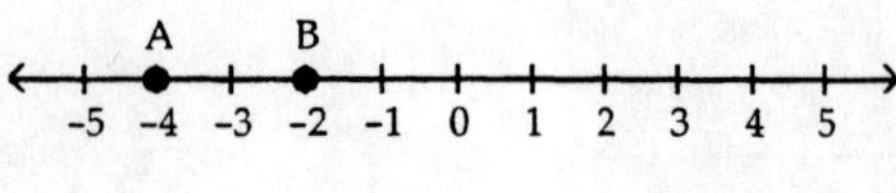

A) 2 B) 3 C) 1 D) -2

3 Evaluate Algebraic Expressions

Find the value of the expression using the given values.

1) $x + 14y$ $x = -4, y = 2$

A) 24 B) -2 C) 10 D) -54

2) $-5xy + 4y - 2$ $x = 5, y = -1$

A) 21 B) 19 C) 23 D) -31

3) $-15x + y$ $x = -4, y = -2$

A) -62 B) -6 C) -17 D) 58

4) $\frac{12x - 6y}{7}$ $x = 8, y = 3$

A) $\frac{114}{7}$ B) $\frac{78}{7}$ C) $\frac{90}{7}$ D) $\frac{12}{7}$

5) $\frac{3x - 6y}{x + 12}$ $x = 10, y = 4$

A) $\frac{24}{11}$ B) 3 C) $\frac{3}{11}$ D) $\frac{3}{8}$

6) $(x + 3y)^2$ $x = 4, y = 4$

A) 16 B) 32 C) 256 D) 49

7) $\frac{2xy + 5}{x}$ $x = 5, y = 7$

A) 19 B) 1 C) 3 D) 15

8) $xyz + y^2$ $x = 5, y = 3, z = 2$

A) -114 B) -108 C) 39 D) 114

9) $|x - y|$ $x = -6, y = 4$

A) 10 B) -10 C) 2 D) -2

10) $|x| + |y|$ $x = -4, y = 2$

A) 2 B) -6 C) -2 D) 6

11) $|6x - 7y|$ $x = 4, y = 7$

A) 73 B) 25 C) -25 D) -73

12) $4|x| + 5|y|$ $x = 9, y = -8$

A) 4 B) -4 C) 76 D) -76

13) $|-9x + 3y|$ $x = -5, y = 1$

A) 42 B) 48 C) 24 D) 6

14) $|-7x| - |3y|$ $x = -1, y = 2$

A) 1 B) 17 C) 13 D) 11

15) z^2 $z = -9$

A) -81 B) 18 C) 81 D) -18

16) w^5 $w = -3$

A) 15 B) 243 C) -15 D) -243

17) $6x^2$ $x = 4$

A) 48 B) 96 C) 576 D) 144

18) $-6x^3y$ $x = 2, y = -3$

A) 108 B) 144 C) 72 D) -144

19) $\frac{3}{7x^2}$ $x = -4$

A) $-\frac{3}{112}$ B) $\frac{3}{112}$ C) $-\frac{3}{56}$ D) $\frac{3}{56}$

Find the value of the expression if x = -2, y = 3, and z = -4.

20) $9x - 2y + 6z$

A) 8 B) -48 C) 7 D) -54

21) $-4z - 9y - 7x$

A) 25 B) 13 C) 23 D) 3

22) $-x^2 - 2z$

A) -4 B) -12 C) 4 D) 8

23) $\frac{7 + z}{y - 3}$

A) 0 B) $-\frac{1}{2}$ C) 3 D) undefined

24) $\frac{4x - 5}{7y + z}$

A) $\frac{13}{17}$ B) $\frac{13}{25}$ C) $-\frac{13}{17}$ D) $-\frac{13}{25}$

4 Determine the Domain of a Variable

Determine which value(s), if any, must be excluded from the domain of the variable in the expression.

1) $\frac{9}{x - 7}$

A) $x = -7$ B) $x = 7$ C) $x = 0$ D) none

2) $\frac{8}{x + 7}$

A) $x = 7$ B) $x = -7$ C) $x = 0$ D) none

3) $\frac{x - 8}{5}$

A) $x = 0$ B) $x = -8$ C) $x = 8$ D) none

4) $\frac{x - 2}{9 - x}$

A) $x = -9$ B) $x = 9, x = 2$ C) $x = 9$ D) none

5) $\frac{4x - 2}{x^2 - 36}$

A) $x = 6$ B) $x = 6, x = -6$ C) $x = \frac{1}{2}$ D) $x = 36$

6) $\frac{x^2 - 81}{x^2 - 5x + 6}$

A) $x = 9, x = -9$ B) $x = 0$ C) $x = 2, x = 3$ D) $x = -2, x = -3$

7) $\frac{25x^2 - 16}{7x - 56}$

A) $x = 8$ B) $x = \frac{4}{5}, x = -\frac{4}{5}$ C) $x = 56$ D) none

8) $\frac{x^3 + 6x^4}{x^2 + 81}$

A) $x = 0, x = -\frac{1}{6}$ B) $x = -9$ C) $x = -81$ D) none

9) $\frac{x^3 + 12x^2 + 35x}{x^2 + 6x}$

A) $x = -7, x = -5$ B) $x = -6$
C) $x = 0, x = -6, x = -7, x = -5$ D) $x = 0, x = -6$

10) $\frac{x^2 + 6x - 40}{x + 10}$

A) $x = 10$ B) $x = -10, x = 4$ C) $x = -10$ D) none

Determine the domain of the variable x in the expression.

11) $\frac{x-5}{x+4}$

A) $\{x \mid x \neq 5\}$ B) $\{x \mid x \neq -4\}$ C) $\{x \mid x \geq -4\}$ D) $\{x \mid x = -4\}$

12) $\frac{3x+4}{x-1}$

A) $\{x \mid x \neq 0, -1\}$ B) $\{x \mid x \neq 1\}$ C) $\{x \mid x \neq 0\}$ D) all real numbers

13) $\frac{4x}{x(x-64)}$

A) $\{x \mid x \neq 64, x \neq 0\}$ B) $\{x \mid x \neq \pm 8\}$ C) $\{x \mid x \neq \pm 64, x \neq 0\}$ D) $\{x \mid x \neq 8\}$

14) $\frac{(x-1)}{(x-3)(x+7)}$

5 Use the Laws of Exponents

Simplify the expression.

1) 5^3

A) 15 B) −125 C) 125 D) −15

2) -2^3

A) 6 B) −8 C) −6 D) 8

3) $6 \cdot 3^3$

A) 54 B) 162 C) 5832 D) 33

4) 2^{-3}

A) $\frac{1}{6}$ B) −8 C) 8 D) $\frac{1}{8}$

5) $(-5)^{-2}$

A) 25 B) $\frac{1}{25}$ C) −25 D) $-\frac{1}{25}$

6) -2^{-3}

A) 8 B) $\frac{1}{6}$ C) −8 D) $-\frac{1}{8}$

7) $(-6)^3$

A) −216 B) 216 C) 18 D) −18

8) -10^1

A) −10 B) 1 C) 10 D) 0

9) $5^{-1} + 6^{-1}$

A) $\frac{30}{11}$ B) $\frac{11}{30}$ C) 2 D) − 1

10) $\left(\frac{5}{7}\right)^2$

A) $\frac{25}{7}$ B) $\frac{7}{25}$ C) $\frac{25}{49}$ D) $\frac{49}{25}$

11) $\left(\frac{1}{7}\right)^2$

A) $\frac{1}{14}$ B) $\frac{1}{9}$ C) $\frac{2}{7}$ D) $\frac{1}{49}$

12) $\left(\frac{1}{5}\right)^{-4}$

A) $\frac{1}{625}$ B) 625 C) -625 D) $\frac{1}{20}$

13) $3^{-8} \cdot 3^6$

A) $\frac{1}{27}$ B) 9 C) 3 D) $\frac{1}{9}$

14) $(3^{-1})^{-2}$

A) 3 B) $\frac{1}{9}$ C) 9 D) $\frac{1}{3}$

Simplify the expression. Express the answer so that all exponents are positive. Whenever an exponent is 0 or negative, we assume that the base is not 0.

15) x^{-2}

A) $x^{1/2}$ B) $-\frac{1}{x^2}$ C) $\frac{1}{x^2}$ D) $-x^2$

16) $\frac{1}{x^{-7}}$

A) $x^{1/7}$ B) $-x^7$ C) x^7 D) $-x^{1/7}$

17) $\frac{x^{-9}}{x^2}$

A) $\frac{1}{x^{18}}$ B) $\frac{1}{x^7}$ C) x^{11} D) $\frac{1}{x^{11}}$

18) $\frac{x^8}{x^{-8}}$

A) $\frac{1}{x^{16}}$ B) x^{64} C) x^{16} D) $\frac{1}{x^{64}}$

19) $-2x^{-3}$

A) $-\frac{1}{8x^3}$ B) $\frac{-8}{x^3}$ C) $\frac{1}{8x^3}$ D) $-\frac{2}{x^3}$

20) $\frac{(xy^7)(x^4y)}{(x^5y^2)^2}$

A) $\frac{y^4}{x^5}$ B) $\frac{y^3}{x^6}$ C) $\frac{x^5}{y^4}$ D) $\frac{x^6}{y^3}$

21) $\left(\frac{5x^{-3}}{8y^{-3}}\right)^{-1}$

A) $\frac{8y^3}{5x^3}$ B) $\frac{8x^3}{5y^3}$ C) $\frac{5x^{31}}{8y^{31}}$ D) $\frac{5x^3}{8y^3}$

22) $\left(\frac{4x^{-2}y^2}{12x^{-4}y^{-1}}\right)^3$

A) $\frac{x^2y^3}{3}$ B) $\frac{x^6y^9}{27}$ C) $\frac{x^8y^6}{9}$ D) $\frac{x^6y^9}{3}$

23) $(3x^2)^3(2x)^{-1}$

A) $6x^5$ B) $6x^4$ C) $\frac{27x^5}{2}$ D) $\frac{9x^5}{2}$

24) $\frac{(4xy^{-2})^{-2}}{2xy^3}$

A) $\frac{y}{32x^3}$ B) $-\frac{8y}{x^3}$ C) $\frac{y}{32}$ D) $-\frac{4}{x^3y^{-7}}$

25) $\frac{-36x^5y^2}{-9x^4y^4}$

26) $(4xy)^3$

A) $\frac{1}{64x^3y^3}$ B) $64x^3y^3$ C) $4x^3y^3$ D) $64xy$

27) $\left(\frac{-8x^7y^{-5}}{7z^2}\right)^{-1}$

A) $\frac{7y^5z^2}{-8x^7}$ B) $\frac{-8x^7}{7y^5z^2}$ C) $\frac{7y^5}{-8x^7z^2}$ D) $\frac{7z^2}{-8x^7y^5}$

28) $(x^{-9}y^6)^{-6}z^5$

A) $\frac{y^{36}z^5}{x^{54}}$ B) $\frac{x^{54}}{y^{36}z^5}$ C) $\frac{y^{36}}{x^{54}z^5}$ D) $\frac{x^{54}z^5}{y^{36}}$

Find the value of the expression using the given values.

29) $(x^{-2}y^{-1})^2$ $x = 4,\ y = -2$

A) -1024 B) $\frac{1}{1024}$ C) $-\frac{1}{1024}$ D) 1024

30) $(9x + 7y)^2$ $x = 4, y = -3$

A) 289 B) 1737 C) 3249 D) 225

6 Evaluate Square Roots

Simplify the expression.

1) $\sqrt{36}$

A) $\frac{1}{36}$ B) 6 C) 1296 D) not a real number

2) $\sqrt{(-5)^2}$

A) 625 B) 5 C) $\frac{1}{25}$ D) not a real number

Find the value of the expression using the given values.

3) $\sqrt{x^2}$ $x = 9$

A) $-3\sqrt{2}$ B) -9 C) $3\sqrt{2}$ D) 9

4) $(\sqrt{x})^2$ $x = 7$

A) 7 B) $\frac{1}{49}$ C) 49 D) $\frac{1}{7}$

5) $\sqrt{x^2 + y^2}$ $x = 15, y = 20$

A) 25 B) 18 C) 20 D) 24

6) $\sqrt{x^2 + y^2}$ $x = 5, y = -3$

A) 8 B) 2 C) $\sqrt{34}$ D) 15

7 Use a Calculator to Evaluate Exponents

Use a calculator to evaluate the expression. Round answers to three decimal places.

1) $(-2.09)^{-5}$

A) -39.878 B) 0.025 C) -0.025 D) 39.878

2) $-(-1.36)^6$

A) -0.158 B) 0.158 C) -6.328 D) 6.328

3) $(-0.52)^3$

A) -0.141 B) -7.112 C) -1.560 D) 7.112

4) $(7.26)^2$

8 Use Scientific Notation

Write the number in scientific notation.

1) 28,888

A) 2.8888×10^5 B) 2.8888×10^1 C) 2.8888×10^4 D) 2.8888×10^{-4}

2) 615.41

A) 6.1541×10^2 B) 6.1541×10^3 C) 6.1541×10^{-2} D) 6.1541×10^{-3}

3) 684.595

A) 6.84595×10^{-3} B) 6.84595×10^2 C) 6.84595×10^3 D) 6.84595×10^{-2}

4) 69,000

A) 6.9×10^5 B) 6.9×10^{-5} C) 6.9×10^4 D) 6.9×10^{-4}

5) 78,000,000

A) 7.8×10^7 B) 7.8×10^6 C) 7.8×10^{-7} D) 7.8×10^{-6}

6) 0.000307

A) 3.07×10^{-5} B) 3.07×10^4 C) 3.07×10^{-3} D) 3.07×10^{-4}

7) 0.00006729

A) 6.729×10^4 B) 6.729×10^{-4} C) 6.729×10^5 D) 6.729×10^{-5}

8) 0.000009421

A) 9.421×10^{-7} B) 9.421×10^{-6} C) 9.421×10^6 D) 9.421×10^{-5}

9) 0.000000228013

A) 2.28013×10^7 B) 2.28013×10^{-6} C) 2.28013×10^{-7} D) 2.28013×10^6

10) 0.000000076509

A) 7.6509×10^{-9} B) 7.6509×10^{-7} C) 7.6509×10^8 D) 7.6509×10^{-8}

11) 580,000,000

A) 5.8×10^{10} B) 5.8×10^9 C) 5.8×10^8 D) 58×10^7

12) The diameter of an atom is about 0.0000001 millimeters (Source: Powers of Ten, Philip and Phylis Morrison). Write this number in scientific notation.

13) In 1995, the United States consumed 17,640,000 barrels of petroleum products per day (Source: 1998 Information Please Almanac). How many barrels on average were consumed in the month of February 1995? Express your answer in scientific notation.

14) Use scientific notation to express the speed of light in miles per hour assuming the speed of light is 186,000 miles per second.

Write the number as a decimal.

15) 8.96×10^5

A) 8,960,000 B) 448 C) 89,600 D) 896,000

16) 1.018×10^5

A) 10,180 B) 50.9 C) 1,018,000 D) 101,800

17) 4.4877×10^5

A) 448,770 B) 224.385 C) 4,487,700 D) 44,877

18) 9.28×10^{-4}

A) 0.000928 B) 0.00928 C) -928,000 D) 0.0000928

19) 1.813×10^{-5}

A) -181,300 B) 0.000001813 C) 0.0001813 D) 0.00001813

20) 3.643×10^{-6}

A) -3,643,000 B) 0.0000003643 C) 0.000003643 D) 0.00003643

21) 3.0156×10^{-7}

A) -301560,000 B) 0.00000030156 C) 0.0000030156 D) 0.000000030156

22) 3.41×10^4

23) The distance from Earth to the Moon is about 4×10^8 meters (Source: Powers of Ten, Philip and Phylis Morrison). Write this distance as a whole number in kilometers. (1 km = 1000 meters)

9 Additional Applications

Insert <, >, or = to make the statement true.

1) -2 ___ -3

A) < B) = C) >

2) 0.8 ___ 1.3

A) > B) = C) <

3) -97 ___ -22

A) = B) > C) <

4) |-10| ___ |9|

A) < B) > C) =

5) |-8| ___ |8|

A) < B) = C) >

6) 10 ___ -10

A) = B) > C) <

7) |-10| ___ 0

A) = B) > C) <

8) 70 ___ 700

A) > B) < C) =

9) $\frac{15}{5}$ ___ $\frac{12}{4}$

A) $<$ B) $>$ C) $=$

10) $\frac{2}{3}$ ___ 0.67

A) $<$ B) $=$ C) $>$

11) 3.14 ___ π

A) $=$ B) $<$ C) $>$

Write the statement as a mathematical statement.

12) z is negative

A) $z < 0$ B) $z > 0$ C) $z \geq 0$ D) $z \leq 0$

13) x is greater than seven

A) $x < 7$ B) $x = 7$ C) $x \geq 7$ D) $x > 7$

14) negative forty-two is less than negative sixteen

A) $-42 < -16$ B) $-42 > -16$ C) $-42 < 16$ D) $-42 \leq -16$

15) nine is greater than or equal to three

A) $9 > 3$ B) $9 \leq 3$ C) $9 \geq 3$ D) $9 = 3$

16) x is less than or equal to thirteen

A) $x \geq 13$ B) $x \leq 13$ C) $x < 13$ D) $x \neq 13$

17) negative eleven is equal to negative eleven

A) $-11 \geq -11$ B) $-11 \leq -11$ C) $-11 \neq -11$ D) $-11 = -11$

18) twenty-five is not equal to negative twenty-five

A) $25 \leq -25$ B) $25 \neq -25$ C) $25 = -25$ D) $25 \geq -25$

19) fifteen is not equal to twenty-one

A) $15 = 21$ B) $15 \neq 21$ C) $15 \leq 21$ D) $15 \geq 21$

20) thirty-three is greater than or equal to thirty-three

A) $33 \geq 33$ B) $33 > 33$ C) $33 \leq 33$ D) $33 = 33$

Use the formula $C = \frac{5}{9}(F - 32)$ for converting degrees Fahrenheit into degrees Celsius to find the Celsius measure of the Fahrenheit temperature.

21) $F = -31°$

A) $-25°$ C B) $-40°$ C C) $-35°$ C D) $-30°$ C

Express the statement as an equation involving the indicated variables.

22) The area A of a rectangle is the product of its length l and its width w.

A) $A = l + w$ B) $A = lw$ C) $A = 2(l + w)$ D) $A = \frac{l}{w}$

23) The perimeter P of a rectangle is twice the sum of its length l and its width w.

A) $P = 2lw$ B) $P = l + w$ C) $P = 2(l + w)$ D) $P = lw$

24) The circumference C of a circle is the product of π and its diameter d.

A) $C = 2\pi d$ B) $C = \pi d$ C) $C = \pi + d$ D) $C = \frac{\pi}{d}$

25) The area A of a triangle is one-half the product of its base b and its height h.

A) $A = bh$ B) $A = \frac{1}{2}(b + h)$ C) $A = \frac{1}{2}bh$ D) $A = 2bh$

26) The volume V of a sphere is $\frac{4}{3}$ times π times the cube of the radius r.

A) $V = \frac{4}{3}\pi r^3$ B) $V = \frac{4}{3}\pi \sqrt[3]{r}$ C) $V = \frac{4}{3}\pi r$ D) $V = \frac{4}{3}\pi r^2$

27) The surface area S of a sphere is 4 times π times the square of the radius r.

A) $S = 4\pi r$ B) $S = 4\pi\sqrt{r}$ C) $S = \pi r^2$ D) $S = 4\pi r^2$

28) The volume V of a cube is the cube of the length x of a side.

A) $V = x^2$ B) $V = x^3$ C) $V = \sqrt[3]{x}$ D) $V = 3x$

29) The surface area S of a cube is 6 times the square of the length x of a side.

A) $S = 6 + x^2$ B) $S = x^2$ C) $S = 6x$ D) $S = 6x^2$

Solve the problem.

30) The weekly production cost C of manufacturing x calendars is given by $C(x) = 28 + 3x$, where the variable C is in dollars. What is the cost of producing 268 calendars?

A) \$804.00 B) \$296.00 C) \$832.00 D) \$7507.00

31) The weekly production cost C of manufacturing x telephones is given by $C = 5000 + 15x$, where the variable C is in dollars. What is the cost of producing 400 telephones?

32) At the beginning of the month, Christopher had a balance of \$284 in his checking account. During the next month, he wrote a check for \$31, deposited \$65, and wrote another check for \$164. What was his balance at the end of the month?

A) -\$24 B) \$24 C) \$154 D) -\$154

33) At the beginning of the month, Ethel had a balance of \$250 in her checking account. During the next month, she wrote a check for \$123, deposited \$56, was assessed a monthly service charge of \$8, wrote two checks for \$50 and \$13, and made another deposit for \$20. What was her balance at the end of the month?

34) The cost of a taxi ride x quarter-miles long is given by the formula $F = 2.85 + 0.65x$, where the variable F is in dollars. What is the cost of a ride $3\frac{1}{4}$ miles long?

35) A breakfast cereal company produces a brand of cereal with a stated net weight of 18 oz. Only boxes with a net weight within 0.02 oz. of this stated amount are acceptable. If x is the net weight of a particular box of cereal brand, a formula describing this situation is $|x - 18| \leq 0.02$. Which of the following cereal boxes is acceptable?

A) a box with a net weight of 18.01 oz
B) a box with a net weight of 18.03 oz
C) a box with a net weight of 17.03 oz
D) a box with a net weight of 17 oz

36) XYZ Dumbbells, Inc. manufactures free weights varying in size from 2-pound weights to 50-pound weights. Only sets of dumbbells that actually weigh within 5% of the stated weight, are considered acceptable. If y is the actual weight of a 30-pound dumbbell that is unacceptable, then this is represented by the expression $|y - 30| > 0.05(30) = 1.5$. Which of the following 30-pound dumbbells is unacceptable?

A) one that actually weighs 31.7 pounds
B) one that actually weighs 28.5 pounds
C) one that actually weighs 28.7 pounds
D) one that actually weighs 30.1 pounds

0.3 Geometry Review

1 Use the Pythagorean Theorem and Its Converse

The lengths of the legs of a right triangle are given. Find the hypotenuse.

1) $a = 6$, $b = 8$

A) 10
B) 9
C) 7
D) 8

Find the hypotenuse. If necessary, round to the nearest tenth.

2)

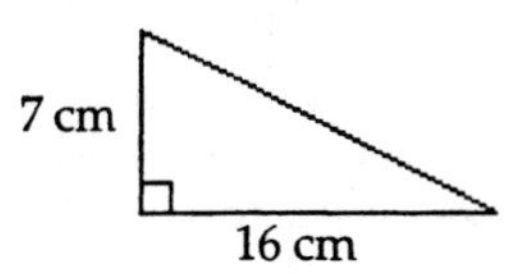

A) 305 cm
B) 17.5 cm
C) 11.5 cm
D) 152.5 cm

The lengths of the sides of a triangle are given. Determine if the triangle is a right triangle. If it is, identify the hypotenuse.

3) 6, 8, 10

A) right triangle; 10
B) not right triangle

4) 12, 16, 20

A) right triangle; 20
B) not a right triangle

5) 12, 13, 5

A) right triangle; 13
B) not a right triangle

6) 5, 11, 10

A) right triangle; 11
B) not a right triangle

7) 1, 2, 3

A) right trianglc; 3
B) not a right triangle

8) 4, 11, 12

A) right triangle; 12

B) not a right triangle

9) $\sqrt{11}, \sqrt{12}, \sqrt{13}$

A) right triangle; $\sqrt{13}$

B) not a right triangle

2 Know Geometry Formulas

Solve the problem.

1) Find the area A of a rectangle with length 8 in and width 9 in.

A) $A = 144 \text{ in}^2$ B) $A = 72 \text{ in}^2$ C) $A = 17 \text{ in}^2$ D) $A = 32 \text{ in}^2$

2) Find the area A of a rectangle with length 2.9 yd and width 4.8 yd.

A) $A = 11.6 \text{ yd}^2$ B) $A = 13.92 \text{ yd}^2$ C) $A = 7.7 \text{ yd}^2$ D) $A = 27.84 \text{ yd}^2$

3) Find the area A of a triangle with height 7 cm and base 5 cm.

A) $A = 35 \text{ cm}^2$ B) $A = \frac{35}{2} \text{ cm}^2$ C) $A = 35 \text{ cm}$ D) $A = \frac{35}{2} \text{ cm}$

4) Find the area A and circumference C of a circle of radius 12 mi.

A) $A = 144\pi \text{ mi}^2$; $C = 24\pi \text{ mi}$

B) $A = 48\pi \text{ mi}^2$; $C = 12\pi \text{ mi}$

C) $A = 576\pi \text{ mi}^2$; $C = 12\pi \text{ mi}$

D) $A = 24\pi \text{ mi}^2$; $C = 24\pi \text{ mi}$

5) Find the area A and circumference C of a circle of diameter 13 mi. Use 3.14 for π. Round the result to the nearest tenth.

A) $A = 81.6 \text{ mi}^2$; $C = 20.4 \text{ mi}$

B) $A = 530.7 \text{ mi}^2$; $C = 20.4 \text{ mi}$

C) $A = 40.8 \text{ mi}^2$; $C = 40.8 \text{ mi}$

D) $A = 132.7 \text{ mi}^2$; $C = 40.8 \text{ mi}$

6) Find the area A and circumference C of a circle of radius 20.5 ft. Use 3.14 for π. Round the result to the nearest tenth.

A) $A = 1319.6 \text{ ft}^2$; $C = 128.7 \text{ ft}$

B) $A = 128.7 \text{ ft}^2$; $C = 128.7 \text{ ft}$

C) $A = 257.5 \text{ ft}^2$; $C = 64.4 \text{ ft}$

D) $A = 5278.3 \text{ ft}^2$; $C = 64.4 \text{ ft}$

7) Find the volume V of a rectangular box with length 4 in, width 9 in, and height 8 in.

A) $V = 128 \text{ in}^3$ B) $V = 324 \text{ in}^3$ C) $V = 288 \text{ in}^3$ D) $V = 576 \text{ in}^3$

8) Find the volume V and surface area S of a sphere of radius 10 centimeters.

A) $V = \frac{1000}{3}\pi \text{ cm}^3$; $S = \frac{4000}{3}\pi \text{ cm}^2$

B) $V = 1000\pi \text{ cm}^3$; $S = 100\pi \text{ cm}^2$

C) $V = 4000\pi \text{ cm}^3$; $S = 25\pi \text{ cm}^2$

D) $V = \frac{4000}{3}\pi \text{ cm}^3$; $S = 400\pi \text{ cm}^2$

9) Find the volume V of a right circular cylinder with radius 17 cm and height 19 cm.

A) $V = \frac{5491}{4}\pi \text{ cm}^3$ B) $V = 5491\pi \text{ cm}^3$ C) $V = \frac{323}{2}\pi \text{ cm}^3$ D) $V = 323\pi \text{ cm}^3$

10) Find the volume V of a sphere of radius 2.5 in. Use 3.14 for π. If necessary, round the result to the nearest tenth.

A) $V = 65.4 \text{ in}^3$ B) $V = 523.3 \text{ in}^3$ C) $V = 36.8 \text{ in}^3$ D) $V = 26.2 \text{ in}^3$

11) Find the volume V of a sphere of diameter 4 cm. Use 3.14 for π. If necessary, round the result to the nearest tenth.

A) $V = 18.8\ cm^3$ B) $V = 33.5\ cm^3$ C) $V = 267.9\ cm^3$ D) $V = 16.7\ cm^3$

12) Find the volume V of a sphere of diameter 1.4 yd. Use 3.14 for π. If necessary, round the result to the nearest tenth.

A) $V = 2.1\ yd^3$ B) $V = 11.5\ yd^3$ C) $V = 0.8\ yd^3$ D) $V = 1.4\ yd^3$

13) Find the area of the shaded region.

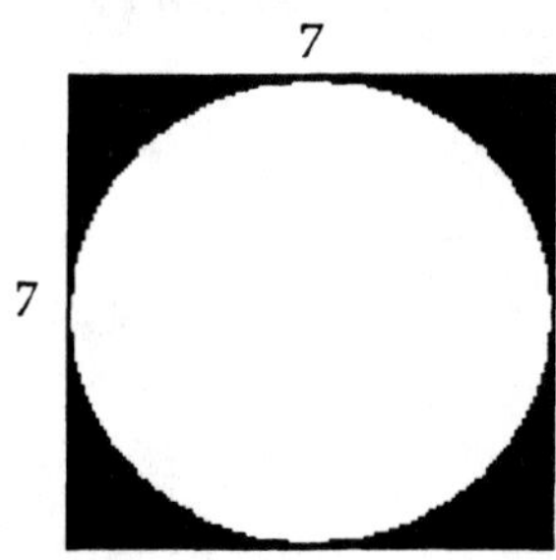

A) $49 - \frac{49}{4}\pi$ square units B) $196 - 49\pi$ square units

C) $\frac{49}{4}\pi + 49$ square units D) $49 - \frac{49}{2}\pi$ square units

14) Find the area of the shaded region in the figure. Round results to the nearest unit.

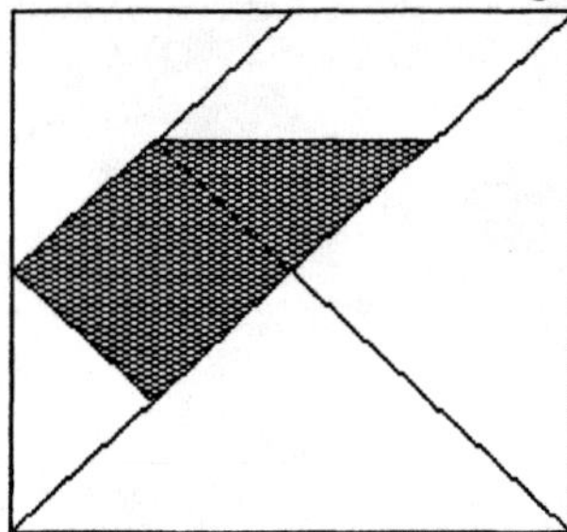

The square portion of the shaded region has sides of length 8.84 m. The hypotenuse of the triangular portion of the shaded region has length 12.5 m. The sides of the outer square have length 25 m.

A) $156\ m^2$ B) $117\ m^2$ C) $94\ m^2$ D) not enough data

15) Find the area of the shaded region in the figure. Round results to the nearest unit.

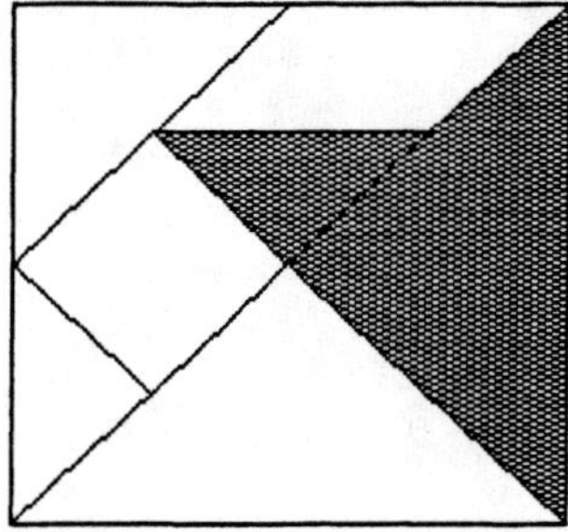

The hypotenuse of the smaller triangular portion of the shaded region has length 11 ft. The sides of the outer square have length 22 ft.

A) $151\ ft^2$ B) $121\ ft^2$ C) $303\ ft^2$ D) not enough data

16) Find the area of the shaded region in the figure. Round results to the nearest unit.

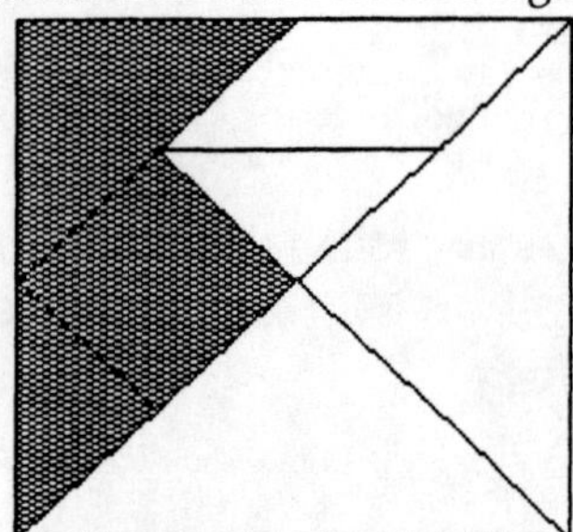

The hypotenuse of the smallest, unshaded triangular region has length 2.13 in. The sides of the outer square have length 6 in. Half the diagonal of the outer square is 4.25 in.

A) 15 in^2 B) 11 in^2 C) 24 in^2 D) not enough data

17) Find the area of the shaded region in the figure. Round results to the nearest unit.

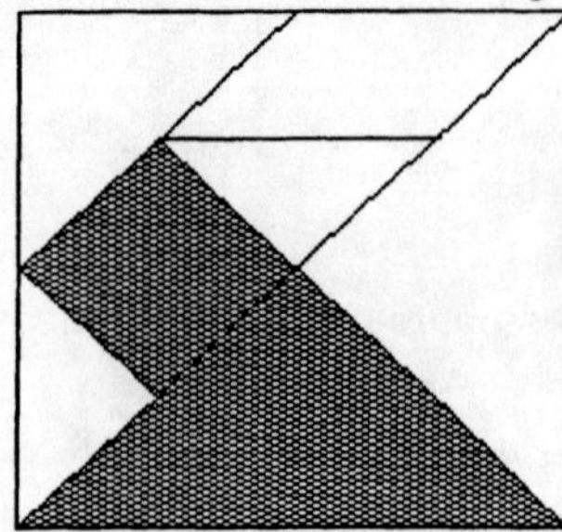

The sides of the square portion of the shaded region have length 1.42 yd. The sides of the outer square have length 4 yd.

A) 8 yd^2 B) 5 yd^2 C) 6 yd^2 D) not enough data

18) Find the area of the shaded region in the figure. Round results to the nearest unit.

The sides of the outer square have length 6 m.

A) 27 m^2 B) 32 m^2 C) 34 m^2 D) not enough data

19) Find the area of the shaded region in the figure. Round results to the nearest unit.

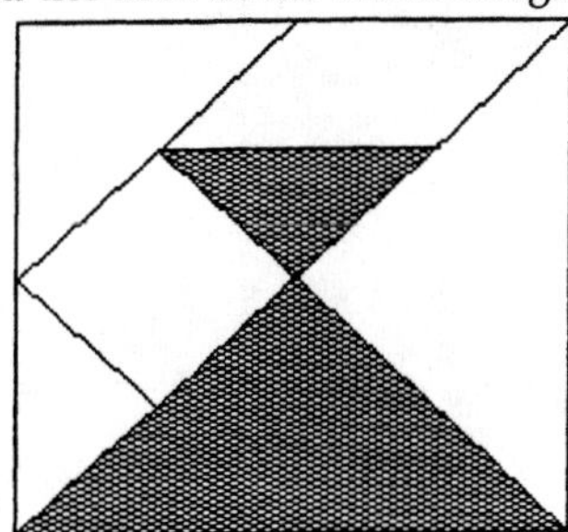

The sides of the outer square have length 22 in. Half the diagonal of the outer square is 15.56 in.

A) 214 in^2 B) 302 in^2 C) 151 in^2 D) not enough data

20) How many inches does a wheel with a diameter of 14 inches travel after 10 revolutions?

21) A rectangular patio has dimensions 10 feet by 15 feet. The patio is surrounded by a border with a uniform width of 3 feet. Find the area of the border.

22) A circular swimming pool, 25 feet in diameter, is enclosed by a circular deck that is 3.5 feet wide. What is the area of the deck?

23) A Norman window consists of a rectangle surmounted by a semicircle. If the width of the window is 5 feet and the total height of the window is 8 feet, how much wood frame is needed to enclose the window?

24) A person who is 5 feet tall is looking out at the Pacific Ocean while standing on an observation deck of a lighthouse that is 175 feet above sea level. To the nearest tenth of a mile, how far can this person see? Assume the radius of the Earth to be 3960 miles. [Note: 1 mile = 5280 feet]

0.4 Polynomials

1 Recognize Monomials

Tell whether the expression is a monomial. If it is, name the variable(s) and the coefficient and give the degree of the monomial.

1) $-5x^8$

A) monomial, variable is x, coefficient is 8, degree is –5

B) monomial, variable is x, coefficient is –5, degree is 1

C) monomial, variable is x, coefficient is –5, degree is 8

D) not a monomial

2) $\frac{24}{x}$

A) monomial, variable is x, coefficient is 1, degree is 24

B) monomial, variable is x, coefficient is 24, degree is 1

C) monomial, variable is x, coefficient is 24, degree is –1

D) not a monomial

3) $-9x^{-3}$

A) monomial, variable is x, coefficient is 9, degree is –3

B) monomial, variable is x, coefficient is –9, degree is –3

C) monomial, variable is x, coefficient is 9, degree is 3

D) not a monomial

4) $-2xy^4$

A) monomial, variables are x and y, coefficient is 2, degree is 4

B) monomial, variables are x and y, coefficient is –2, degree is 5

C) monomial, variables are x and y, coefficient is –2, degree is 4

D) not a monomial

5) $6x^8y^3$

A) monomial, variables are x and y, coefficient is 6, degree is 3

B) monomial, variables are x and y, coefficient is 6, degree is 8

C) monomial, variables are x and y, coefficient is 6, degree is 11

D) not a monomial

6) $\frac{5x^6}{y^8}$

A) monomial, variables are x and y, coefficient is 5, degree is 8

B) monomial, variables are x and y, coefficient is 5, degree is 6

C) monomial, variables are x and y, coefficient is 5, degree is 14

D) not a monomial

7) $2x^5 + 3$

A) monomial, variable is x, coefficient is 2, degree is 5

B) monomial, variable is x, coefficient is 3, degree is 5

C) monomial, variable is x, coefficient is 2, degree is 3

D) not a monomial

2 Recognize Polynomials

Tell whether the expression is a polynomial. If it is, give its degree.

1) $-8x^2 + 4$

A) polynomial, degree –8

B) polynomial, degree 2

C) polynomial, degree 4

D) not a polynomial

2) –4

A) polynomial, degree 1

B) polynomial, degree –4

C) polynomial, degree 0

D) not a polynomial

3) 9π

A) polynomial, degree 1

B) polynomial, degree 9

C) polynomial, degree 0

D) not a polynomial

4) $-9x^4 - \frac{9}{x}$

A) polynomial, degree 4
B) polynomial, degree 3
C) polynomial, degree -1
D) not a polynomial

5) $6y^3 - \sqrt{7}$

A) polynomial, degree 7
B) polynomial, degree 6
C) polynomial, degree 3
D) not a polynomial

6) $5z^3 + z$

A) polynomial, degree 5
B) polynomial, degree 3
C) polynomial, degree 1
D) not a polynomial

7) $\frac{x^8 - 2}{x^9 - 2}$

A) polynomial, degree -1
B) polynomial, degree 17
C) polynomial, degree 8
D) not a polynomial

3 Add and Subtract Polynomials

Perform the indicated operations. Express the answer as a single polynomial in standard form.

1) $(4x^6 + 17x^4 + 13) - (7x^6 + 2x^4 - 10)$

A) $-3x^6 + 15x^4 + 23$
B) $35x^{10}$
C) $-3x^6 + 24x^4 + 3$
D) $-3x^6 + 15x^4 + 3$

2) $(9x^4 - 2x^3 - 4x^2 - 7) + (4x^4 - 8x^3 + 8x^2 + 2)$

A) $13x^8 - 10x^6 + 4x^4 - 5$
B) $-15x^4 - 15x^3 + 11x^2 + 2$
C) $13x^4 - 10x^3 + 4x^2 - 5$
D) $7x^{18} - 5$

3) $(3x^7 + 2x^9 - 6 - 9x^8) - (4 + 4x^8 + 7x^9 - 7x^7)$

A) $-5x^9 - 5x^8 - 4x^7 - 2$
B) $9x^9 - 5x^8 - 4x^7 - 10$
C) $9x^9 - 5x^8 - 4x^7 - 2$
D) $-5x^9 - 13x^8 + 10x^7 - 10$

4) $(5x^8 + 10x^6 - 8x^3 + 12) - (12x^8 - 6x^5 + 9x^3 - 2)$

A) $-7x^8 + 10x^6 + 6x^5 - 17x^3 + 14$
B) $7x^8 + 10x^6 - 6x^5 - 17x^3 + 14$
C) $7x^8 + 10x^6 + 6x^5 - 17x^3 + 14$
D) $-7x^8 + 10x^6 - 6x^5 - 17x^3 + 14$

5) $(28y^2 + 5) - (-6y^4 - 14y^2 + 5)$

A) $-6y^4 + 14y^2 + 10$
B) $48y^6$
C) $6y^4 + 42y^2 - 10$
D) $6y^4 + 42y^2$

6) $(11x^2 - xy - y^2) + (x^2 + 10xy + 7y^2)$

A) $12x^2 + 11xy + 8y^2$
B) $10x^2 - 11xy - 8y^2$
C) $12x^2 + 9xy + 6y^2$
D) $11x^2 + 10xy + 7y^2$

7) $8z - (10 - 3z)$

A) $5z + 10$
B) $11z - 10$
C) $5z - 10$
D) $11z + 10$

8) $[(1.2x^3 + 7.7x^2 + 4.3) + (6.3x - 2.9)] - (3.4x^2 - x - 9.9)$

A) $1.2x^3 + 4.3x^2 + 7.3x + 11.3$
B) $1.2x^3 + 4.3x^2 + 6.3x + 11.3$
C) $1.2x^3 + 11.1x^2 + 5.3x - 8.5$
D) $12.8x^6 + 11.3$

4 Multiply Polynomials

Perform the indicated operations. Express the answer as a single polynomial in standard form.

1) $-5x(12x + 9)$

A) $-105x^2$
B) $-60x^2 + 9x$
C) $12x^2 - 45x$
D) $-60x^2 - 45x$

2) $-11x^3(8x - 11)$

A) $-88x^4 - 11$
B) $-88x + 121$
C) $33x^3$
D) $-88x^4 + 121x^3$

3) $-9x^4(6x^3 - 7)$

A) $-54x^7 - 7$
B) $-54x^3 + 63$
C) $9x^4$
D) $-54x^7 + 63x^4$

4) $11x^6(-6x^6 + 10x^4)$

A) $44x^6$
B) $-66x^{12} + 110x^{10}$
C) $-66x^{12} + 10x^4$
D) $44x^{12} + 44x^{10}$

5) $-10x^5(-3x^5 - 12x^4 - 2)$

A) $30x^{10} + 120x^9 + 20x^5$
B) $30x^{10} + 120x^9$
C) $30x^5 + 120x^4 + 20$
D) $30x^{10} - 12x^4 - 2$

6) $(x - 12)(x^2 + 4x - 3)$

A) $x^3 - 8x^2 - 51x + 36$
B) $x^3 + 16x^2 + 51x + 36$
C) $x^3 + 16x^2 + 45x - 36$
D) $x^3 - 8x^2 - 45x - 36$

7) $(x + 12)(x^3 + 4x - 5)$

A) $x^4 + 12x^3 + 4x^2 + 53x + 60$
B) $x^4 + 4x^2 - 5x + 12$
C) $x^3 + 16x^2 + 43x - 60$
D) $x^4 + 12x^3 + 4x^2 + 43x - 60$

8) $(9y + 11)(5y^2 - 2y - 3)$

A) $45y^3 + 37y^2 - 49y - 33$
B) $100y^2 - 40y - 60$
C) $45y^3 - 18y^2 - 27y + 11$
D) $45y^3 + 73y^2 + 49y + 33$

9) $(2 - z)(8 - 3z + 4z^2)$

A) $4z^3 + 11z^2 + 14z + 16$
B) $4z^3 + 11z^2 - 14z + 16$
C) $-4z^3 + 11z^2 - 15z + 16$
D) $-4z^3 + 11z^2 - 14z + 16$

Multiply the polynomials using the FOIL method. Express the answer as a single polynomial in standard form.

10) $(x - 12)(x - 12)$

A) $x^2 - 25x + 144$
B) $x^2 + 144x - 24$
C) $x^2 - 24x + 144$
D) $x^2 - 24x - 24$

11) $(-2x - 3)(x + 5)$

A) $-2x^2 - 13x - 13$
B) $-2x^2 - 13x - 15$
C) $-2x^2 - 15x - 13$
D) $-2x^2 - 15x - 15$

12) $(4x - 1)(5x + 8)$

A) $9x^2 + 27x - 8$
B) $20x^2 + 27x + 27$
C) $9x^2 + 27x + 27$
D) $20x^2 + 27x - 8$

13) $(x + 9y)(x - 12y)$

A) $x^2 - 3xy - 108y^2$ B) $x^2 - 6xy - 108y^2$ C) $x^2 - 3xy - 3y^2$ D) $x - 3xy - 108y$

14) $(x + 7y)(4x + 6y)$

A) $4x^2 + 34xy + 34y^2$ B) $4x^2 + 34xy + 42y^2$ C) $x^2 + 34xy + 34y^2$ D) $x^2 + 34xy + 42y^2$

15) $(4x + 2y)(3x + 12y)$

A) $12x^2 + 54xy + 54y^2$ B) $12x^2 + 6xy + 24y^2$ C) $12x^2 + 54xy + 24y^2$ D) $12x^2 + 48xy + 24y^2$

16) $(1 + x)(5 + x)$

A) $x^2 + 5x + 5$ B) $x^2 + 6x + 6$ C) $x^2 + 6x + 5$ D) $x^2 + 5x + 6$

5 Know Formulas for Special Products

Multiply the polynomials. Express the answer as a single polynomial in standard form.

1) $(x + 3)^2$

A) $9x^2 + 6x + 9$ B) $x^2 + 9$ C) $x^2 + 6x + 9$ D) $x + 9$

2) $(x - 15)^2$

A) $225x^2 - 30x + 225$ B) $x + 225$ C) $x^2 + 225$ D) $x^2 - 30x + 225$

3) $(6 + y)^2$

A) $y^2 + 12y + 36$ B) $y^2 + 36$ C) $y + 36$ D) $36y^2 + 12y + 36$

4) $(z - x)^2$

A) $z^2 - 2zx + x^2$ B) $z^2 - 2zx - x^2$ C) $z^2 - zx + x^2$ D) $z^2 + 2zx + x^2$

5) $(9x + 5)^2$

A) $81x^2 + 25$ B) $81x^2 + 90x + 25$ C) $9x^2 + 25$ D) $9x^2 + 90x + 25$

6) $(5x - 4)^2$

A) $5x^2 + 16$ B) $25x^2 - 40x + 16$ C) $25x^2 + 16$ D) $5x^2 - 40x + 16$

7) $(3x + 11y)^2$

A) $3x^2 + 66xy + 121y^2$ B) $9x^2 + 66xy + 121y^2$ C) $3x^2 + 121y^2$ D) $9x^2 + 121y^2$

8) $(7x - 6y)^2$

A) $7x^2 - 84xy + 36y^2$ B) $49x^2 - 84xy + 36y^2$ C) $7x^2 + 36y^2$ D) $49x^2 + 36y^2$

9) $(x + 10)(x - 10)$

A) $x^2 + 20x - 100$ B) $x^2 - 20$ C) $x^2 - 100$ D) $x^2 - 20x - 100$

10) $(9 + y)(9 - y)$

A) $81 + 18y - y^2$ B) $81 - 18y - y^2$ C) $81 - y^2$ D) $18 - y^2$

11) $(6z + 11)(6z - 11)$

A) $z^2 - 121$ B) $36z^2 + 132z - 121$ C) $36z^2 - 121$ D) $36z^2 - 132z - 121$

12) $(3 - 10x)(3 + 10x)$

A) $9 - 100x^2$ B) $9 + 60x - 100x^2$ C) $9 - 60x - 100x^2$ D) $9 - 10x^2$

13) $(x + 4y)(x - 4y)$

A) $x^2 - 16y^2$ B) $x^2 + 8xy - 16y^2$ C) $x^2 - 8y^2$ D) $x^2 - 8xy - 16y^2$

14) $(12y + x)(12y - x)$

A) $144y^2 - 24xy - x^2$ B) $24y^2 - x^2$ C) $144y^2 - x^2$ D) $144y^2 + 24xy - x^2$

15) $(x + \frac{3}{8})(x - \frac{3}{8})$

A) $x^2 - 9$ B) $x^2 - \frac{3}{4}x - \frac{9}{64}$ C) $x^2 - \frac{9}{64}$ D) $x^2 + \frac{3}{4}x - \frac{9}{64}$

16) $(\frac{2}{3}y - 3)(\frac{2}{3}y + 3)$

A) $\frac{4}{9}y^2 - 6$ B) $\frac{4}{9}y^2 - 4y - 9$ C) $\frac{4}{9}y^2 + 4y - 9$ D) $\frac{4}{9}y^2 - 9$

17) $(4 + \frac{4}{7}z)(4 - \frac{4}{7}z)$

A) $16 - \frac{8}{7}z^2$ B) $16 - \frac{16}{49}z^2$ C) $16 + \frac{32}{7}z - \frac{16}{49}z^2$ D) $16 - \frac{32}{7}z - \frac{16}{49}z^2$

18) $(x - 3)^3$

A) $x^3 - 9x^2 + 27x - 27$ B) $x^3 - 9x^2 + 15x - 27$ C) $x^3 - 3x^2 + 15x - 27$ D) $x^3 - 9x^2 + 9x - 27$

19) $(2x + 5)^3$

A) $8x^3 + 60x^2 + 60x + 125$ B) $4x^6 + 10x^3 + 15{,}625$

C) $8x^3 + 60x^2 + 150x + 125$ D) $4x^2 + 20x + 25$

20) $(x + 2y)^3$

A) $3(x + 2y)$ B) $x^3 + 6x^2y + 12xy^2 + 8y^3$

C) $x^3 + 8y^3$ D) $x^3 + 2x^2y + 4xy + 4xy^2 + 8y^2 + 8y^3$

0.5 Factoring Polynomials

1 Factor the Difference of Two Squares and the Sum and the Difference of Two Cubes

Factor completely. If the polynomial cannot be factored, say it is prime.

1) $x^2 - 121$

A) $(x - 11)^2$ B) $(x + 11)(x - 11)$ C) $(x + 11)^2$ D) prime

2) $25x^2 - 36$

A) $(5x - 6)^2$ B) prime C) $(5x + 6)(5x - 6)$ D) $(5x + 6)^2$

3) $x^3 + 216$

A) $(x + 6)(x^2 - 6x + 36)$
B) $(x - 216)(x^2 - 1)$
C) $(x + 6)(x^2 + 36)$
D) $(x - 6)(x^2 + 6x + 36)$

4) $216 - x^3$

A) $(6 - x)(36 + 6x + x^2)$
B) $(6 + x)(36 - x^2)$
C) $(6 - x)(36 + x^2)$
D) $(6 + x)(36 - 6x + x^2)$

5) $x^3 - 64$

A) $(x + 4)(x^2 - 4x + 16)$
B) $(x + 64)(x^2 - 1)$
C) $(x - 4)(x^2 + 4x + 16)$
D) $(x - 4)(x^2 + 16)$

6) $1000x^3 - 1$

A) $(10x - 1)(100x^2 + 1)$
B) $(10x + 1)(100x^2 - 10x + 1)$
C) $(10x - 1)(100x^2 + 10x + 1)$
D) $(1000x - 1)(x^2 + 10x + 1)$

7) $512x^3 - 729$

A) $(8x - 9)(64x^2 + 81)$
B) $(8x + 9)(64x^2 - 72x + 81)$
C) $(8x - 9)(64x^2 + 72x + 81)$
D) $(512x - 9)(x^2 + 72x + 81)$

2 Factor Perfect Squares

Factor completely. If the polynomial cannot be factored, say it is prime.

1) $x^2 + 16x + 64$

A) $(x + 8)(x - 8)$
B) $(x - 8)^2$
C) $(x + 8)^2$
D) prime

2) $x^2 - 16xy + 64y^2$

A) $(x - 8y)^2$
B) $(x - 8y)(x + 8y)$
C) $(x + 8y)^2$
D) prime

3) $x^2 + 6x + 9$

A) $(x + 3)^2$
B) $(x - 3)^2$
C) $(x + 3)(x - 3)$
D) $(x + 6)(x - 6)$

4) $49x^2 - 56x + 16$

A) $(49x + 1)(x + 16)$
B) $(7x - 4)^2$
C) $(7x - 4)(7x + 4)$
D) $(7x + 4)^2$

5) $25x^2 - 40xy + 16y^2$

A) $(5x - 4y)(5x + 4y)$
B) $(5x - 4y)^2$
C) $(25x + y)(x + 16y)$
D) $(5x + 4y)^2$

3 Factor a Second-degree Polynomial: $x^2 + Bx + C$

Factor completely. If the polynomial cannot be factored, say it is prime.

1) $x^2 - x - 56$

A) $(x + 1)(x - 56)$
B) $(x + 7)(x - 8)$
C) $(x + 8)(x - 7)$
D) prime

2) $x^2 + 8x - 48$

A) $(x - 12)(x + 1)$
B) $(x - 12)(x + 4)$
C) $(x + 12)(x - 4)$
D) prime

3) $x^2 + 7x + 10$

A) $(x - 5)(x + 1)$
B) $(x + 5)(x + 2)$
C) $(x - 5)(x + 2)$
D) prime

4) $x^2 - x - 45$

A) $(x + 5)(x - 9)$ B) $(x - 45)(x + 1)$ C) $(x - 5)(x + 9)$ D) prime

4 Factor by Grouping

Factor completely. If the polynomial cannot be factored, say it is prime.

1) $6x^2 + 8x + 15x + 20$

A) $(6x + 5)(x + 4)$ B) $(2x - 5)(3x - 4)$ C) $(6x - 5)(x - 4)$ D) $(2x + 5)(3x + 4)$

2) $8x^6 + 10x^3 - 12x^3 - 15$

A) $(2x^6 - 3)(4x + 5)$ B) $(2x^3 - 3)(4x^3 + 5)$ C) $(2x^3 + 3)(4x^3 - 5)$ D) $(8x^3 + 3)(x^3 - 5)$

3) $15x^2 + 12xy - 10xy - 8y^2$

A) $(3x - 2)(5x + 4)$ B) $(3x - 2y)(5x + 4y)$ C) $(15x - 2y)(x + 4y)$ D) $(3x + 2y)(5x + 4y)$

4) $12x^3 - 20x^2y + 15xy^2 - 25y^3$

A) $(4x^2 - 5y^2)(3x + 5y)$ B) $(12x^2 + 5y^2)(x - 5y)$ C) $(4x^2 + 5y^2)(3x - 5y)$ D) $(4x^2 + 5y)(3x - 5y)$

5) $64x^3 - 48x^2 - 4x + 3$

A) $(4x - 1)^2(4x - 3)$ B) $(4x + 1)(4x - 1)(4x - 3)$

C) $x(64x^2 - 48x - 4x) + 3$ D) $16x^2(4x - 3) + (-4x + 3)$

6) $x^3 + 2x^2 + x + 2$

7) $4x^2 + 4x + 1 - 16y^2$

8) $x^4 + 2x^3 + x^2 - x^2y^2$

An expression that occurs in calculus is given. Factor completely.

9) $6(x + 8)(x - 7)^3 + (x + 8)^2 \cdot 4(x - 7)^2$

A) $(x + 8)(x - 7)^2(5x - 5)$ B) $(x + 8)(x - 7)^2(10x - 10)$

C) $2(x + 7)(x - 8)^2(5x - 5)$ D) $2(x + 8)(x - 7)^2(5x - 5)$

10) $(2x + 1)^2 + 3(2x + 1) - 4$

11) $3(x + 5)^2(2x - 1)^2 + 4(x + 5)^3(2x - 1)$

5 Factor a Second-degree Polynomial: $Ax^2 + Bx + C$

Factor completely. If the polynomial cannot be factored, say it is prime.

1) $15x^2 + 22x + 8$

A) $(3x - 2)(5x - 4)$ B) $(3x + 2)(5x + 4)$ C) $(15x + 2)(x + 4)$ D) prime

2) $12y^2 + 17y + 6$

A) $(12y + 2)(y + 3)$ B) $(3y - 2)(4y - 3)$ C) $(3y + 2)(4y + 3)$ D) prime

3) $20z^2 + 7z - 6$

A) $(20z + 3)(z - 2)$ B) $(4z - 3)(5z + 2)$ C) $(4z + 3)(5z - 2)$ D) prime

4) $8z^2 - 6z - 9$

A) $(2z - 3)(4z + 3)$　B) $(8z - 3)(z + 3)$　C) $(2z + 3)(4z - 3)$　D) prime

5) $35g^2 + 44g + 12$

A) $(7g - 6)(5g - 2)$　B) $(7g - 6)(5g + 2)$　C) $(7g + 6)(5g + 2)$　D) $(7g + 6)(5g - 2)$

6 Additional Applications

Factor the polynomial by removing the common monomial factor.

1) $50x + 10$

A) $5(10x + 2)$　B) $10(5x)$　C) $10(5x + 1)$　D) $2(25x + 5)$

2) $35x^4y + 40xy^5$

A) $5y(7x^4 + 8xy^4)$　B) $5x(7x^3y + 8y^5)$　C) $xy(35x^3 + 40y^4)$　D) $5xy(7x^3 + 8y^4)$

3) $21x + 7y - 7$

A) $7(3x + y - 1)$　B) $7x(3 + y - 1)$　C) $7(3x + y)$　D) no common factors

4) $27y^3 - 9y^2 + 15y$

A) $3y(9y^2 - 3y + 5)$　B) $3y(9y^3 - 3y^2 + 5y)$　C) $3(9y^3 - 3y^2 + 5y)$　D) $y(27y^2 - 9y + 15)$

Factor completely. If the polynomial cannot be factored, say it is prime.

5) $12x^2 - 52x - 40$

A) $4(3x - 2)(x + 5)$　B) $(12x + 8)(x - 5)$　C) $4(3x + 2)(x - 5)$　D) prime

6) $16x^2 - 56x - 32$

A) $(16x - 8)(x + 4)$　B) $8(2x + 1)(x - 4)$　C) $8(2x - 1)(x + 4)$　D) prime

7) $12y^2 + 54y - 30$

A) $(12y - 6)(y + 5)$　B) $6(2y + 1)(y - 5)$　C) $6(2y - 1)(y + 5)$　D) prime

8) $42x^9 + 12x^7 + 36x^3$

A) $6x^3(7x^6 + 2x^4 + 6)$　B) $x^3(42x^6 + 12x^4 + 36)$　C) $6(7x^9 + 2x^7 + 6x^3)$　D) prime

9) $27x^4y - 147y^3$

A) $3y(3x^2 + 7y)(3x^2 - 7y)$　B) $(9x + 21y)(3xy - 7y^2)$

C) $3y(3x + 7y)^2$　D) $3y(3x - 7y)^2$

10) $49 - (x + 3y)^2$

A) $(7 + x + y)(7 - x - y)$　B) $(7 + x + 3y)(7 - x - 3y)$

C) $(7 + 3xy)(7 - 3xy)$　D) $(7 + x - 3y)(7 - x - 3y)$

11) $49x^3y - 112x^2y^2 + 64xy^3$

A) $xy(7x - 8y)^2$　B) $xy(7x - 8y)(7x + 8y)$　C) $xy(7x + 8y)^2$　D) prime

12) $x^2 + 119x + 120$

A) $(x + 120)(x - 1)$　B) $(x - 12)(x + 10)$　C) $(x + 12)(x - 10)$　D) prime

13) $-75x^2 - 70x - 15$

A) $(3x + 1)(-25x - 15)$ B) $(-5x - 5)(5x + 3)$ C) $-5(3x + 1)(5x + 3)$ D) $-5(3x - 1)(5x - 3)$

14) $x^2 - 4x + 16$

A) $(x + 4)(x - 4)$ B) $(x - 4)^2$ C) $(x + 4)^2$ D) prime

15) $x^3 - 25x + 3x^2 - 75$

A) $(x + 5)(x - 5)(x + 3)$ B) $(x - 5)^2(x + 3)$ C) $(x^2 - 25)(x + 3)$ D) prime

16) $x^4 - 9x^2 - 400$

A) $(x + 5)(x - 5)(x^2 + 16)$ B) $(x - 5)^2(x^2 + 16)$

C) $(x + 5)(x - 5)(x + 4)(x - 4)$ D) $(x^2 - 25)(x^2 + 16)$

17) $x^4 + 5x^3 + 216x + 1080$

A) $(x + 5)(x^3 + 216)$ B) $(x + 5)(x + 6)(x^2 - 6x + 36)$

C) $(x + 5)(x - 6)(x^2 + 6x + 36)$ D) prime

0.6 Polynomial Division; Synthetic Division

1 Divide Polynomials Using Long Division

Find the quotient and the remainder.

1) $21x^8 - 35x^5$ divided by $7x$

A) $3x^9 - 5x^6$; remainder 0 B) $21x^7 - 35x^4$; remainder 0

C) $3x^7 - 5x^4$; remainder 0 D) $3x^8 - 5x^5$; remainder 0

2) $35x^2 + 45x - 12$ divided by $5x$

A) $7x^2 + 9x - \frac{12}{5}$; remainder 0 B) $35x + 45$; remainder -12

C) $7x - 3$; remainder 0 D) $7x + 9$; remainder -12

3) $x^2 + 9x + 14$ divided by $x + 7$

A) $x - 7$; remainder 0 B) $x^2 + 2$; remainder 0 C) $x^3 - 7$; remainder 0 D) $x + 2$; remainder 0

4) $8x^2 + 14x - 4$ divided by $x + 2$

A) $x - 2$; remainder 0 B) $8x + 2$; remainder 0 C) $8x - 2$; remainder 3 D) $8x - 2$; remainder 0

5) $x^2 + 2x - 22$ divided by $x + 6$

A) $x - 4$; remainder 2 B) $x - 2$; remainder 4 C) $x + 4$; remainder 2 D) $x - 4$; remainder 0

6) $x^2 + 15x + 47$ divided by $x + 6$

A) $x + 9$; remainder -7 B) $x + 9$; remainder 0 C) $x + 10$; remainder 0 D) $x + 9$; remainder 7

7) $3x^3 + 5x^2 - 9x + 9$ divided by $x + 3$

A) $x^2 + 5x + 6$; remainder 0 B) $x^2 + 4x + 3$; remainder 0

C) $3x^2 - 4x + 3$; remainder 0 D) $3x^2 + 4x + 3$; remainder 0

8) $-8x^3 - 10x^2 + 15x - 5$ divided by $2x + 5$

A) $-4x^2 + 5x - 5$; remainder 0

B) $-4x^2 + 5x - 5$; remainder 23

C) $-4x^2 + 5x - 5$; remainder 20

D) $x^2 - 5$; remainder 5

9) $5x^3 - 7x^2 + 7x - 8$ divided by $5x - 2$

A) $x^2 + x - 1$; remainder -6

B) $x^2 - x + 1$; remainder 6

C) $x^2 - x + 1$; remainder -6

D) $x^2 - x + 1$; remainder 10

10) $x^4 + 16$ divided by $x - 2$

A) $x^3 + 2x^2 + 4x + 8$; remainder 16

B) $x^3 + 2x^2 + 4x + 8$; remainder 32

C) $x^3 - 2x^2 + 4x - 8$; remainder 32

D) $x^3 + 2x^2 + 4x + 8$; remainder 0

11) $8k^4 + 12k^3 - 2k$ divided by $2k^2 + k$

A) $4k^2 + 4k$; remainder $-6k$

B) $4k^2 + 8k + 4$; remainder $2k$

C) $4k^2 + 4k - 2$; remainder 0

D) $4k^2 + 6k$; remainder $-2k$

2 Divide Polynomials Using Synthetic Division

Use synthetic division to find the quotient and remainder.

1) $2x^4 - 4x^3 - 12x - 14$ divided by $x - 3$

A) $2x^2 + 2x + 6$; remainder 6

B) $2x^3 + 2x^2 + 6x$; remainder 6

C) $2x^3 + 2x^2 + 6x + 6$; remainder 4

D) $2x^2 + 2x + 6$; remainder 4

2) $5x^4 - 13x^2 - 6x + 9$ divided by $x - 3$

A) $5x^3 - 15x^2 + 32x - 102$; remainder -297

B) $5x^3 + 15x^2 + 32x + 90$; remainder 279

C) $5x^3 + 8x^2 + 18x$; remainder 63

D) $5x^3 + 8x^2 + 18x + 63$; remainder 0

3) $6x^5 - 5x^4 + x - 4$ divided by $x + \frac{1}{2}$

A) $6x^4 - 8x^3 + 5$; remainder $-\frac{13}{2}$

B) $6x^4 - 2x^3 + x^2 - \frac{1}{2}x + \frac{5}{4}$; remainder $-\frac{37}{8}$

C) $6x^4 - 2x^3 - x^2 + \frac{1}{2}x + \frac{5}{4}$; remainder $-\frac{27}{8}$

D) $6x^4 - 8x^3 + 4x^2 - 2x + 2$; remainder -5

Use synthetic division to find the quotient and remainder when f(x) is divided by g(x).

4) $f(x) = x^3 - 2x^2 + x + 1$; $g(x) = x - 1$

5) $f(x) = x^5 - 3x^3 + 2$; $g(x) = x + \frac{1}{2}$

Use synthetic division to determine whether x – c is a factor of the given polynomial.

6) $3x^4 - 10x^3 - 59x^2 + 146x + 120$; $x + 5$

A) Yes

B) No

7) $x^4 + 7x^3 - 19x^2 - 103x + 210$; $x - 3$

A) Yes

B) No

8) $2x^4 - 4x^3 - 12x - 14; \quad x - 3$

A) Yes　　B) No

9) $6x^5 - 5x^4 + x - 4; \quad x + \frac{1}{2}$

A) Yes　　B) No

10) $3x^4 - 2x^2 + x - 1; \quad x + 1$

A) Yes　　B) No

11) $2x^6 - 18x^4 + x^2 - 9; \quad x + 3$

A) Yes　　B) No

0.7 Rational Expressions

1 Reduce a Rational Expression to Lowest Terms

Reduce the rational expression to lowest terms.

1) $\frac{4 - x}{x - 4}$

A) $-x$　　B) 1　　C) -1　　D) $\frac{4 - x}{x - 4}$

2) $\frac{5x - 10}{12 - 6x}$

A) $-\frac{5}{6}$　　B) -1　　C) 1　　D) $\frac{5}{6}$

3) $\frac{3x + 2}{6x^2 + 19x + 10}$

A) $\frac{3x}{2x + 5}$　　B) $\frac{1}{2x + 5}$　　C) $\frac{3x + 2}{6x^2 + 19x + 10}$　　D) $\frac{3x + 2}{2x + 19}$

4) $\frac{y^2 + 8y + 12}{y^2 + 9y + 18}$

A) $\frac{8y + 12}{9y + 18}$　　B) $\frac{y + 2}{y + 3}$　　C) $-\frac{y^2 + 8y + 12}{y^2 + 9y + 18}$　　D) $\frac{8y + 2}{9y + 3}$

5) $\frac{4x^2 - 38x + 48}{x - 8}$

A) $4x^2 - 44$　　B) $4x - 6$　　C) $\frac{4x^2 - 38x + 48}{x - 8}$　　D) $\frac{1}{x - 8}$

6) $\frac{9x^2 + 27x^3}{7x + 21x^2}$

A) $\frac{9x}{7}$　　B) $\frac{9}{7}$　　C) $\frac{9 + 27x^3}{7x + 21}$　　D) $\frac{9x^2 + 27x^3}{7x + 21x^2}$

7) $\dfrac{x^2 - xy + 9x - 9y}{x + 9}$

A) $\dfrac{x^2 - xy + 9x - 9y}{x + 9}$ B) $x - y$ C) $x - 2y + 1$ D) $\dfrac{1}{x + 9}$

8) $\dfrac{y^3 - 216}{y - 6}$

A) $\dfrac{y^3 - 216}{y - 6}$ B) $y^2 + 6y + 36$ C) $\dfrac{1}{y - 6}$ D) $y^2 - 36$

9) $\dfrac{-x^2 - 2x + 8}{2 - x}$

A) $(x - 4)$ B) $-(x - 4)$ C) $-(x + 4)$ D) $(x + 4)$

10) $\dfrac{x^2 - 4x - 21}{x^2 - 49}$

An expression that occurs in calculus is given. Reduce the expression to lowest terms.

11) $\dfrac{(x^2 + 7) \cdot 6 - (6x + 7) \cdot 8x}{(x^2 + 7)^3}$

A) $\dfrac{-48x^2 - 56x + 6}{(x^2 + 7)^2}$ B) $\dfrac{42x^2 + 56x - 42}{(x^2 + 7)^3}$ C) $\dfrac{-42x^2 - 56x + 42}{(x^2 + 7)^3}$ D) $\dfrac{-54x^2 - 56x + 42}{(x^2 + 7)^3}$

12) $\dfrac{(2x + 5)4x - 2x^2(2)}{(2x + 5)^2}$

2 Multiply and Divide Rational Expressions

Perform the indicated operations and simplify the result. Leave the answer in factored form.

1) $\dfrac{2x}{4x + 2} \cdot \dfrac{6x + 3}{4}$

A) $\dfrac{3}{4}$ B) $\dfrac{x}{4}$ C) $\dfrac{3x}{4}$ D) $\dfrac{3x}{8}$

2) $\dfrac{2x - 2}{x} \cdot \dfrac{9x^2}{8x - 8}$

A) $\dfrac{16x^2 + 32x + 16}{9x^3}$ B) $\dfrac{18x^3 - 18x^2}{8x^2 - 8x}$ C) $\dfrac{9x}{4}$ D) $\dfrac{4}{9x}$

3) $\dfrac{x^3 + 1}{x^3 - x^2 + x} \cdot \dfrac{3x}{-18x - 18}$

A) $-\dfrac{x^3 + 1}{6(x + 1)}$ B) $\dfrac{x + 1}{6(-x - 1)}$ C) $-\dfrac{x^2 + 1}{6}$ D) $-\dfrac{1}{6}$

4) $\dfrac{x^2+8x+12}{x^2+13x+42} \cdot \dfrac{x^2+15x+56}{x^2+10x+16}$

A) $\dfrac{x+2}{x+7}$ B) $\dfrac{x+7}{x+8}$ C) 1 D) $\dfrac{1}{x+8}$

5) $\dfrac{x^2-7x+6}{x^2-14x+45} \cdot \dfrac{x^2-15x+50}{x^2-17x+66}$

A) $\dfrac{(x+1)(x+10)}{(x+9)(x+11)}$ B) $\dfrac{(x-1)(x-10)}{(x-9)(x-11)}$

C) $\dfrac{(x^2-7x+6)(x^2-15x+50)}{(x^2-14x+45)(x^2-17x+66)}$ D) $\dfrac{(x-1)}{(x-11)}$

6) $\dfrac{x^2+12x+27}{x^2+17x+72} \cdot \dfrac{x^2+8x}{x^2-3x-18}$

A) $\dfrac{1}{x-6}$ B) $\dfrac{x}{x^2+17x+72}$ C) $\dfrac{x(x+8)}{x-6}$ D) $\dfrac{x}{x-6}$

7) $\dfrac{9x^4-72x}{3x^2-12} \cdot \dfrac{x^2+x-2}{4x^3+8x^2+16x}$

A) $\dfrac{3x(x-1)}{4}$ B) $\dfrac{3(x-1)}{4}$ C) $\dfrac{3x(x+1)}{4}$ D) $\dfrac{3x(x-1)(x-2)^2}{4(x+2)^2}$

8) $\dfrac{x^2+6x+8}{x^2+7x+10} \cdot \dfrac{x^2+5x}{x^2+13x+36}$

A) $\dfrac{x}{x^2+7x+10}$ B) $\dfrac{1}{x+9}$ C) $\dfrac{x^2+5x}{x+9}$ D) $\dfrac{x}{x+9}$

9) $\dfrac{10x^2-3x-4}{5x^2-x-4} \cdot \dfrac{15x^2+12x}{1-4x^2}$

A) $\dfrac{3x(5x-4)}{(1-2x)(x-1)}$ B) $\dfrac{(5x+4)(5x-4)}{(1-2x)(x-1)}$ C) $\dfrac{3x(5x+4)}{(1-2x)(x-1)}$ D) $\dfrac{3x}{(1-2x)(x-1)}$

10)

$$\dfrac{\dfrac{5x-5}{x}}{\dfrac{9x-9}{6x^2}}$$

A) $\dfrac{45(x+1)^2}{6x^3}$ B) $\dfrac{30x^2(x-1)}{9x(x-1)}$ C) $\dfrac{3}{10x}$ D) $\dfrac{10x}{3}$

11)

$$\frac{\frac{x^2+8x+15}{x^2+11x+24}}{\frac{x^2+5x}{x^2+13x+40}}$$

A) $\frac{x}{(x+8)(x+3)}$ B) $\frac{x+5}{x(x+8)}$ C) $\frac{x+5}{x}$ D) $x+5$

12)

$$\frac{\frac{x^2-10x+25}{7x-35}}{\frac{4x-20}{28}}$$

A) 1 B) 28 C) $\frac{(x-5)^2}{49}$ D) $\frac{x^2-10x+25}{(x-5)^2}$

13)

$$\frac{\frac{12x^2+8xy-15y^2}{14x^2+29xy+12y^2}}{\frac{42x^2-47xy+10y^2}{35x^2+32xy-12y^2}}$$

A) $\frac{(7x-4y)}{(5x-6y)}$ B) $\frac{5x+6y}{7x+4y}$ C) $\frac{(6x+5y)^2}{(5x-6y)(7x-4y)}$ D) $\frac{(6x-5y)^2}{(5x+6y)(7x+4y)}$

14)

$$\frac{\frac{(y-2)^2}{11}}{\frac{11y-22}{121}}$$

A) $\frac{(y-2)^3}{121}$ B) $y-2$ C) $\frac{1}{y-2}$ D) $\frac{11(y-2)^2}{11y-22}$

15)

$$\frac{\frac{x^2-4x+xy-4y}{3x^2-3y^2}}{\frac{x-4}{5x-5y}}$$

A) $\frac{5(x^2-4x+xy-4y)}{3(x+y)(x-4)}$ B) $\frac{(x-4)^2}{15(x-y)^2}$ C) $\frac{5}{3}$ D) 1

16)
$$\frac{\frac{1}{x+2}}{\frac{5}{x^2-4}}$$

A) $x - 2$ B) $\frac{5}{x-2}$ C) $\frac{x+2}{5}$ D) $\frac{x-2}{5}$

17)
$$\frac{\frac{z^2+5z-6}{z^2+9z+18}}{\frac{z^2-1}{z^2+7z+12}}$$

3 Add and Subtract Rational Expressions

Perform the indicated operations and simplify the result. Leave the answer in factored form.

1) $\frac{6}{11x} + \frac{2}{11x}$

A) $\frac{8}{22x}$ B) 1 C) $\frac{11x}{8}$ D) $\frac{8}{11x}$

2) $\frac{2x+5}{2} - \frac{2x-5}{2}$

A) $2x$ B) 5 C) 25 D) 0

3) $\frac{3x}{x-6} - \frac{18}{x-6}$

A) $\frac{3(x+6)}{x-6}$ B) 0 C) $\frac{3}{x-6}$ D) 3

4) $\frac{9y^2}{y-1} - \frac{9y}{y-1}$

A) 0 B) $\frac{9y(y+1)}{y-1}$ C) $\frac{9y}{y-1}$ D) $9y$

5) $\frac{15}{x-8} - \frac{2}{x-8}$

A) $\frac{13}{x-8}$ B) $\frac{13}{x}$ C) $\frac{17}{x-8}$ D) $\frac{15(x-8)}{2(x-8)}$

6) $\frac{x^2-7x}{x-2} + \frac{10}{x-2}$

A) $x - 5$ B) $x + 5$ C) $x + 2$ D) $x - 2$

7) $\frac{8}{x+5} + \frac{3}{x-5}$

4 Use the Least Common Multiple Method

Find the LCM of the given polynomials.

1) x, $x + 7$

A) $x + 7$ B) $x^2(x + 7)$ C) x D) $x(x + 7)$

2) $15xy$, $12x^2$

A) $60x^3y$ B) $60xy^3$ C) $60xy^2$ D) $60x^2y$

3) $4x + 20$, $x^2 + 5x$

A) $4x(x + 5)$ B) $4x^2 + 20$ C) $4x^2 + 5$ D) $4x + 5$

4) $x^2 + 8x + 16$, $x^2 + 4x$

A) $x(x + 1)(x + 4)$ B) $(x + 4)^2$ C) $x(x + 4)$ D) $x(x + 4)^2$

5) $x^2 + 2x$, $x^2 + 5x + 6$

A) $(x - 1)^2$ B) $x(x - 1)(x + 3)$ C) $x(x - 1)^2$ D) $x(x + 2)(x + 3)$

6) $x^2 + 3x - 4$, $-2x + 2$

A) $-2(x + 4)(x + 1)$ B) $-2(x + 4)(x - 1)$ C) $-2(x - 4)(x - 1)$ D) $-2(x - 4)(x + 1)$

7) $x^2 - 8x + 15$, $x^2 - 2x - 3$

A) $(x - 3)(x + 1)$ B) $(x + 5)(x + 3)(x + 1)$ C) $(x - 5)(x - 3)$ D) $(x - 5)(x - 3)(x + 1)$

8) $x - 14$, $14 - x$

A) $(x - 14)$ or $(14 - x)$ B) $(x - 14)(14 - x)$ C) $x + 14$ D) -1

9) $15x - 14$, $15x + 14$

A) $(15x + 14)$ B) $(15x - 14)$

C) $(15x - 14)(15x + 14)$ D) $(15x - 14)$ or $(15x + 14)$

Perform the indicated operations and simplify the result. Leave the answer in factored form.

10) $\frac{2}{x} + \frac{9}{x - 6}$

A) $\frac{11x - 12}{x(x - 6)}$ B) $\frac{12x - 11}{x(x - 6)}$ C) $\frac{12x - 11}{x(6 - x)}$ D) $\frac{11x - 12}{x(6 - x)}$

11) $-\frac{3}{2} - \frac{8}{3x}$

A) $\frac{-9x - 16}{6x}$ B) $\frac{-9x + 16}{6x}$ C) $\frac{9x - 16}{6x}$ D) $\frac{-17}{6 - 3x}$

12) $\frac{x}{x^2 - 64} - \frac{8}{64 - x^2}$

A) $\frac{x + 8}{x - 8}$ B) $\frac{1}{x + 8}$ C) $\frac{1}{x - 8}$ D) $\frac{x - 8}{x + 8}$

13) $\dfrac{x + 1}{x^2 + 3x - 4} + \dfrac{2x + 3}{x^2 + 6x - 7}$

A) $3x + 4$ B) $\dfrac{3x + 4}{2x^2 + 9x - 11}$ C) $\dfrac{3x^2 + 19x + 19}{(x - 1)(x + 4)(x + 7)}$ D) $\dfrac{3x^2 + 19x + 19}{(x + 1)(x - 4)(x - 7)}$

14) $\dfrac{3}{x^2 - 3x + 2} + \dfrac{7}{x^2 - 1}$

A) $\dfrac{10x - 11}{(x - 1)(x - 2)}$ B) $\dfrac{10x - 11}{(x - 1)(x + 1)(x - 2)}$ C) $\dfrac{11x - 10}{(x - 1)(x + 1)(x - 2)}$ D) $\dfrac{42x - 11}{(x - 1)(x + 1)(x - 2)}$

15) $\dfrac{x}{x^2 - 16} - \dfrac{6}{x^2 + 5x + 4}$

A) $\dfrac{x^2 - 5x + 24}{(x - 4)(x + 4)(x + 1)}$ B) $\dfrac{x^2 - 5}{(x - 4)(x + 4)(x + 1)}$ C) $\dfrac{x^2 + 5x + 24}{(x - 4)(x + 4)(x + 1)}$ D) $\dfrac{x^2 - 5x + 24}{(x - 4)(x + 4)}$

16) $\dfrac{12}{x - 8} - \dfrac{9}{8 - x}$

A) $\dfrac{21}{8 - x}$ B) $\dfrac{168 - 21x}{(x - 8)(8 - x)}$ C) $\dfrac{3}{x - 8}$ D) $\dfrac{21}{x - 8}$

17) $\dfrac{x + 3}{x^2 + x - 12} - \dfrac{x + 4}{x^2 - 9}$

18) $\dfrac{x - 3}{x^2 + 7x + 12} + \dfrac{x - 5}{x^2 - 16}$

A) $\dfrac{2x^2 - x - 27}{(x + 4)(x - 4)(x + 3)}$ B) $\dfrac{2x^2 - 9x - 27}{(x + 4)(x - 4)(x + 3)}$ C) $\dfrac{2x^2 - x - 3}{(x + 4)(x - 4)(x + 3)}$ D) $\dfrac{2x^2 - 9x - 3}{(x + 4)(x - 4)(x + 3)}$

19) $\dfrac{3x}{x^2 - 5x - 36} - \dfrac{x - 1}{x^2 - 16} + \dfrac{1}{x^2 - 13x + 36}$

A) $\dfrac{2x^2 - 21x + 13}{(x - 9)(x + 4)(x - 4)}$ B) $\dfrac{2x - 1}{(x - 9)(x - 4)}$ C) $\dfrac{2x^2 - x - 5}{(x - 9)(x + 4)(x - 4)}$ D) $\dfrac{2x - 1}{(x - 9)(x + 4)}$

5 Simplify Mixed Quotients

Perform the indicated operations and simplify the result. Leave the answer in factored form.

1)

$$\dfrac{\dfrac{1}{x} + 1}{\dfrac{1}{x} - 1}$$

A) $\dfrac{x}{(1 - x)(1 + x)}$ B) $(1 - x)(1 + x)$ C) $\dfrac{1 + x}{1 - x}$ D) 1

2)

$$\frac{4+\frac{2}{x}}{\frac{x}{3}+\frac{1}{6}}$$

A) 1 B) $\frac{12}{x}$ C) $\frac{x}{12}$ D) 12

3)

$$\frac{x-\frac{1}{x^2}}{x-\frac{1}{x^3}}$$

A) $\frac{x}{(x-1)(x+1)}$ B) $\frac{x(x^2+x+1)}{(x+1)(x^2+1)}$ C) $\frac{1}{x+1}$ D) $\frac{x(x^2-x+1)}{(x+1)(x^2+1)}$

4)

$$\frac{\frac{1}{9}-\frac{1}{5}}{\frac{1}{6}+\frac{1}{8}}$$

A) $-\frac{7}{270}$ B) $-\frac{270}{7}$ C) $-\frac{105}{32}$ D) $-\frac{32}{105}$

5)

$$\frac{\frac{64x^2-49y^2}{xy}}{\frac{8}{y}-\frac{7}{x}}$$

A) $\frac{8x+7y}{xy}$ B) $8x+7y$ C) $\frac{xy}{7x+8y}$ D) $7x+8y$

6)

$$2+\frac{y}{2+\frac{2}{2+y}}$$

7)

$$\frac{\frac{x+2}{2x-1}+\frac{x+1}{x-1}}{\frac{2}{x}-\frac{x}{2x+3}}$$

Solve the problem.

8) The focal length f of a lens with index of refraction n is $\frac{1}{f} = (n - 1)\left[\frac{1}{R_1} + \frac{1}{R_2}\right]$ where R_1 and R_2 are the radii of curvature of the front and back surfaces of the lens. Express f as a rational expression.

9) An electrical circuit contains three resistors connected in parallel. If the resistance of each is R_1, R_2, and R_3 ohms, respectively, their combined resistance R is given by the formula $\frac{1}{R} = \frac{1}{R_1} + \frac{1}{R_2} + \frac{1}{R_3}$. Express R as a rational expression.

0.8 nth Roots; Rational Exponents

1 Work with nth Roots

Simplify the expression. Assume that all variables are positive when they appear.

1) $\sqrt[3]{125}$

A) 11 B) ±5 C) 25 D) 5

2) $\sqrt[3]{-1000}$

A) 32 B) −10 C) ±10 D) 100

3) $\sqrt[4]{16}$

A) 16 B) −2 C) 2 D) not a real number

2 Simplify Radicals

Simplify the expression. Assume that all variables are positive when they appear.

1) $\sqrt{15}$

A) 3 B) $5\sqrt{3}$ C) $3\sqrt{5}$ D) $\sqrt{15}$

2) $\sqrt{325}$

A) 65 B) $25\sqrt{13}$ C) $5\sqrt{13}$ D) $\sqrt{325}$

3) $\sqrt{\frac{100}{9}}$

A) $\frac{\sqrt{10}}{\sqrt{3}}$ B) 3 C) $\frac{\sqrt{10}}{3}$ D) $\frac{10}{3}$

4) $\sqrt[3]{\frac{1}{27}}$

A) $\frac{1}{\sqrt[3]{3}}$ B) 3 C) $\frac{1}{9}$ D) $\frac{1}{3}$

5) $\sqrt[3]{40}$

A) $2\sqrt[3]{10}$ B) 10 C) 2 D) $2\sqrt[3]{5}$

6) $\sqrt[4]{16}$

A) 24 B) 8 C) 2 D) 16

7) $\sqrt{100y^{18}}$

A) $100y^9$ B) $10y^{16}$ C) $10y^{18}$ D) $10y^9$

8) $\sqrt{125x^2}$

A) $5x\sqrt{5}$ B) $125x$ C) $5\sqrt{5x}$ D) $5x^2\sqrt{5}$

9) $\sqrt{\frac{18x^2y}{25}}$

A) $9x\sqrt{2y}$ B) $\frac{3x\sqrt{2y}}{5}$ C) $x\sqrt{\frac{18y}{5}}$ D) $\frac{3\sqrt{2x^2y}}{5}$

10) $\sqrt{y^5}$

A) $y^4\sqrt{y}$ B) $\sqrt{y^5}$ C) $y\sqrt{y^3}$ D) $y^2\sqrt{y}$

11) $\sqrt[3]{-343x^{12}}$

A) $-7x^{12}$ B) $7x^4$ C) $-7x^4$ D) $-7x^9$

12) $\sqrt[3]{p^{29}}$

A) p^{11} B) $\sqrt[3]{p^{29}}$ C) $p\sqrt[3]{p^{26}}$ D) $p^9\sqrt[3]{p^2}$

13) $\sqrt[3]{\frac{3}{y^{33}}}$

A) $\frac{3}{y^{11}}$ B) $\frac{\sqrt[3]{3}}{y^{30}}$ C) $\sqrt[3]{\frac{3}{y^{33}}}$ D) $\frac{\sqrt[3]{3}}{y^{11}}$

14) $\sqrt[3]{\frac{320}{x^{24}}}$

A) $\frac{\sqrt[3]{320}}{x^8}$ B) $\sqrt[3]{\frac{320}{x^{24}}}$ C) $\frac{4\sqrt[3]{5}}{x^{21}}$ D) $\frac{4\sqrt[3]{5}}{x^8}$

15) $\sqrt{5} \cdot \sqrt{7}$

A) $\sqrt{12}$ B) $\sqrt{35}$ C) 35 D) $\sqrt{5+7}$

16) $\sqrt{2} \cdot \sqrt{4}$

A) $\sqrt{8}$ B) 2 C) $\sqrt{4}$ D) $2\sqrt{2}$

17) $\sqrt{18} \cdot \sqrt{32}$

A) $12\sqrt{2}$ B) 12 C) 32 D) 24

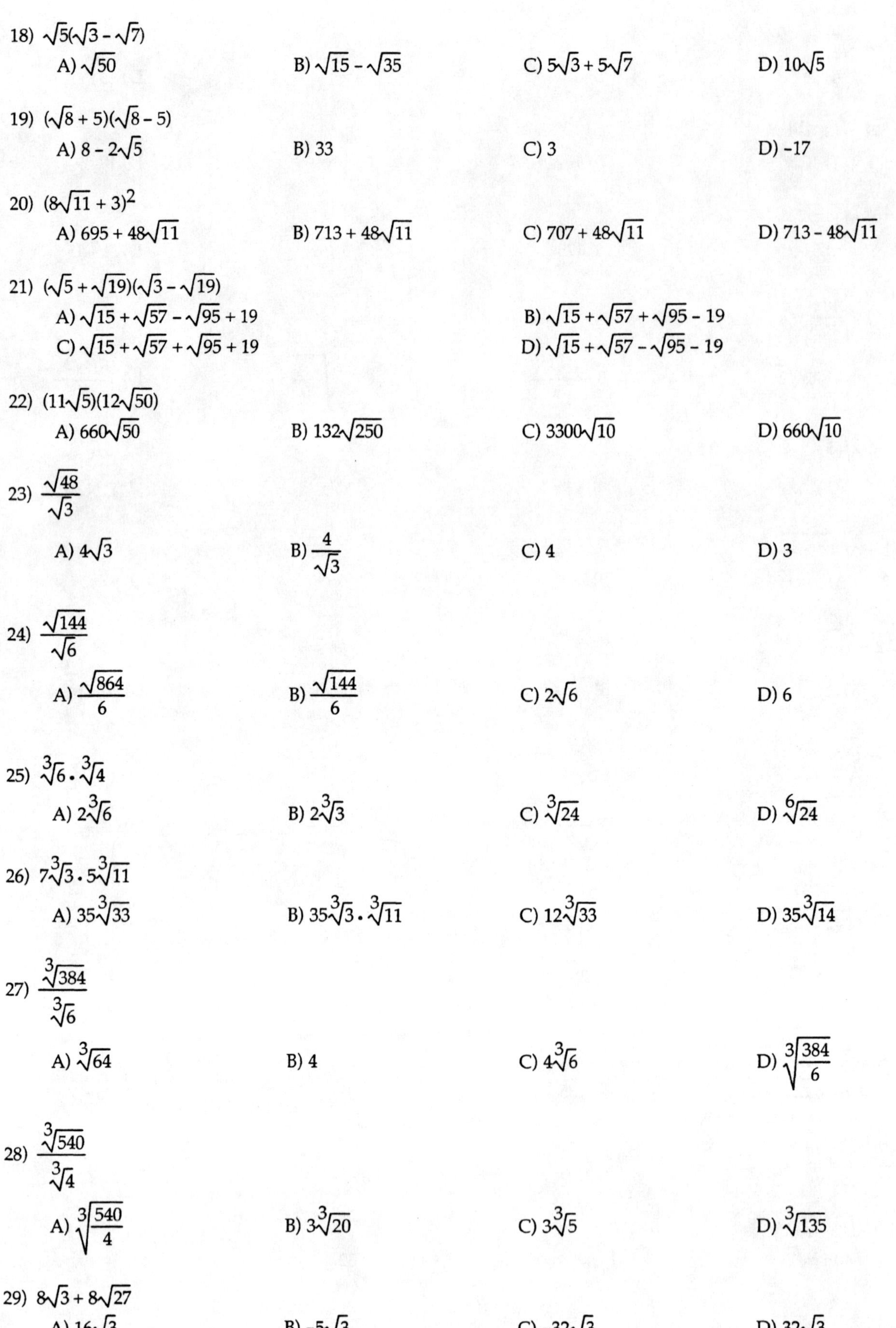

18) $\sqrt{5}(\sqrt{3} - \sqrt{7})$

A) $\sqrt{50}$ B) $\sqrt{15} - \sqrt{35}$ C) $5\sqrt{3} + 5\sqrt{7}$ D) $10\sqrt{5}$

19) $(\sqrt{8} + 5)(\sqrt{8} - 5)$

A) $8 - 2\sqrt{5}$ B) 33 C) 3 D) -17

20) $(8\sqrt{11} + 3)^2$

A) $695 + 48\sqrt{11}$ B) $713 + 48\sqrt{11}$ C) $707 + 48\sqrt{11}$ D) $713 - 48\sqrt{11}$

21) $(\sqrt{5} + \sqrt{19})(\sqrt{3} - \sqrt{19})$

A) $\sqrt{15} + \sqrt{57} - \sqrt{95} + 19$ B) $\sqrt{15} + \sqrt{57} + \sqrt{95} - 19$
C) $\sqrt{15} + \sqrt{57} + \sqrt{95} + 19$ D) $\sqrt{15} + \sqrt{57} - \sqrt{95} - 19$

22) $(11\sqrt{5})(12\sqrt{50})$

A) $660\sqrt{50}$ B) $132\sqrt{250}$ C) $3300\sqrt{10}$ D) $660\sqrt{10}$

23) $\dfrac{\sqrt{48}}{\sqrt{3}}$

A) $4\sqrt{3}$ B) $\dfrac{4}{\sqrt{3}}$ C) 4 D) 3

24) $\dfrac{\sqrt{144}}{\sqrt{6}}$

A) $\dfrac{\sqrt{864}}{6}$ B) $\dfrac{\sqrt{144}}{6}$ C) $2\sqrt{6}$ D) 6

25) $\sqrt[3]{6} \cdot \sqrt[3]{4}$

A) $2\sqrt[3]{6}$ B) $2\sqrt[3]{3}$ C) $\sqrt[3]{24}$ D) $\sqrt[6]{24}$

26) $7\sqrt[3]{3} \cdot 5\sqrt[3]{11}$

A) $35\sqrt[3]{33}$ B) $35\sqrt[3]{3} \cdot \sqrt[3]{11}$ C) $12\sqrt[3]{33}$ D) $35\sqrt[3]{14}$

27) $\dfrac{\sqrt[3]{384}}{\sqrt[3]{6}}$

A) $\sqrt[3]{64}$ B) 4 C) $4\sqrt[3]{6}$ D) $\sqrt[3]{\dfrac{384}{6}}$

28) $\dfrac{\sqrt[3]{540}}{\sqrt[3]{4}}$

A) $\sqrt[3]{\dfrac{540}{4}}$ B) $3\sqrt[3]{20}$ C) $3\sqrt[3]{5}$ D) $\sqrt[3]{135}$

29) $8\sqrt{3} + 8\sqrt{27}$

A) $16\sqrt{3}$ B) $-5\sqrt{3}$ C) $-32\sqrt{3}$ D) $32\sqrt{3}$

30) $-5\sqrt{200} + 7\sqrt{18}$

A) $71\sqrt{2}$
B) $29\sqrt{2}$
C) $-71\sqrt{2}$
D) $-29\sqrt{2}$

31) $-6\sqrt{32} - 5\sqrt{200} - 6\sqrt{72}$

A) $12\sqrt{2}$
B) $-110\sqrt{2}$
C) $-6\sqrt{2}$
D) $-12\sqrt{2}$

32) $\sqrt{5x^2} - 6\sqrt{80x^2} - 3\sqrt{80x^2}$

A) $-9x\sqrt{5}$
B) $-35x\sqrt{5}$
C) $-35x\sqrt{50}$
D) $-9x\sqrt{50}$

33) $18\sqrt[3]{2} - 3\sqrt[3]{54}$

A) $-9\sqrt[3]{2}$
B) $9\sqrt[3]{2}$
C) $18\sqrt[3]{2} - 3\sqrt[3]{54}$
D) $15\sqrt[3]{2}$

34) $\sqrt[3]{8y} - \sqrt[3]{128y}$

A) $4\sqrt[3]{2y} - 2\sqrt[3]{y}$
B) $2\sqrt[3]{y} - 4\sqrt[3]{2y}$
C) $2 - 4\sqrt[3]{2}$
D) $6\sqrt[3]{y}$

35) $2\sqrt[3]{8x} + 2\sqrt[3]{64x}$

A) $2\sqrt[3]{72x}$
B) $6\sqrt[3]{x}$
C) $12\sqrt[3]{x}$
D) $12x$

36) $2\sqrt[3]{256x} - 4\sqrt[3]{32x}$

A) $8\sqrt[3]{x} - 4\sqrt[3]{32x}$
B) $24\sqrt[3]{x}$
C) 0
D) cannot simplify

37) $\sqrt[3]{125} + 14 - \sqrt[3]{12}$

A) $19 - 2\sqrt[3]{6}$
B) $19 - \sqrt[3]{12}$
C) $\sqrt[3]{125} + 14 - \sqrt[3]{12}$
D) $19\sqrt[3]{12}$

38) $\sqrt{3x^2} - \sqrt[3]{250} + \sqrt{300x^2}$

A) $11x\sqrt{3} - 5\sqrt[3]{2}$
B) $11\sqrt{3x^2} - \sqrt[3]{250}$
C) $11x\sqrt{3} - 5\sqrt{10}$
D) $10x\sqrt{3} - 5\sqrt[3]{2}$

39) $(\sqrt{12} + \sqrt{z})(\sqrt{12} - \sqrt{z})$

A) $12 - 2\sqrt{12z}$
B) $12z$
C) $12 - z$
D) $12 - 2\sqrt{z}$

40) $\dfrac{\sqrt{56x^5y^6}}{\sqrt{2y^4}}$

A) $4x^2y\sqrt{7x}$
B) $2x^2y\sqrt{7x}$
C) $28xy\sqrt{x}$
D) $2x^4y^2\sqrt{7xy}$

41) $\dfrac{\sqrt{600x^{11}}}{\sqrt{6x}}$

A) $10x^{10}$
B) $10x^5\sqrt{6}$
C) $10\sqrt{x^{11}}$
D) $10x^5$

42) $\dfrac{\sqrt{162x^7}}{\sqrt{3x^5}}$

A) $x\sqrt{54}$
B) $3x\sqrt{6}$
C) $3x^6\sqrt{6}$
D) $3x\sqrt{18}$

Solve the problem.

43) A gas station stores its gasoline in an underground tank that is a right circular cylinder lying on its side. The volume V of gasoline in the tank (in gallons) is given by the formula $V = 40h^2\sqrt{\frac{96}{h} - 0.608}$ where h is the height of the gasoline (in inches) as measured on a depth stick. Approximately how many gallons are in the tank if the gas is to a level of 10 inches?

44) The final velocity v of an object in feet per second (ft/sec) after it slides down a frictionless inclined plane of height h feet is $v = \sqrt{64h + v_0^2}$ where v_0 is the initial velocity (in ft/sec) of the object. What is the final velocity v of an object that slides down a frictionless plane of height 4 feet if the initial velocity is 8 ft/sec?

45) The period T, in seconds, of a pendulum of length l feet may be approximated using the formula $T = 2\pi\sqrt{\frac{l}{32}}$. Find the period T of a pendulum whose length is 8 feet.

3 Rationalize Denominators

Rationalize the denominator of the expression. Assume that all variables are positive when they appear.

1) $\frac{\sqrt{9}}{\sqrt{11}}$

A) $3\sqrt{11}$ B) $\frac{3\sqrt{11}}{11}$ C) 124 D) $\frac{9\sqrt{11}}{11}$

2) $\frac{8}{\sqrt{5}}$

A) $\frac{8\sqrt{5}}{5}$ B) $\frac{8}{5}$ C) $\frac{64\sqrt{5}}{5}$ D) $8\sqrt{5}$

3) $\frac{2}{7 - \sqrt{2}}$

A) $\frac{2}{7} - \frac{2}{\sqrt{2}}$ B) $\frac{14 + 2\sqrt{2}}{-5}$ C) $\frac{14 + 2\sqrt{2}}{47}$ D) $\frac{14 - 2\sqrt{2}}{47}$

4) $\frac{\sqrt{7}}{\sqrt{6} + 7}$

A) $\frac{\sqrt{42} - 7\sqrt{7}}{13}$ B) $\frac{\sqrt{42} + 7\sqrt{7}}{-43}$ C) $\frac{\sqrt{42} - 7\sqrt{7}}{-43}$ D) $\frac{3\sqrt{42} + 6\sqrt{42}}{7}$

5) $\frac{3 - \sqrt{2}}{3 + \sqrt{2}}$

A) $\frac{7 - 6\sqrt{2}}{11}$ B) 1 C) $\frac{11 - 6\sqrt{2}}{7}$ D) $\frac{11 + 6\sqrt{2}}{7}$

6) $\frac{8}{9 - \sqrt{2}}$

A) $\frac{72 + 8\sqrt{2}}{79}$ B) $\frac{72 + 8\sqrt{2}}{-7}$ C) $\frac{72 - 8\sqrt{2}}{79}$ D) $\frac{8}{9} - \frac{8}{\sqrt{2}}$

7) $\frac{4}{\sqrt{x+h}-\sqrt{x}}$

A) $\frac{4(\sqrt{x+h}+\sqrt{x})}{h}$
B) $\frac{4\sqrt{h}}{h}$
C) $\frac{4\sqrt{x+h}+\sqrt{x}}{h}$
D) $\frac{4(\sqrt{x+h}-\sqrt{x})}{h}$

4 Simplify Expressions with Rational Exponents

Simplify the expression.

1) $100^{1/2}$

A) 40
B) 5
C) 20
D) 10

2) $625^{1/4}$

A) 100
B) 5
C) 3125
D) 20

3) $-32^{1/5}$

A) 16
B) -8
C) 32
D) -2

4) $\left(\frac{1}{343}\right)^{1/3}$

A) 7
B) $\frac{1}{7}$
C) -7
D) $\frac{-1}{7}$

5) $64^{4/3}$

A) 4096
B) 256
C) 16,384
D) 1024

6) $16^{5/4}$

A) 256
B) 32
C) 512
D) 128

7) $243^{4/5}$

A) 19,683
B) 81
C) 6561
D) 2187

8) $\left(\frac{8}{27}\right)^{2/3}$

A) $\frac{4}{13}$
B) $\frac{4}{9}$
C) $-\frac{4}{9}$
D) $\frac{3}{4}$

9) $(-64)^{4/3}$

A) 4096
B) not a real number
C) 256
D) -256

10) $36^{-3/2}$

A) $-\frac{1}{216}$
B) 216
C) $\frac{1}{216}$
D) -216

11) $8^{-4/3}$

A) $\frac{1}{16}$
B) 16
C) $-\frac{1}{16}$
D) not a real number

12) $16^{-5/4}$

A) $\frac{1}{32}$ B) not a real number C) 32 D) $-\frac{1}{32}$

13) $32^{-4/5}$

A) $-\frac{1}{16}$ B) 16 C) not a real number D) $\frac{1}{16}$

14) $-8^{-4/3}$

A) $\frac{1}{16}$ B) $-\frac{1}{16}$ C) 16 D) not a real number

15) $-243^{-4/5}$

A) $-\frac{1}{81}$ B) not a real number C) $\frac{1}{81}$ D) 81

16) $125^{-1/3}$

A) $\frac{1}{25}$ B) 5 C) $\frac{1}{5}$ D) 25

Simplify the expression. Express the answer so that only positive exponents occur. Assume that all variables are positive when they appear.

17) $(8x^{12})^{1/3}$

A) $2x^{12}$ B) $2x^4$ C) $2\sqrt[4]{x}$ D) $8x^4$

18) $(-243x^{25})^{1/5}$

A) $-3x^{25}$ B) $-243x^5$ C) $-3x^5$ D) $-3\sqrt[5]{x}$

19) $x^{1/8} \cdot x^{7/8}$

A) $\frac{1}{x}$ B) $x^{7/8}$ C) x D) $x^{7/64}$

20) $\frac{y^{3/4}}{y^{1/4}}$

A) $y^{3/4}$ B) $\frac{1}{y}$ C) $y^{1/2}$ D) y

21) $(x^5)^{2/5}$

A) $x^{1/2}$ B) $x^{7/5}$ C) $x^{2/25}$ D) x^2

22) $z^{-2/7} \cdot z^{3/7}$

A) $z^{1/7}$ B) $z^{7/6}$ C) $z^{6/7}$ D) $z^{-1/7}$

23) $(9x^{1/5} \cdot y^{1/5})^2$

A) $81x^2y^2$ B) $81x^{1/10}y^{1/10}$ C) $81x^{1/25}y^{1/25}$ D) $81x^{2/5}y^{2/5}$

24) $y^{5/9}(y^{3/9} - 8y^{2/9})$

A) $y^{5/3} - 8y^{5/2}$ B) $y^{8/9} - 8y^{7/9}$ C) $y^{2/9} - 8y^{3/9}$ D) $y^{15/81} - 8y^{10/81}$

25) $(-3x^{3/5} + 2x^{4/5})(-3x^{3/5} + 2x^{4/5})$

A) $-6x^{6/5} - 6x^{4/5} + 4x^{8/5}$ B) $9x^{6/5} - 12x^{7/5} + 4x^{8/5}$

C) $9x^{9/25} + 7x$ D) $9x^{6/5} - 6x^{7/5} + 4x^{8/5}$

26) $\dfrac{x^{4/7} \cdot x^{6/5}}{x^{5/6}}$

A) $x^{197/210}$ B) $\dfrac{1}{x^{197/210}}$ C) $x^{547/210}$ D) $\dfrac{1}{x^{547/210}}$

27) $\dfrac{(3x^{5/2})^5}{x^{-6/5}}$

A) $243x^{113/10}$ B) $3x^{137/10}$ C) $3x^{113/10}$ D) $243x^{137/10}$

28) $(x^5y^2)^{3/4}$

A) $x^{20/3}y^{8/3}$ B) $x^{3/2}y^{15/4}$ C) $x^{23/4}y^{11/4}$ D) $x^{15/4}y^{3/2}$

29) $(x^{-2}y^8z^{-4})^{-5/2}$

A) $\dfrac{x^5y^{20}}{z^{10}}$ B) $\dfrac{x^5z^{10}}{y^{20}}$ C) $\dfrac{y^{20}}{x^5z^{10}}$ D) $\dfrac{y^{20}z^{10}}{x^5}$

30) $(16x^4y^{-2})^{5/2}$

A) $\dfrac{16x^{10}}{y^5}$ B) $\dfrac{1024}{x^{10}y^5}$ C) $1024x^{10}y^5$ D) $\dfrac{1024x^{10}}{y^5}$

5 Additional Applications

An expression that occurs in calculus is given. Write the expression as a single quotient in which only positive exponents and/or radicals appear.

1) $\dfrac{(49 - x^2)^{1/2} + 3x^2(49 - x^2)^{-1/2}}{49 - x^2}$

A) $\dfrac{3x^2}{(49 - x^2)^{3/2}}$ B) $\dfrac{49 + 2x^2}{(49 - x^2)^{3/2}}$ C) $\dfrac{49 - 4x^2}{(49 - x^2)^{3/2}}$ D) $\dfrac{3x^2}{\sqrt{49 - x^2}}$

2) $6x^{3/2}(x^3 + x^2) - 8x^{5/2} - 8x^{3/2}$

A) $6x^{3/2}(x + 1)(3x^2 - 4)$ B) $2x^{3/2}(x + 1)(3x^2 - 4)$

C) $6x^{3/2}(x^3 - 1) + 6x^3 - 2x^{3/2}(x - 1)$ D) $x^{3/2}(x - 1)[6(x^2 - 1) - 2] + 6x^3$

3) $(x^2 - 1)^{1/2} \cdot \frac{3}{2}(2x + 5)^{1/2} \cdot 2 + (2x + 5)^{3/2} \cdot \frac{1}{2}(x^2 - 1)^{-1/2} \cdot 2x$

4) $\dfrac{(x^2 + 2)^{1/2} \cdot \frac{3}{2}(2x + 1)^{1/2} \cdot 2 - (2x + 1)^{3/2} \cdot \frac{1}{2}(x^2 + 2)^{-1/2} \cdot 2x}{x^2 + 2}$

Factor the expression.

5) $x^{6/7} + 3x^{4/7}$

A) $x^{1/7}(x^{6/7} + 3)$ B) $x^{1/7}(x^{2/7} + 3)$ C) $x^{4/7}(x^{2/7} + 3)$ D) $x^{4/7}(x^{-2/7} + 3)$

6) $11x^{5/7} - 4x^{-6/7}$

A) $x^{1/3}(11 - 4x^{11/4})$ B) $x^{-6/7}(11x^{11/7} - 4x)$

C) $x^{-6/7}(11x^{11/7} - 4)$ D) $x^{5/7}(11x^{11/7} - 4x^{11/4})$

7) $x^{3/2} + x^{5/4}$

A) $x^{5/4}(x^{1/4} - 1)$ B) $x^{5/4}(x^{1/4})$ C) $x^{5/4}(x^{1/4} + 1)$ D) $x^{5/4}(x^{1/4} + x)$

8) $x^{6/7} - x^{5/7}$

A) $x^{5/7}(x^{1/7} - 1)$ B) $x^{5/7}(x^{1/7} - x)$ C) $x^{5/7}(x^{1/7})$ D) $x^{5/7}(x^{1/7} + 1)$

Ch. 0 (Chapter R) Review
Answer Key

0.1 Real Numbers

1 Classify Numbers

1) C
2) A
3) B
4) D
5) A
6) B
7) A
8) D
9) B
10) C
11) B
12) D
13) D
14) rational numbers, real numbers

2 Evaluate Numerical Expressions

1) C
2) A
3) C
4) C
5) B
6) C
7) D
8) C
9) B
10) A
11) D
12) C
13) C
14) A
15) $\frac{1}{6}$

3 Work with Properties of Real Numbers

1) D
2) C
3) C
4) B
5) B
6) D
7) D
8) D
9) D
10) C
11) B
12) C
13) B
14) C
15) D
16) C
17) B

18) B
19) A
20) C
21) C
22) A

4 Additional Applications

1) C
2) (a) 2.556 (b) 2.555
3) D
4) 2.61
5) B
6) C
7) C
8) A
9) C
10) C
11) C

0.2 Algebra Review

1 Graph Inequalities

1) C
2) D
3) A
4) C

2 Find Distance on the Real Number Line

1) A

3 Evaluate Algebraic Expressions

1) A
2) B
3) D
4) B
5) C
6) C
7) D
8) C
9) A
10) D
11) B
12) C
13) B
14) A
15) C
16) D
17) B
18) B
19) B
20) B
21) D
22) C
23) D
24) C

4 Determine the Domain of a Variable

1) B
2) B
3) D
4) C

5) B
6) C
7) A
8) D
9) D
10) C
11) B
12) B
13) A
14) $\{x \mid x \neq -7, x \neq 3\}$

5 Use the Laws of Exponents

1) C
2) B
3) B
4) D
5) B
6) D
7) A
8) A
9) B
10) C
11) D
12) B
13) D
14) C
15) C
16) C
17) D
18) C
19) D
20) A
21) B
22) B
23) C
24) A
25) $\frac{4x}{y^2}$
26) B
27) A
28) D
29) B
30) D

6 Evaluate Square Roots

1) B
2) B
3) D
4) A
5) A
6) C

7 Use a Calculator to Evaluate Exponents

1) C
2) C
3) A
4) 52.708

8 **Use Scientific Notation**

1) C
2) A
3) B
4) C
5) A
6) D
7) D
8) B
9) C
10) D
11) C
12) 1×10^{-7} mm
13) 4.9392×10^{8} barrels in February 1995
14) 6.696×10^{8} mi/hr
15) D
16) D
17) A
18) A
19) D
20) C
21) B
22) 34,100
23) 400,000 km

9 **Additional Applications**

1) C
2) C
3) C
4) B
5) B
6) B
7) B
8) B
9) C
10) A
11) B
12) A
13) D
14) A
15) C
16) B
17) D
18) B
19) B
20) A
21) C
22) B
23) C
24) B
25) C
26) A
27) D
28) B
29) D

30) C
31) $11,000
32) C
33) $132
34) $11.30
35) A
36) A

0.3 Geometry Review

1 Use the Pythagorean Theorem and Its Converse

1) A
2) B
3) A
4) A
5) A
6) B
7) B
8) B
9) B

2 Know Geometry Formulas

1) B
2) B
3) B
4) A
5) D
6) A
7) C
8) D
9) B
10) A
11) B
12) D
13) A
14) B
15) A
16) B
17) C
18) B
19) C
20) $140\pi \approx 439.8$ inches
21) 186 sq ft
22) $99.75\pi \approx 313.4$ sq ft
23) $16 + 2.5\pi \approx 23.9$ ft
24) 16.4 mi

0.4 Polynomials

1 Recognize Monomials

1) C
2) D
3) D
4) B
5) C
6) D
7) D

2 Recognize Polynomials

1) B

2) C
3) C
4) D
5) C
6) B
7) D

3 **Add and Subtract Polynomials**

1) A
2) C
3) D
4) A
5) D
6) C
7) B
8) A

4 **Multiply Polynomials**

1) D
2) D
3) D
4) B
5) A
6) A
7) D
8) A
9) D
10) C
11) B
12) D
13) A
14) B
15) C
16) C

5 **Know Formulas for Special Products**

1) C
2) D
3) A
4) A
5) B
6) B
7) B
8) B
9) C
10) C
11) C
12) A
13) A
14) C
15) C
16) D
17) B
18) A
19) C
20) B

0.5 Factoring Polynomials

1 Factor the Difference of Two Squares and the Sum and the Difference of Two Cubes

1) B
2) C
3) A
4) A
5) C
6) C
7) C

2 Factor Perfect Squares

1) C
2) A
3) A
4) B
5) B

3 Factor a Second-degree Polynomial: $x^2 + Bx + C$

1) B
2) C
3) B
4) D

4 Factor by Grouping

1) D
2) B
3) B
4) C
5) B
6) $(x + 2)(x^2 + 1)$
7) $(2x - 4y + 1)(2x + 4y + 1)$
8) $x^2(x - y + 1)(x + y + 1)$
9) D
10) $2x(2x + 5)$
11) $(x + 5)^2(2x - 1)(10x + 17)$

5 Factor a Second-degree Polynomial: $Ax^2 + Bx + C$

1) B
2) C
3) C
4) A
5) C

6 Additional Applications

1) C
2) D
3) A
4) A
5) C
6) B
7) C
8) A
9) A
10) B
11) A
12) D
13) C
14) D
15) A

16) A
17) B

0.6 Polynomial Division; Synthetic Division

1 Divide Polynomials Using Long Division

1) C
2) D
3) D
4) D
5) A
6) A
7) C
8) C
9) C
10) B
11) C

2 Divide Polynomials Using Synthetic Division

1) C
2) B
3) D
4) $x^2 - x$; remainder 1
5) $x^4 - \frac{1}{2}x^3 - \frac{11}{4}x^2 + \frac{11}{8}x - \frac{11}{16}$; remainder $\frac{75}{32}$
6) B
7) A
8) B
9) B
10) B
11) A

0.7 Rational Expressions

1 Reduce a Rational Expression to Lowest Terms

1) C
2) A
3) B
4) B
5) B
6) A
7) B
8) B
9) D
10) $\frac{x + 3}{x + 7}$
11) C
12) $\frac{4x(x + 5)}{(2x + 5)^2}$

2 Multiply and Divide Rational Expressions

1) C
2) C
3) D
4) C
5) B
6) D
7) B
8) D

9) A
10) D
11) C
12) A
13) B
14) B
15) C
16) D
17) $\frac{z+4}{z+1}$

3 Add and Subtract Rational Expressions

1) D
2) B
3) D
4) D
5) A
6) A
7) $\frac{11x - 25}{x^2 - 25}$

4 Use the Least Common Multiple Method

1) D
2) D
3) A
4) D
5) D
6) B
7) D
8) A
9) C
10) A
11) A
12) C
13) C
14) B
15) A
16) D
17) $\frac{-2x - 7}{(x + 4)(x + 3)(x - 3)}$

18) D
19) C

5 Simplify Mixed Quotients

1) C
2) B
3) B
4) D
5) B
6) $\frac{y^2 + 6y + 12}{2y + 6}$

7) $\frac{x(2x + 3)(3x^2 + 2x - 3)}{(x - 1)(2x - 1)(-x^2 + 4x + 6)}$

8) $f = \frac{R_1 R_2}{(n - 1)(R_1 + R_2)}$

9) $R = \dfrac{R_1 R_2 R_3}{(R_2 R_3 + R_1 R_3 + R_1 R_2)}$

0.8 nth Roots; Rational Exponents

1 Work with nth Roots

1) D
2) B
3) C

2 Simplify Radicals

1) D
2) C
3) D
4) D
5) D
6) C
7) D
8) A
9) B
10) D
11) C
12) D
13) D
14) D
15) B
16) D
17) D
18) B
19) D
20) B
21) D
22) D
23) C
24) C
25) B
26) A
27) B
28) C
29) D
30) D
31) B
32) B
33) B
34) B
35) C
36) C
37) B
38) A
39) C
40) B
41) D
42) B
43) 11,994.67 gal
44) $8\sqrt{5} \approx 17.89$ ft/sec
45) $\pi \approx 3.14$ sec

3 Rationalize Denominators

1) B
2) A
3) C
4) C
5) C
6) A
7) A

4 Simplify Expressions with Rational Exponents

1) D
2) B
3) D
4) B
5) B
6) B
7) B
8) B
9) C
10) C
11) A
12) A
13) D
14) B
15) A
16) C
17) B
18) C
19) C
20) C
21) D
22) A
23) D
24) B
25) B
26) A
27) D
28) D
29) B
30) D

5 Additional Applications

1) B
2) B
3) $\dfrac{(2x + 5)^{1/2}(5x^2 + 5x - 3)}{(x^2 - 1)^{1/2}}$
4) $\dfrac{(2x + 1)^{1/2}(x^2 - x + 6)}{(x^2 + 2)^{3/2}}$
5) C
6) C
7) C
8) A

Ch. 1 Equations and Inequalities

1.1 Linear Equations

1 Solve a Linear Equation

Solve the equation.

1) $2x = -2$

A) {2} B) {-2} C) {1} D) {-1}

2) $4x = 2$

A) $\{-\frac{1}{2}\}$ B) {2} C) $\{\frac{1}{2}\}$ D) {- 2}

3) $4x + 12 = 0$

A) {-4} B) {3} C) {-3} D) {4}

4) $11x - 7 = 0$

A) $\{-\frac{7}{11}\}$ B) $\{-\frac{11}{7}\}$ C) $\{\frac{7}{11}\}$ D) $\{\frac{11}{7}\}$

5) $8x + 5 = 0$

A) $\{\frac{8}{5}\}$ B) $\{-\frac{5}{8}\}$ C) $\{-\frac{8}{5}\}$ D) $\{\frac{5}{8}\}$

6) $\frac{8}{9}x = \frac{1}{7}$

A) $\{\frac{9}{7}\}$ B) $\{-\frac{9}{56}\}$ C) $\{\frac{56}{9}\}$ D) $\{\frac{9}{56}\}$

7) $12x = 1 + 11x$

A) {1} B) {2} C) {-1} D) {12}

8) $13x - 18 = 4x - 45$

A) {5} B) {-3} C) {3} D) {-5}

9) $-3x + 66 = 3x + 18$

A) {11} B) {-8} C) {-11} D) {8}

10) $-30 - 8x = -5 - 3x$

A) {-6} B) {6} C) {-5} D) {5}

11) $5x - (4x - 1) = 2$

A) $\{-\frac{1}{9}\}$ B) $\{\frac{1}{9}\}$ C) {- 1} D) {1}

12) $4(3x - 1) = 16$

A) $\{\frac{5}{4}\}$ B) $\{\frac{17}{12}\}$ C) {1} D) $\{\frac{5}{3}\}$

13) $13(6x - 5) = 3x - 4$

A) $\{\frac{61}{81}\}$ B) $\{\frac{23}{25}\}$ C) $\{-\frac{61}{75}\}$ D) $\{\frac{61}{75}\}$

14) $6(x + 5) = 7[x - (3 - x)]$

A) $\{\frac{15}{4}\}$ B) $\{\frac{51}{8}\}$ C) $\{-\frac{51}{8}\}$ D) $\{-\frac{15}{4}\}$

15) $3(x + 3) = 4(x - 2)$

A) $\{-17\}$ B) $\{1\}$ C) $\{4\}$ D) $\{17\}$

16) $5(2x - 3) = 9(x + 3)$

A) $\{-42\}$ B) $\{17\}$ C) $\{-12\}$ D) $\{42\}$

17) $3(x + 2) = (3x + 6)$

A) $\{12\}$ B) $\{0\}$ C) all real numbers D) no solution

18) $6x - 4 - 6(x + 1) = 3x - 1$

A) $\{\frac{3}{8}\}$ B) $\{-3\}$ C) $\{\frac{1}{10}\}$ D) $\{3\}$

19) $-8x + 1 + 6x = -2x + 6$

A) $\{-1\}$ B) $\{5\}$ C) all real numbers D) no solution

20) $\frac{x}{6} - 3 = 1$

A) $\{24\}$ B) $\{12\}$ C) $\{-24\}$ D) $\{-12\}$

21) $\frac{x}{5} - \frac{1}{5} = -3$

A) $\{16\}$ B) $\{-14\}$ C) $\{14\}$ D) $\{-16\}$

22) $\frac{2x}{5} = 2 + \frac{x}{3}$

A) $\{-60\}$ B) $\{30\}$ C) $\{60\}$ D) $\{-30\}$

23) $\frac{x}{4} + 1 = \frac{x}{5} + 6$

A) $\{-100\}$ B) $\{\frac{1}{4}\}$ C) $\{-\frac{1}{4}\}$ D) $\{100\}$

24) $\frac{1}{4} - \frac{x}{5} = \frac{19}{20}$

A) $\{-\frac{14}{5}\}$ B) $\{-\frac{7}{2}\}$ C) $\{\frac{14}{5}\}$ D) $\{\frac{7}{2}\}$

25) $-3.1x + 1.1 = -16.9 - 1.3x$

A) $\{6.2\}$ B) $\{-20\}$ C) $\{10\}$ D) $\{5.8\}$

26) $\frac{-5x + 9}{6} + \frac{4}{3} = -\frac{5x}{3}$

A) $\{-\frac{17}{5}\}$ B) $\{\frac{17}{15}\}$ C) $\{-\frac{1}{5}\}$ D) $\{\frac{1}{5}\}$

Solve the equation. The letters a, b, and c are constants.

27) $ax - b = c, a \neq 0$

A) $x = \frac{c - b}{a}$ B) $x = \frac{b + c}{a}$ C) $x = -\frac{b + c}{a}$ D) $x = \frac{b - c}{a}$

28) $\frac{x}{a} + \frac{x}{b} = c, a \neq 0, b \neq 0, a \neq -b$

A) $x = \frac{a + b}{abc}$ B) $x = \frac{c}{ab}$ C) $x = abc$ D) $x = \frac{abc}{a + b}$

2 Solve Equations That Lead to Linear Equations

Solve the equation.

1) $1 - \frac{3}{4x} = \frac{3}{2}$

A) $\{-\frac{3}{2}\}$ B) $\{-\frac{5}{2}\}$ C) $\{\frac{3}{2}\}$ D) $\{-\frac{12}{5}\}$

2) $\frac{5}{x} + \frac{2}{5} = \frac{7}{x}$

A) $\{-5\}$ B) $\{2\}$ C) $\{5\}$ D) $\{-2\}$

3) $(x + 6)(x - 1) = (x + 1)^2$

A) $\{\frac{7}{6}\}$ B) $\{2\}$ C) $\{7\}$ D) $\{\frac{7}{3}\}$

4) $x(6x - 1) = (6x + 1)(x - 4)$

A) $\{2\}$ B) $\{-\frac{2}{11}\}$ C) $\{-\frac{4}{23}\}$ D) $\{-\frac{4}{5}\}$

5) $x(1 + 3x) = (3x - 1)(x - 3)$

A) $\{-\frac{3}{121}\}$ B) $\{\frac{3}{11}\}$ C) $\{\frac{3}{121}\}$ D) $\{-\frac{3}{11}\}$

6) $x(x^2 + 2) = 7 + x^3$

A) $\{2\}$ B) $\{\frac{7}{2}\}$ C) $\{7\}$ D) $\{\frac{2}{7}\}$

7) $\frac{3}{x + 5} + \frac{2}{2x + 1} = \frac{4}{x - 2}$

A) $\{\frac{46}{47}\}$ B) $\{-\frac{47}{46}\}$ C) $\{-\frac{46}{47}\}$ D) $\{\frac{47}{46}\}$

8) $\frac{2x}{x^2 - 16} = \frac{2}{x^2 - 16} - \frac{1}{x + 4}$

A) $\{-2\}$ B) $\{\frac{1}{2}\}$ C) $\{-\frac{2}{3}\}$ D) $\{2\}$

9) $\frac{5 - x}{x} + \frac{3}{4} = \frac{7}{x}$

A) $\{-\frac{8}{7}\}$ B) $\{-8\}$ C) $\{-4\}$ D) $\{8\}$

10) $\frac{1}{x} + \frac{1}{x - 6} = \frac{x - 5}{x - 6}$

A) $\{-6\}$ B) $\{-1\}$ C) $\{6\}$ D) $\{1\}$

11) $\frac{6}{2x - 3} = \frac{4}{2x + 5}$

A) $\{\frac{21}{2}\}$ B) $\{-\frac{21}{2}\}$ C) $\{\frac{2}{21}\}$ D) $\{-\frac{2}{21}\}$

12) $\frac{x - 2}{x - 5} = \frac{x + 6}{x - 7}$

A) $\{-\frac{15}{7}\}$ B) $\{\frac{22}{5}\}$ C) $\{\frac{8}{5}\}$ D) $\{-4\}$

13) $\frac{8x + 4}{2x - 7} = \frac{12x - 6}{3x - 3}$

A) $\{-\frac{5}{18}\}$ B) $\{\frac{5}{14}\}$ C) $\{\frac{9}{14}\}$ D) $\{-\frac{1}{2}\}$

14) $\frac{5}{4x} - \frac{1}{x+1} = \frac{1}{x(3x + 3)}$

A) $\{-11\}$ B) $\{-\frac{11}{12}\}$ C) $\{-\frac{11}{3}\}$ D) no solution

15) $\frac{4}{x + 2} - \frac{6}{x - 2} = \frac{4}{x^2 - 4}$

A) $\{12\}$ B) $\{2\sqrt{6}\}$ C) $\{-12\}$ D) $\{24\}$

16) $\frac{6x - 9}{2x - 6} = \frac{9x - 1}{3x + 8}$

A) $\{-\frac{6}{7}\}$ B) $\{\frac{78}{77}\}$ C) $\{-\frac{78}{35}\}$ D) $\{\frac{66}{35}\}$

17) $\frac{7}{x} + \frac{4}{x} = -2$

A) $\{\frac{2}{11}\}$ B) $\{-\frac{11}{2}\}$ C) $\{\frac{11}{2}\}$ D) $\{-\frac{2}{11}\}$

Solve the equation. The letters a, b, and c are constants.

18) $\frac{a}{x} + \frac{b}{x} = c, c \neq 0$

A) $x = \frac{c}{ab}$ B) $x = \frac{a + b}{c}$ C) $x = \frac{c}{a + b}$ D) $x = \frac{ab}{c}$

3 Solve Applied Problems Involving Linear Equations

Solve the formula for the indicated variable.

1) $PV = nRT$ for R

A) $R = \frac{PVT}{n}$ B) $R = \frac{PV}{T}$ C) $R = \frac{nPV}{T}$ D) $R = \frac{PV}{nT}$

2) $S = 2\pi rh + 2\pi r^2$ for h

A) $h = 2\pi(S - r)$ B) $h = \frac{S}{2\pi r} - 1$ C) $h = S - r$ D) $h = \frac{S - 2\pi r^2}{2\pi r}$

3) $F = \frac{9}{5}C + 32$ for C

A) $C = \frac{9}{5}(F - 32)$ B) $C = \frac{5}{9}(F - 32)$ C) $C = \frac{F - 32}{9}$ D) $C = \frac{5}{F - 32}$

4) $A = P(1 + rt)$ for t

A) $t = \frac{A + P}{rP}$ B) $t = \frac{A - P}{rP}$ C) $t = -\frac{A + P}{rP}$ D) $t = \frac{P - A}{rP}$

5) $P - \frac{3Q}{7} = \frac{P + 5}{2} + 1$ for P

A) $P = \frac{49 + 6Q}{7}$ B) $P = \frac{21 - 6Q}{7}$ C) $P = \frac{49 - 6Q}{7}$ D) $P = \frac{21 + 6Q}{7}$

Solve the problem.

6) Mary and her brother John collect foreign coins. Mary has four times the number of coins that John has. Together they have 200 foreign coins. Find how many coins Mary has.

A) 160 coins B) 152 coins C) 40 coins D) 32 coins

7) Center City East Parking Garage has a capacity of 258 cars more than Center City West Parking Garage. If the combined capacity for the two garages is 1236 cars, find the capacity for each garage.

A) Center City East: 747 cars
Center City West: 489 cars

B) Center City East: 757 cars
Center City West: 479 cars

C) Center City East: 479 cars
Center City West: 757 cars

D) Center City East: 489 cars
Center City West: 747 cars

8) During an intramural basketball game, Team A scored 19 fewer points than Team B. Together, both teams scored a total of 147 points. How many points did Team A score during the game?

A) 83 points B) 64 points C) 73 points D) 65 points

9) An auto repair shop charged a customer $289 to repair a car. The bill listed $64 for parts and the remainder for labor. If the cost of labor is $45 per hour, how many hours of labor did it take to repair the car?

A) 5 hr B) 5.5 hr C) 6 hr D) 4 hr

10) Going into the final exam, which will count as three tests, Jerome has test scores of 61, 72, 59, 75, and 77. What score does Jerome need on the final in order to earn a C, which requires an average of 70?

11) After a 16% price reduction, a boat sold for $29,400. What was the boat's price before the reduction? (Round to the nearest cent, if necessary.)

A) $35,000 B) $183,750.00 C) $4704.00 D) $34,104.00

12) Inclusive of a 6.7% sales tax, a diamond ring sold for $1813.90. Find the price of the ring before the tax was added. (Round to the nearest cent, if necessary.)

A) $121.53 B) $1935.43 C) $1700 D) $1692.37

13) It costs $33 per hour plus a flat fee of $17 for a plumber to make a house call. After writing an equation for this situation, suppose the total cost to have a plumber come to a house is $149. How many hours did the plumber work?

A) 3 hr B) 11 hr C) 12 hr D) 4 hr

14) A rectangular carpet has a perimeter of 258 inches. The length of the carpet is 79 inches more than the width. What are the dimensions of the carpet?

A) 77 by 102 in. B) 116.5 by 129 in. C) 104 by 25 in. D) 104 by 129 in.

15) The perimeter of a triangle is 65 centimeters. Find the lengths of its sides, if the longest side is 7 centimeters longer than the shorter side, and the remaining side is 4 centimeters longer than the shorter side.

4 Additional Applications

Solve the problem.

1) x represents the number of textbooks purchased at $53 per book, and y represents the total amount of money spent on textbooks. What is an equation of the form $y = ax$ for this situation?

A) $y = 27x$ B) $y = 159x$ C) $y = 53x$ D) $y = 106x$

2) It costs $18 per hour plus a flat fee of $15 for a plumber to make a house call. What is an equation of the form $y = ax + b$ for this situation?

A) $y = 15x + 18$ B) $y = 15x$ C) $y = 18x$ D) $y = 18x + 15$

3) Using a phone card to make a long distance call costs a flat fee of $0.68 plus $0.33 per minute starting with the first minute. What is an equation of the form $y = ax + b$ for this situation?

A) $y = 0.68x$ B) $y = 0.68x + 0.33$ C) $y = 0.33x$ D) $y = 0.33x + 0.68$

4) x represents the number of textbooks purchased at $27 per book, and y represents the total amount of money spent on textbooks. After writing an equation for this situation, what is the cost of 4 textbooks?

A) $54.00 B) $6.75 C) $108 D) $216

5) x represents the number of cassette tapes sold at $8.75 per tape, and y represents the total cost of the cassette tapes. After writing an equation for this situation, what is the total cost of 20 cassettes?

A) $175 B) $350 C) $0.44 D) $87.50

6) It costs \$26 per hour plus a flat fee of \$31 for a plumber to make a house call. After writing an equation for this situation, what is the total cost to have a plumber come to a house for 10 hours?

A) \$291 B) \$260 C) \$816 D) \$336

1.2 Quadratic Equations

1 Solve a Quadratic Equation by Factoring

Solve the equation by factoring.

1) $15x^2 - 7x = 0$

A) $\{\frac{7}{15}, 0\}$ B) $\{\frac{7}{15}, -\frac{7}{15}\}$ C) $\{-\frac{7}{15}, 0\}$ D) $\{0\}$

2) $14x^2 + 20x = 0$

A) $\{0\}$ B) $\{\frac{10}{7}, -\frac{10}{7}\}$ C) $\{-\frac{10}{7}, 0\}$ D) $\{\frac{10}{7}, 0\}$

3) $x^2 - 36 = 0$

A) $\{-6\}$ B) $\{36\}$ C) $\{6\}$ D) $\{6, -6\}$

4) $x^2 - x = 72$

5) $x^2 - 6x + 8 = 0$

A) $\{2, 4\}$ B) $\{-2, 4\}$ C) $\{-2, -4\}$ D) $\{2, -4\}$

6) $x^2 - 3x - 18 = 0$

A) $\{-3, -6\}$ B) $\{-3, 6\}$ C) $\{3, 6\}$ D) $\{3, -6\}$

7) $5x^2 + 13x - 6 = 0$

A) $\{-\frac{2}{5}, -3\}$ B) $\{\frac{2}{5}, -3\}$ C) $\{-\frac{2}{5}, 3\}$ D) $\{\frac{2}{5}, 3\}$

8) $5x^2 - 35 = 0$

A) $\{-7, 7\}$ B) $\{17.5\}$ C) $\{8\}$ D) $\{-\sqrt{7}, \sqrt{7}\}$

9) $x(x - 6) + 5 = 0$

A) $\{-1, 5\}$ B) $\{1, -5\}$ C) $\{1, 5\}$ D) $\{-1, -5\}$

10) $25x^2 - 40x + 16 = 0$

A) $\{\frac{5}{4}\}$ B) $\{-\frac{4}{5}\}$ C) $\{\frac{4}{5}\}$ D) $\{-\frac{5}{4}\}$

11) $12x^2 - 5x - 25 = 0$

A) $\{-\frac{5}{4}, -\frac{5}{3}\}$ B) $\{-\frac{5}{4}, \frac{5}{3}\}$ C) $\{\frac{5}{4}, \frac{5}{3}\}$ D) $\{\frac{5}{4}, -\frac{5}{3}\}$

12) $7x - 41 = \frac{6}{x}$

A) $\{-7, 6\}$ B) $\{-\frac{1}{7}, 7\}$ C) $\{\frac{1}{41}, -\frac{1}{7}\}$ D) $\{-\frac{1}{7}, 6\}$

13) $24x + \frac{6}{x} = -24$

A) $\{24, 2\}$ B) $\{-2, \frac{1}{2}\}$ C) $\{-\frac{1}{2}, -\frac{1}{2}\}$ D) $\{\frac{1}{2}, \frac{1}{2}\}$

14) $\frac{x-4}{x} = \frac{15}{x+4}$

A) $\{16, -1\}$ B) $\{4, -1\}$ C) $\{16, 1\}$ D) $\{4, 1\}$

2 Know How to Complete the Square

What number should be added to complete the square of the expression?

1) $x^2 + 14x$

A) 7 B) 49 C) 25 D) 98

2) $x^2 - 14x$

A) -7 B) 25 C) 98 D) 49

3) $x^2 + \frac{1}{3}x$

A) $\frac{1}{9}$ B) $\frac{1}{36}$ C) $\frac{1}{18}$ D) $\frac{1}{6}$

4) $x^2 - \frac{3}{4}x$

A) $\frac{9}{64}$ B) $-\frac{1}{4}$ C) $\frac{9}{16}$ D) $-\frac{9}{32}$

Solve the problem.

5) Find the number that should be added to complete the square of the expression, and factor the perfect square trinomial: $x^2 - x$.

6) Find the term that must be added to both sides of the equation so that the equation can be solved by the method of completing the square: $x^2 + 2x = 3$.

A) 2 B) 4 C) 1 D) -3

3 Solve a Quadratic Equation by Completing the Square

Solve the equation by completing the square.

1) $x^2 - 4x + 1 = 0$

2) $x^2 + 8x = 7$

A) $\{-4 - 2\sqrt{23}, -4 + 2\sqrt{23}\}$ B) $\{4 + \sqrt{23}\}$
C) $\{-1 - \sqrt{23}, -1 + \sqrt{23}\}$ D) $\{-4 - \sqrt{23}, -4 + \sqrt{23}\}$

3) $x^2 + 2x - 24 = 0$

A) {-4, 6}
B) {-18, -6}
C) $\{\sqrt{5}, -1\}$
D) {4, -6}

4) $x^2 + 18x + 67 = 0$

A) $\{-9 - \sqrt{14}, -9 + \sqrt{14}\}$
B) $\{-18 + \sqrt{67}\}$
C) $\{9 + \sqrt{14}\}$
D) $\{9 - \sqrt{67}, 9 + \sqrt{67}\}$

5) $x^2 + 5x - 5 = 0$

A) $\{-5 - 3\sqrt{5}, -5 + 3\sqrt{5}\}$
B) $\{\frac{-5 - 3\sqrt{5}}{2}\}$
C) $\{\frac{-5 - 3\sqrt{5}}{2}, \frac{-5 + 3\sqrt{5}}{2}\}$
D) $\{\frac{5 + 3\sqrt{5}}{2}\}$

6) $x^2 + 8x - 7 = 0$

A) $\{-1 - \sqrt{23}, -1 + \sqrt{23}\}$
B) $\{-4 - \sqrt{23}, -4 + \sqrt{23}\}$
C) $\{4 + \sqrt{23}\}$
D) $\{-4 - 2\sqrt{23}, -4 + 2\sqrt{23}\}$

7) $x^2 - 4x - 7 = 0$

A) $\{4 - \sqrt{23}, 4 + \sqrt{23}\}$
B) $\{2 - \sqrt{11}, 2 + \sqrt{11}\}$
C) $\{2 - \sqrt{7}, 2 + \sqrt{7}\}$
D) $\{-2 - \sqrt{11}, -2 + \sqrt{11}\}$

8) $\frac{1}{4}x^2 + \frac{1}{16}x - \frac{1}{8} = 0$

A) $\{\frac{\sqrt{33} - 1}{8}, \frac{\sqrt{33} + 1}{8}\}$
B) $\{\frac{\sqrt{33}}{8}, -\frac{\sqrt{33}}{8}\}$
C) $\{\frac{\sqrt{33} - 1}{8}, -\frac{\sqrt{33} + 1}{8}\}$
D) $\{\frac{1}{8}, -\frac{1}{8}\}$

9) $5x^2 - 2x - 1 = 0$

A) $\{-1, \frac{7}{5}\}$
B) $\{\frac{5 - \sqrt{6}}{25}, \frac{5 + \sqrt{6}}{25}\}$
C) $\{\frac{-1 - \sqrt{6}}{5}, \frac{-1 + \sqrt{6}}{5}\}$
D) $\{\frac{1 - \sqrt{6}}{5}, \frac{1 + \sqrt{6}}{5}\}$

10) $25x^2 + 20x + 3 = 0$

A) $\{-\frac{3}{25}, \frac{6}{25}\}$
B) $\{-\frac{1}{25}, -\frac{3}{25}\}$
C) $\{\frac{1}{5}, \frac{3}{5}\}$
D) $\{-\frac{1}{5}, -\frac{3}{5}\}$

11) $x^2 + \frac{5}{7}x + \frac{4}{49} = 0$

A) $\{\frac{1}{7}, -\frac{4}{7}\}$
B) $\{-\frac{1}{7}, \frac{4}{7}\}$
C) $\{-\frac{1}{7}, -\frac{4}{7}\}$
D) $\{\frac{1}{7}, \frac{4}{7}\}$

4 Solve a Quadratic Equation Using the Quadratic Formula

Find the real solutions, if any, of the equation. Use the quadratic formula.

1) $x^2 + 8x - 7 = 0$

A) $\{4 + \sqrt{23}\}$
B) $\{-1 - \sqrt{23}, -1 + \sqrt{23}\}$
C) $\{-4 - 2\sqrt{23}, -4 + 2\sqrt{23}\}$
D) $\{-4 - \sqrt{23}, -4 + \sqrt{23}\}$

2) $x^2 - 4x - 15 = 0$

A) $\{2 + \sqrt{15}, 2 - \sqrt{15}\}$
B) $\{2 + \sqrt{19}, 2 - \sqrt{19}\}$
C) $\{4 + \sqrt{19}, 4 - \sqrt{19}\}$
D) $\{-2 + \sqrt{19}, -2 - \sqrt{19}\}$

3) $2x^2 + 3x - 9 = 0$

A) $\{\frac{3}{2}, 3\}$
B) $\{-\frac{3}{2}, 3\}$
C) $\{\frac{3}{2}, -3\}$
D) $\{-\frac{3}{2}, -3\}$

4) $8x^2 - x + 5 = 0$

A) $\{\frac{-1 - \sqrt{161}}{16}, \frac{1 + \sqrt{161}}{16}\}$
B) $\{\frac{-1 - \sqrt{161}}{16}, \frac{-1 + \sqrt{161}}{16}\}$
C) $\{\frac{-1 + \sqrt{161}}{16}, \frac{1 + \sqrt{161}}{16}\}$
D) no real solution

5) $x^2 + 7x + 3 = 0$

A) $\{\frac{-7 - \sqrt{61}}{2}, \frac{-7 + \sqrt{61}}{2}\}$
B) $\{\frac{-7 - \sqrt{37}}{14}, \frac{-7 + \sqrt{37}}{14}\}$
C) $\{\frac{7 - \sqrt{37}}{2}, \frac{7 + \sqrt{37}}{2}\}$
D) $\{\frac{-7 - \sqrt{37}}{2}, \frac{-7 + \sqrt{37}}{2}\}$

6) $5x^2 + x - 3 = 0$

A) $\{\frac{-1 - \sqrt{61}}{2}, \frac{-1 + \sqrt{61}}{2}\}$
B) $\{\frac{-1 - \sqrt{61}}{10}, \frac{-1 + \sqrt{61}}{10}\}$
C) $\{\frac{1 - \sqrt{61}}{10}, \frac{1 + \sqrt{61}}{10}\}$
D) no real solution

7) $2x^2 + 10x + 2 = 0$

A) $\{\frac{-10 - \sqrt{21}}{2}, \frac{-10 + \sqrt{21}}{2}\}$
B) $\{\frac{-5 - \sqrt{21}}{2}, \frac{-5 + \sqrt{21}}{2}\}$
C) $\{\frac{-5 - \sqrt{21}}{4}, \frac{-5 + \sqrt{21}}{4}\}$
D) $\{\frac{-5 - \sqrt{29}}{2}, \frac{-5 + \sqrt{29}}{2}\}$

8) $14x = 8x^2$

A) $\{0\}$
B) $\{\frac{7}{4}, -\frac{7}{4}\}$
C) $\{0, \frac{7}{4}\}$
D) $\{0, -\frac{7}{4}\}$

9) $9x^2 + -48x + 64 = 0$

A) $\{-\frac{8}{3}\}$
B) $\{\frac{8}{3}, -24\}$
C) $\{\frac{8}{3}\}$
D) no real solution

10) $36x^2 + 5 = -36x$

A) $\{-\frac{1}{36}, -\frac{5}{36}\}$
B) $\{\frac{1}{6}, \frac{5}{6}\}$
C) $\{-\frac{1}{6}, -\frac{5}{6}\}$
D) $\{-\frac{5}{36}, \frac{5}{18}\}$

11) $6x^2 + 12x = -2$

A) $\{\frac{-12 - \sqrt{6}}{3}, \frac{-12 + \sqrt{6}}{3}\}$

B) $\{\frac{-3 - \sqrt{6}}{3}, \frac{-3 + \sqrt{6}}{3}\}$

C) $\{\frac{-3 - \sqrt{3}}{3}, \frac{-3 + \sqrt{3}}{3}\}$

D) $\{\frac{-3 - \sqrt{6}}{12}, \frac{-3 + \sqrt{6}}{12}\}$

12) $2x^2 = -8x - 1$

A) $\{\frac{-4 - \sqrt{2}}{2}, \frac{-4 + \sqrt{2}}{2}\}$

B) $\{\frac{-8 - \sqrt{14}}{2}, \frac{-8 + \sqrt{14}}{2}\}$

C) $\{\frac{-4 - \sqrt{14}}{4}, \frac{-4 + \sqrt{14}}{4}\}$

D) $\{\frac{-4 - \sqrt{14}}{2}, \frac{-4 + \sqrt{14}}{2}\}$

13) $7x = 1 + \frac{-9}{x}$

A) $\{\frac{1 - \sqrt{253}}{14}\}$

B) $\{\frac{1 - \sqrt{253}}{14}, \frac{1 + \sqrt{253}}{14}\}$

C) $\{-\frac{1 + \sqrt{253}}{14}, \frac{1 - \sqrt{253}}{14}\}$

D) no real solution

Find the real solutions, if any, of the equation. Use the quadratic formula and a calculator. Express any solutions rounded to two decimal places.

14) $x^2 + 2\sqrt{5}x - 20 = 0$

A) {-7.24, 2.76}

B) {-7.24, 7.24}

C) {-7.24, -2.76}

D) {-2.76, 7.24}

Find the real solutions, if any, of the equation. Use the quadratic formula and a calculator. Express any solutions rounded to two decimal places. Use 3.14 to approximate π.

15) $\pi x^2 + \pi x - 8 = 0$

A) {-2.17, -1.17}

B) {-2.17, 1.17}

C) {1.17, 2.17}

D) {-1.17, 2.17}

Use the discriminant to determine whether the quadratic equation has two unequal real solutions, a repeated real solution, or no real solution without solving the equation.

16) $x^2 + 5x + 4 = 0$

A) repeated real solution

B) two unequal real solutions

C) no real solution

17) $x^2 + 12x + 36 = 0$

A) repeated real solution

B) two unequal real solutions

C) no real solution

18) $x^2 + 2x + 8 = 0$

A) repeated real solution

B) two unequal real solutions

C) no real solution

Solve using the quadratic formula. Round any answers to two decimal places.

19) $\frac{1}{4}x^2 - 2\sqrt{3}x = 3$

A) {0.21, -14.67}

B) {0.82, -14.67}

C) {-0.21, 14.67}

D) {-0.82, 14.67}

5 Solve Applied Problems Involving Quadratic Equations

Solve the problem.

1) The length of a vegetable garden is 3 feet longer than its width. If the area of the garden is 154 square feet, find its dimensions.

A) 11 ft by 14 ft B) 10 ft by 13 ft C) 12 ft by 15 ft D) 10 ft by 15 ft

2) The area of a circle is found by the equation $A = \pi r^2$. If the area A of a certain circle is 9π square centimeters, find its radius r.

A) {3 cm, -3 cm} B) $3\sqrt{\pi}$ cm C) 3 cm D) 3π cm

3) A 35-inch-square TV is on sale at the local electronics store. If 35 inches is the measure of the diagonal of the screen, use the Pythagorean theorem to find the length of the side of the screen.

A) $\frac{\sqrt{35}}{2}$ in. B) $\frac{1225}{2}$ in. C) $\sqrt{35}$ in. D) $\frac{35\sqrt{2}}{2}$ in.

4) The surface area A of a right circular cylinder is $A = 2\pi r^2 + 2\pi rh$, where r is the radius and h is the height. Find the radius of a right circular cylinder whose surface area is 95.36π square inches and whose height is 11.7 inches.

5) An open box is to be constructed from a square sheet of plastic by removing a square of side 4 inches from each corner, and then turning up the sides. If the box must have a volume of 1600 cubic inches, find the length of one side of the open box.

A) 20 in. B) 19 in. C) 28 in. D) 24 in.

6) A ball is thrown vertically upward from the top of a building 96 feet tall with an initial velocity of 80 feet per second. The distance s (in feet) of the ball from the ground after t seconds is $s = 96 + 80t - 16t^2$. After how many seconds does the ball strike the ground?

A) 5 sec B) 8 sec C) 6 sec D) 97 sec

7) As part of a physics experiment, Ming drops a baseball from the top of a 305-foot building. To the nearest tenth of a second, for how many seconds will the baseball fall? (Hint: Use the formula $h = 16t^2$, which gives the distance h, in feet, that a free-falling object travels in t seconds.)

A) 1.1 sec B) 19.1 sec C) 76.3 sec D) 4.4 sec

8) The net income y (in millions of dollars) of Pet Products Unlimited from 1997 to 1999 is given by the equation $y = 9x^2 + 15x + 52$, where x represents the number of years after 1997. Assume this trend continues and predict the year in which Pet Products Unlimited's net income will be $916 million.

A) 2006 B) 2005 C) 2008 D) 2007

9) The formula $A = P(1 + r)^2$ is used to find the amount of money, A, in an account after P dollars have been invested in the account paying an annual interest rate, r, for 2 years. Find the interest rate r if $500 grows to $845 in 2 years.

A) 3% B) 69% C) 230% D) 30%

10) A circular pool measures 10 feet across. One cubic yard of concrete is to be used to create a circular border of uniform width around the pool. If the border is to have a depth of 1 inch, how wide will the border be? Use 3.14 to approximate π. Express your solution rounded to two decimal places. (1 cubic yard = 27 cubic feet)

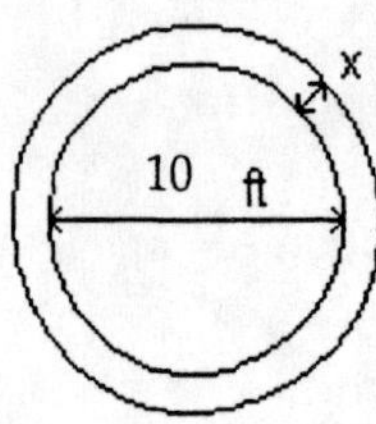

A) 12.91 ft B) 6.32 ft C) 6.96 ft D) 3.78 ft

11) If a polygon, of n sides has $\frac{1}{2}n(n - 3)$ diagonals, how many sides will a polygon with 275 diagonals have?

A) 24 sides B) 25 sides C) 26 sides D) 27 sides

6 Additional Applications

Solve the equation by the Square Root Method.

1) $x^2 = 169$

A) {13, −13} B) {14, −14} C) {84.5} D) {13}

2) $(x - 3)^2 = 16$

A) {7, −1} B) {−1, −7} C) {19} D) {4, −4}

3) $(2x - 1)^2 = 49$

A) {3, −4} B) {8, −6} C) {4, −3} D) {6, −8}

4) $x^2 = 11$

A) $\{\sqrt{11}, -\sqrt{11}\}$ B) $\{\sqrt{11}\}$ C) {11, −11} D) no real solution

5) $(x + 5)^2 = 10$

A) $\{\sqrt{10}, -\sqrt{10}\}$ B) $\{5 + \sqrt{10}, 5 - \sqrt{10}\}$
C) {5} D) $\{-5 + \sqrt{10}, -5 - \sqrt{10}\}$

Find the real solutions, if any, of the quadratic equation.

6) $36x^2 + 72x + 35 = 0$

A) $\{-\frac{7}{6}, \frac{7}{6}\}$ B) $\{-\frac{5}{6}, -\frac{7}{6}\}$ C) $\{\frac{5}{6}, \frac{7}{6}\}$ D) $\{-\frac{5}{36}, -\frac{7}{36}\}$

7) $12 = x + x^2$

A) {4, −3} B) {4, 3} C) {−4, 3} D) no real solution

Solve using the quadratic formula. Round any answers to two decimal places.

8) $2.758x^2 + 5.7 = 3.116x$

A) {0.56 + 1.32i} B) {1.88, −0.76}
C) {0.56 + 1.32i, 0.56 − 1.32i} D) {0.56 + 1.32i, 0.56 − 1.32i, 1.88, −0.76}

Solve the problem.

9) Find a positive value of k such that the equation $x^2 + kx + 9 = 0$ has a repeated real solution.

A) 7 B) 4 C) 5 D) 6

1.3 Quadratic Equations in the Complex Number System

1 Add, Subtract, Multiply, and Divide Complex Numbers

Write the expression in the standard form a + bi.

1) $(8 - 9i) + (2 + 6i)$

A) $-10 + 3i$ B) $6 + 15i$ C) $10 + 3i$ D) $10 - 3i$

2) $(6 + 7i) - (-5 + i)$

A) $11 - 6i$ B) $11 + 6i$ C) $1 + 8i$ D) $-11 - 6i$

3) $9i(5 - 4i)$

A) $36 + 45i$ B) $45i + 36i^2$ C) $45i - 36i^2$ D) $45i - 36$

4) $(7 + 2i)(3 - 9i)$

A) $39 - 57i$ B) $3 + 69i$ C) $-18i^2 - 57i + 21$ D) $39 + 57i$

5) $(4 + 3i)(4 - 3i)$

A) $16 - 9i^2$ B) $16 - 9i$ C) 7 D) 25

6) $\dfrac{4}{3i}$

A) $-\dfrac{4}{3}$ B) $\dfrac{4}{3}i$ C) $-\dfrac{4}{3}i$ D) $\dfrac{4}{3}$

7) $\dfrac{4 - i}{2i}$

A) $\dfrac{1}{2} + 2i$ B) $-\dfrac{1}{2} - 2i$ C) $\dfrac{1}{2} - 2i$ D) $-\dfrac{1}{2} + 2i$

8) $\dfrac{5}{6 + 8i}$

A) $-\dfrac{15}{14} + \dfrac{10}{7}i$ B) $\dfrac{3}{10} - \dfrac{2}{5}i$ C) $\dfrac{3}{10} + \dfrac{2}{5}i$ D) $-\dfrac{15}{14} - \dfrac{10}{7}i$

9) $\dfrac{9}{8 + i}$

A) $\dfrac{72}{65} + \dfrac{9}{65}i$ B) $\dfrac{8}{7} - \dfrac{1}{7}i$ C) $\dfrac{72}{65} - \dfrac{9}{65}i$ D) $\dfrac{8}{7} + \dfrac{1}{7}i$

10) $\dfrac{10 + 24i}{3 + 2i}$

A) $4 - 6i$ B) $4 + 6i$ C) $6 + 4i$ D) $6 - 4i$

11) $\dfrac{9+9i}{7-2i}$

A) $\dfrac{45}{53}+\dfrac{81}{53}i$ B) $\dfrac{81}{53}-\dfrac{45}{53}i$ C) $1-\dfrac{9}{5}i$ D) $\dfrac{9}{5}-\dfrac{9}{5}i$

12) $\left(\dfrac{\sqrt{2}}{2}+\dfrac{\sqrt{2}}{2}i\right)^2$

A) $-i$ B) $\dfrac{i}{2}$ C) $-\dfrac{i}{2}$ D) i

13) $(5+6i)^2$

A) -11 B) $-11+60i$ C) $61+60i$ D) $25+60i+36i^2$

14) i^{28}

A) i B) -1 C) $-i$ D) 1

15) i^{21}

A) i B) -1 C) $-i$ D) 1

16) i^{-19}

A) i B) 1 C) -1 D) $-i$

17) i^{27}

A) i B) 1 C) $-i$ D) -1

18) $2i^{15}-i^7$

A) i B) -1 C) $-i$ D) 1

19) $5i^5(1+i^3)$

A) $5-5i$ B) $-5-5i$ C) $-5+5i$ D) $5+5i$

20) $(1+i)^7$

A) $8+i$ B) $8+8i$ C) $-8-8i$ D) $8-8i$

21) $i^6+i^4+i^2+1$

A) 1 B) 0 C) -1 D) i

Perform the indicated operations and express your answer in the form a + bi.

22) $\sqrt{-4}$

A) $\{2, -2\}$ B) $-2i$ C) $-\sqrt{2}i$ D) $2i$

23) $\sqrt{(5i-12)(12+5i)}$

A) $-13i$ B) $\sqrt{119}i$ C) 13 D) $13i$

24) $\sqrt{(2+i)(i-2)}$

A) $-\sqrt{5}i$ B) $-\sqrt{5}$ C) $\sqrt{5}i$ D) $\sqrt{5}$

2 Solve Quadratic Equations in the Complex Number System

Solve the equation in the complex number system.

1) $x^2 + 100 = 0$

A) $\{-10i, 10i\}$ B) $\{10\}$ C) $\{10i\}$ D) $\{-10, 10\}$

2) $x^2 - 14x + 58 = 0$

A) $\{7 + 3i, 7 - 3i\}$ B) $\{10, 4\}$ C) $\{7 - 9i, 7 + 9i\}$ D) $\{7 + 3i\}$

3) $x^2 + x + 4 = 0$

A) $\{\frac{1}{2} - \frac{\sqrt{15}}{2}i, \frac{1}{2} + \frac{\sqrt{15}}{2}i\}$ B) $\{-\frac{1}{2} - \frac{\sqrt{15}}{2}i, -\frac{1}{2} + \frac{\sqrt{15}}{2}i\}$

C) $\{\frac{-1 - \sqrt{15}}{2}, \frac{-1 + \sqrt{15}}{2}\}$ D) $\{\frac{1 - \sqrt{15}}{2}, \frac{1 + \sqrt{15}}{2}\}$

4) $8x^2 + 1 = 5x$

A) $\{\frac{5}{16} - \frac{\sqrt{7}}{16}i, -\frac{5}{16} + \frac{\sqrt{7}}{16}i\}$ B) $\{-\frac{5}{16} - \frac{\sqrt{7}}{16}i, \frac{5}{16} + \frac{\sqrt{7}}{16}i\}$

C) $\{\frac{5}{16} - \frac{\sqrt{7}}{16}i, \frac{5}{16} + \frac{\sqrt{7}}{16}i\}$ D) $\{-\frac{5}{16} - \frac{\sqrt{7}}{16}i, -\frac{5}{16} + \frac{\sqrt{7}}{16}i\}$

5) $x^3 - 729 = 0$

A) $\{9, -\frac{9}{2} - \frac{9\sqrt{3}}{2}i, -\frac{9}{2} + \frac{9\sqrt{3}}{2}i\}$ B) $\{9\}$

C) $\{9, -\frac{9}{2} - \frac{9\sqrt{3}}{2}, -\frac{9}{2} + \frac{9\sqrt{3}}{2}\}$ D) $\{9, -9i, 9i\}$

6) $x^4 = 16$

A) $\{2\}$ B) $\{-2, 2, -2i, 2i\}$ C) $\{-2, 2\}$ D) $\{-2, 2, 2i\}$

7) $x^4 - 2x^2 - 3 = 0$

A) $\{\sqrt{3}, 3\}$ B) $\{-\sqrt{3}i, -i\}$ C) $\{-\sqrt{3}, \sqrt{3}, i, -i\}$ D) $\{\sqrt{3}i, i\}$

Without solving, determine the character of the solutions of the equation in the complex number system.

8) $x^2 - 6x + 8 = 0$

A) two complex solutions that are conjugates of each other

B) a repeated real solution

C) two unequal real solutions

9) $x^2 - 4x + 4 = 0$

A) two complex solutions that are conjugates of each other

B) two unequal real solutions

C) a repeated real solution

10) $x^2 - 5x + 8 = 0$

A) a repeated real solution

B) two unequal real solutions

C) two complex solutions that are conjugates of each other

11) $x^2 - 3x - 3 = 0$

A) two complex solutions that are conjugates of each other

B) two unequal real solutions

C) a repeated real solution

Solve the problem.

12) 2 - i is a solution of a quadratic equation with real coefficients. Find the other solution.

A) -2 + i   B) 2 + i   C) -2 - i   D) 2 - i

Write the expression in the standard form a + bi.

13) If $z = 3 - 6i$ and $w = 1 + 5i$, evaluate $\overline{\overline{z + w}}$.

1.4 Radical Equations; Equations Quadratic in Form; Factorable Equations

1 Solve Radical Equations

Find the real solutions of the equation.

1) $\sqrt{x + 2} = 9$

A) {121}   B) {79}   C) {81}   D) {83}

2) $\sqrt{10x - 9} = 9$

A) $\{\frac{36}{5}\}$ B) $\{\frac{81}{10}\}$ C) {9}   D) {81}

3) $\sqrt{4x + 5} = 5$

A) $\{\frac{15}{2}\}$ B) $\{\frac{25}{4}\}$ C) {25}   D) {5}

4) $\sqrt[3]{4x - 5} + 7 = 8$

5) $\sqrt[3]{3x + 2} = 2$

A) $\{\frac{2}{3}\}$ B) $\{\frac{8}{3}\}$ C) $\{\frac{5}{3}\}$ D) {2}

6) $\sqrt[5]{1 - x} = -2$

A) {32}   B) {33}   C) {-33}   D) {-32}

7) $\sqrt[3]{1 + x} = -1$

A) {-2}   B) {1}   C) {2}   D) {-1}

8) $x = 7\sqrt{x}$

A) {0, 7}   B) {0, 49}   C) {-49, 49}   D) {-7, 7}

9) $\sqrt{5x + 36} = x$

A) {9} B) {- 9} C) {-4, 9} D) no real solution

10) $\sqrt{24x - 24} = x + 5$

A) {-6} B) {8} C) {7} D) {-7}

11) $\sqrt{x + 7} + \sqrt{x} = 3$

A) {4} B) $\{\frac{1}{9}\}$ C) {1} D) no real solution

12) $\sqrt{x^2 - 3x + 4} = x - 1$

A) {2} B) {3} C) {-3} D) $\{-\frac{3}{2}\}$

13) $\sqrt{x^2 - 2x + 34} = x + 2$

A) {3} B) {-5} C) {5} D) {7}

14) $\sqrt{x^2 + 2} - \sqrt{2x + 5} = 0$

A) {-3, 1} B) {3, -1} C) {3} D) no real solution

15) $\sqrt{2x + 3} - \sqrt{x + 1} = 1$

A) {3, -1} B) {-3, -1} C) {3} D) no real solution

16) $\sqrt{2x + 5} - \sqrt{x - 2} = 3$

A) {2, 38} B) {2} C) {-2} D) {3, 8}

17) $\sqrt{3x + 10} - \sqrt{x + 2} = 2$

A) {-2} B) {2} C) {-2, 2} D) {-3}

18) $\sqrt{2 - 3\sqrt{x}} = 6$

A) $\{\sqrt{-\frac{34}{3}}\}$ B) $\{\sqrt{\frac{34}{3}}\}$ C) $\{\sqrt{\frac{2}{3}}\}$ D) no real solution

19) $(3x - 7)^{1/2} = 4$

A) $\{\frac{11}{3}\}$ B) $\{\frac{22}{3}\}$ C) $\{\frac{23}{3}\}$ D) no real solution

20) $(3x + 2)^{1/2} = 3$

A) {3} B) {6} C) $\{-\frac{2}{3}\}$ D) $\{\frac{7}{3}\}$

21) $\sqrt[3]{4x - 5} + 7 = 8$

22) $(4x + 1)^{1/3} = 3$

A) $\{\frac{27}{4}\}$ B) {2} C) {23} D) $\{\frac{13}{2}\}$

23) $(x + 3)^{1/3} = -2$

A) {-9} B) {-11} C) {1} D) no real solution

24) $(x^2 - 6)^{1/2} = 11$

A) {17} B) $\{\sqrt{127}, -\sqrt{127}\}$ C) $\{\sqrt{17}\}$ D) {127, -127}

25) $x^{3/2} - 8x^{1/2} = 0$

A) {0, 8} B) $\{\sqrt{2}\}$ C) $\{2\sqrt{2}\}$ D) {0, 64}

26) $x^{5/4} - 4x^{1/4} = 0$

A) {-4, 0, 4} B) {0, 2} C) {0} D) {0, 4}

2 Solve Equations Quadratic in Form

Find the real solutions of the equation.

1) $x^4 - 26x^2 + 25 = 0$

A) {-26, 26} B) {-5, 5} C) {-25, 25} D) {-1, 1, -5, 5}

2) $5x^4 + 7x^2 - 6 = 0$

A) $\{-\frac{3}{5}, \frac{3}{5}\}$ B) $\{-\sqrt{\frac{2}{5}}, \sqrt{\frac{2}{5}}\}$ C) $\{-\sqrt{2}, \sqrt{2}\}$ D) $\{-\sqrt{\frac{3}{5}}, \sqrt{\frac{3}{5}}\}$

3) $x^6 - 26x^3 - 27 = 0$

A) {-3, 1} B) {27} C) {3, -1} D) {3}

4) $3(x+1)^2 + 17(x + 1) + 10 = 0$

A) $\{-\frac{2}{3}, -6\}$ B) {- 1, -6} C) $\{\frac{1}{3}, 4\}$ D) {- 1, -5}

5) $(2x - 7)^2 + 2(2x - 7) - 24 = 0$

A) $\{\frac{3}{2}, -\frac{13}{2}\}$ B) $\{-\frac{11}{2}, -\frac{1}{2}\}$ C) $\{\frac{11}{2}, \frac{1}{2}\}$ D) $\{-\frac{3}{7}, \frac{13}{2}\}$

6) $(x - 4)^2 + 3(x - 4) - 18 = 0$

A) {-1, 6} B) {-2, 7} C) {-6, 1} D) {-7, 2}

7) $x + \sqrt{x} = 90$

A) {10} B) {100} C) {81} D) {9}

8) $x + 8x^{1/2} + 7 = 0$

A) {324} B) {14, - 22} C) {196, 484} D) no real solution

9) $x^{1/2} - 9x^{1/4} + 18 = 0$

A) {36, 9} B) {1296, 81} C) {-6, -3} D) {6, 3}

10) $x^2 + 5x - \sqrt{x^2 + 5x} = 42$

A) {5}

B) {25, -25}

C) $\{\frac{-5 + \sqrt{221}}{2}, \frac{-5 - \sqrt{221}}{2}\}$

D) {6}

11) $\frac{1}{(x - 2)^2} - \frac{2}{x - 2} = 3$

A) $\{1, \frac{1}{3}\}$

B) $\{1, \frac{7}{3}\}$

C) $\{-1, \frac{7}{3}\}$

D) $\{-1, \frac{1}{3}\}$

12) $2 + \frac{5}{3x - 1} = \frac{-2}{(3x - 1)^2}$

A) $\{-\frac{1}{3}, 0\}$

B) $\{-\frac{1}{3}, \frac{1}{6}\}$

C) $\{-2, -\frac{1}{2}\}$

D) $\{-\frac{1}{3}, -\frac{1}{6}\}$

13) $4x^{-2} - 19x^{-1} - 5 = 0$

A) $\{-4, \frac{1}{5}\}$

B) $\{4, \frac{1}{5}\}$

C) $\{-\frac{1}{4}, 5\}$

D) $\{-\frac{1}{4}, -5\}$

14) $x^{2/3} - 8x^{1/3} + 15 = 0$

A) {3, 5}

B) {-5, -3}

C) {-125, -27}

D) {27, 125}

15) $x^{2/3} + 2x^{1/3} - 3 = 0$

A) {-1, 3}

B) {-1, 27}

C) {-27, 1}

D) {-3, 1}

Find the real solutions of the equation. Use a calculator to express the solutions rounded to two decimal places.

16) $\pi(1 + x)^2 - 5 = 2(1 + x)$

A) {0.62, -1.98}

B) {-0.62, -0.18}

C) {-0.62, 0.82}

D) {-1.62, -0.62}

17) $x^{2/5} - 3x^{1/5} - 4 = 0$

A) {-1, 1.32}

B) {-1, 4}

C) {-1}

D) {-1, 1024}

Solve the problem.

18) If $k = \frac{x + 3}{x - 1}$ and $k^2 - 7k = 8$, find x.

A) $\{\frac{11}{8}, -1\}$

B) $\{\frac{11}{7}, \frac{1}{2}\}$

C) $\{\frac{8}{7}, -1\}$

D) $\{\frac{11}{7}, -1\}$

3 Solve Equations by Factoring

Find the real solutions of the equation by factoring.

1) $x^3 - 9x = 0$

A) {0, -3}

B) {0, 3}

C) {0, 3, -3}

D) {0, 9}

2) $5x^5 = 405x^3$

A) {-9, 9}

B) $\{-9\sqrt{5}, 0, 9\sqrt{5}\}$

C) {0}

D) {-9, 0, 9}

3) $x^3 + 5x^2 - 4x - 20 = 0$

A) {4, -5}
B) {-2, 2, -5}
C) {2, -5}
D) {-2, 2, 5}

4) $x^3 + 4x^2 + 9x + 36 = 0$

A) {-3, 3, -4}
B) {-4}
C) {4}
D) no real solution

4 Additional Applications

Solve the problem.

1) The square root of the sum of a number and 9 is the same as one half of the sum of the number and 1. Find the number.

2) The quotient of a number and the sum of the number and 2 is equal to the number. Find the number.

3) The formula for the lateral surface area of a right circular cone is $A = \pi r\sqrt{r^2 + h^2}$, where r and h represent the radius and height of the cone, respectively. Find the height of such a cone if its lateral surface area is 15π square inches and its radius is 3 inches.

4) Solve for c in the quadratic formula $x = \dfrac{-b \pm \sqrt{b^2 - 4ac}}{2a}$.

5) Solve for n in the formula $\dfrac{x}{1-n} = kn$.

A) $n = \dfrac{k+1}{x}$
B) $n = \sqrt{\dfrac{x}{k}}$
C) $n = \dfrac{\sqrt{k^2 - 4kx}}{2}$
D) $n = \dfrac{k + \sqrt{k^2 - 4kx}}{2k}$

6) Solve for r in the quadratic formula $P = \dfrac{A}{(1+r)^2}$.

A) $r = \dfrac{P + \sqrt{P^2 + AP}}{2}$
B) $r = \dfrac{-P + \sqrt{AP}}{P}$
C) $r = \dfrac{-P + \sqrt{P^2 - AP}}{P}$
D) $r = \dfrac{P + \sqrt{AP}}{2P}$

7) The length (in centimeters) of one leg of a right triangle is twice the square of the length (in centimeters) of the other leg, and the length of the hypotenuse is $3\sqrt{37}$ centimeters. Let x be the length of one leg; then $2x^2$ is the length of the other leg, and by the Pythagorean Theorem, $(x)^2 + (2x^2)^2 = (3\sqrt{37})^2$. Solve this equation to find the length of the legs of the triangle.

1.5 Solving Inequalities

1 Use Interval Notation

Express the graph shown using interval notation. Also express it as an inequality involving x.

1)

-15 -14 -13 -12 -11 -10 -9 -8 -7 -6 -5 -4 -3 -2 -1 0 1 2

A) (-9, -4]
$-9 < x \leq -4$

B) [-9, -4]
$-9 \leq x \leq -4$

C) [-9, -4)
$-9 \leq x < -4$

D) (-9, -4)
$-9 < x < -4$

2)

-6 -5 -4 -3 -2 -1 0 1 2

A) $(-2, \infty)$ $x > -2$

B) $(-\infty, -2)$ $x < -2$

C) $[-2, \infty)$ $x \geq -2$

D) $(-\infty, -2]$ $x \leq -2$

3)

-11 -10 -9 -8 -7 -6 -5 -4 -3

A) $(-7, \infty)$ $x > -7$

B) $(-\infty, -7)$ $x < -7$

C) $(-\infty, -7]$ $x \leq -7$

D) $[-7, \infty)$ $x \geq -7$

4)

-1 0 1 2 3 4 5 6 7 8 9 10 11 12 13 14

A) $[3, 10)$ $3 \leq x < 10$

B) $[3, 10]$ $3 \leq x \leq 10$

C) $(3, 10]$ $3 < x \leq 10$

D) $(3, 10)$ $3 < x < 10$

5)

-13 -12 -11 -10 -9 -8 -7 -6 -5 -4

A) $(-10, -7]$ $-10 < x \leq -7$

B) $[-10, -7]$ $-10 \leq x \leq -7$

C) $(-10, -7)$ $-10 < x < -7$

D) $[-10, -7)$ $-10 \leq x < -7$

6)

A) $(-\infty, 6)$ $x < 6$

B) $[-7, 6)$ $-7 \leq x < 6$

C) $[-7, 6]$ $-7 \leq x \leq 6$

D) $(-7, 6]$ $-7 < x \leq 6$

Write the inequality using interval notation, and illustrate the inequality using the real number line.

7) $-1 < x < 2$

A) $[-1, 2]$

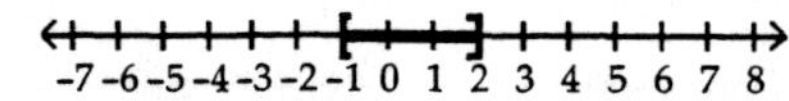

B) $(-1, 2)$

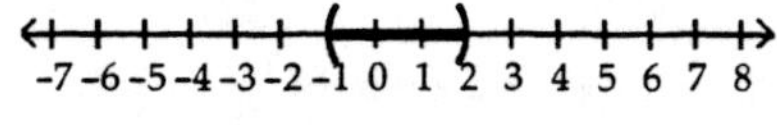

C) $(-1, 2)$

-7 -6 -5 -4 -3 -2 -1 0 1 2 3 4 5 6 7 8

D) $[-1, 2)$

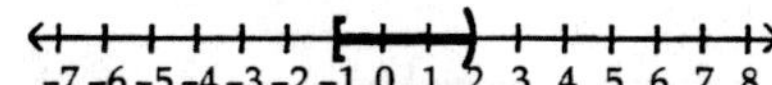

8) $1 \le x \le 2$

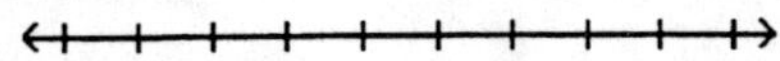

A) (1, 2]

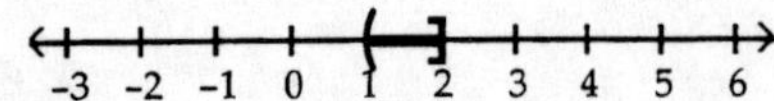

B) [1, 2]

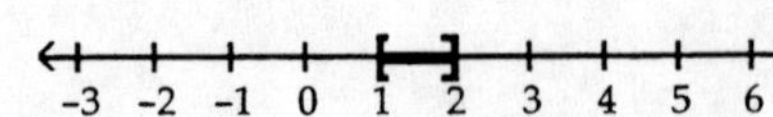

C) [1, 2]

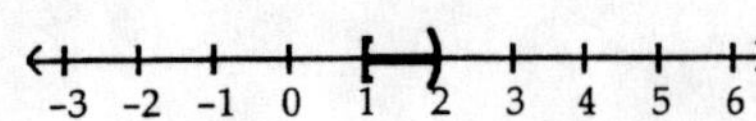

D) (1, 2)

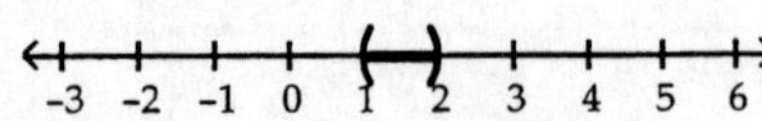

9) $-1 \le x < 4$

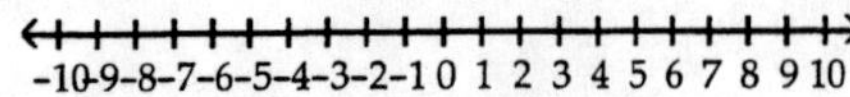

A) (-1, 4]

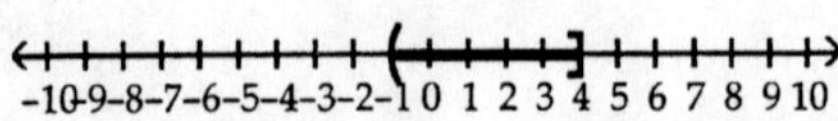

B) [-1, 4)

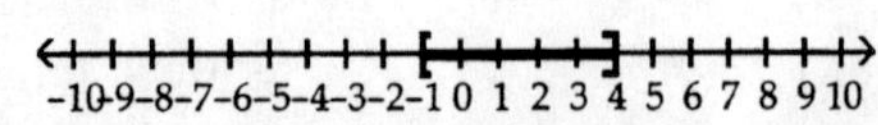

C) $(-\infty, 4)$

D) [-1, 4)

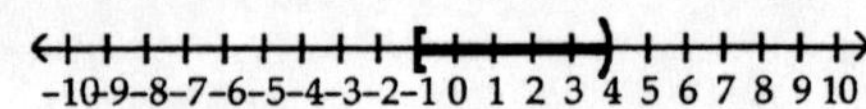

10) $t \ge 1$

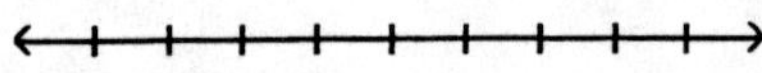

A) $[1, \infty]$

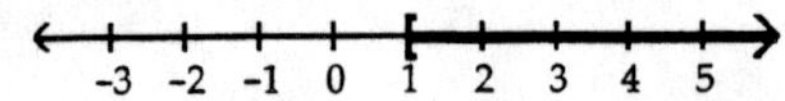

B) $(1, \infty]$

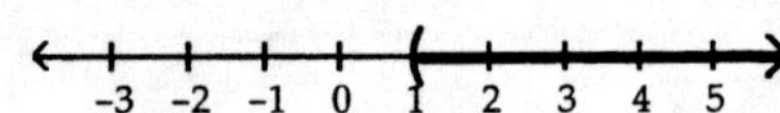

C) $(1, \infty)$

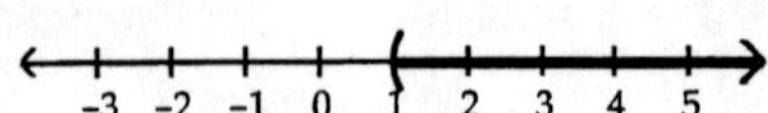

D) $[1, \infty)$

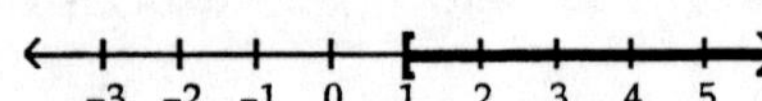

11) $y < -2$

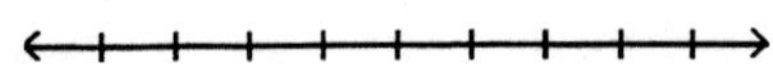

A) $[-\infty, -2]$

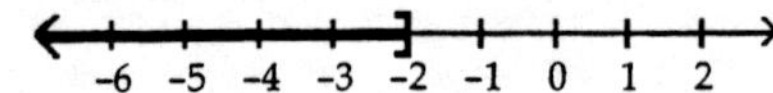

B) $(-\infty, -2]$

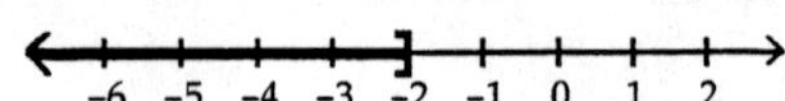

C) $[-\infty, -2)$

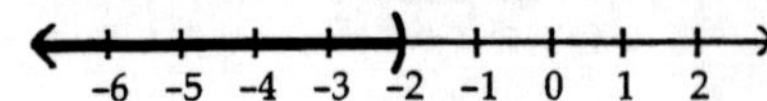

D) $(-\infty, -2)$

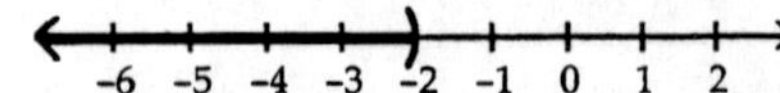

Write the interval as an inequality involving x, and illustrate the inequality using the real number line.

12) [-6, 2)

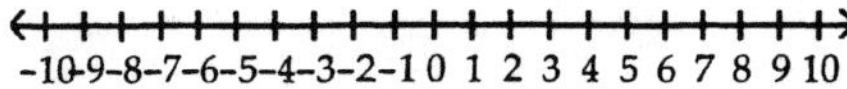

A) $-6 < x \le 2$

B) $x < 2$

C) $-6 \le x < 2$

D) $-6 \le x < 2$

13) [-6, 3]

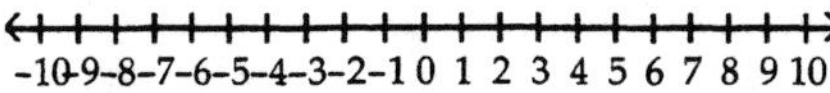

A) $-6 \le x < 3$

B) $-6 < x \le 3$

C) $-6 < x < 3$

D) $-6 \le x \le 3$

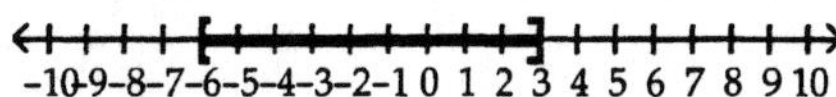

14) (-8, ∞)

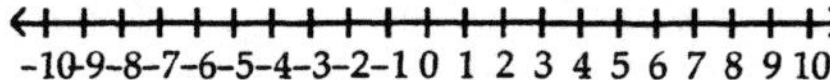

A) $x > -8$

B) $x > -8$

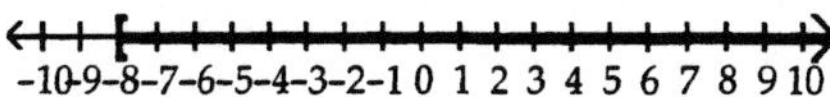

C) $x \ge -8$

D) $x \ge -8$

15) [8, ∞)

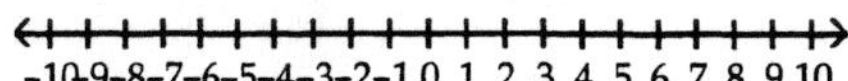

A) $x > 8$

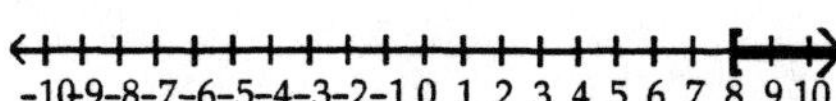

B) $x \ge 8$

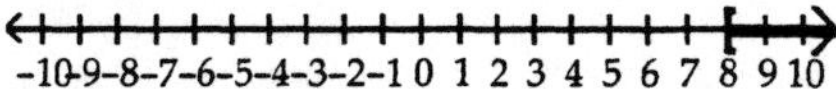

C) $x > 8$

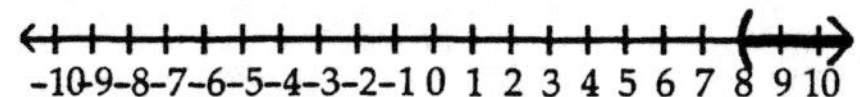

D) $x \ge 8$

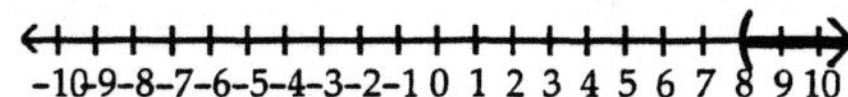

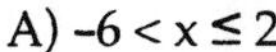

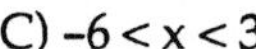

16) $(-\infty, -7)$

A) $x < -7$

B) $x < -7$

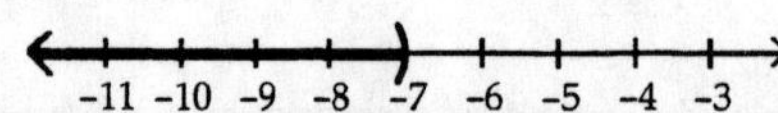

C) $x \leq -7$

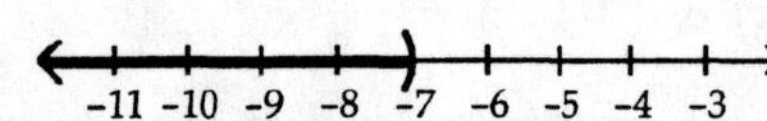

D) $x \leq -7$

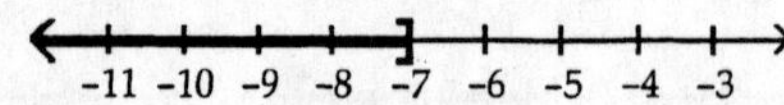

2 Use Properties of Inequalities

Write the inequality obtained by performing the indicated operation on the given inequality.

1) Add 4 to each side of the inequality $1 + 2x > -1$.

A) $5 + 2x > 3$ B) $5 + 6x < 3$ C) $5 + 2x < 3$ D) $5 + 6x > 3$

2) Multiply each side of the inequality $5 + 4x > -5$ by -3.

A) $-15 - 12x > 15$ B) $-15 - 12x < 15$ C) $-15 + 4x > 15$ D) $-15 + 4x < 15$

Solve the problem.

3) Which statement is true if $x > -0.7$?

A) $x + 0.2 < -0.5$ B) $x - 0.2 < -0.9$ C) $x - 0.2 > -0.9$ D) none of these

Fill in the blank with the correct inequality symbol.

4) If $x < 5$, then $x - 5$ ____ 0.

A) $>$ B) $<$ C) $\leq$ D) $\geq$

5) If $x < -11$, then $x + 11$ ____ 0.

A) $<$ B) $\geq$ C) $\leq$ D) $>$

6) If $x > -3$, then $3x$ ____ -9.

A) $<$ B) $>$ C) $\leq$ D) $\geq$

7) If $x < 5$, then $-4x$ ____ -20.

A) $>$ B) $\geq$ C) $<$ D) $\leq$

8) If $x > -8$, then $-3x$ ____ 24.

A) $<$ B) $\leq$ C) $\geq$ D) $>$

9) If $3x > -6$, then x _____ -2.

A) $>$ B) $<$ C) $\leq$ D) $\geq$

Solve the inequality. Express your answer using interval notation. Graph the solution set.

1) $x - 6 < -11$

A) $(-\infty, -17]$

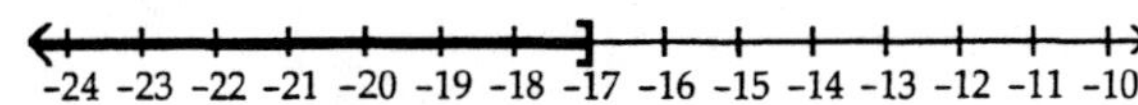

B) $(-\infty, -17)$

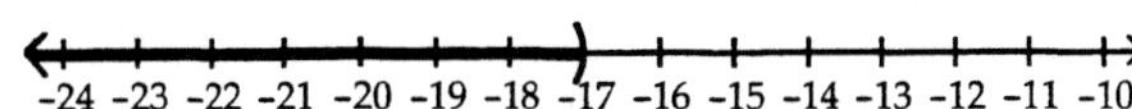

C) $(-\infty, -5)$

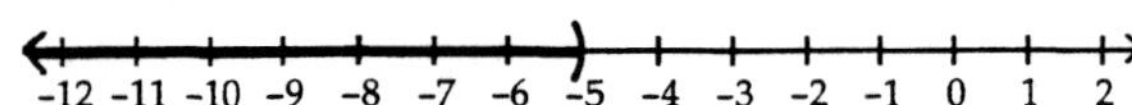

D) $(-5, \infty)$

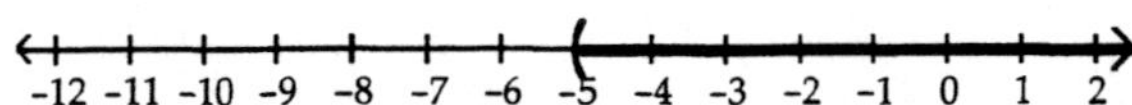

2) $x - 1 < -3$

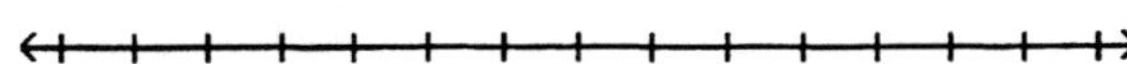

A) $(-2, \infty)$

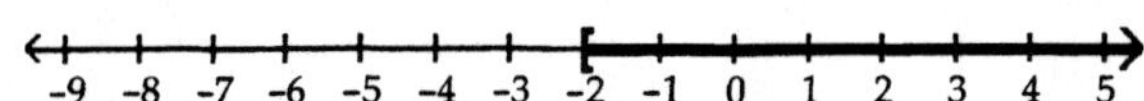

B) $(-\infty, -2)$

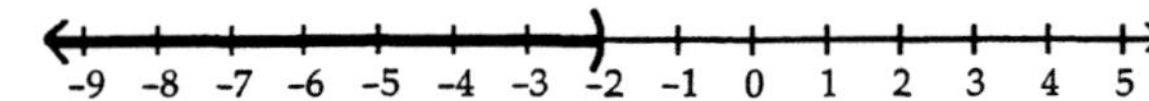

C) $(-\infty, -4)$

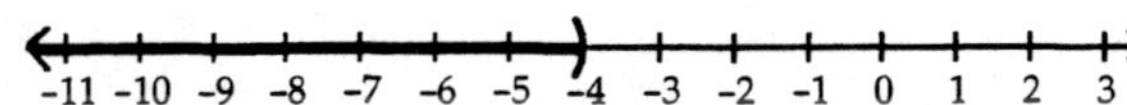

D) $(-\infty, -2]$

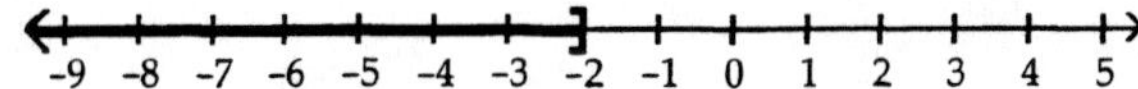

3) $2x + 3 < 17$

-10 -9 -8 -7 -6 -5 -4 -3 -2 -1 0 1 2 3 4 5 6 7 8 9 10

A) $(-\infty, 7]$

-10 -9 -8 -7 -6 -5 -4 -3 -2 -1 0 1 2 3 4 5 6 7 8 9 10

B) $(7, \infty)$

-10 -9 -8 -7 -6 -5 -4 -3 -2 -1 0 1 2 3 4 5 6 7 8 9 10

C) $[7, \infty)$

-10 -9 -8 -7 -6 -5 -4 -3 -2 -1 0 1 2 3 4 5 6 7 8 9 10

D) $(-\infty, 7)$

-10 -9 -8 -7 -6 -5 -4 -3 -2 -1 0 1 2 3 4 5 6 7 8 9 10

4) $5x - 2 > 4x + 5$

A) $(-\infty, 7]$

0 1 2 3 4 5 6 7 8 9 10 11 12 13 14

B) $[7, \infty)$

0 1 2 3 4 5 6 7 8 9 10 11 12 13 14

C) $(3, \infty)$

-4 -3 -2 -1 0 1 2 3 4 5 6 7 8 9 10

D) $(7, \infty)$

0 1 2 3 4 5 6 7 8 9 10 11 12 13 14

5) $5x - 7 \leq 4x - 14$

A) $(-\infty, -7)$

-14 -13 -12 -11 -10 -9 -8 -7 -6 -5 -4 -3 -2 -1 0

B) $(-\infty, -7]$

-14 -13 -12 -11 -10 -9 -8 -7 -6 -5 -4 -3 -2 -1 0

C) $[-7, \infty)$

-14 -13 -12 -11 -10 -9 -8 -7 -6 -5 -4 -3 -2 -1 0

D) $[-21, \infty)$

-28 -27 -26 -25 -24 -23 -22 -21 -20 -19 -18 -17 -16 -15 -14

6) $-4x - 5 \geq -5x - 11$

A) $(-\infty, -6]$

-13 -12 -11 -10 -9 -8 -7 -6 -5 -4 -3 -2 -1 0 1

B) $[-6, \infty)$

-13 -12 -11 -10 -9 -8 -7 -6 -5 -4 -3 -2 -1 0 1

C) $(-16, \infty)$

-23 -22 -21 -20 -19 -18 -17 -16 -15 -14 -13 -12 -11 -10 -9

D) $(-\infty, -6)$

-13 -12 -11 -10 -9 -8 -7 -6 -5 -4 -3 -2 -1 0 1

7) $-6x + 3 > -7x + 6$

A) $[3, \infty)$

-4 -3 -2 -1 0 1 2 3 4 5 6 7 8 9 10

B) $(-\infty, 3]$

-4 -3 -2 -1 0 1 2 3 4 5 6 7 8 9 10

C) $(3, \infty)$

-4 -3 -2 -1 0 1 2 3 4 5 6 7 8 9 10

D) $(9, \infty)$

2 3 4 5 6 7 8 9 10 11 12 13 14 15 16

8) $4 - 2(1 - x) \leq -14$

-10 -8 -6 -4 -2 0 2 4 6 8 10

A) $(-\infty, -8]$

-10 -8 -6 -4 -2 0 2 4 6 8 10

B) $(-\infty, -8)$

-10 -8 -6 -4 -2 0 2 4 6 8 10

C) $(-\infty, -7]$

-10 8 -6 -4 -2 0 2 4 6 8 10

D) $[-8, \infty)$

-10 -8 -6 -4 -2 0 2 4 6 8 10

9) $-16x - 12 \leq -4(3x + 7)$

A) $(-\infty, 4]$

B) $[4, \infty)$

C) $[4, \infty)$

D) $(-\infty, 4)$

10) $-5(6x + 8) < -35x - 25$

A) $(-\infty, 13]$

B) $(-\infty, 3]$

C) $(-\infty, 3)$

D) $(3, \infty)$

11) $\frac{x}{3} \geq 2 + \frac{x}{9}$

A) $[9, \infty)$

B) $(9, \infty)$

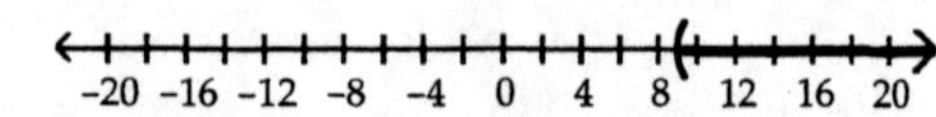

C) $[-9, \infty)$

D) $(-\infty, 9]$

12) $(3x + 5)^{-1} < 0$

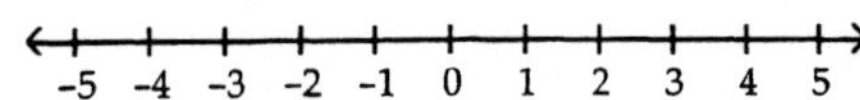

A) $(-\frac{5}{3}, \infty)$

B) $(-\infty, -\frac{5}{3}]$

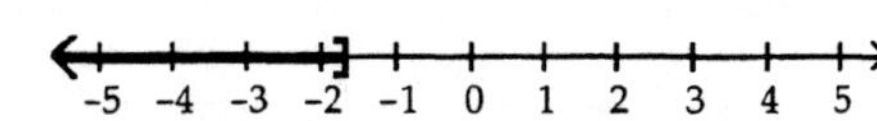

C) $(-\infty, -\frac{5}{3})$

D) $[-\frac{5}{3}, \infty)$

13) $\left(4 - \frac{1}{2}x\right)^{-1/2} > 0$

A) $(-\infty, 8]$

B) $(-\infty, 8)$

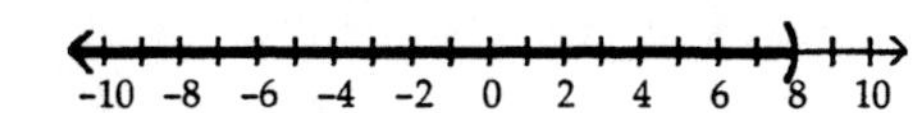

C) $(8, \infty)$

D) $[8, \infty)$

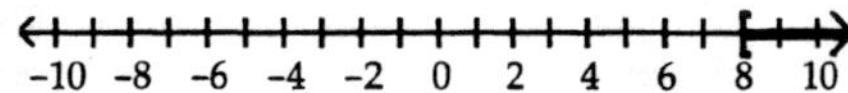

14) $x(25x - 7) \leq (5x + 6)^2$

A) $[-\frac{36}{67}, \infty)$

B) $(-\infty, -\frac{36}{67}]$

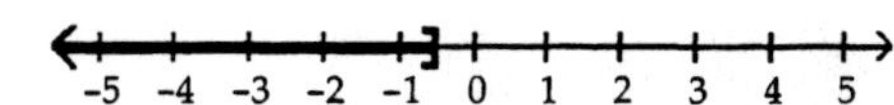

C) $(-\infty, -\frac{36}{67}]$

D) $[\frac{36}{67}, \infty)$

Solve the inequality. Express your answer using interval notation. Graph the solution set.

1) $6 < 3x \leq 18$

A) (2, 6]

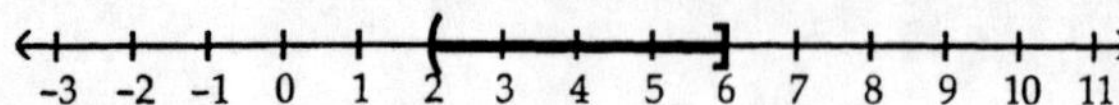

B) [-6, -2)

C) [2, 6)

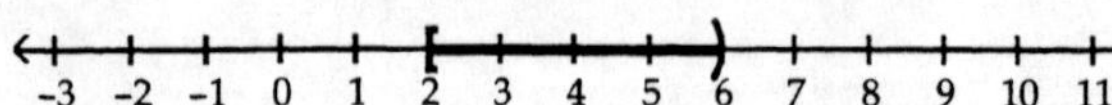

D) (-6, -2]

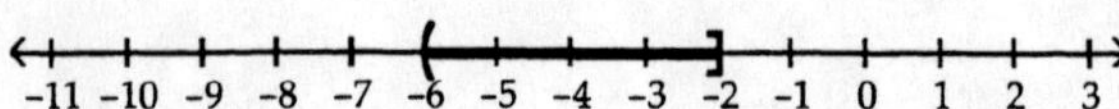

2) $11 \leq 5x + 1 \leq 26$

A) (-5, -2)

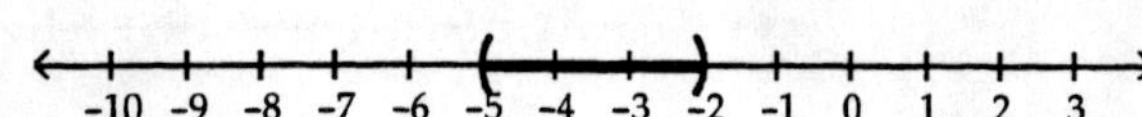

B) (2, 5)

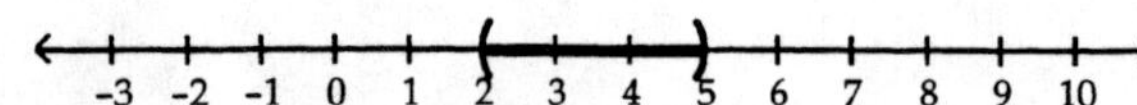

C) [2, 5]

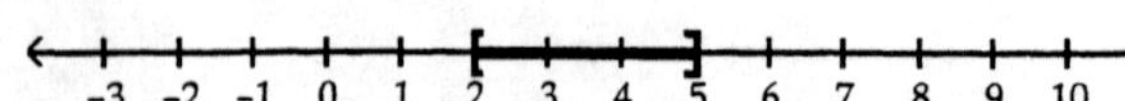

D) [-5, -2]

3) $-27 \leq -4x + 1 < -11$

A) [3, 7)

-2 -1 0 1 2 3 4 5 6 7 8 9 10 11 12

B) [-7, -3)

-12 -11 -10 -9 -8 -7 -6 -5 -4 -3 -2 -1 0 1 2

C) (-7, -3]

-12 -11 -10 -9 -8 -7 -6 -5 -4 -3 -2 -1 0 1 2

D) (3, 7]

-2 -1 0 1 2 3 4 5 6 7 8 9 10 11 12

4) $-23 \leq -3x - 5 \leq -17$

A) (-6, -4)

-12 -11 -10 -9 -8 -7 -6 -5 -4 -3 -2 -1 0 1 2

B) (4, 6)

-2 -1 0 1 2 3 4 5 6 7 8 9 10 11 12

C) [4, 6]

-2 -1 0 1 2 3 4 5 6 7 8 9 10 11 12

D) [-6, -4]

-12 -11 -10 -9 -8 -7 -6 -5 -4 -3 -2 -1 0 1 2

5) $0 \le \frac{3x + 2}{3} < 4$

A) $(-\frac{2}{3}, \frac{10}{3}]$

B) $(-\frac{2}{3}, \frac{10}{3})$

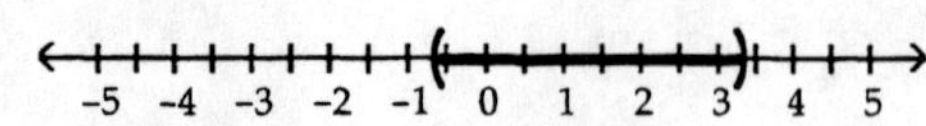

C) $[-\frac{2}{3}, \frac{10}{3})$

D) $[-\frac{2}{3}, \frac{10}{3}]$

6) $11 \le \frac{2}{3}x + 9 < 13$

A) $(3, 4]$

B) $(3, 6]$

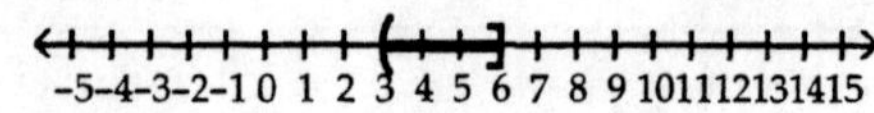

C) $[3, 4)$

D) $[3, 6)$

7) $0 < \frac{1}{x} < \frac{2}{5}$

A) $(\frac{2}{5}, \infty)$

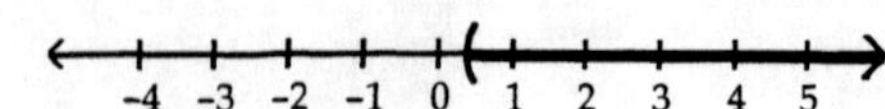

B) $(-\infty, \frac{5}{2})$

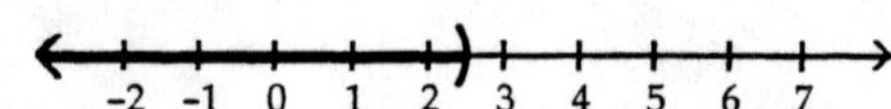

C) $(-\infty, \frac{2}{5})$

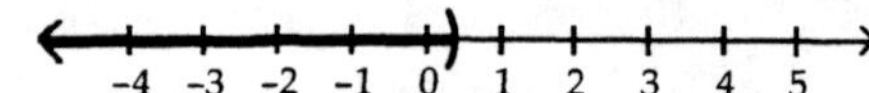

D) $(\frac{5}{2}, \infty)$

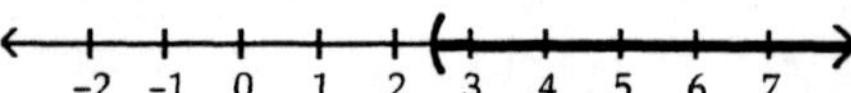

8) $0 < (2x - 8)^{-1} < \frac{1}{2}$

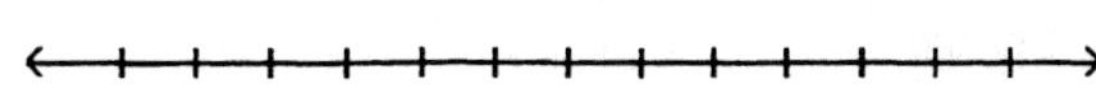

A) $[5, \infty)$

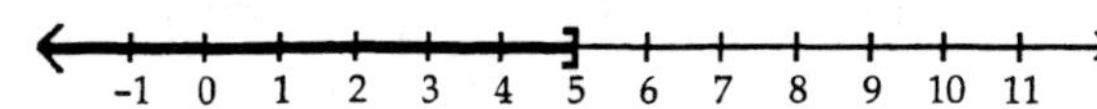

B) $(-\infty, 5]$

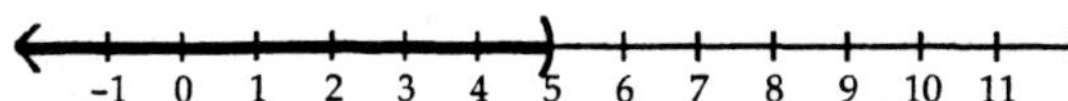

C) $(-\infty, 5)$

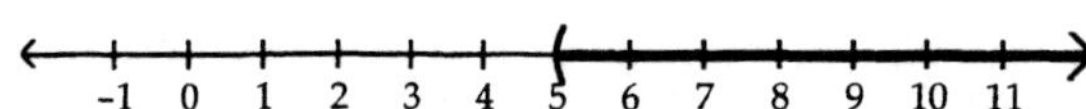

D) $(5, \infty)$

-1 0 1 2 3 4 5 6 7 8 9 10 11

Find a and b.

9) If $-4 < x < -1$, then $a < \frac{2}{3}x < b$.

A) $a = \frac{8}{3}, b = -\frac{2}{3}$ B) $a = \frac{8}{3}, b = \frac{2}{3}$ C) $a = -\frac{8}{3}, b = \frac{2}{3}$ D) $a = -\frac{8}{3}, b = -\frac{2}{3}$

10) If $0 < 2x < 6$, then $a < x^2 < b$.

A) a = 0, b = 36 B) a = 0, b = 16 C) a = 0, b = 9 D) a = 0, b = 3

5 Additional Applications

Solve the problem.

1) What is the domain of the variable in the expression $\sqrt{4x - 12}$?

2) In one city, the local cable TV company charges $1.24 for each pay-per-view movie watched. In addition, each monthly bill contains a basic customer charge of $19.50. If last month's bills ranged from a low of $24.46 to a high of $34.38, over what range did customers watch pay-per-view movies?

A) movies watched varied from 5 to 13 inclusive B) movies watched varied from 3 to 13 inclusive
C) movies watched varied from 4 to 12 inclusive D) movies watched varied from 3 to 11 inclusive

3) During the first five months of the year, Len earned commissions of $3000, $3480, $2720, $3790, and $2800. If Len must have average monthly earnings of at least $3130 in order to qualify for retirement benefits, what must he earn in the sixth month in order to qualify for benefits?

A) at least $2990 B) at least $3153 C) at least $3158 D) at least $3130

4) A real estate agent agrees to sell an office building according to the following commission schedule: $26,000 plus 15% of the selling price in excess of $900,000. Assuming that the office building will sell at some price between $900,000 and $1,100,000, inclusive, over what range does the agent's commission vary?

A) The commission will vary between $161,000 and $191,000, inclusive.

B) The commission will vary between $27,000 and$56,000, inclusive.

C) The commission will vary between $26,000 and $181,000, inclusive.

D) The commission will vary between $26,000 and $56,000, inclusive.

5) Jim has gotten scores of 74 and 90 on his first two tests. What score must he get on his third test to keep an average of 85 or better?

A) at least 82　　B) at least 91　　C) at least 89　　D) at least 83

6) In his algebra class, Rob has scores of 79, 85, 81, and 65 on his first four tests. To get a grade of C, the average of the first five tests must be greater than or equal to 70 and less than 80. Solve an inequality to find the range of scores that Rob can earn on the fifth test to get a C.

7) Marianne is planning a shopping trip to buy birthday gifts for her son. She estimates that the total price of the items she plans to purchase will be between $350 and $400 inclusive. If sales are taxed at a rate of 8.375% in her area, what is the range of the amount of sales tax she should expect to pay on her purchases? If Marianne's budget for the shopping trip is $425, will she necessarily be able to buy all the gifts that she has planned?

8) At Bargain Car Rental, the cost of renting an economy car for one day is $19.95 plus 20 cents per mile. At Best Deal Car Rental, the cost of renting a similar car for one day is $24.95 plus 15 cents per mile. Solve the inequality $24.95 + 0.15x < 19.95 + 0.20x$ to find the range of miles driven such that Best Deal is a better deal than Bargain.

1.6 Equations and Inequalities Involving Absolute Value

1 Solve Equations Involving Absolute Value

Solve the equation.

1) $|x| = 10$

A) $\{-10, 10\}$　　B) $\{-10\}$　　C) $\{100\}$　　D) $\{10\}$

2) $|x| = -3$

A) $\{3, -3\}$　　B) $\{-3\}$　　C) $\{3\}$　　D) no solution

3) $|x| - 10 = -18$

A) $\{8, -8\}$　　B) $\{-8\}$　　C) $\{8\}$　　D) no solution

4) $|x - 2| = 9$

A) $\{-11\}$　　B) $\{-7, 11\}$　　C) $\{7, 11\}$　　D) no solution

5) $|5x + 9| = 7$

A) $\{\frac{2}{5}, \frac{16}{5}\}$　　B) $\{-\frac{2}{9}, -\frac{16}{9}\}$　　C) $\{-\frac{2}{5}, -\frac{16}{5}\}$　　D) no solution

6) $|x + 5| = 0$

A) $\{-5\}$　　B) $\{5\}$　　C) $\{5, -5\}$　　D) no solution

7) $3|x - 3| = 18$

A) {9, -3}　B) {3}　C) {3, -9}　D) no solution

8) $|5x + 8| + 9 = 12$

A) $\{-\frac{5}{8}, -\frac{11}{8}\}$　B) $\{1, \frac{11}{5}\}$　C) $\{-1, -\frac{11}{5}\}$　D) no solution

9) $\left|\frac{11x + 33}{3}\right| = 11$

A) {6, 0}　B) {-6, 6}　C) {-6, 0}　D) no solution

10) $|3(x + 1) + 6| = 12$

A) {-7, 0}　B) {-5, 0}　C) {-5, 3}　D) {-7, 1}

11) $|x^2 - 1x| = 0$

A) {-1, 0, 1}　B) {0, 1}　C) {-1, 0}　D) no solution

12) $|x^2 - 4x - 4| = 8$

A) {-2, 2, -6}　B) {2, 6}　C) {-2, 2}　D) {-2, 2, 6}

13) $|2x^2 - x - 1| = 3$

A) $\{\frac{1 - \sqrt{33}}{4}, -\frac{1 + \sqrt{33}}{4}\}$　B) $\{-\frac{1 - \sqrt{33}}{4}, -\frac{1 + \sqrt{33}}{4}\}$

C) $\{\frac{1 - \sqrt{33}}{4}, \frac{1 + \sqrt{33}}{4}\}$　D) no solution

14) $|x^2 - 4x + 4| = 2$

A) $\{2 - \sqrt{2}\}$　B) $\{2 - \sqrt{2}, 2 + \sqrt{2}\}$　C) $\{2 + \sqrt{2}\}$　D) no solution

2 Solve Inequalities Involving Absolute Value

Solve the inequality. Express your answer using interval notation. Graph the solution set.

1) $|x| < 9$

A) [0, 9]　B) (-∞, -9) and (9, ∞)　C) (-9, 9)　D) (-∞, -9) or (9, ∞)

2) $|x| < -6$

A) (-6, 6)　B) {-6}　C) (-∞, ∞)　D) no solution

3) $|x| > -9$

A) {9}　B) (-9, 9)　C) (-∞, ∞)　D) no solution

4) $|4x - 3| - 9 > -17$

A) $(-\infty, -\frac{5}{4})$ or $(\frac{11}{4}, \infty)$　B) $(-\frac{5}{4}, \frac{11}{4})$　C) (-∞, ∞)　D) no solution

5) $|x + 8| - 4 \le 5$

A) [17, 5]　B) (-∞, -17) or (1, ∞)　C) [-17, 1]　D) no solution

6) $|3x - 7| - 6 < 0$

A) $(-\infty, \frac{1}{3})$ B) $(-\infty, \frac{1}{3})$ or $(\frac{13}{3}, \infty)$ C) $(\frac{1}{3}, \frac{13}{3})$ D) no solution

7) $|5x - 1| \geq 5$

A) $(-\infty, -\frac{4}{5}]$ or $[\frac{6}{5}, \infty)$ B) $[-\frac{4}{5}, \frac{6}{5}]$ C) $(-\frac{4}{5}, \frac{6}{5})$ D) $(-\infty, -\frac{4}{5})$ or $(\frac{6}{5}, \infty)$

8) $|5x + 2| > 3$

A) $(-\infty, -1)$ or $(\frac{1}{5}, \infty)$ B) $(-\infty, -1]$ or $[\frac{1}{5}, \infty)$ C) $(-1, \frac{1}{5})$ D) $[-1, \frac{1}{5}]$

9) $|5 - 7x| > 9$

A) $(-\frac{4}{7}, 2)$ B) $(-\infty, -\frac{4}{7})$ or $(2, \infty)$ C) $(-\infty, -\frac{4}{7})$ or $(\frac{4}{7}, \infty)$ D) $(-\frac{4}{7}, \infty)$ or $(2, \infty)$

10) $|8x + 9| + |-3| \leq 10$

A) $[\frac{1}{4}, 2]$ B) $(-\infty, \frac{1}{4}]$ or $[2, \infty)$ C) $(-\infty, -2]$ or $[-\frac{1}{4}, \infty)$ D) $[-2, -\frac{1}{4}]$

11) $|x + 4| < 0$

A) $(4, \infty)$ B) $(4, -4)$ C) $(-\infty, -4)$ D) no solution

12) $|x + 9| \leq 0$

A) $(-\infty, -9)$ B) $\{-9\}$ C) $\{9\}$ D) no solution

Solve the problem.

13) Express that x differs from -7 by more than 3 as an inequality involving absolute value. Solve for x.

14) A landscaping company sells 40-pound bags of top soil. The actual weight x of a bag, however, may differ from the advertised weight by as much as 0.75 pound. Write an inequality involving absolute value that expresses the relationship between the actual weight x of a bag and 40 pounds. Over what range may the weight of a 40-pound bag of to soil vary?

15) Chi is assigned to construct a triangle with the measure b of the base and the measure h of the height differing by no more than 0.2 centimeters. Express the relationship between b and h as an inequality involving absolute value.

Find a and b.

16) If $|x + 1| < 5$, then $a < x + 3 < b$.

A) $a = -3, b = 7$ B) $a = -8, b = 2$ C) $a = -6, b = 4$ D) $a = -2, b = 8$

17) If $|x + 1| \leq 2$, then $a \leq \frac{1}{x + 5} \leq b$.

A) $a = \frac{1}{6}, b = \frac{1}{8}$ B) $a = \frac{1}{2}, b = \frac{1}{6}$ C) $a = \frac{1}{8}, b = \frac{1}{6}$ D) $a = \frac{1}{6}, b = \frac{1}{2}$

1.7 Applications: Interest, Mixture, Uniform Motion, Constant Rate Jobs

1 Translate Verbal Descriptions into Mathematical Expressions

Translate the sentence into a mathematical equation. Be sure to identify the meaning of all symbols.

1) The surface area of a sphere is 4π times the square of the radius.

A) If S represents the surface area and r the radius, then $4\pi S = r^2$.

B) If S represents the surface area and r the radius, then $S = 4\pi r^2$.

C) If S represents the surface area and r the radius, then $S = \pi r^2$.

D) If S represents the surface area and r the radius, then $S = 4\pi r$.

2) The volume of a right prism is the area of the base times the height of the prism.

A) If V represents the volume, B the area of the base, and h the height, then $V = B + h$.

B) If V represents the volume, B the area of the base, and h the height, then $V = \frac{1}{2}Bh$.

C) If V represents the volume, B the area of the base, and h the height, then $V = Bh$.

D) If V represents the volume, B the area of the base, and h the height, then $V = \frac{B}{h}$.

3) Speed is measured by distance divided by time.

A) If S represents speed, d distance, and t time, then $d = \frac{S}{t}$.

B) If S represents speed, d distance, and t time, then $t = \frac{S}{d}$.

C) If S represents speed, d distance, and t time, then $S = \frac{t}{d}$.

D) If S represents speed, d distance, and t time, then $S = \frac{d}{t}$.

4) Momentum is the product of the mass of an object and its velocity.

A) If M represents momentum, m mass, and v velocity, then $M = mv$.

B) If M represents momentum, m mass, and v velocity, then $M = m + v$.

C) If M represents momentum, m mass, and v velocity, then $M = \frac{1}{2}mv$.

D) If M represents momentum, m mass, and v velocity, then $M = \frac{m}{v}$.

5) The force of gravity between two objects is the gravitational constant times the product of their masses divided by the square of the distance between them.

A) If F is the force of gravity, G the gravitational constant, m_1 the mass of one object, m_2 the mass of the second, and d the distance between them, then $F = G\frac{m_1 m_2}{d^2}$.

B) If F is the force of gravity, G the gravitational constant, m_1 the mass of one object, m_2 the mass of the second, and d the distance between them, then $F = G\frac{m_1 m_2}{d}$.

C) If F is the force of gravity, G the gravitational constant, m_1 the mass of one object, m_2 the mass of the second, and d the distance between them, then $F = G\frac{m_1 + m_2}{d^2}$.

D) If F is the force of gravity, G the gravitational constant, m_1 the mass of one object, m_2 the mass of the second, and d the distance between them, then $FG = \frac{m_1 m_2}{d^2}$.

6) The total cost of producing refrigerators in one production line is $3900 plus $150 per unit produced.

A) If C is the total cost and x is the number of units produced, then $C = \frac{3900}{150x}$.

B) If C is the total cost and x is the number of units produced, then $C = 3900x + 150$.

C) If C is the total cost and x is the number of units produced, then $C = 3900 + 150x$.

D) If C is the total cost and x is the number of units produced, then $C = (3900 + 150)x$.

7) The profit derived from the sale of x video cameras is $380 per unit less the sum of $2700 costs plus $160 per unit.

A) If P is profit and x the units sold, then $P = 380x + 2700 - 160x$ or $P = 220x + 2700$.

B) If P is profit and x the units sold, then $P = 380x - (2700 + 160x)$ or $P = 220x - 2700$.

C) If P is profit and x the units sold, then $P = 380x - (2700 - 160x)$ or $P = 540x - 2700$.

D) If P is profit and x the units sold, then $P = \frac{380}{x} - (2700 + \frac{160}{x})$ or $P = \frac{220}{x} - 2700$.

2 Solve Interest Problems

Solve the problem.

1) Don James wants to invest $54,000 to earn $6230 per year. He can invest in B-rated bonds paying 14% per year or in a Certificate of Deposit (CD) paying 7% per year. How much money should be invested in each to realize exactly $6230 in interest per year?

A) $35,000 in B-rated bonds and $19,000 in a CD
B) $18,000 in B-rated bonds and $36,000 in a CD
C) $19,000 in B-rated bonds and $35,000 in a CD
D) $36,000 in B-rated bonds and $18,000 in a CD

2) A bank loaned out $51,000, part of it at the rate of 13% per year and the rest at a rate of 8% per year. If the interest received was $5780, how much was loaned at 13%?

A) $35,000
B) $16,000
C) $34,000
D) $17,000

3) A loan officer at a bank has $94,000 to lend and is required to obtain an average return of 16% per year. If he can lend at the rate of 17% or the rate of 14%, how much can he lend at the 14% rate and still meet his required return?

A) $1,034,000.00 B) $31,333.33 C) $5529.41 D) $3032.26

4) A college student earned $7400 during summer vacation working as a waiter in a popular restaurant. The student invested part of the money at 10% and the rest at 8%. If the student received a total of $632 in interest at the end of the year, how much was invested at 10%?

A) $5400 B) $925 C) $2000 D) $3700

5) Susan purchased some municipal bonds yielding 7% annually and some certificates of deposit yielding 9% annually. If Susan's investment amounts to $19,000 and the annual income is $1590, how much money is invested in bonds and how much is invested in certificates of deposit?

A) $6000 in bonds; $13,000 in certificates of deposit B) $13,000 in bonds; $6000 in certificates of deposit
C) $13,500 in bonds; $5500 in certificates of deposit D) $5500 in bonds; $13,500 in certificates of deposit

6) Martin purchased some municipal bonds yielding 8% annually and some certificates of deposit yielding 11% annually. If Martin's investment amounts to $23,000 and the annual income is $2230, how much money is invested in bonds and how much is invested in certificates of deposit?

7) Kevin invested part of his $10,000 bonus in a certificate of deposit that paid 6% annual simple interest, and the remainder in a mutual fund that paid 11% annual simple interest. If his total interest for that year was $700, how much did Kevin invest in the mutual fund?

A) $8000 B) $2000 C) $3000 D) $1000

3 Solve Mixture Problems

Solve the problem.

1) The manager of a coffee shop has one type of coffee that sells for $7 per pound and another type that sells for $15 per pound. The manager wishes to mix 90 pounds of the $15 coffee to get a mixture that will sell for $12 per pound. How many pounds of the $7 coffee should be used?

A) 72 lb B) 27 lb C) 144 lb D) 54 lb

2) The owners of a candy store want to sell, for $6 per pound, a mixture of chocolate-covered raisins, which usually sells for $3 per pound, and chocolate-covered macadamia nuts, which usually sells for $8 per pound. They have a 70-pound barrel of the raisins. How many pounds of the nuts should they mix with the barrel of raisins so that they hit their target value of $6 per pound for the mixture?

A) 91 lb B) 98 lb C) 105 lb D) 112 lb

3) How much pure acid should be mixed with 7 gallons of a 50% acid solution in order to get an 80% acid solution?

A) 3.5 gal B) 17.5 gal C) 28 gal D) 10.5 gal

4) A chemist needs 130 milliliters of a 79% solution but has only 29% and 94% solutions available. Find how many milliliters of each that should be mixed to get the desired solution.

A) 90 ml of 29%; 40 ml of 94% B) 40 ml of 29%; 90 ml of 94%
C) 30 ml of 29%; 100 ml of 94% D) 100 ml of 29%; 30 ml of 94%

Solve the problem.

1) A boat heads upstream a distance of 30 miles on the Mississippi river, whose current is running at 5 miles per hour. If the trip back takes an hour less, what was the speed of the boat in still water? Give the answer rounded to two decimal places, if necessary.

A) 18.03 mph B) 15 mph C) 6 mph D) 16.58 mph

2) Two cars start from the same point and travel in the same direction. If one car is traveling 60 miles per hour and the other car is traveling at 52 miles per hour, how far apart will they be after 6.7 hours?

A) 53.6 mi B) 750.4 mi C) 348.4 mi D) 402 mi

3) Two trains leave a train station at the same time. One travels east at 11 miles per hour. The other train travels west at 10 miles per hour. In how many hours will the two trains be 176.4 miles apart?

A) 4.2 hr B) 16.8 hr C) 8.4 hr D) 8.9 hr

4) Ken and Kara are 29 miles apart on a calm lake paddling toward each other. Ken paddles at 4 miles per hour, while Kara paddles at 7 miles per hour. How long will it take them to meet?

A) $2\frac{1}{8}$ hr B) $9\frac{2}{3}$ hr C) $2\frac{7}{11}$ hr D) 18 hr

5) A freight train leaves a station traveling at 32 km/h. Two hours later, a passenger train leaves the same station traveling in the same direction at 52 km/h. How long does it takes the passenger train to catch up to the freight train?

A) 2.2 hr B) 5.2 hr C) 4.2 hr D) 3.2 hr

6) Five friends drove at an average rate of 60 miles per hour to a weekend retreat. On the way home, they took the same route but averaged 65 miles per hour. What was the distance between home and the retreat if the round trip took 10 hours?

A) 312 mi B) $5\frac{1}{5}$ mi C) 624 mi D) 7800 mi

7) Gary can hike on level ground 3 miles an hour faster than he can on uphill terrain. Yesterday, he hiked 31 miles, spending 2 hours on level ground and 5 hours on uphill terrain. Find his average speed on level ground.

A) $6\frac{4}{7}$ mph B) $3\frac{4}{7}$ mph C) 7 mph D) $4\frac{3}{7}$ mph

8) During a hurricane evacuation from the east coast of Georgia, a family traveled 230 miles west. For part of the trip, they averaged 50 mph, but as the congestion got bad, they had to slow to 30 mph. If the total time of travel was 8 hours, how many miles did they drive at the reduced speed?

A) 250 mi B) 255 mi C) 260 mi D) 265 mi

9) Richard works for a company that pays $24 per hour for the first forty hours and $36 per hour for each hour in the week worked above the 40 hours. If he earned $1284 this week, how many overtime hours did he work?

A) 9 hr B) 8 hr C) 49 hr D) 13.5 hr

10) Jamie sells handcrafted dolls at local art fairs. She sells small dolls for $20 and large dolls for $45. At the end of the Little Town Art Fair, she determined that the total amount she made by selling 17 dolls was $490. Determine the number of small and the number of large dolls that she sold.

A) 6 small, 11 large B) 11 small, 6 large C) 10 small, 7 large D) 7 small, 10 large

5 Solve Constant Rate Job Problems

Solve the problem.

1) An experienced bank auditor can check a bank's deposits twice as fast as a new auditor. Working together it takes the auditors 22 hours to do the job. How long would it take the experienced auditor working alone?

A) 44 hr B) 66 hr C) 22 hr D) 33 hr

2) BJ can overhaul a boat's diesel inboard engine in 20 hours. His apprentice takes 60 hours to do the same job. How long would it take them working together assuming no gain or loss in efficiency?

A) 15 hr B) 6 hr C) 12 hr D) 80 hr

3) Tracy can wallpaper 2 rooms in a new house in 10 hours. Together with her trainee they can wallpaper the 2 rooms in 7 hours. How long would it take the trainee working by herself to do the job?

A) 10 hr B) 30 hr C) 40 hr D) 20 hr

4) Brandon can paint a fence in 12 hours and Elaine can paint the same fence in 11 hours. How long will they take to paint the fence if they work together?

A) $5\frac{17}{23}$ hr B) $5\frac{13}{24}$ hr C) $5\frac{3}{4}$ hr D) $11\frac{1}{2}$ hr

5) Sue can sew a precut dress in 3 hours. Helen can sew the same dress in 2 hours. If they work together, how long will it take them to complete sewing that dress? Give your answer rounded to one decimal place, if necessary.

A) 2.5 hr B) 1.8 hr C) 5 hr D) 1.2 hr

6) Two pumps can fill a water tank in 45 minutes when working together. Alone, the second pump takes 3 times longer than the first to fill the tank. How long does it take the first pump alone to fill the tank?

Ch. 1 Equations and Inequalities
Answer Key

1.1 Linear Equations
1 Solve a Linear Equation

1) D
2) C
3) C
4) C
5) B
6) D
7) A
8) B
9) D
10) C
11) D
12) D
13) D
14) B
15) D
16) D
17) C
18) B
19) D
20) A
21) B
22) B
23) D
24) B
25) C
26) A
27) B
28) D

2 Solve Equations That Lead to Linear Equations

1) A
2) C
3) D
4) B
5) B
6) B
7) C
8) D
9) B
10) D
11) B
12) B
13) C
14) C
15) C
16) B
17) B
18) B

3 Solve Applied Problems Involving Linear Equations

1) D
2) D

3) B
4) B
5) A
6) A
7) A
8) B
9) A
10) 72
11) A
12) C
13) D
14) C
15) 18 cm, 22 cm, 25 cm

4 Additional Applications

1) C
2) D
3) D
4) C
5) A
6) A

1.2 Quadratic Equations

1 Solve a Quadratic Equation by Factoring

1) A
2) C
3) D
4) {-8, 9}
5) A
6) B
7) B
8) D
9) C
10) C
11) B
12) D
13) C
14) A

2 Know How to Complete the Square

1) B
2) D
3) B
4) A
5) $x^2 - x + \frac{1}{4} = (x - \frac{1}{2})^2$
6) C

3 Solve a Quadratic Equation by Completing the Square

1) $\{2 + \sqrt{3}, 2 - \sqrt{3}\}$
2) D
3) D
4) A
5) C
6) B
7) B
8) C
9) D

10) D
11) C

4 Solve a Quadratic Equation Using the Quadratic Formula

1) D
2) B
3) C
4) D
5) D
6) B
7) B
8) C
9) C
10) C
11) B
12) D
13) D
14) A
15) B
16) B
17) A
18) C
19) D

5 Solve Applied Problems Involving Quadratic Equations

1) A
2) C
3) D
4) 3.2 in.
5) A
6) C
7) D
8) A
9) D
10) B
11) B

6 Additional Applications

1) A
2) A
3) C
4) A
5) D
6) B
7) C
8) C
9) D

1.3 Quadratic Equations in the Complex Number System

1 Add, Subtract, Multiply, and Divide Complex Numbers

1) D
2) B
3) A
4) A
5) D
6) C
7) B
8) B
9) C

10) C
11) A
12) D
13) B
14) D
15) A
16) A
17) C
18) C
19) D
20) D
21) B
22) D
23) D
24) C

2 Solve Quadratic Equations in the Complex Number System

1) A
2) A
3) B
4) C
5) A
6) B
7) C
8) C
9) C
10) C
11) B
12) B
13) 4 - i

1.4 Radical Equations; Equations Quadratic in Form; Factorable Equations

1 Solve Radical Equations

1) B
2) C
3) D
4) $\{\frac{3}{2}\}$
5) D
6) B
7) A
8) B
9) A
10) C
11) B
12) B
13) C
14) B
15) A
16) A
17) C
18) D
19) C
20) D
21) $\{\frac{3}{2}\}$

22) D
23) B
24) B
25) A
26) D

2 Solve Equations Quadratic in Form

1) D
2) D
3) C
4) B
5) C
6) B
7) C
8) D
9) B
10) C
11) B
12) B
13) A
14) D
15) C
16) A
17) D
18) D

3 Solve Equations by Factoring

1) C
2) D
3) B
4) B

4 Additional Applications

1) 7
2) -1 or 0
3) 4 in.
4) $c = -ax^2 - bx$
5) D
6) B
7) 3 cm and 18 cm

1.5 Solving Inequalities

1 Use Interval Notation

1) C
2) D
3) A
4) D
5) B
6) B
7) B
8) B
9) D
10) D
11) D
12) D
13) D
14) A
15) B

16) B

2 Use Properties of Inequalities

1) A
2) B
3) C
4) B
5) A
6) B
7) A
8) A
9) A

3 Solve Inequalities

1) C
2) B
3) D
4) D
5) B
6) B
7) C
8) A
9) B
10) C
11) A
12) C
13) B
14) A

4 Solve Combined Inequalities

1) A
2) C
3) D
4) C
5) C
6) D
7) D
8) D
9) D
10) C

5 Additional Applications

1) $x \geq 3$
2) C
3) A
4) D
5) B
6) $40 \leq x < 90$, where x represents Bob's score on the fifth test
7) $\$29.31 \leq x \leq \$33.50$, where x represents the amount of sales tax; no
8) $x > 100$ mi

1.6 Equations and Inequalities Involving Absolute Value

1 Solve Equations Involving Absolute Value

1) A
2) D
3) D
4) B
5) C
6) A
7) A

8) C
9) C
10) D
11) B
12) D
13) C
14) B

2 Solve Inequalities Involving Absolute Value

1) C
2) D
3) C
4) C
5) C
6) C
7) A
8) A
9) B
10) D
11) D
12) B
13) $|x+7| > 3$; $x < -10$ or $x > -4$
14) $|x-40| \le 0.75$; $39.25 \le x \le 40.75$
15) $|b-h| \le 0.2$
16) A
17) D

1.7 Applications: Interest, Mixture, Uniform Motion, Constant Rate Jobs

1 Translate Verbal Descriptions into Mathematical Expressions

1) B
2) C
3) D
4) A
5) A
6) C
7) B

2 Solve Interest Problems

1) A
2) C
3) B
4) C
5) A
6) $10,000 in bonds; $13,000 in certificates of deposit
7) B

3 Solve Mixture Problems

1) D
2) C
3) D
4) C

4 Solve Uniform Motion Problems

1) A
2) A
3) C
4) C
5) D
6) A
7) A

8) B
9) A
10) B

5 Solve Constant Rate Job Problems

1) D
2) A
3) D
4) A
5) D
6) 60 min

Ch. 2 Graphs

2.1 Rectangular Coordinates

1 Use the Distance Formula

Find the distance $d(P_1, P_2)$ between the points P_1 and P_2.

1) $P_1 = (3, 2)$; $P_2 = (-2, -2)$

A) 20 B) $\sqrt{41}$ C) 3 D) 1

2) $P_1 = (4, -2)$; $P_2 = (6, -6)$

A) $12\sqrt{3}$ B) 6 C) $2\sqrt{5}$ D) 12

3) $P_1 = (-6, -5)$; $P_2 = (6, 1)$

A) $108\sqrt{3}$ B) 6 C) 108 D) $6\sqrt{5}$

4) $P_1 = (-3, 2)$; $P_2 = (-7, 5)$

A) 10 B) 5 C) 6 D) 25

5) $P_1 = (0, 0)$; $P_2 = (-3, 8)$

A) $2\sqrt{2}$ B) $\sqrt{73}$ C) 5 D) 73

6) $P_1 = (2, 2)$; $P_2 = (2, -5)$

A) 8 B) 7 C) $\sqrt{7}$ D) 6

7) $P_1 = (-0.8, -0.7)$; $P_2 = (-2.8, -1.3)$ Round to three decimal places, if necessary.

A) 2.088 B) 6.603 C) 13 D) 2.188

Decide whether or not the points are the vertices of a right triangle.

8) (3, 9), (10, 9), (10, 13)

A) No B) Yes

9) (-6, 2), (-4, 6), (-2, 5)

A) No B) Yes

10) (7, -3), (13, -1), (12, -6)

A) No B) Yes

11) (1, 10), (7, 12), (13, 5)

A) No B) Yes

Solve the problem.

12) Find all values of k so that the given points are $\sqrt{29}$ units apart.
(-5, 5), (k, 0)

A) 3, 7 B) -3, -7 C) -7 D) 7

13) Find the area of the right triangle ABC with A = (-2, 7), B = (7, -1), C = (3, 9).

A) $\frac{\sqrt{58}}{2}$ square units B) 29 square units C) 58 square units D) $\frac{\sqrt{29}}{2}$ square units

14) Find the length of each side of the triangle determined by the three points P_1, P_2, and P_3. State whether the triangle is an isosceles triangle, a right triangle, neither of these, or both.
$P_1 = (-5, -4)$, $P_2 = (-3, 4)$, $P_3 = (0, -1)$

A) $d(P_1, P_2) = 2\sqrt{17}$; $d(P_2, P_3) = \sqrt{34}$; $d(P_1, P_3) = 5\sqrt{2}$
right triangle

B) $d(P_1, P_2) = 2\sqrt{17}$; $d(P_2, P_3) = \sqrt{34}$; $d(P_1, P_3) = \sqrt{34}$
both

C) $d(P_1, P_2) = 2\sqrt{17}$; $d(P_2, P_3) = \sqrt{34}$; $d(P_1, P_3) = \sqrt{34}$
isosceles triangle

D) $d(P_1, P_2) = 2\sqrt{17}$; $d(P_2, P_3) = \sqrt{34}$; $d(P_1, P_3) = 5\sqrt{2}$
neither

15) Find all the points having an x-coordinate of 9 whose distance from the point (3, -2) is 10.

A) (9, -12), (9, 8) B) (9, 2), (9, -4) C) (9, 13), (9, -7) D) (9, 6), (9, -10)

16) Find the length of the line segment. Assume that the endpoints of the line segment have integer coordinates. If necessary, round to two decimal places.

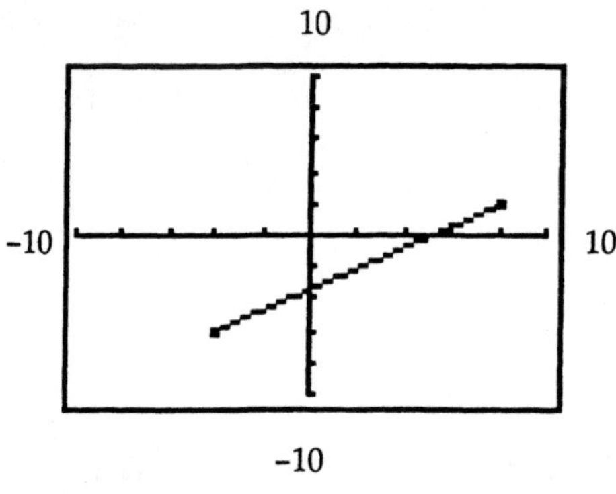

A) 14.42 B) 21.63 C) 13.41 D) 8.94

17) Find the length of the line segment.

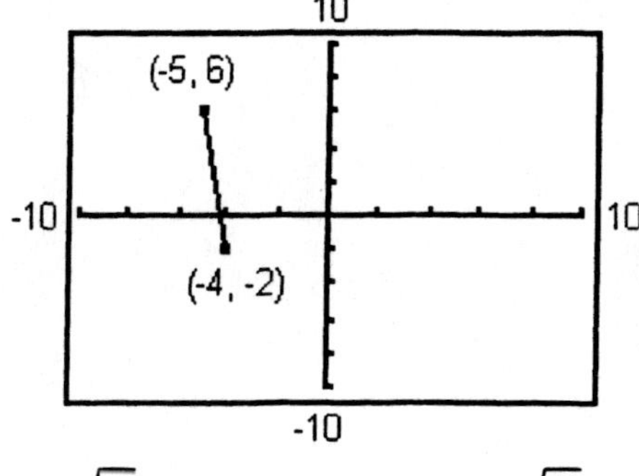

A) $\sqrt{65}$ B) $\sqrt{97}$ C) 8 D) $\sqrt{63}$

18) A middle school's baseball playing field is a square, 65 feet on a side. How far is it directly from home plate to second base (the diagonal of the square)? If necessary, round to the nearest foot.

A) 99 feet B) 92 feet C) 91 feet D) 93 feet

19) A motorcycle and a car leave an intersection at the same time. The motorcycle heads north at an average speed of 20 miles per hour, while the car heads east at an average speed of 48 miles per hour. Find an expression for their distance apart in miles at the end of t hours.

A) $52\sqrt{t}$ miles B) $2t\sqrt{13}$ miles C) 52t miles D) $t\sqrt{68}$ miles

20) A rectangular city park has a jogging loop that goes along a length, width, and diagonal of the park. To the nearest yard, find the length of the jogging loop, if the length of the park is 125 yards and its width is 75 yards.

A) 346 yards B) 146 yards C) 145 yards D) 345 yards

2 Use the Midpoint Formula

Find the midpoint of the line segment joining the points P_1 and P_2.

1) $P_1 = (6, 3)$; $P_2 = (8, 5)$

A) (4, 7) B) (-2, -2) C) (7, 4) D) (14, 8)

2) $P_1 = (-6, 8)$; $P_2 = (-5, 4)$

A) $(-\frac{11}{2}, 6)$ B) (-11, 12) C) $(-\frac{1}{2}, 2)$ D) (-1, 4)

3) $P_1 = (7, 1)$; $P_2 = (-16, -16)$

A) (-9, -15) B) $(\frac{23}{2}, \frac{17}{2})$ C) $(-\frac{9}{2}, -\frac{15}{2})$ D) (9, 15)

4) $P_1 = (0.8, 0.1)$; $P_2 = (1.6, -1.9)$

A) (1.2, -0.9) B) (0.4, -1) C) (-0.9, 1.2) D) (-1, 0.4)

5) $P_1 = (y, 5)$; $P_2 = (0, 6)$

A) $(\frac{y}{2}, \frac{11}{2})$ B) (y, 11) C) $(y, \frac{11}{2})$ D) $(-\frac{y}{2}, 1)$

6) $P_1 = (9b, 4)$; $P_2 = (10b, 5)$

A) (19b, 9) B) (b, 1) C) $(\frac{9b}{2}, \frac{19}{2})$ D) $(\frac{19b}{2}, \frac{9}{2})$

Solve the problem.

7) Find the midpoint of the line segment shown on the graphing utility.

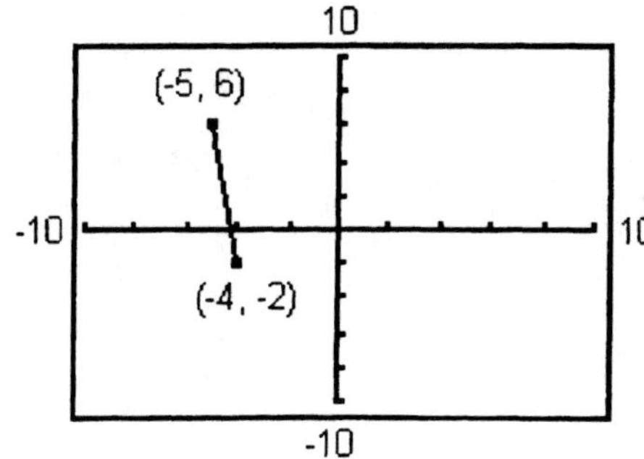

A) (−1, 8) B) $(-\frac{9}{2}, 2)$ C) (−9, 4) D) $(\frac{9}{2}, 2)$

8) The medians of a triangle intersect at a point. The distance from the vertex to the point is exactly two–thirds of the distance from the vertex to the midpoint of the opposite side. Find the exact distance of that point from the vertex A(3, 4) of a triangle, given that the other two vertices are at (0, 0) and (8, 0).

A) $\frac{\sqrt{17}}{3}$ B) 2 C) $\frac{2\sqrt{17}}{3}$ D) $\frac{8}{3}$

9) If (−4, −3) is the endpoint of a line segment, and (−1, −2) is its midpoint, find the other endpoint.

A) (−2, 3) B) (2, −1) C) (−10, −5) D) (2, −4)

10) If (2, 2) is the endpoint of a line segment, and (−3, −1) is its midpoint, find the other endpoint.

A) (−4, −8) B) (−8, −4) C) (−8, 5) D) (12, 8)

11) If (9, 4) is the endpoint of a line segment, and (10, 0) is its midpoint, find the other endpoint.

A) (11, 8) B) (11, −4) C) (1, 6) D) (7, 12)

12) If (−2, −7) is the endpoint of a line segment, and (−7, −5) is its midpoint, find the other endpoint.

A) (−12, −3) B) (8, −11) C) (−12, −9) D) (2, −17)

3 Additional Applications

Name the quadrant in which the point is located.

1) (14, 2)

A) I B) II C) III D) IV

2) (−16, 5)

A) I B) II C) III D) IV

3) (−17, −18)

A) I B) II C) III D) IV

4) (15, −20)

A) I B) II C) III D) IV

Identify the points in the graph for the ordered pairs.

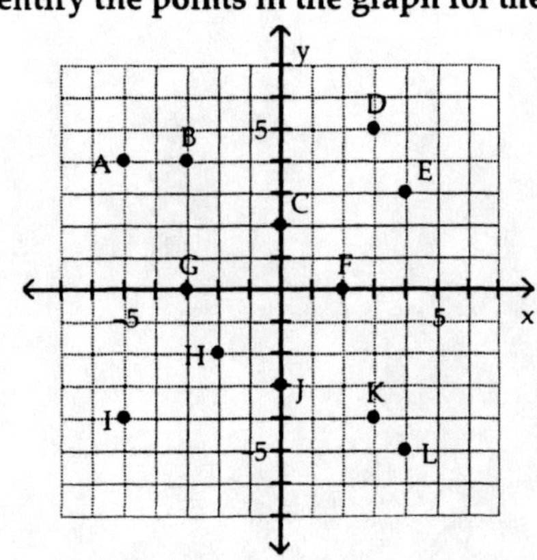

5) (0, 2), (4, 3)

A) B and C B) C and E C) C and K D) F and E

6) (-5, -4), (0, -3)

A) A and G B) A and J C) I and J D) G and I

7) (-3, 4), (2, 0), (4, -5)

A) A, B, and F B) B, C, and L C) F, K, and L D) B, F, and L

8) (3, 5), (-3, 0)

A) L and J B) I and G C) D and G D) D and J

Give the coordinates of the points shown on the graph.

9)

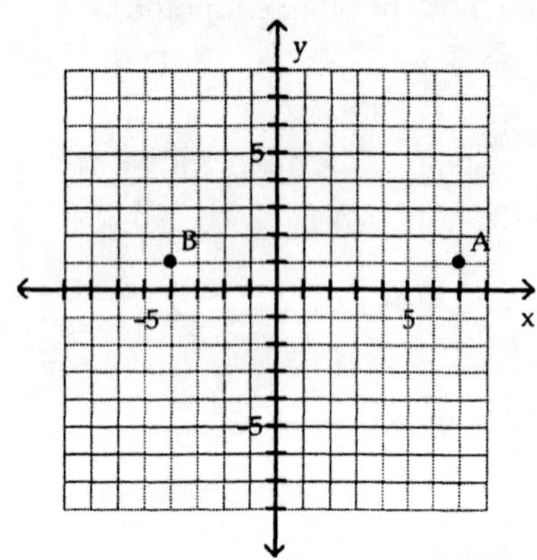

A) A = (7, 1), B = (1, 1) B) A = (7, 1), B = (1, -4)

C) A = (7, 1), B = (-4, 1) D) A = (1, 30), B = (1, -4)

10)

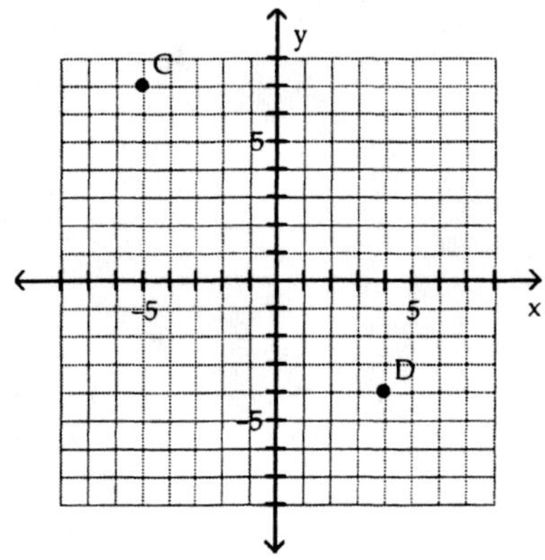

A) C = (-5, 7), D = (4, -4)

B) C = (-5, -4), D = (7, -4)

C) C = (-5, 7), D = (-4, 4)

D) C = (7, -5), D = (-4, 4)

11)

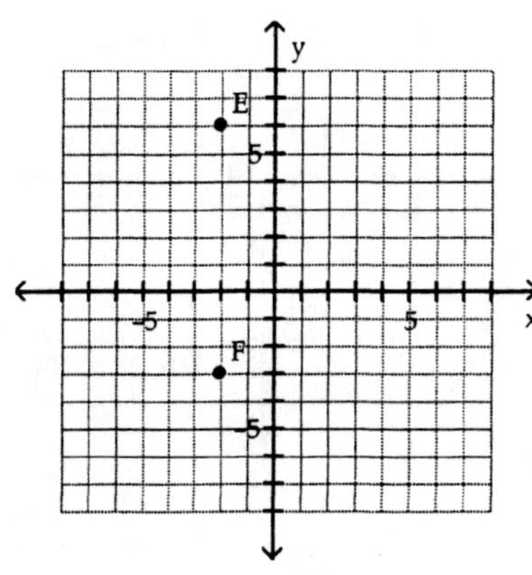

A) E = (6, -2), F = (-3, -2)

B) E = (-2, 6), F = (-2, -3)

C) E = (-2, -3), F = (6, -3)

D) E = (-2, -3) , F = (-2, 6)

12)

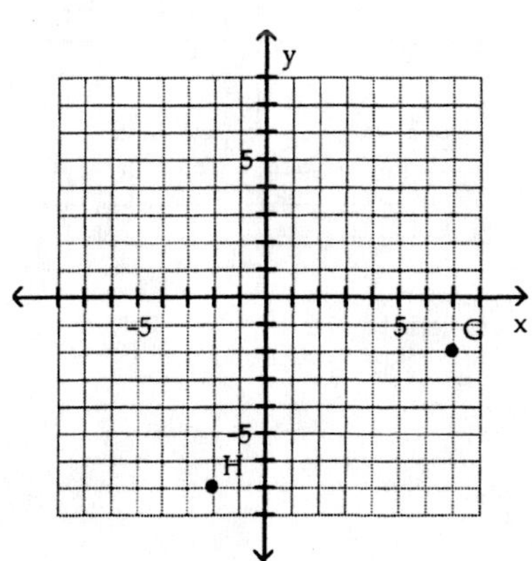

A) G = (7, -2), H = (-2, -7)

B) G = (7, -2), H = (-7, -2)

C) G = (-2, 7), H = (-7, -2)

D) G = (7, -7), H = (-2, -7)

Plot the point in the xy-plane. Tell in which quadrant or on what axis the point lies.

13) (3, 4)

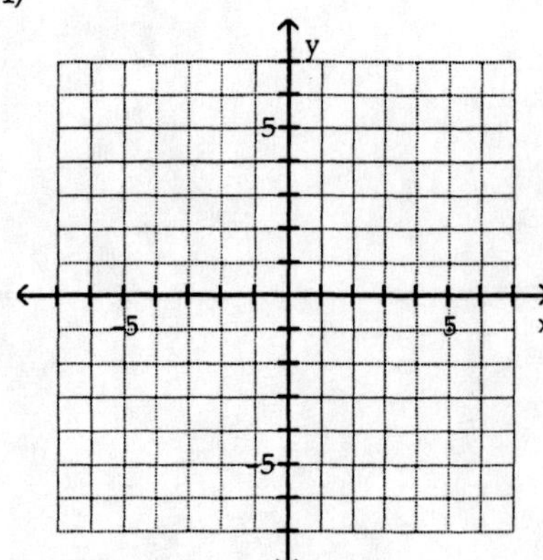

A)

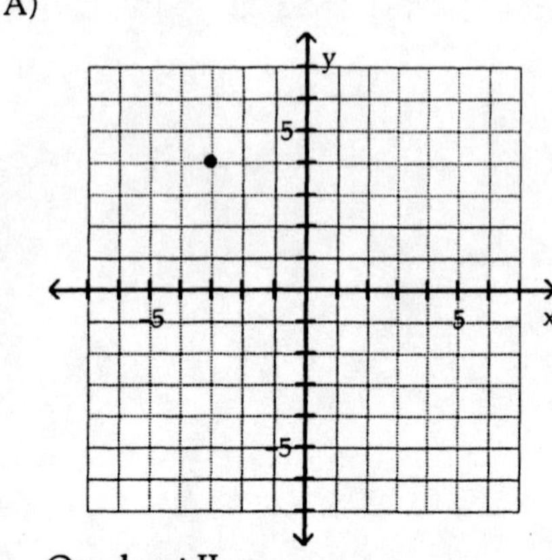

Quadrant II

B)

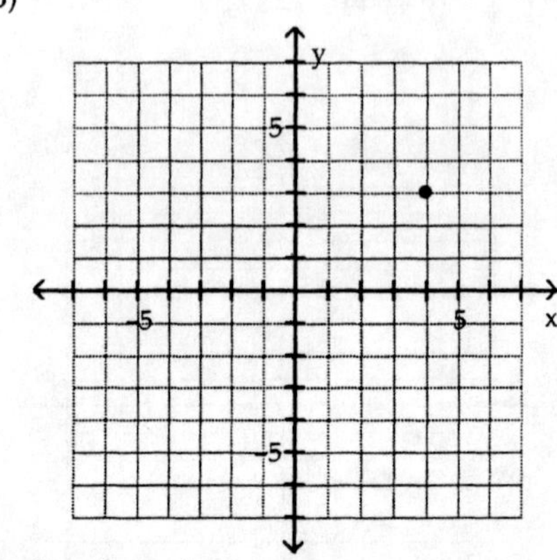

Quadrant I

C)

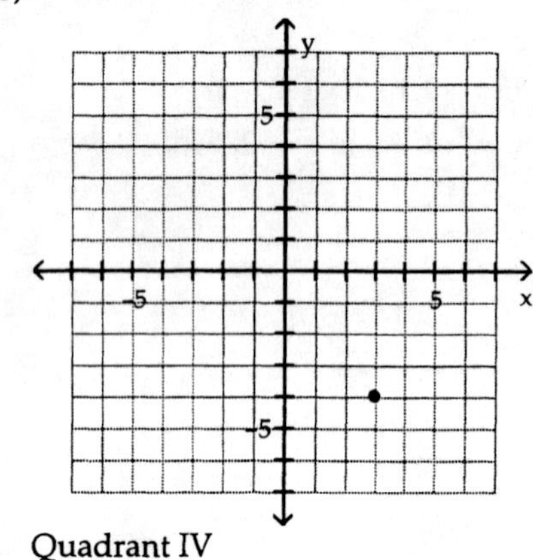

Quadrant IV

D)

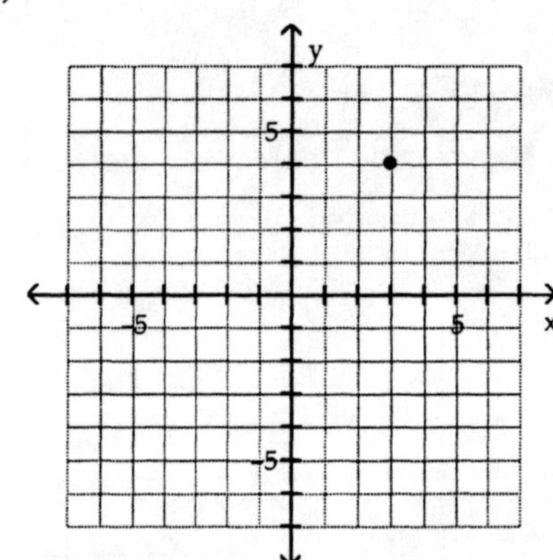

Quadrant I

14) (-4, 6)

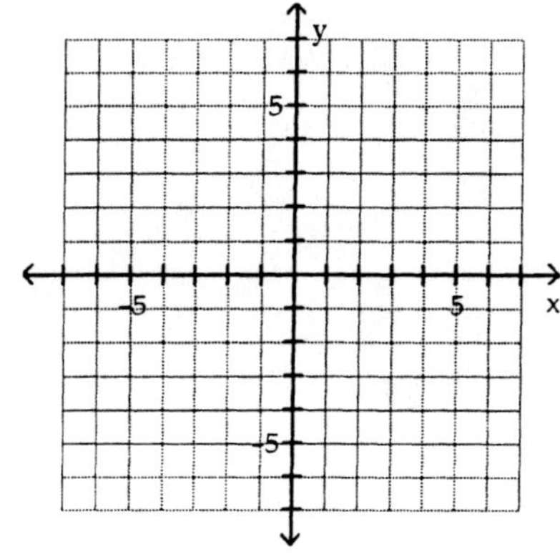

A)

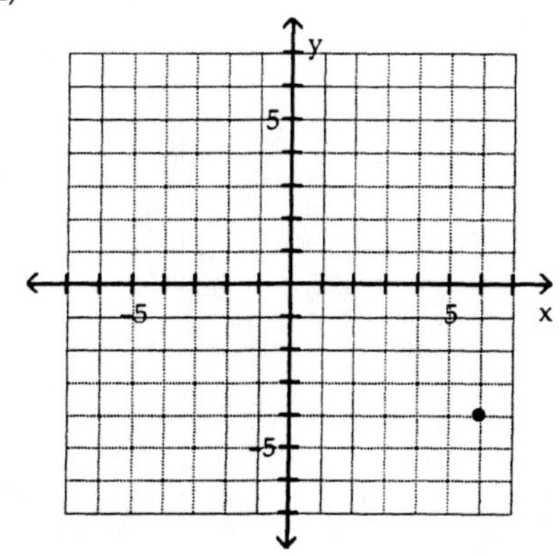

Quadrant IV

B)

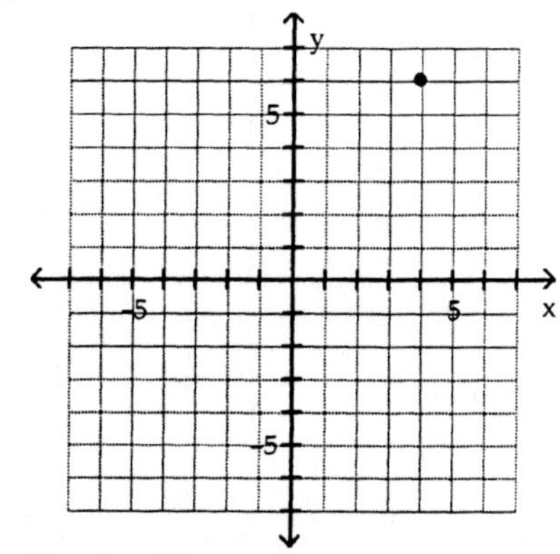

Quadrant I

C)

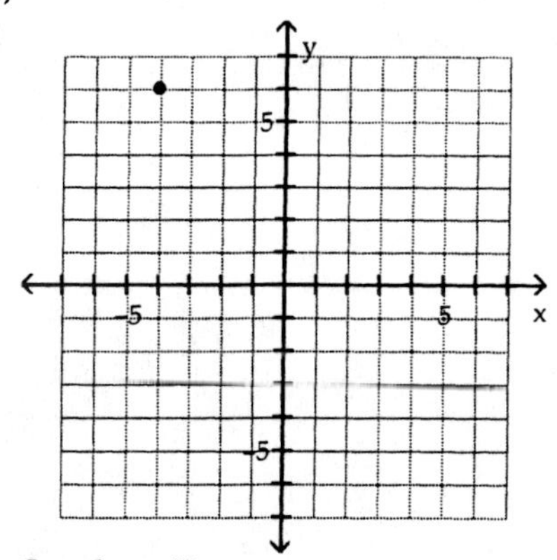

Quadrant II

D)

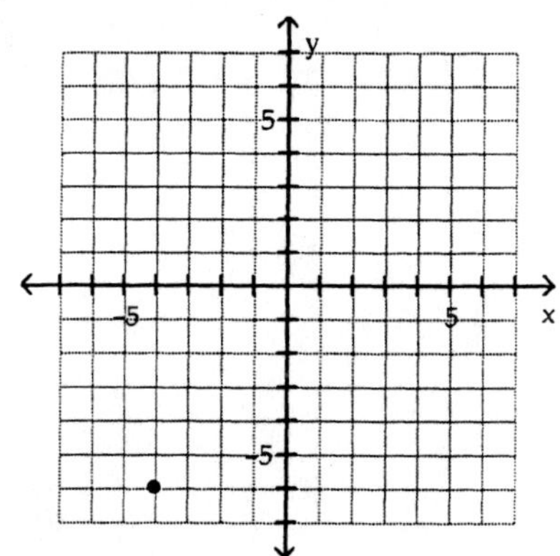

Quadrant III

15) (6, –5)

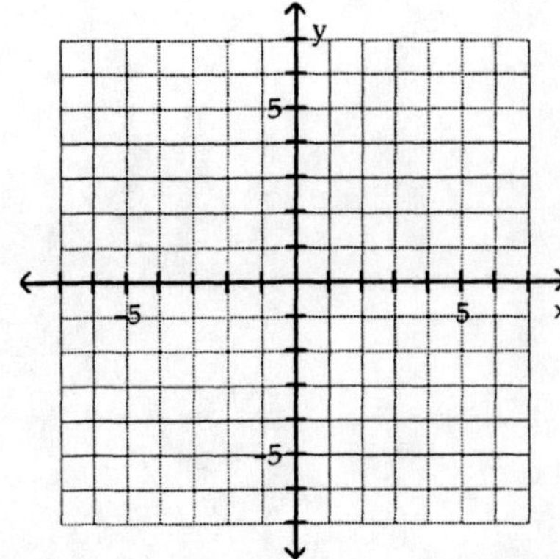

A)

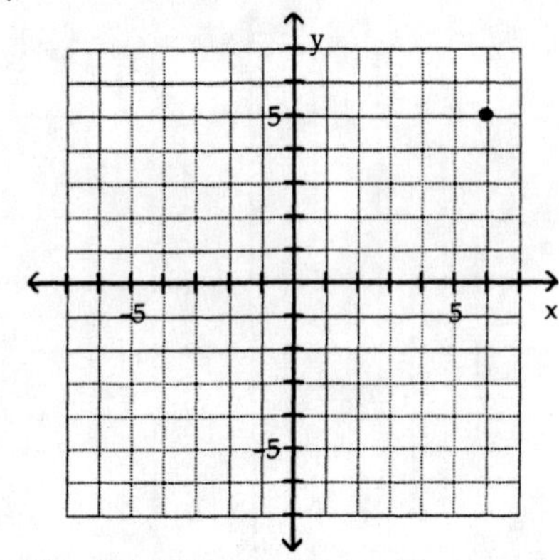

Quadrant I

B)

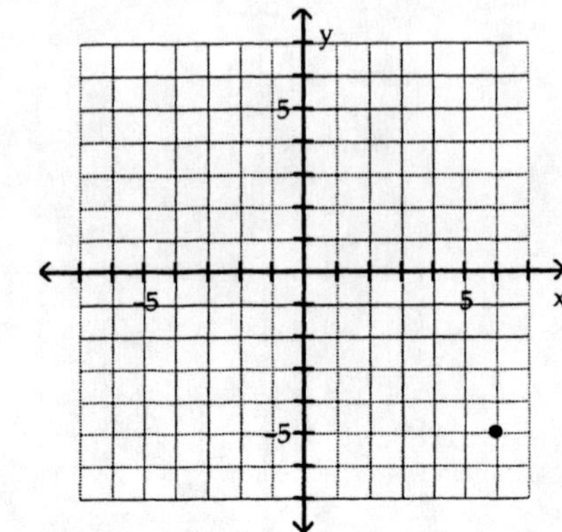

Quadrant IV

C)

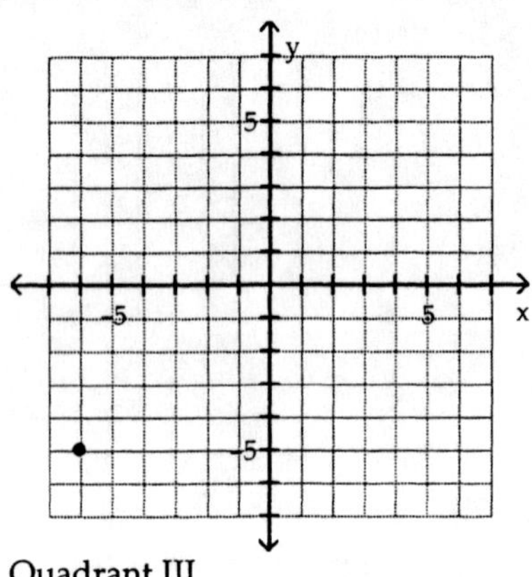

Quadrant III

D)

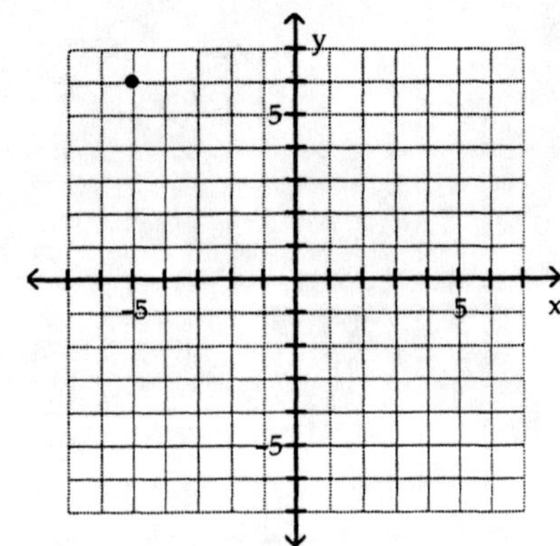

Quadrant II

16) (-3, -2)

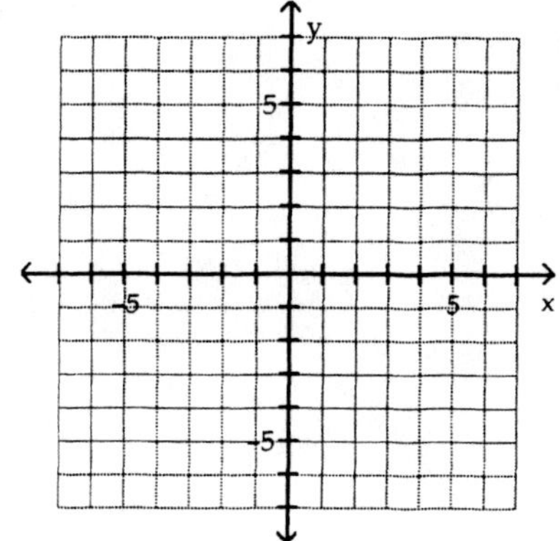

A)

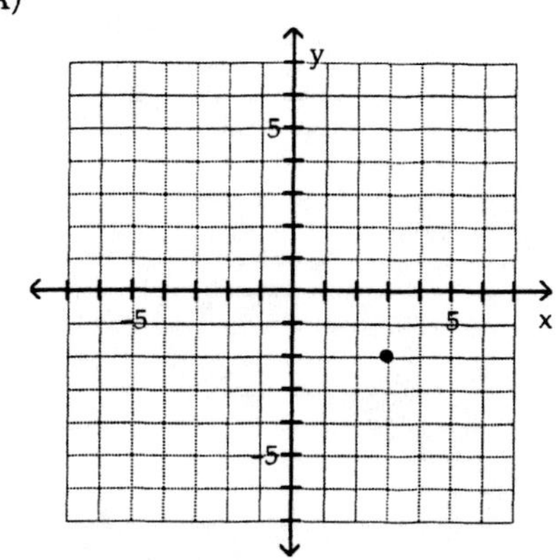

Quadrant IV

B)

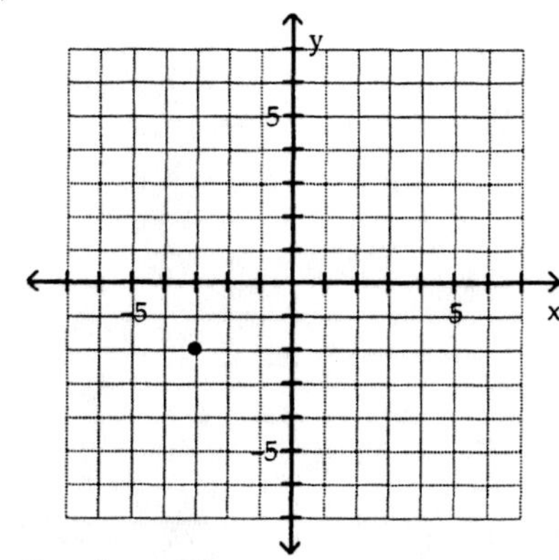

Quadrant III

C)

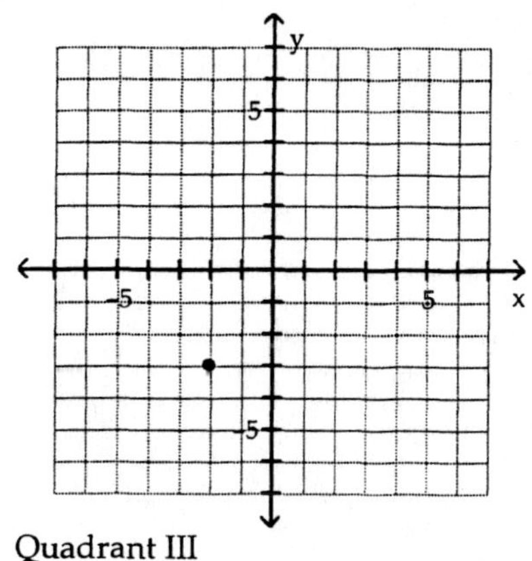

Quadrant III

D)

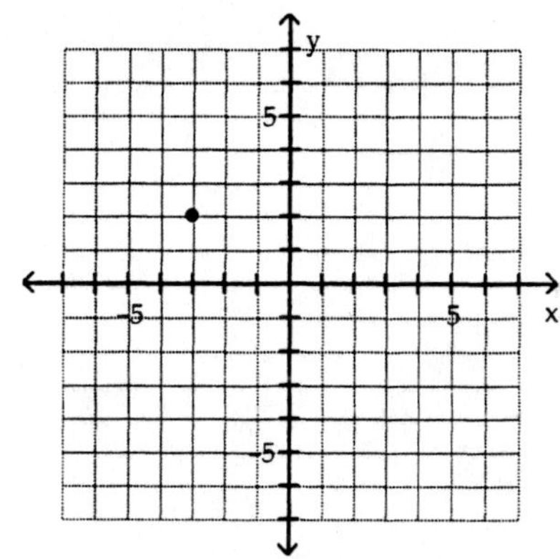

Quadrant II

17) (0, -1)

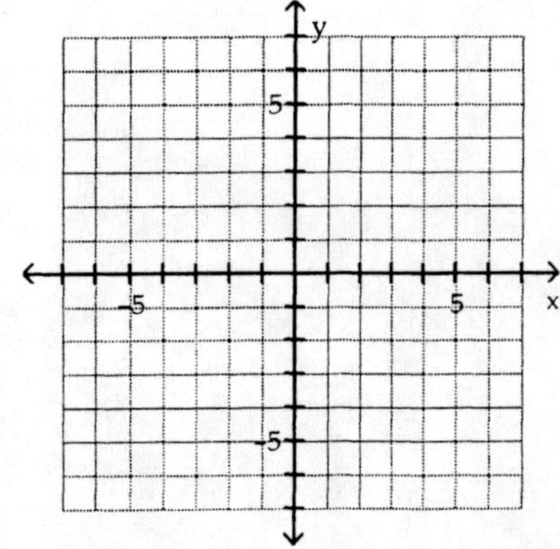

A)

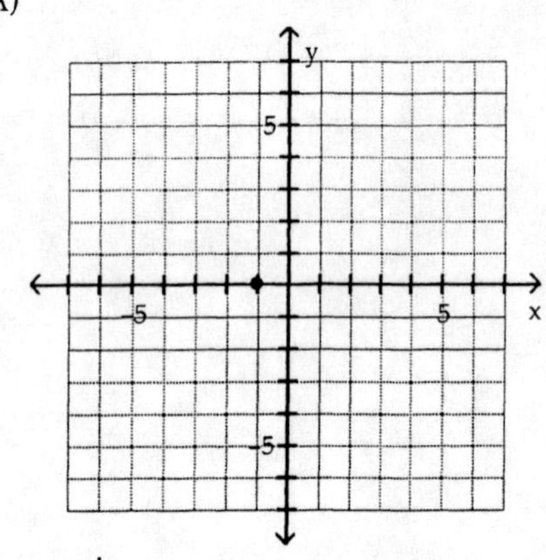

x–axis

B)

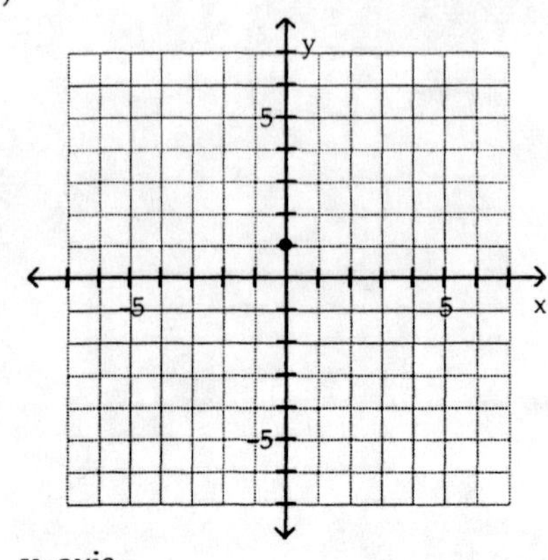

y–axis

C)

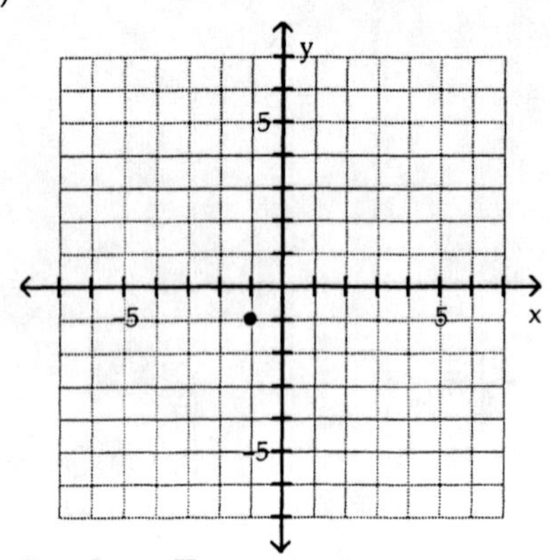

Quadrant II

D)

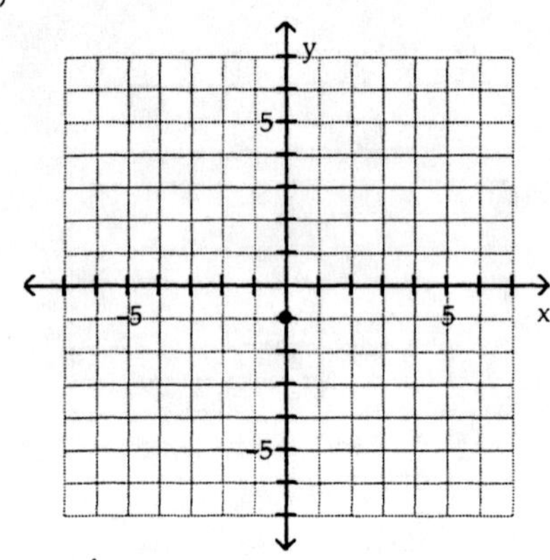

y–axis

18) (4, 0)

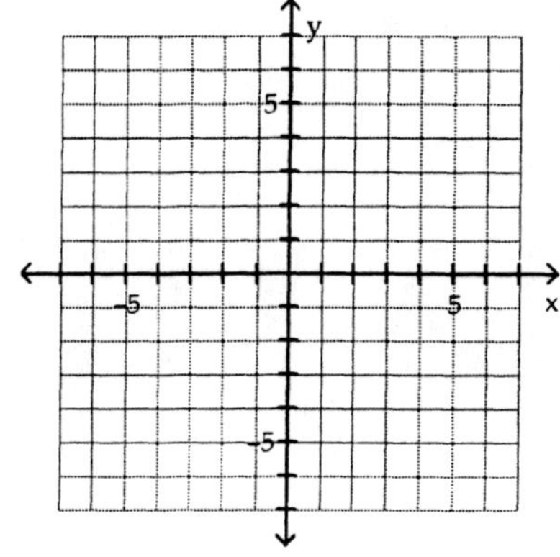

A)

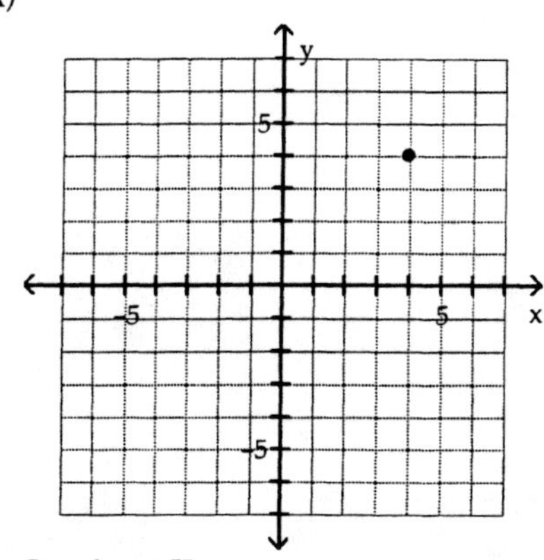

Quadrant II

B)

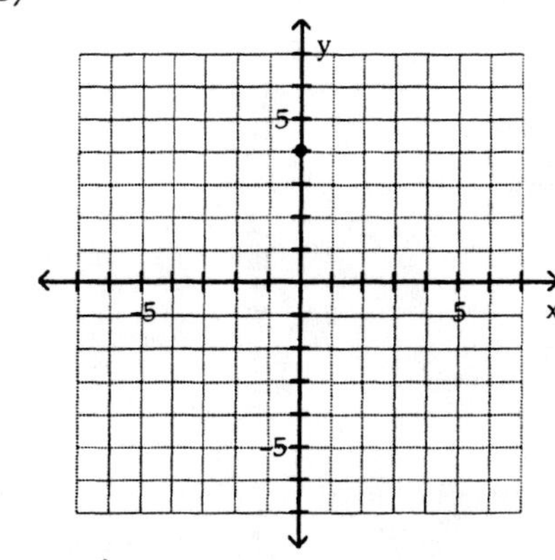

y–axis

C)

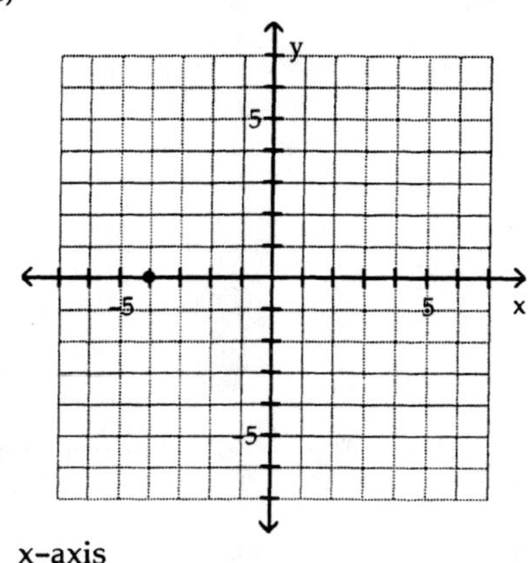

x–axis

D)

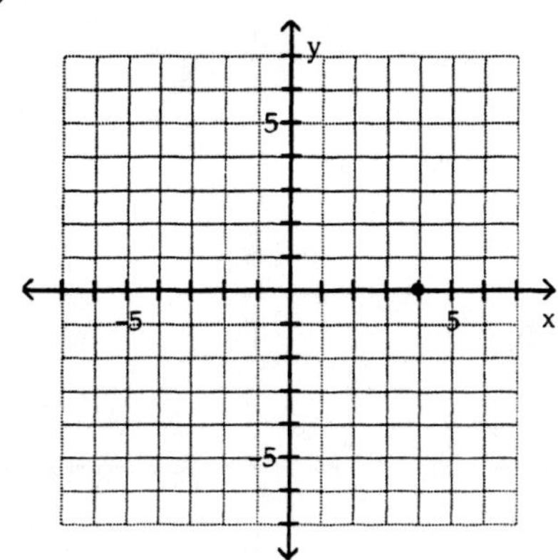

x–axis

2.2 Graphs of Equations

1 Graph Equations by Plotting Points

Graph the equation by plotting points.

1) $y = x - 5$

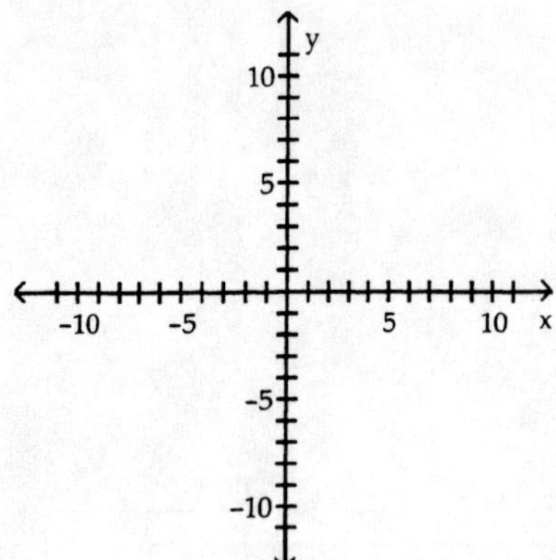

A)

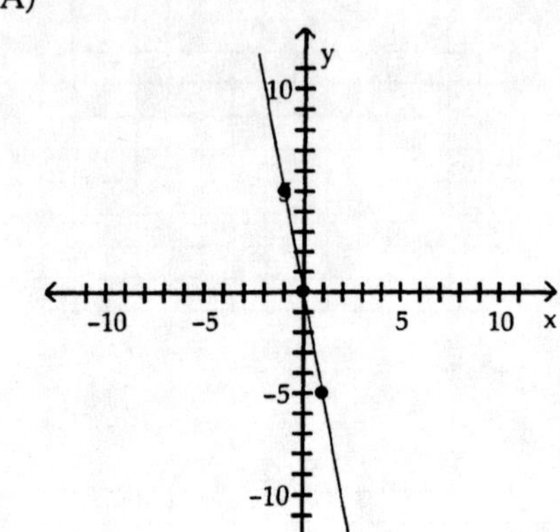

B)

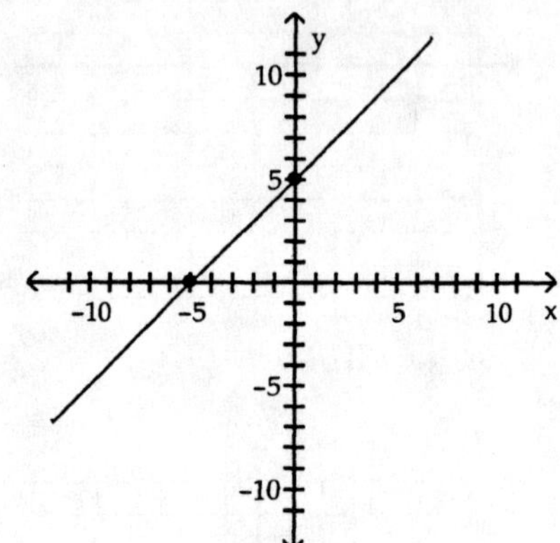

C)

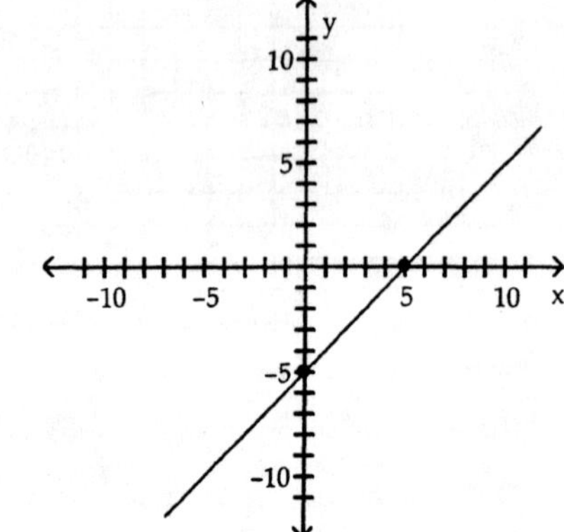

D)

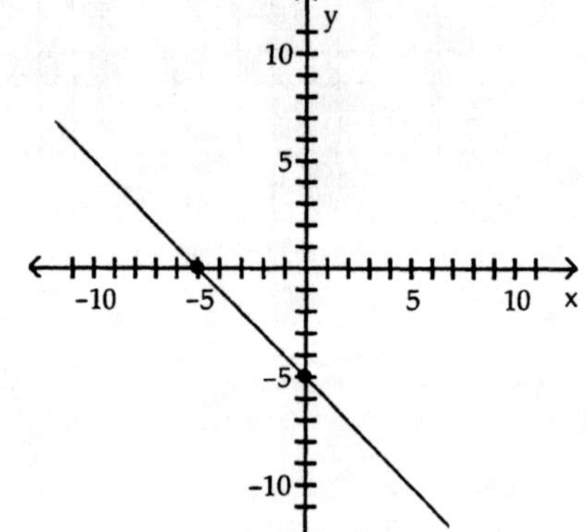

2) $y = -5x$

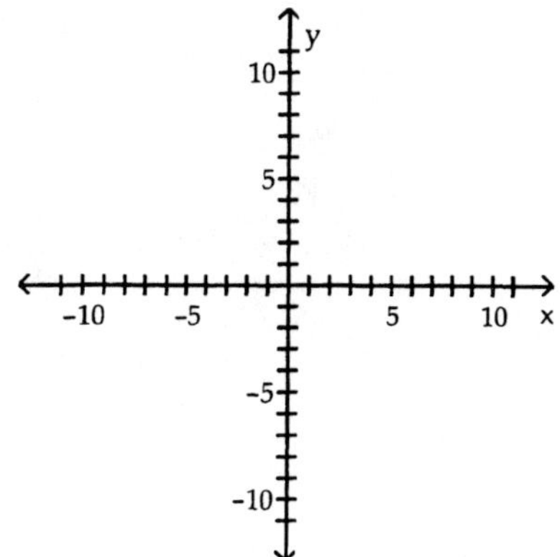

A)

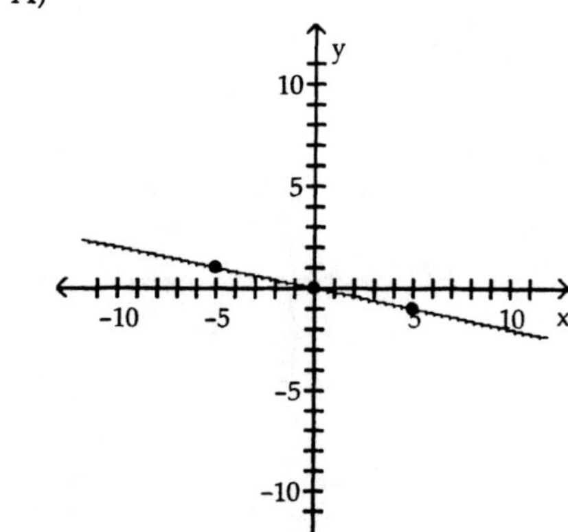

B)

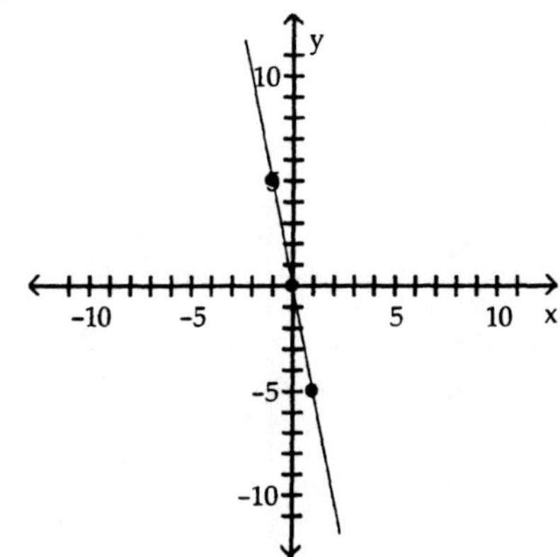

C)

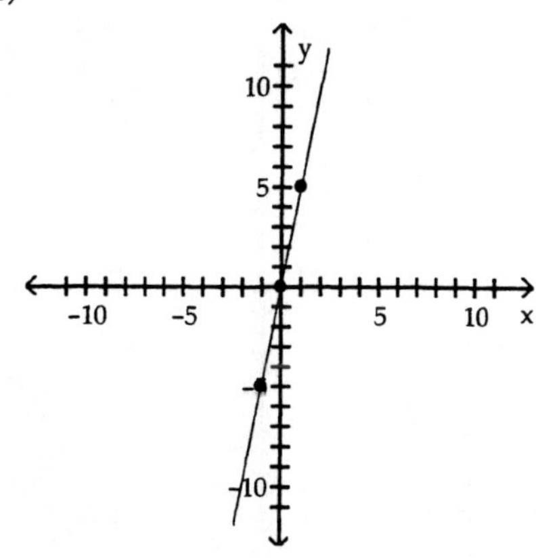

D)

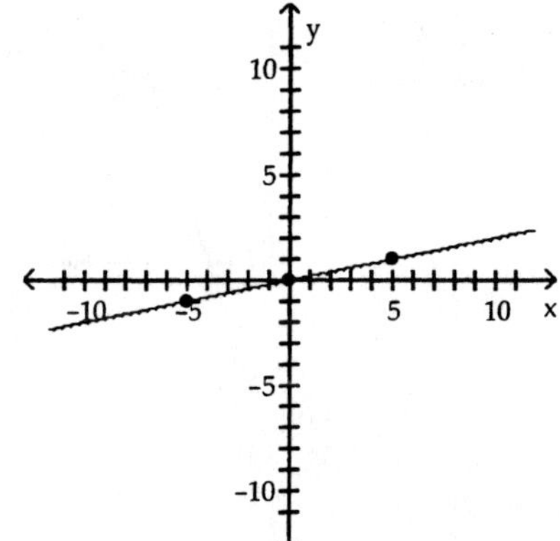

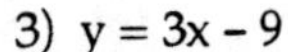
3) $y = 3x - 9$

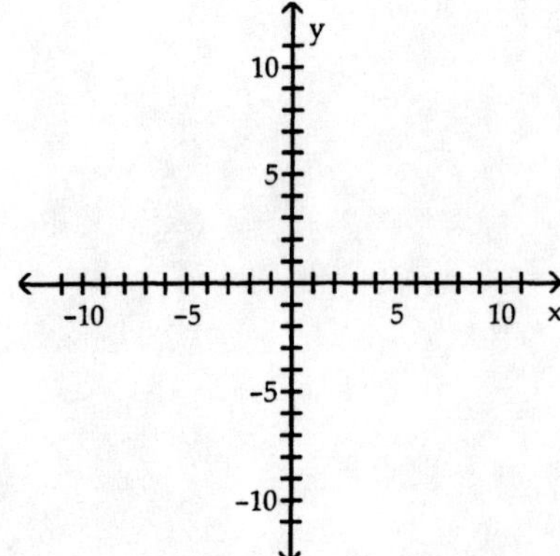

y
10
5
-10
-5
5
10
x
-5
-10

y
10
5
-10
-5
5
10
x
-5
-10

y
10
5
-10
-5
5
10
x
-5
-10

y
10
5
-10
-5
5
10
x
-5
-10

y
10
5
-10
-5
5
10
x
-5
-10

A)

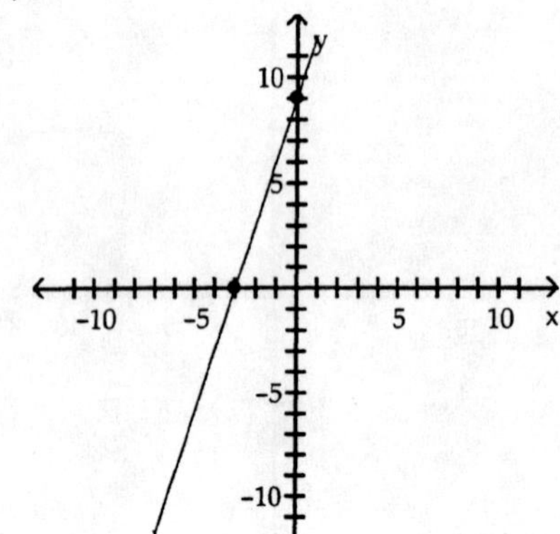

B)

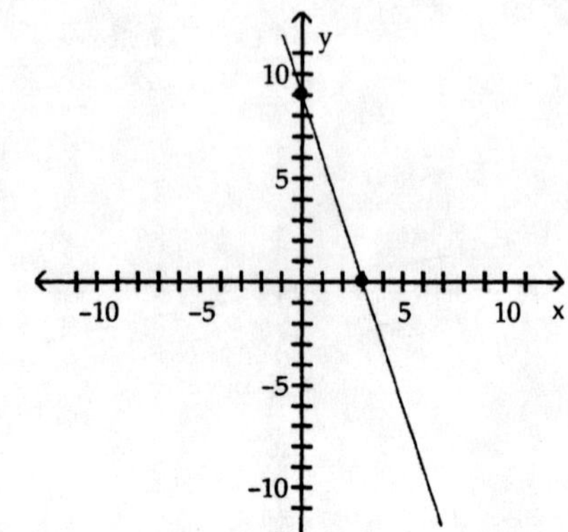

C)

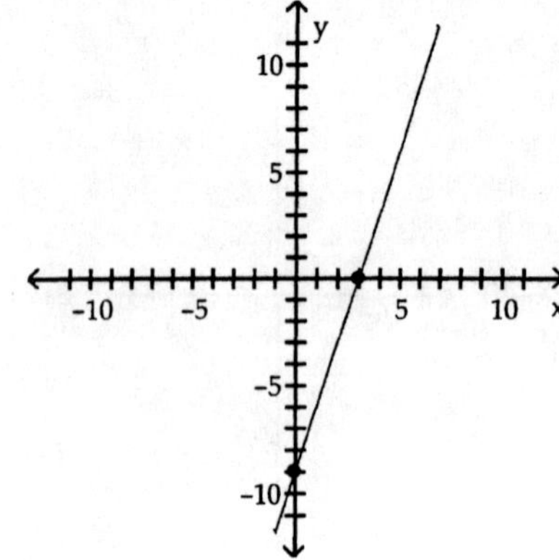

D)

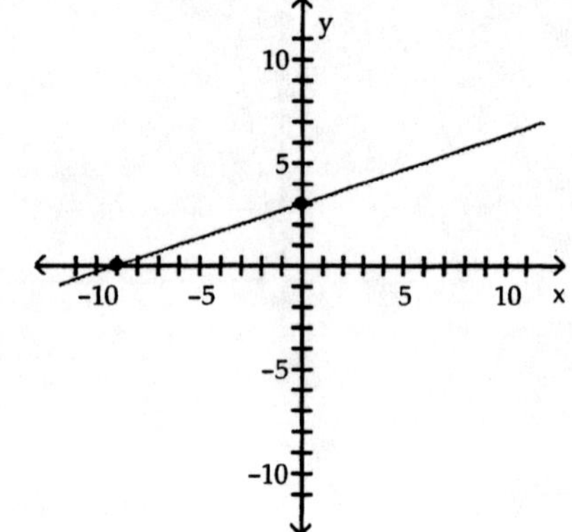

4) $y = x^2$

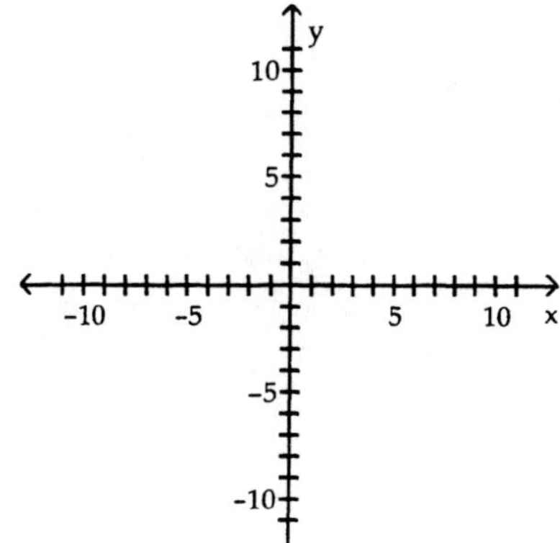

A)

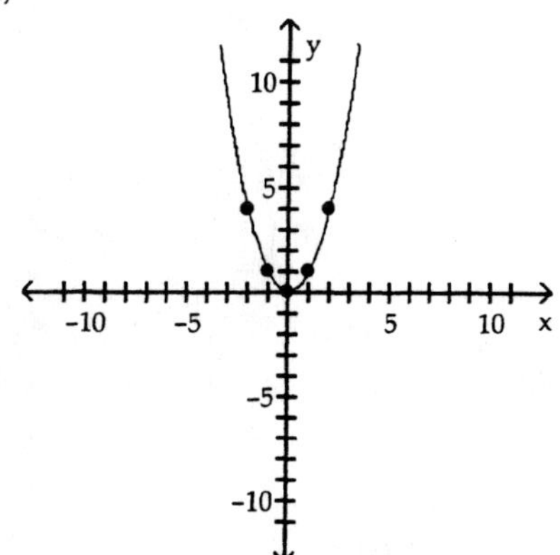

B)

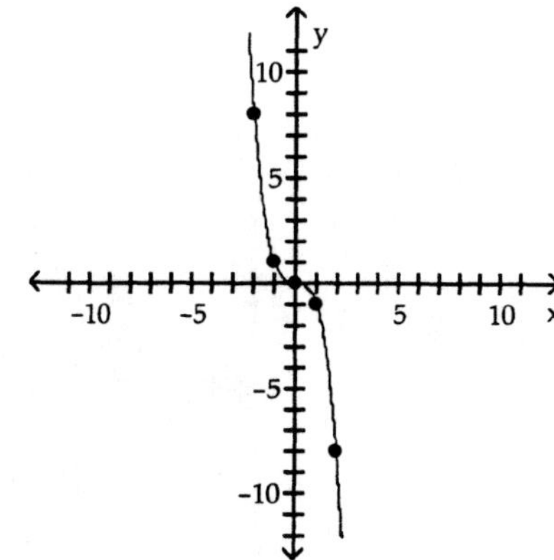

C)

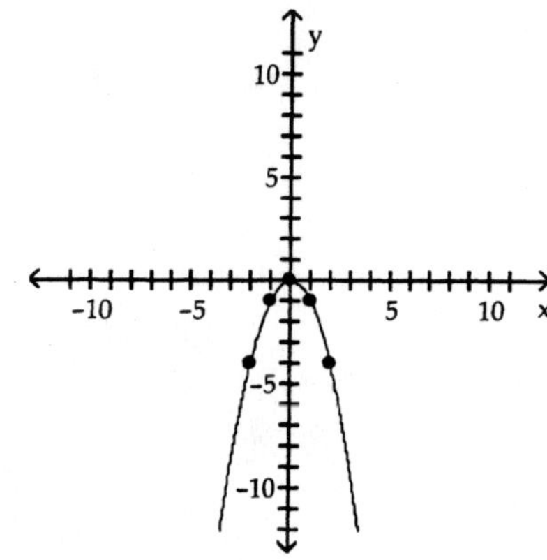

D)

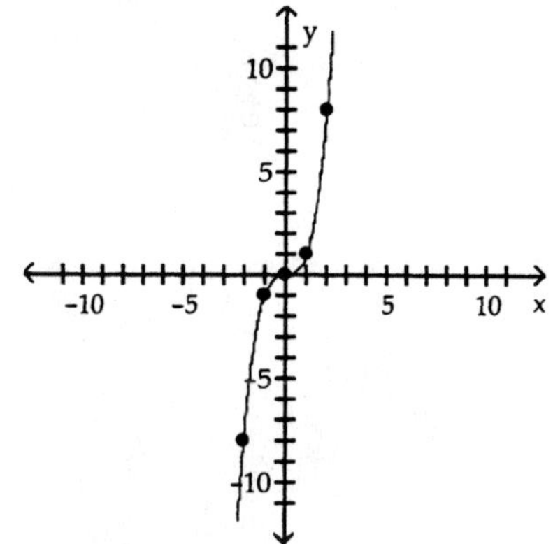

5) $y = x^3$

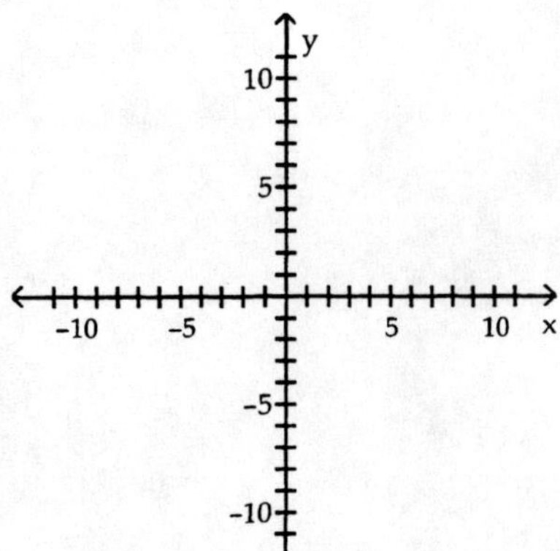

A)

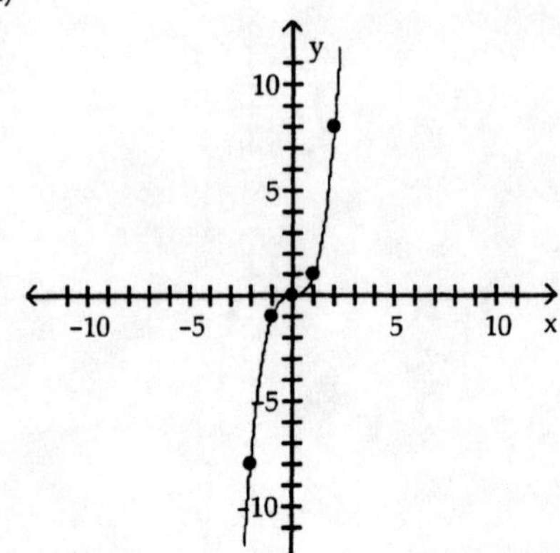

B)

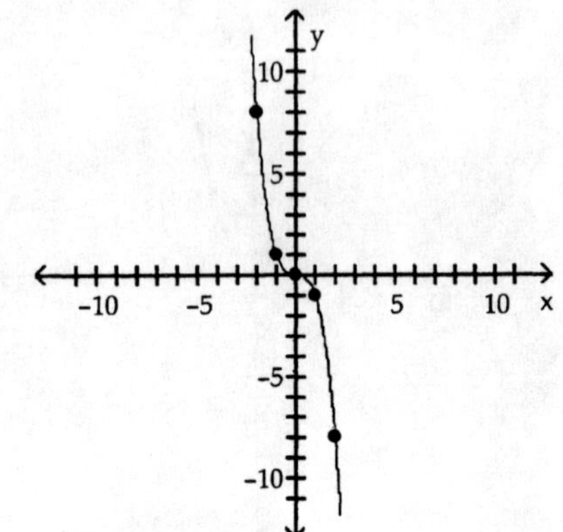

C)

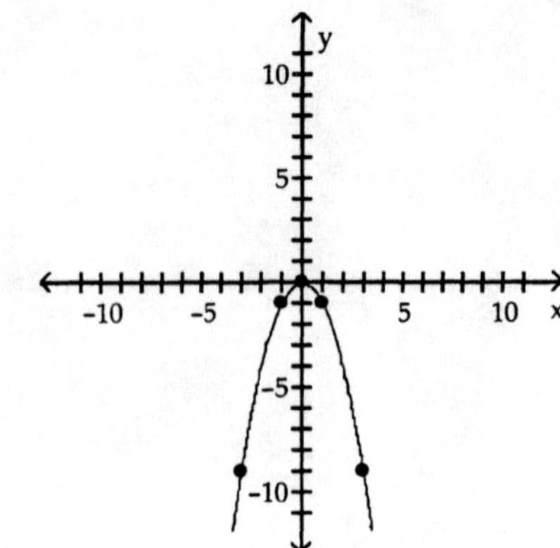

D)

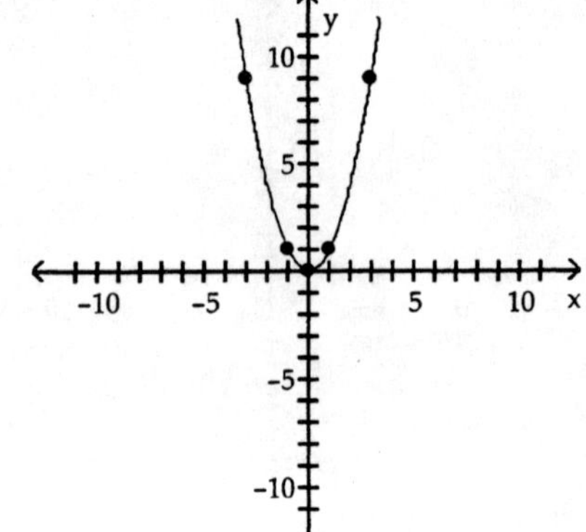

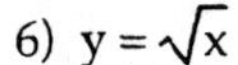

6) $y = \sqrt{x}$

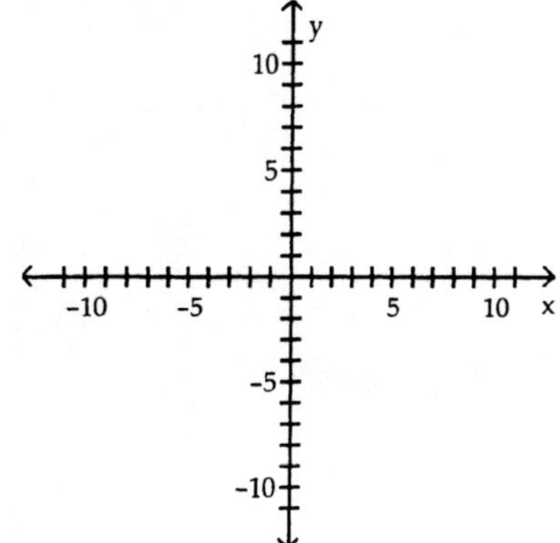

A)

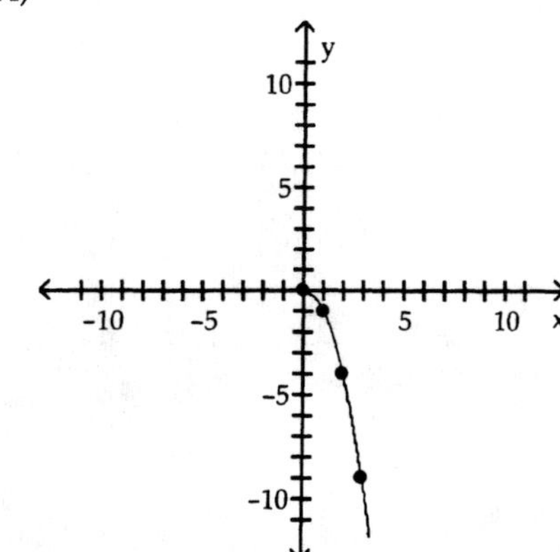

B)

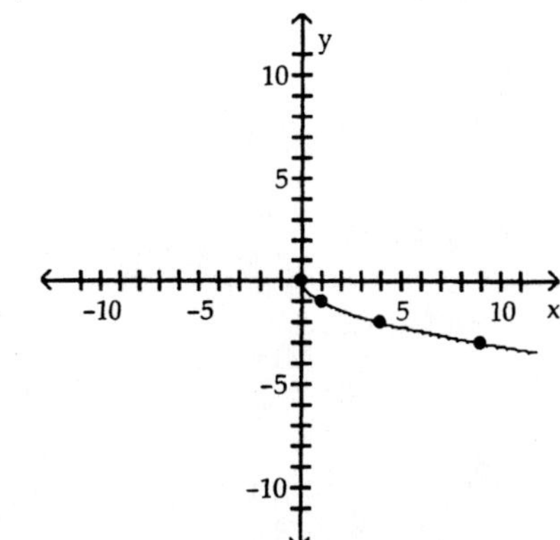

C)

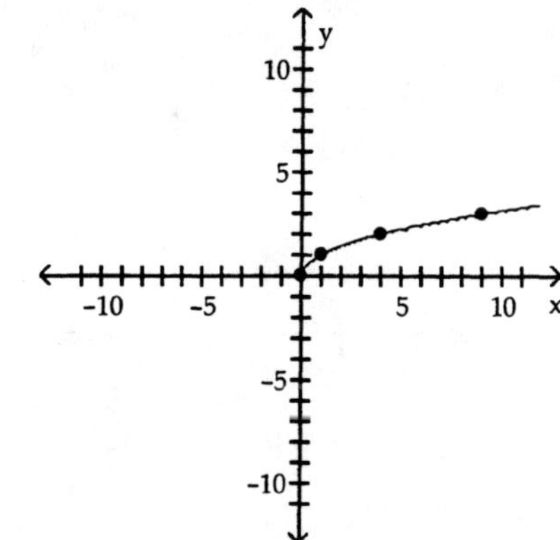

D)

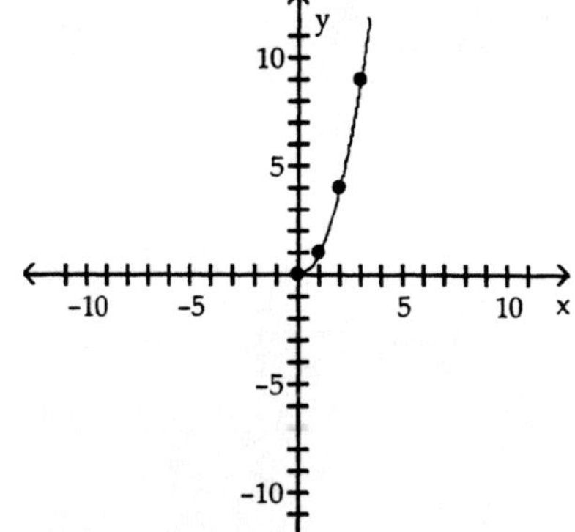

7) $y = \frac{1}{x}$

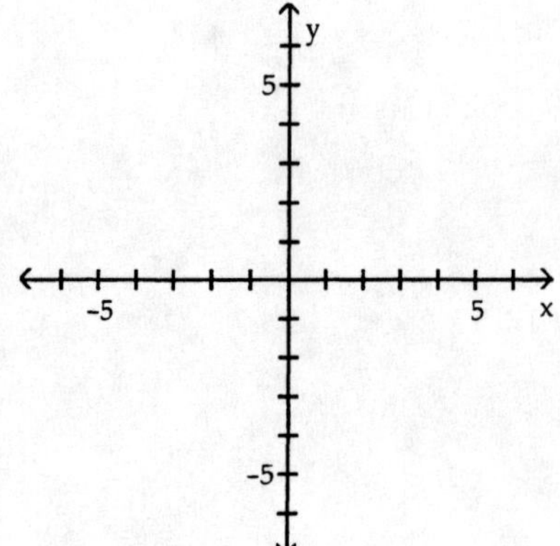

A)

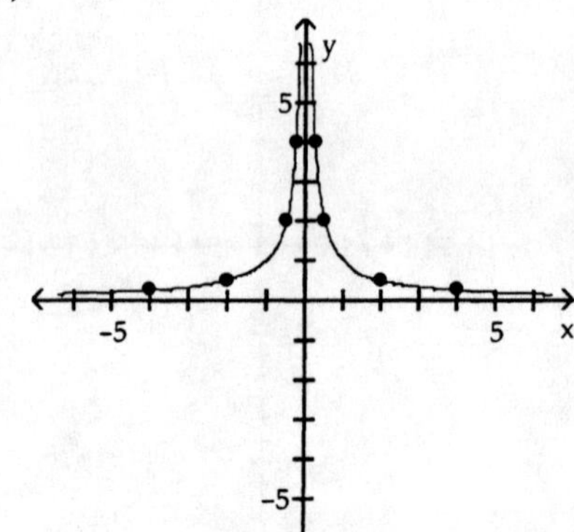

B)

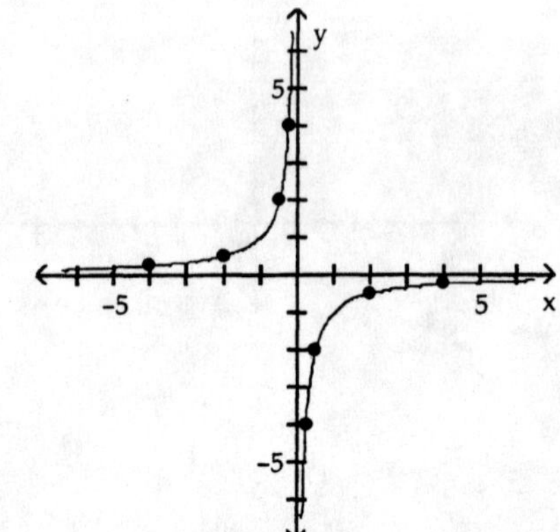

C)

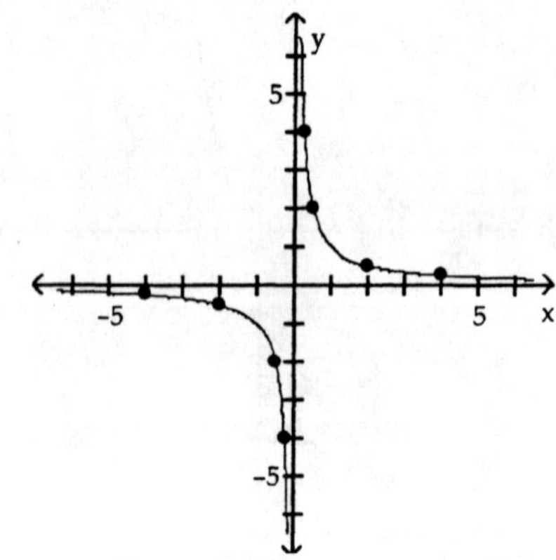

D)

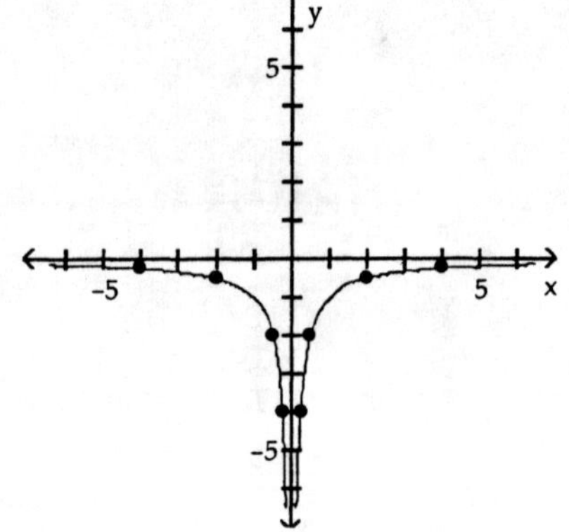

8) $x = y^2$

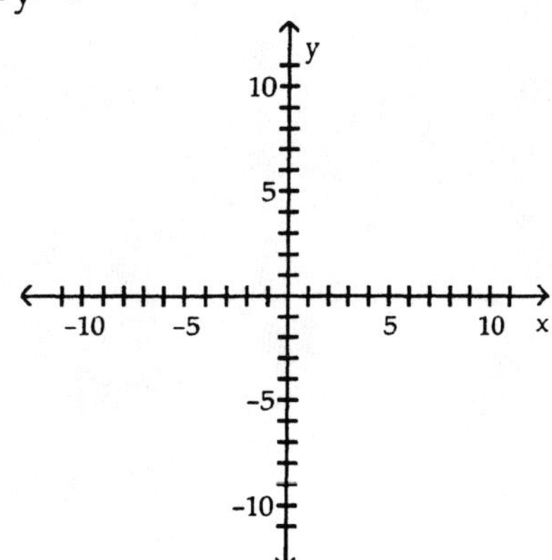

A)

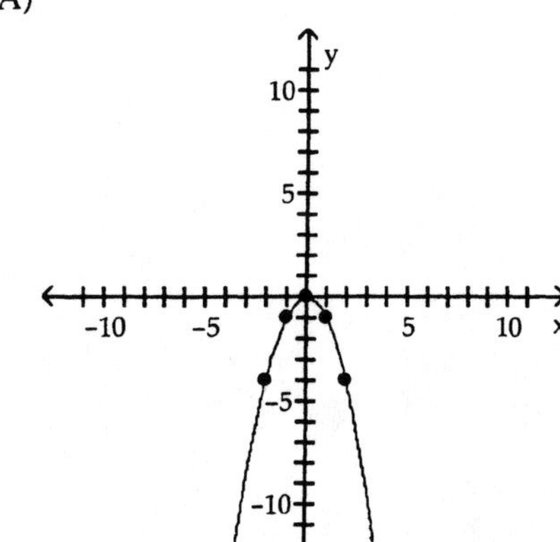

B)

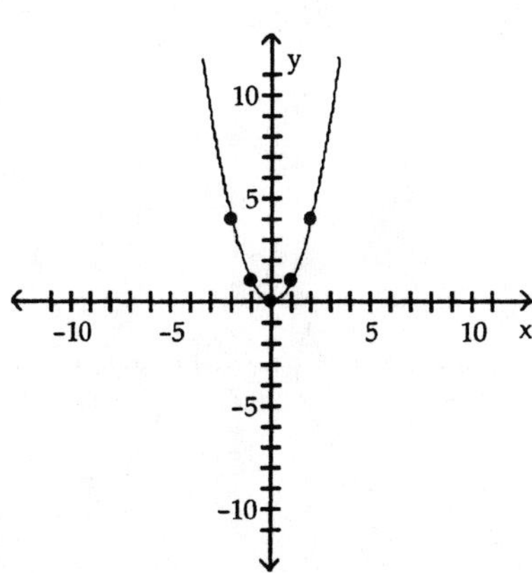

C)

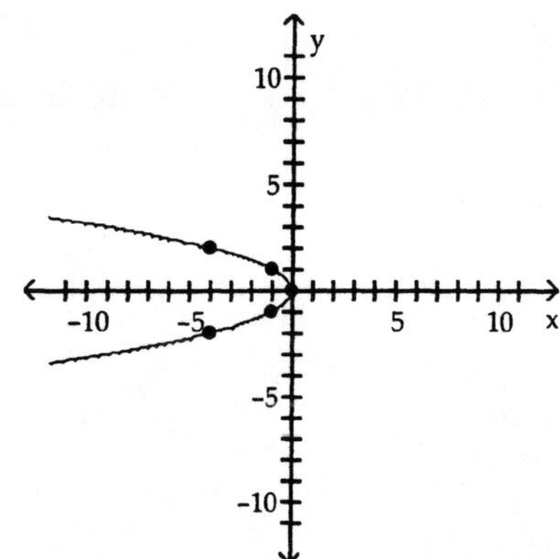

D)

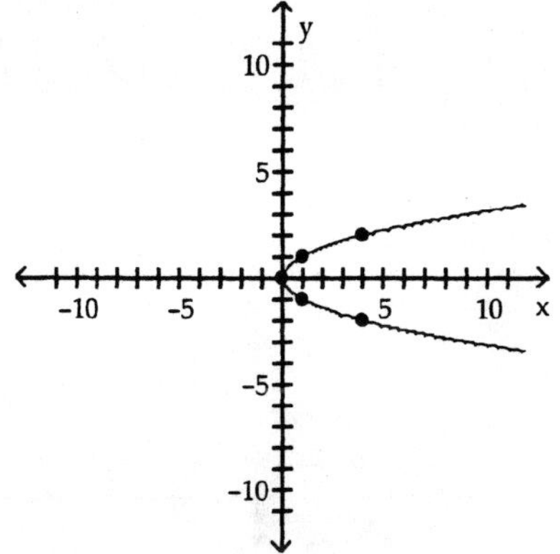

Determine whether the given point is on the graph of the equation.

9) Equation: $y = x^4 - \sqrt{x}$
Point: (0, 0)

A) Yes

B) No

10) Equation: $x^2 - y^2 = 4$
Point: (0, -2)

A) No

B) Yes

Solve the problem.

11) If (a, 3) is a point on the graph of $y = 2x - 5$, what is a?

A) -1 B) 1 C) -4 D) 4

12) If (3, b) is a point on the graph of $3x - 2y = 17$, what is b?

A) $\frac{23}{3}$ B) -4 C) 4 D) $\frac{11}{3}$

13) The height of a baseball (in feet) at time t (in seconds) is given by $y = -16x^2 + 80x + 5$. Which one of the following points is not on the graph of the equation?

A) (4, 69) B) (3, 101) C) (1, 69) D) (2, 117)

2 Find Intercepts from a Graph

List the intercepts of the graph.

1)

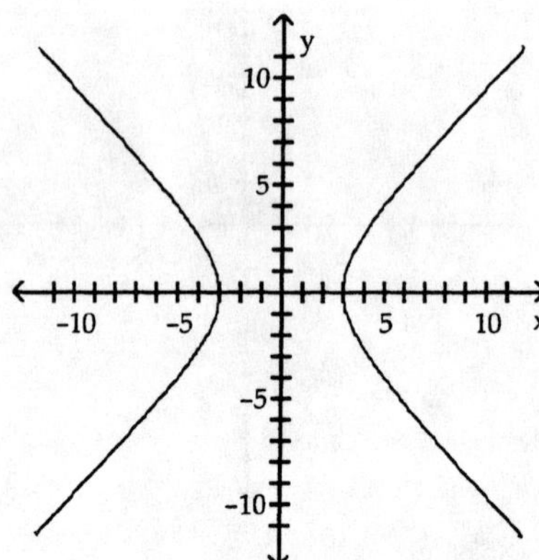

A) (0, -3), (3, 0) B) (-3, 0), (3, 0) C) (-3, 0), (0, 3) D) (0, -3), (0, 3)

2)

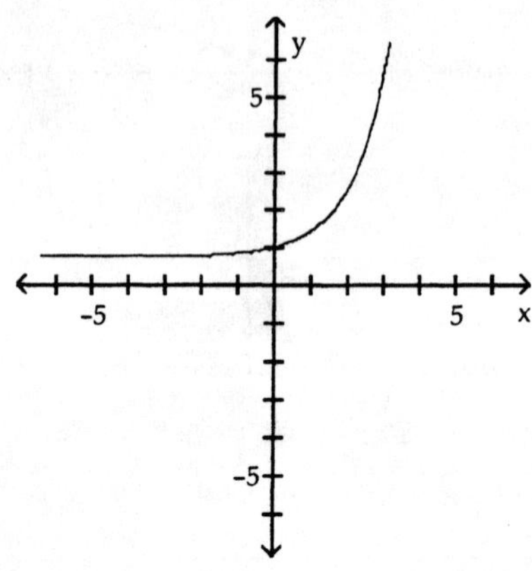

A) (0, 0) B) (1, 0) C) (0, 1) D) (1, 1)

3)

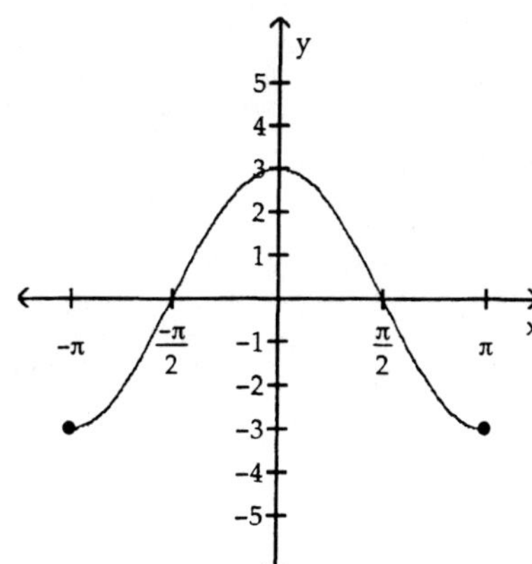

A) $(-\frac{\pi}{2}, 0)$, (3, 0), $(\frac{\pi}{2}, 0)$

B) $(0, -\frac{\pi}{2})$, (0, 3), $(0, \frac{\pi}{2})$

C) $(-\frac{\pi}{2}, 0)$, (0, 3), $(\frac{\pi}{2}, 0)$

D) $(0, -\frac{\pi}{2})$, (3, 0), $(0, \frac{\pi}{2})$

4)

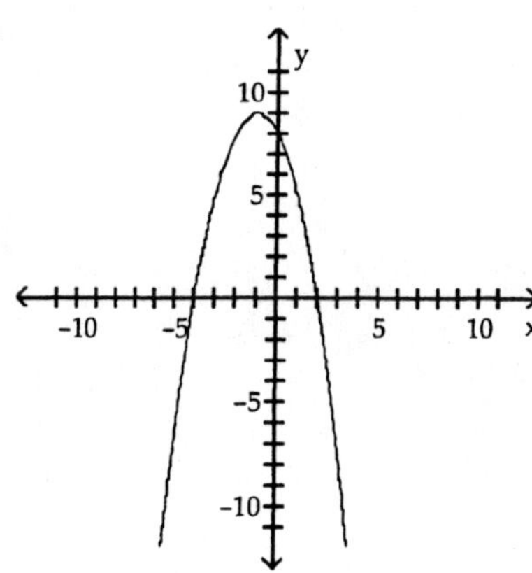

A) (0, -4), (0, 8), (2, 0)

B) (-4, 0), (0, 8), (2, 0)

C) (-4, 0), (0, 8), (0, 2)

D) (0, -4), (8, 0), (0, 2)

5)

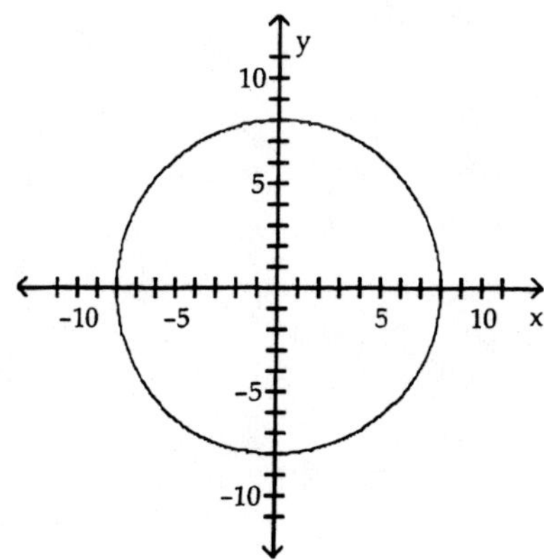

A) (-8, 0), (0, -8), (0, 8), (8, 0)

B) (0, 8), (8, 0)

C) (-8, 0), (0, -8), (0, 0), (0, 8), (8, 0)

D) (-8, 0), (0, 8)

6)

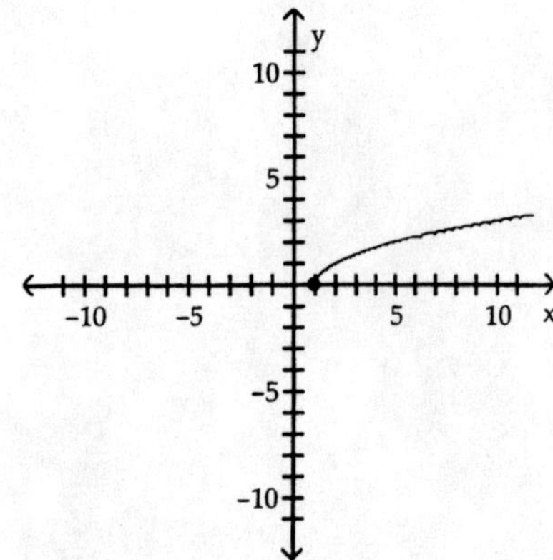

A) (0, 1) B) (1, 0) C) (–1, 0) D) (0, –1)

7)

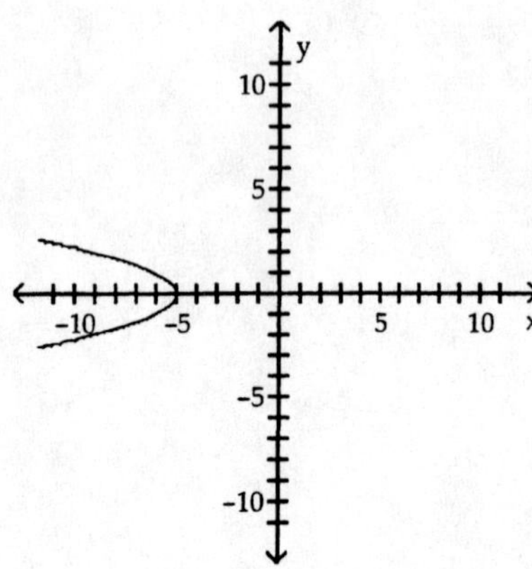

A) (–5, 0) B) (5, 0) C) (0, –5) D) (0, 5)

8)

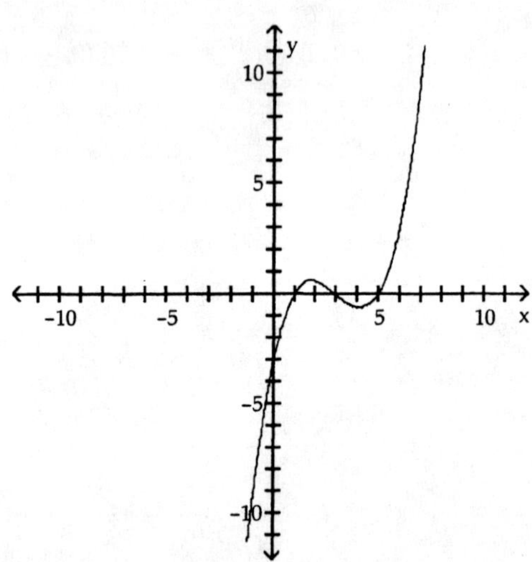

A) (3, 0), (1, 0) (5, 0), (0, –3) B) (–3, 0), (1, 0), (–5, 0), (0, –3)

C) (–3, 0), (0, –3), (0, 1), (0, –5) D) (–3, 0), (0, 3), (0, 1), (0, 5)

9)

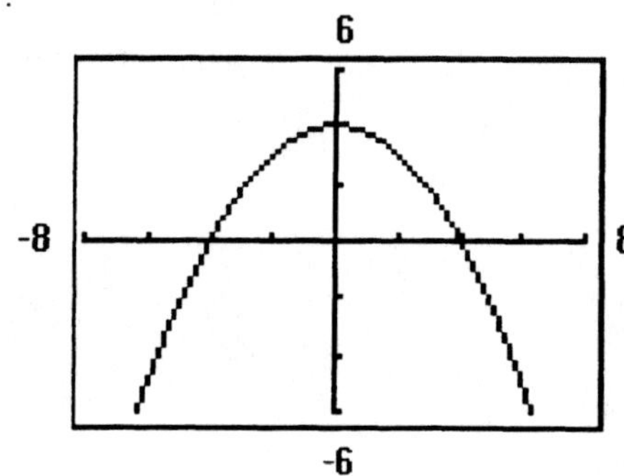

A) (−4, 0), (0, 4), (4, 0) B) (−2, 0), (2, 0) C) (−2, 0), (0, 4), (2, 0) D) (−2, 0), (0, 2), (2, 0)

3 Find Intercepts from an Equation

List the intercepts for the graph of the equation.

1) $y = x - 6$

A) (6, 0), (0, 6) B) (−6, 0), (0, 6) C) (6, 0), (0, −6) D) (−6, 0), (0, −6)

2) $y = -5x$

A) (0, −5) B) (−5, −5) C) (−5, 0) D) (0, 0)

3) $x^2 = y$

A) (−1, 0), (1, 0) B) (0, 0) C) (−1, 0), (0, −1), (1, 0) D) (0, 1)

4) $y^2 = x$

A) (0, 0) B) (−1, 0), (0, 1), (0, −1) C) (0, 1), (0, −1) D) (1, 0)

5) $x^2 + y - 36 = 0$

A) (−6, 0), (0, 36), (6, 0) B) (−6, 0), (0, −36), (6, 0)

C) (6, 0), (0, 36), (0, −36) D) (0, −6), (36, 0), (0, 6)

6) $y^2 - x - 64 = 0$

A) (−8, 0), (0, −64), (8, 0) B) (0, −8), (64, 0), (0, 8)

C) (8, 0), (0, 64), (0, −64) D) (0, −8), (−64, 0), (0, 8)

7) $4x^2 + 16y^2 = 64$

A) (−4, 0), (0, −2), (0, 2), (4, 0) B) (−16, 0), (0, −4), (0, 4), (16, 0)

C) (−2, 0), (−4, 0), (4, 0), (2, 0) D) (−4, 0), (−16, 0), (16, 0), (4, 0)

8) $16x^2 + y^2 = 16$

A) (−1, 0), (0, −16), (0, 16), (1, 0) B) (−16, 0), (0, −1), (0, 1), (16, 0)

C) (−4, 0), (0, −1), (0, 1), (4, 0) D) (−1, 0), (0, −4), (0, 4), (1, 0)

9) $y = x^2 + 10x + 25$

A) (5, 0), (5, 0), (0, 25) B) (0, 5), (0, 5), (25, 0)

C) (0, −5), (0, −5), (25, 0) D) (−5, 0), (−5, 0), (0, 25)

10) $y = \frac{7x}{x^2 + 49}$

A) (-49, 0), (0, 0), (49, 0)
B) (0, -7), (0, 0), (0, 7)
C) (0, 0)
D) (-7, 0), (0, 0), (7, 0)

11) $y = \frac{x^2 - 16}{4x^4}$

A) (-16, 0), (0, 0), (16, 0)
B) (-4, 0), (4, 0)
C) (0, 0)
D) (0, -4), (0, 4)

12) $y = x^2 + 4$

A) (4, 0)
B) (4, 0), (0, -2), (0, 2)
C) (0, 4), (-2, 0), (2, 0)
D) (0, 4)

13) $y = x^3 - 125$

A) (-125, 0), (0, 5)
B) (0, -125), (5, 0)
C) (0, -5), (-5, 0)
D) (0, -5), (0, 5)

14) $y = (x - 2)^2 - 1$

A) (0, 3), (0, 1), (3, 0)
B) (0, 3), (3, 0) (1, 0)
C) (0, 3), (2, 0)
D) (0, 2), (3, 0)

4 Test an Equation for Symmetry with Respect to the (a) x-axis, (b) y-axis, and (c) Origin

Determine whether the graph is symmetric with respect to the x-axis, the y-axis, and/or the origin.

1)

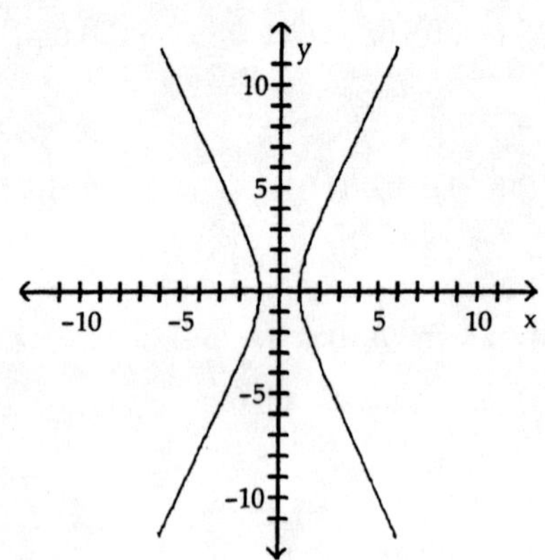

A) y-axis
B) x-axis
C) origin
D) x-axis, y-axis, origin
E) none

2)

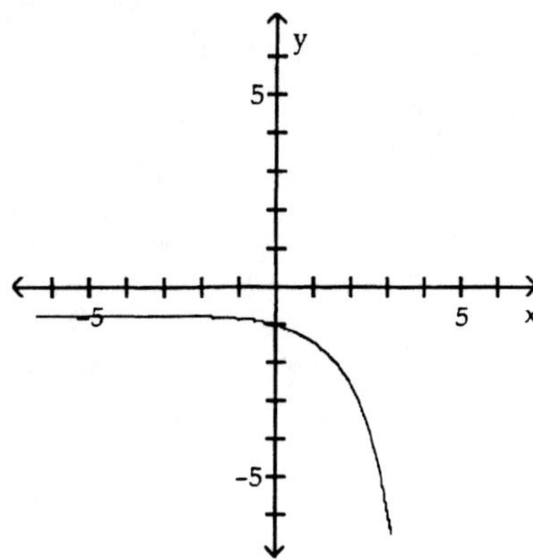

A) y-axis
B) origin
C) x-axis
D) x-axis, y-axis, origin
E) none

3)

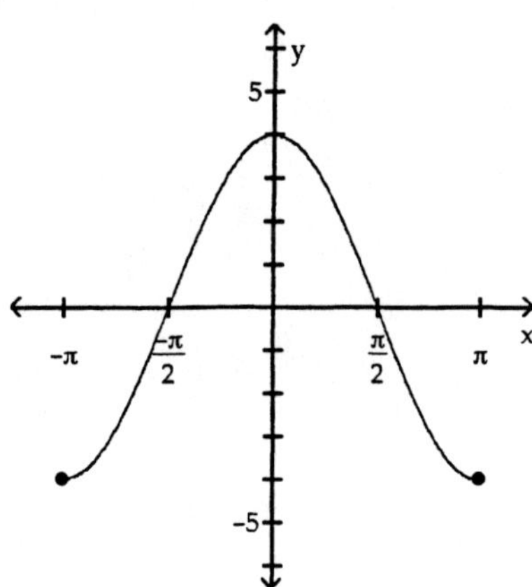

A) x-axis
B) y-axis
C) origin
D) x-axis, y-axis, origin
E) none

4)

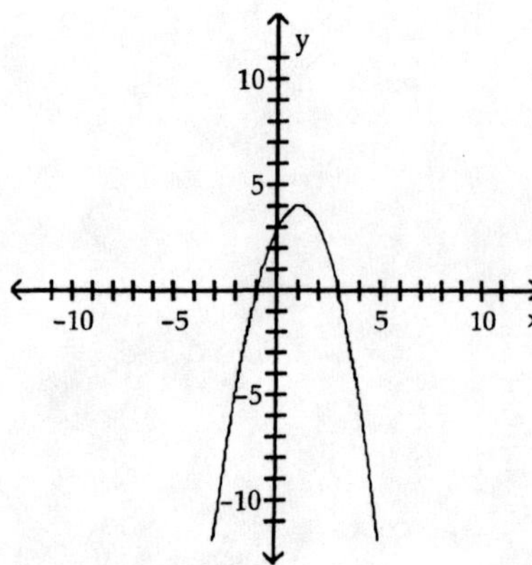

A) y-axis

B) x-axis

C) origin

D) x-axis, y-axis, origin

E) none

5)

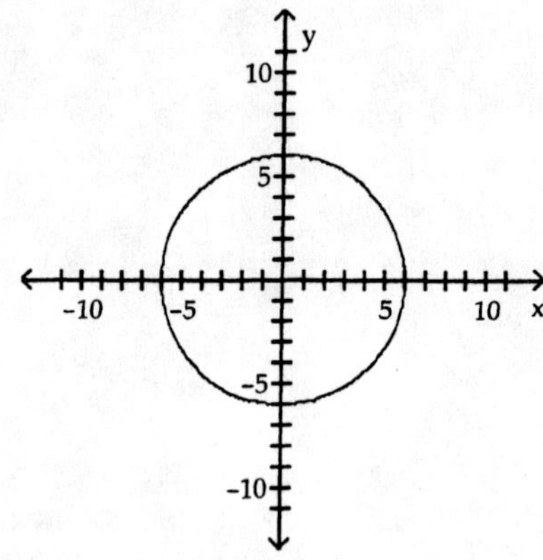

A) y-axis

B) x-axis

C) origin

D) x-axis, y-axis, origin

E) none

6)

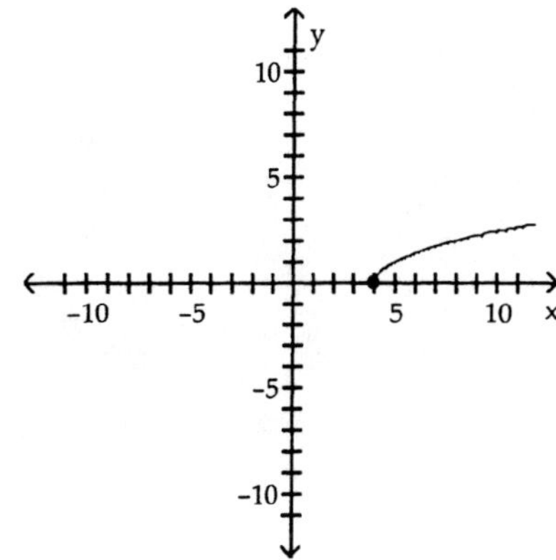

A) origin

B) x–axis

C) y–axis

D) x–axis, y–axis, origin

E) none

7)

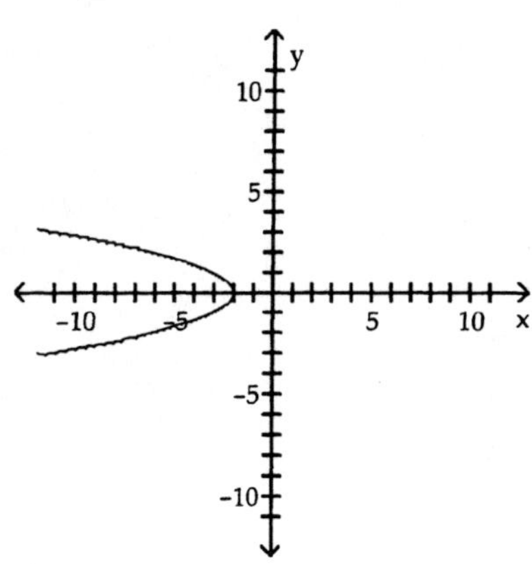

A) y–axis

B) origin

C) x–axis

D) x–axis, y–axis, origin

E) none

Determine whether the graph of the equation is symmetric with respect to the x–axis, the y–axis, and/or the origin.

8) $y = x + 1$

A) origin

B) y–axis

C) x–axis

D) x–axis, y–axis, origin

E) none

9) $y = -4x$

A) y-axis

B) x-axis

C) origin

D) x-axis, y-axis, origin

E) none

10) $x^2 = y$

A) x-axis

B) origin

C) y-axis

D) x-axis, y-axis, origin

E) none

11) $y^2 = x$

A) origin

B) x-axis

C) y-axis

D) x-axis, y-axis, origin

E) none

12) $x^2 + y - 9 = 0$

A) y-axis

B) origin

C) x-axis

D) x-axis, y-axis, origin

E) none

13) $y^2 - x - 25 = 0$

A) y-axis

B) x-axis

C) origin

D) x-axis, y-axis, origin

E) none

14) $9x^2 + 16y^2 = 144$

A) y-axis

B) x-axis

C) origin

D) x-axis, y-axis, origin

E) none

15) $16x^2 + y^2 = 16$

A) y-axis

B) origin

C) x-axis

D) x-axis, y-axis, origin

E) none

16) $y = x^2 + 7x + 10$

A) y-axis

B) x-axis

C) origin

D) x-axis, y-axis, origin

E) none

17) $y = \frac{8x}{x^2 + 64}$

A) origin

B) y-axis

C) x-axis

D) x-axis, y-axis, origin

E) none

18) $y = \frac{x^2 - 64}{8x^4}$

A) y-axis

B) origin

C) x-axis

D) x-axis, y-axis, origin

E) none

19) $y = 4x^2 + 5$

A) x-axis

B) y-axis

C) origin

D) x-axis, y-axis, origin

E) none

20) $y = (x + 6)(x + 5)$

A) y-axis

B) x-axis

C) origin

D) x-axis, y-axis, origin

E) none

21) $y = -7x^3 + 6x$

A) x–axis

B) y–axis

C) origin

D) x–axis, y–axis, origin

E) none

22) $y = -1x^4 + 3x - 9$

A) y–axis

B) x–axis

C) origin

D) x–axis, y–axis, origin

E) none

Solve the problem.

23) Find the equation whose graph is symmetric with respect to the y–axis.

A) $x = 2y^2$ B) $3x^2 = y$ C) $y = 5x$ D) $y = 2x^3$

24) If a graph is symmetric with respect to the y–axis and it contains the point (5, –6), which of the following points is also on the graph?

A) (–6, 5) B) (–5, –6) C) (–5, 6) D) (5, –6)

25) If a graph is symmetric with respect to the origin and it contains the point (–4, 7), which of the following points is also on the graph?

A) (4, 7) B) (4, –7) C) (7, –4) D) (–4, –7)

5 **Additional Applications**

List the intercepts and type(s) of symmetry, if any.

1) $y^2 - 1 = x$

A) intercepts: (–1, 0), (0, 1), (0, –1)
symmetric with respect to x–axis

B) intercepts: (0, –1), (1, 0), (–1, 0)
symmetric with respect to y–axis

C) intercepts: (1, 0), (0, 1), (0, –1)
symmetric with respect to x–axis

D) intercepts: (0, 1), (1, 0), (–1, 0)
symmetric with respect to y–axis

2) $16x^2 + y^2 = 16$

A) intercepts: (4, 0), (–4, 0), (0, 1), (0, –1)
symmetric with respect to the origin

B) intercepts: (4, 0), (–4, 0), (0, 1), (0, –1)
symmetric with respect to x–axis and y–axis

C) intercepts: (1, 0), (–1, 0), (0, 4), (0, –4)
symmetric with respect to x–axis and y–axis

D) intercepts: (1, 0), (–1, 0), (0, 4), (0, –4)
symmetric with respect to x–axis, y–axis, and origin

3) $y = \frac{-x^3}{x^2 - 2}$

A) intercept: (0, 0)
symmetric with respect to origin

B) intercepts: $(\sqrt{2}, 0)$, $(-\sqrt{2}, 0)$, (0, 0)
symmetric with respect to origin

C) intercept: (0, 0)
symmetric with respect to y-axis

D) intercept: (0, 0)
symmetric with respect to x-axis

List the intercepts of the graph. Tell whether the graph is symmetric with respect to the x-axis, y-axis, origin, or none of these.

4)

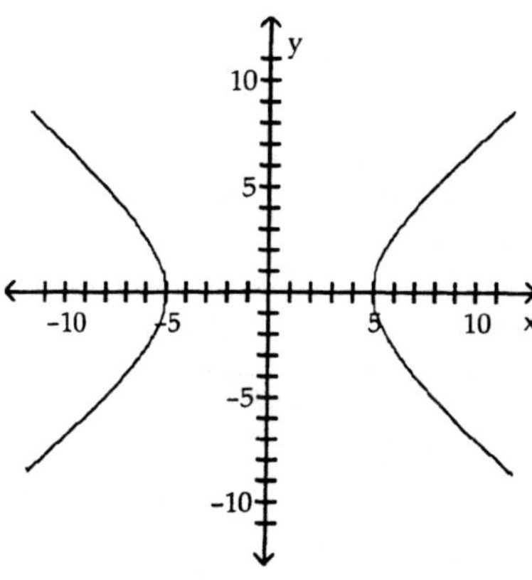

A) intercepts: (-5, 0) and (5, 0)
symmetric with respect to origin

B) intercepts: (0, -5) and (0, 5)
symmetric with respect to x-axis, y-axis, and origin

C) intercepts: (0, -5) and (0, 5)
symmetric with respect to y-axis

D) intercepts: (-5, 0) and (5, 0)
symmetric with respect to x-axis, y-axis, and origin

5)

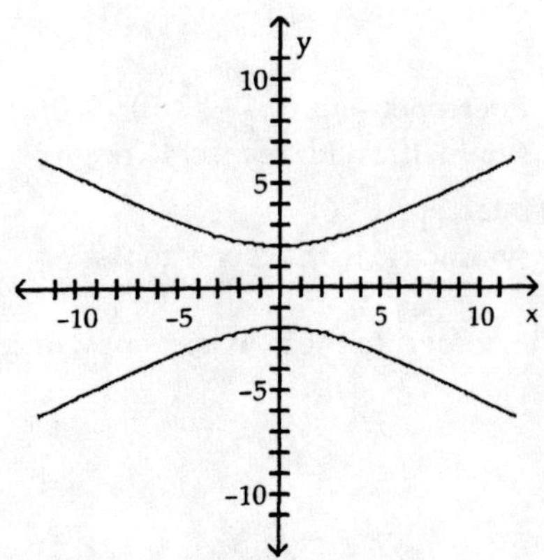

A) intercepts: (0, 2) and (0, -2)
symmetric with respect to x-axis, y-axis, and origin

B) intercepts: (0, 2) and (0, -2)
symmetric with respect to origin

C) intercepts: (2, 0) and (-2, 0
symmetric with respect to y-axis

D) intercepts: (2, 0) and (-2, 0)
symmetric with respect to x-axis, y-axis, and origin

6)

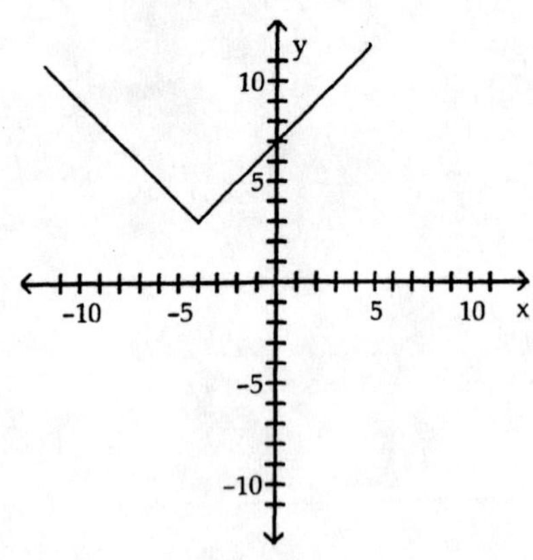

A) intercept: (7, 0)
symmetric with respect to y-axis

B) intercept: (0, 7)
no symmetry

C) intercept: (7, 0)
no symmetry

D) intercept: (0, 7)
symmetric with respect to x-axis

7)

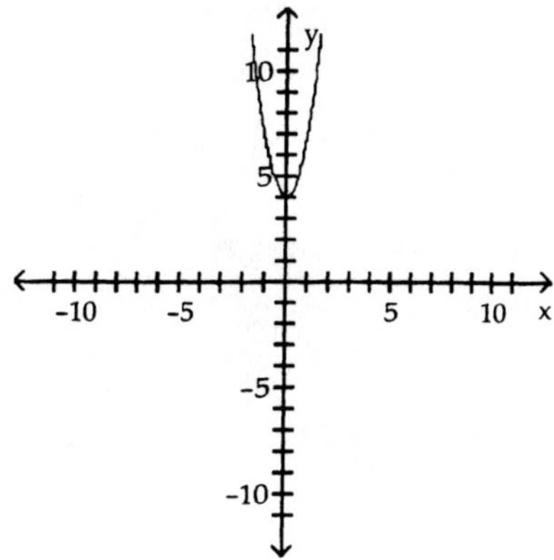

A) intercept: (0, 4)
symmetric with respect to origin

B) intercept: (4, 0)
symmetric with respect to x-axis

C) intercept: (4, 0)
symmetric with respect to y-axis

D) intercept: (0, 4)
symmetric with respect to y-axis

8)

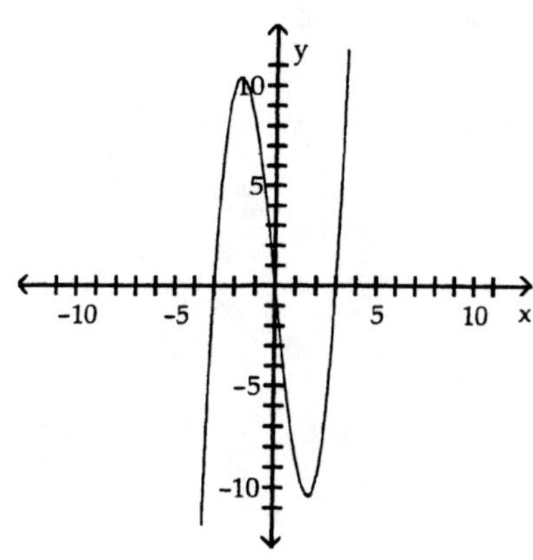

A) intercepts: (-3, 0), (0, 0), (3, 0)
symmetric with respect to origin

B) intercepts: (-3, 0), (0, 0), (3, 0)
symmetric with respect to x-axis, y-axis, and origin

C) intercepts: (-3, 0), (0, 0), (3, 0)
symmetric with respect to y-axis

D) intercepts: (-3, 0), (0, 0), (3, 0)
symmetric with respect to x-axis

2.3 Circles

1 Write the Standard Form of the Equation of a Circle

Write the standard form of the equation of the circle.

1)

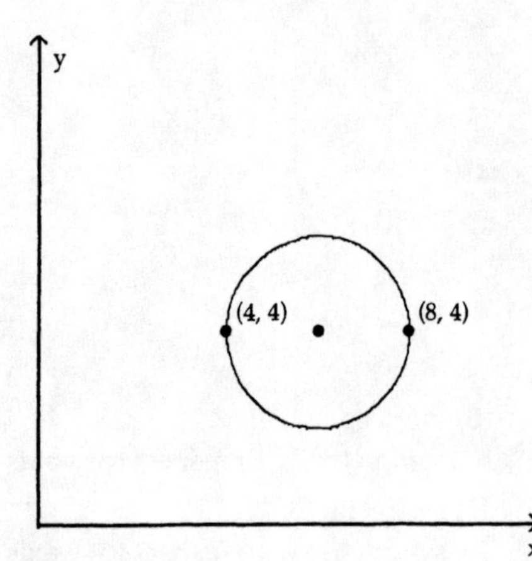

A) $(x - 6)^2 + (y - 4)^2 = 2$

B) $(x + 6)^2 + (y + 4)^2 = 4$

C) $(x - 6)^2 + (y - 4)^2 = 4$

D) $(x + 6)^2 + (y + 4)^2 = 2$

2)

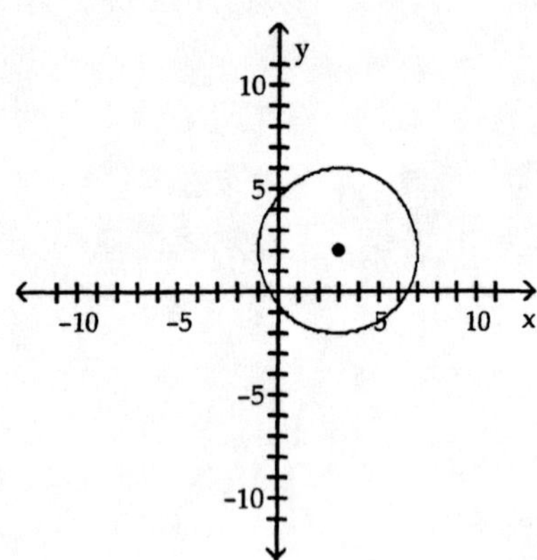

A) $(x - 3)^2 + (y - 2)^2 = 16$

B) $(x - 2)^2 + (y - 3)^2 = 16$

C) $(x + 3)^2 + (y + 2)^2 = 16$

D) $(x + 2)^2 + (y + 3)^2 = 16$

Write the standard form of the equation of the circle with radius r and center (h, k).

3) $r = 3$; $(h, k) = (0, 0)$

A) $(x - 3)^2 + (y - 3)^2 = 9$

B) $x^2 + y^2 = 9$

C) $(x - 3)^2 + (y - 3)^2 = 3$

D) $x^2 + y^2 = 3$

4) $r = 10$; $(h, k) = (-6, 5)$

A) $(x - 6)^2 + (y + 5)^2 = 10$

B) $(x + 6)^2 + (y - 5)^2 = 10$

C) $(x + 6)^2 + (y - 5)^2 = 100$

D) $(x - 6)^2 + (y + 5)^2 = 100$

5) $r = 5$; $(h, k) = (-9, 0)$

A) $x^2 + (y - 9)^2 = 5$

B) $(x - 9)^2 + y^2 = 25$

C) $x^2 + (y + 9)^2 = 5$

D) $(x + 9)^2 + y^2 = 25$

6) $r = 7$; $(h, k) = (0, 6)$

A) $x^2 + (y + 6)^2 = 7$ B) $(x - 6)^2 + y^2 = 49$ C) $x^2 + (y - 6)^2 = 49$ D) $(x + 6)^2 + y^2 = 49$

7) $r = \sqrt{10}$; $(h, k) = (5, -1)$

A) $(x + 1)^2 + (y - 5)^2 = 100$ B) $(x + 5)^2 + (y - 1)^2 = 10$
C) $(x - 5)^2 + (y + 1)^2 = 10$ D) $(x - 1)^2 + (y + 5)^2 = 100$

8) $r = \sqrt{11}$; $(h, k) = (0, 3)$

A) $(x + 3)^2 + y^2 = 121$ B) $x^2 + (y + 3)^2 = 11$ C) $(x - 3)^2 + y^2 = 121$ D) $x^2 + (y - 3)^2 = 11$

9) $r = 4\sqrt{5}$; $(h, k) = (-9, -8)$

A) $(x + 9)^2 + (y + 8)^2 = 16\sqrt{5}$ B) $(x - 9)^2 + (y - 8)^2 = 16\sqrt{5}$
C) $(x - 9)^2 + (y - 8)^2 = 80$ D) $(x + 9)^2 + (y + 8)^2 = 80$

Solve the problem.

10) Find the equation of a circle in standard form where C(6, –2) and D(–4, 4) are endpoints of a diameter.

A) $(x + 1)^2 + (y + 1)^2 = 34$ B) $(x - 1)^2 + (y - 1)^2 = 34$
C) $(x - 1)^2 + (y - 1)^2 = 136$ D) $(x + 1)^2 + (y + 1)^2 = 136$

11) Find the equation of a circle in standard form with center at the point (–3, 2) and tangent to the line y = 4.

A) $(x + 3)^2 + (y - 2)^2 = 16$ B) $(x + 3)^2 + (y - 2)^2 = 4$
C) $(x - 3)^2 + (y + 2)^2 = 4$ D) $(x - 3)^2 + (y + 2)^2 = 16$

12) Find the equation of a circle in standard form that is tangent to the line x = –3 at (–3, 5) and also tangent to the line x = 9.

A) $(x - 3)^2 + (y + 5)^2 = 36$ B) $(x + 3)^2 + (y + 5)^2 = 36$
C) $(x - 3)^2 + (y - 5)^2 = 36$ D) $(x + 3)^2 + (y - 5)^2 = 36$

Find the center (h, k) and radius r of the circle with the given equation.

13) $x^2 + y^2 = 16$

A) $(h, k) = (0, 0)$; $r = 4$ B) $(h, k) = (4, 4)$; $r = 4$
C) $(h, k) = (0, 0)$; $r = 16$ D) $(h, k) = (4, 4)$; $r = 16$

14) $(x + 8)^2 + (y + 1)^2 = 49$

A) $(h, k) = (-1, -8)$; $r = 49$ B) $(h, k) = (-8, -1)$; $r = 7$
C) $(h, k) = (-1, -8)$; $r = 7$ D) $(h, k) = (-8, -1)$; $r = 49$

15) $(x - 3)^2 + y^2 = 81$

A) $(h, k) = (3, 0)$; $r = 9$ B) $(h, k) = (0, 3)$; $r = 81$
C) $(h, k) = (0, 3)$; $r = 9$ D) $(h, k) = (3, 0)$; $r = 81$

16) $x^2 + (y + 7)^2 = 9$

A) $(h, k) = (0, -7)$; $r = 3$ B) $(h, k) = (0, -7)$; $r = 9$
C) $(h, k) = (-7, 0)$; $r = 3$ D) $(h, k) = (-7, 0)$; $r = 9$

Find the center (h, k) and radius r of the circle.

17) $3(x - 3)^2 + 3(y + 6)^2 = 45$

A) $(h, k) = (3, -6)$; $r = 3\sqrt{15}$

B) $(h, k) = (-3, 6)$; $r = \sqrt{15}$

C) $(h, k) = (3, -6)$; $r = \sqrt{15}$

D) $(h, k) = (-3, 6)$; $r = 3\sqrt{15}$

Solve the problem.

18) Find the standard form of the equation of the circle. Assume that the center has integer coordinates and the radius is an integer.

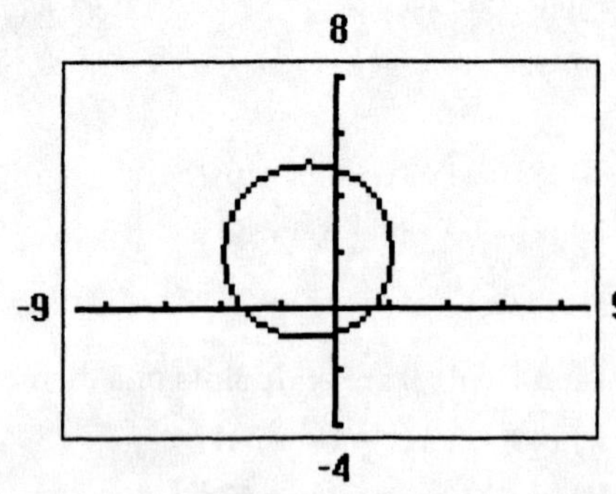

A) $x^2 + y^2 - 2x + 4y - 4 = 0$

B) $x^2 + y^2 + 2x - 4y - 4 = 0$

C) $(x - 1)^2 + (y + 2)^2 = 9$

D) $(x + 1)^2 + (y - 2)^2 = 9$

2 Graph a Circle

Graph the circle with radius r and center (h, k).

1) $r = 4$; $(h, k) = (0, 0)$

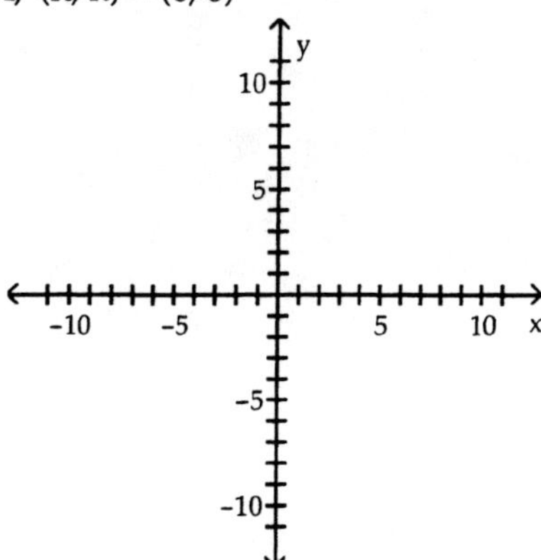

A)

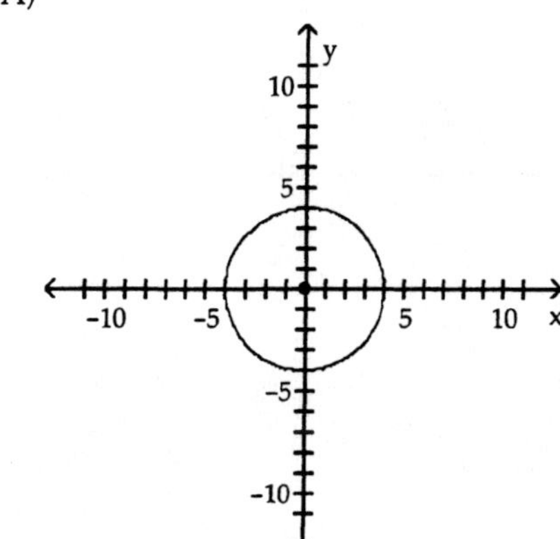

B)

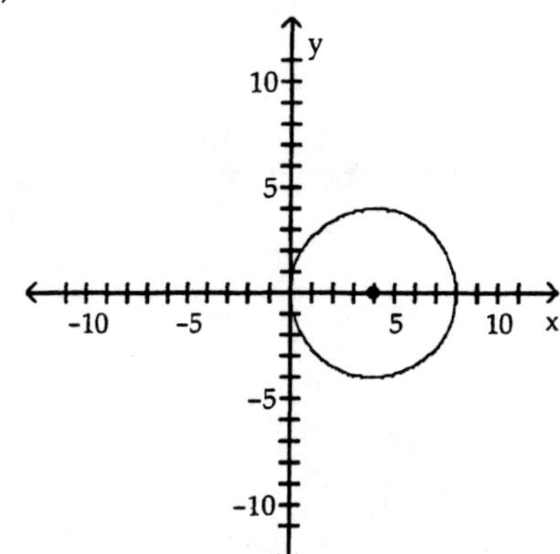

C)

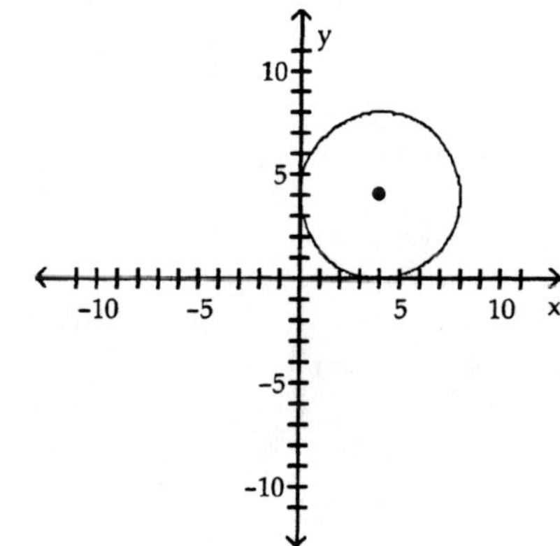

D)

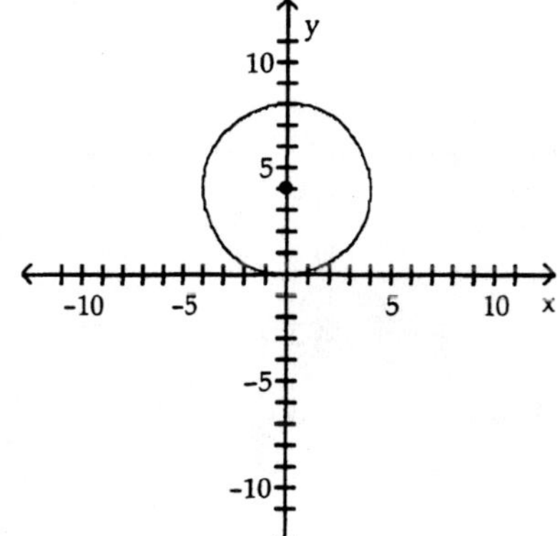

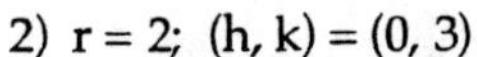
2) $r = 2$; $(h, k) = (0, 3)$

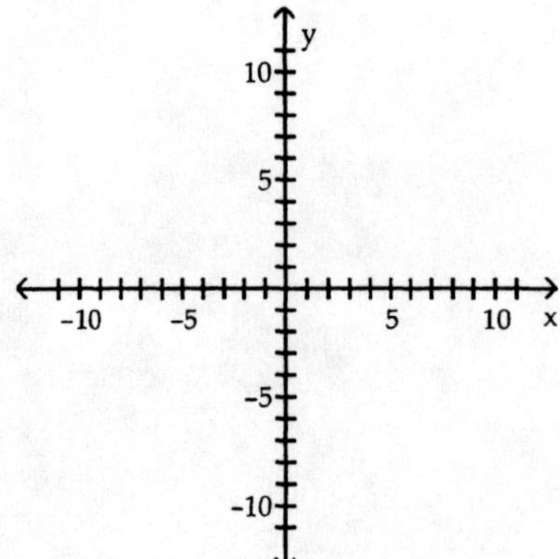

A)

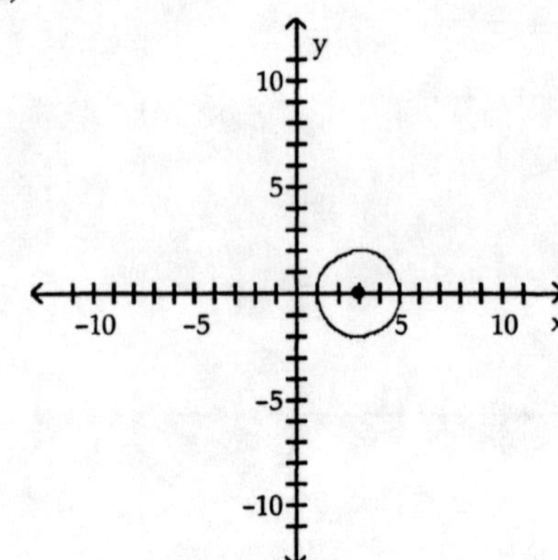

B)

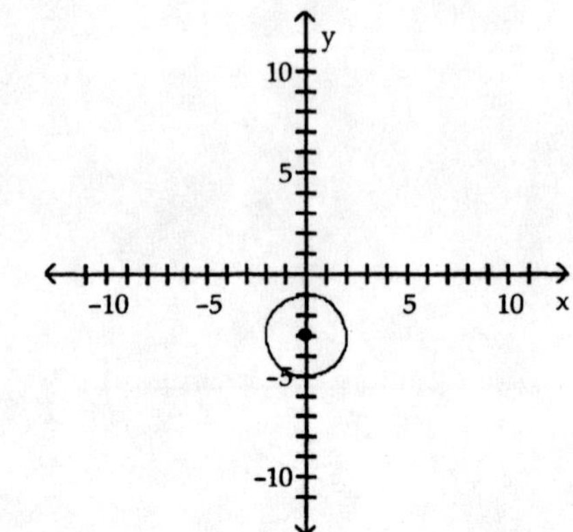

C)

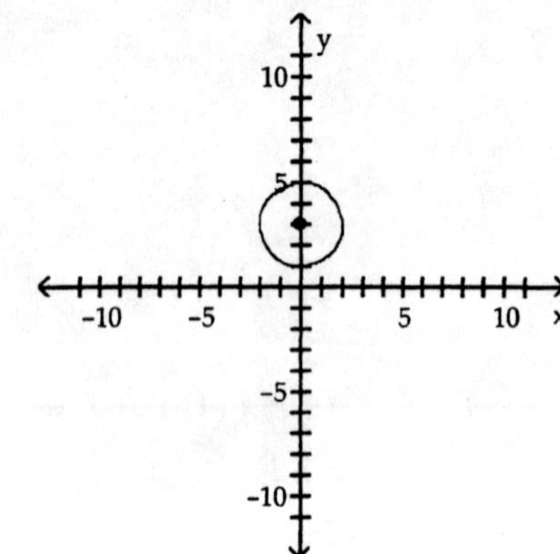

D)

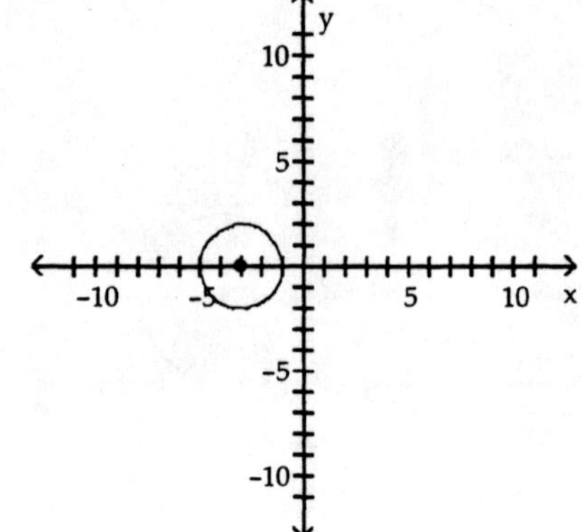

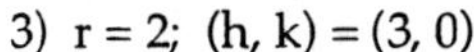

3) $r = 2$; $(h, k) = (3, 0)$

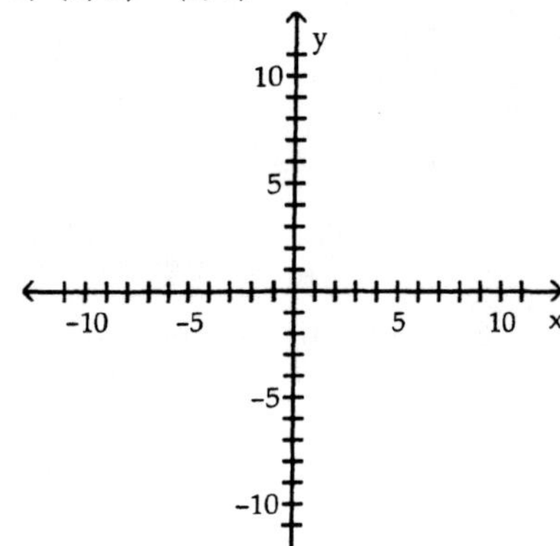

A)

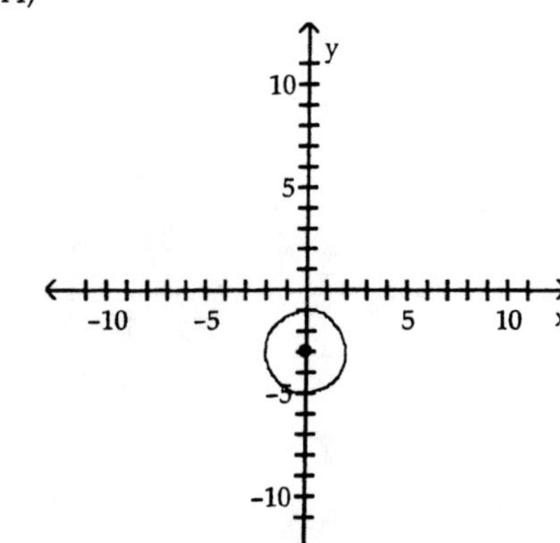

B)

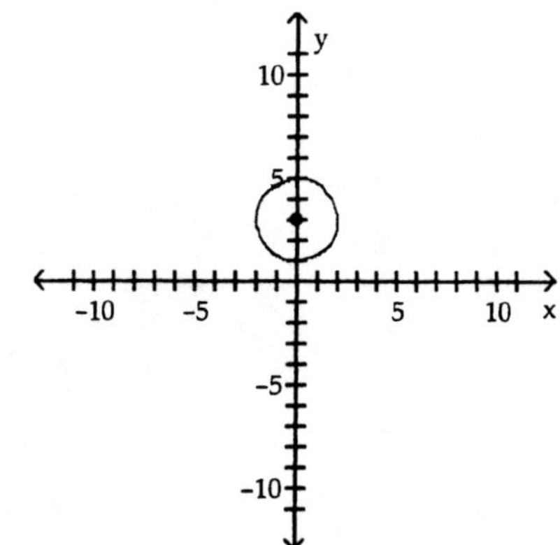

C)

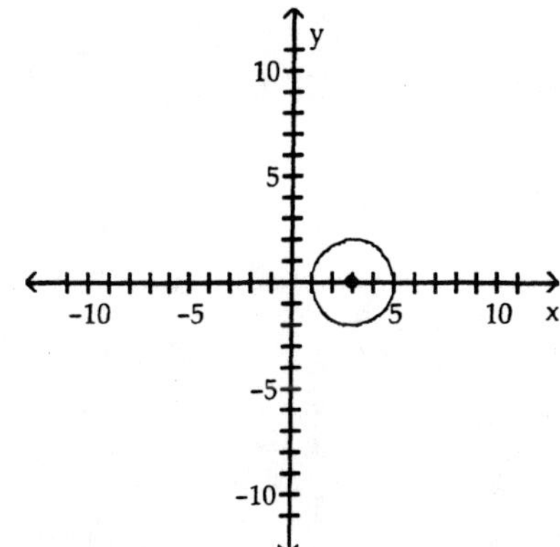

D)

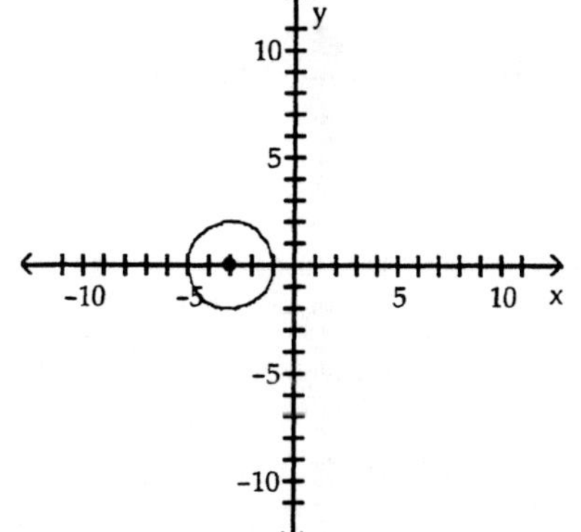

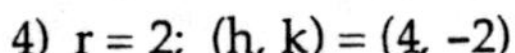

4) $r = 2$; $(h, k) = (4, -2)$

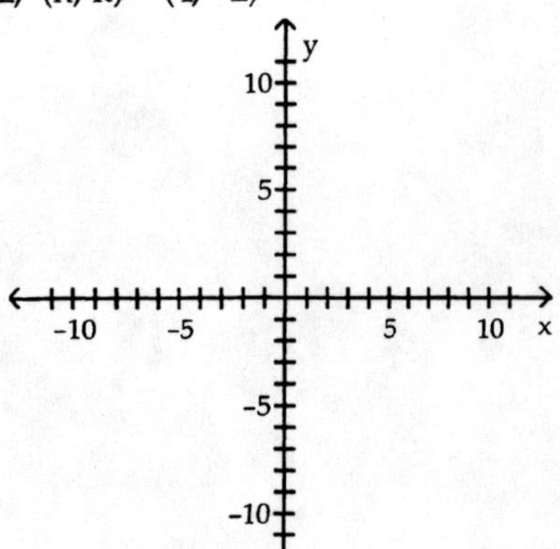

A)

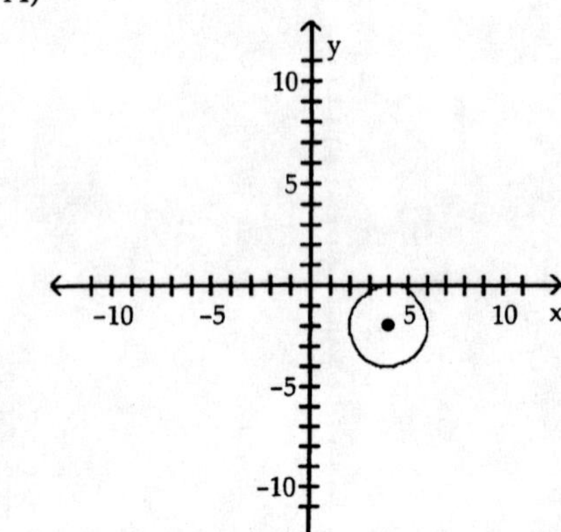

B)

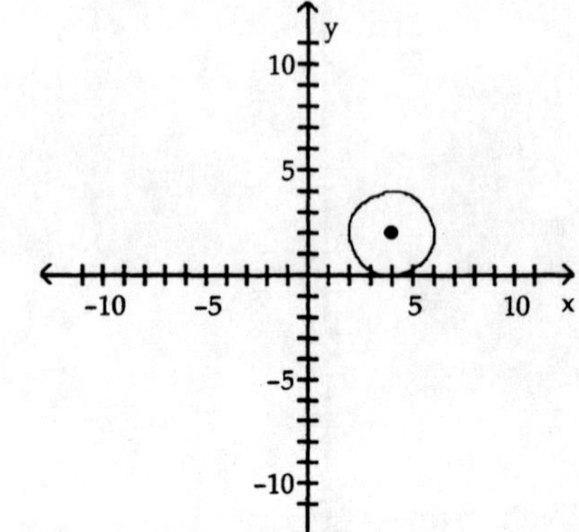

C)

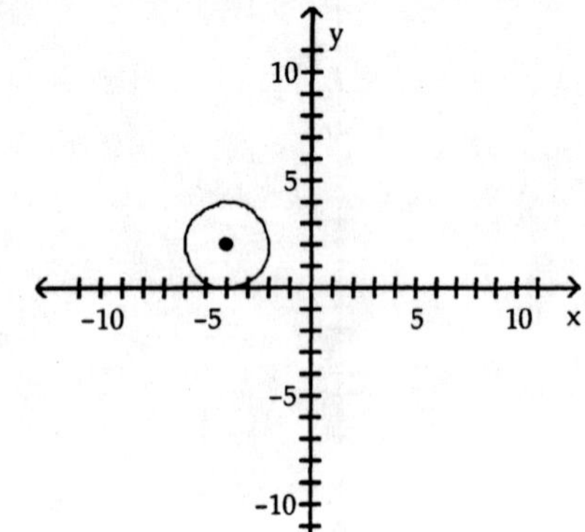

D)

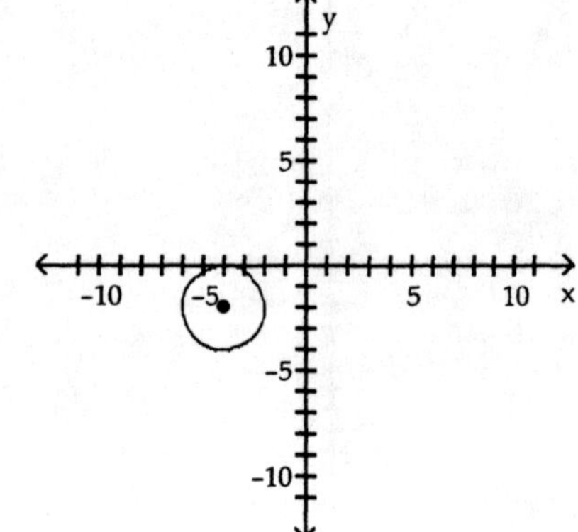

Graph the equation.

5) $x^2 + y^2 = 9$

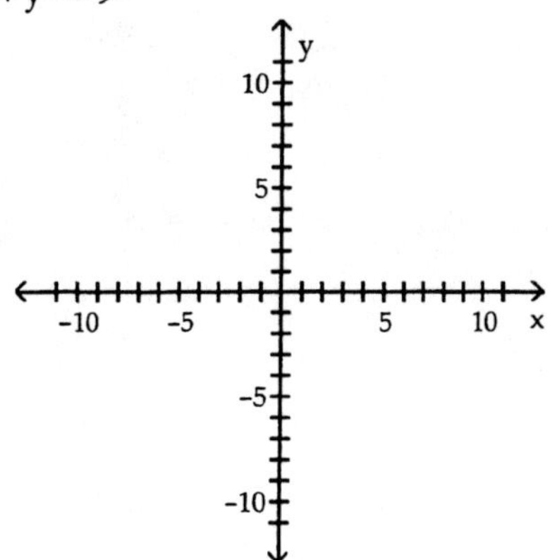

A)

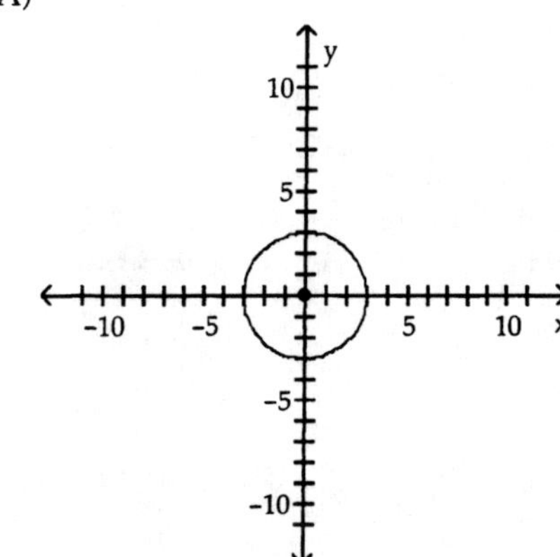

B)

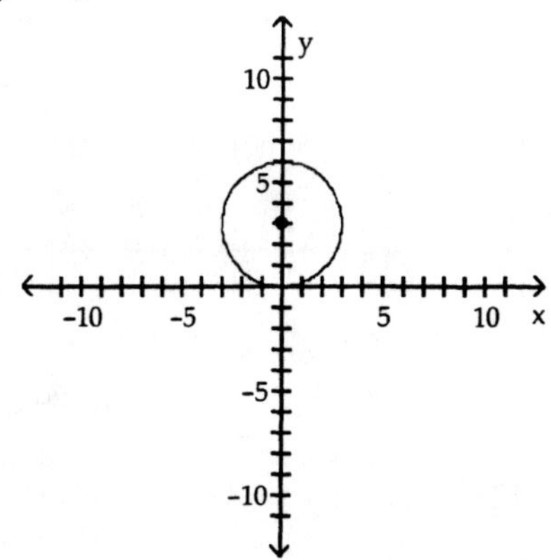

C)

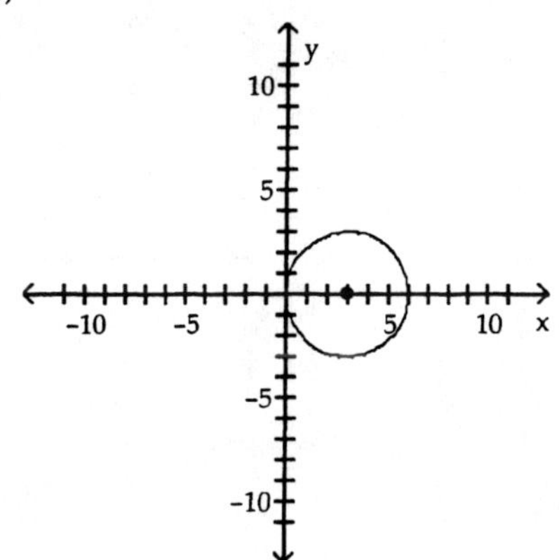

D)

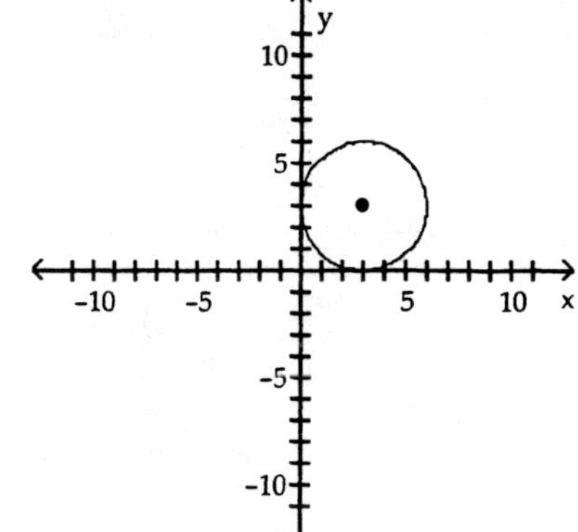

6) $(x + 2)^2 + (y + 5)^2 = 16$

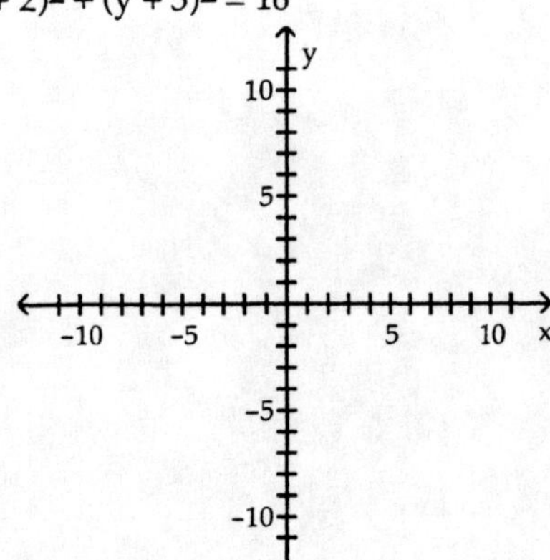

A)

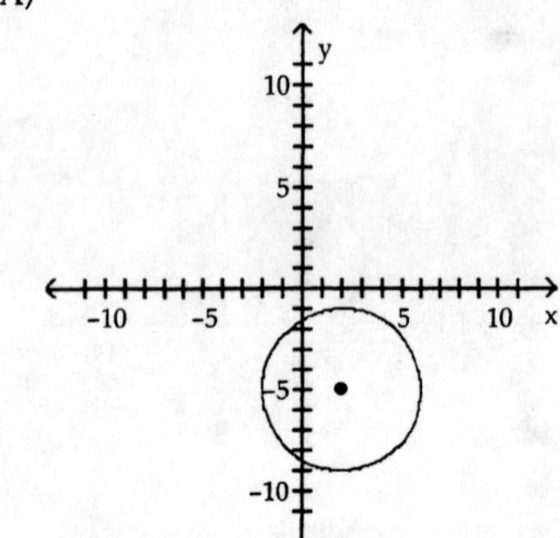

B)

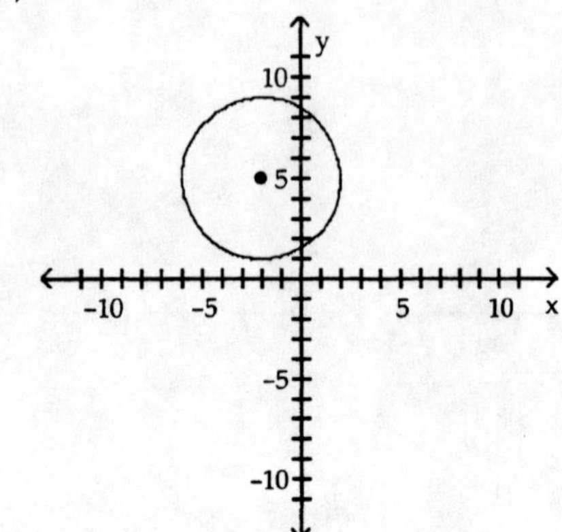

C)

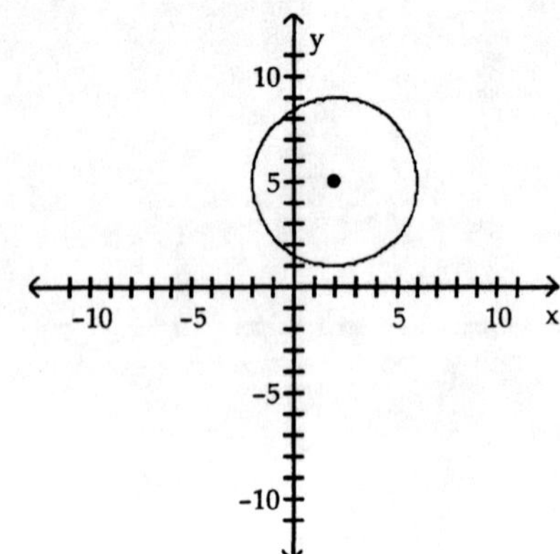

D)

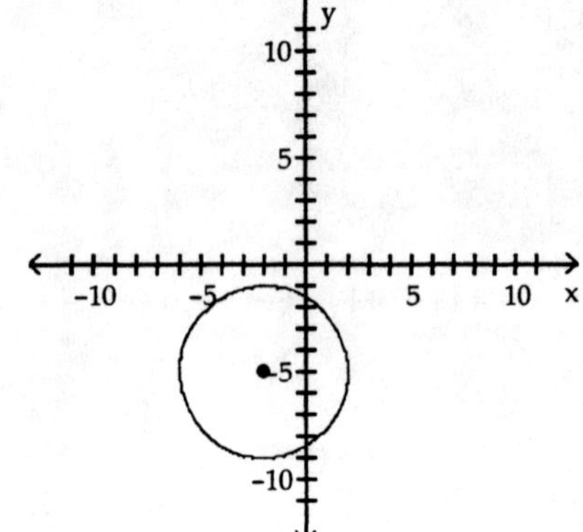

7) $x^2 + (y - 5)^2 = 4$

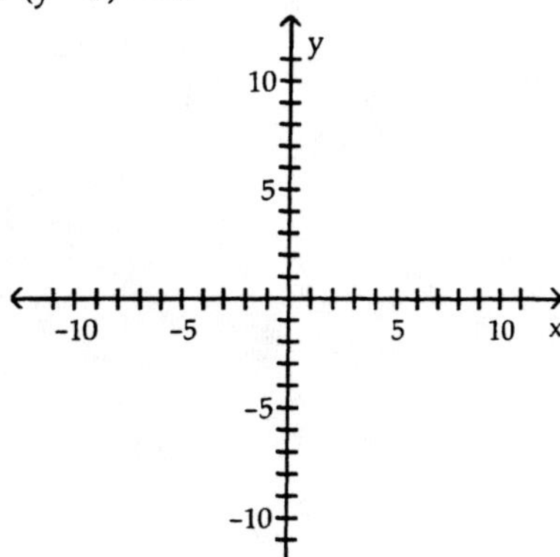

A)

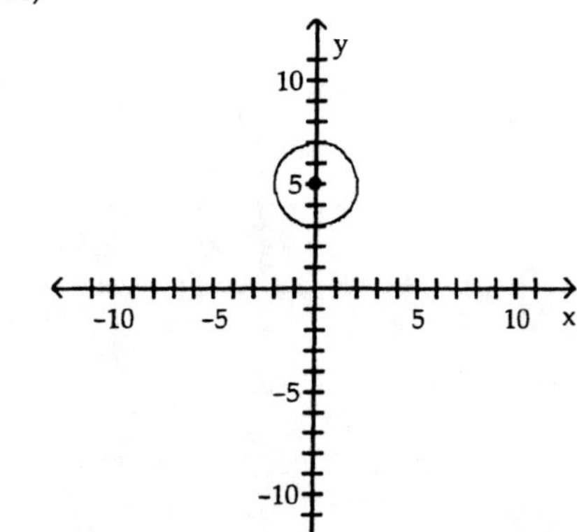

B)

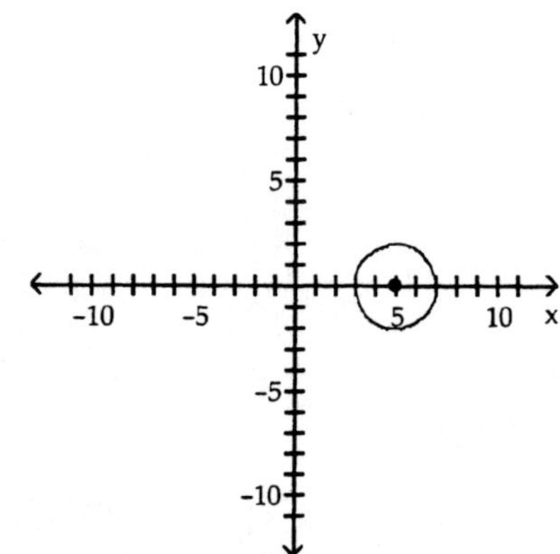

C)

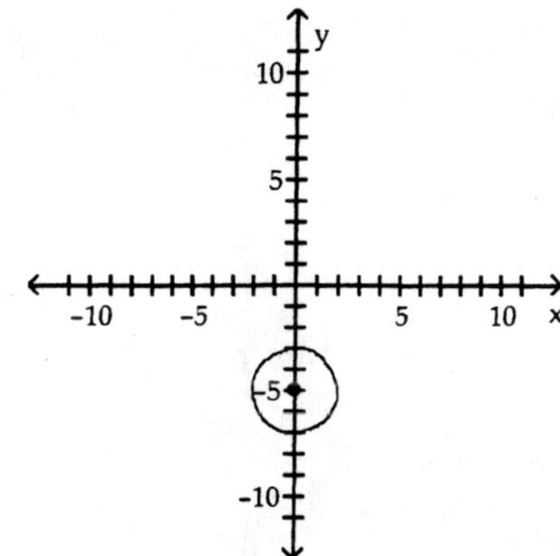

D)

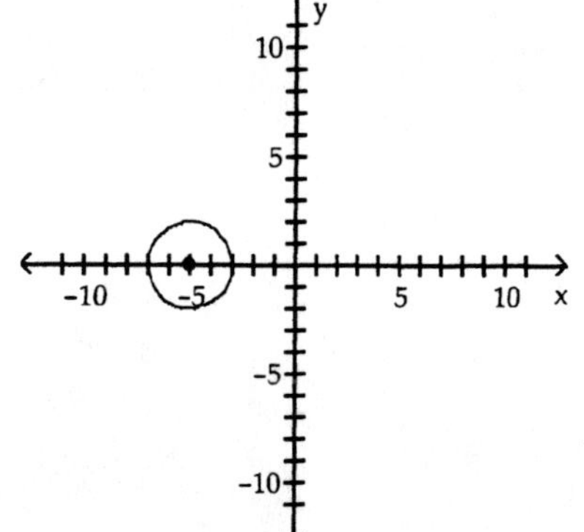

8) $(x - 4)^2 + y^2 = 9$

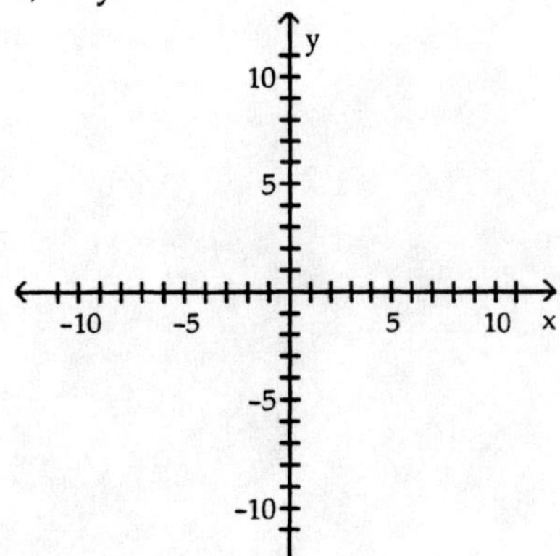

A)

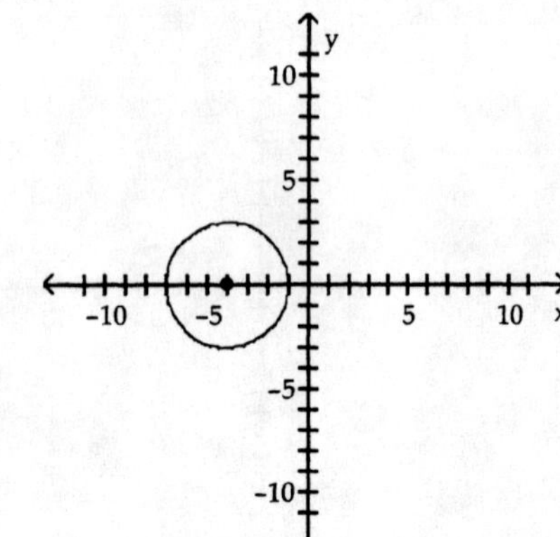

B)

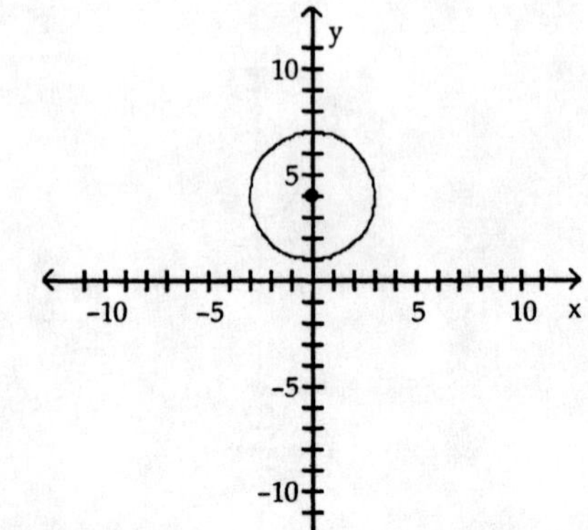

C)

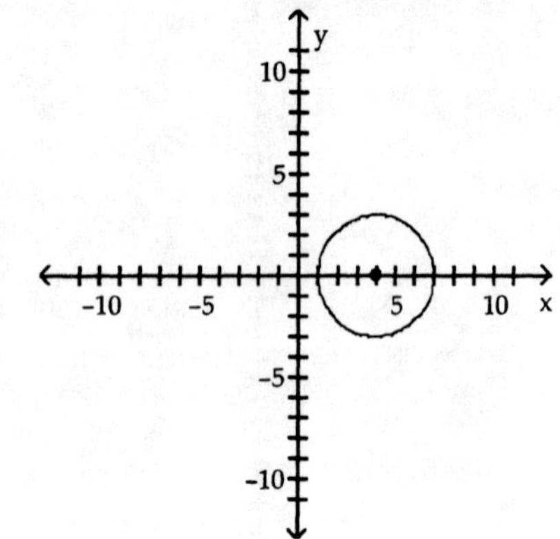

D)

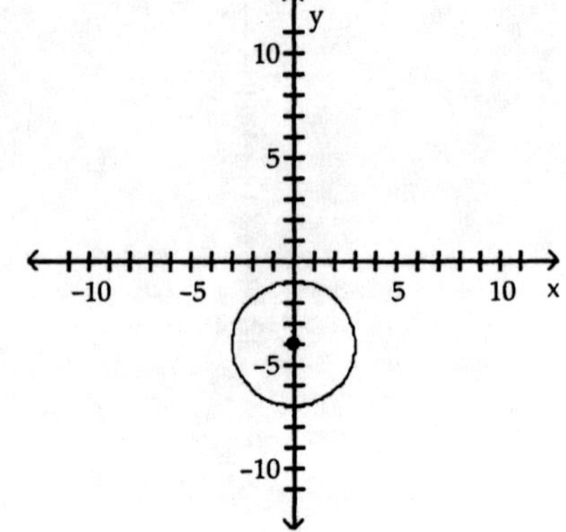

3 Find the Center and Radius of a Circle in General Form and Graph It

Find the center (h, k) and radius r of the circle. Graph the circle.

1) $x^2 + y^2 - 6x - 6y - 7 = 0$

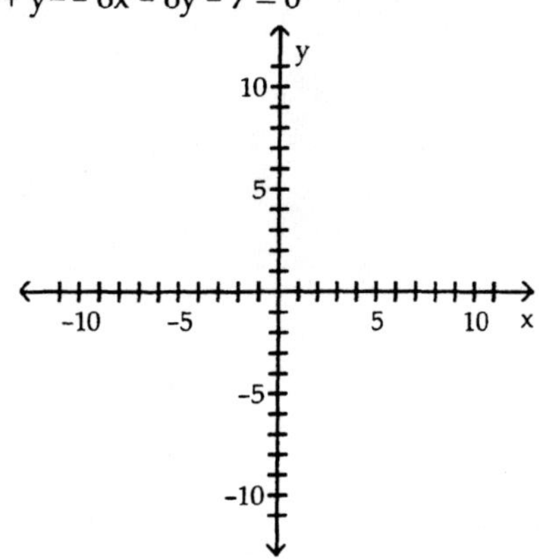

A) $(h, k) = (3, -3)$; $r = 5$

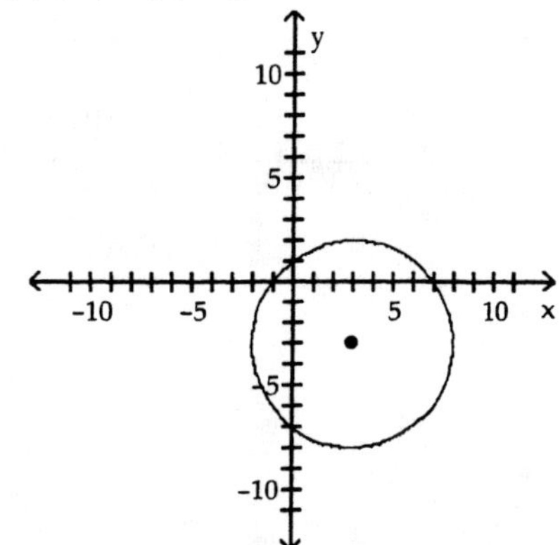

B) $(h, k) = (3, 3)$; $r = 5$

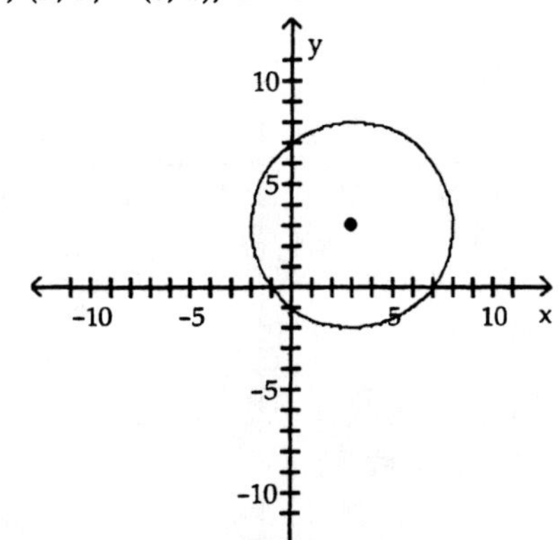

C) $(h, k) = (-3, 3)$; $r = 5$

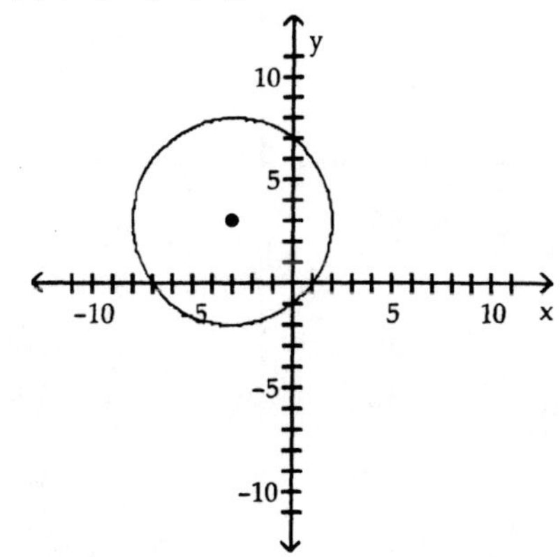

D) $(h, k) = (-3, -3)$; $r = 5$

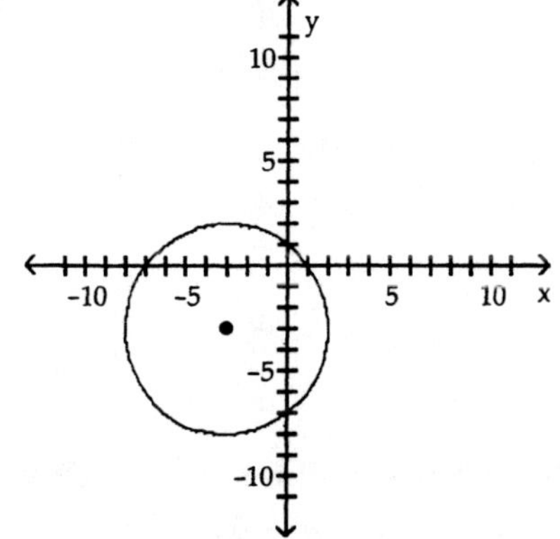

2) $x^2 + y^2 + 4x + 2y - 31 = 0$

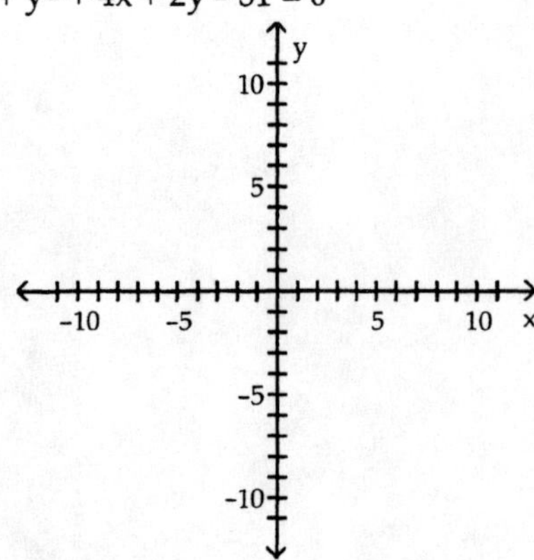

A) $(h, k) = (2, -1);\ r = 6$

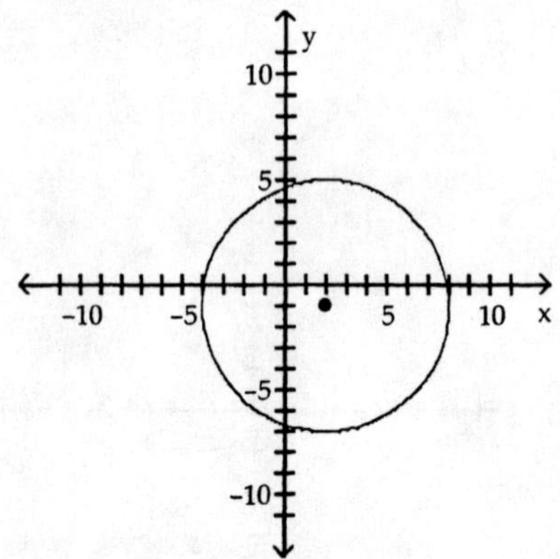

B) $(h, k) = (2, 1);\ r = 6$

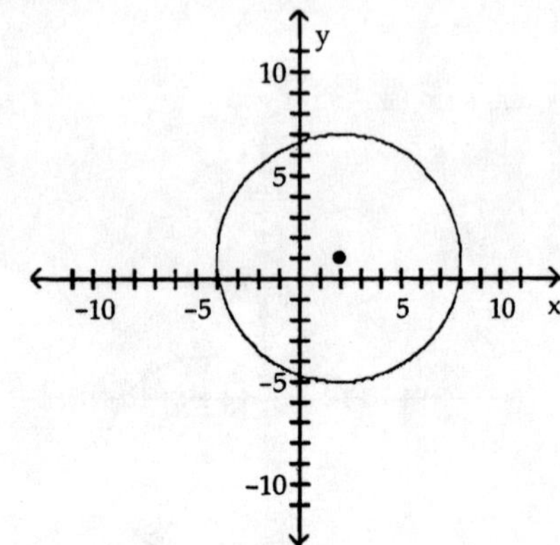

C) $(h, k) = (-2, -1);\ r = 6$

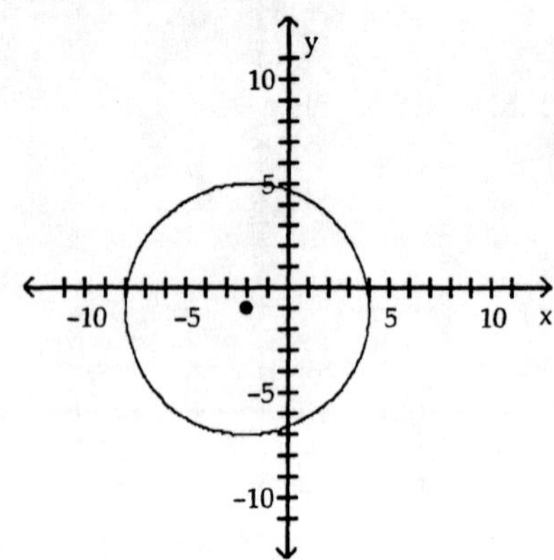

D) $(h, k) = (-2, 1);\ r = 6$

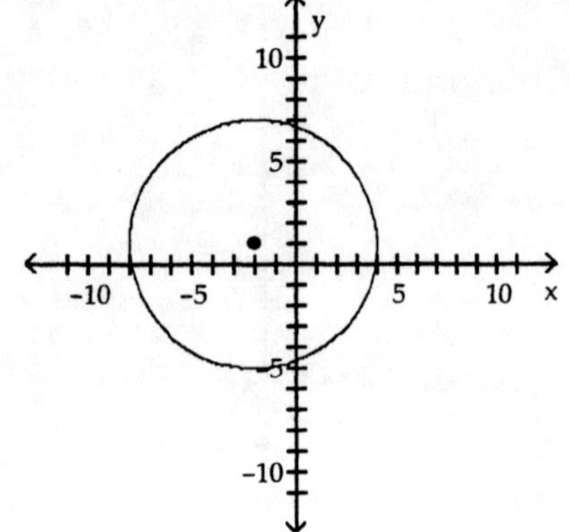

Find the center (h, k) and radius r of the circle with the given equation.

3) $4x^2 + 4y^2 - 12x + 16y - 5 = 0$

A) $(h, k) = (\frac{3}{2}, -2);\ r = \frac{3\sqrt{5}}{2}$

B) $(h, k) = (-\frac{3}{2}, 2);\ r = \frac{3\sqrt{5}}{2}$

C) $(h, k) = (-\frac{3}{2}, 2);\ r = \frac{\sqrt{30}}{2}$

D) $(h, k) = (\frac{3}{2}, -2);\ r = \frac{\sqrt{30}}{2}$

4) $x^2 - 8x + 16 + (y - 3)^2 = 49$

A) $(h, k) = (-4, -3);\ r = 49$

B) $(h, k) = (3, 4);\ r = 7$

C) $(h, k) = (-3, -4);\ r = 49$

D) $(h, k) = (4, 3);\ r = 7$

5) $x^2 + 12x + 36 + y^2 - 6y + 9 = 64$

A) $(h, k) = (-6, 3);\ r = 8$

B) $(h, k) = (-3, 6);\ r = 64$

C) $(h, k) = (6, -3);\ r = 64$

D) $(h, k) = (3, -6);\ r = 8$

6) $x^2 + y^2 + 6x + 8y + 25 = 4$

A) $(h, k) = (-3, -4);\ r = 2$

B) $(h, k) = (-4, -3);\ r = 2$

C) $(h, k) = (4, 3);\ r = 4$

D) $(h, k) = (3, 4);\ r = 4$

7) $x^2 + y^2 + 14x + 16y = -32$

A) $(h, k) = (-7, -8);\ r = 9$

B) $(h, k) = (7, 8);\ r = 81$

C) $(h, k) = (-8, -7);\ r = 9$

D) $(h, k) = (8, 7);\ r = 81$

Find the general form of the equation of the the circle.

8) Center at the point $(-4, -3)$; containing the point $(-3, 3)$

A) $x^2 + y^2 - 6x + 6y - 12 = 0$

B) $x^2 + y^2 + 6x + 8y - 17 = 0$

C) $x^2 + y^2 + 6x - 6y - 17 = 0$

D) $x^2 + y^2 + 8x + 6y - 12 = 0$

9) Center at the point $(2, -3)$; containing the point $(5, -3)$

A) $x^2 + y^2 + 4x - 6y + 22 = 0$

B) $x^2 + y^2 + 4x - 6y + 4 = 0$

C) $x^2 + y^2 - 4x + 6y + 22 = 0$

D) $x^2 + y^2 - 4x + 6y + 4 = 0$

10) Center at the point $(3, 5)$; tangent to x-axis

A) $x^2 + y^2 - 6x - 10y + 9 = 0$

B) $x^2 + y^2 - 6x - 10y + 25 = 0$

C) $x^2 + y^2 - 6x - 10y + 59 = 0$

D) $x^2 + y^2 + 6x + 10y + 9 = 0$

2.4 Lines

1 Calculate and Interpret the Slope of a Line

Find the slope of the line through the points and interpret the slope.

1)

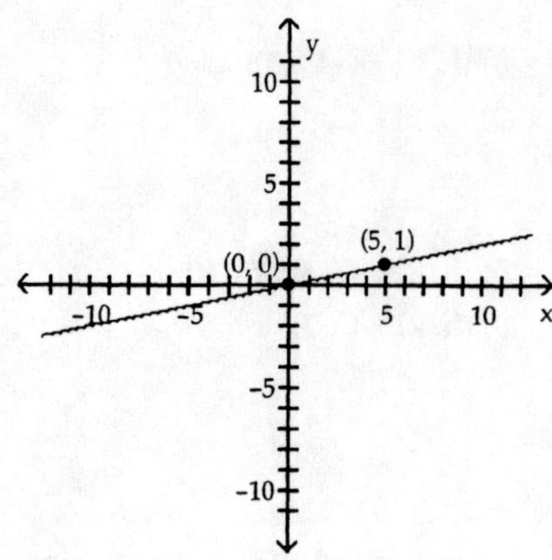

A) $-\frac{1}{5}$; for every 5-unit increase in x, y will decrease by 1 unit

B) $\frac{1}{5}$; for every 5-unit increase in x, y will increase by 1 unit

C) -5; for every 1-unit increase in x, y will decrease by 5 units

D) 5; for every 1-unit increase in x, y will increase by 5 units

2) $(1, -3)$; $(7, 8)$

A) $\frac{6}{11}$; for every 11-unit increase in x, y will increase by 6 units

B) $-\frac{11}{6}$; for every 6-unit increase in x, y will decrease by 11 units

C) $-\frac{6}{11}$; for every 11-unit increase in x, y will decrease by 6 units

D) $\frac{11}{6}$; for every 6-unit increase in x, y will increase by 11 units

Find the slope of the line.

3)

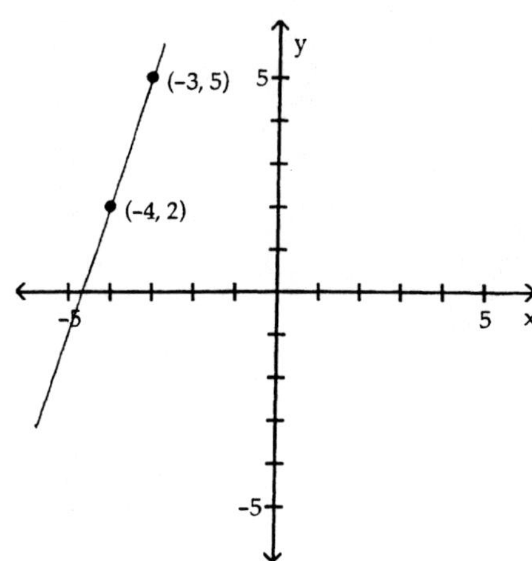

A) $\frac{1}{3}$

B) $-\frac{1}{3}$

C) 3

D) -3

4)

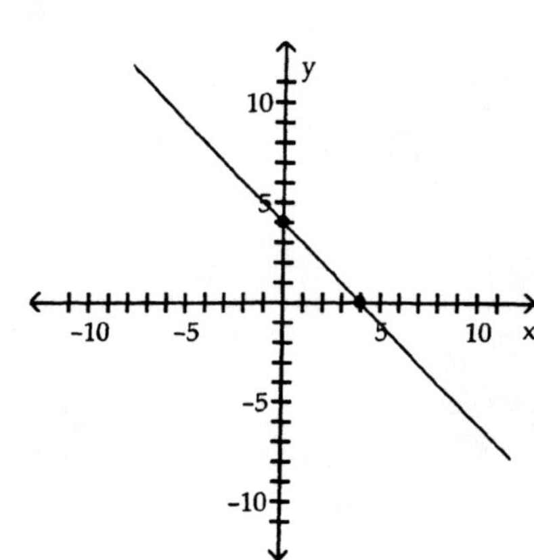

A) -4

B) 1

C) 4

D) -1

5)

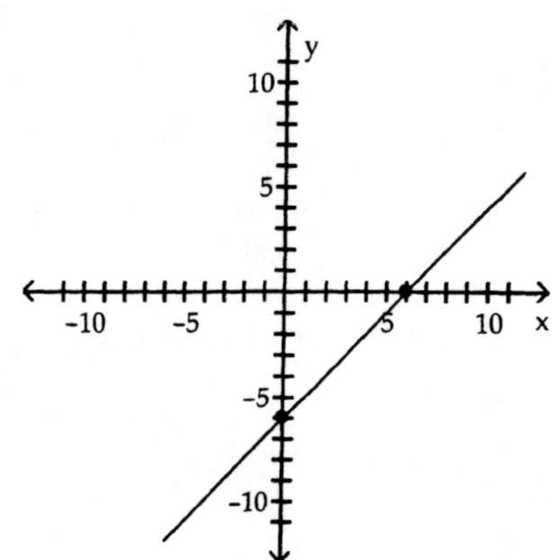

A) 6

B) -1

C) -6

D) 1

6)

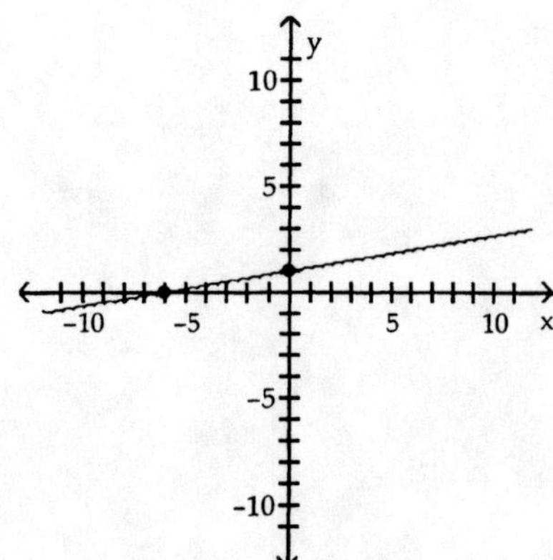

A) 6 B) $\frac{1}{6}$ C) -6 D) $-\frac{1}{6}$

Find the slope of the line containing the two points.

7) (3, 0); (-5, 9)

A) $-\frac{9}{8}$ B) $\frac{8}{9}$ C) $-\frac{8}{9}$ D) $\frac{9}{8}$

8) (5, 0); (0, 4)

A) $\frac{4}{5}$ B) $-\frac{4}{5}$ C) $\frac{5}{4}$ D) $-\frac{5}{4}$

9) (-5, -1); (6, 2)

A) $-\frac{3}{11}$ B) $\frac{3}{11}$ C) $\frac{11}{3}$ D) $-\frac{11}{3}$

10) (-9, -1); (-9, -2)

A) 0 B) 1 C) -1 D) undefined

11) (-9, 9); (-1, 9)

A) $-\frac{1}{8}$ B) 0 C) 8 D) undefined

2 Graph Lines Given a Point and the Slope

Graph the line containing the point P and having slope m.

1) $P = (-5, 6);\ m = -\frac{10}{7}$

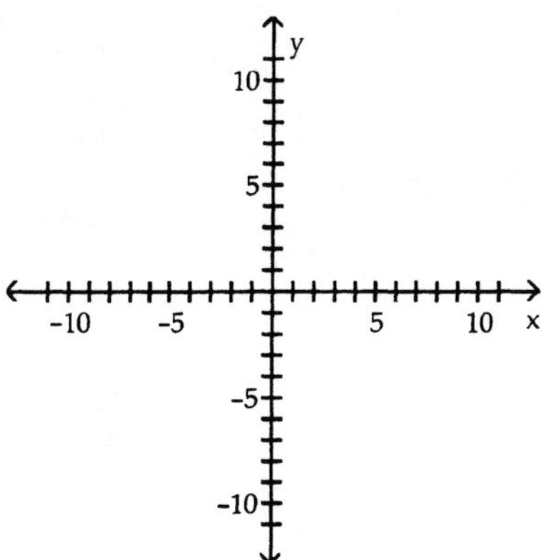

A)

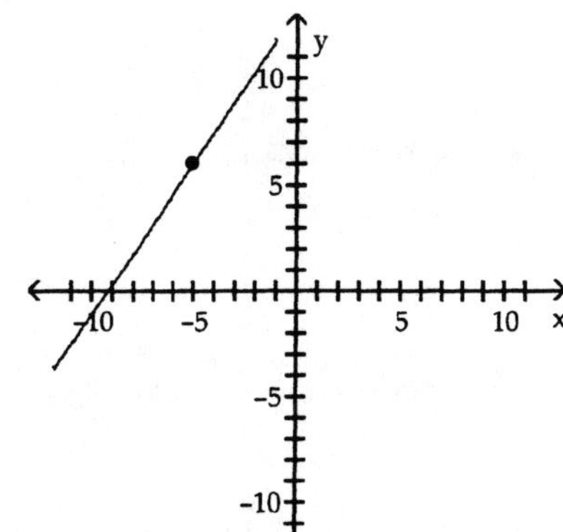

B)

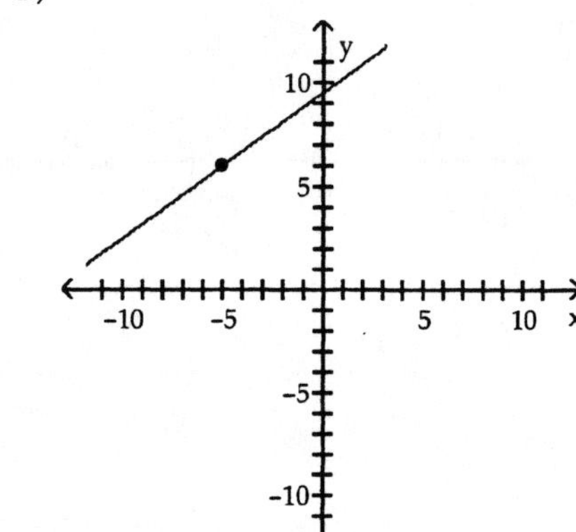

C)

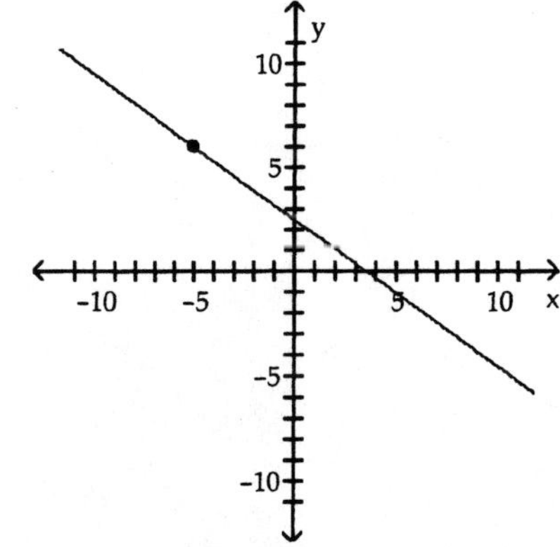

D)

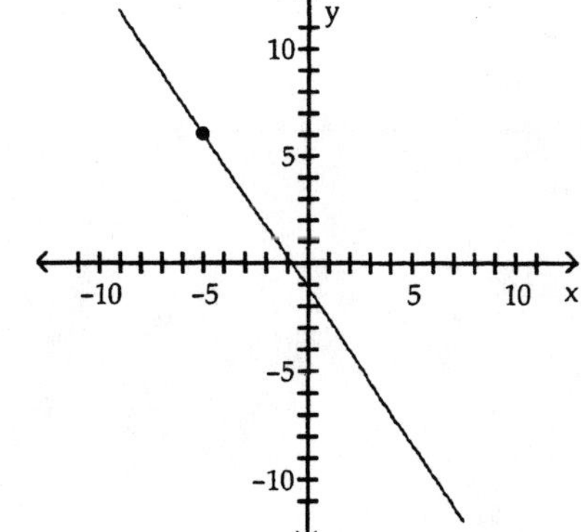

2) $P = (-5, -9)$; $m = \frac{2}{3}$

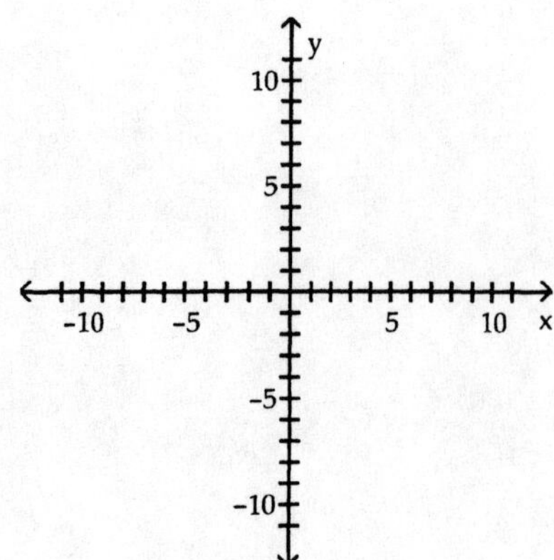

A)

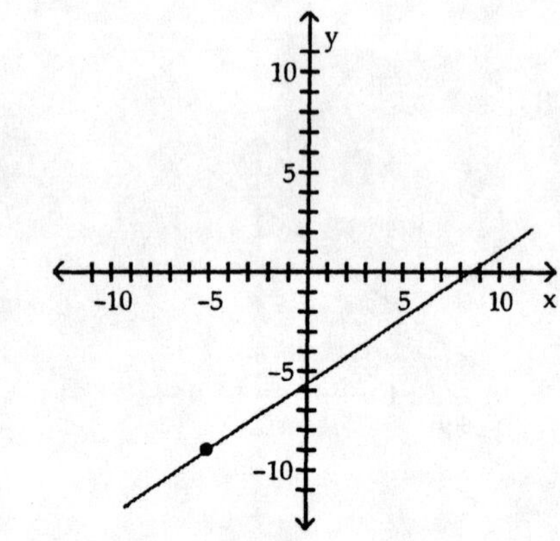

B)

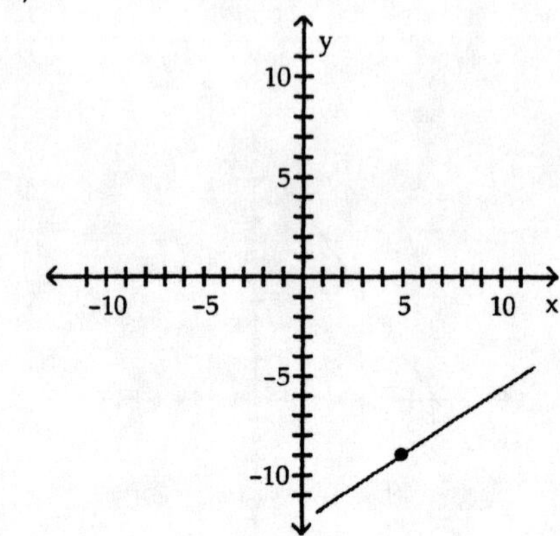

C)

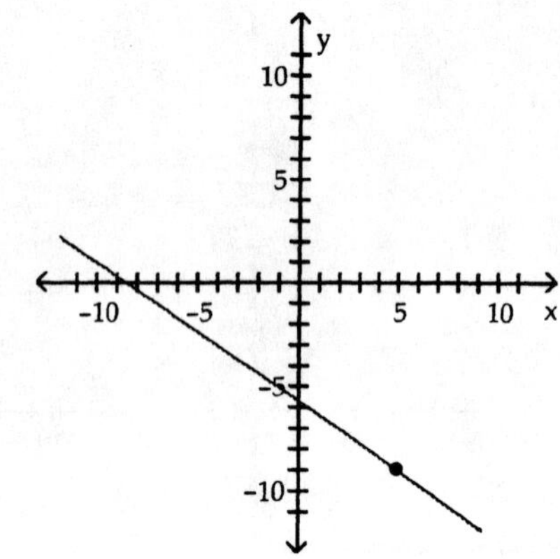

D)

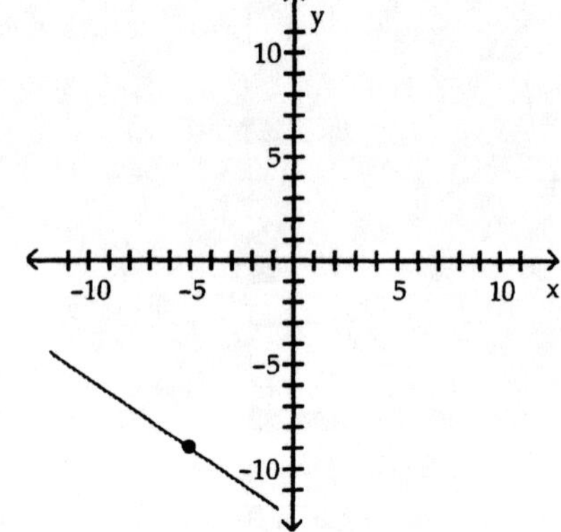

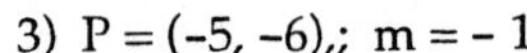
3) P = (-5, -6),; m = - 1

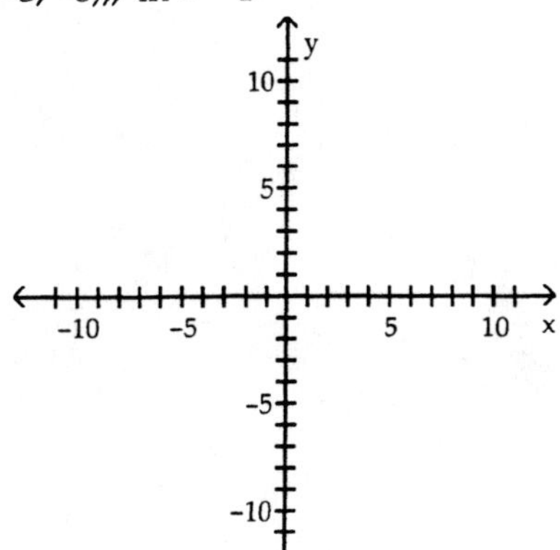

A)

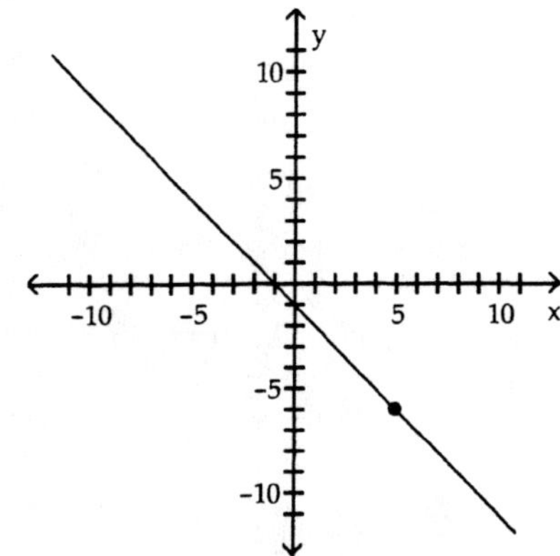

B)

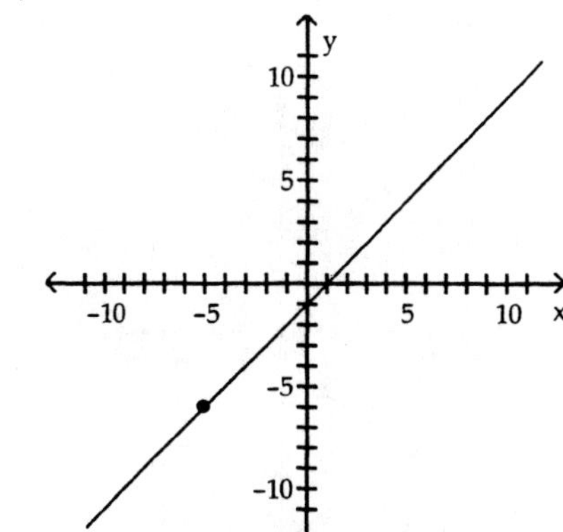

C)

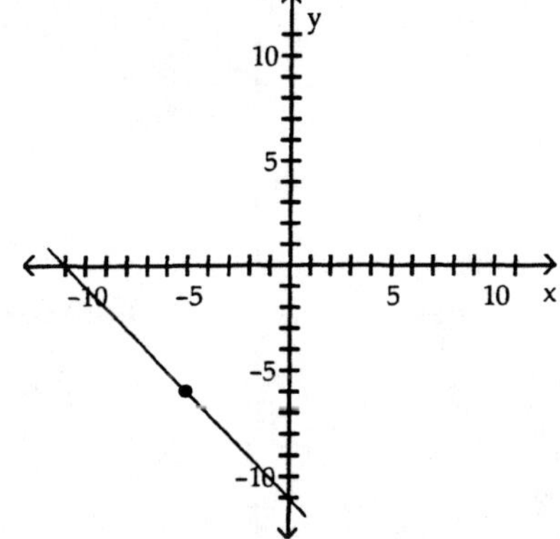

D)

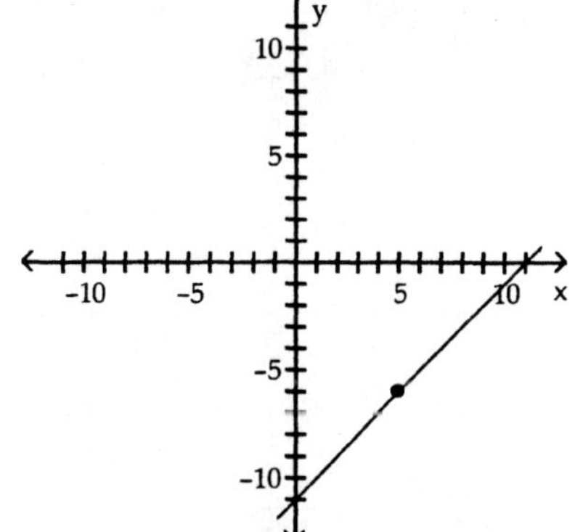

4) $P = (0, 3);\ m = \frac{3}{5}$

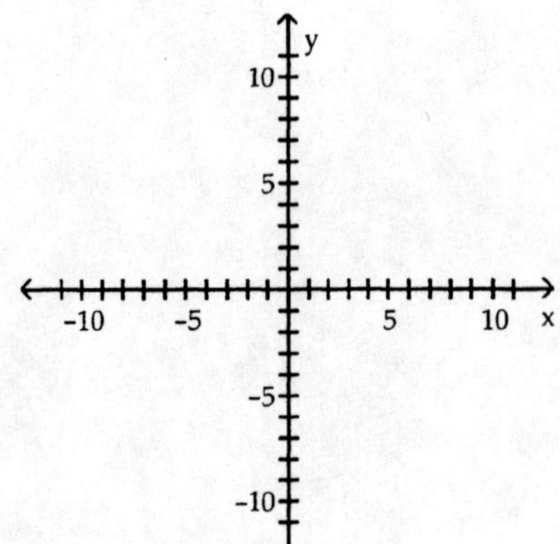

A)

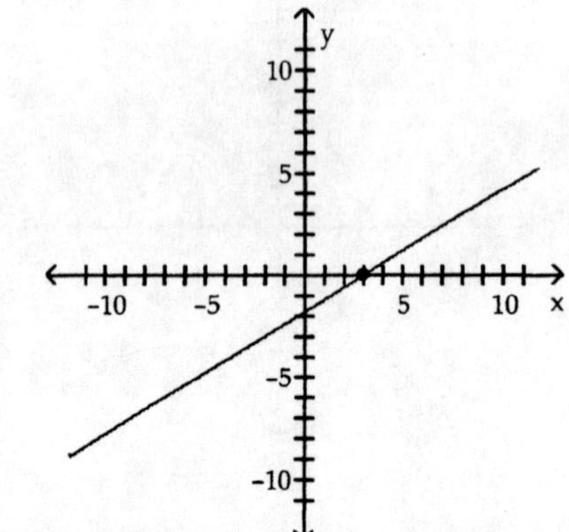

B)

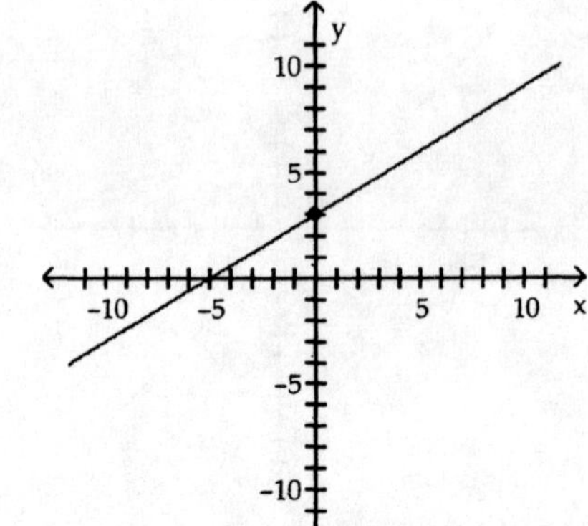

C)

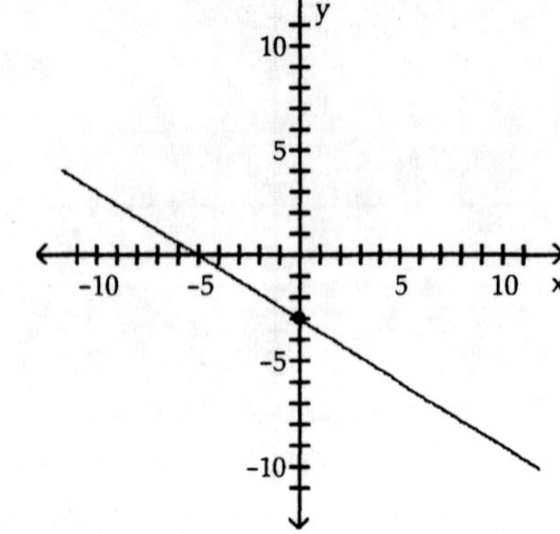

D)

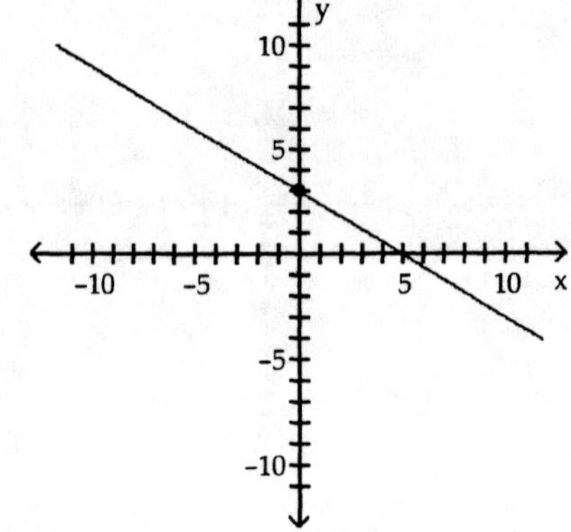

5) $P = (0, 3)$; $m = -\frac{4}{3}$

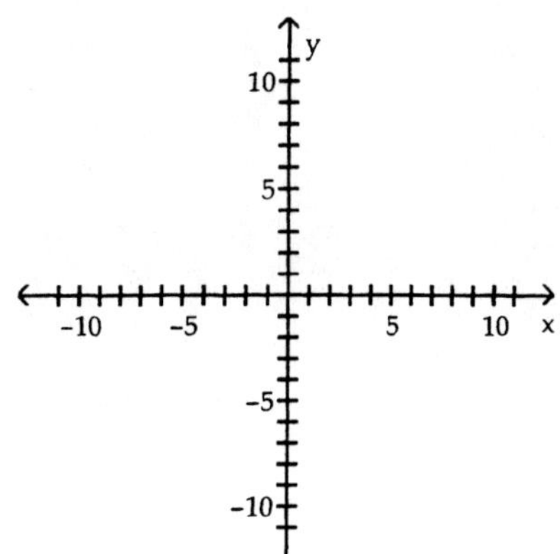

A)

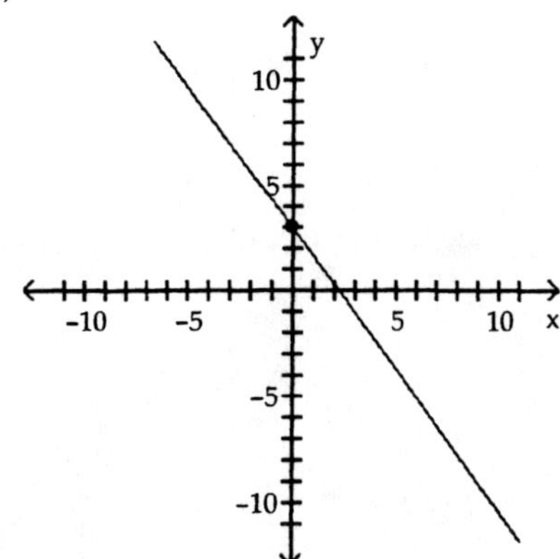

B)

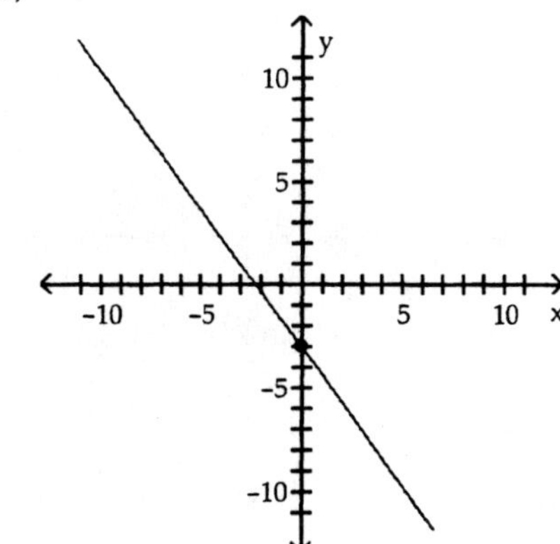

C)

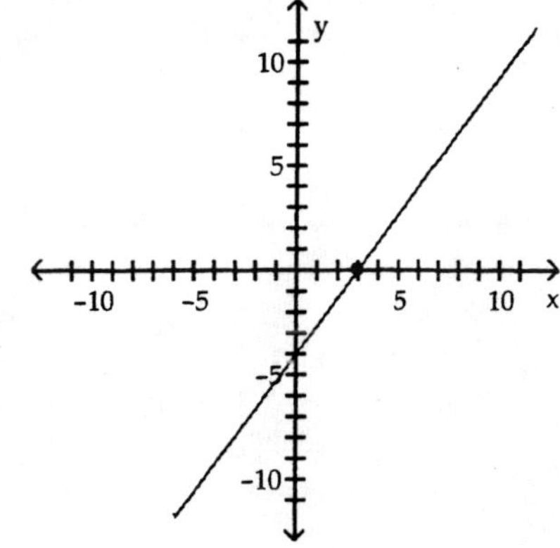

D)

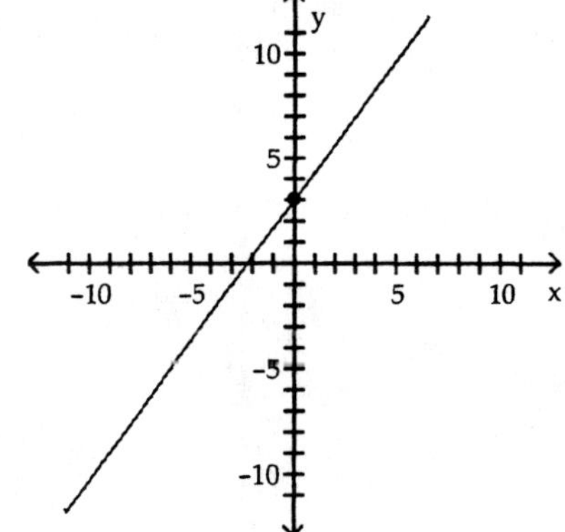

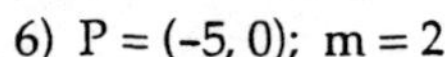

6) $P = (-5, 0)$; $m = 2$

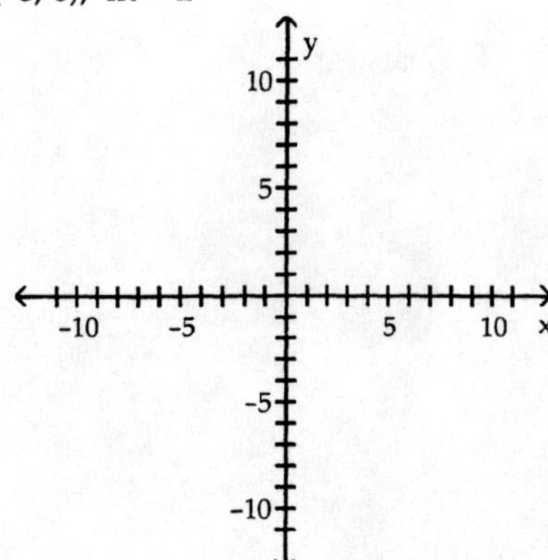

A)

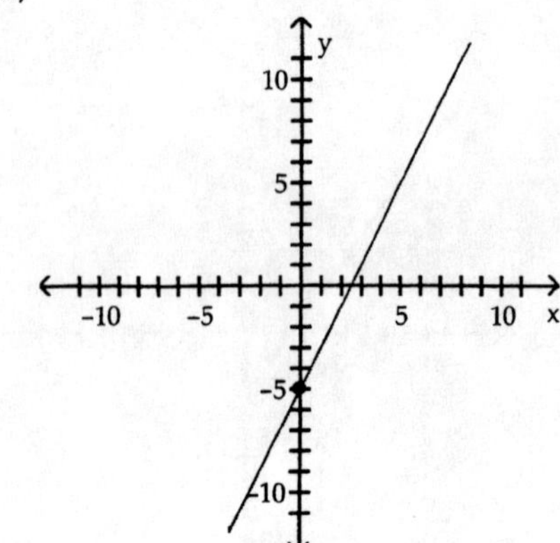

B)

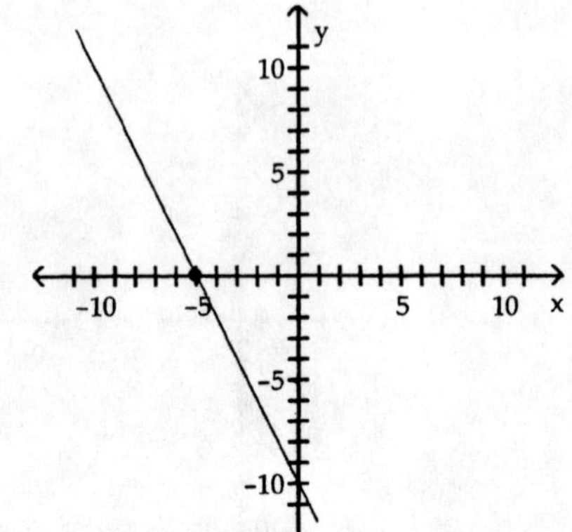

C)

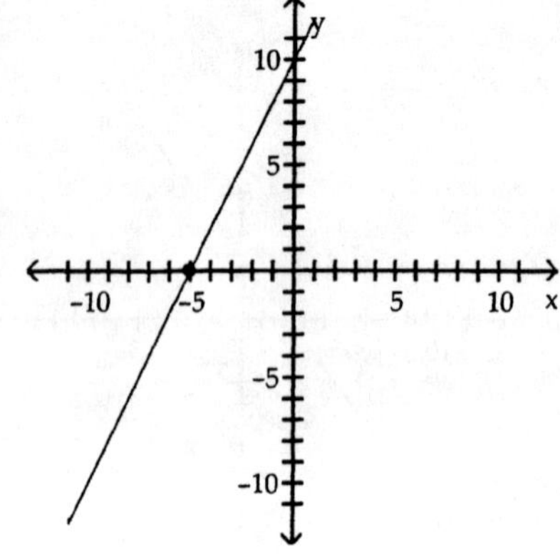

D)

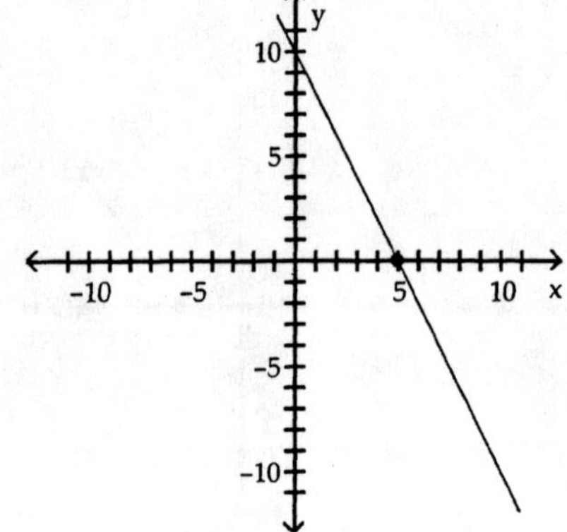

7) $P = (5, 0);\ m = -\frac{4}{5}$

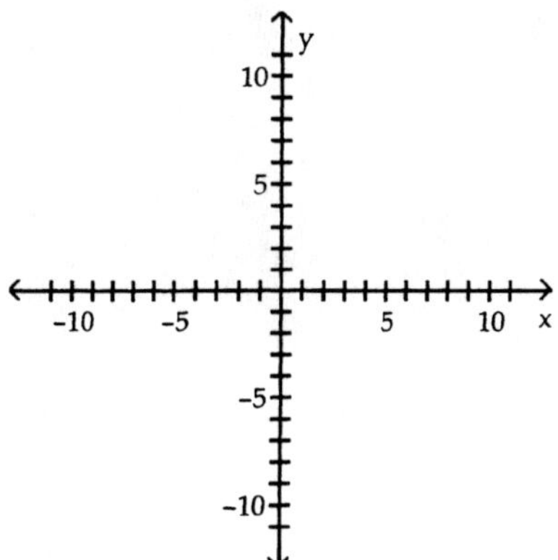

A)

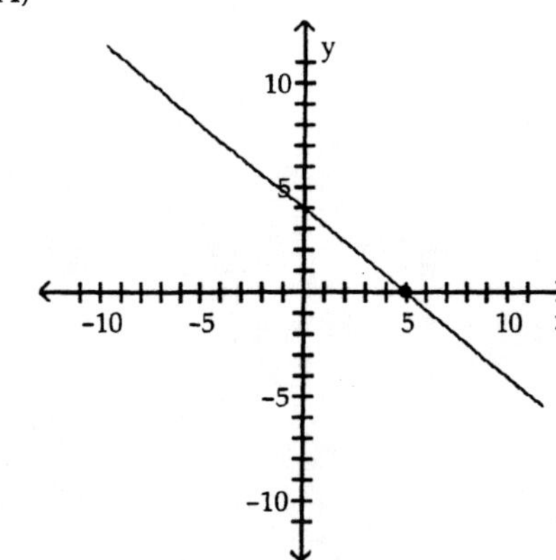

B)

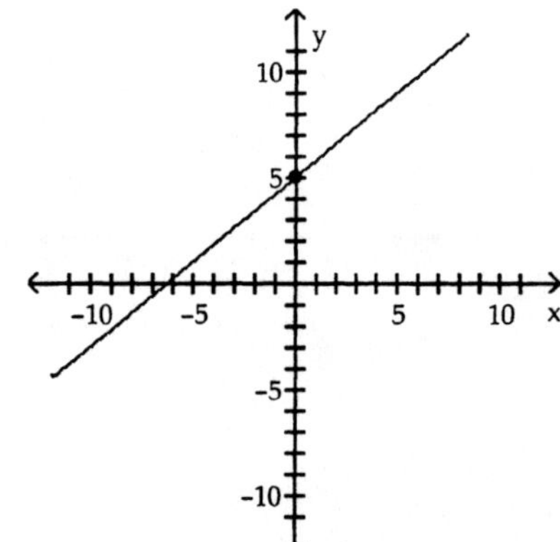

C)

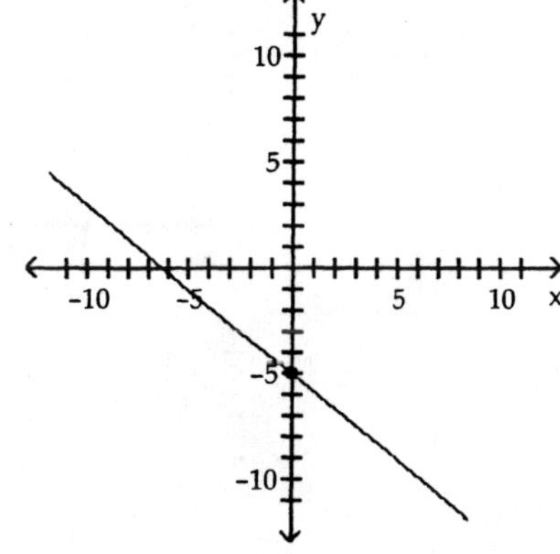

D)

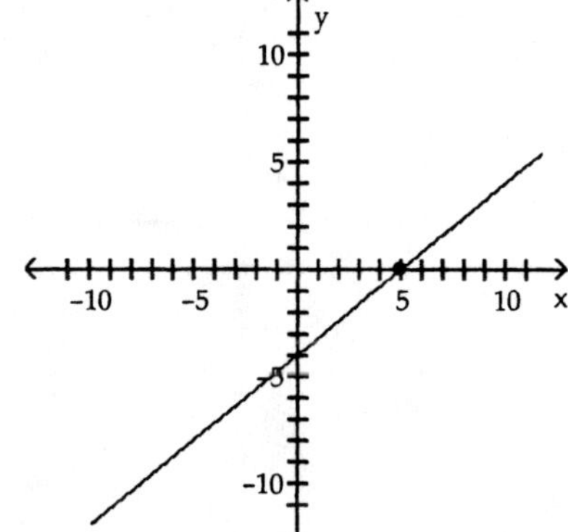

8) $P = (2, -7)$; $m = 0$

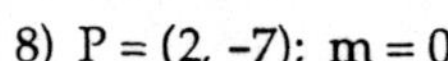

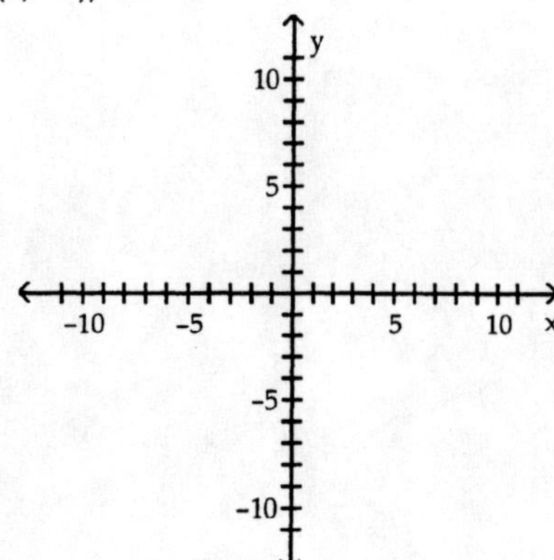

A)

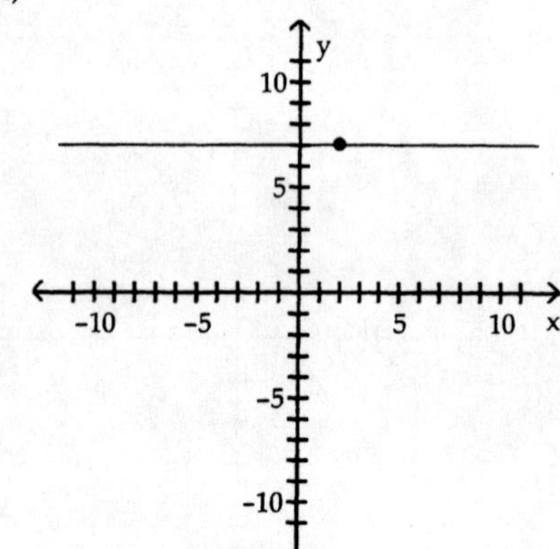

B)

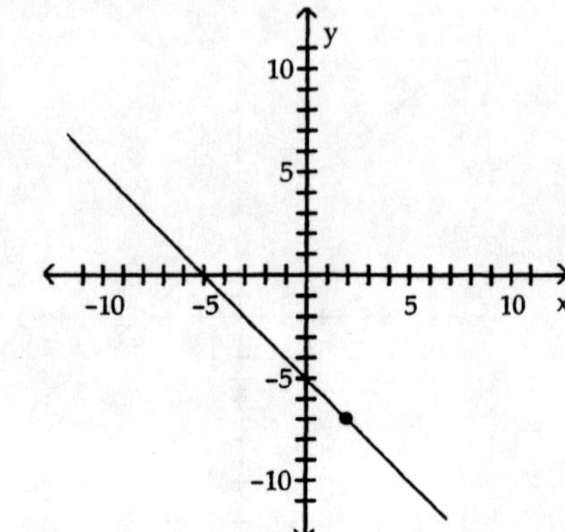

C)

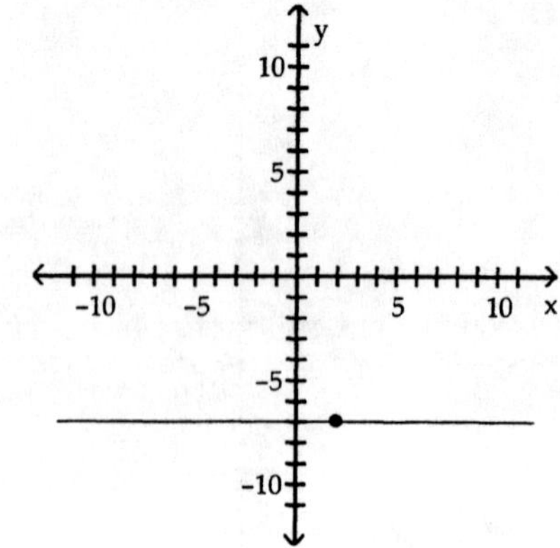

D)

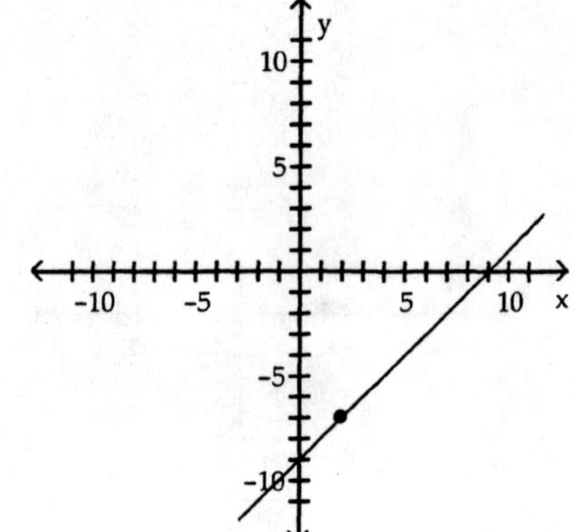

9) P = (2, 1); slope undefined

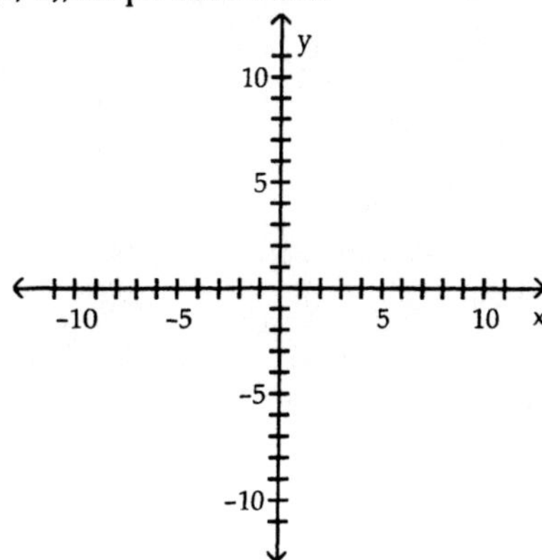

A)

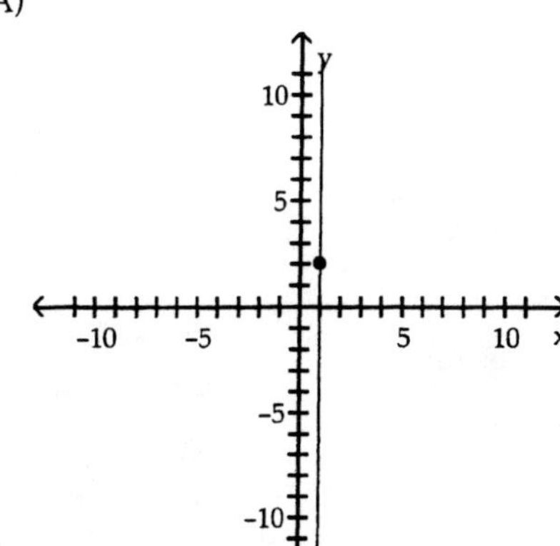

B)

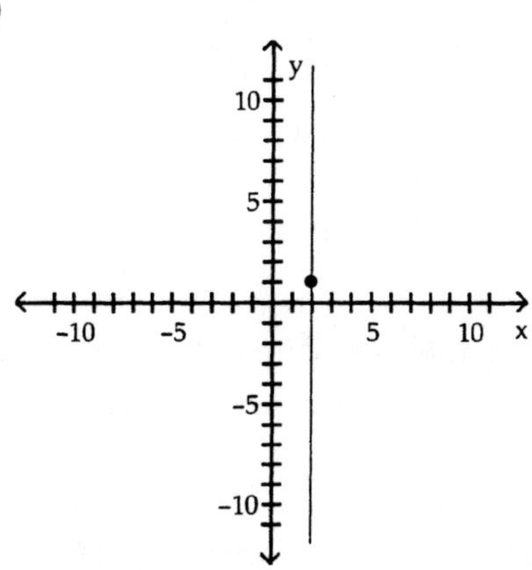

C)

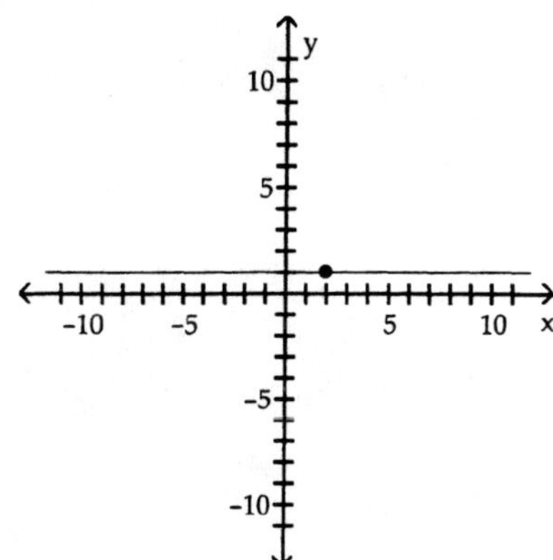

D)

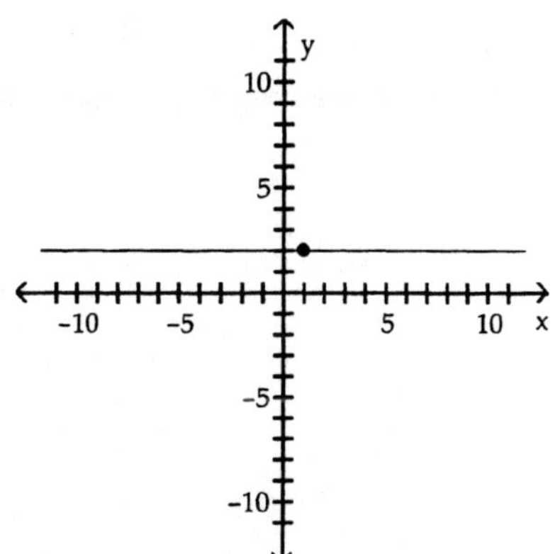

3 Find the Equation of a Vertical Line

Solve the problem.

1) Find an equation of the line with slope undefined and containing the point (–10, 8).

A) x = 8 B) y = 8 C) y = –10 D) x = –10

2) Find an equation of the vertical line containing the point (–5, –9).

A) y = –5 B) y = –9 C) x = –9 D) x = –5

3) Find an equation of the y–axis.

A) x = 0 B) y = 1 C) y = 0 D) x = 1

4) Find an equation of the line with slope undefined and containing the point $(-\frac{7}{8}, 4)$.

A) $y = 4$ B) $y = -\frac{7}{8}$ C) $x = -\frac{7}{8}$ D) $x = 4$

5) Find an equation of the vertical line through the point (-9.0, -8.1).

A) $x = -9.0$ B) $x = 0$ C) $x = -8.1$ D) $x = 17.1$

6) Find an equation of the line through the point $(-\frac{2}{9}, 9)$ with undefined slope.

A) $x = \frac{2}{9}$ B) $x = -\frac{2}{9}$ C) $x = 0$ D) $x = 9$

4 Use the Point-Slope Form of a Line; Identify Horizontal Lines

Find the point-slope equation for the line with the given properties.

1) slope $= \frac{5}{3}$; containing the point (-3, -5)

A) $y - 5 = \frac{5}{3}(x - 3)$ B) $y + 3 = \frac{5}{3}(x + 5)$ C) $y + 5 = \frac{5}{3}(x + 3)$ D) $x - 5 = \frac{5}{3}(y - 3)$

Find the slope-intercept form of the equation of the line with the given properties.

2) horizontal; containing the point (-9, -10)

A) $y = -10$ B) $y = -9$ C) $x = -9$ D) $x = -10$

3) slope = 0; containing the point (9, -8)

A) $y = -8$ B) $x = 9$ C) $y = 9$ D) $x = -8$

4) horizontal; containing the point (-5.4, 3.4)

A) $y = 3.4$ B) $y = 0$ C) $y = 2$ D) $y = -5.4$

5) horizontal; containing the point $(-\frac{1}{4}, 9)$

A) $y = -\frac{1}{4}$ B) $y = 0$ C) $y = -9$ D) $y = 9$

5 Find the Equation of a Line Given Two Points

Find the equation of the line in slope-intercept form.

1)

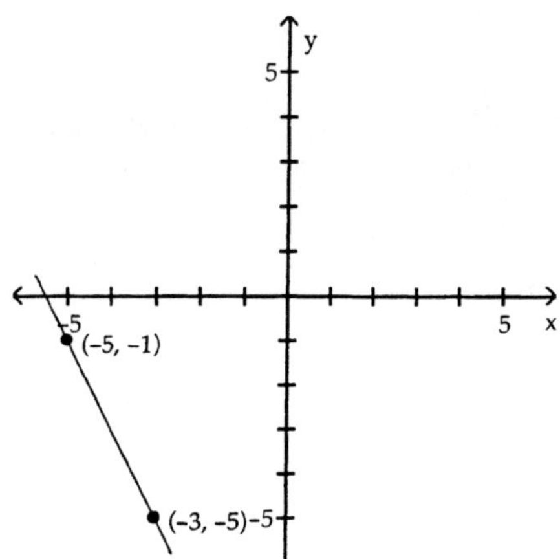

A) $y = -2x + \frac{21}{2}$ B) $y = -2x - 11$ C) $y = -2x - \frac{9}{2}$ D) $y = -\frac{1}{2}x - \frac{9}{22}$

Solve the problem.

2) Find the slope-intercept form of the equation of the line containing the points (7, -4) and (-7, -6).

A) $y = \frac{1}{7}x - 5$ B) $y + 4 = \frac{1}{7}(x - 7)$ C) $y = mx - 5$ D) $y = -\frac{1}{7}x - 5$

3) Find the general form of the equation of the line containing the points (7, -5) and (-2, 5).

A) $-12x + 7y = -11$ B) $10x + 9y = 25$ C) $-10x + 9y = 25$ D) $12x - 7y = -11$

4) Find the point-slope form of the equation of the line containing the points (-3, -1) and (1, -2). Use (-3, -1) as the point (x_1, y_1).

A) $y - 3 = -\frac{1}{4}(x - 1)$ B) $y + 3 = -\frac{1}{4}(x + 1)$ C) $y + 1 = -\frac{1}{4}(x + 3)$ D) $y - 1 = -\frac{1}{4}(x - 3)$

5) Find an equation of the line containing the points (8, -6), and (8, -4).

A) $y = -6$ B) $y = \frac{7}{8}x - 13$ C) $y = \frac{7}{8}x$ D) $x = 8$

6) Find an equation of the line containing the points (-2, 4), and (-6, 4).

A) $y = -4$ B) $x = -4$ C) $y = 4$ D) $x = 4$

6 Write the Equation of a Line in Slope-Intercept Form

Find the slope-intercept form of the equation of the line with the given properties.

1) slope = -7; y-intercept = 10

A) $y = -7x - 10$ B) $y = 10x - 7$ C) $y = -7x + 10$ D) $y = 10x + 7$

2) slope = 2; containing the point (-5, -6)

A) $y = -2x - 4$ B) $y = 2x + 4$ C) $y = 2x - 4$ D) $y = -2x + 4$

3) Containing the points (-6, -7) and (4, -5)

A) $y = 5x - \frac{29}{5}$ B) $y + 7 = \frac{1}{5}(x + 6)$ C) $y = -\frac{1}{5}x - \frac{29}{5}$ D) $y = \frac{1}{5}x - \frac{29}{5}$

4) x-intercept = 5; y-intercept = 6

A) $y = -\frac{5}{6}x + 5$ B) $y = \frac{6}{5}x + 6$ C) $y = -\frac{6}{5}x + 5$ D) $y = -\frac{6}{5}x + 6$

5) horizontal; containing the point (6, -7)

A) $y = -7$ B) $x = 6$ C) $x = -7$ D) $y = 6$

6) slope = 0; containing the point (9, -3)

A) $y = -3$ B) $y = 9$ C) $x = 9$ D) $x = -3$

Write the equation in slope-intercept form.

7) $5x - 6y = 2$

A) $y = \frac{5}{6}x - \frac{1}{3}$ B) $-6y = -5x + 2$ C) $y = -5x + 8$ D) $x = \frac{6}{5}y + \frac{2}{5}$

8) $6x + 5y = -2$

A) $y = -\frac{5}{6}x - \frac{1}{3}$ B) $y = -\frac{6}{5}x - \frac{2}{5}$ C) $y = \frac{6}{5}x - \frac{2}{5}$ D) $y = \frac{5}{6}x - \frac{1}{3}$

9) $16x + 5y = 9$

A) $y = 16x - 9$ B) $y = \frac{16}{5}x + \frac{9}{5}$ C) $y = \frac{16}{5}x - \frac{9}{5}$ D) $y = -\frac{16}{5}x + \frac{9}{5}$

10) $8x - 7y = 9$

A) $y = 8x - 9$ B) $y = \frac{7}{8}x + \frac{9}{8}$ C) $y = \frac{8}{7}x + \frac{9}{7}$ D) $y = \frac{8}{7}x - \frac{9}{7}$

11) $x = 7y + 5$

A) $y = x - \frac{5}{7}$ B) $y = \frac{1}{7}x - 5$ C) $y = 7x - 5$ D) $y = \frac{1}{7}x - \frac{5}{7}$

12) $2x + 5y = 9x + 8$

A) $y = \frac{5}{7}x - \frac{8}{7}$ B) $y = 7x + 11$ C) $y = \frac{7}{5}x + \frac{8}{5}$ D) $y = \frac{11}{5}x + \frac{8}{5}$

7 Identify the Slope and y-Intercept of a Line from Its Equation

Find the slope and y-intercept of the line.

1) $y = \frac{7}{9}x - \frac{7}{2}$

A) slope = $-\frac{7}{9}$; y-intercept = $\frac{7}{2}$ B) slope = $\frac{7}{9}$; y-intercept = $-\frac{7}{2}$

C) slope = $-\frac{7}{2}$; y-intercept = $\frac{7}{9}$ D) slope = $\frac{9}{7}$; y-intercept = $\frac{7}{2}$

2) $x + y = 4$

A) slope = 1; y-intercept = 4

B) slope = 0; y-intercept = 4

C) slope = -1; y-intercept = 4

D) slope = -1; y-intercept = -4

3) $11x + y = 7$

A) slope = $\frac{11}{7}$; y-intercept = $\frac{1}{7}$

B) slope = -11; y-intercept = 7

C) slope = $-\frac{1}{11}$; y-intercept = $\frac{7}{11}$

D) slope = 11; y-intercept = 7

4) $6x - 12y = 72$

A) slope = $\frac{1}{2}$; y-intercept = -6

B) slope = 6; y-intercept = 72

C) slope = 2; y-intercept = 12

D) slope = $-\frac{1}{2}$; y-intercept = 6

5) $x + 10y = 1$

A) slope = 1; y-intercept = 1

B) slope = -10; y-intercept = 10

C) slope = $-\frac{1}{10}$; y-intercept = $\frac{1}{10}$

D) slope = $\frac{1}{10}$; y-intercept = $\frac{1}{10}$

6) $-x + 9y = 63$

A) slope = 9; y-intercept = -63

B) slope = -1; y-intercept = 63

C) slope = $-\frac{1}{9}$; y-intercept = 7

D) slope = $\frac{1}{9}$; y-intercept = 7

7) $y = 1$

A) slope = 1; y-intercept = 0

B) slope = 1; y-intercept = 1

C) slope = 0; y-intercept = 1

D) slope = 0; no y-intercept

8) $x = 4$

A) slope undefined; y-intercept = 4

B) slope undefined; no y-intercept

C) slope = 4; y-intercept = 0

D) slope = 0; y-intercept = 4

9) $y = -6x$

A) slope = 0; y-intercept = -6

B) slope = $-\frac{1}{6}$; y-intercept = 0

C) slope = -6; y-intercept = 0

D) slope = 6; y-intercept = 0

10) $6x - 2y = -24$

A) slope = 3; y-intercept = 12

B) slope = -3; y-intercept = -12

C) slope = 18; y-intercept = $\frac{1}{3}$

D) slope = -12; y-intercept = $-\frac{1}{3}$

11) $12x + 4y = -96$

A) slope = 3; y-intercept = -24
B) slope = -3; y-intercept = 24
C) slope = -3; y-intercept = -24
D) slope = 3; y-intercept = 24

8 Write the Equation of a Line in General Form

Find the general form of the equation for the line with the given properties.

1) slope = $\frac{3}{5}$; y-intercept = $\frac{11}{5}$

A) $y = \frac{3}{5}x + \frac{11}{5}$
B) $3x - 5y = -11$
C) $3x + 5y = -11$
D) $y = \frac{3}{5}x - \frac{11}{5}$

2) Containing the points (12, 0) and (0, –4)

A) $4x - 12y = 48$
B) $y = -\frac{1}{3}x - 4$
C) $4x + 12y = 48$
D) $y = -\frac{1}{3}x + 12$

3) Containing the points (–2, 4) and (–9, 1)

A) $6x + 10y = -64$
B) $-3x + 7y = 34$
C) $-6x - 10y = -64$
D) $3x + 7y = 34$

4) slope = 1; y-intercept = –4

A) $x - y = 4$
B) $x + y = 4$
C) $-x + y = 4$
D) $-x - y = 4$

5) slope = 4; containing the point (3, –5)

A) $4x - y = -17$
B) $4x + y = 17$
C) $4x + y = -17$
D) $4x - y = 17$

6) slope = $-\frac{3}{4}$; containing the point (2, 3)

A) $3x + 4y = 18$
B) $4x + 3y = -18$
C) $3x - 4y = 18$
D) $3x + 4y = -18$

7) slope = $-\frac{3}{7}$; containing the point (0, 5)

A) $3x + 7y = 35$
B) $7x + 3y = -35$
C) $3x + 7y = -35$
D) $3x - 7y = 35$

8) slope = $\frac{5}{7}$; containing (0, 5)

A) $7x - 5y = -35$
B) $-5x - 7y = 35$
C) $-5x + 7y = -35$
D) $-5x + 7y = 35$

9) Containing the points (–3, 4) and (0, –6)

A) $-7x + 6y = -36$
B) $10x - 3y = 18$
C) $-10x - 3y = 18$
D) $7x - 6y = -36$

10) Containing the points (9, 0) and (5, 3)

A) $3x + 4y = 27$
B) $-9x - 2y = -39$
C) $9x + 2y = -39$
D) $-3x + 4y = 27$

11) Containing the points (6, 9) and (–3, –7)

A) $-16x + 9y = -15$
B) $3x - 4y = -37$
C) $16x + 9y = -15$
D) $-3x + 4y = -37$

12) Containing the points (3, 6) and (3, 9)

A) $6x - 9y = 0$
B) $x = 3$
C) $y = 3$
D) $9x + 6y = 0$

Solve the problem.

1) The relationship between Celsius (°C) and Fahrenheit (°F) degrees of measuring temperature is linear. Find an equation relating °C and °F if 10°C corresponds to 50°F and 30°C corresponds to 86°F. Use the equation to find the Celsius measure of 24° F.

A) $C = \frac{5}{9}F - \frac{160}{9}$; $-\frac{40}{9}$ °C
B) $C = \frac{9}{5}F - 80$; $-\frac{184}{5}$ °C
C) $C = \frac{5}{9}F - 10$; $\frac{10}{3}$ °C
D) $C = \frac{5}{9}F + \frac{160}{9}$; $\frac{280}{9}$ °C

2) A truck rental company rents a moving truck one day by charging $31 plus $0.07 per mile. Write a linear equation that relates the cost C, in dollars, of renting the truck to the number x of miles driven. What is the cost of renting the truck if the truck is driven 180 miles?

A) $C = 31x + 0.07$; $5580.07
B) $C = 0.07x + 31$; $32.26
C) $C = 0.07x + 31$; $43.60
D) $C = 0.07x - 31$; $18.40

3) Each week a soft drink machine sells x cans of soda for $0.75/soda. The cost to the owner of the soda machine for each soda is $0.10. The weekly fixed cost for maintaining the soda machine is $25/week. Write an equation that relates the weekly profit, P, in dollars to the number of cans sold each week. Then use the equation to find the weekly profit when 92 cans of soda are sold in a week.

4) Each day the commuter train transports x passengers to or from the city at $1.75/passenger. The daily fixed cost for running the train is $1200. Write an equation that relates the daily profit, P, in dollars to the number of passengers each day. Then use the equation to find the daily profit when the train has 920 passengers in a day.

5) Each month a beauty salon gives x manicures for $12.00/manicure. The cost to the owner of the beauty salon for each manicure is $7.35. The monthly fixed cost to maintain a manicure station is $120.00. Write an equation that relates the monthly profit, in dollars, to the number of manicures given each month. Then use the equation to find the monthly profit when 200 manicures are given in a month.

6) Each week a maid service cleans x houses at $60/house. The cost to the maid service for cleaning each house is $40.00. Write an equation that relates the monthly profit, in dollars, to the number of houses that are cleaned each month. Then use the equation to find the monthly profit when 60 houses are cleaned in a month.

7) Each month a gas station sells x gallons of gas at $1.92/gallon. The cost to the owner of the gas station for each gallon of gas is $1.32. The monthly fixed cost for running the gas station is $37,000. Write an equation that relates the monthly profit, in dollars, to the number of gallons of gasoline sold. Then use the equation to find the monthly profit when 75,000 gallons of gas are sold in a month.

8) A business purchasing an item for business purposes may use straight-line depreciation to obtain a tax deduction. The formula for the present value, P, after t years is $P = C - \left(\frac{C - s}{L}\right)t$, where C is the cost and s is the scrap value after L years. The number L is called the useful life of the item. If a certain piece of equipment costs $20,000 and has a scrap value of $5000 after 6 years, find the present value of the equipment after 7 years.

A) $7142.86
B) $2500
C) $17,500
D) $14,133.67

9) Find an equation for the line graphed on a graphing utility.

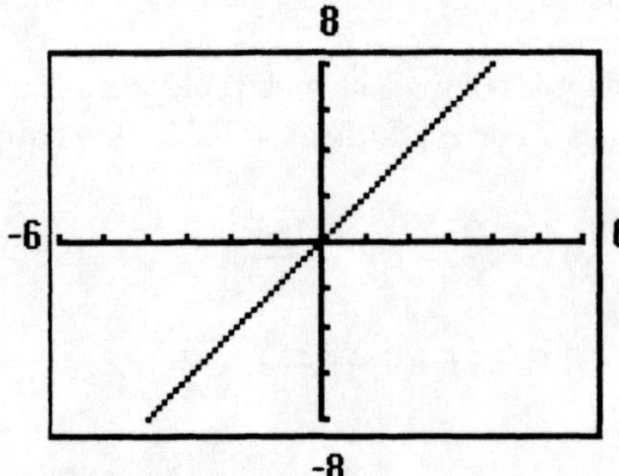

A) $y = 2x$ B) $y = \frac{3}{4}x$ C) $y = \frac{4}{3}x$ D) $y = x$

10) Find an equation in slope-intercept form for the line graphed on a graphing utility.

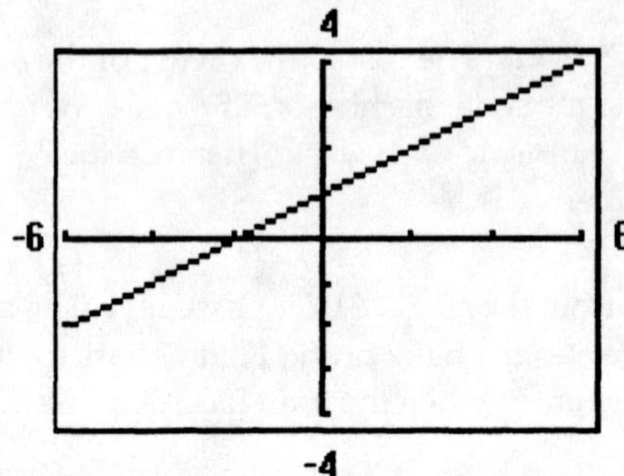

A) $x - 2y = -2$ B) $y = \frac{1}{2}x + 1$ C) $y = x + 1$ D) $y = 2x + 1$

11) Find an equation in general form for the line graphed on a graphing utility.

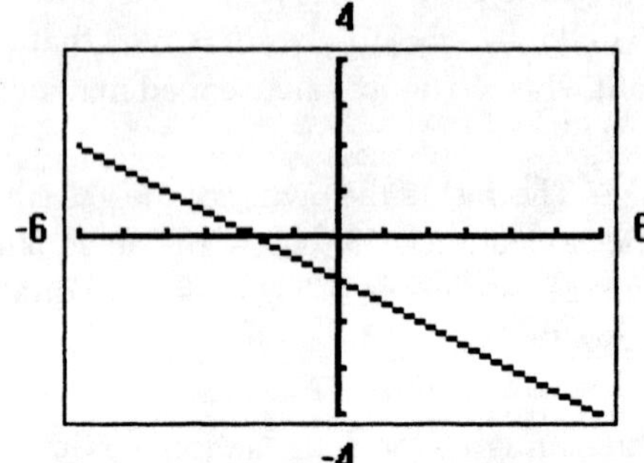

A) $x + 2y = -2$ B) $y = -2x - 1$ C) $2x + y = -1$ D) $y = -\frac{1}{2}x - 1$

Find the slope of the line and sketch its graph.

12) $3x + 4y = 17$

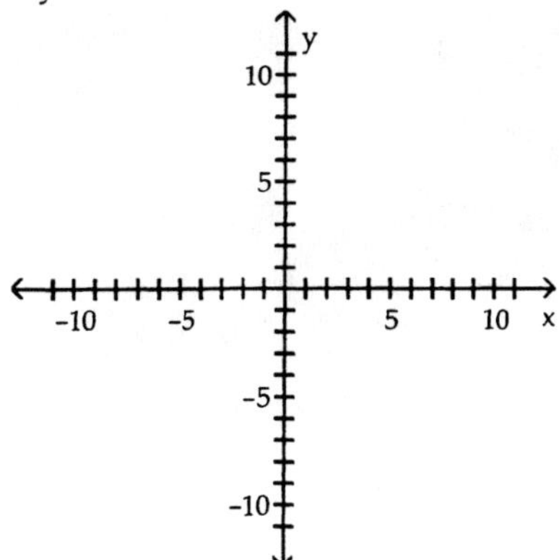

A) slope $= \frac{4}{3}$

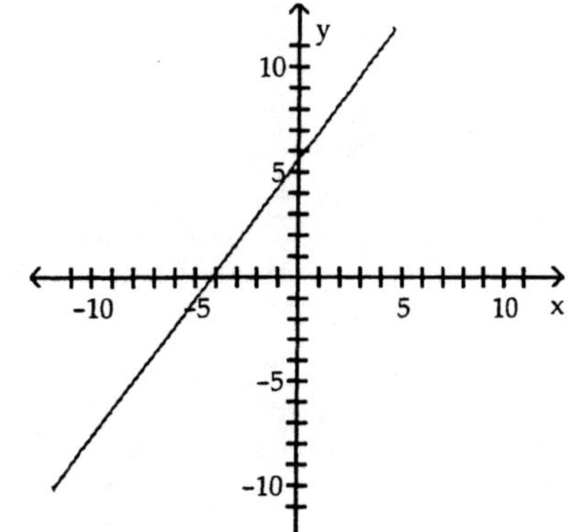

B) slope $= -\frac{4}{3}$

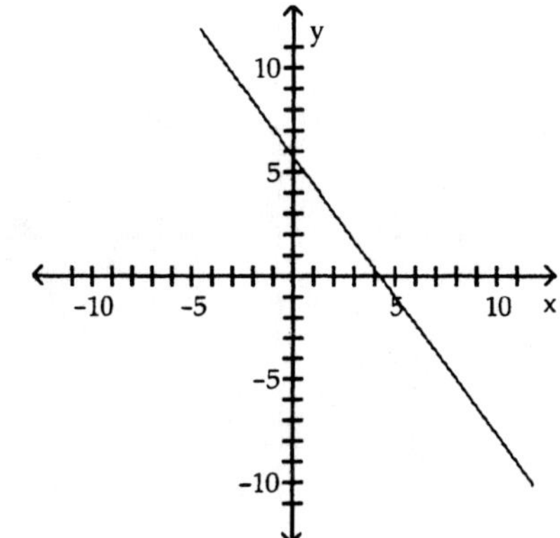

C) slope $= -\frac{3}{4}$

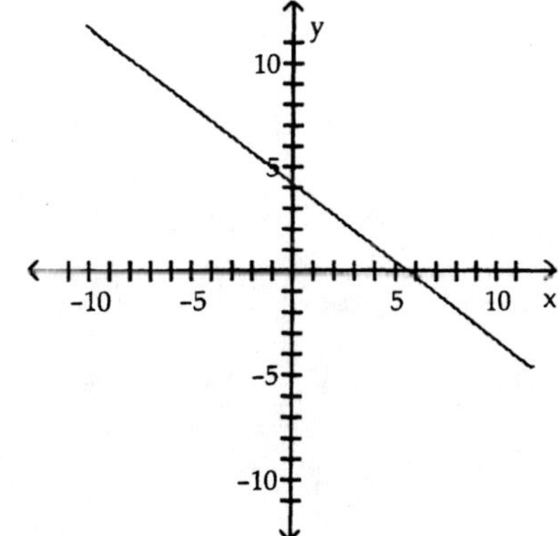

D) slope $= \frac{3}{4}$

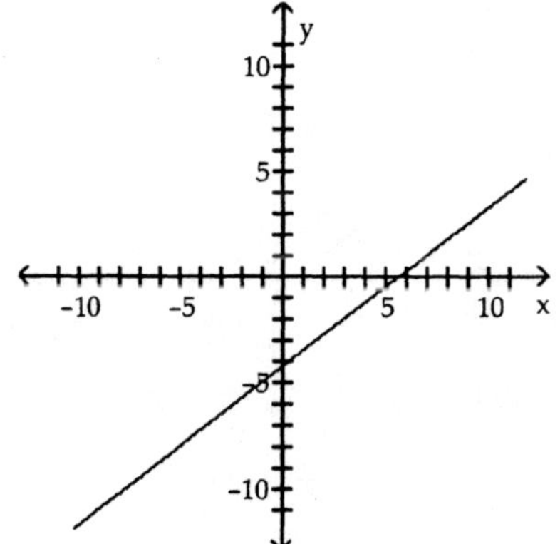

13) $3x - 4y = -2$

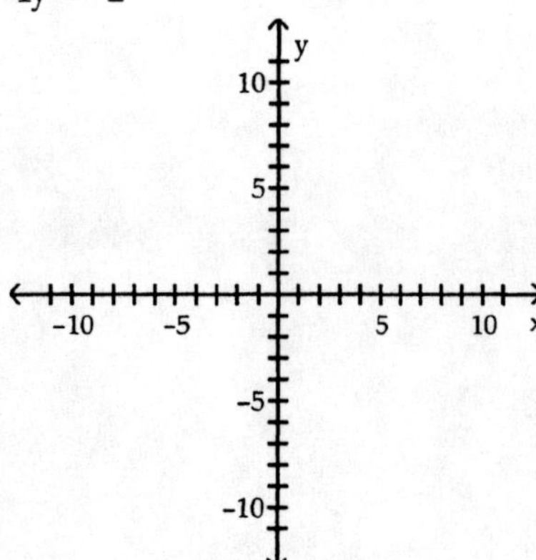

A) slope $= \frac{4}{3}$

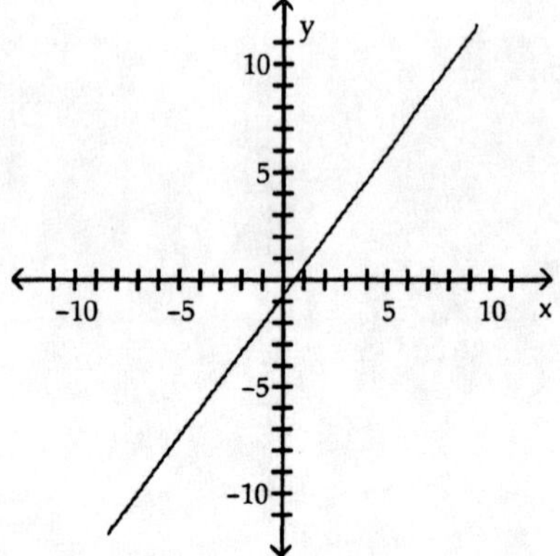

B) slope $= -\frac{4}{3}$

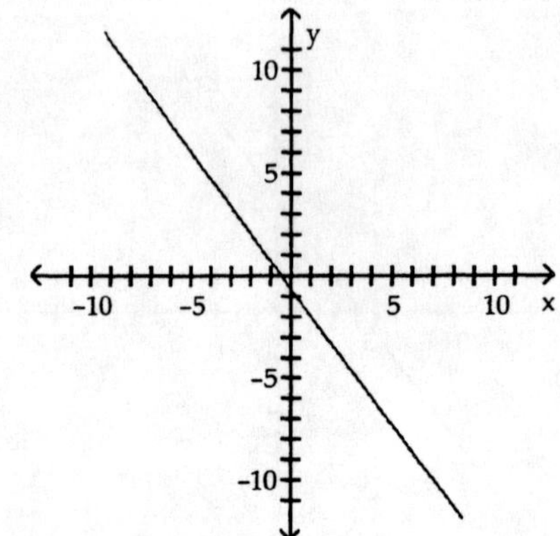

C) slope $= -\frac{3}{4}$

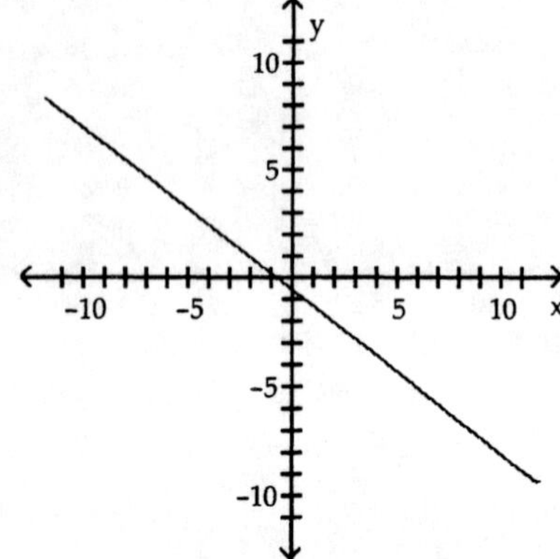

D) slope $= \frac{3}{4}$

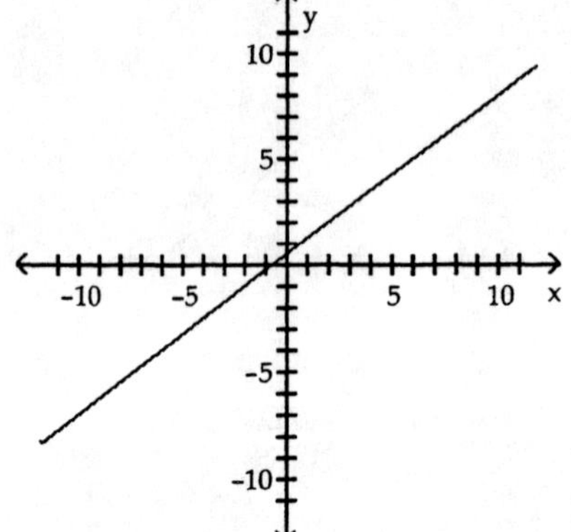

14) $-4y = -3x + 1$

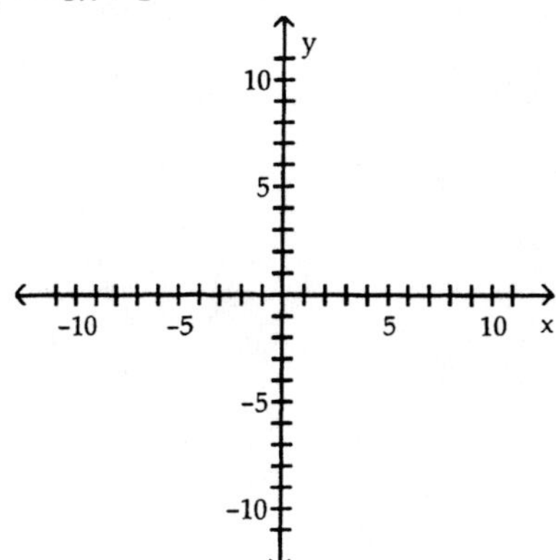

A) slope $= -\frac{3}{4}$

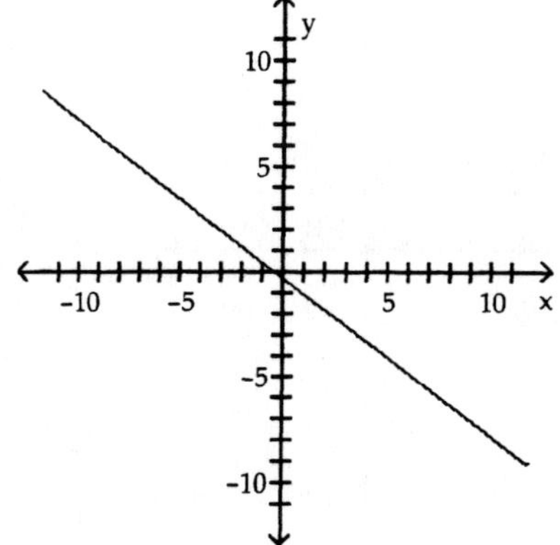

B) slope $= \frac{3}{4}$

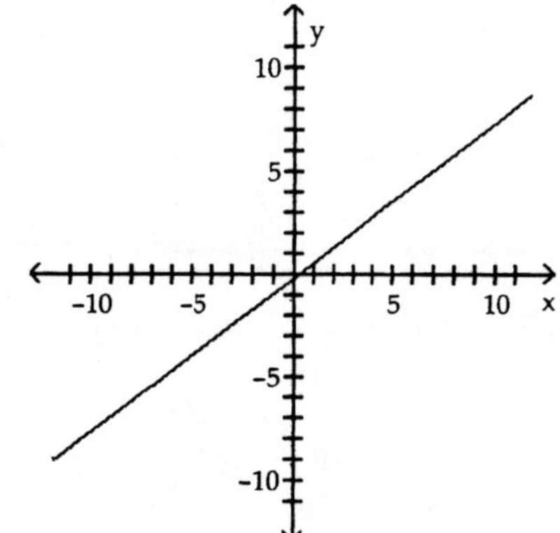

C) slope $= -\frac{4}{3}$

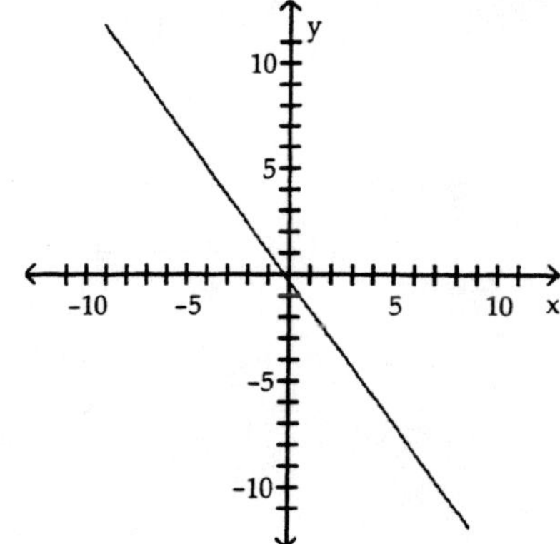

D) slope $= \frac{4}{3}$

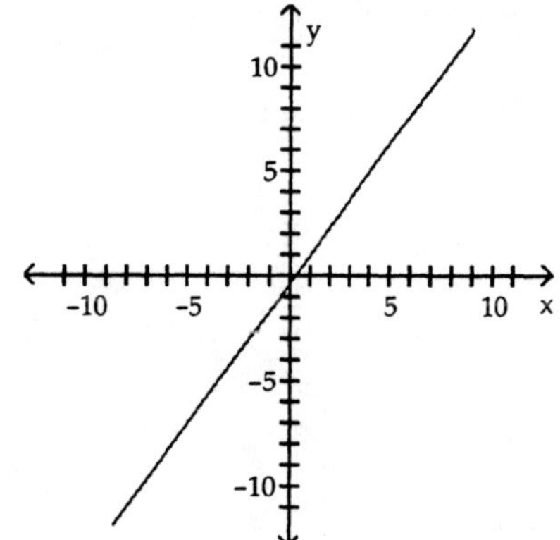

15) $y + 4 = 0$

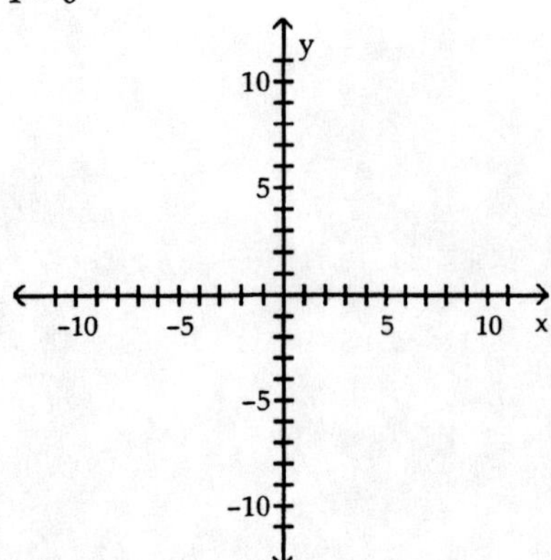

A) slope = 0

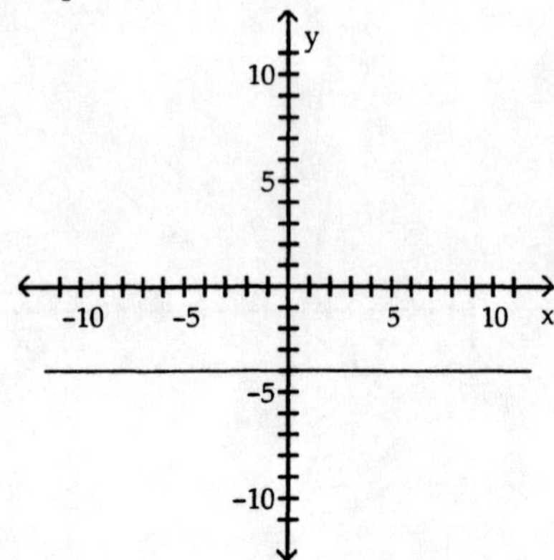

B) slope is undefined

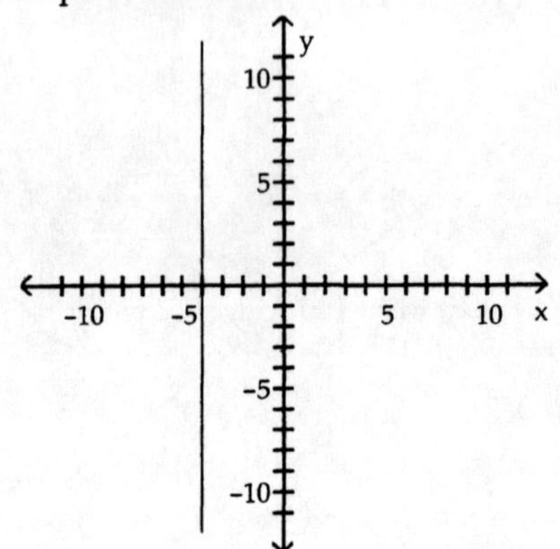

C) slope = $-\frac{1}{4}$

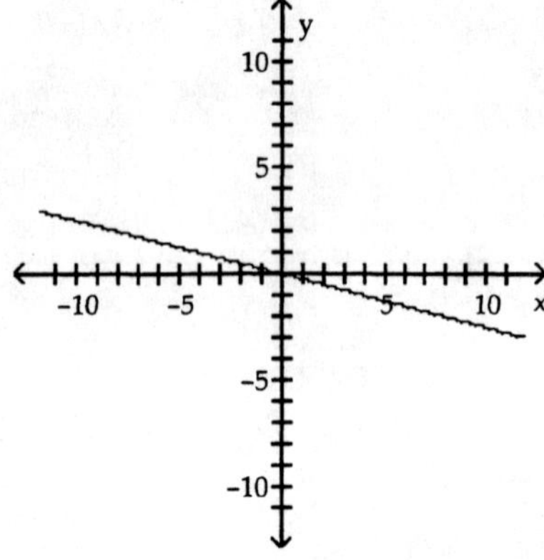

D) slope = -4

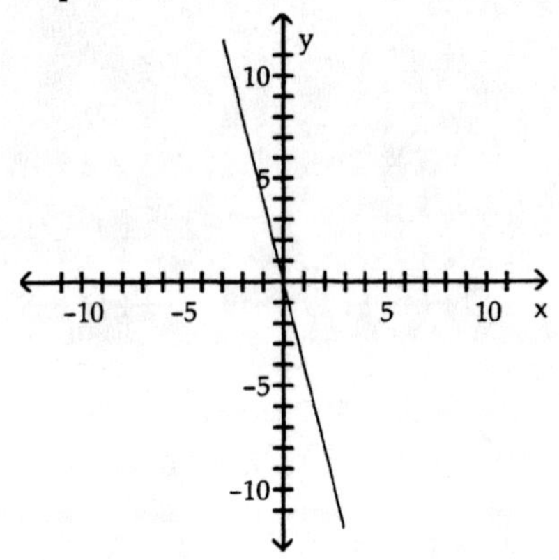

16) $y = -2x$

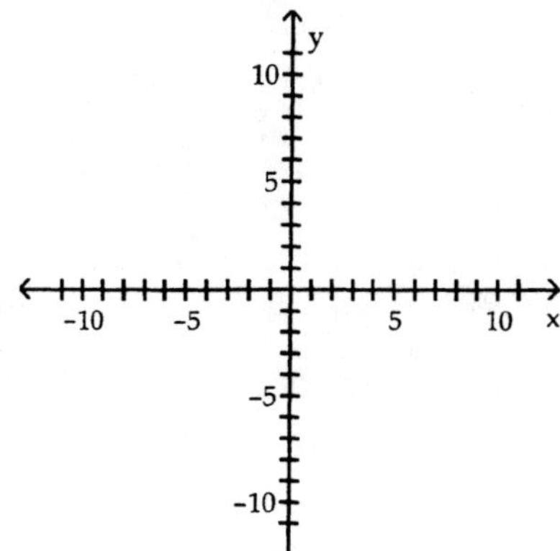

A) slope = -2

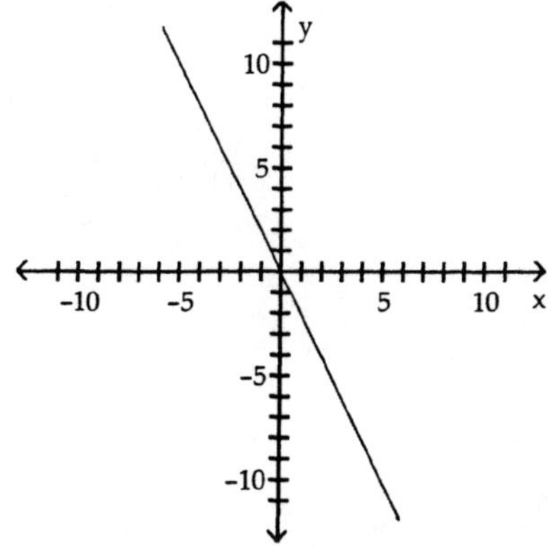

B) slope = $-\frac{1}{2}$

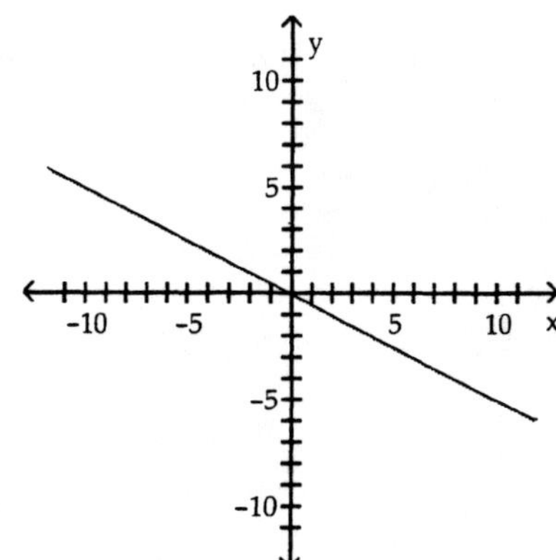

C) slope = 0

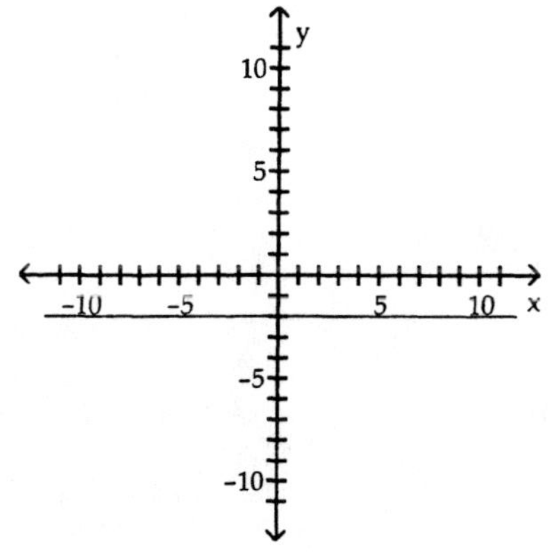

D) slope = 2

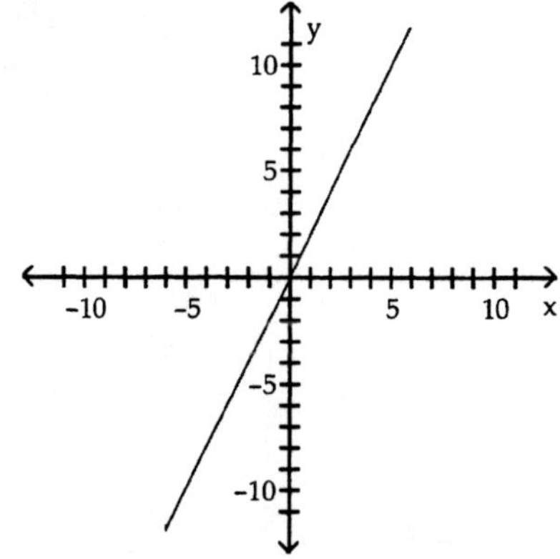

2.5 Parallel and Perpendicular Lines

1 Define Parallel Lines

Complete the statement.

1) Two nonvertical lines are parallel if and only if ________.

A) their slopes are equal and they have the same y–intercepts

B) the product of their slopes is –1

C) their slopes are equal and they have different y–intercepts

D) the product of their slopes is 1

Solve the problem.

2) Find the slope of a line parallel to the line containing the points (2, 1) and (–8, –2).

A) $-\frac{10}{3}$ B) $\frac{10}{3}$ C) $\frac{3}{10}$ D) $-\frac{3}{10}$

The equation of a line L is given. Find the slope of a line that is parallel to L.

3) $y = 5x$

A) 5 B) -5 C) $\frac{1}{5}$ D) $-\frac{1}{5}$

4) $y = \frac{1}{5}x - 2$

A) 5 B) -2 C) $\frac{1}{5}$ D) $-\frac{1}{5}$

5) $x + y = 3$

A) 3 B) -3 C) -1 D) 1

6) $4x + y = -2$

A) –4 B) 4 C) $-\frac{1}{4}$ D) -2

7) $2x - 8y = 16$

A) 2 B) $\frac{1}{4}$ C) $-\frac{1}{4}$ D) 4

8) $x + 13y = 1$

A) 1 B) $\frac{1}{13}$ C) $-\frac{1}{13}$ D) –13

9) $-x + 4y = 44$

A) 11 B) $\frac{1}{4}$ C) -1 D) 4

10) $y = -1$

A) 1 B) 0 C) -1 D) undefined

11) $x = -1$

A) 0 B) 1 C) -1 D) undefined

2 Find Equations of Parallel Lines

Find an equation for the line with the given properties.

1) The solid line L contains the point (3, 2) and is parallel to the dotted line whose equation is $y = 2x$. Give the equation for the line L in slope-intercept form.

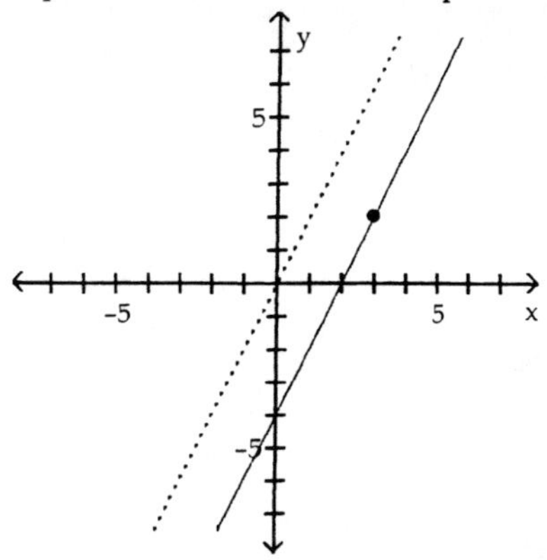

A) $y - 2 = 2(x - 3)$ B) $y = 2x - 1$ C) $y = 2x + b$ D) $y = 2x - 4$

2) Parallel to the line $y = -3x$; containing the point (8, 4)

A) $y = -3x$ B) $y = -3x - 28$ C) $y - 4 = -3x - 8$ D) $y = -3x + 28$

3) Parallel to the line $x + 4y = 6$; containing the point (0, 0)

A) $y = \frac{1}{4}x$ B) $y = -\frac{1}{4}x$ C) $y = \frac{5}{4}$ D) $y = -\frac{1}{4}x + 6$

4) Parallel to the line $3x - y = 5$; containing the point (0, 0)

A) $y = -\frac{1}{3}x + 5$ B) $y = -\frac{1}{3}x$ C) $y = 3x$ D) $y = \frac{1}{3}x$

5) Parallel to the line $y = 5$; containing the point (8, 7)

A) $y = 5$ B) $y = 8$ C) $y = 7$ D) $y = -7$

6) Parallel to the line $x = -5$; containing the point (8, 7)

A) $y = -5$ B) $x = 8$ C) $x = 7$ D) $y = 7$

7) Parallel to the line $3x + 4y = 60$; containing the point (8, 15)

A) $8x + 4y = 60$ B) $4x + 3y = 15$ C) $3x - 4y = 84$ D) $3x + 4y = 84$

8) Parallel to the line $-5x + 6y = 2$; x-intercept = 5

A) $6x + 5y = 30$ B) $6x + 5y = 25$ C) $-5x + 6y = 30$ D) $-5x + 6y = -25$

9) Parallel to the line $y = -4x - 1$; containing the point (2, 6)

A) $y = 4x - 14$ B) $y = -4x + 14$ C) $y = -4x + 26$ D) $y = 4x - 26$

10) Parallel to the line y = -6x + 1; containing the point (6, 0)

A) $y = 6x - 36$ B) $y = \frac{1}{6}x - 1$ C) $y = -6x + 36$ D) $y = -\frac{1}{6}x + 1$

3 Define Perpendicular Lines

Complete the statement.

1) Two nonvertical lines are perpendicular if and only if ___________.

A) the product of their slopes is –1

B) their slopes are equal and they have the same y-intercepts

C) the product of their slopes is 1

D) their slopes are equal and they have different y-intercepts

Solve the problem.

2) Find the slope of a line perpendicular to the line containing the points (–5, –5) and (–1, –2).

A) $-\frac{4}{3}$ B) $\frac{3}{4}$ C) $-\frac{3}{4}$ D) $\frac{4}{3}$

The equation of a line L is given. Find the slope of a line that is perpendicular to L.

3) $y = -12x$

A) -12 B) $-\frac{1}{12}$ C) $\frac{1}{12}$ D) 12

4) $y = \frac{1}{4}x + 1$

A) $\frac{1}{4}$ B) $-\frac{1}{4}$ C) 4 D) - 4

5) $x + y = 10$

A) -10 B) 1 C) 10 D) -1

6) $2x + y = -12$

A) $-\frac{1}{6}$ B) $\frac{1}{2}$ C) -2 D) $-\frac{1}{2}$

7) $2x - 9y = 18$

A) $\frac{9}{2}$ B) $-\frac{9}{2}$ C) $\frac{2}{9}$ D) $-\frac{2}{9}$

8) $x + 7y = 1$

A) $\frac{1}{7}$ B) -7 C) 7 D) $-\frac{1}{7}$

9) $-x + 2y = 10$

A) 2 B) -2 C) $\frac{1}{2}$ D) $-\frac{1}{2}$

10) $y = -10$

A) -10 B) $\frac{1}{10}$ C) 0 D) undefined

11) $x = -5$

A) -5 B) $\frac{1}{5}$ C) 0 D) undefined

4 Find Equations of Perpendicular Lines

Find an equation for the line with the given properties.

1) The solid line L contains the point (3, 4) and is perpendicular to the dotted line whose equation is $y = 2x$. Give the equation of line L in slope-intercept form.

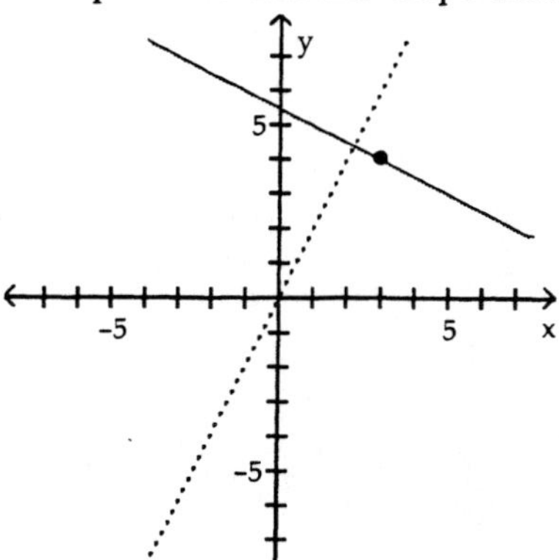

A) $y = \frac{1}{2}x + \frac{11}{2}$ B) $y - 4 = 2(x - 3)$ C) $y = -\frac{1}{2}x + \frac{11}{2}$ D) $y - 4 = -\frac{1}{2}(x - 3)$

2) Perpendicular to the line $y = 2x - 4$; containing the point (2, 3)

A) $y = -2x + 4$ B) $y = 2x + 4$ C) $y = -\frac{1}{2}x + 4$ D) $y = \frac{1}{2}x + 4$

3) Perpendicular to the line $y = \frac{1}{3}x + 6$; containing the point (4, -5)

A) $y = -3x + 7$ B) $y = -\frac{1}{3}x - \frac{7}{3}$ C) $y = -3x - 7$ D) $y = 3x - 7$

4) Perpendicular to the line $-2x - y = 3$; containing the point $(0, -\frac{3}{2})$

A) $y = -1$ B) $y = \frac{1}{2}x + 3$ C) $y = \frac{1}{2}x - \frac{3}{2}$ D) $y = -\frac{1}{2}x - \frac{3}{2}$

5) Perpendicular to the line $x - 7y = 3$; containing the point (5, 2)

A) $y = 7x - 37$ B) $y = -7x - 37$ C) $y = -7x + 37$ D) $y = -\frac{1}{7}x - \frac{37}{7}$

6) Perpendicular to the line $y = -9$; containing the point (4, 3)

A) $y = 4$ B) $x = 4$ C) $y = 3$ D) $x = 3$

7) Perpendicular to the line $x = 3$; containing the point (2, 5)

A) $x = 5$ B) $x = 2$ C) $y = 2$ D) $y = 5$

8) Perpendicular to the line $8x - 7y = -25$; containing the point (3, -7)

A) $3x + 7y = -25$ B) $7x + 8y = -35$ C) $7x - 8y = -35$ D) $8x + 7 = 8$

9) Perpendicular to the line $8x - 9y = 88$; containing the point (2, -14)

A) $-9x + 8y = 88$ B) $-9x - 8y = 94$ C) $-9x + 8y = 94$ D) $8x + 9y = 94$

10) Perpendicular to the line $4x - 3y = -3$; y-intercept = -1

A) $4x - 3y = 3$ B) $4x - 3y = -4$ C) $-3x - 4y = 4$ D) $-3x - 4y = 3$

11) Perpendicular to the line $y = -4x - 5$; containing the point (2, 1)

A) $-4x + y = -7$ B) $x - 4y = -2$ C) $-x + 4y = 2$ D) $4x - y = 7$

12) Perpendicular to the line $6x - 3y = 8$; containing the point (0, 8)

A) $y = 2x + 8$ B) $y = -2x + 8$ C) $y = \frac{1}{2}x + 8$ D) $y = -\frac{1}{2}x + 8$

5 Additional Applications

Decide whether the pair of lines is parallel, perpendicular, or neither.

1) $3x - 8y = 2$
$32x + 12y = -13$

A) parallel B) perpendicular C) neither

2) $3x - 2y = -19$
$2x + 3y = 3$

A) parallel B) perpendicular C) neither

3) $6x + 2y = 8$
$24x + 8y = 35$

A) parallel B) perpendicular C) neither

4) The line through (3, -5) and (-1, 7) and the line through (6, -13) and (-2, 11)

A) parallel B) perpendicular C) neither

5) The line through (-20, 5) and (-4, 7) and the line through (-5, 5) and (7, 4)

A) parallel B) perpendicular C) neither

Solve the problem.

6) Find the two remaining vertices of square ABCD that has A = (-2, 1), B = (6, -5), and a vertex in quadrant III.

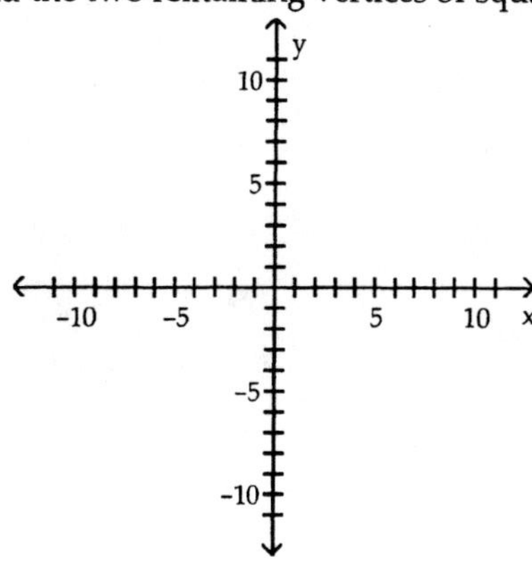

7) Find the slope of the sides of rectangle ABCD having vertices A = (1, 4) and B = (-5, 3).

8) Which of the following 4 points, forms the vertices of a parallelogram?

A) (1, 2), (2, 0), (3, 5), (4, 4)
B) (1, 3), (3, 6), (4, 4), (6, 7)
C) (2, 3), (3, 5), (5, 2), (6, 5)
D) (0, 3), (3, 0), (5, 6), (8, 4)

2.6 Scatter Diagrams; Linear Curve Fitting

1 Draw and Interpret Scatter Diagrams

Plot and interpret the appropriate scatter diagram.

1) The table shows the study times and test scores for a number of students. Draw a scatter plot of score versus time treating time as the independent variable.

Study Time (min)	9	16	21	26	33	36	40	47
Test Score	59	61	64	65	73	74	78	78

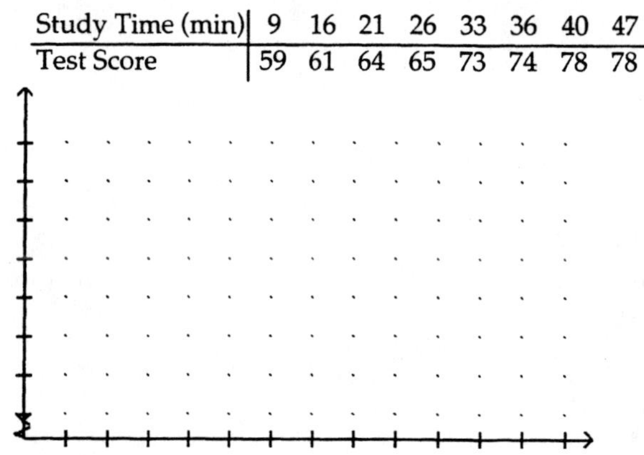

2) The table gives the times spent watching TV and the grades of several students.

Weekly TV (h)	6	12	18	24	30	36
Grade (%)	92.5	87.5	72.5	77.5	62.5	57.5

Which scatter diagram describes the data and the relationship, if any?

A)

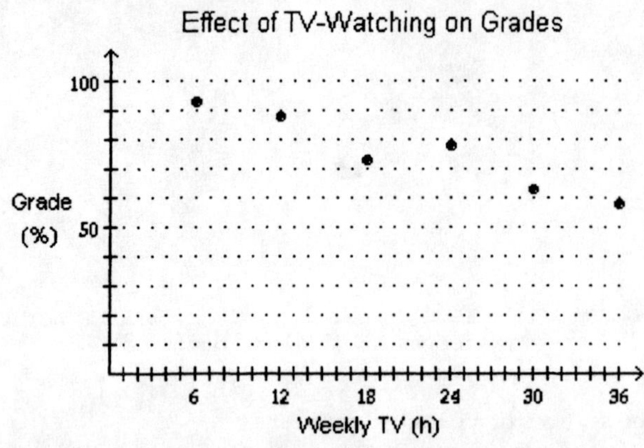

More hours spent watching TV may increase grades.

B)

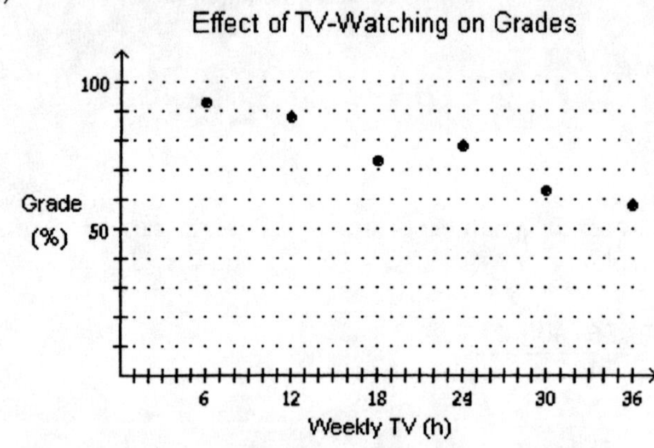

More hours spent watching TV may reduce grades.

C)

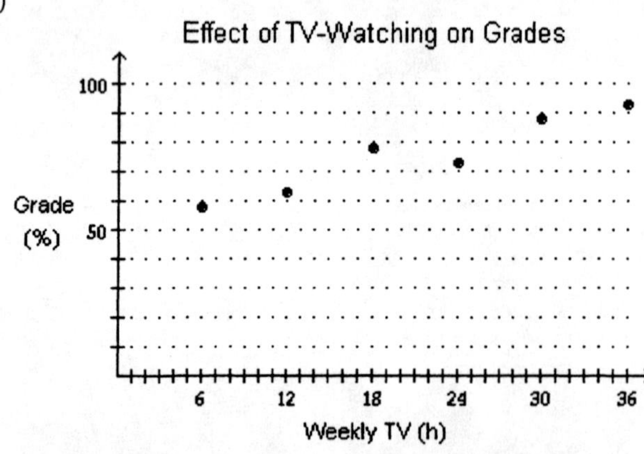

More hours spent watching TV may reduce grades.

D) none of these

Plot a scatter diagram.

3)

x:	24	−11	18	−10	−1	8	10	22	8	-7
y:	64	23	39	−9	−1	38	7	67	-5	5

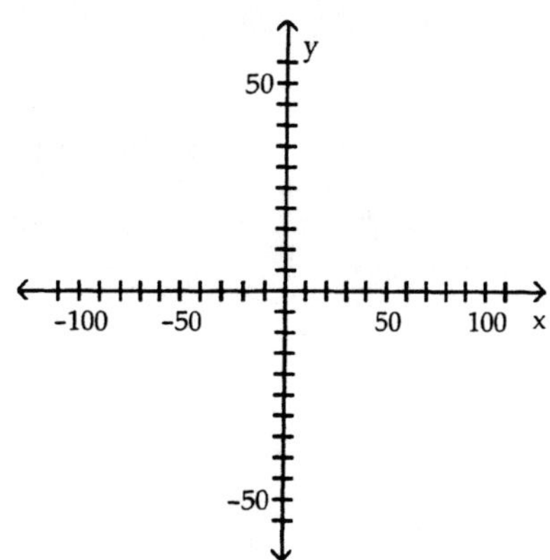

A)

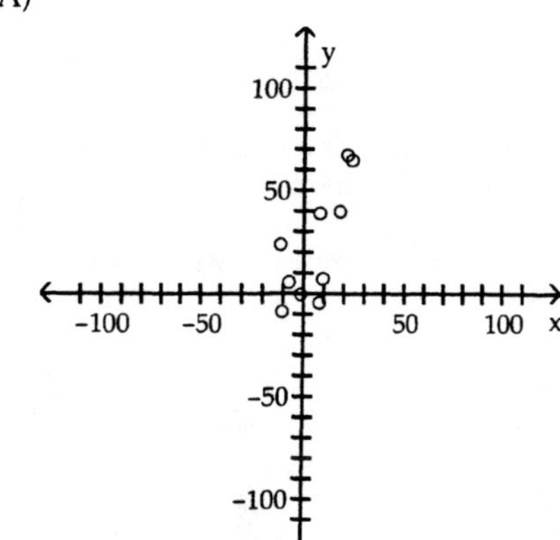

B)

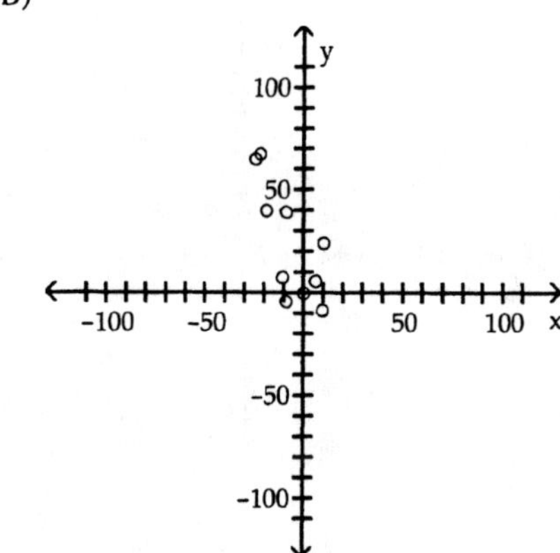

C)

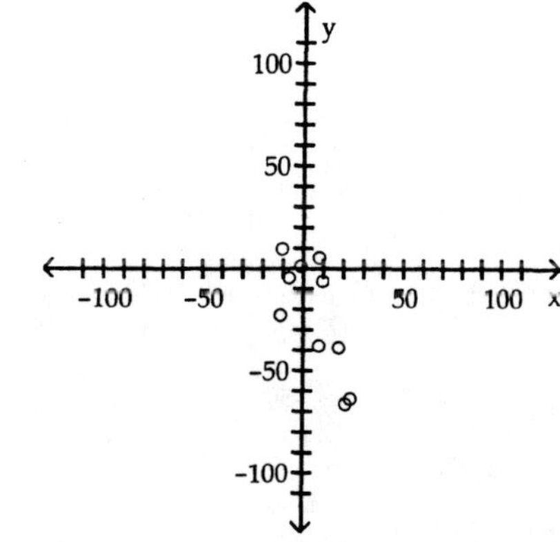

D)

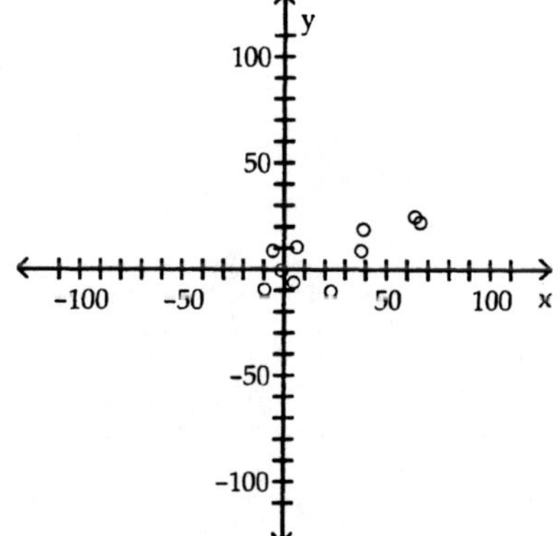

4)

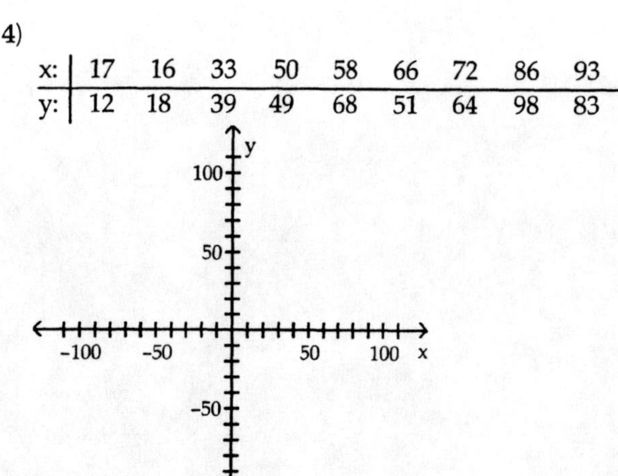

x:	17	16	33	50	58	66	72	86	93
y:	12	18	39	49	68	51	64	98	83

A)

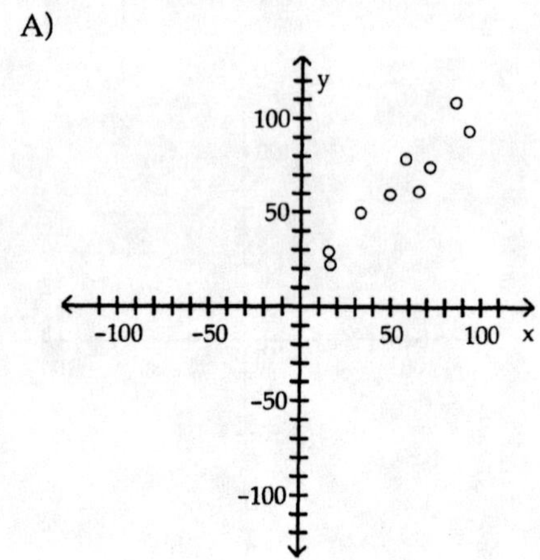

B)

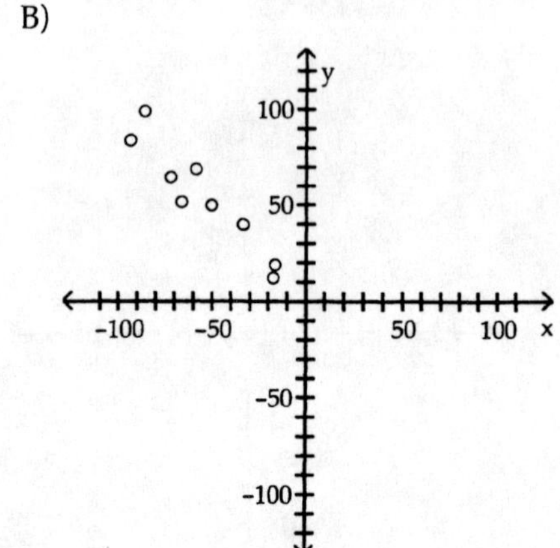

C)

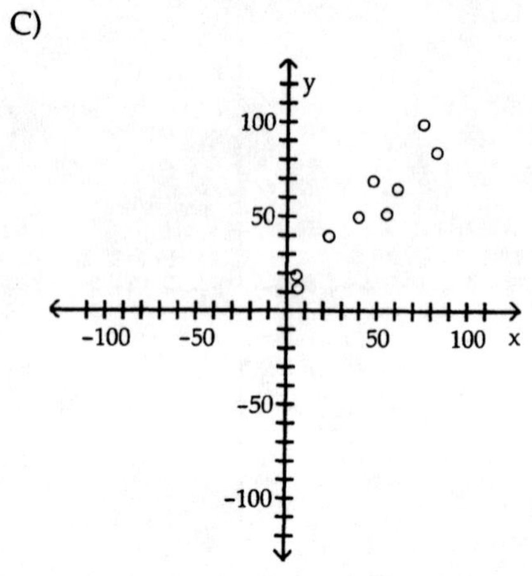

D)

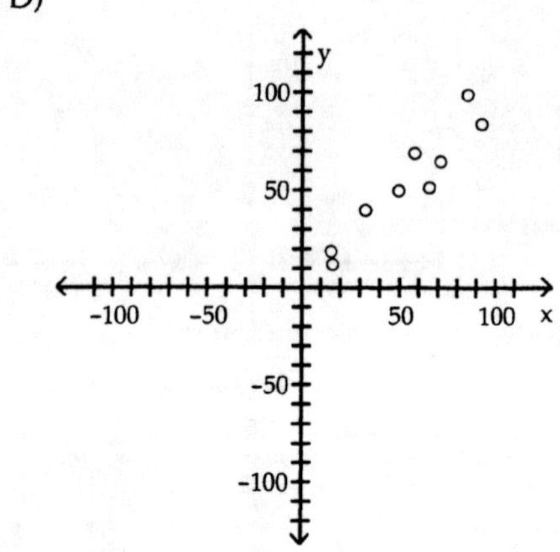

2 Distinguish Between Linear and Nonlinear Relations

Determine if the type of relation is linear, nonlinear, or none.

1)

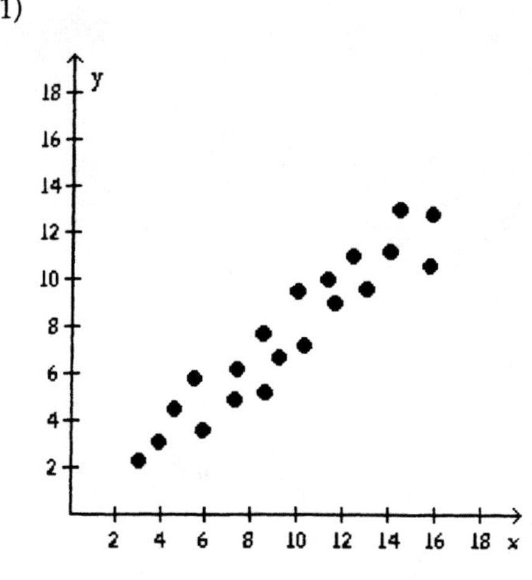

A) none B) nonlinear C) linear

2)

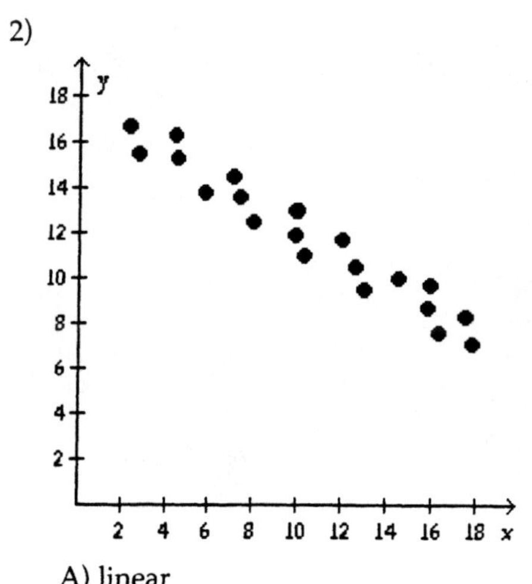

A) linear B) none C) nonlinear

3)

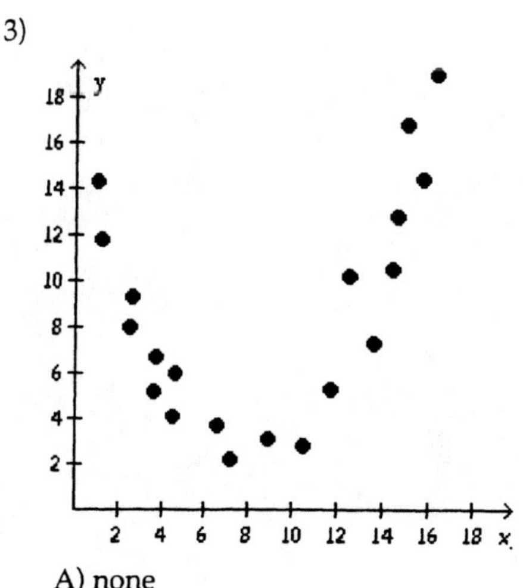

A) none B) nonlinear C) linear

4)

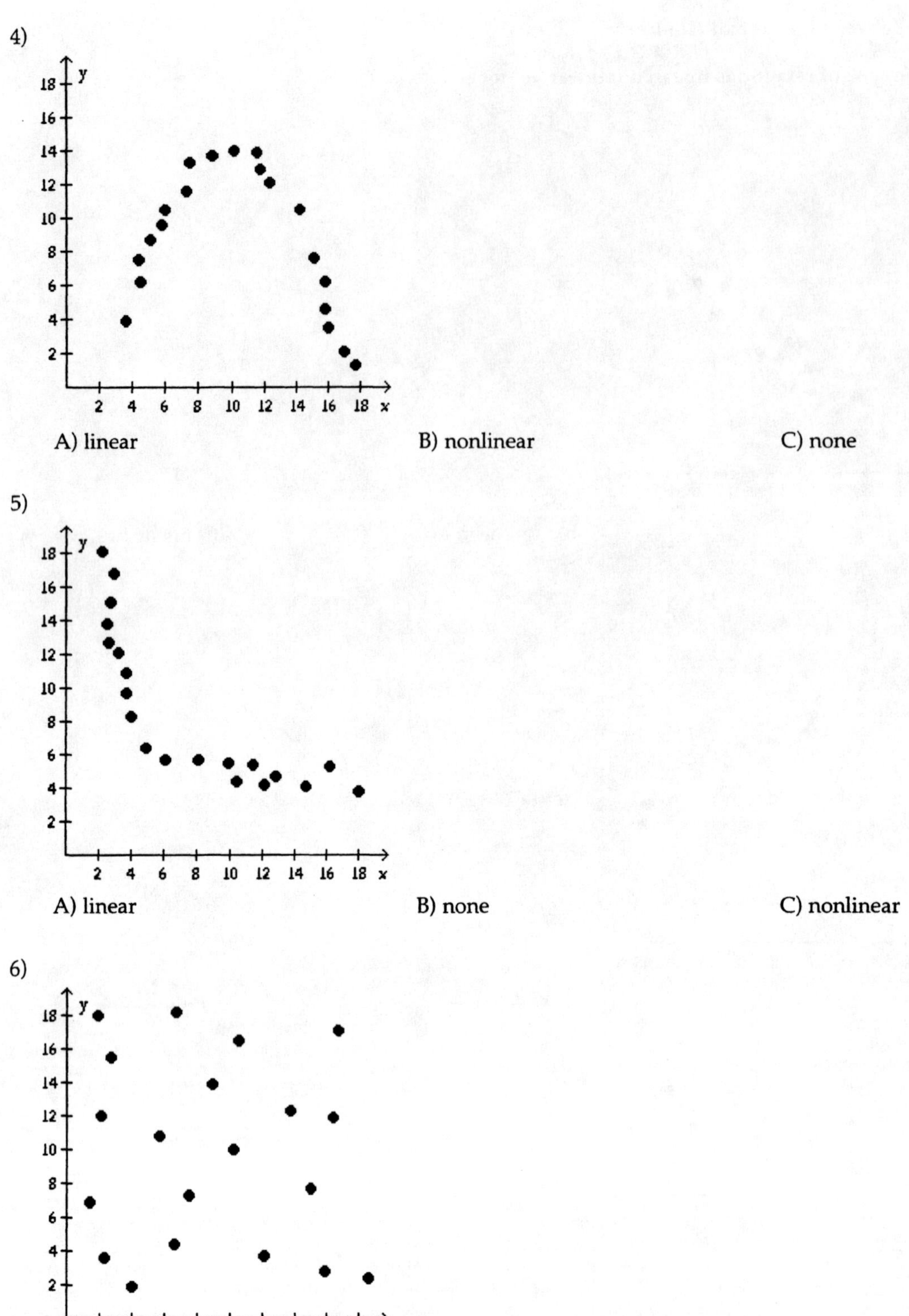

A) linear
B) nonlinear
C) none

5)

A) linear
B) none
C) nonlinear

6)

A) linear
B) nonlinear
C) none

Solve the problem.

7) Identify the scatter diagram of the relation that appears linear.

A)

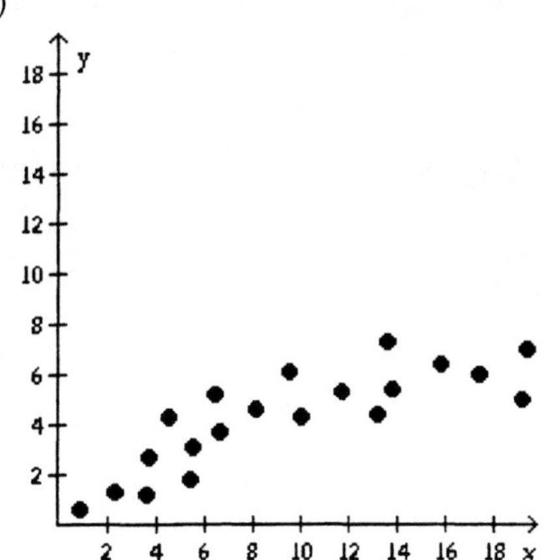

B)

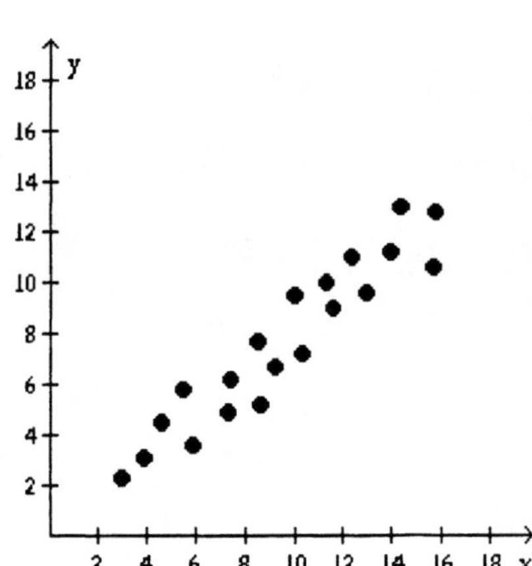

C)

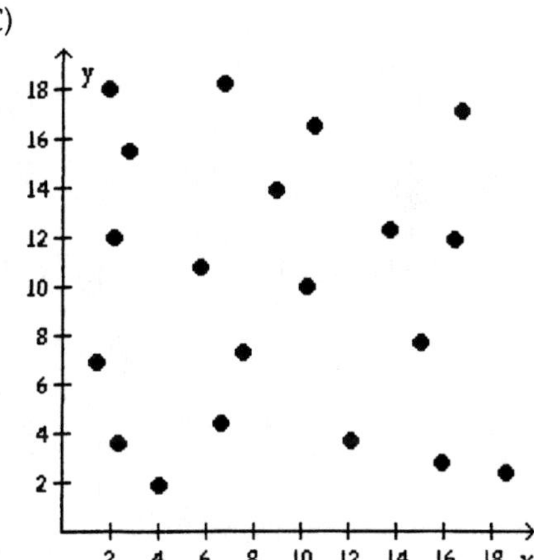

D)

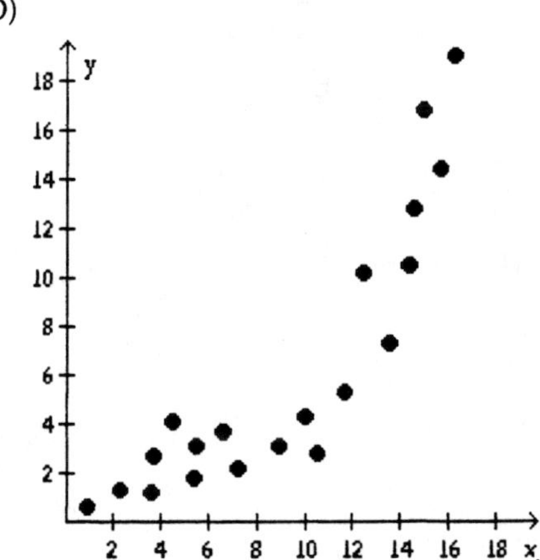

3 Use a Graphing Utility to Find the Line of Best Fit

Use a graphing utility to find the equation of the line of best fit. Round to two decimal places, if necessary.

1)

x	2	4	5	6
y	7	11	13	20

A) $y = 2.8x + 0.15$ B) $y = 2.8x$ C) $y = 3x$ D) $y = 3x + 0.15$

2)

x	6	8	20	28	36
y	2	4	13	20	30

A) $y = 0.95x - 2.79$ B) $y = 0.90x - 3.79$ C) $y = 0.80x - 3.79$ D) $y = 0.85x - 2.79$

3)

x	1	3	5	7	9
y	143	116	100	98	90

A) $y = -6.2x + 140.4$ B) $y = -6.8x + 150.7$ C) $y = 6.2x - 140.4$ D) $y = 6.8x - 150.7$

4)

x	1	2	3	4	5	6
y	17	20	19	22	21	24

A) $y = 1.03x + 18.9$ B) $y = 1.17x + 16.4$ C) $y = 1.17x + 18.9$ D) $y = 1.03x + 16.4$

5)

x	0	3	4	5	12
y	8	2	6	9	12

A) $y = 0.53x + 4.88$ B) $y = 0.73x + 4.98$ C) $y = 0.43x + 4.98$ D) $y = 0.63x + 4.88$

6)

x	3	5	7	15	16
y	8	11	7	14	20

A) $y = 0.75x + 5.07$ B) $y = 0.95x + 3.07$ C) $y = 0.75x + 4.07$ D) $y = 0.85x + 3.07$

7)

x	24	26	28	30	32
y	15	13	20	16	24

A) $y = 0.95x + 11.8$ B) $y = 1.05x + 11.8$ C) $y = 0.95x - 11.8$ D) $y = 1.05x - 11.8$

8)

x	2	4	6	8	10
y	15	37	60	75	94

A) $y = 9x - 3$ B) $y = 9.8x - 2.6$ C) $y = 10x - 3$ D) $y = 9.2x - 2.1$

9)

x	1.2	1.4	1.6	1.8	2.0
y	54	53	55	54	56

A) $y = 2.5x + 50.4$ B) $y = 55.3$ C) $y = 54$ D) $y = 3x + 50$

10)

x	10	20	30	40	50
y	3.9	4.6	5.4	6.9	8.3

A) $y = x - 8$ B) $y = 0.17x + 2.11$ C) $y = 0.11x + 2.49$ D) $y = 0.5x - 2$

11)

x	2	3	7	8	10
y	3	4	4	5	6

A) $y = 0.32x + 2.57$ B) $y = 0.30x + 2.57$ C) $y = 0.32x + 4.29$ D) $y = 0.30x + 4.29$

12)

x	2	3	7	8	10
y	2	4	4	6	6

A) $y = 1.79x + 0.43$ B) $y = -1.86x + 1.79$ C) $y = 0.43x + 1.79$ D) $y = 1.79x - 1.86$

Solve the problem.

13) A drug company establishes that the most effective dose of a new drug relates to body weight as shown below. Let body weight be the independent variable and drug dosage be the dependent variable. Use a graphing utility to draw a scatter diagram and to find the line of best fit. What is the most effective dosage for a person weighing 140 lbs?

Body Weight (lbs)	Drug Dosage (mg)
50	10
100	10
150	14
200	19
250	19

A) 10.07 mg B) 13.86 mg C) 12 mg D) 17.75 mg

14) A marina owner wishes to estimate a linear function that relates boat length in feet and its draft (depth of boat below water line) in feet. He collects the following data. Let boat length represent the independent variable and draft represent the dependent variable. Use a graphing utility to draw a scatter diagram and to find the line of best fit. What is the draft for a boat 60 ft in length (to the nearest tenth)?

Boat Length (ft)	Draft (ft)
25	2.5
25	2
30	3
30	3.5
45	6
45	7
50	7
50	8

A) 10.5 B) 10.3 C) 15.7 D) 9.7

15) A survey of the interest rates earned by Certificates of Deposit (CDs) showed the following percents for the length of time (in years) for holding the CD. Let length of time represent the independent variable and interest rate represent the dependent variable. Use a graphing utility to draw a scatter diagram and to find the line of best fit. What is the estimate of the interest rate for a CD held for 30 years (to the nearest thousandth)?

CD Maturity (yrs)	Interest rate (%)
5	8.458
10	8.470
15	8.496
20	8.580
25	8.625

A) 8.669 B) 8.675 C) 8.874 D) 9.064

16) Super Sally, a truly amazing individual, picks up a rock and throws it as hard as she can. The table below displays the relationship between the rock's horizontal distance, d (in feet) from Sally and the initial speed with which she throws.

Initial speed(in ft/sec), v	10	15	20	25	30
Horizontal distance of the rock (in feet), d	9.9	14.8	19.1	24.5	28.2

Assume that the horizontal distance travelled varies linearly with the speed with which the rock is thrown. Using a graphing utility, find the line of best fit, and estimate, rounded to two decimal places, the horizontal distance of the rock if the initial speed is 33 ft/sec.

A) 34.76 feet B) 26.67 feet C) 31.34 feet D) 31.33 feet

17) The following data represents the amount of money Tom is saving each month since he graduated from college.

month	1	2	3	4	5	6	7
savings	$52	$70	$81	$91	$102	$118	$132

Using the line of best fit for the data set, predict the amount he will save in the 24th month after graduating from college.

18) The following data represents the amount of money Tom is saving each month since he graduated from college.

month	1	2	3	4	5	6	7
savings	$52	$70	$81	$91	$102	$118	$132

Find the slope of the line of best fit for the data set and interpret it.

19) The following data represents the Olympic winning time in Women's 100 m Freestyle.

year	1972	1976	1980	1984	1988	1992	1996
time	58.59	55.65	54.79	55.92	54.93	54.65	54.50

Using the line of best fit (with slope correct to 5 decimal places) for the data set, predict the Olympic winning time in 2000.

20) The following data represents the Olympic winning time in Women's 100 m Freestyle.

year	1972	1976	1980	1984	1988	1992	1996
time	58.59	55.65	54.79	55.92	54.93	54.65	54.50

Find the slope of the line of best fit for the data set and interpret it.

21) The following data represents the number of employees at a company at the start of each year since the company began.

month	1	2	3	4	5	6	7
number	3	172	403	571	823	1061	1194

Using the line of best fit for the data set, predict the number of employees at the start of the 10th year.

22) The following data represents the number of employees at a company at the start of each year since the company began.

month	1	2	3	4	5	6	7
number	3	172	403	571	823	1061	1194

Find the slope of the line of best fit for the data set and interpret it.

2.7 Variation

1 Construct a Model Using Direct Variation

Write a general formula to describe the variation.

1) v varies directly with t; $v = 9$ when $t = 14$

A) $v = \frac{9}{14}t$ B) $v = \frac{9}{14t}$ C) $v = \frac{14}{9}t$ D) $v = \frac{14}{9t}$

2) A varies directly with t^2; $A = 147$ when $t = 7$

A) $A = 21t^2$ B) $A = 3t^2$ C) $A = \frac{3}{t^2}$ D) $A = \frac{21}{t^2}$

3) z varies directly with the sum of the squares of x and y; $z = 15$ when $x = 9$ and $y = 12$

A) $z = \frac{1}{225}(x^2 + y^2)$ B) $z = \frac{1}{15}(x^2 + y^2)$ C) $z^2 = x^2 + y^2$ D) $z = \frac{1}{30}(x^2 + y^2)$

4) The volume V of a right circular cone varies directly with the square of its base radius r and its height h. The constant of proportionality is $\frac{1}{3}\pi$.

A) $V = \frac{1}{3}\pi r^2 h^2$ B) $V = \frac{1}{3}\pi r^2 h$ C) $V = \frac{1}{3}r^2 h$ D) $V = \frac{1}{3}\pi r h$

5) The surface area S of a right circular cone varies directly as the radius r times the square root of the sum of the squares of the base radius r and the height h. The constant of proportionality is π.

A) $S = \pi r\sqrt{r^2 h}$ B) $S = \pi r\sqrt{r^2 h^2}$ C) $S = \pi r\sqrt{r^2 + h^2}$ D) $S = \pi\sqrt{r^2 + h^2}$

Solve the problem.

6) In simplified form, the period of vibration P for a pendulum varies directly as the square root of its length L. If P is 0.5 sec. when L is 4 in., what is the period when the length is 49 in.?

A) 12.25 sec. B) 28 sec. C) 196 sec. D) 1.75 sec.

7) The amount a spring will stretch, S, varies directly with the force (or weight), F, attached to the spring. If a spring stretches 0.6 inches with 20 pounds attached, how far will it stretch with 90 pounds attached? Round to two decimal places, if necessary.

A) 0.33 in. B) 54 in. C) 0.03 in. D) 2.70 in .

8) On planet X, an object falls 18 feet in 3 seconds. Knowing the distance it falls varies directly with the square of the time of fall, how long does it take an object to fall 100 feet? Round to three decimal places, if necessary.

9) Sonya's earnings are directly proportional to the number of hours she tutors math. If she earns \$422.50 when she tutors 13 hours, what will she earn if she tutors 28 hours?

2 Construct a Model Using Inverse Variation

Write a general formula to describe the variation.

1) A varies inversely with x^2; $A = 8$ when $x = 4$

A) $A = 32x^2$ B) $A = \frac{128}{x^2}$ C) $A = \frac{32}{x^2}$ D) $A = \frac{1}{2}x^2$

Solve the problem.

2) The price per person of renting a bus varies inversely with the number of people renting the bus. It costs \$26 per person if 51 people rent the bus. How much will it cost per person if 96 people rent the bus?

A) \$13.81 B) \$188.31 C) \$48.94 D) \$16.90

3) The intensity, I, of light received from a source varies inversely as the square of the distance, d, from the source. If the light intensity is 2 foot-candles at 11 feet, find the light intensity at 19 feet. Round your answer to the nearest hundredth if necessary.

4) The length of time it takes Brad to get to work varies inversely with his speed of travel. If it takes him 2.5 hours when traveling 15 miles per hour, how long does it take when traveling 50 miles per hour?

5) The amount of time Jesse works on his homework each day varies inversely with the amount of time he spends watching TV that day. If he spends 3 hours working on homework when he watches 0.5 hours of TV, how much time does he spend on homework when he watches 2.5 hours of TV.

3 Construct a Model Using Joint or Combined Variation

Write a general formula to describe the variation.

1) The square of G varies directly with the cube of x and inversely with the square of y; G = 5 when x = 3 and y = 4

A) $G = \frac{25}{432}(x^3 + y^2)$ B) $G = \frac{20}{3}\frac{x^3}{y^2}$ C) $G = \frac{400}{27}\frac{x^3}{y^2}$ D) $G = \frac{675}{16}\frac{y^3}{x^2}$

2) R varies directly with g and inversely with the square of h; R = 3 when g = 3 and h = 5.

3) z varies jointly as the cube root of x and the cube of y; z = 50 when x = 8 and y = 5.

A) $z = 3125\frac{\sqrt[3]{x}}{y^3}$ B) $z = \frac{1}{5}\sqrt[3]{x}y^3$ C) $z = 5\sqrt[3]{x}y^3$ D) $z = \frac{1}{3125}\frac{\sqrt[3]{x}}{y^3}$

4) The centrifugal force F of an object speeding around a circular course varies directly as the product of the object's mass m and the square of it's velocity v and inversely as the radius of the turn r.

A) $F = \frac{km^2v}{r}$ B) $F = \frac{kmv^2}{r}$ C) $F = \frac{kmr}{v^2}$ D) $F = \frac{kmv}{r}$

5) The safety load λ of a beam with a rectangular cross section that is supported at each end varies directly as the product of the width W and the square of the depth D and inversely as the length L of the beam between the supports.

A) $\lambda = \frac{kWD}{L}$ B) $\lambda = \frac{kL}{WD^2}$ C) $\lambda = \frac{kWD^2}{L}$ D) $\lambda = \frac{k(W + D^2)}{L}$

6) The illumination I produced on a surface by a source of light varies directly as the candlepower c of the source and inversely as the square of the distance d between the source and the surface.

A) $I = kcd^2$ B) $I = \frac{kc^2}{d^2}$ C) $I = \frac{kc}{d^2}$ D) $I = \frac{kd^2}{c}$

Solve the problem.

7) The volume V of a given mass of gas varies directly as the temperature T and inversely as the pressure P. A measuring device is calibrated to give V = 696 in^3 when T = 580° and P = 10 lb/in^2. What is the volume on this device when the temperature is 260° and the pressure is 20 lb/in^2?

A) V = 13 in^3 B) V = 126 in^3 C) V = 186 in^3 D) V = 156 in^3

8) The time in hours it takes a satellite to complete an orbit around the earth varies directly as the radius of the orbit (from the center of the earth) and inversely as the orbital velocity. If a satellite completes an orbit 770 miles above the earth in 17 hours at a velocity of 30,000 mph, how long would it take a satellite to complete an orbit if it is at 1800 miles above the earth at a velocity of 39,000 mph? (Use 3960 miles as the radius of the earth.)

A) 159.25 hours B) 15.92 hours C) 30.57 hours D) 4.98 hours

9) The wattage rating of an appliance, W, varies jointly as the square of the current, I, and the resistance, R. If the wattage is 12 watts when the current is 0.2 ampere and the resistance is 300 ohms, find the wattage when the current is 0.3 ampere and the resistance is 250 ohms.

A) 22.5 watts B) 18,750 watts C) 150 watts D) 75 watts

10) The area of the projected rectangular image varies jointly with the length and width of the image being projected. The area of the projected image of an 8.5-inch by 11-inch image is 374 square inches. Find the area of the projected image of an 11-inch by 17-inch image.

11) The volume of an ideal gas varies directly with the temperature T and inversely with the Pressure P. A cylinder contains 15 liters of oxygen at a temperature of 240 K and a pressure of 80 atmospheres. If a piston is lowered into the cylinder decreasing the volume occupied by the gas to 10 liters and the temperature remains constant, what is the pressure?

Ch. 2 Graphs
Answer Key

2.1 Rectangular Coordinates
1 Use the Distance Formula

1) B
2) C
3) D
4) B
5) B
6) B
7) A
8) B
9) B
10) A
11) A
12) B
13) B
14) B
15) D
16) A
17) A
18) B
19) C
20) A

2 Use the Midpoint Formula

1) C
2) A
3) C
4) A
5) A
6) D
7) B
8) C
9) B
10) B
11) B
12) A

3 Additional Applications

1) A
2) B
3) C
4) D
5) B
6) C
7) D
8) C
9) C
10) A
11) B
12) A
13) D
14) C
15) B
16) B

17) D
18) D

2.2 Graphs of Equations

1 Graph Equations by Plotting Points

1) C
2) B
3) C
4) A
5) A
6) C
7) C
8) D
9) A
10) A
11) D
12) B
13) D

2 Find Intercepts from a Graph

1) B
2) C
3) C
4) B
5) A
6) B
7) A
8) A
9) A

3 Find Intercepts from an Equation

1) C
2) D
3) B
4) A
5) A
6) D
7) A
8) D
9) D
10) C
11) B
12) D
13) B
14) B

4 Test an Equation for Symmetry with Respect to the (a) x-axis, (b) y-axis, and (c) Origin

1) D
2) E
3) B
4) E
5) D
6) E
7) C
8) E
9) C
10) C
11) B
12) A

13) B
14) D
15) D
16) E
17) A
18) A
19) B
20) E
21) C
22) E
23) B
24) B
25) B

5 Additional Applications

1) A
2) D
3) A
4) D
5) A
6) B
7) D
8) A

2.3 Circles

1 Write the Standard Form of the Equation of a Circle

1) C
2) A
3) B
4) C
5) D
6) C
7) C
8) D
9) D
10) B
11) B
12) C
13) A
14) B
15) A
16) A
17) C
18) D

2 Graph a Circle

1) A
2) C
3) C
4) A
5) A
6) D
7) A
8) C

3 Find the Center and Radius of a Circle in General Form and Graph It

1) B
2) C
3) D

4) D
5) A
6) A
7) A
8) D
9) D
10) A

2.4 Lines

1 Calculate and Interpret the Slope of a Line

1) B
2) D
3) C
4) D
5) D
6) B
7) A
8) B
9) B
10) D
11) B

2 Graph Lines Given a Point and the Slope

1) D
2) A
3) C
4) B
5) A
6) C
7) A
8) C
9) B

3 Find the Equation of a Vertical Line

1) D
2) D
3) A
4) C
5) A
6) B

4 Use the Point–Slope Form of a Line; Identify Horizontal Lines

1) C
2) A
3) A
4) A
5) D

5 Find the Equation of a Line Given Two Points

1) B
2) A
3) B
4) C
5) D
6) C

6 Write the Equation of a Line in Slope–Intercept Form

1) C
2) B
3) D
4) D

5) A
6) A
7) A
8) B
9) D
10) D
11) D
12) C

7 Identify the Slope and y–Intercept of a Line from Its Equation

1) B
2) C
3) B
4) A
5) C
6) D
7) C
8) B
9) C
10) A
11) C

8 Write the Equation of a Line in General Form

1) B
2) A
3) B
4) A
5) D
6) A
7) A
8) D
9) C
10) A
11) A
12) B

9 Additional Applications

1) A
2) C
3) $P = 0.65x - 25$; \$34.80
4) $P = 1.75x - 1200$; \$410
5) $P = 4.65x - 120$; \$810
6) $P = 20x$; \$1200
7) $P = 0.60x - 37{,}000$; \$8000
8) B
9) A
10) B
11) A
12) C
13) D
14) B
15) A
16) A

2.5 Parallel and Perpendicular Lines

1 Define Parallel Lines

1) C
2) C
3) A

4) C
5) C
6) A
7) B
8) C
9) B
10) B
11) D

2 Find Equations of Parallel Lines

1) D
2) D
3) B
4) C
5) C
6) B
7) D
8) D
9) B
10) C

3 Define Perpendicular Lines

1) A
2) A
3) C
4) D
5) B
6) B
7) B
8) C
9) B
10) D
11) C

4 Find Equations of Perpendicular Lines

1) C
2) C
3) A
4) C
5) C
6) B
7) D
8) B
9) B
10) C
11) B
12) D

5 Additional Applications

1) B
2) B
3) A
4) A
5) C
6) C = (0, −13), D = (−8, −7)
7) Slopes of AB and CD are $\frac{1}{6}$ and the slopes of BC and AD are −6.

8) B

2.6 Scatter Diagrams; Linear Curve Fitting

1 Draw and Interpret Scatter Diagrams

1)

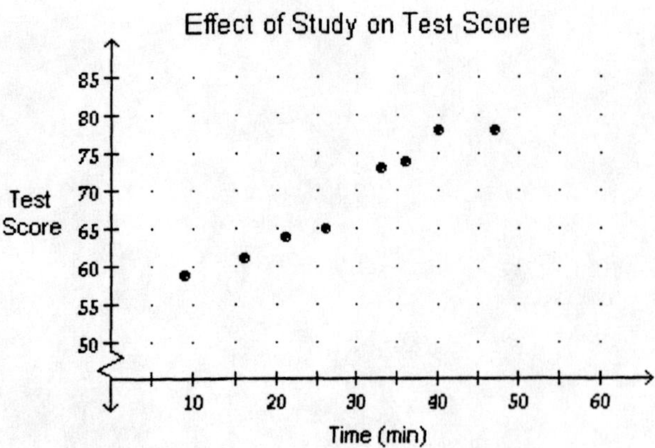

More time spent studying may increase test scores.

2) B
3) A
4) D

2 Distinguish Between Linear and Nonlinear Relations

1) C
2) A
3) B
4) B
5) C
6) C
7) B

3 Use a Graphing Utility to Find the Line of Best Fit

1) C
2) B
3) A
4) B
5) A
6) A
7) D
8) B
9) A
10) C
11) B
12) C
13) B
14) D
15) A
16) C
17) $347.29
18) The slope is 12.75 which means that the amount Tom saves increases $12.75 each month.
19) 53.56
20) The slope is about -0.12616 which means that the winning time is decreasing by 0.12616 of a second each year.
21) 1840
22) The slope is about 206.1 which means that the number of employees is increasing by about 206 employees each year.

2.7 Variation

1 Construct a Model Using Direct Variation

1) A
2) B
3) B
4) B
5) C
6) D
7) D
8) 7.071 sec.
9) $910

2 Construct a Model Using Inverse Variation

1) B
2) A
3) 0.67 foot-candles
4) 0.75 hours
5) 0.6 hours

3 Construct a Model Using Joint or Combined Variation

1) C
2) $R = 25\frac{g}{h^2}$
3) B
4) B
5) C
6) C
7) D
8) B
9) A
10) 748 square inches
11) 120 atmospheres

Ch. 3 Functions and Their Graphs

3.1 Functions

1 Determine Whether a Relation Represents a Function

Determine whether the relation represents a function. If it is a function, state the domain and range.

1)

4 →	20
9 →	45
14 →	70
19 →	95

A) function
domain: {4, 9, 14, 19}
range: {20, 45, 70, 95}

B) function
domain:{20, 45, 70, 95}
range: {4, 9, 14, 19}

C) not a function

2)

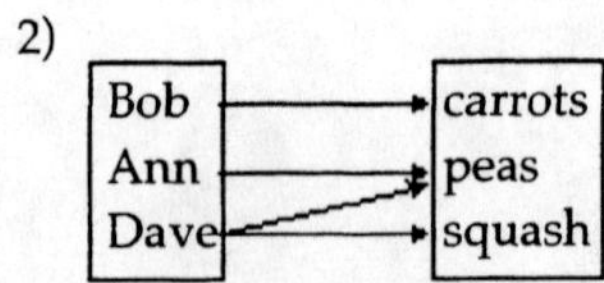

A) function
domain: {carrots, peas, squash}
range: {Bob, Ann, Dave}

B) function
domain: {Bob, Ann, Dave}
range: {carrots, peas, squash}

C) not a function

3)

Bob → Ms. Lee
Ann → Mr. Bar
Dave → Mr. Bar

A) function
domain: {Bob, Ann, Dave}
range: {Ms. Lee, Mr. Bar}

B) function
domain: {Ms. Lee, Mr. Bar}
range: {Bob, Ann, Dave}

C) not a function

4) $\{(-2, 6), (2, 4), (4, 0), (6, -2)\}$

A) function
domain: $\{-2, 2, 4, 6\}$
range: $\{6, 4, 0, -2\}$

B) function
domain: $\{6, 4, 0, -2\}$
range: $\{-2, 2, 4, 6\}$

C) not a function

5) $\{(41, -3), (5, -2), (5, 0), (9, 2), (21, 4)\}$

A) function
domain: $\{-3, -2, 0, 2, 4\}$
range: $\{41, 9, 5, 21\}$

B) function
domain: $\{41, 9, 5, 21\}$
range: $\{-3, -2, 0, 2, 4\}$

C) not a function

6) {(−2, 0), (−1, −3), (0, −4), (1, −3), (3, 5)}

A) function
domain: {0, −3, −4, 5}
range: {−2, −1, 0, 1, 3}

B) function
domain: {−2, −1, 0, 1, 3}
range: {0, −3, −4, 5}

C) not a function

7) $\{(9.88, 13.98), (9.888, -14), (\frac{5}{3}, 0), (1.67, -9)\}$

A) function
domain: {13.98, −14, 0, −9}
range: $\{9.88, 9.888, \frac{5}{3}, 1.67\}$

B) function
domain: $\{9.88, 9.888, \frac{5}{3}, 1.67\}$
range: {13.98, −14, 0, −9}

C) not a function

Determine whether the equation is a function.

8) $y = x^3$

A) function B) not a function

9) $y = \frac{1}{x}$

A) function B) not a function

10) $y = |x|$

A) function B) not a function

11) $y^2 = 5 - x^2$

A) function B) not a function

12) $y = \pm\sqrt{1 - 4x}$

A) function B) not a function

13) $x = y^2$

A) function B) not a function

14) $y^2 + x = 7$

A) function B) not a function

15) $y = 3x^2 - 3x + 8$

A) function B) not a function

16) $y = \frac{3x - 2}{x - 3}$

A) function B) not a function

17) $x^2 + 5y^2 = 1$

A) function B) not a function

18) $x - 8y = 6$

A) function B) not a function

19) $-5x + x^2 - 22 = y$

A) function

B) not a function

2 Find the Value of a Function

Find the value for the function.

1) Find f(4) when $f(x) = x^2 - 5x - 6$.

A) 42 B) −10 C) 2 D) 30

2) Find f(1) when $f(x) = \frac{x^2 - 7}{x - 2}$.

A) − 8 B) 6 C) − 2 D) − 1

3) Find f(−9) when $f(x) = |x| - 6$.

A) −15 B) −3 C) 3 D) 15

4) Find f(4) when $f(x) = \sqrt{x^2 + 6x}$.

A) $2\sqrt{13}$ B) $2\sqrt{10}$ C) $\sqrt{42}$ D) $\sqrt{22}$

5) Find f(−x) when $f(x) = 3x^2 - 3x + 2$.

A) $-3x^2 + 3x - 2$ B) $-3x^2 + 3x + 2$ C) $3x^2 + 3x - 2$ D) $3x^2 + 3x + 2$

6) Find f(−x) when $f(x) = \frac{x}{x^2 + 5}$.

A) $\frac{x}{-x^2 + 5}$ B) $\frac{-x}{x^2 + 5}$ C) $\frac{-x}{-x^2 + 5}$ D) $\frac{-x}{x^2 - 5}$

7) Find −f(x) when $f(x) = -2x^2 + 5x - 3$.

A) $-2x^2 - 5x - 3$ B) $-2x^2 - 5x + 3$ C) $2x^2 - 5x - 3$ D) $2x^2 - 5x + 3$

8) Find −f(x) when $f(x) = |x| - 9$.

A) $-|x| + 9$ B) $|-x| + 9$ C) $-|x| - 9$ D) $|-x| - 9$

9) Find f(x − 1) when $f(x) = 2x^2 - 4x + 5$.

A) $2x^2 - 8x + 3$ B) $2x^2 - 8x + 11$ C) $-8x^2 + 2x + 11$ D) $2x^2 + 6x + 3$

10) Find f(x + 1) when $f(x) = \frac{x^2 - 7}{x - 5}$.

A) $\frac{x^2 - 6}{x - 4}$ B) $\frac{x^2 + 2x - 6}{x + 6}$ C) $\frac{x^2 + 2x - 6}{x - 4}$ D) $\frac{x^2 + 2x + 8}{x - 4}$

11) Find f(−x) when $f(x) = 2x^2 - 2x - 1$.

A) $2x^2 + 2x + 1$ B) $2x^2 + 2x - 1$ C) $-2x^2 + 2x + 1$ D) $-2x^2 + 2x - 1$

12) Find f(2x) when $f(x) = \sqrt{7x^2 - 4x}$.

A) $\sqrt{14x^2 - 16x}$ B) $\sqrt{14x^2 - 8x}$ C) $\sqrt{28x^2 - 8x}$ D) $2\sqrt{7x^2 - 4x}$

13) Find $f(x + h)$ when $f(x) = -3x^2 - 4x + 5$.

A) $-3x^2 - 3xh - 3h^2 - 4x - 4h + 5$
B) $-3x^2 - 3h^2 - 4x - 4h + 5$
C) $-3x^2 - 6xh - 3h^2 - 4x - 4h + 5$
D) $-3x^2 - 3h^2 - 10x - 10h + 5$

14) Find $f(x + h)$ when $f(x) = \frac{7x + 5}{5x - 9}$.

A) $\frac{7x + 7h + 5}{5x - 9}$
B) $\frac{7x + 7h + 5}{5x + 5h - 9}$
C) $\frac{7x + 5h}{5x - 9h}$
D) $\frac{7x + 12h}{5x - 4h}$

Solve the problem.

15) If $f(x) = 8x^3 + 5x^2 - x + C$ and $f(2) = 1$, what is the value of C?

A) $C = 87$
B) $C = -81$
C) $C = -13$
D) $C = 47$

16) If $f(x) = \frac{x - B}{x - A}$, $f(-1) = 0$, and $f(5)$ is undefined, what are the values of A and B?

A) $A = -5, B = 1$
B) $A = 5, B = -1$
C) $A = -1, B = 5$
D) $A = 1, B = -5$

17) If $f(x) = \frac{x - 5A}{-15x + 3}$ and $f(-15) = -10$, what is the value of A?

A) $A = -231$
B) $A = -453$
C) $A = 231$
D) $A = 453$

18) If a rock falls from a height of 60 meters on Earth, the height H (in meters) after x seconds is approximately
$H(x) = 60 - 4.9x^2$.
What is the height of the rock when $x = 1.7$ seconds? Round to the nearest hundredth, if necessary.

A) 45.84 m
B) 74.16 m
C) 46.13 m
D) 51.67 m

19) If a rock falls from a height of 50 meters on Earth, the height H (in meters) after x seconds is approximately
$H(x) = 50 - 4.9x^2$.
When does the rock strike the ground? Round to the nearest hundredth, if necessary.

A) 2.08 sec
B) 3.19 sec
C) 1.44 sec
D) 10.2 sec

3 Find the Domain of a Function

Find the domain of the function.

1) $f(x) = 9x - 7$

A) $\{x \mid x \geq 7\}$
B) $\{x \mid x > 0\}$
C) $\{x \mid x \neq 0\}$
D) all real numbers

2) $f(x) = x^2 + 8$

A) $\{x \mid x > -8\}$
B) $\{x \mid x \neq -8\}$
C) $\{x \mid x \geq -8\}$
D) all real numbers

3) $f(x) = \frac{x}{x^2 + 20}$

A) all real numbers
B) $\{x \mid x \neq 0\}$
C) $\{x \mid x \neq -20\}$
D) $\{x \mid x > -20\}$

4) $g(x) = \frac{2x}{x^2 - 1}$

A) all real numbers
B) $\{x \mid x \neq 0\}$
C) $\{x \mid x \neq -1, 1\}$
D) $\{x \mid x > 1\}$

5) $h(x) = \dfrac{x - 3}{x^3 - 9x}$

A) $\{x \mid x \neq -3, 0, 3\}$ B) $\{x \mid x \neq 0\}$ C) all real numbers D) $\{x \mid x \neq 3\}$

6) $f(x) = \sqrt{4 - x}$

A) $\{x \mid x \neq 4\}$ B) $\{x \mid x \neq 2\}$ C) $\{x \mid x \leq 4\}$ D) $\{x \mid x \leq 2\}$

7) $\dfrac{x}{\sqrt{x - 10}}$

A) $\{x \mid x \geq 10\}$ B) $\{x \mid x > 10\}$ C) $\{x \mid x \neq 10\}$ D) all real numbers

4 Form the Sum, Difference, Product and Quotient of Two Functions

For the given functions f and g, find the requested function and state its domain.

1) $f(x) = 2 - 3x$; $g(x) = -6x + 3$
Find $f + g$.

A) $(f + g)(x) = -9x + 5$; all real numbers
B) $(f + g)(x) = 3x + 5$; $\{x \mid x \neq \frac{5}{3}\}$
C) $(f + g)(x) = -4x$; all real numbers
D) $(f + g)(x) = -6x + 2$; $\{x \mid x \neq \frac{1}{3}\}$

2) $f(x) = 2x - 4$; $g(x) = 4x - 8$
Find $f - g$.

A) $(f - g)(x) = 2x - 4$; all real numbers
B) $(f - g)(x) = 6x - 12$; $\{x \mid x \neq 1\}$
C) $(f - g)(x) = -2x + 4$; all real numbers
D) $(f - g)(x) = -2x - 12$; $\{x \mid x \neq -6\}$

3) $f(x) = 8x - 5$; $g(x) = 9x - 2$
Find $f \cdot g$.

A) $(f \cdot g)(x) = 72x^2 + 10$; $\{x \mid x \neq 10\}$
B) $(f \cdot g)(x) = 72x^2 - 47x + 10$; $\{x \mid x \neq 10\}$
C) $(f \cdot g)(x) = 72x^2 - 61x + 10$; all real numbers
D) $(f \cdot g)(x) = 17x^2 - 61x - 7$; all real numbers

4) $f(x) = 4x + 5$; $g(x) = 6x - 5$
Find $\frac{f}{g}$.

A) $(\frac{f}{g})(x) = \dfrac{4x + 5}{6x - 5}$; $\{x \mid x \neq \frac{5}{6}\}$
B) $(\frac{f}{g})(x) = \dfrac{6x - 5}{4x + 5}$; $\{x \mid x \neq \frac{5}{6}\}$
C) $(\frac{f}{g})(x) = \dfrac{6x - 5}{4x + 5}$; $\{x \mid x \neq -\frac{5}{4}\}$
D) $(\frac{f}{g})(x) = \dfrac{4x + 5}{6x - 5}$; $\{x \mid x \neq -\frac{5}{4}\}$

5) $f(x) = 16 - x^2$; $g(x) = 4 - x$
Find $f + g$.

A) $(f + g)(x) = -x^2 - x + 20$; $\{x \mid x \neq 4, x \neq -5\}$
B) $(f + g)(x) = 4 + x$; $\{x \mid x \neq -4\}$
C) $(f + g)(x) = -x^2 + x + 12$; all real numbers
D) $(f + g)(x) = x^3 - 4x^2 - 16x + 64$; all real numbers

6) $f(x) = x + 9$; $g(x) = 9x^2$
Find $f - g$.

A) $(f - g)(x) = -9x^2 + x + 9$; $\{x \mid x \neq -9\}$
B) $(f - g)(x) = 9x^2 - x - 9$; all real numbers
C) $(f - g)(x) = -9x^2 + x + 9$; all real numbers
D) $(f - g)(x) = 9x^2 + x + 9$; all real numbers

7) $f(x) = 5x^3 - 3$; $g(x) = 2x^2 + 3$
Find $f \cdot g$.
A) $(f \cdot g)(x) = 10x^5 + 15x^3 - 6x^2 - 9$; $\{x \mid x \neq 0\}$
B) $(f \cdot g)(x) = 10x^5 + 15x^3 - 6x^2 - 9$; all real numbers
C) $(f \cdot g)(x) = 10x^6 + 15x^3 - 6x^2 - 9$; all real numbers
D) $(f \cdot g)(x) = 5x^3 + 2x^2 - 9$; all real numbers

8) $f(x) = \sqrt{x}$; $g(x) = 3x - 2$
Find $\frac{f}{g}$.
A) $(\frac{f}{g})(x) = \frac{\sqrt{x}}{3x - 2}$; $\{x \mid x \neq \frac{2}{3}\}$
B) $(\frac{f}{g})(x) = \frac{\sqrt{x}}{3x - 2}$; $\{x \mid x \neq 0\}$
C) $(\frac{f}{g})(x) = \frac{3x - 2}{\sqrt{x}}$; $\{x \mid x \geq 0\}$
D) $(\frac{f}{g})(x) = \frac{\sqrt{x}}{3x - 2}$; $\{x \mid x \geq 0, x \neq \frac{2}{3}\}$

9) $f(x) = \sqrt{7 - x}$; $g(x) = \sqrt{x - 2}$
Find $f \cdot g$.
A) $(f \cdot g)(x) = \sqrt{-x^2 - 14}$; $\{x \mid x \neq 14\}$
B) $(f \cdot g)(x) = \sqrt{(7 - x)(x - 2)}$; $\{x \mid 2 \leq x \leq 7\}$
C) $(f \cdot g)(x) = \sqrt{(7 - x)(x - 2)}$; $\{x \mid x \geq 0\}$
D) $(f \cdot g)(x) = \sqrt{(7 - x)(x - 2)}$; $\{x \mid x \neq 2, x \neq 7\}$

10) $f(x) = \frac{9x + 7}{7x - 5}$; $g(x) = \frac{2x}{7x - 5}$
Find $f + g$.
A) $(f + g)(x) = \frac{7x - 7}{7x - 5}$; $\{x \mid x \neq \frac{5}{7}\}$
B) $(f + g)(x) = \frac{11x + 7}{7x - 5}$; $\{x \mid x \neq \frac{5}{7}\}$
C) $(f + g)(x) = \frac{11x + 7}{7x - 5}$; $\{x \mid x \neq \frac{5}{7}, x \neq -\frac{7}{11}\}$
D) $(f + g)(x) = \frac{11x + 7}{7x - 5}$; $\{x \mid x \neq 0\}$

11) $f(x) = \sqrt{x + 7}$; $g(x) = \frac{2}{x}$
Find $f \cdot g$.
A) $(f \cdot g)(x) = \frac{\sqrt{2x + 14}}{x}$; $\{x \mid x \geq -7, x \neq 0\}$
B) $(f \cdot g)(x) = \sqrt{\frac{9}{x}}$; $\{x \mid x \neq 0\}$
C) $(f \cdot g)(x) = \sqrt{\frac{2x + 14}{x}}$; $\{x \mid x \geq -7, x \neq 0\}$
D) $(f \cdot g)(x) = \frac{2\sqrt{x + 7}}{x}$; $\{x \mid x \geq -7, x \neq 0\}$

Solve the problem.

12) Given $f(x) = \frac{1}{x}$ and $(\frac{f}{g})(x) = \frac{x + 6}{x^2 - 5x}$, find the function g.
A) $g(x) = \frac{x + 5}{x - 6}$
B) $g(x) = \frac{x + 6}{x - 5}$
C) $g(x) = \frac{x - 6}{x + 5}$
D) $g(x) = \frac{x - 5}{x + 6}$

5 Additional Applications

Solve the problem.

1) Express the gross salary G of a person who earns $45 per hour as a function of the number x of hours worked.
A) $G(x) = 45x^2$
B) $G(x) = 45x$
C) $G(x) = \frac{45}{x}$
D) $G(x) = 45 + x$

2) Jacey, a commissioned salesperson, earns \$450 base pay plus \$31 per item sold. Express Jacey's gross salary G as a function of the number x of items sold.

A) $G(x) = 450(x + 31)$ B) $G(x) = 450x + 31$ C) $G(x) = 31x + 450$ D) $G(x) = 31(x + 450)$

Find and simplify the difference quotient of f, $\frac{f(x + h) - f(x)}{h}$, $h \neq 0$, for the function.

3) $f(x) = 3x + 7$

A) $3 + \frac{14}{h}$ B) 0 C) 3 D) $3 + \frac{6(x + 7)}{h}$

4) $f(x) = x^2 + 7x + 6$

A) $2x + h + 6$ B) $2x + h + 7$

C) 1 D) $\frac{2x^2 + 2x + 2xh + h^2 + h + 12}{h}$

5) $f(x) = \frac{1}{8x}$

A) $\frac{-1}{x(x + h)}$ B) $\frac{-1}{8x(x + h)}$ C) $\frac{1}{8x}$ D) 0

Solve the problem.

6) Suppose that P(x) represents the percentage of income spent on automobile insurance in year x and I(x) represents income in year x. Determine a function A that represents total automobile insurance expenditures in year x.

A) $A(x) = (P \cdot I)(x)$ B) $A(x) = (\frac{I}{P})(x)$ C) $A(x) = (I - P)(x)$ D) $A(x) = (P + I)(x)$

3.2 The Graph of a Function

1 Identify the Graph of a Function

Determine whether the graph is that of a function. If it is, use the graph to find its domain and range, the intercepts, if any, and any symmetry with respect to the x-axis, the y-axis, or the origin.

1)

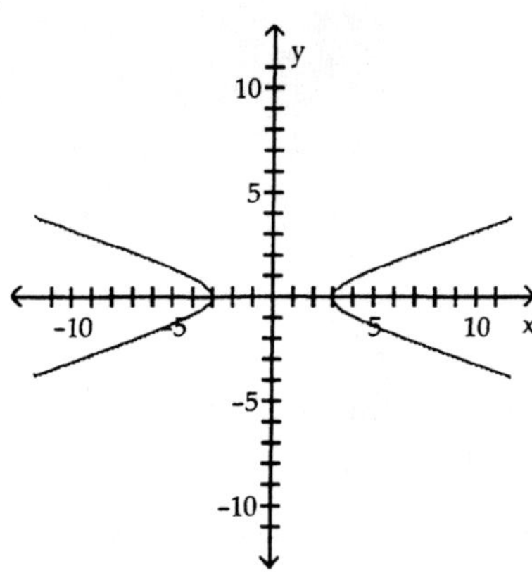

A) function
domain: all real numbers
range: $\{y \mid y \le -3 \text{ or } y \ge 3\}$
intercepts: $(-3, 0)$, $(3, 0)$
symmetry: y-axis

B) function
domain: $\{x \mid -3 \le x \le 3\}$
range: all real numbers
intercepts: $(-3, 0)$, $(3, 0)$
symmetry: x-axis, y-axis

C) function
domain: $\{x \mid x \le -3 \text{ or } x \ge 3\}$
range: all real numbers
intercepts: $(-3, 0)$, $(3, 0)$
symmetry: x-axis, y-axis, origin

D) not a function

2)

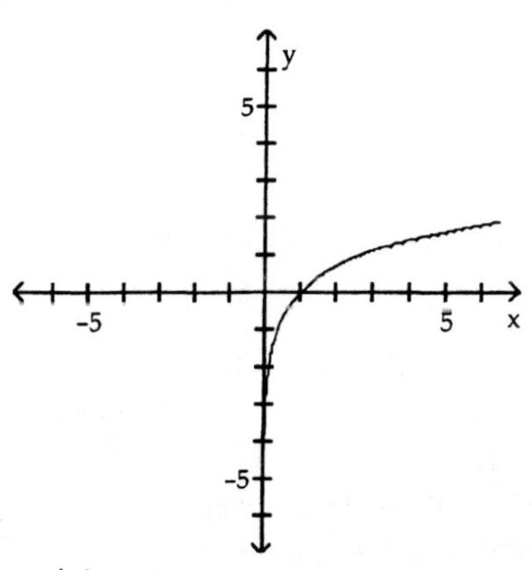

A) function
domain: all real numbers
range: $\{y \mid y > 0\}$
intercept: $(1, 0)$
symmetry: none

B) function
domain: $\{x \mid x > 0\}$
range: all real numbers
intercept: $(0, 1)$
symmetry: origin

C) function
domain: $\{x \mid x > 0\}$
range: all real numbers
intercept: $(1, 0)$
symmetry: none

D) not a function

3)

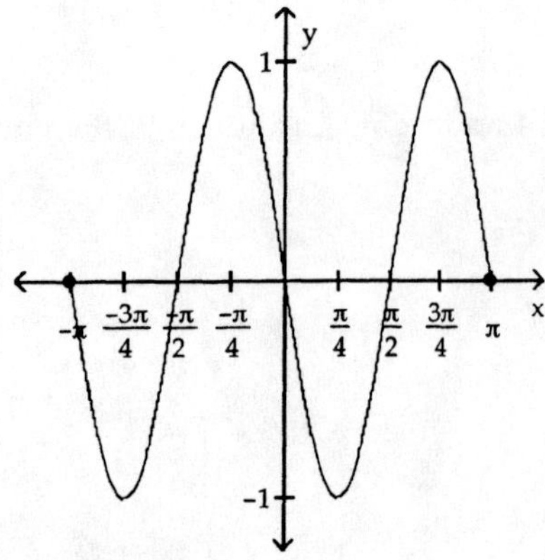

A) function
domain: $\{x \mid -\pi \le x \le \pi\}$
range: $\{y \mid -1 \le y \le 1\}$
intercepts: $(-\pi, 0)$, $(-\frac{\pi}{2}, 0)$, $(0, 0)$, $(\frac{\pi}{2}, 0)$, $(\pi, 0)$
symmetry: origin

B) function
domain: all real numbers
range: $\{y \mid -1 \le y \le 1\}$
intercepts: $(-\pi, 0)$, $(-\frac{\pi}{2}, 0)$, $(0, 0)$, $(\frac{\pi}{2}, 0)$, $(\pi, 0)$
symmetry: origin

C) function
domain: $\{x \mid -1 \le x \le 1\}$
range: $\{y \mid -\pi \le y \le \pi\}$
intercepts: $(-\pi, 0)$, $(-\frac{\pi}{2}, 0)$, $(0, 0)$, $(\frac{\pi}{2}, 0)$, $(\pi, 0)$
symmetry: none

D) not a function

4)

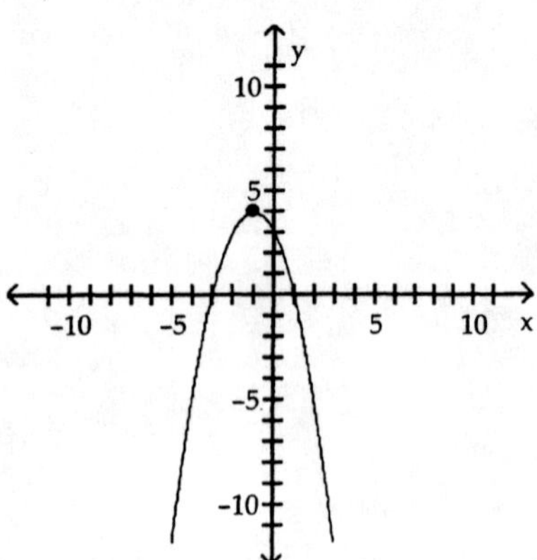

A) function
domain: all real numbers
range: $\{y \mid y \le 4\}$
intercepts: $(0, -3)$, $(3, 0)$, $(0, 1)$
symmetry: none

B) function
domain: all real numbers
range: $\{y \mid y \le 4\}$
intercepts: $(-3, 0)$, $(0, 3)$, $(1, 0)$
symmetry: none

C) function
domain: $\{x \mid x \le 4\}$
range: all real numbers
intercepts: $(-3, 0)$, $(0, 3)$, $(1, 0)$
symmetry: y-axis

D) not a function

5)

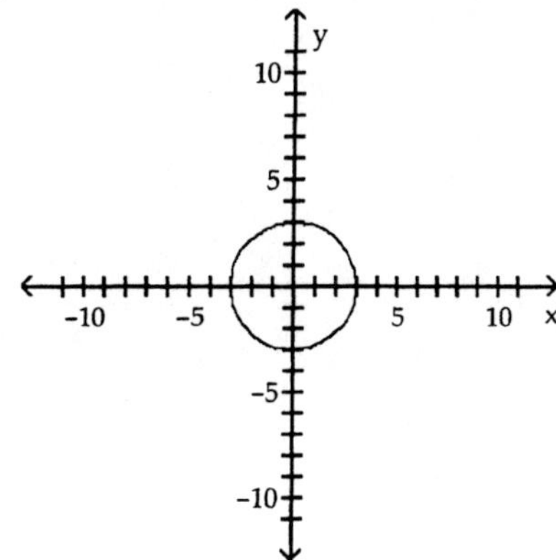

A) function
domain: $\{x \mid -3 \le x \le 3\}$
range: $\{y \mid -3 \le y \le 3\}$
intercepts: (-3, 0), (0, -3), (0, 3), (3, 0)
symmetry: x-axis, y-axis

B) function
domain: $\{x \mid -3 \le x \le 3\}$
range: $\{y \mid -3 \le y \le 3\}$
intercepts: (-3, 0), (0, -3), (0, 0), (0, 3), (3, 0)
symmetry: origin

C) function
domain: $\{x \mid -3 \le x \le 3\}$
range: $\{y \mid -3 \le y \le 3\}$
intercepts: (-3, 0), (0, -3), (0, 3), (3, 0)
symmetry: x-axis, y-axis, origin

D) not a function

6)

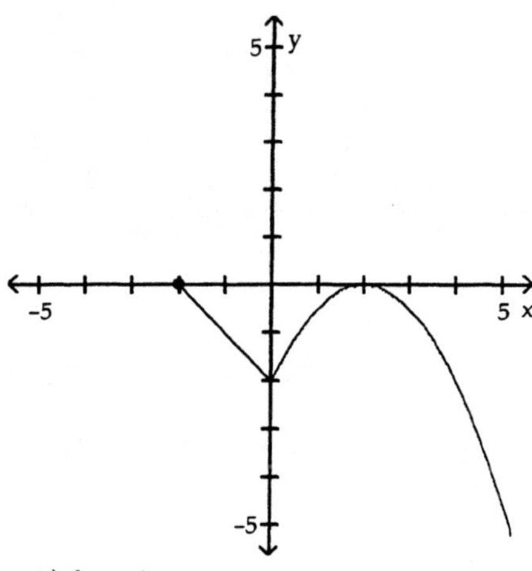

A) function
domain: $\{x \mid x \le 0\}$
range: $\{y \mid y \ge -2\}$
intercepts: (-2, 0), (0, -2), (2, 0)
symmetry: y-axis

B) function
domain: $\{x \mid x \ge -2\}$
range: $\{y \mid y \le 0\}$
intercepts: (-2, 0), (0, -2), (2, 0)
symmetry: none

C) function
domain: all real numbers
range: all real numbers
intercepts: (-2, 0), (0, -2), (2, 0)
symmetry: none

D) not a function

7)

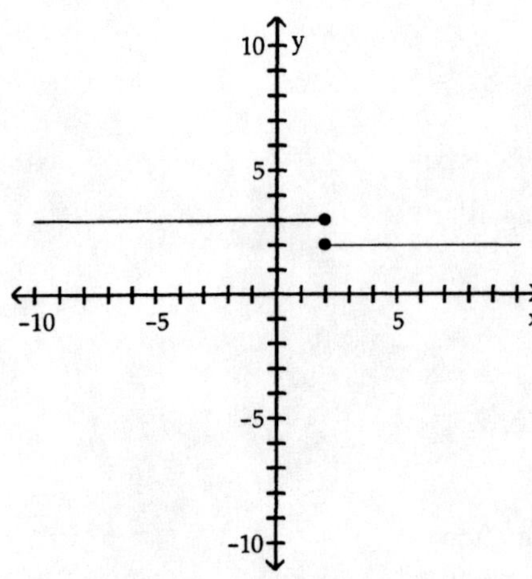

A) function
domain: {x | x = 2 or x = 3}
range: all real numbers
intercept: (3, 0)
symmetry: x-axis

B) function
domain: all real numbers
range: {y | y = 2 or y = 3}
intercept: (0, 3)
symmetry: none

C) function
domain: all real numbers
range: all real numbers
intercept: (0, 3)
symmetry: none

D) not a function

2 Obtain Information from or about the Graph of a Function

The graph of a function f is given. Use the graph to answer the question.

1) Use the graph of f given below to find f(-5).

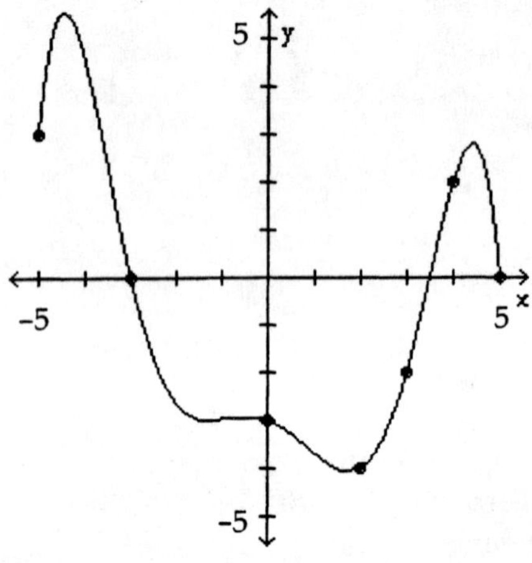

A) 0

B) -5

C) 3

D) 8

2) Is f(−5) positive or negative?

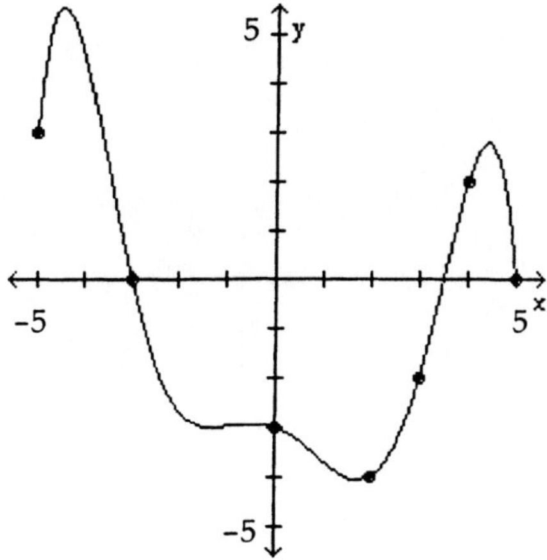

A) positive

B) negative

3) Is f(30) positive or negative?

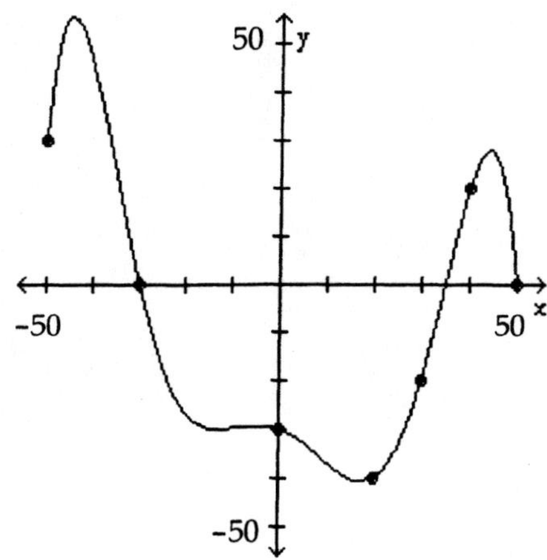

A) positive

B) negative

4) For what numbers x is $f(x) = 0$?

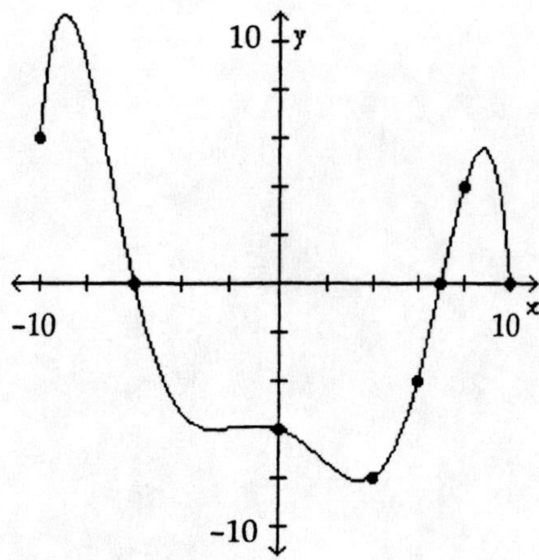

A) (-6, 7) B) -6, 7, 10 C) -6 D) (-10, -6), (7, 10)

5) For what numbers x is $f(x) > 0$?

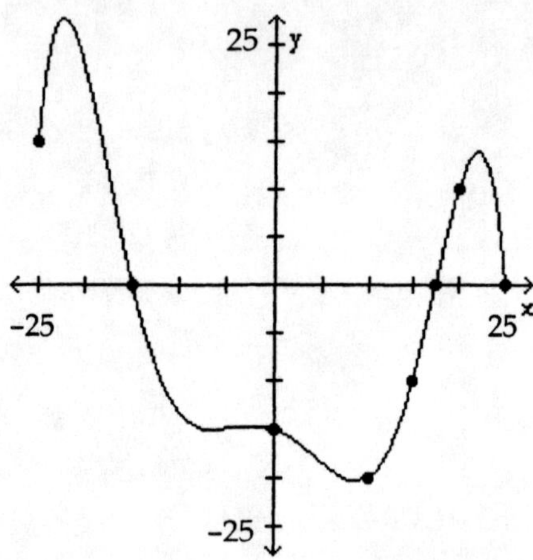

A) (-15, ∞) B) (-∞ -15) C) (-15, 17.5) D) [-25, -15), (17.5, 25)

6) For what numbers x is $f(x) < 0$?

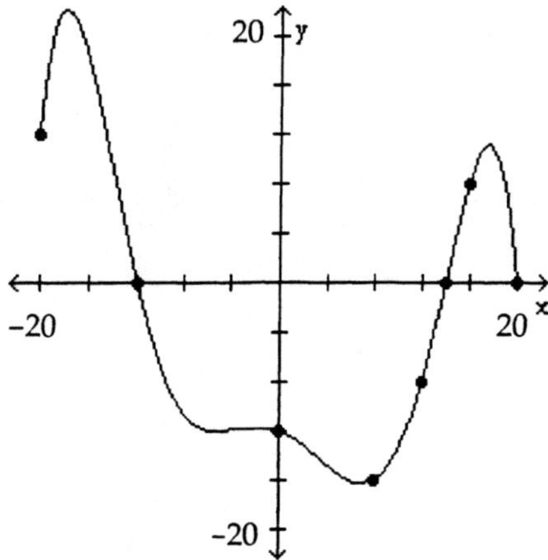

A) $[-20, -12), (14, 20)$ B) $(-12, \infty)$ C) $(-12, 14)$ D) $(-\infty, -12)$

7) What is the domain of f?

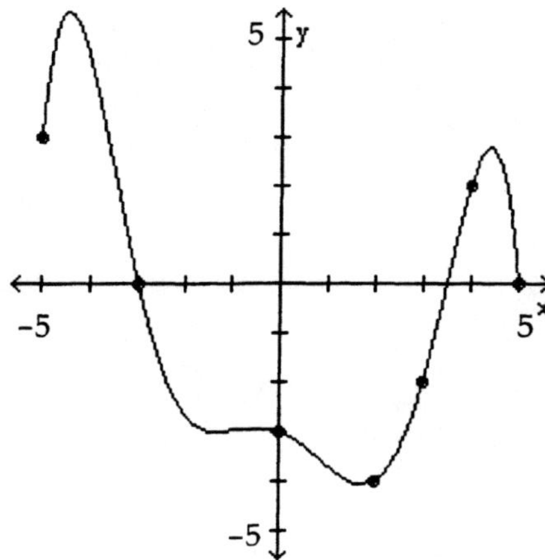

A) $\{x \mid x \geq 0\}$ B) $\{x \mid -5 \leq x \leq 5\}$ C) $\{x \mid -4 \leq x \leq 5.5\}$ D) all real numbers

8) What are the x-intercepts?

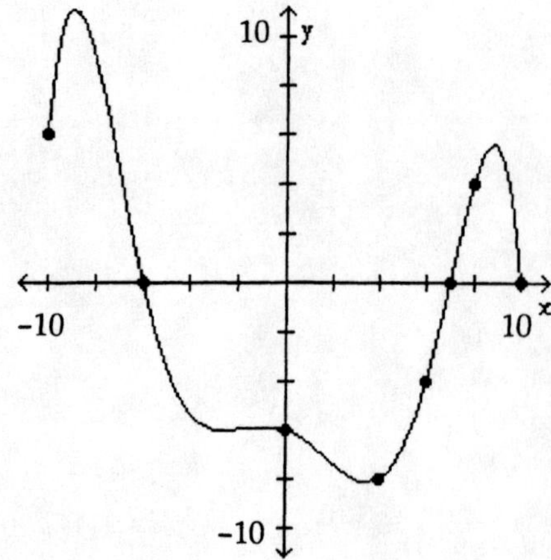

A) -6 B) -6, 7, 10 C) -6, 7 D) -10, -6, 7, 10

9) What is the y-intercept?

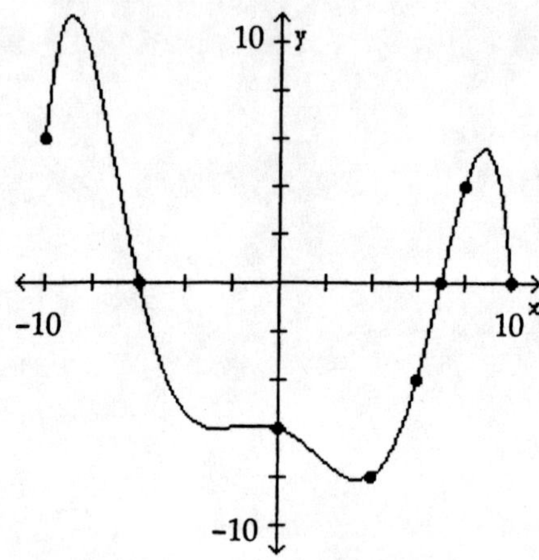

A) 7 B) 10 C) -8 D) -6

10) How often does the line y = -50 intersect the graph?

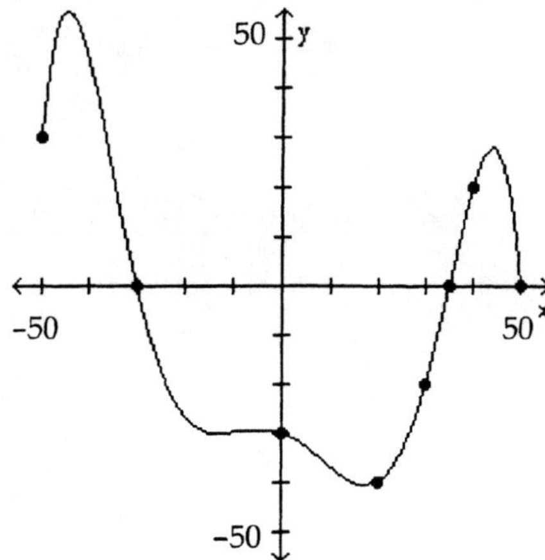

A) once B) twice C) three times D) does not intersect

11) How often does the line y = 5 intersect the graph?

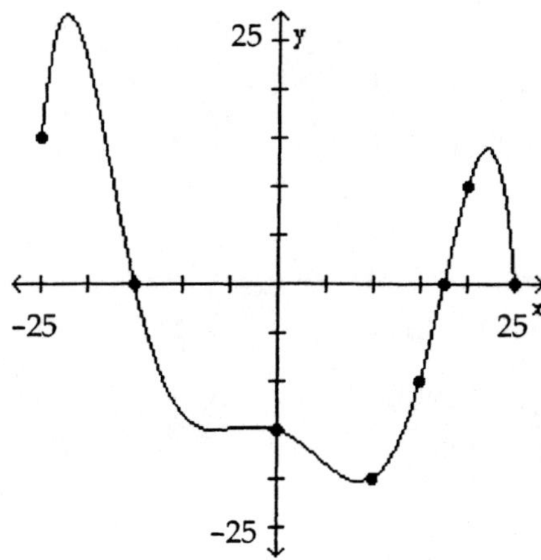

A) once B) twice C) three times D) does not intersect

12) For which of the following values of x does f(x) = -20?

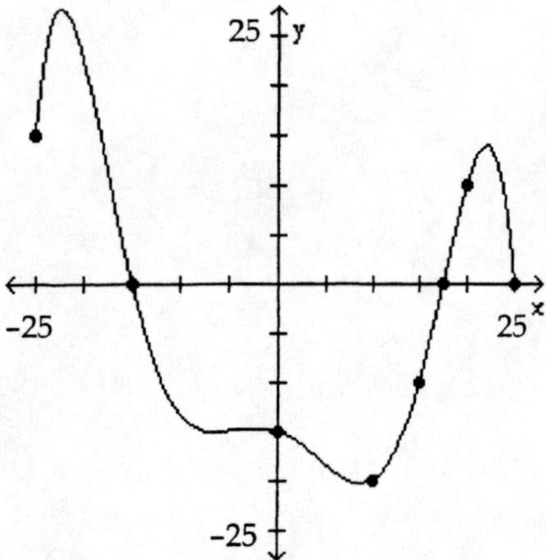

A) 15 B) 10 C) -20 D) 0

Answer the question about the given function.

13) Given the function $f(x) = -2x^2 - 4x + 4$, is the point (-1, 6) on the graph of f?

A) Yes B) No

14) Given the function $f(x) = 7x^2 - 14x - 3$, is the point (2, 11) on the graph of f?

A) Yes B) No

15) Given the function $f(x) = -6x^2 + 12x - 3$, if x = 1, what is f(x)? What point is on the graph of f?

A) 3; (1, 3) B) 3; (3, 1) C) -21; (-21, 1) D) -21; (1, -21)

16) Given the function $f(x) = -2x^2 + 4x - 3$, what is the domain of f?

A) all real numbers B) $\{x \mid x \ge 1\}$ C) $\{x \mid x \le 1\}$ D) $\{x \mid x \ge -1\}$

17) Given the function $f(x) = x^2 + 6x - 16$, list the x-intercepts, if any, of the graph of f.

A) (-8, 0), (1, 0) B) (8, 0), (2, 0) C) (8, 0), (-2, 0) D) (-8, 0), (2, 0)

18) Given the function $f(x) = 4x^2 + 8x - 3$, list the y-intercept, if there is one, of the graph of f.

A) -3 B) 5 C) 9 D) -7

19) Given the function $f(x) = \frac{x^2 - 3}{x + 2}$, is the point $(1, -\frac{2}{3})$ on the graph of f?

A) Yes B) No

20) Given the function $f(x) = \frac{x^2 - 9}{x - 3}$, is the point $(-1, -\frac{5}{2})$ on the graph of f?

A) Yes B) No

21) Given the function $f(x) = \frac{x^2 - 2}{x + 3}$, if x = 2, what is f(x)? What point is on the graph of f?

A) $\frac{2}{5}$; $(2, \frac{2}{5})$ B) $\frac{6}{5}$; $(\frac{6}{5}, 2)$ C) $\frac{2}{5}$; $(\frac{2}{5}, 2)$ D) $\frac{6}{5}$; $(2, \frac{6}{5})$

22) Given the function $f(x) = \frac{x^2 + 7}{x + 8}$, what is the domain of f?

A) $\{x \mid x \neq -8\}$ B) $\{x \mid x \neq 7\}$ C) $\{x \mid x \neq 8\}$ D) $\{x \mid x \neq -\frac{7}{8}\}$

23) Given the function $f(x) = \frac{x^2 + 7}{x + 4}$, list the x-intercepts, if any, of the graph of f.

A) $(-\sqrt{7}, 0)$ B) $(-4, 0)$ C) $(7, 0), (-7, 0)$ D) none

24) Given the function $f(x) = \frac{x^2 + 7}{x - 9}$, list the y-intercept, if there is one, of the graph of f.

A) $(0, 9)$ B) $(0, -\frac{7}{9})$ C) $(0, -7)$ D) $(-\frac{7}{9}, 0)$

Solve the problem.

25) If an object weighs m pounds at sea level, then its weight W (in pounds) at a height of h miles above sea level is given approximately by $W(h) = m\left(\frac{4000}{4000 + h}\right)^2$. How much will a man who weighs 165 pounds at sea level weigh on the top of a mountain which is 14,494 feet above sea level? Round to the nearest hundredth of a pound, if necessary.

A) 165.23 pounds B) 165 pounds C) 7.72 pounds D) 164.77 pounds

Match the function with the graph that best describes the situation.

26) The amount of rainfall as a function of time, if the rain fell more and more softly.

A)

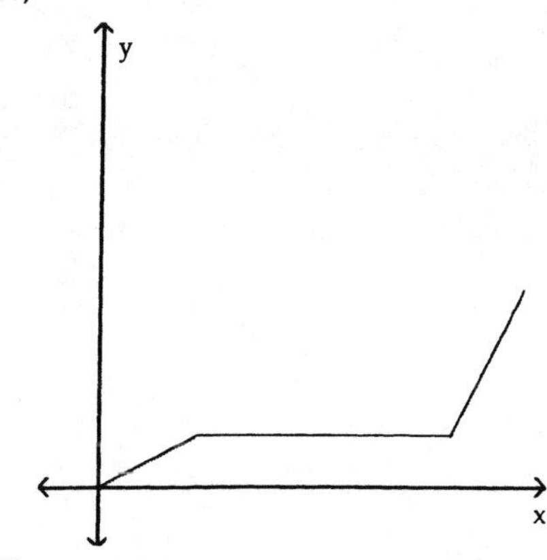

B)

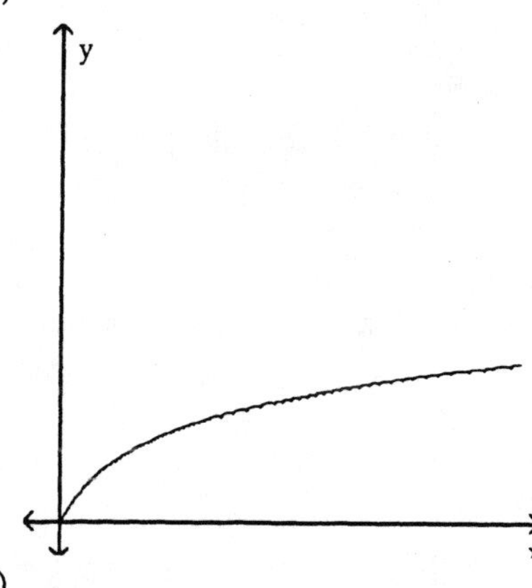

C)

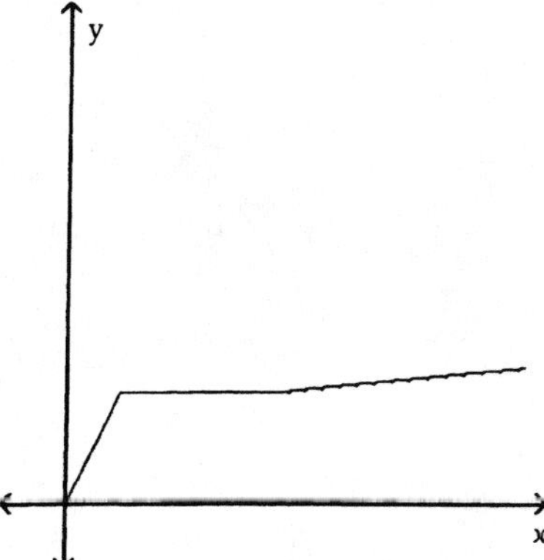

D)

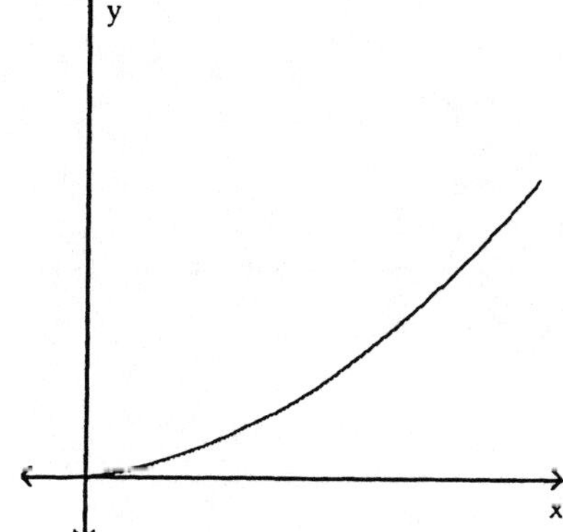

27) The height of an animal as a function of time.

A)

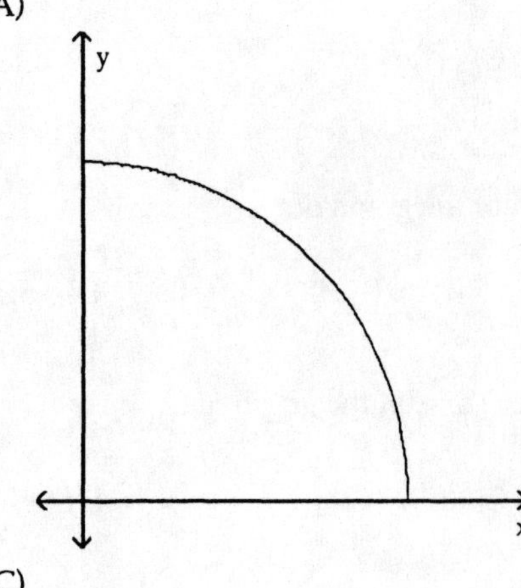

B)

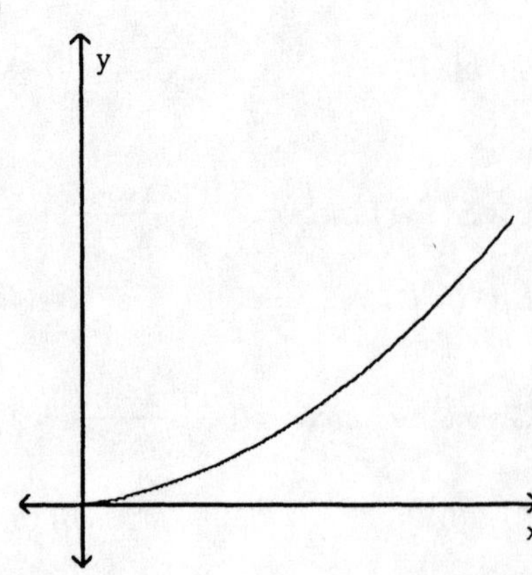

C)

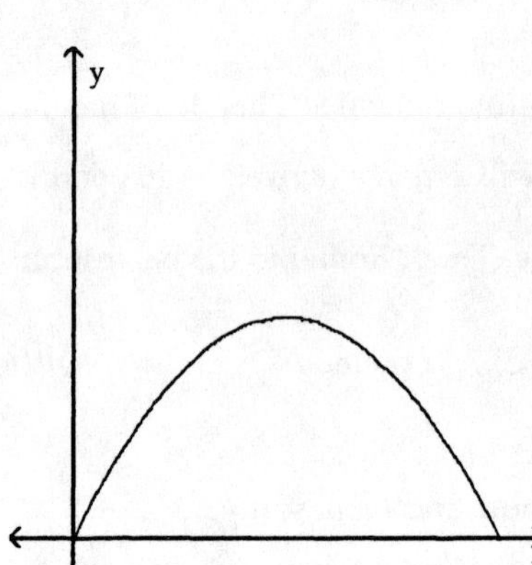

D)

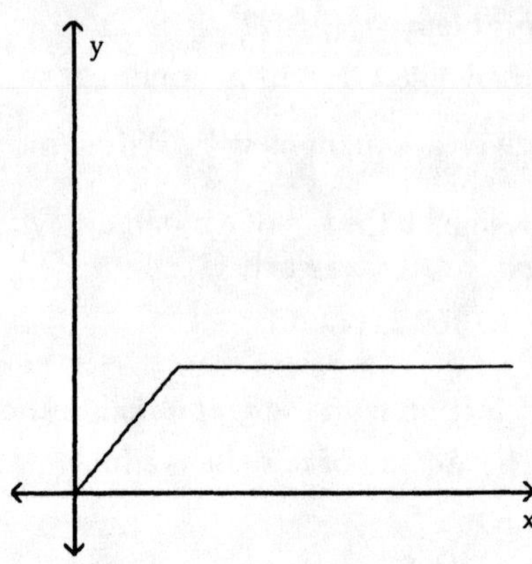

Solve the problem.

28) Michael decides to walk to the mall to do some errands. He leaves home, walks 5 blocks in 15 minutes at a constant speed, and realizes that he forgot his wallet at home. So Michael runs back in 11 minutes. At home, it takes him 2 minutes to find his wallet and close the door. Michael walks 2 blocks in 9 minutes and then decides to jog to the mall. It takes him 11 minutes to get to the mall which is 5 blocks away. Draw a graph of Michael's distance from home (in blocks) as a function of time.

y

x

3.3 Properties of Functions

1 Determine Even and Odd Functions from a Graph

The graph of a function is given. Decide whether it is even, odd, or neither.

1)

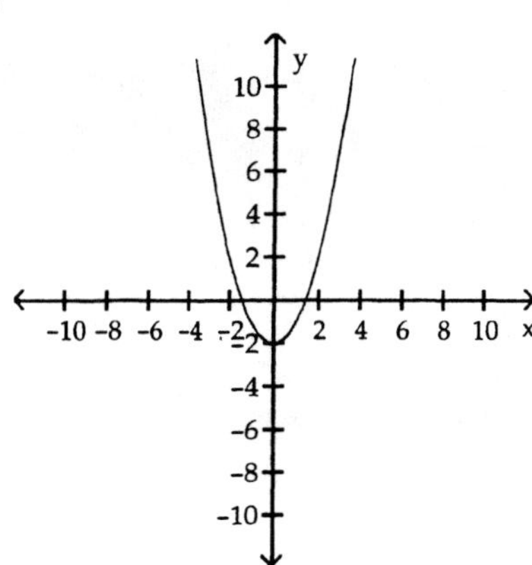

A) even B) odd C) neither

2)

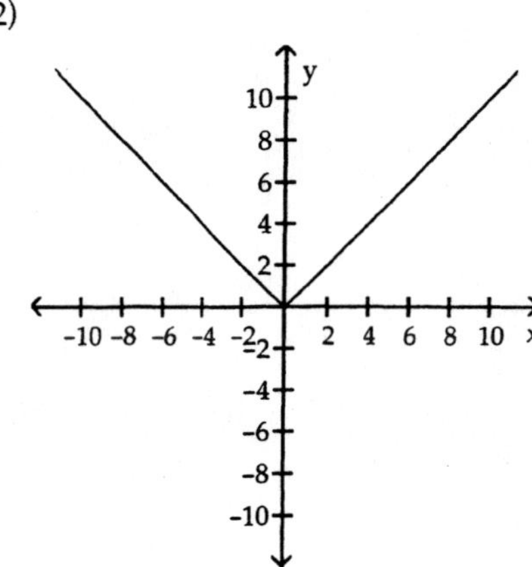

A) even B) odd C) neither

3)

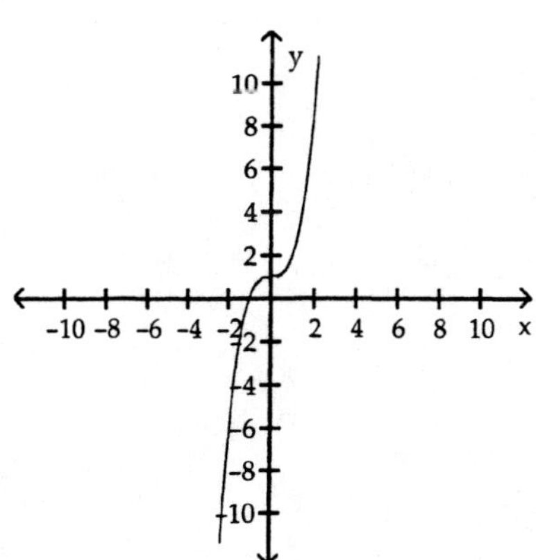

A) even B) odd C) neither

4)

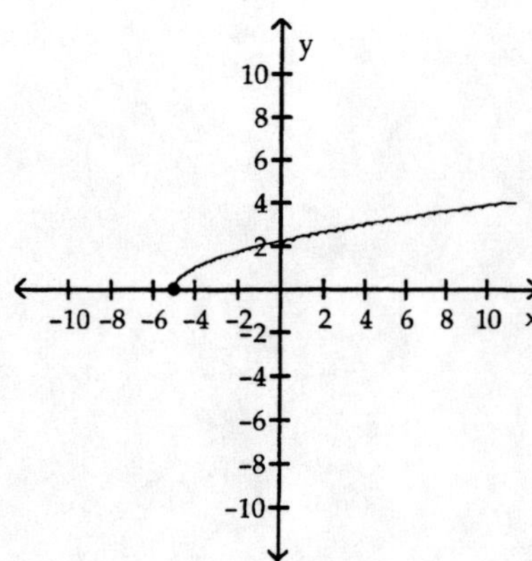

A) even

B) odd

C) neither

5)

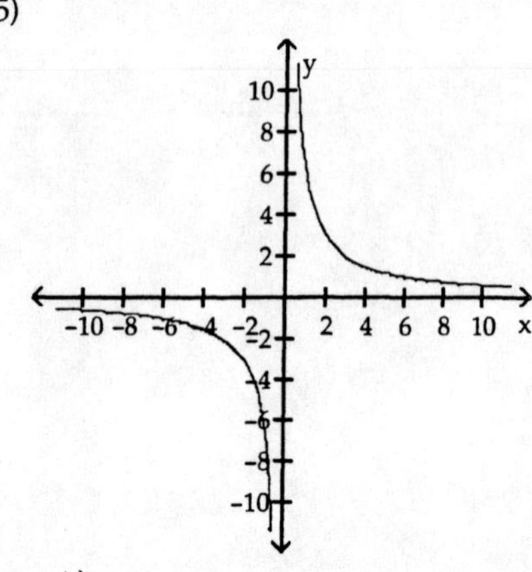

A) even

B) odd

C) neither

6)

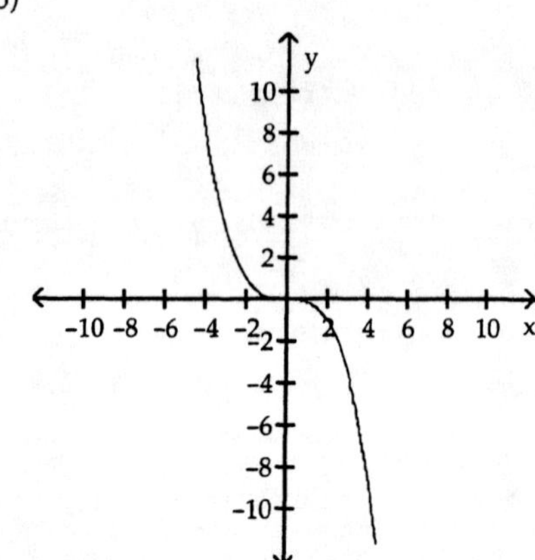

A) even

B) odd

C) neither

7)

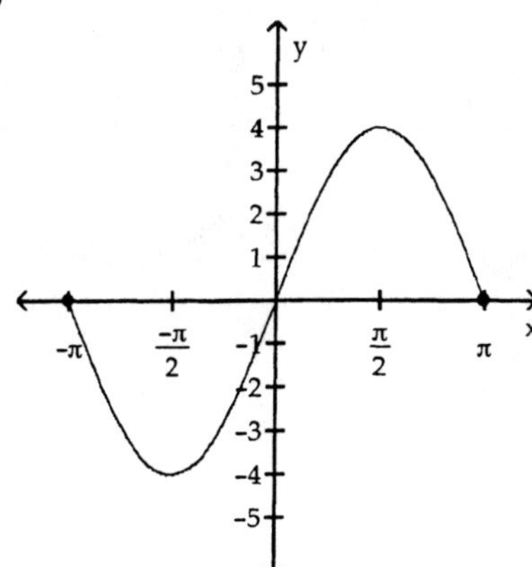

A) even B) odd C) neither

8)

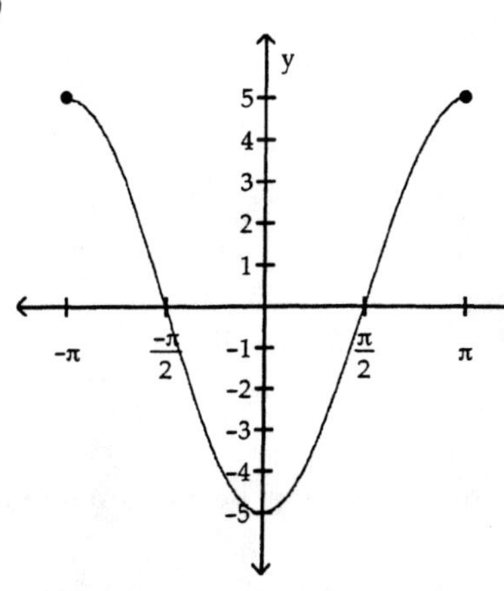

A) even B) odd C) neither

2 Identify Even and Odd Functions from the Equation

Determine algebraically whether the function is even, odd, or neither.

1) $f(x) = 3x^3$

A) even B) odd C) neither

2) $f(x) = 5x^4 - x^2$

A) even B) odd C) neither

3) $f(x) = -3x^2 - 4$

A) even B) odd C) neither

4) $f(x) = 9x^3 + 5$

A) even B) odd C) neither

5) $f(x) = \sqrt[3]{x}$

A) even B) odd C) neither

6) $f(x) = \sqrt{x}$

A) even B) odd C) neither

7) $\sqrt[3]{3x^2 + 8}$

A) even B) odd C) neither

8) $f(x) = \dfrac{1}{x^2}$

A) even B) odd C) neither

9) $f(x) = \dfrac{x}{x^2 - 5}$

A) even B) odd C) neither

10) $f(x) = \dfrac{-x^3}{5x^2 - 7}$

A) even B) odd C) neither

11) $f(x) = \dfrac{4x}{|x|}$

A) even B) odd C) neither

3 Use a Graph to Determine Where a Function Is Increasing, Is Decreasing, or Is Constant

The graph of a function is given. Determine whether the function is increasing, decreasing, or constant on the given interval.

1) $(-2, -1)$

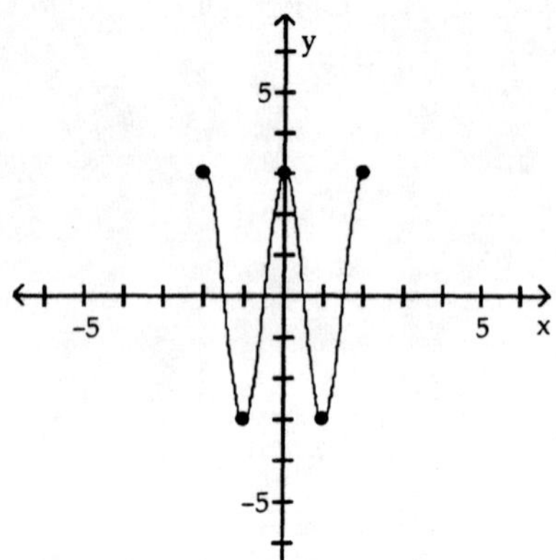

A) constant B) increasing C) decreasing

2) $(-1, 0)$

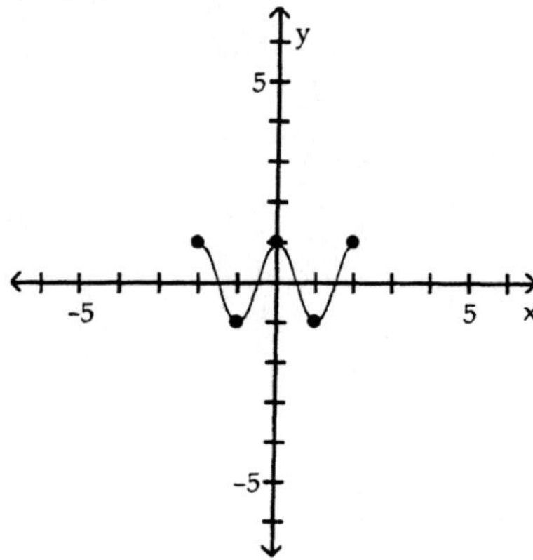

A) increasing

B) constant

C) decreasing

3) $(0, 1)$

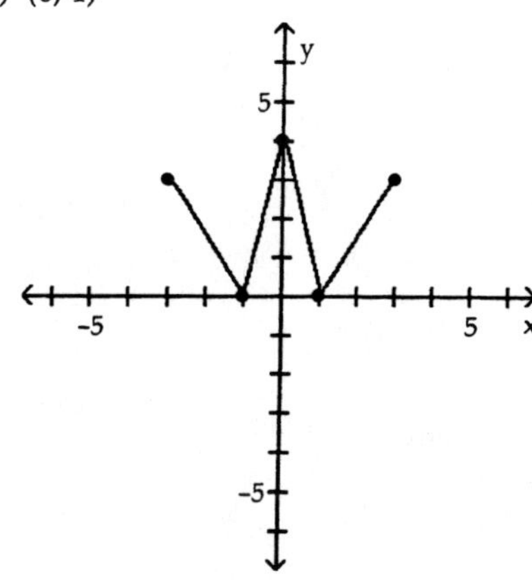

A) decreasing

B) constant

C) increasing

4) $(1, 4)$

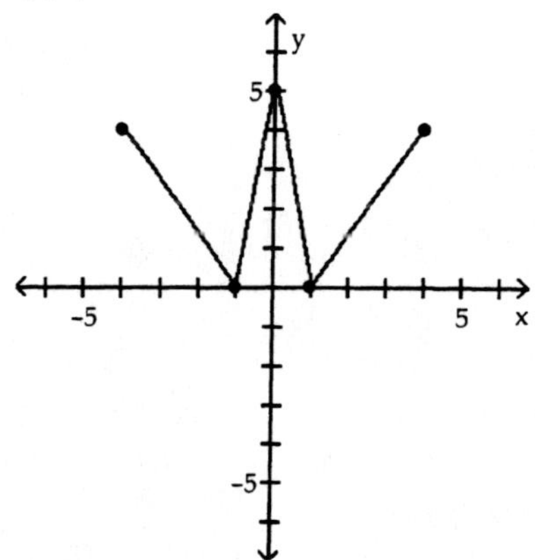

A) decreasing

B) increasing

C) constant

5) 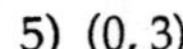 (0, 3)

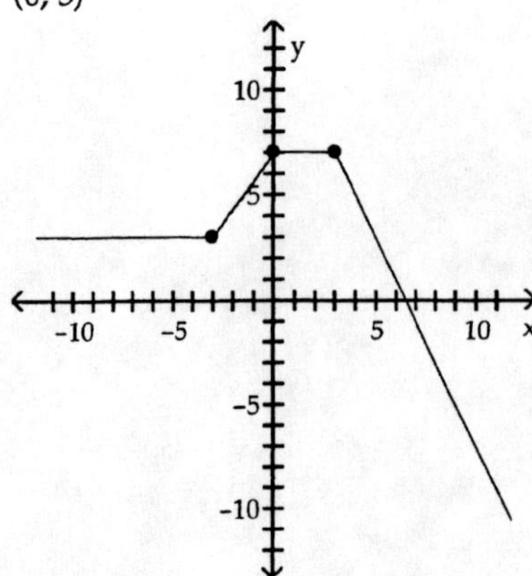

A) increasing B) decreasing C) constant

6) (-2, 0)

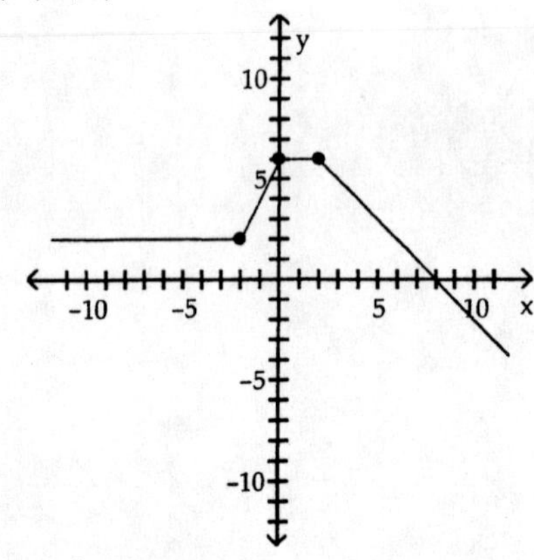

A) decreasing B) constant C) increasing

7) (2, ∞)

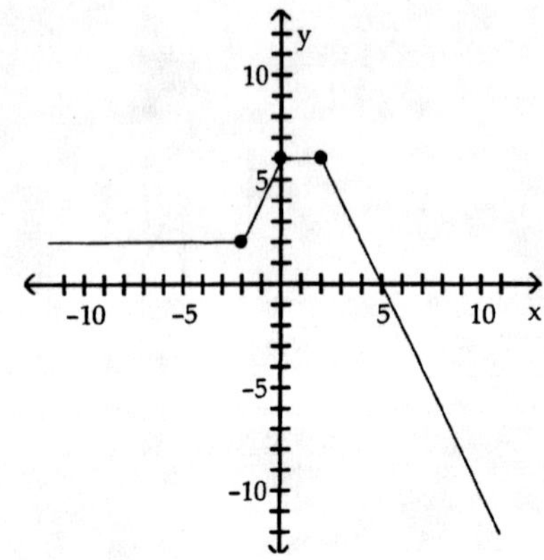

A) constant B) increasing C) decreasing

8) $(2, \infty)$

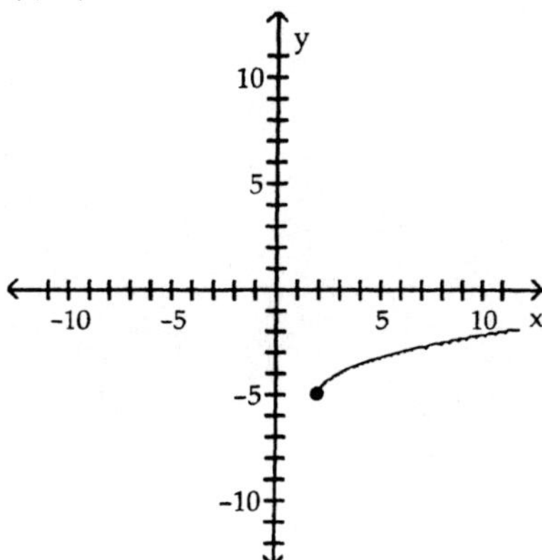

A) constant B) increasing C) decreasing

9) $(-1, 0)$

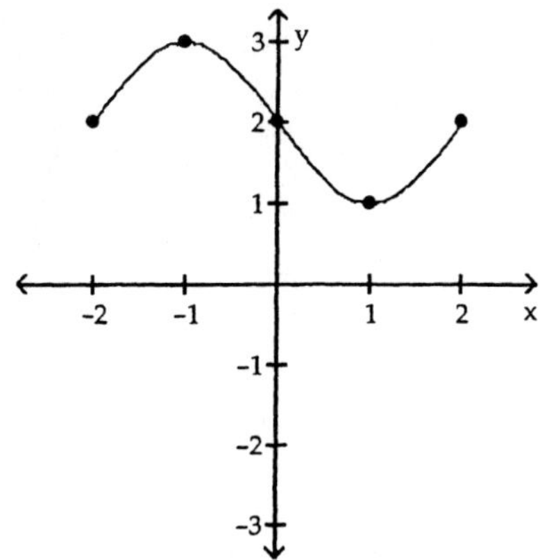

A) increasing B) constant C) decreasing

10) $(0.5, 3.5)$

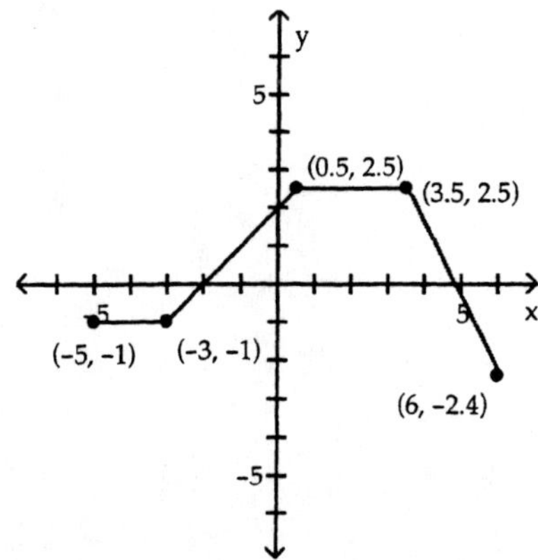

A) decreasing B) increasing C) constant

11) (0, 2)

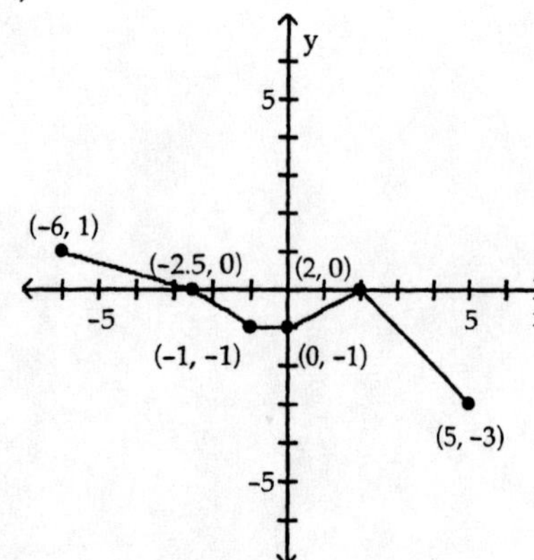

A) increasing B) decreasing C) constant

12) $(5, \infty)$

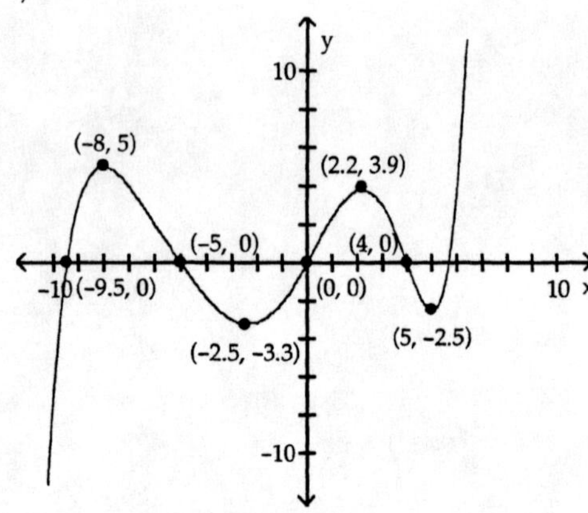

A) decreasing B) increasing C) constant

4 Use a Graph to Locate Local Maxima and Minima

The graph of a function f is given. Use the graph to answer the question.

1) Find the numbers, if any, at which f has a local maximum. What are the local maxima?

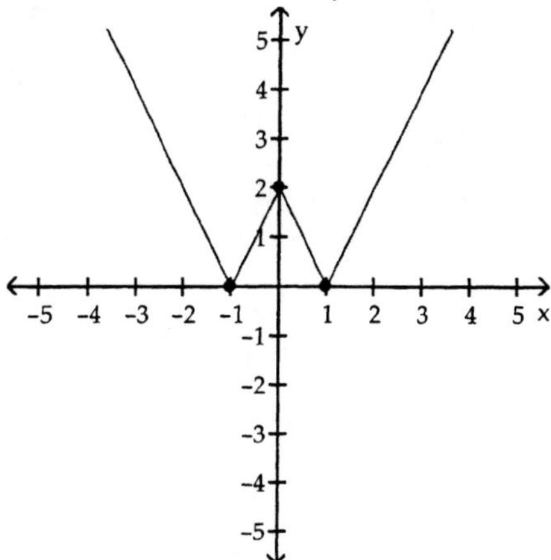

A) f has a local maximum at x = -1 and 1; the local maximum is 0

B) f has a local maximum at x = 1; the local maximum is 2

C) f has a local maximum at x = 0; the local maximum is 2

D) f has no local maximum

2) Find the numbers, if any, at which f has a local minimum. What are the local minima?

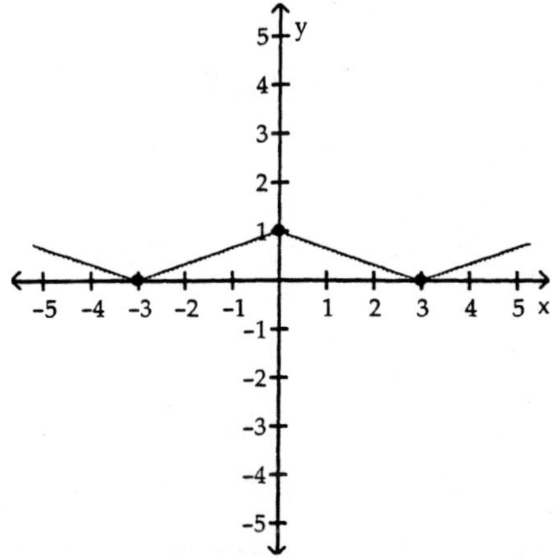

A) f has a local minimum at x = -3; the local minimum is 0

B) f has a local minimum at x = -3 and 3; the local minimum is 0

C) f has a local minimum at x = 0; the local minimum is 1

D) f has no local minimum

3) Find the numbers, if any, at which f has a local maximum. What are the local maxima?

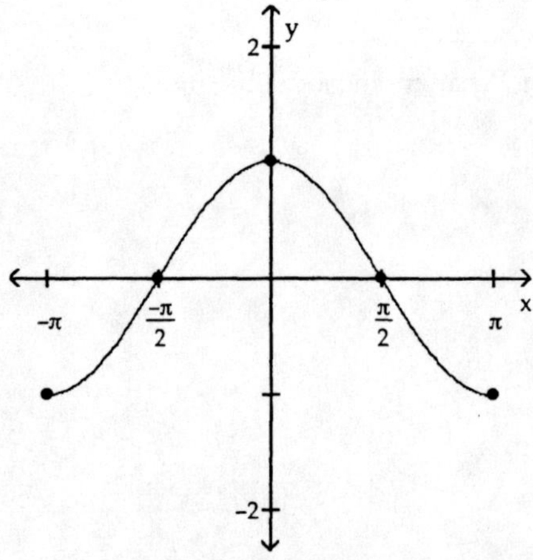

A) f has a local maximum at $-\pi$; the local maximum is 1

B) f has no local maximum

C) f has a local maximum at $x = 0$; the local maximum is 1

D) f has a local maximum at $x = -\pi$ and π; the local maximum is -1

4) Find the numbers, if any, at which f has a local minimum. What are the local minima?

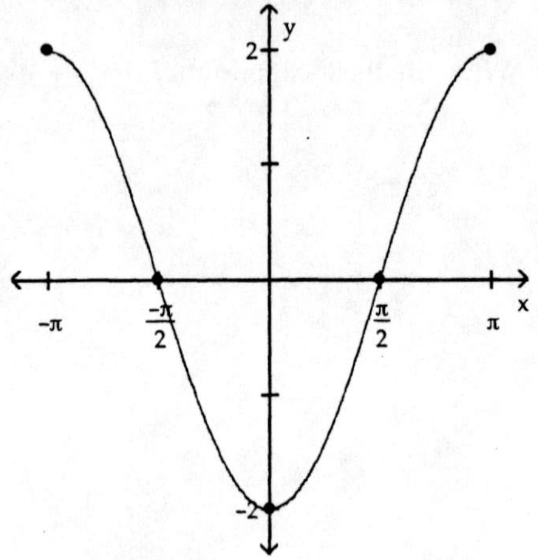

A) f has no local minimum

B) f has a local minimum at $x = 0$; the local minimum is -2

C) f has a local minimum at $x = -\pi$ and π; the local minimum is 2

D) f has a local minimum at $x = -\pi$; the local minimum is -2

5)

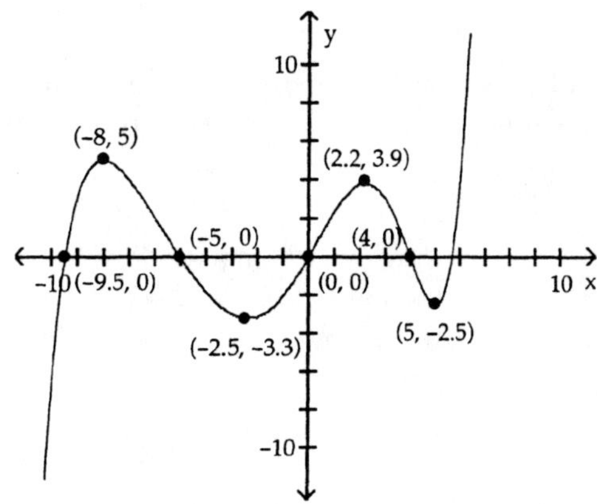

Find the numbers, if any, at which f has a local minimum. What are the local maxima?

A) f has a local minimum at x = -2.5 and 5; the local minimum at -2.5 is -3.3; the local minimum at 5 is -2.5

B) f has a local minimum at x = -3.3 and -2.5; the local minimum at -3.3 is -2.5; the local minimum at -2.5 is 5

C) f has a local maximum at x = -3.3 and -2.5; the local maximum at -3.3 is -2.5; the local maximum at -2.5 is 5

D) f has a local maximum at x = -2.5 and 5; the local maximum at -2.5 is -3.3; the local maximum at 5 is -2.5

Solve the problem.

6) The height s of a ball (in feet) thrown with an initial velocity of 90 feet per second from an initial height of 3 feet is given as a function of time t (in seconds) by $s(t) = -16t^2 + 90t + 3$. What is the maximum height? Round to the nearest hundredth, if necessary.

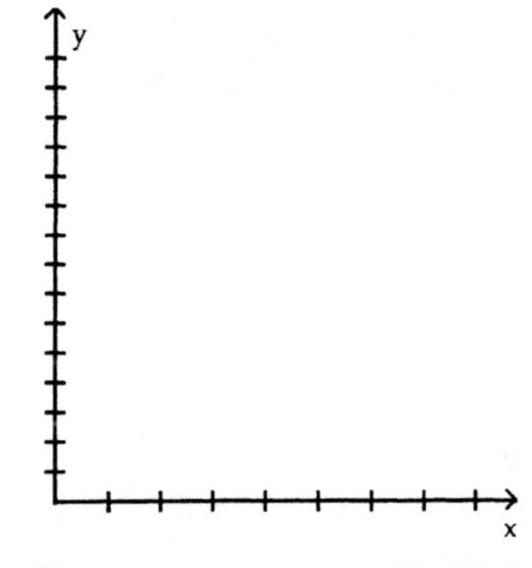

A) 126.75 ft B) -98.25 ft C) 129.56 ft D) 143.63 ft

5 Use a Graphing Utility to Approximate Local Maxima and Minima and to Determine Where a Function Is Increasing or Decreasing

Use a graphing utility to graph the function over the indicated interval and approximate any local maxima and local minima. Determine where the function is increasing and where it is decreasing. If necessary, round answers to two decimal places.

1) $f(x) = x^3 - 3x^2 + 1$, (-1, 3)

A) local maximum at (2, -3)
local minimum at (0, 1)
increasing on (-1, 0)
decreasing on (0, 2)

B) local maximum at (0, 1)
local minimum at (2, -3)
increasing on (-1, 0) and (2, 3)
decreasing on (0, 2)

C) local maximum at (0, 1)
local minimum at (2, -3)
increasing on (0, 2)
decreasing on (-1, 0) and (2, 3)

D) local maximum at (2, -3)
local minimum at (0, 1)
increasing on (-1, 0) and (2, 3)
decreasing on (0, 2)

2) $f(x) = x^3 - 4x^2 + 6$; (-1, 4)

3) $f(x) = x^5 - x^2$; (-2, 2)

4) $f(x) = -0.3x^3 + 0.2x^2 + 4x - 5$; (-4, 5)

5) $f(x) = 0.15x^4 + 0.3x^3 - 0.8x^2 + 5$; (-4, 2)

Use a graphing utility to graph the function over the indicated interval and approximate any local maxima and local minima. If necessary, round answers to two decimal places.

6) $f(x) = x^2 + 2x - 3$; (-5, 5)

A) local minimum at (1, 4)
B) local maximum at (1, -4)
C) local maximum at (-1, 4)
D) local minimum at (-1, -4)

7) $f(x) = 2 + 8x - x^2$; (-5, 5)

A) local minimum at (4, 50)
B) local maximum at (-4, 50)
C) local maximum at (4, 18)
D) local minimum at (-4, 18)

8) $f(x) = x^3 - 3x^2 + 1$; (-5, 5)

A) local minimum at (2, -3)

B) local maximum at (0, 1)
local minimum at (2, -3)

C) local minimum at (0, 1)
local maximum at (2, -3)

D) none

9) $f(x) = x^3 - 12x + 2$; (-5, 5)

A) local maximum at (-2, 18)
local minimum at (0, 0)
local minimum at (2, -14)

B) local minimum at (0, 0)

C) local maximum at (-2, 18)
local minimum at (2, -14)

D) none

10) $f(x) = x^4 - 5x^3 + 3x^2 + 9x - 3$; $(-5, 5)$

A) local minimum at $(-1, -6)$
local maximum at $(1, 6)$
local minimum at $(3, -3)$

B) local minimum at $(-0.61, -5.64)$
local maximum at $(1.41, 6.12)$
local minimum at $(3, -3)$

C) local minimum at $(-0.57, -6.12)$
local maximum at $(1.32, 5.64)$
local minimum at $(3, -3)$

D) local minimum at $(-3, -3)$
local maximum at $(-1.32, 5.64)$
local minimum at $(0.57, -6.12)$

6 Find the Average Rate of Change of a Function

For the function, find the average rate of change of f from 1 to x:
$\dfrac{f(x) - f(1)}{x - 1}, x \neq 1$

1) $f(x) = 7x$

A) 6　B) $\dfrac{7}{x - 1}$　C) 7　D) 0

2) $f(x) = x^2 - 2x$

A) 1　B) $\dfrac{x^2 - 2x - 1}{x - 1}$　C) $x - 1$　D) $x + 1$

3) $f(x) = \dfrac{6}{x + 5}$

A) $-\dfrac{1}{x + 5}$　B) $\dfrac{6}{(x - 1)(x + 5)}$　C) $\dfrac{1}{x + 5}$　D) $\dfrac{6}{x(x + 5)}$

4) $f(x) = \sqrt{x + 15}$

A) $\dfrac{\sqrt{x + 15} - 4}{x - 1}$　B) $\dfrac{\sqrt{x + 15} + 4}{x - 1}$　C) $\dfrac{\sqrt{x + 15} + 4}{x + 1}$　D) $\dfrac{\sqrt{x + 15} - 4}{x + 1}$

Find the average rate of change for the function between the given values.

5) $f(x) = -2x + 6$; from 1 to 3

A) -6　B) 6　C) -2　D) 2

6) $f(x) = x^2 + 4x$; from 1 to 5

A) 10　B) $\dfrac{45}{4}$　C) 9　D) 8

7) $f(x) = 8x^3 + 4x^2 + 5$; from -9 to 3

A) 480　B) $\dfrac{257}{3}$　C) $\dfrac{257}{12}$　D) 1920

8) $f(x) = \sqrt{2x}$; from 2 to 8

A) $-\dfrac{3}{10}$　B) 7　C) 2　D) $\dfrac{1}{3}$

9) $f(x) = \frac{3}{x-2}$; from 4 to 7

A) $\frac{1}{3}$ B) 7 C) 2 D) $-\frac{3}{10}$

10) $f(x) = 4x^2$; from 0 to $\frac{7}{4}$

A) 2 B) $\frac{1}{3}$ C) $-\frac{3}{10}$ D) 7

11) $f(x) = -3x^2 - x$; from 5 to 6

A) $\frac{1}{2}$ B) $-\frac{1}{6}$ C) -2 D) -34

12) $f(x) = x^3 + x^2 - 8x - 7$; from 0 to 2

A) $\frac{1}{2}$ B) -28 C) $-\frac{1}{6}$ D) -2

13) $f(x) = \sqrt{2x - 1}$; from 1 to 5

A) -28 B) $\frac{1}{2}$ C) -2 D) $-\frac{1}{6}$

14) $f(x) = \frac{3}{x+2}$; from 1 to 4

A) $\frac{1}{2}$ B) -28 C) $-\frac{1}{6}$ D) -2

Find an equation of the secant line containing (1, f(1)) and (2, f(2)).

15) $f(x) = x^2 - 2x$

A) $y = -x - 2$ B) $y = x - 2$ C) $y = x + 2$ D) $y = -x + 2$

16) $f(x) = \frac{9}{x+8}$

A) $y = \frac{1}{10}x + \frac{11}{9}$ B) $y = \frac{1}{10}x + \frac{9}{10}$ C) $y = \frac{9}{10}x + \frac{1}{10}$ D) $y = -\frac{1}{10}x + \frac{11}{10}$

17) $f(x) = \sqrt{x + 24}$

A) $y = (\sqrt{26} - 5)x + \sqrt{26} - 10$ B) $y = (\sqrt{26} - 5)x - \sqrt{26} + 10$

C) $y = (-\sqrt{26} + 5)x + \sqrt{26} - 10$ D) $y = (-\sqrt{26} - 5)x - \sqrt{26} + 10$

3.4 Library of Functions; Piecewise-defined Functions

1 Graph the Functions Listed in the Library of Functions

Match the graph to the function listed whose graph most resembles the one given.

1)

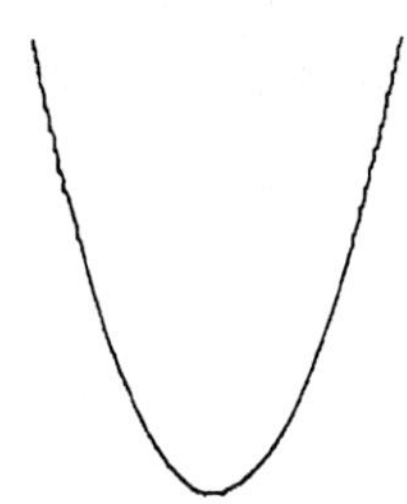

A) cube function
B) square function
C) reciprocal function
D) absolute value function

2)

A) reciprocal function
B) absolute value function
C) constant function
D) linear function

3)

A) square root function
B) cube function
C) cube root function
D) square function

4)

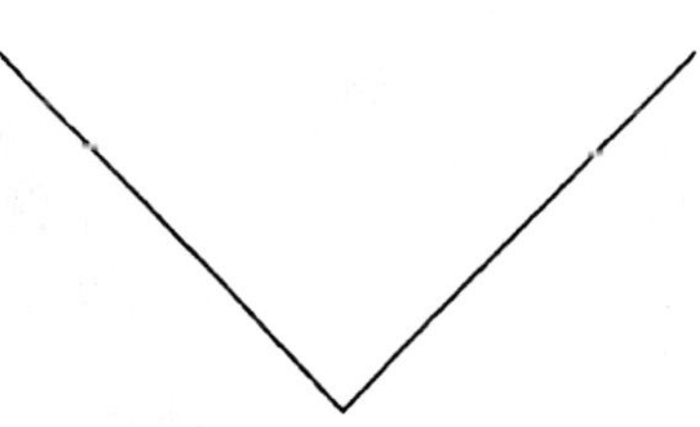

A) square function
B) linear function
C) absolute value function
D) reciprocal function

5)

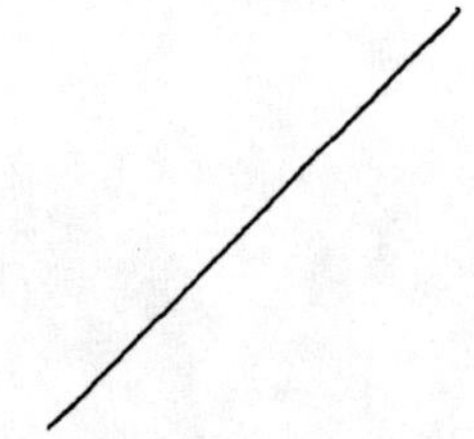

A) reciprocal function
B) constant function
C) linear function
D) absolute value function

6)

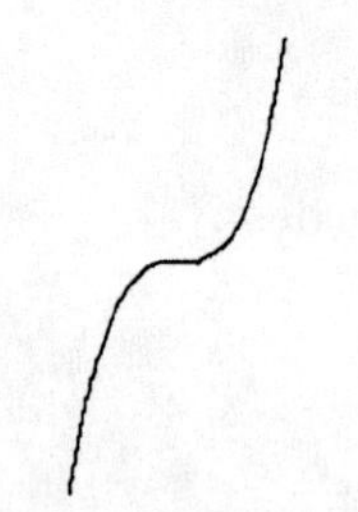

A) cube function
B) square root function
C) square function
D) cube root function

7)

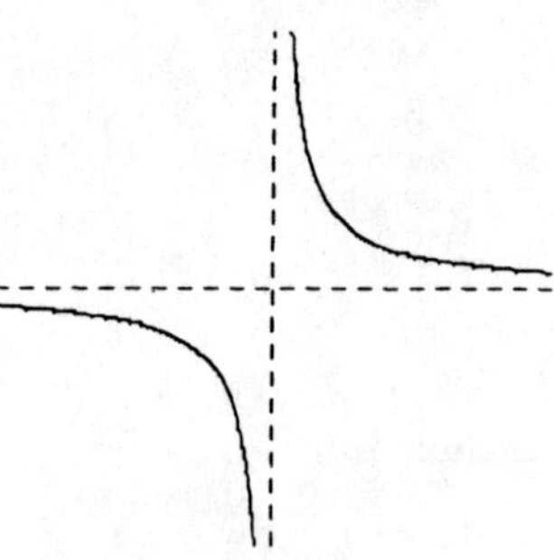

A) square function
B) absolute value function
C) reciprocal function
D) square root function

8)

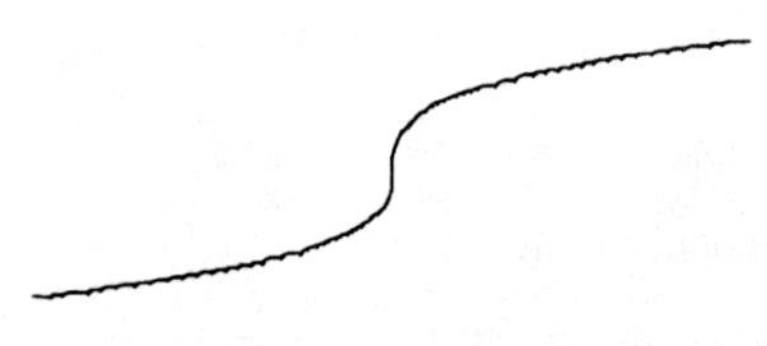

A) square function
B) cube function
C) cube root function
D) square root function

Graph the function.

9) $f(x) = x$

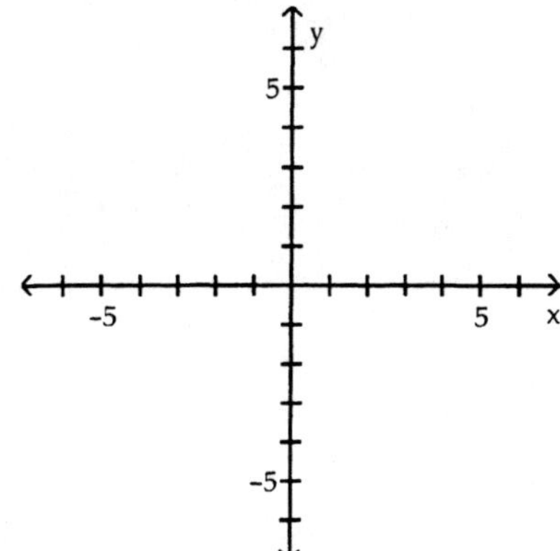

A)

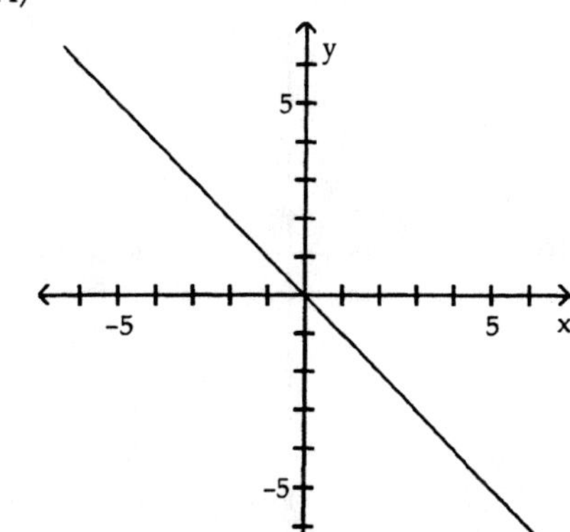

B)

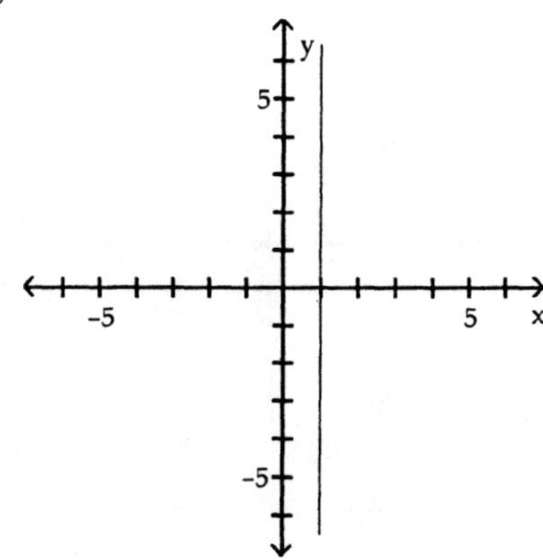

C)

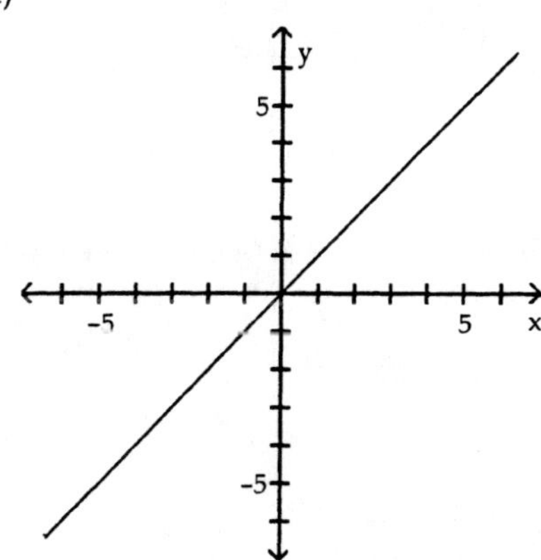

D)

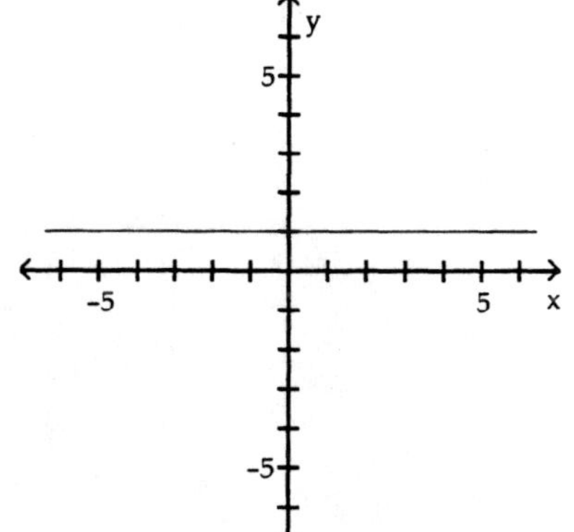

10) $f(x) = x^2$

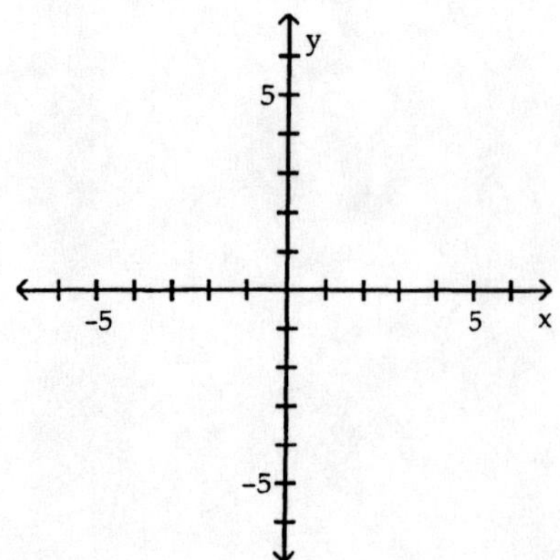

A)

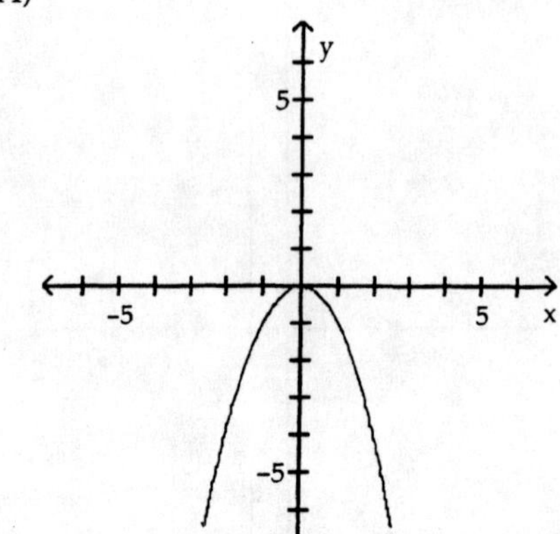

B)

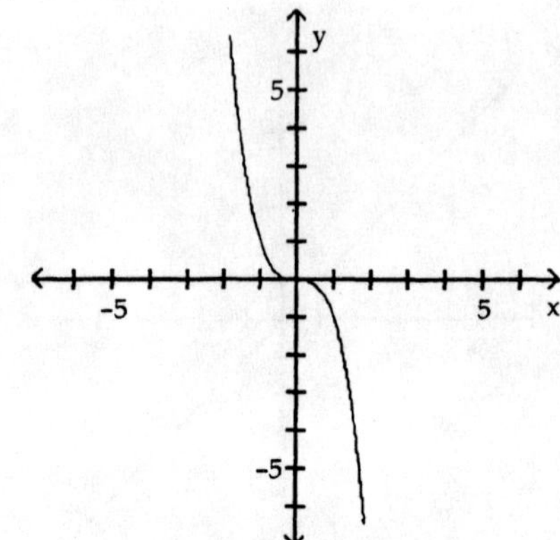

C)

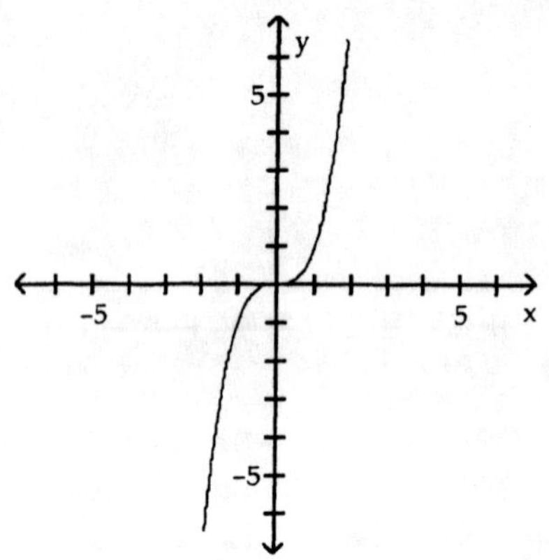

D)

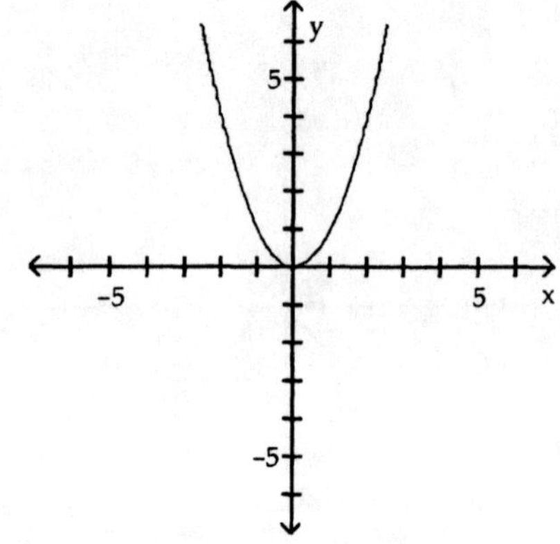

11) $f(x) = x^3$

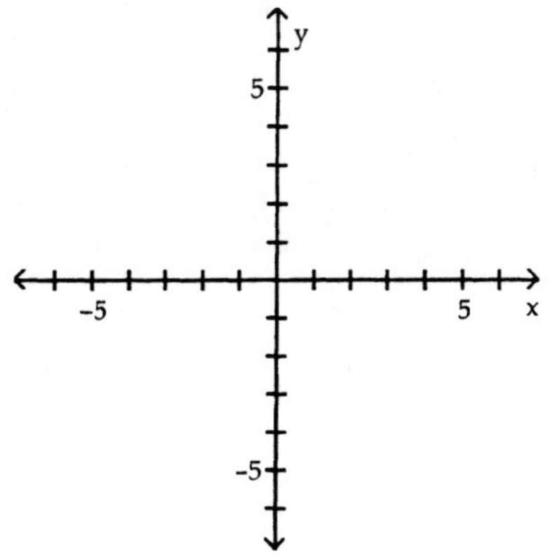

A)

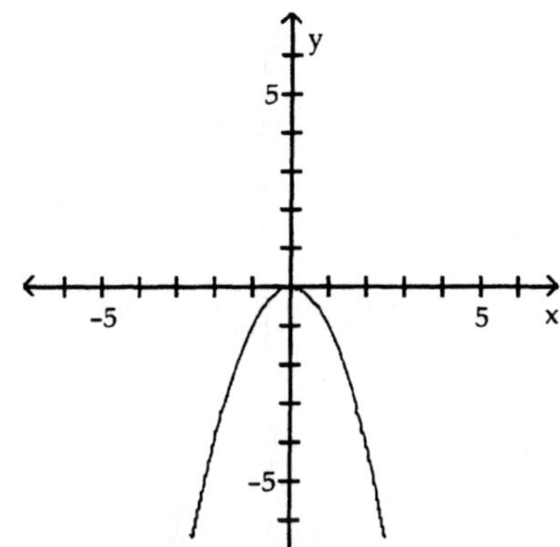

B)

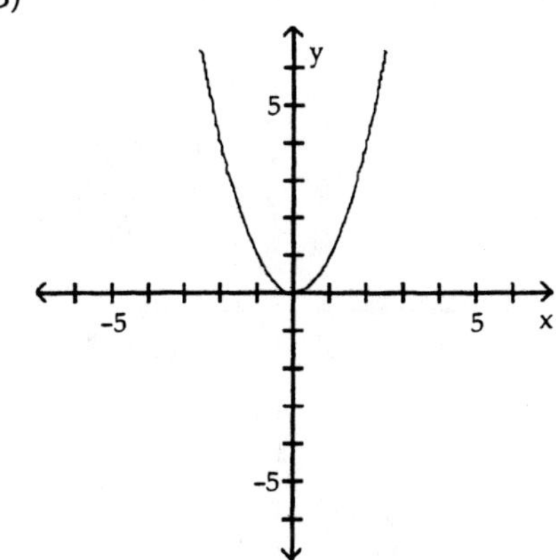

C)

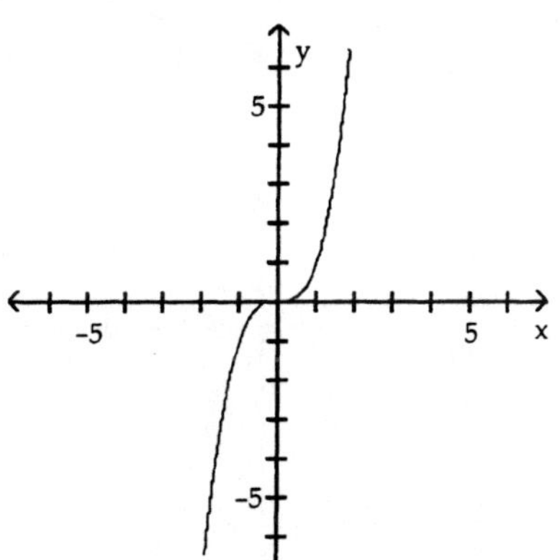

D)

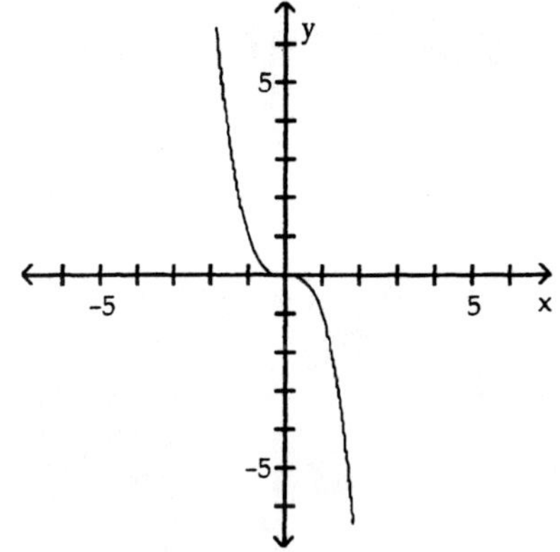

12) $f(x) = \sqrt{x}$

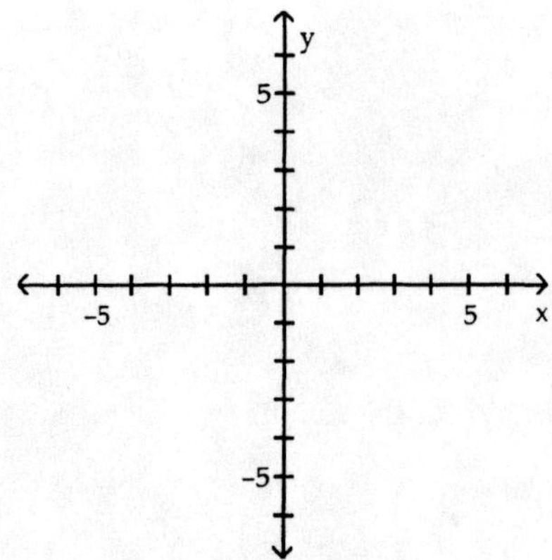

A)

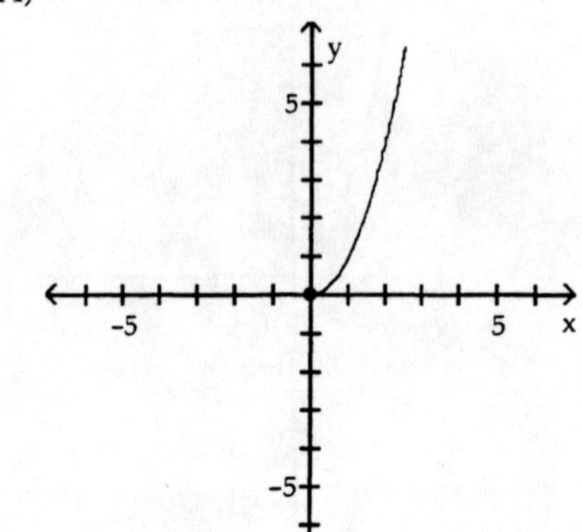

B)

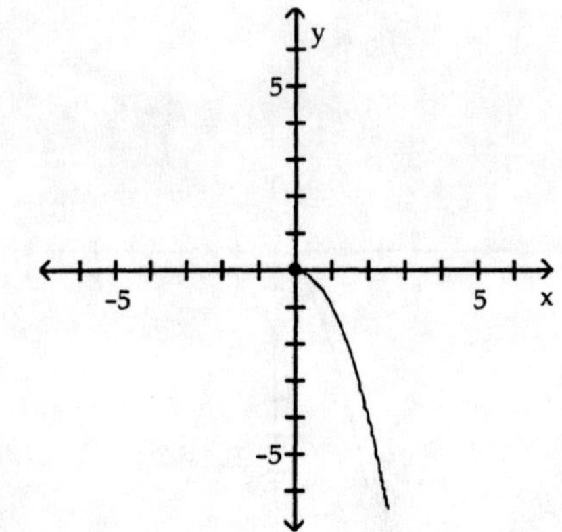

C)

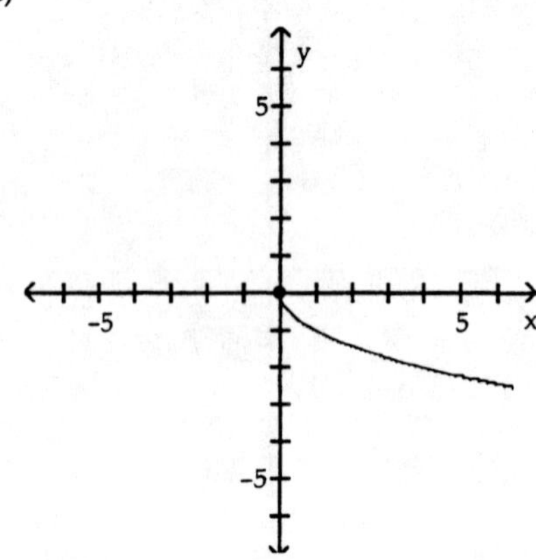

D)

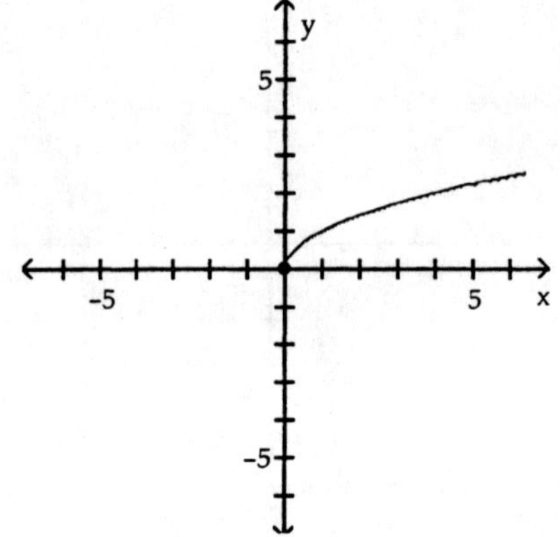

13) $f(x) = \frac{1}{x}$

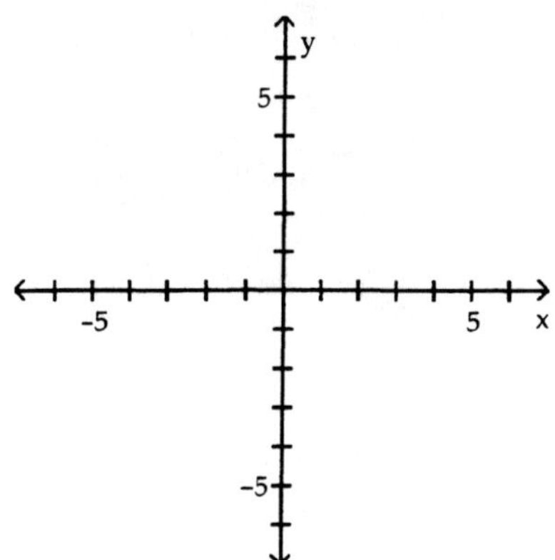

A)

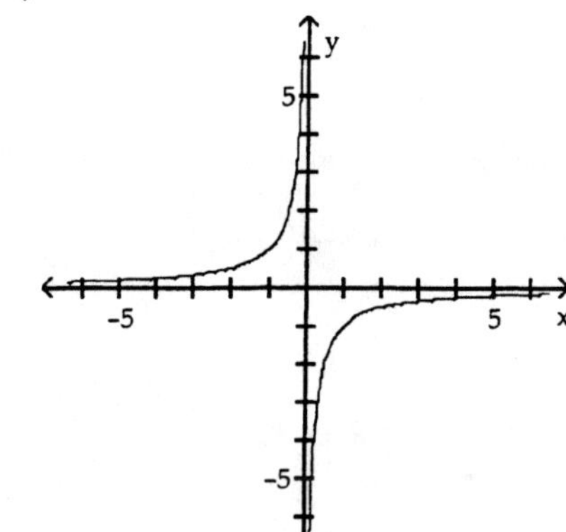

B)

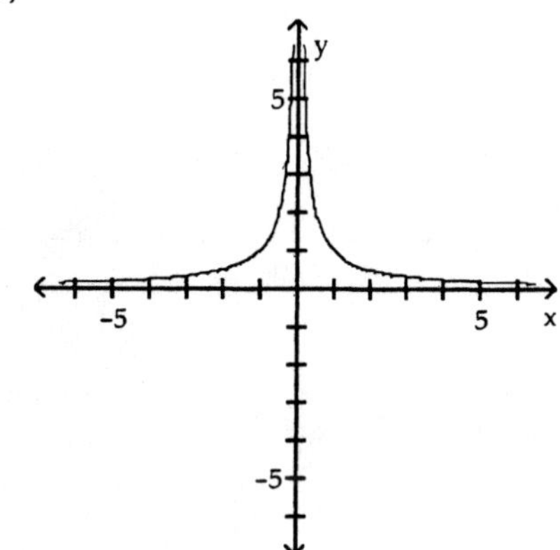

C)

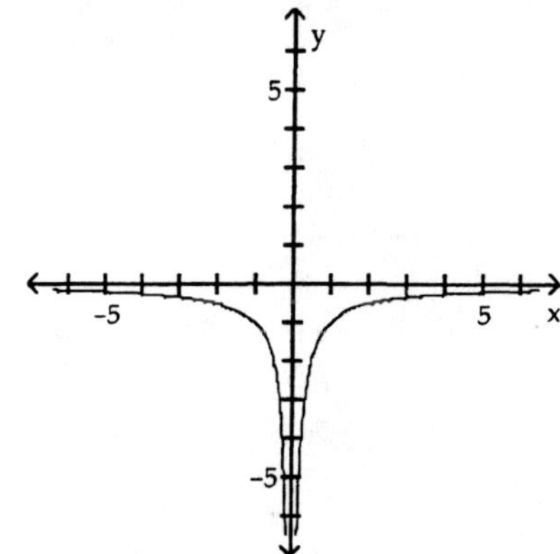

D)

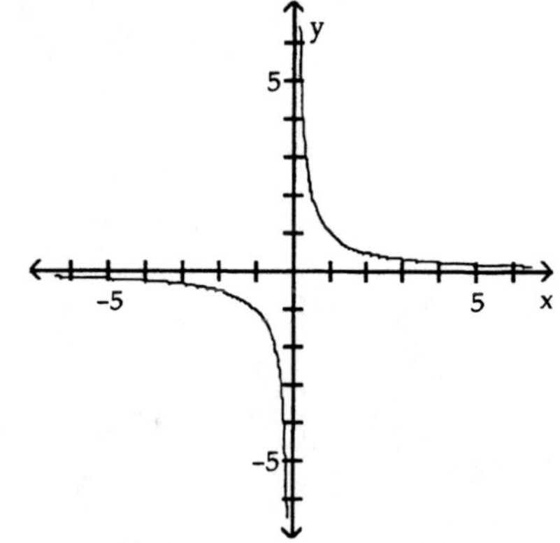

14) $f(x) = |x|$

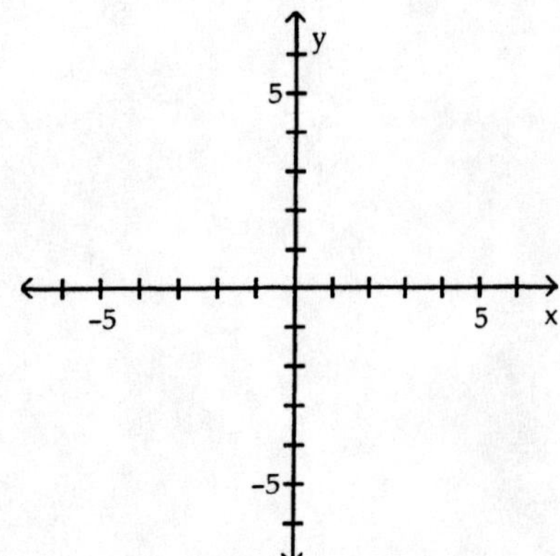

A)

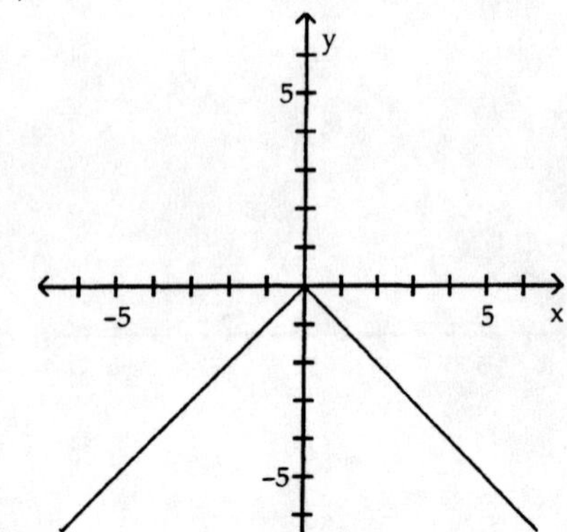

B)

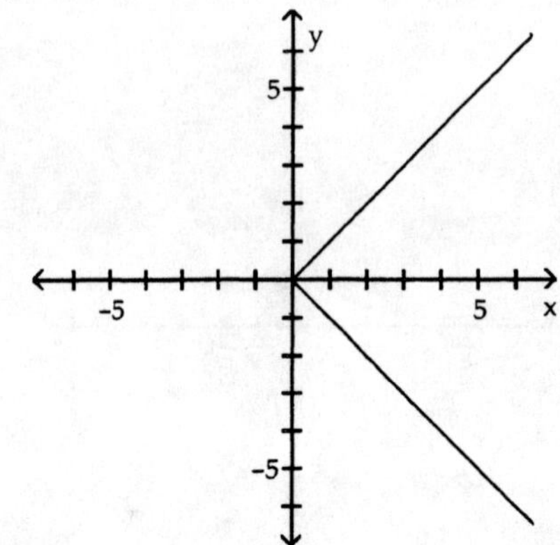

C)

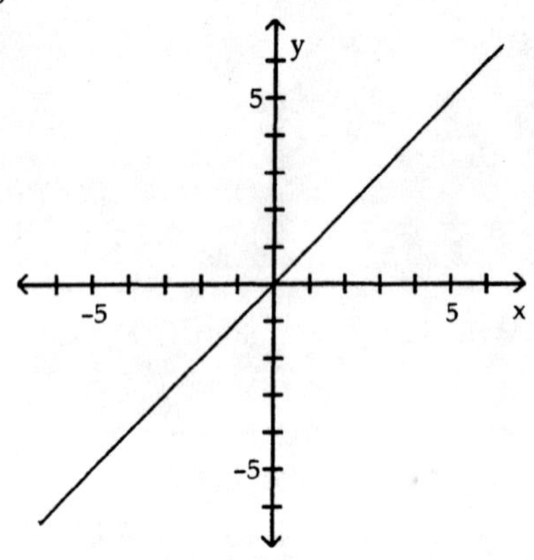

D)

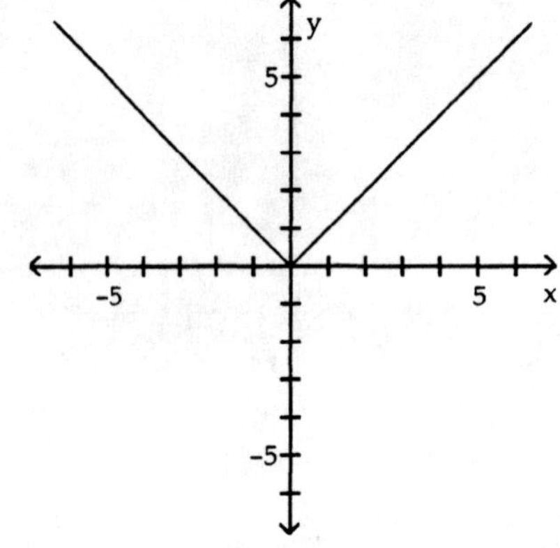

15) $f(x) = \sqrt[3]{x}$

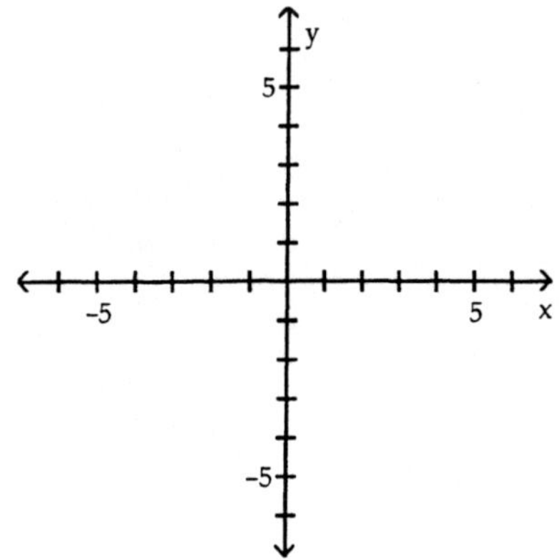

A)

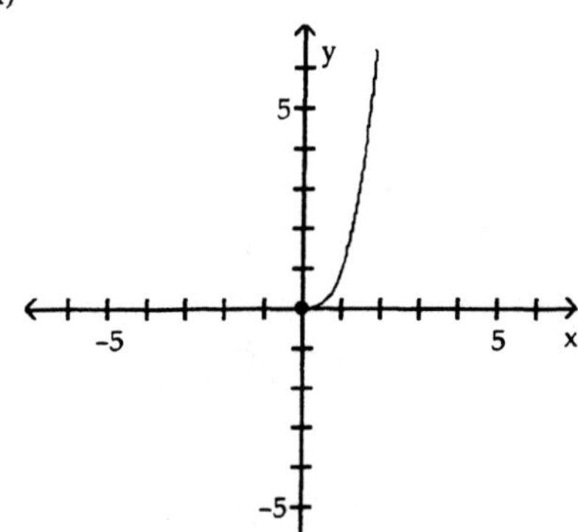

B)

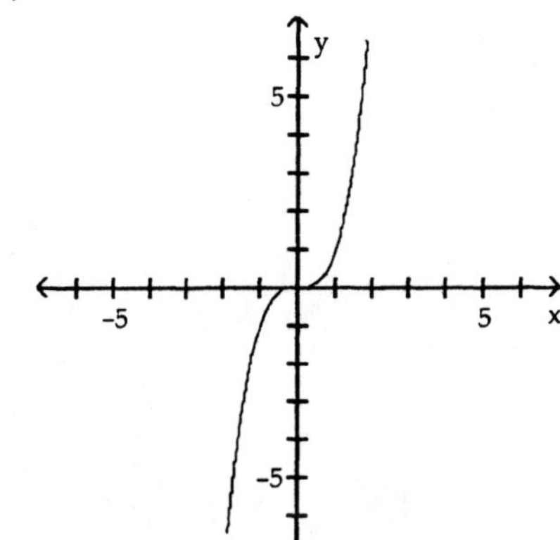

C)

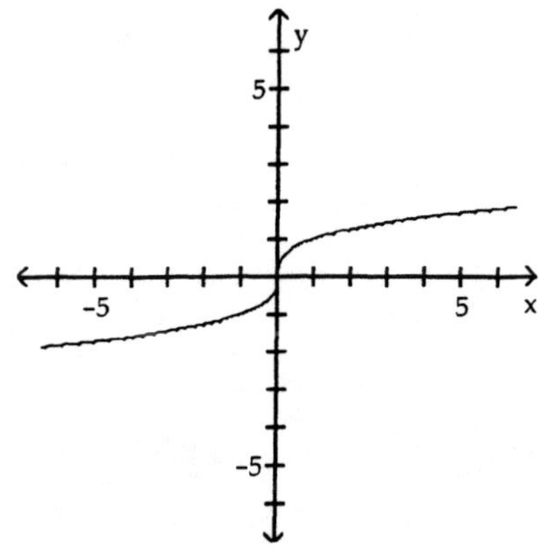

D)

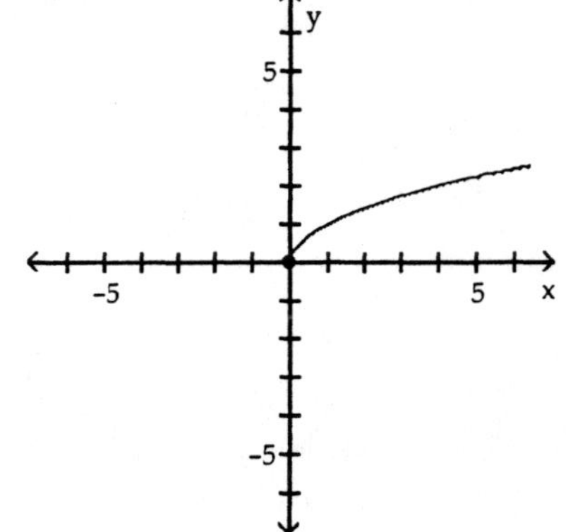

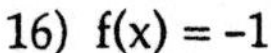
16) $f(x) = -1$

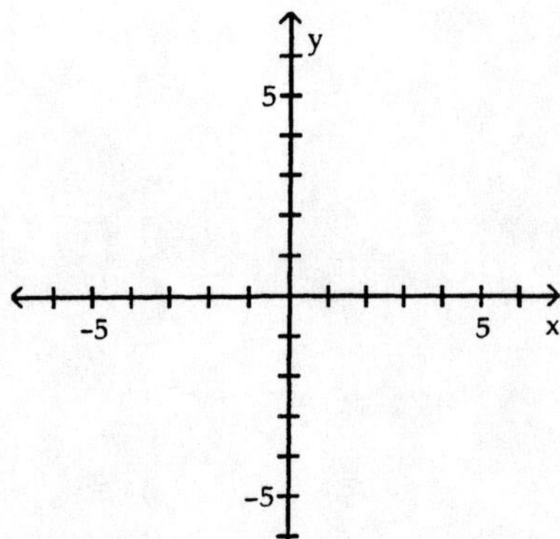

A)

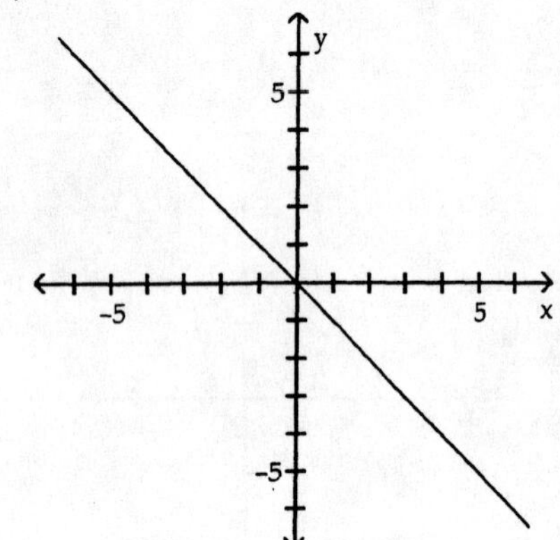

B)

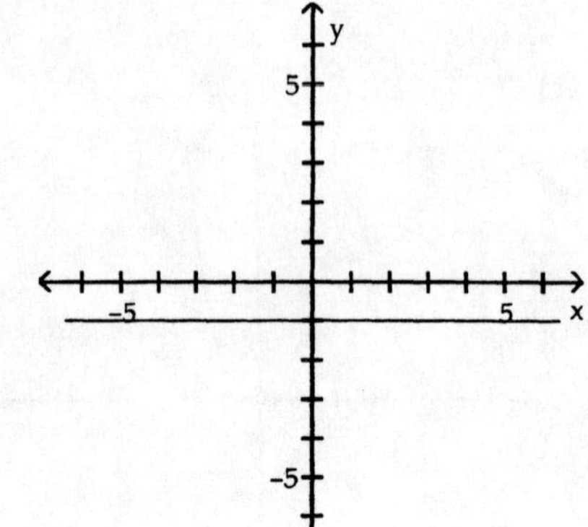

C)

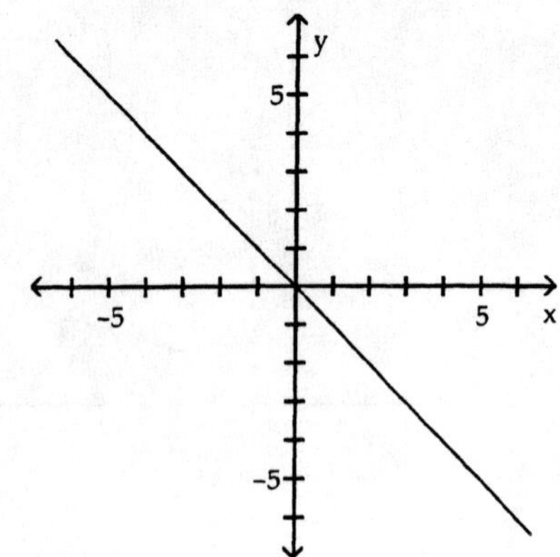

D)

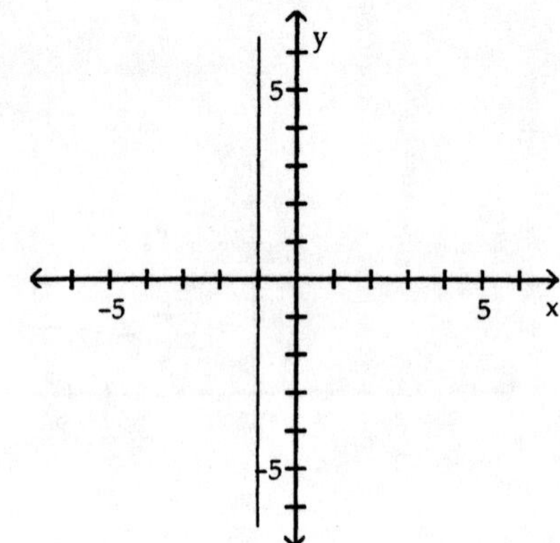

2 Graph Piecewise-defined Functions

Graph the function.

1)

$$f(x) = \begin{cases} x + 2 & \text{if } x < 1 \\ -3 & \text{if } x \geq 1 \end{cases}$$

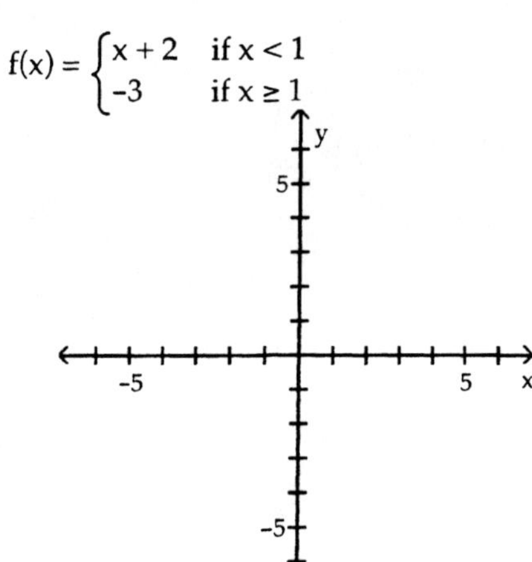

A)

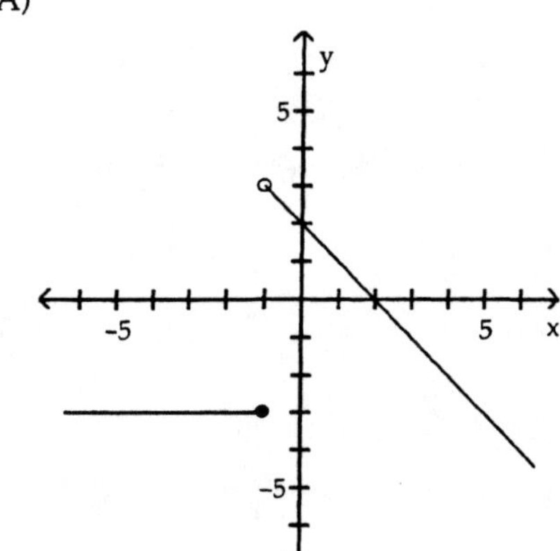

B)

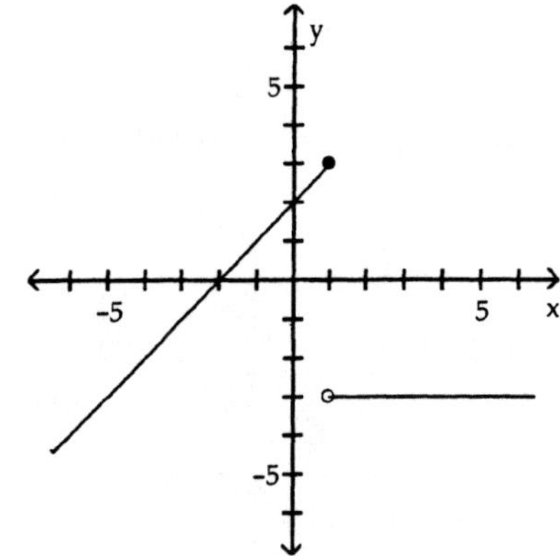

C)

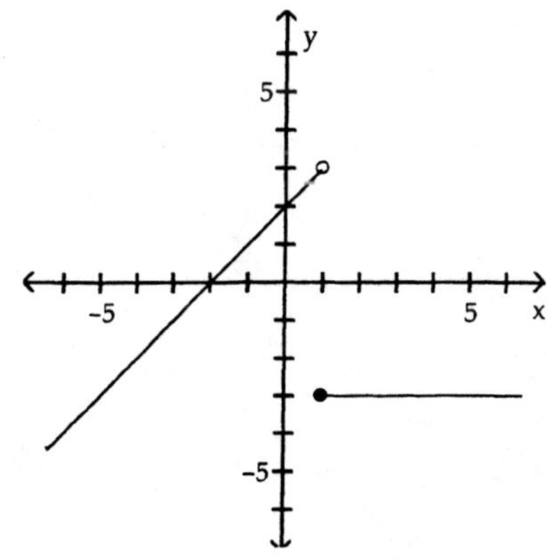

D)

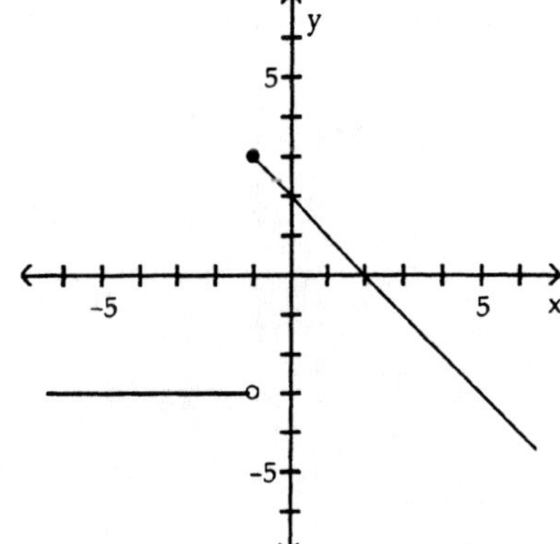

2)

$$f(x) = \begin{cases} -x + 3 & \text{if } x < 2 \\ 2x - 3 & \text{if } x \geq 2 \end{cases}$$

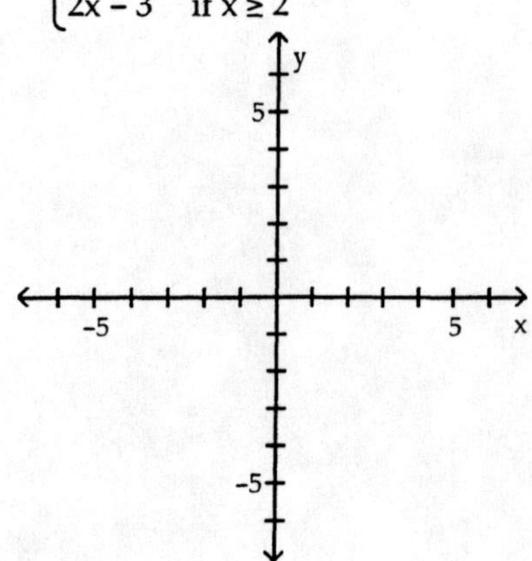

A)

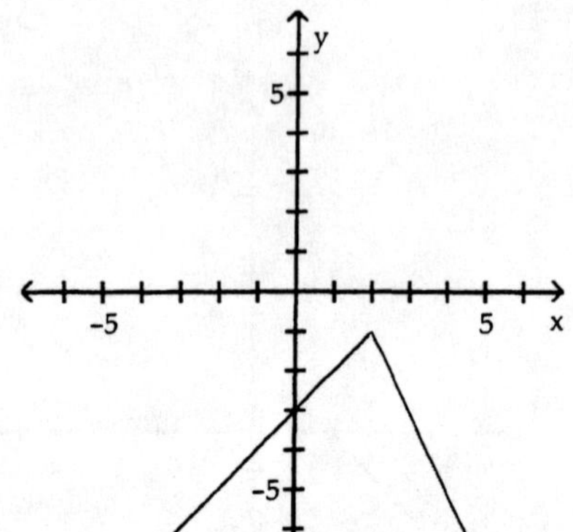

B)

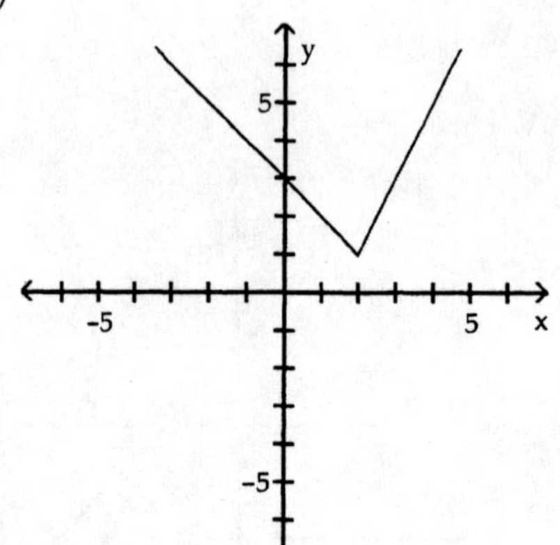

C)

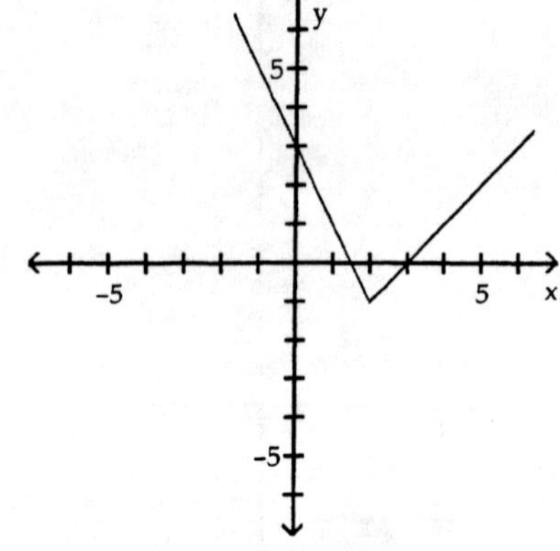

D)

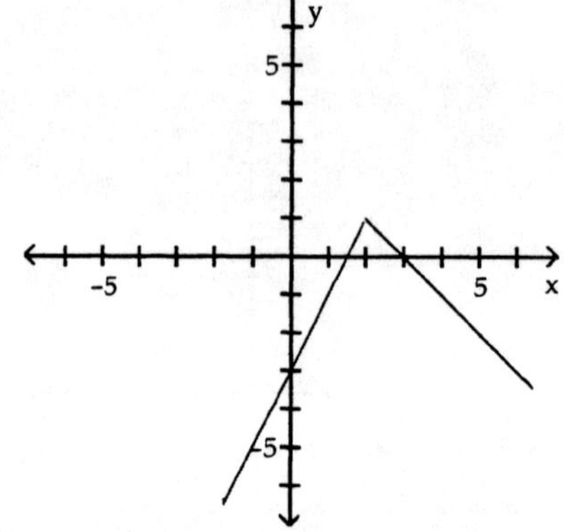

3)

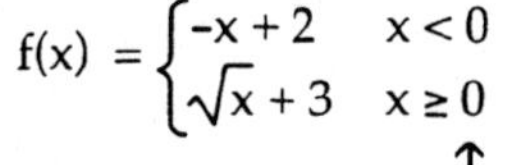

$$f(x) = \begin{cases} -x + 2 & x < 0 \\ \sqrt{x} + 3 & x \geq 0 \end{cases}$$

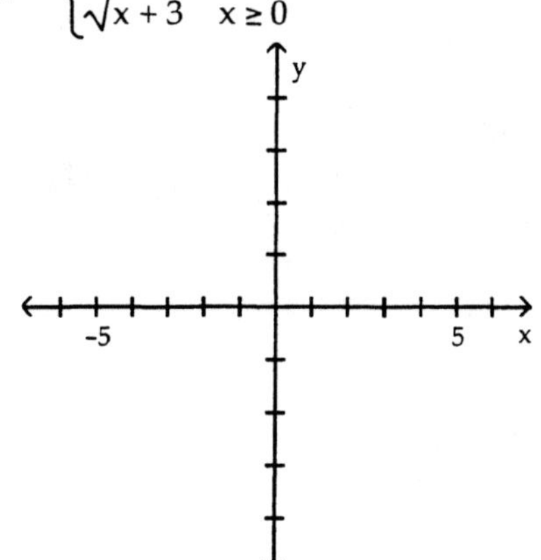

A)

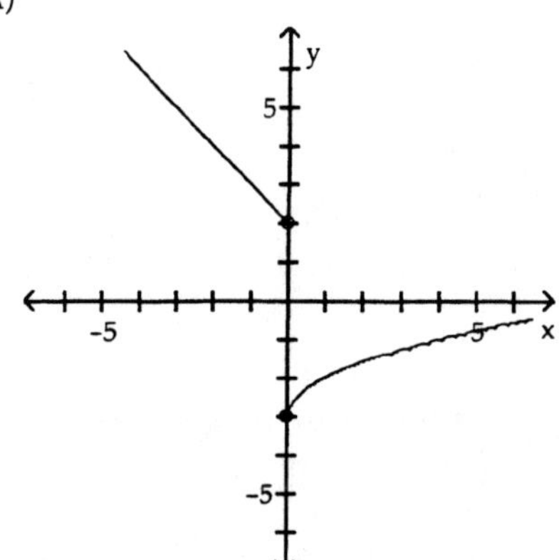

B)

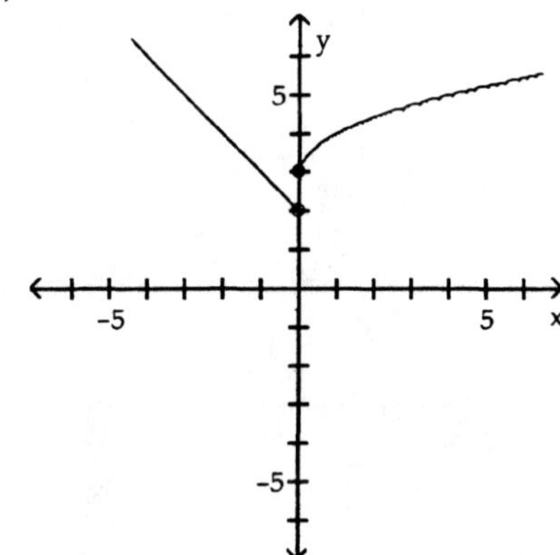

C)

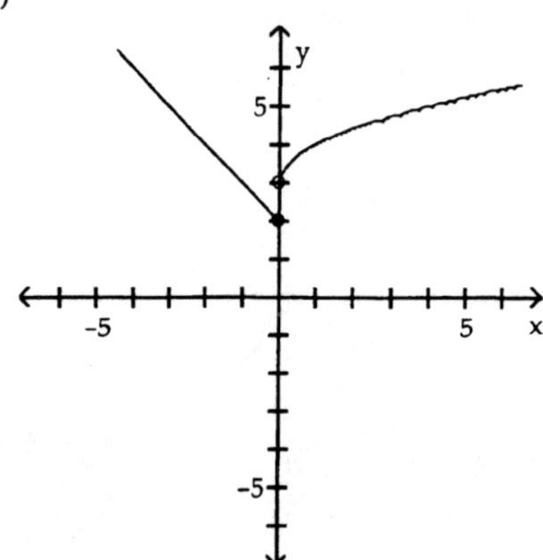

D)

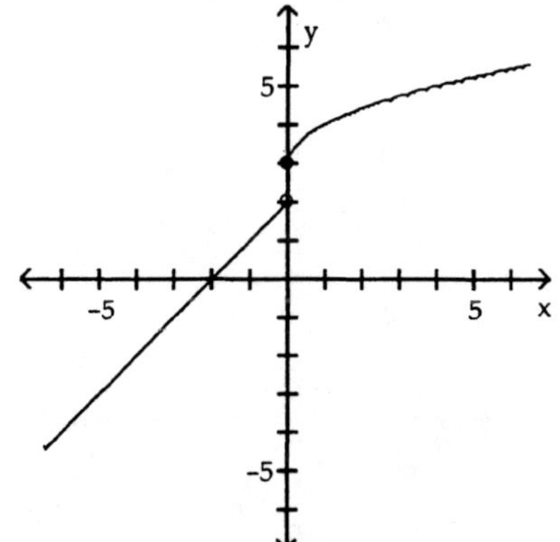

4)

$$f(x) = \begin{cases} x + 5 & \text{if } -7 \le x < -2 \\ -5 & \text{if } x = -2 \\ -x + 6 & \text{if } x > -2 \end{cases}$$

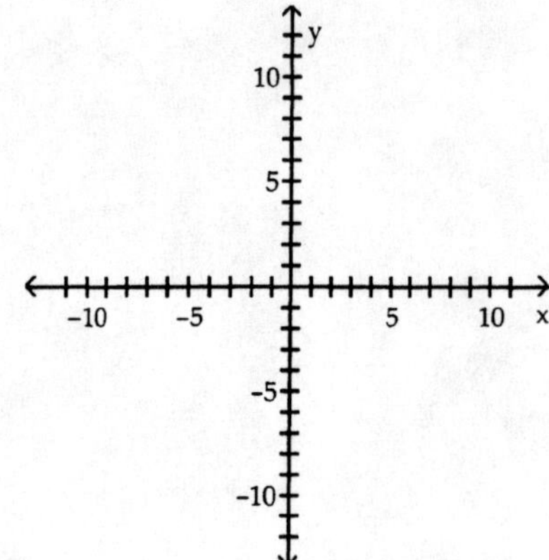

A)

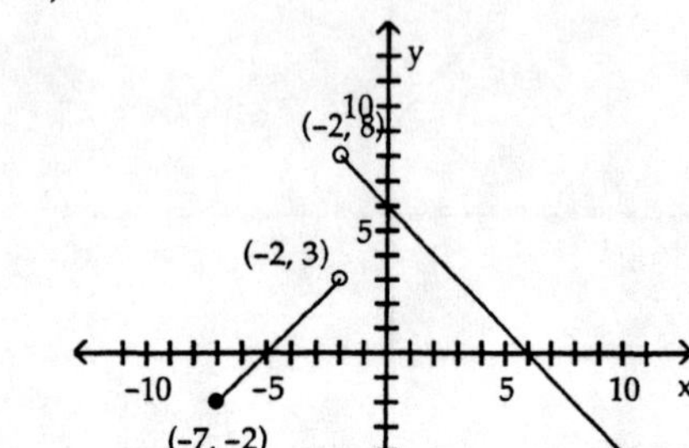

B)

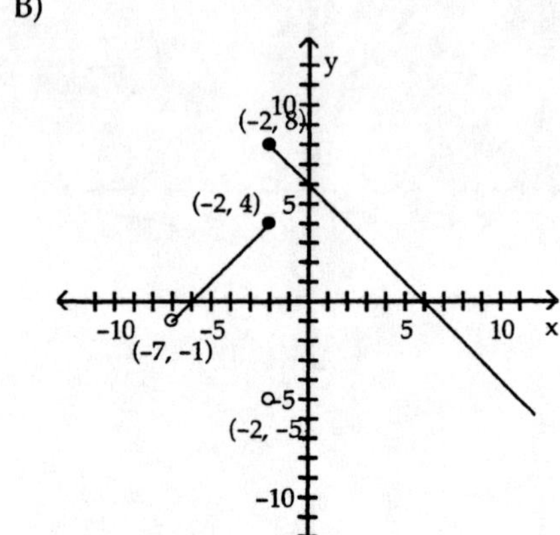

C)

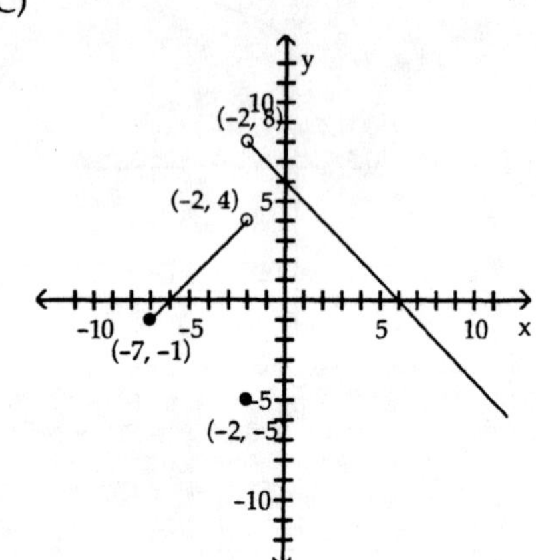

D)

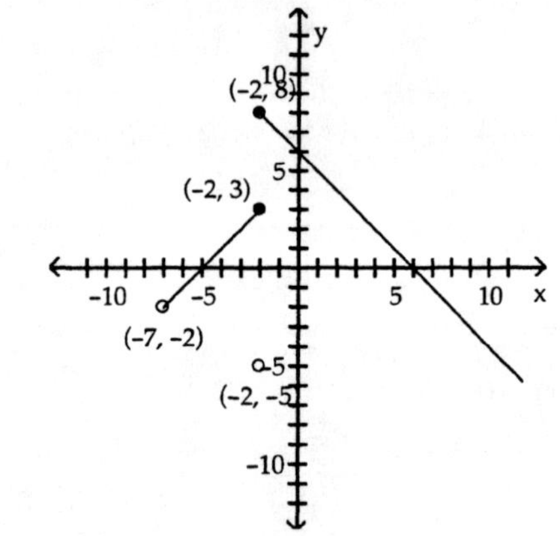

5)

$$f(x) = \begin{cases} 1 & \text{if } -3 \le x < 6 \\ |x| & \text{if } 6 \le x < 8 \\ \sqrt[3]{x} & \text{if } 8 \le x \le 11 \end{cases}$$

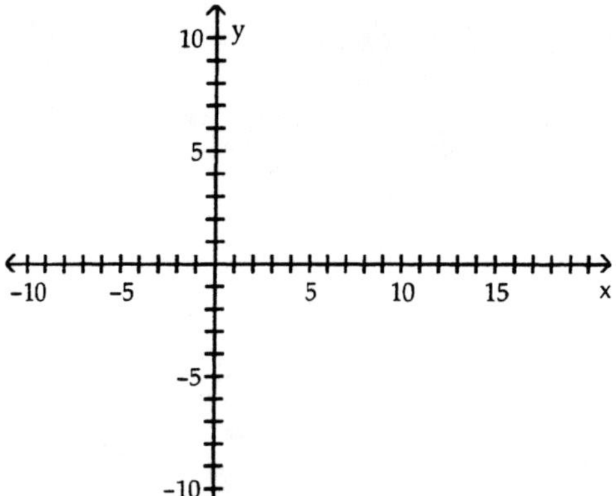

A)

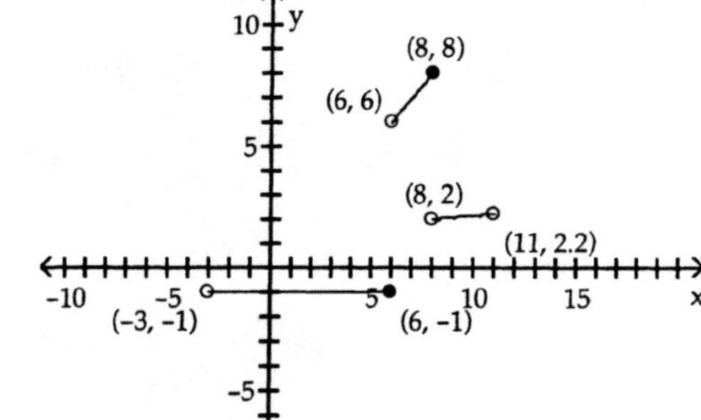

B)

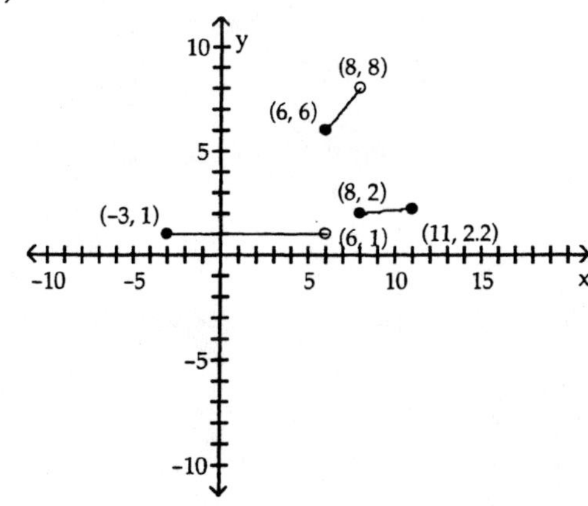

C)

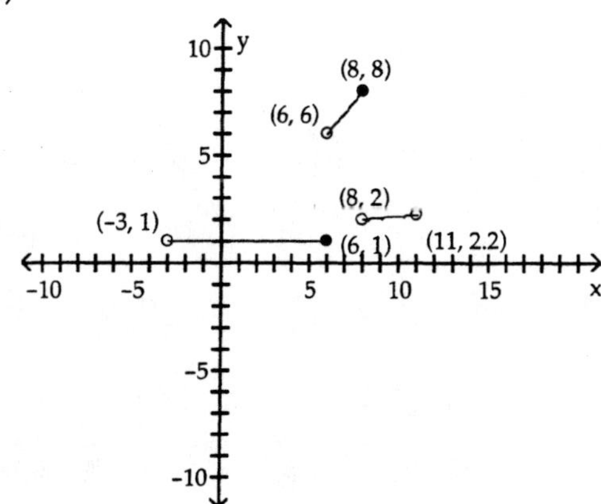

D)

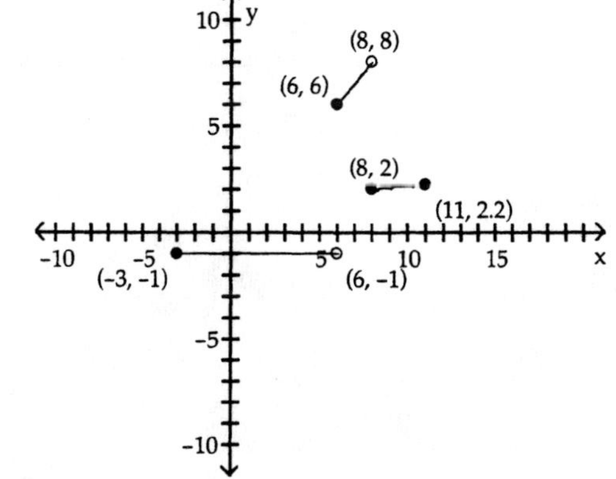

Find the domain of the function.

6)

$$f(x) = \begin{cases} -3x & \text{if } x \ne 0 \\ -5 & \text{if } x = 0 \end{cases}$$

A) $\{x \mid x \le 0\}$ B) $\{0\}$ C) $\{x \mid x \ne 0\}$ D) all real numbers

7)

$$f(x) = \begin{cases} 1 & \text{if } -7 \le x < -3 \\ |x| & \text{if } -3 \le x < 7 \\ \sqrt[3]{x} & \text{if } 7 \le x \le 30 \end{cases}$$

A) $\{x \mid -7 \le x < 7 \text{ or } 7 < x \le 30\}$

B) $\{x \mid x \ge -7\}$

C) $\{x \mid 7 \le x \le 30\}$

D) $\{x \mid -7 \le x \le 30\}$

Locate any intercepts of the function.

8)

$$f(x) = \begin{cases} -6x + 9 & \text{if } x < 1 \\ 9x - 6 & \text{if } x \ge 1 \end{cases}$$

A) $(0, -6), (\frac{3}{2}, 0), (\frac{2}{3}, 0)$

B) $(0, 9)$

C) $(0, 9), (\frac{3}{2}, 0), (\frac{2}{3}, 0)$

D) $(0, -6)$

9)

$$f(x) = \begin{cases} 1 & \text{if } -4 \le x < -1 \\ |x| & \text{if } -1 \le x < 4 \\ \sqrt{x} & \text{if } 4 \le x \le 25 \end{cases}$$

A) $(0, 0), (0, 1)$

B) $(0, 0)$

C) $(0, 0), (1, 0)$

D) none

Based on the graph, find the range of $y = f(x)$.

10)

$$f(x) = \begin{cases} -\frac{1}{2}x & \text{if } x \ne 0 \\ -8 & \text{if } x = 0 \end{cases}$$

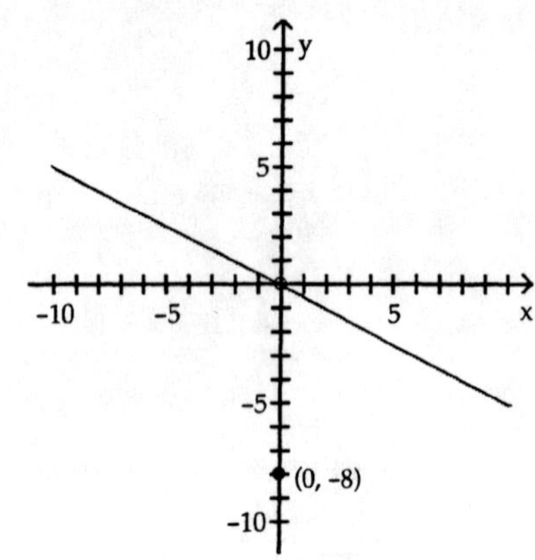

A) $(-\infty, 0)$ or $(0, \infty)$

B) $(-\infty, 0)$ or $\{0\}$ or $(0, \infty)$

C) $(-10, 10)$

D) $(-\infty, \infty)$

11)

$$f(x) = \begin{cases} 4 & \text{if } -6 \le x < -2 \\ |x| & \text{if } -2 \le x < 8 \\ \sqrt{x} & \text{if } 8 \le x \le 13 \end{cases}$$

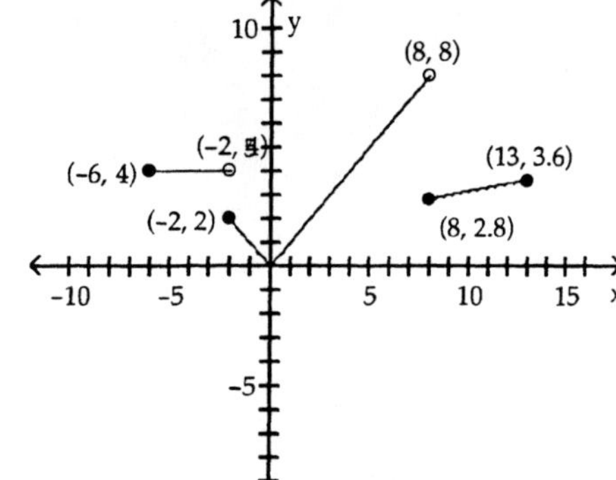

A) $[0, \sqrt{13}]$ B) $[0, 8]$ C) $[0, 8)$ D) $[0, \infty)$

The graph of a piecewise-defined function is given. Write a definition for the function.

12)

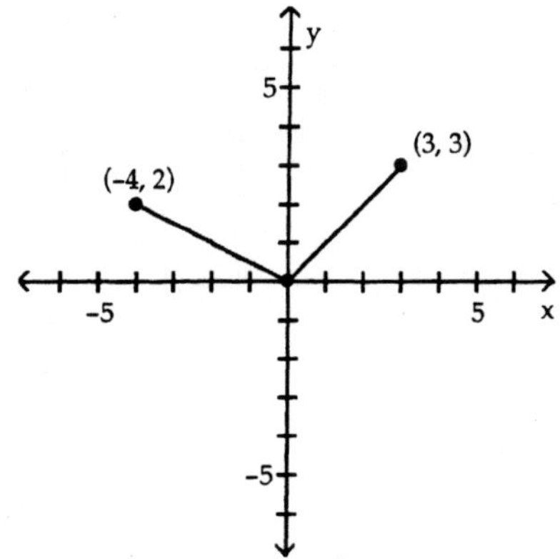

A)
$$f(x) = \begin{cases} -2x & \text{if } -4 \le x \le 0 \\ x & \text{if } 0 < x \le 3 \end{cases}$$

B)
$$f(x) = \begin{cases} -\frac{1}{2}x & \text{if } -4 \le x \le 0 \\ x & \text{if } 0 < x \le 3 \end{cases}$$

C)
$$f(x) = \begin{cases} -\frac{1}{2}x & \text{if } -4 < x < 0 \\ x & \text{if } 0 < x < 3 \end{cases}$$

D)
$$f(x) = \begin{cases} \frac{1}{2}x & \text{if } -4 < x < 0 \\ x & \text{if } 0 < x < 3 \end{cases}$$

13)

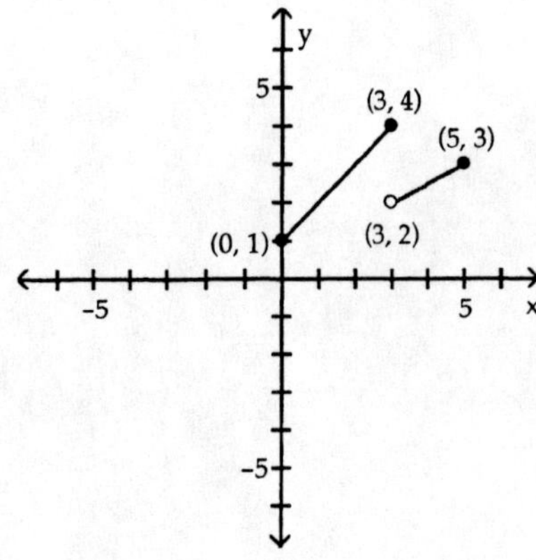

A)

$$f(x) = \begin{cases} x + 1 & \text{if } 0 \le x \le 3 \\ \frac{1}{2}x + \frac{1}{2} & \text{if } 3 < x \le 5 \end{cases}$$

B)

$$f(x) = \begin{cases} x + 1 & \text{if } 0 \le x \le 3 \\ \frac{1}{2}x - \frac{1}{2} & \text{if } 3 < x \le 5 \end{cases}$$

C)

$$f(x) = \begin{cases} x + 1 & \text{if } 0 \le x \le 3 \\ \frac{1}{2}x & \text{if } 3 < x \le 5 \end{cases}$$

D)

$$f(x) = \begin{cases} x + 1 & \text{if } 0 \le x \le 3 \\ \frac{1}{2}x + 2 & \text{if } 3 < x \le 5 \end{cases}$$

14)

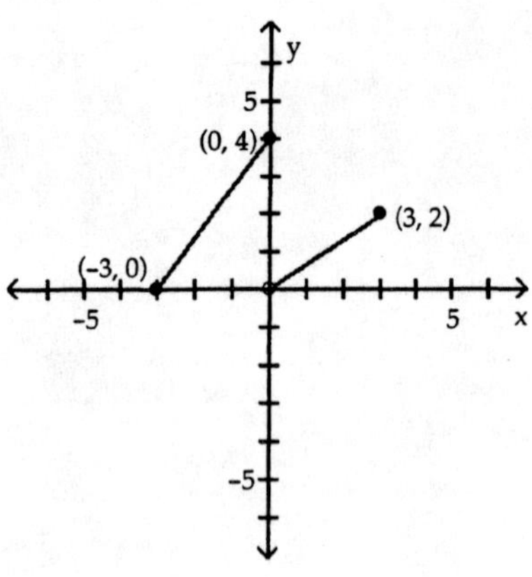

A)

$$f(x) = \begin{cases} \frac{3}{4}x + 4 & \text{if } -3 \le x \le 0 \\ \frac{3}{2}x & \text{if } 0 < x \le 3 \end{cases}$$

B)

$$f(x) = \begin{cases} \frac{4}{3}x + 4 & \text{if } -3 \le x \le 0 \\ \frac{2}{3}x + 2 & \text{if } 0 < x \le 3 \end{cases}$$

C)

$$f9x) = \begin{cases} \frac{4}{3}x - 4 & \text{if } -3 \le x \le 0 \\ \frac{2}{3}x & \text{if } 0 \le x \le 3 \end{cases}$$

D)

$$f(x) = \begin{cases} \frac{4}{3}x + 4 & \text{if } -3 \le x \le 0 \\ \frac{2}{3}x & \text{if } 0 < x \le 3 \end{cases}$$

15)

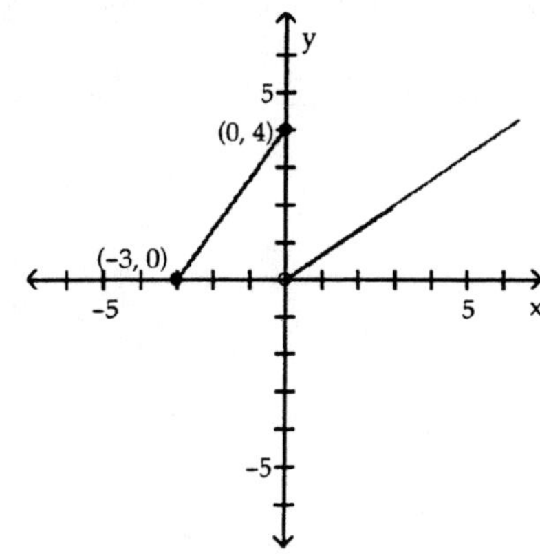

A) $f(x) = \begin{cases} \frac{4}{3}x + 4 & \text{if } -3 \le x \le 0 \\ \frac{2}{3}x & \text{if } 0 < x \le 3 \end{cases}$

B) $f(x) = \begin{cases} \frac{4}{3}x + 4 & \text{if } -3 \le x \le 0 \\ \frac{2}{3}x & \text{if } x > 0 \end{cases}$

C) $f(x) = \begin{cases} \frac{3}{4}x + 4 & \text{if } -3 \le x \le 0 \\ \frac{3}{2}x & \text{if } x \ge 0 \end{cases}$

D) $f(x) = \begin{cases} \frac{3}{4}x + 4 & \text{if } -3 \le x \le 0 \\ \frac{3}{2}x & \text{if } x > 0 \end{cases}$

3 Additional Applications

Solve the problem.

1) If $f(x) = \text{int}(4x)$, find $f(1.6)$.

A) 6 B) 2 C) 7 D) 1

2) A gas company has the following rate schedule for natural gas usage in single-family residences:

Monthly service charge	\$8.80
Per therm service charge	
1st 25 therms	\$0.6686/therm
Over 25 therms	\$0.85870/therm

What is the charge for using 25 therms in one month?
What is the charge for using 45 therms in one month?
Construct a function that gives the monthly charge C for x therms of gas.

3) An electric company has the following rate schedule for electricity usage in single-family residences:

Monthly service charge	\$4.93
Per kilowatt service charge	
1st 300 kilowatts	\$0.11589/kW
Over 300 kilowatts	\$0.13321/kW

What is the charge for using 300 kilowatts in one month?
What is the charge for using 375 kilowatts in one month?
Construct a function that gives the monthly charge C for x kilowatts of electricity.

4) One Internet service provider has the following rate schedule for high-speed Internet service:

Monthly service charge	\$18.00
1st 50 hours of use	free
Next 50 hours of use	\$0.25/hour
Over 100 hours of use	\$1.00/hour

What is the charge for 50 hours of high-speed Internet use in one month?
What is the charge for 75 hours of high-speed Internet use in one month?
What is the charge for 135 hours of high-speed Internet use in one month?

5) The wind chill factor represents the equivalent air temperature at a standard wind speed that would produce the same heat loss as the given temperature and wind speed. One formula for computing the equivalent temperature is

$$W(t) = \begin{cases} t & \text{if } 0 \le v < 1.79 \\ 33 - \dfrac{(10.45 + 10\sqrt{v} - v)(33 - t)}{22.04} & \text{if } 1.79 \le v < 20 \\ 33 - 1.5958(33 - t) & \text{if } v \ge 20 \end{cases}$$

where v represents the wind speed (in meters per second) and t represents the air temperature (°C). Compute the wind chill for an air temperature of 15°C and a wind speed of 12 meters per second. (Round the answer to one decimal place.)

6) A cellular phone plan had the following schedule of charges:

Basic service, including 100 minutes of calls	\$20.00 per month
2nd 100 minutes of calls	\$0.075 per minute
Additional minutes of calls	\$0.10 per minute

What is the charge for 200 minutes of calls in one month?
What is the charge for 250 minutes of calls in one month?
Construct a function that relates the monthly charge C for x minutes of calls.

3.5 Graphing Techniques: Transformations

1 Graph Functions Using Horizontal and Vertical Shifts

Graph the function by starting with the graph of the basic function and then using the techniques of shifting, compressing, stretching, and/or reflecting.

1) $f(x) = x^2 + 2$

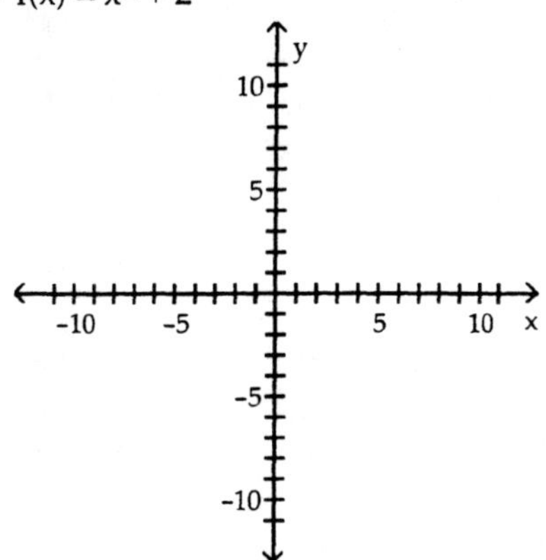

A)

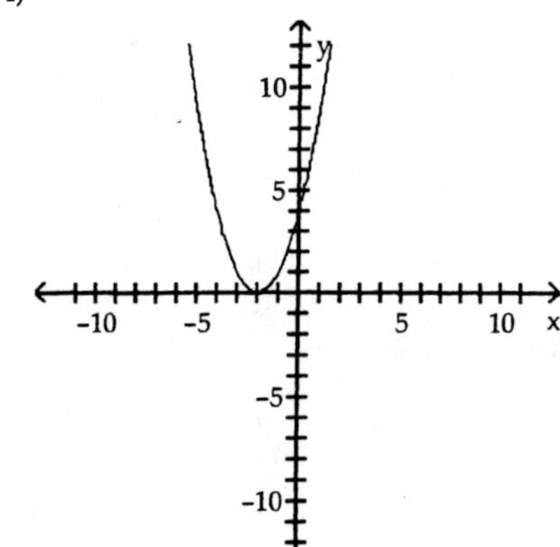

B)

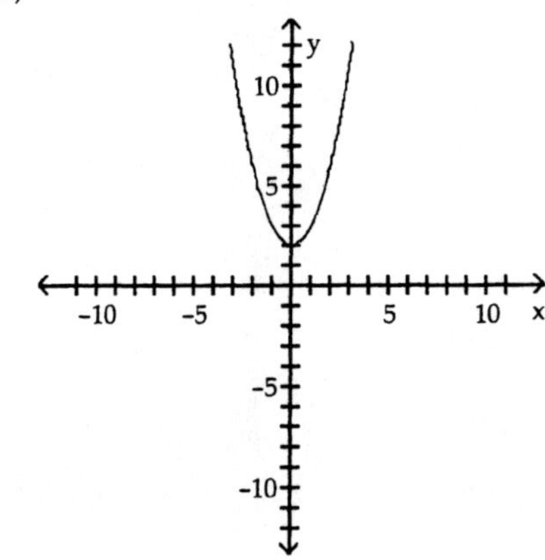

C)

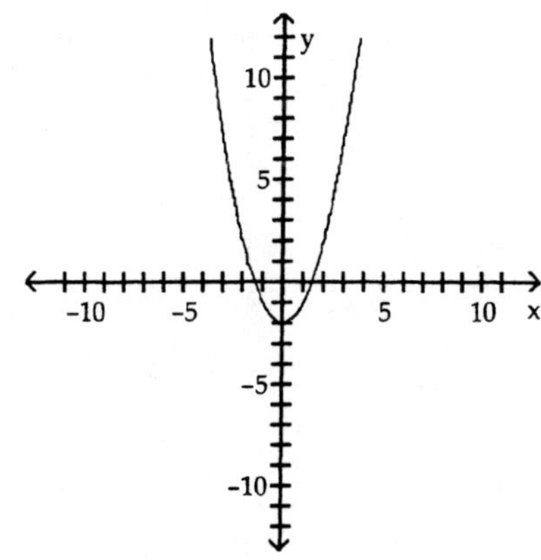

D)

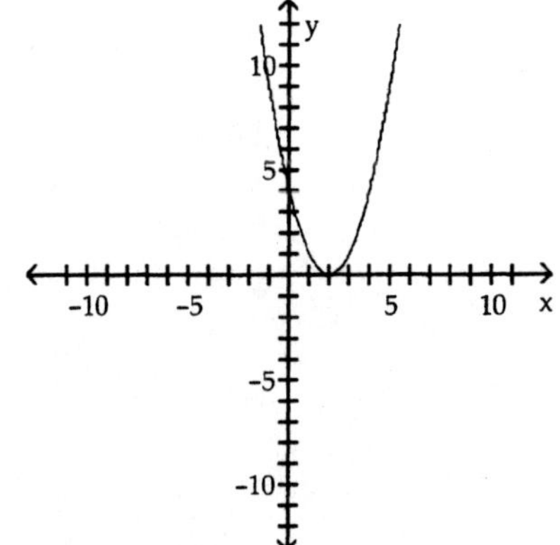

2) $f(x) = (x - 3)^2$

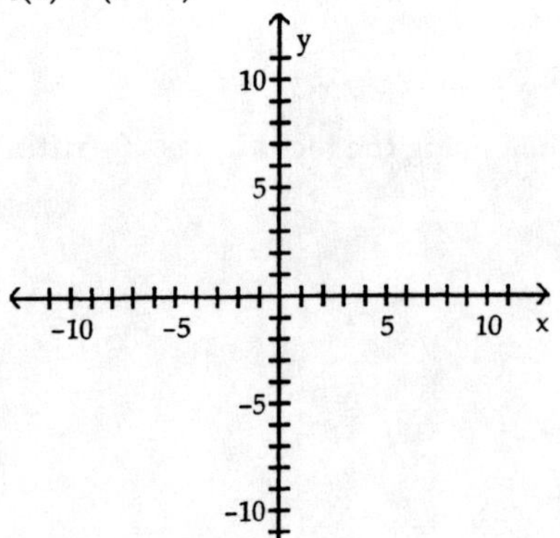

A)

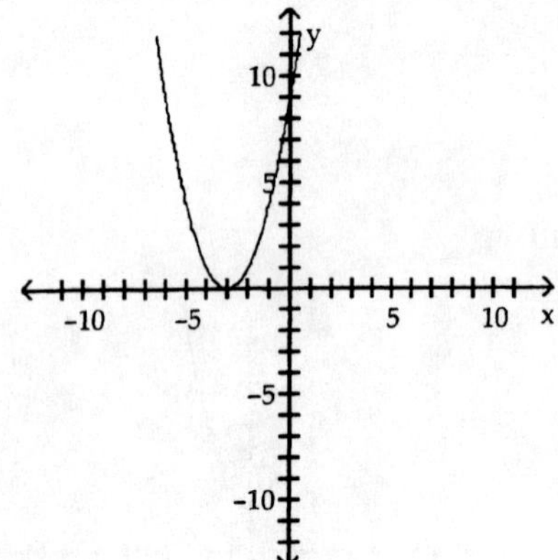

B)

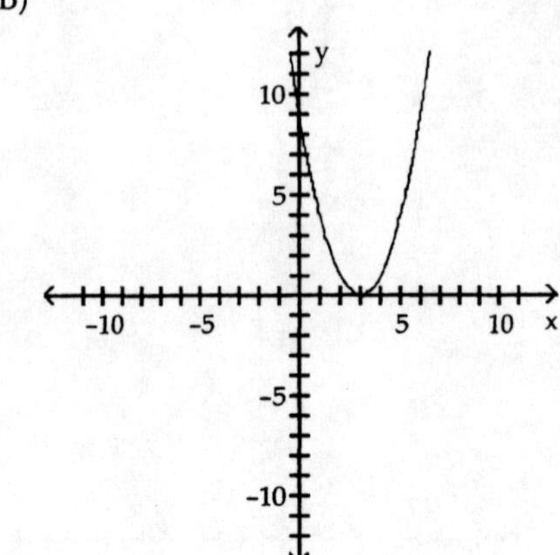

C)

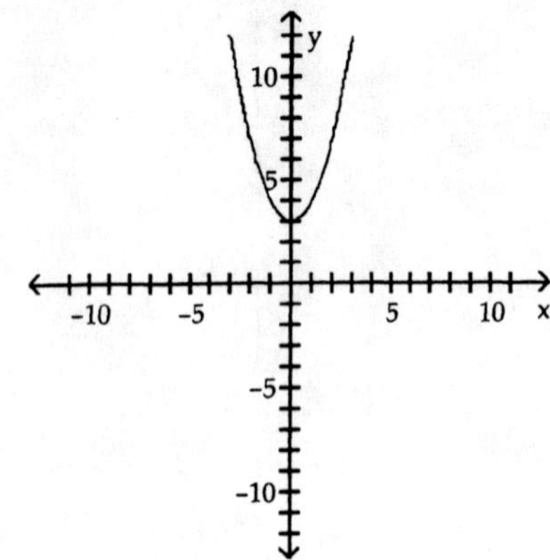

D)

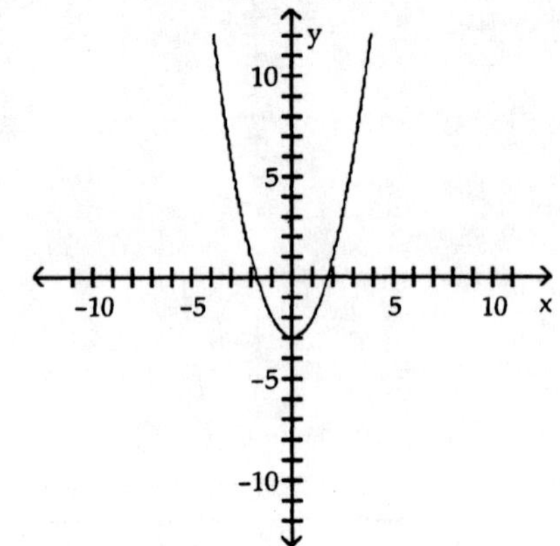

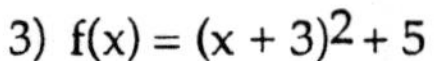

3) $f(x) = (x + 3)^2 + 5$

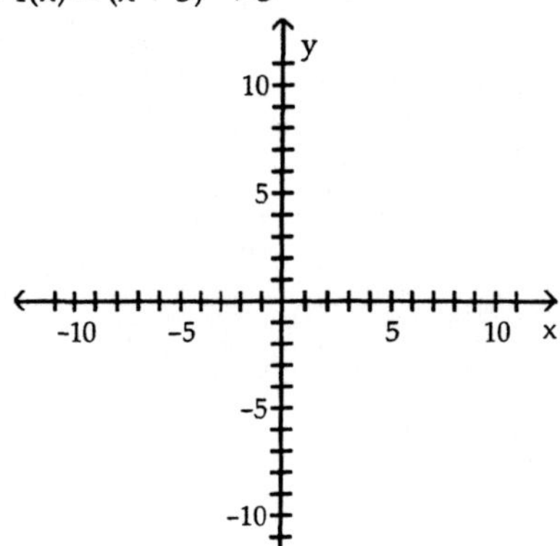

A)

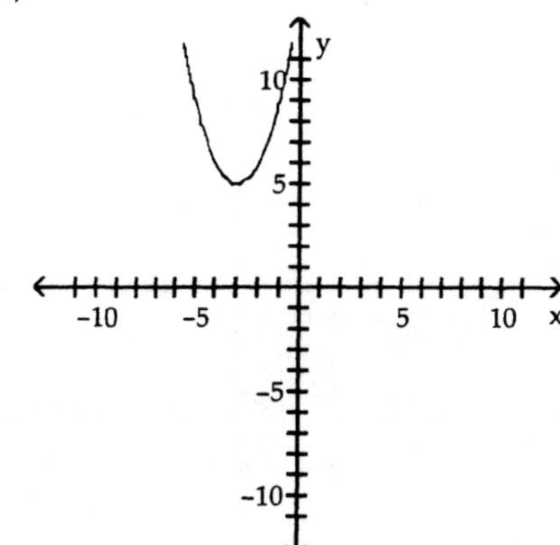

B)

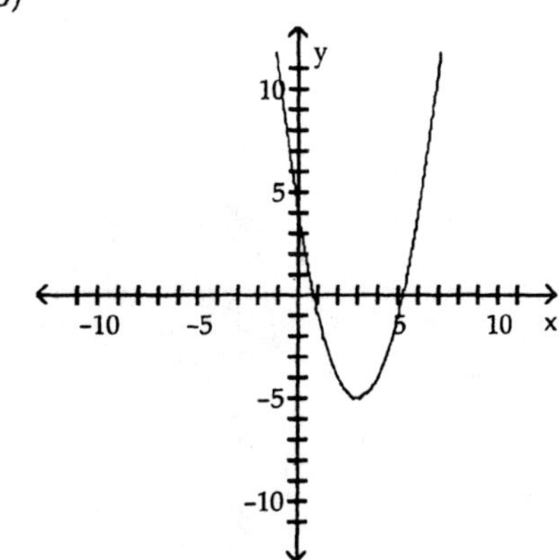

C)

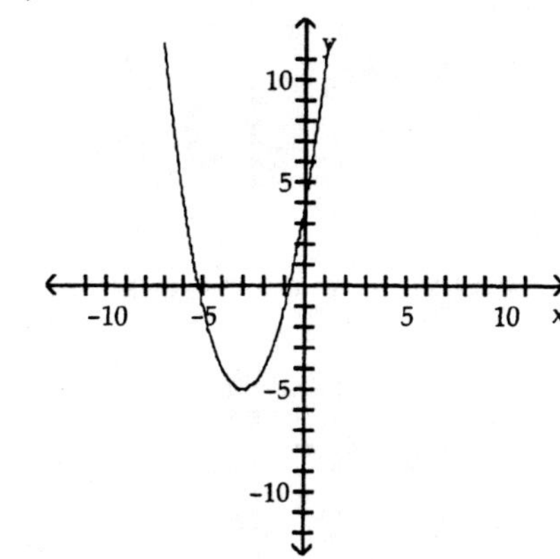

D)

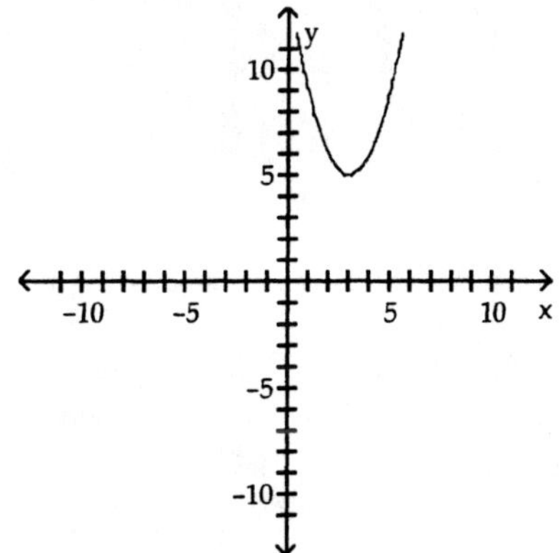

4) $f(x) = x^3 - 3$

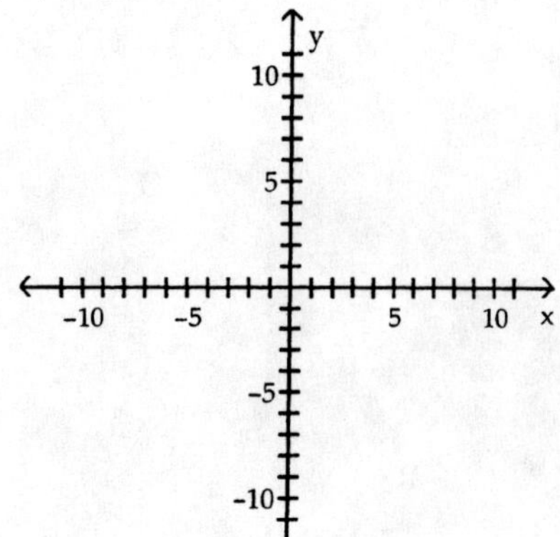

A)

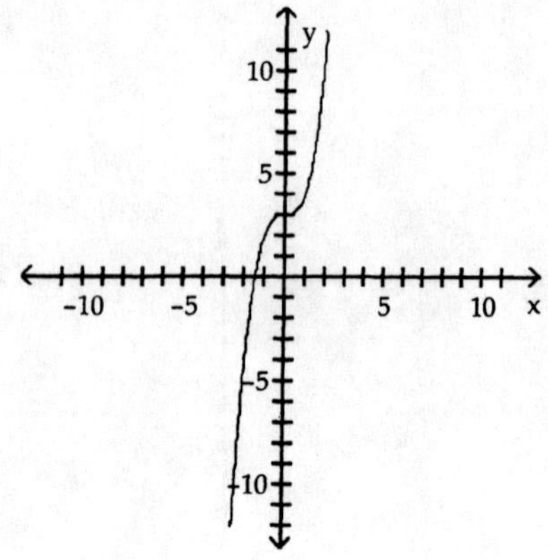

B)

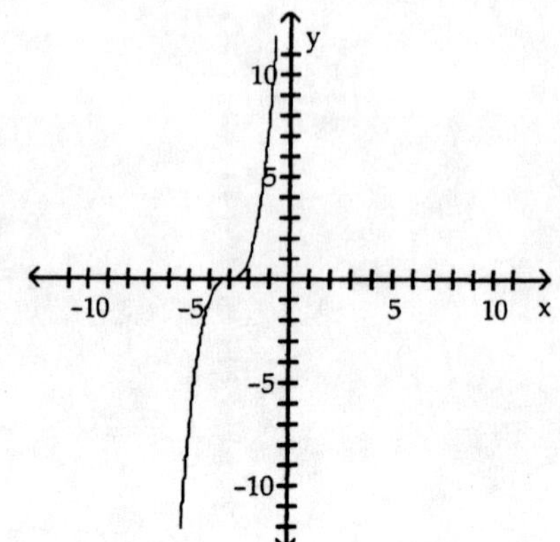

C)

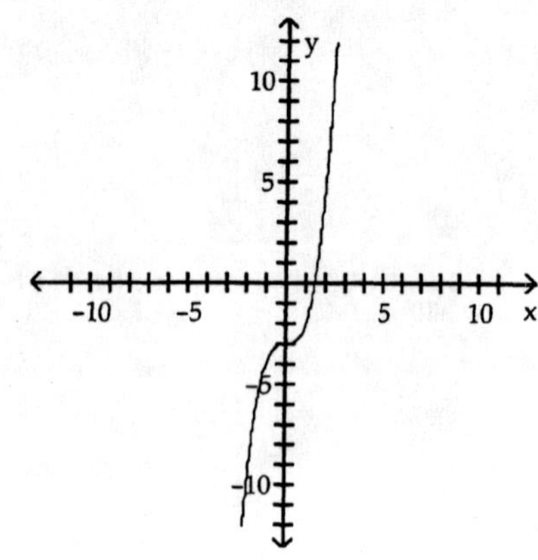

D)

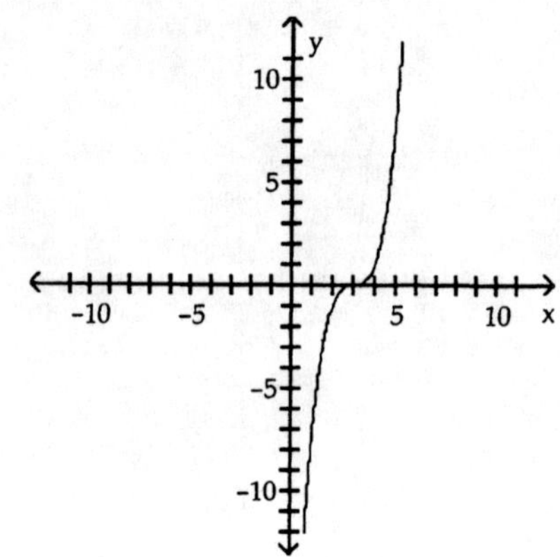

5) $f(x) = (x - 3)^3$

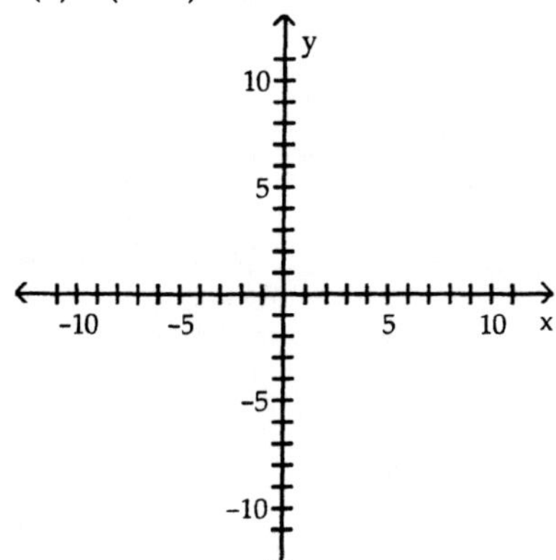

A)

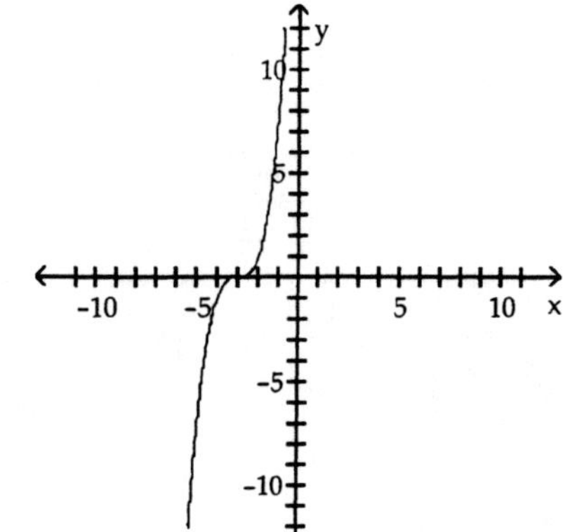

B)

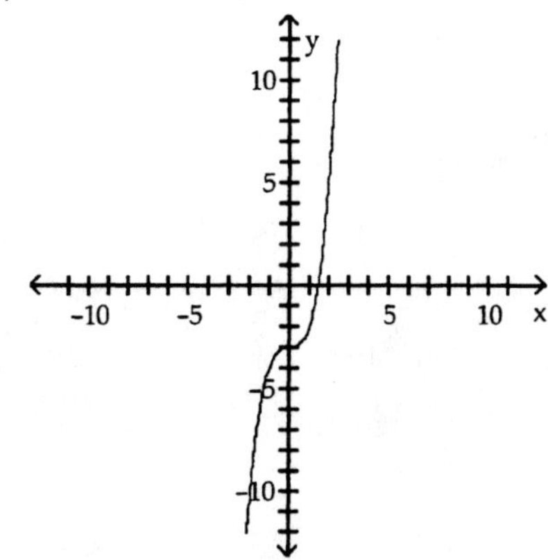

C)

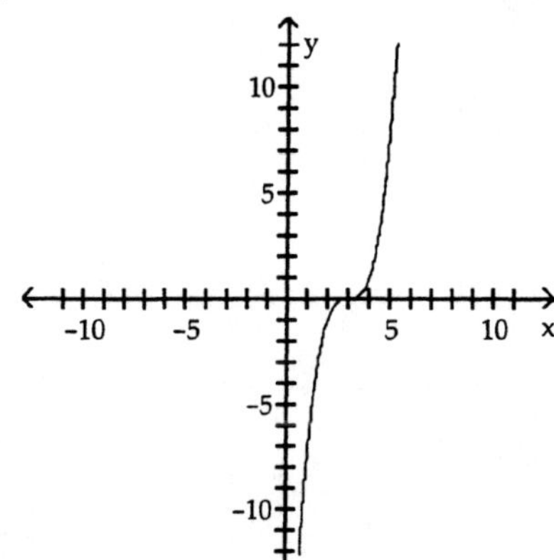

D)

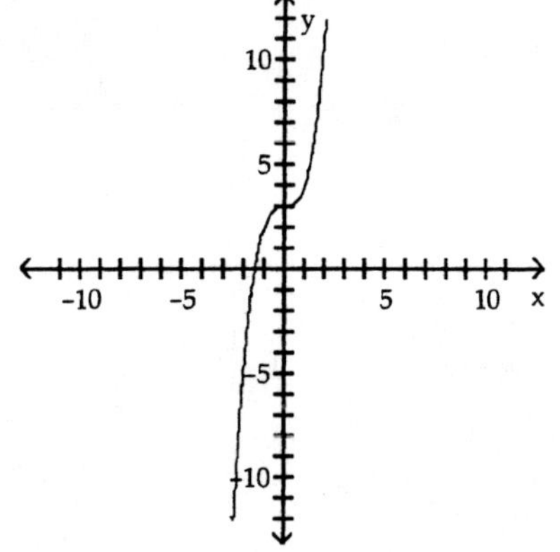

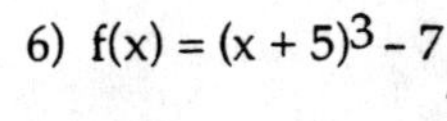

6) $f(x) = (x + 5)^3 - 7$

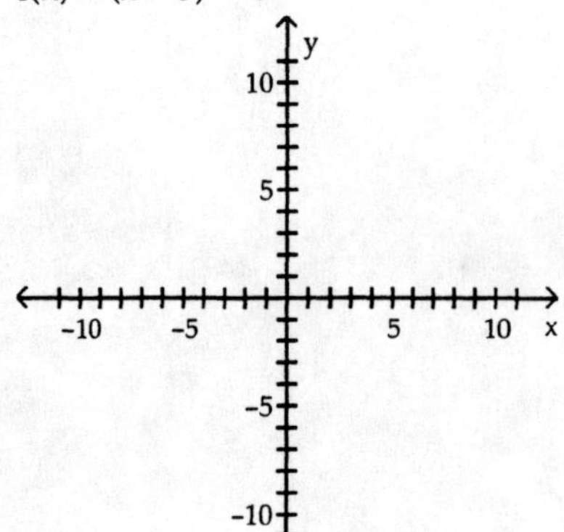

A)

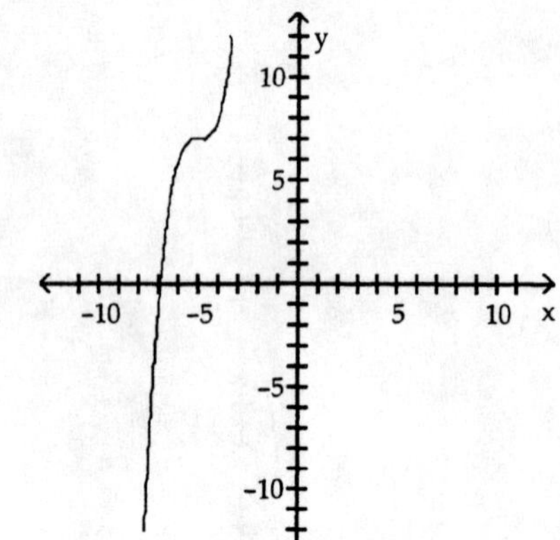

B)

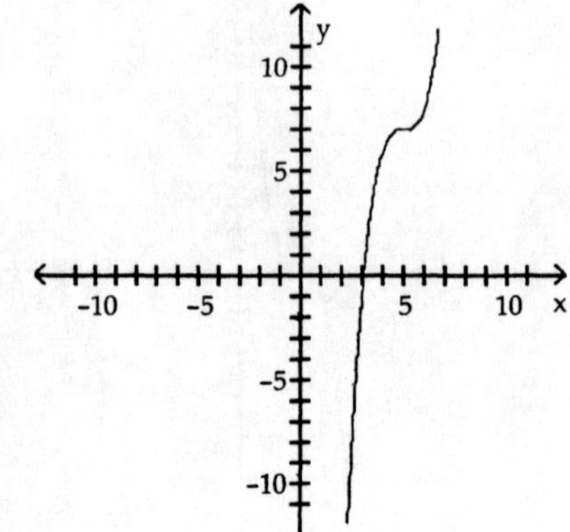

C)

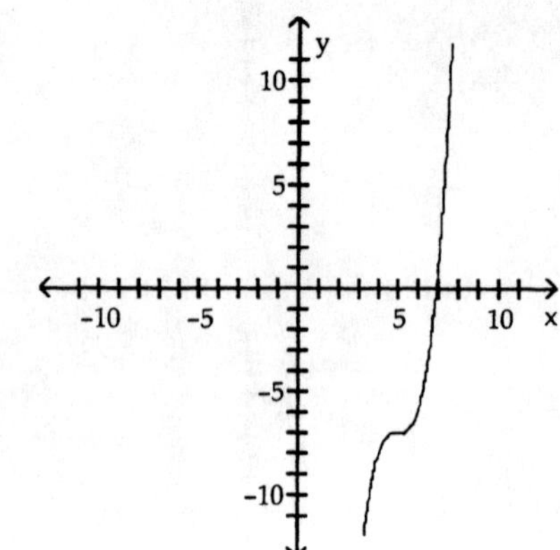

D)

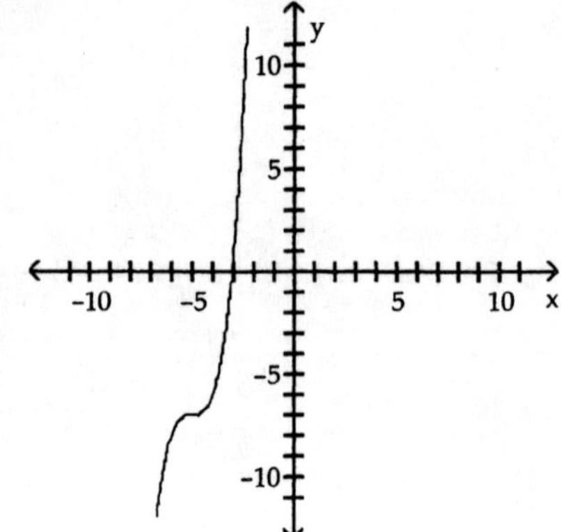

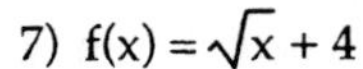
7) $f(x) = \sqrt{x} + 4$

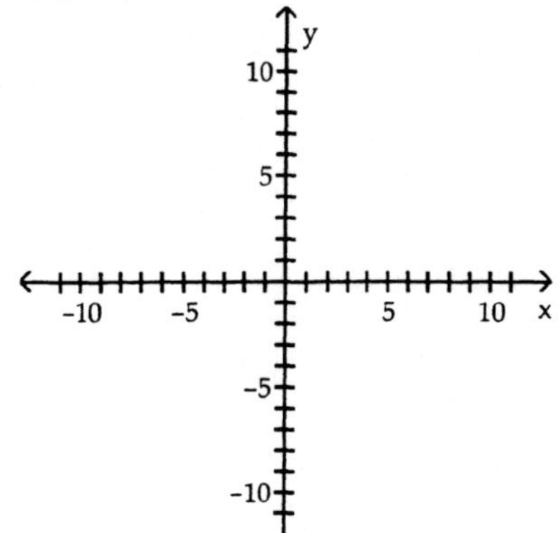

A)

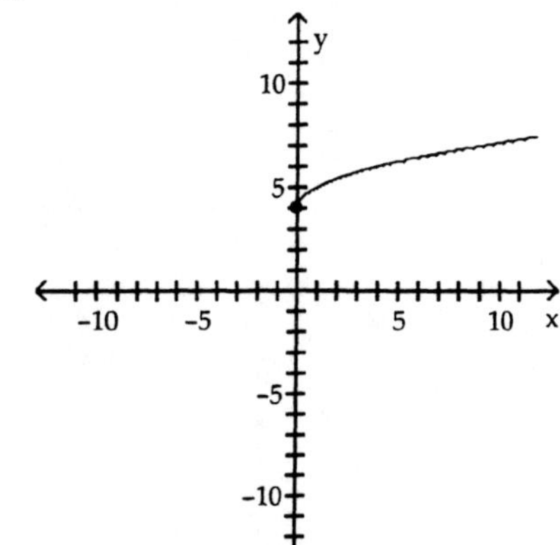

B)

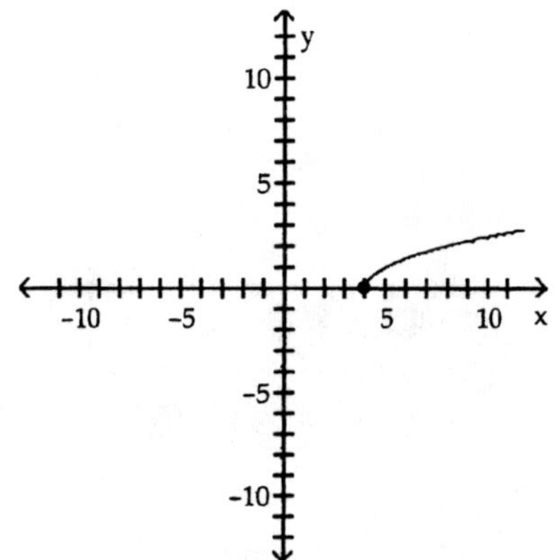

C)

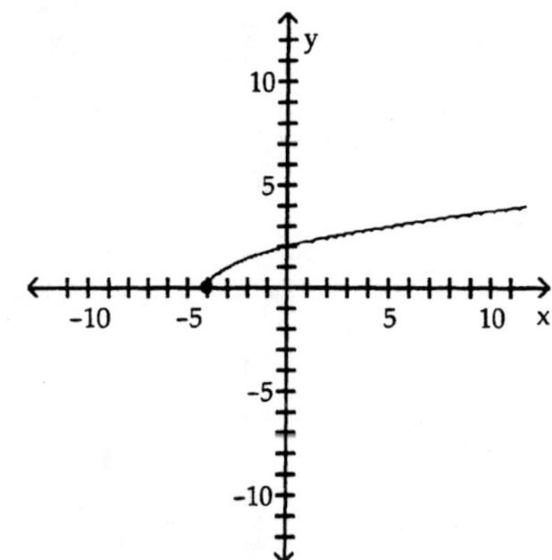

D)

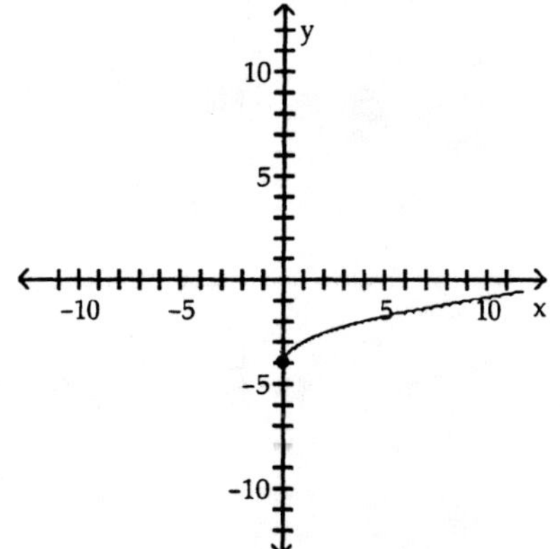

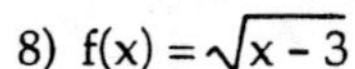
8) $f(x) = \sqrt{x - 3}$

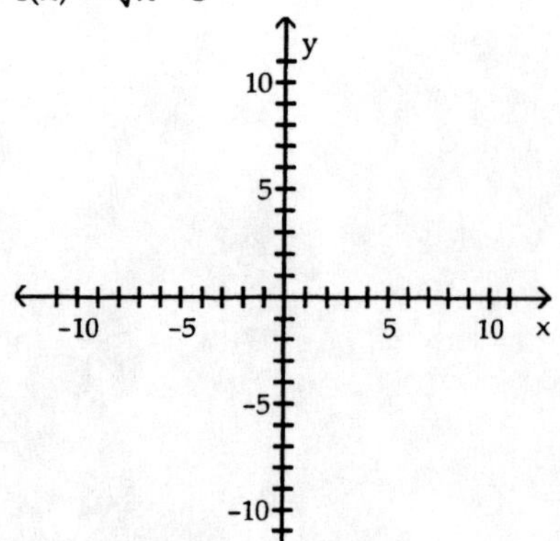

A)

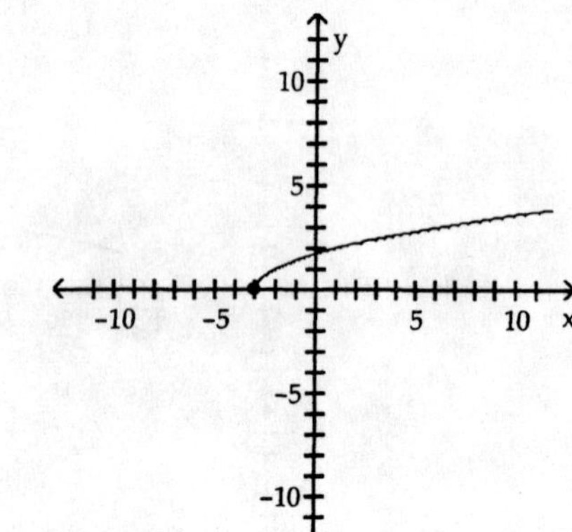

B)

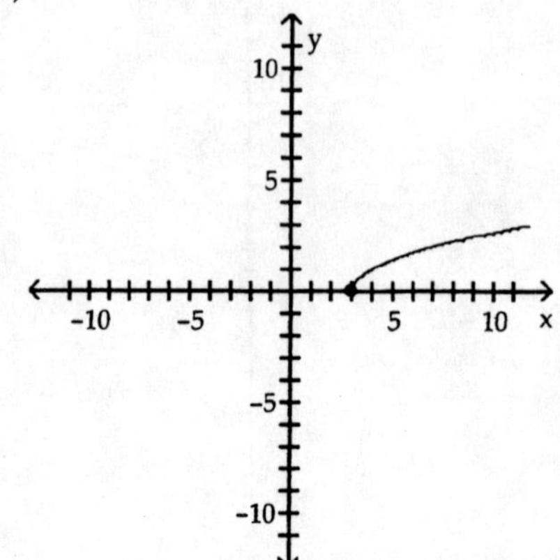

C)

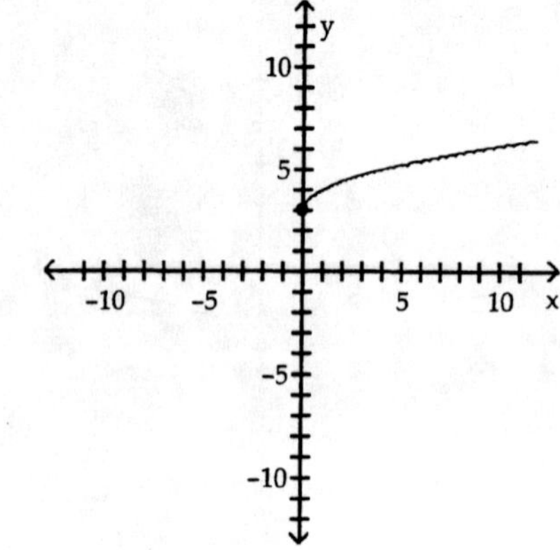

D)

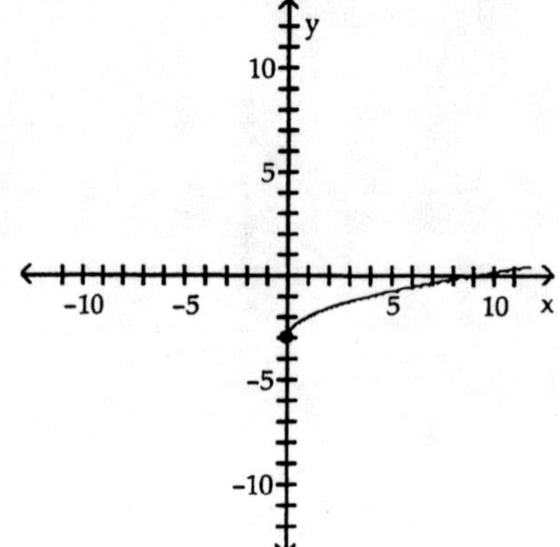

9) $f(x) = \sqrt{x + 6} - 7$

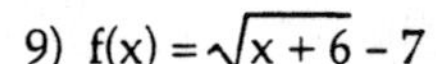

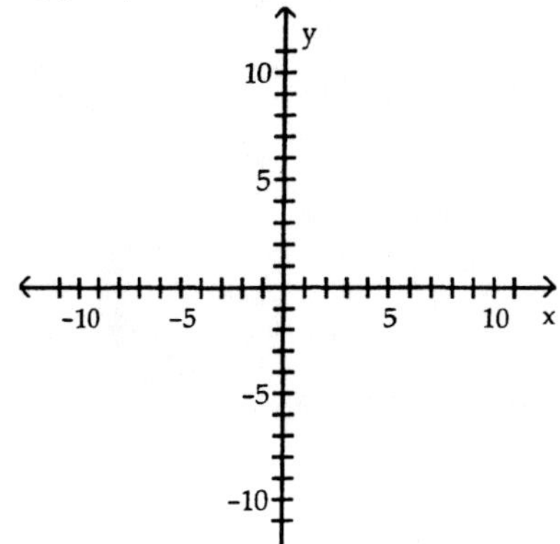

A)

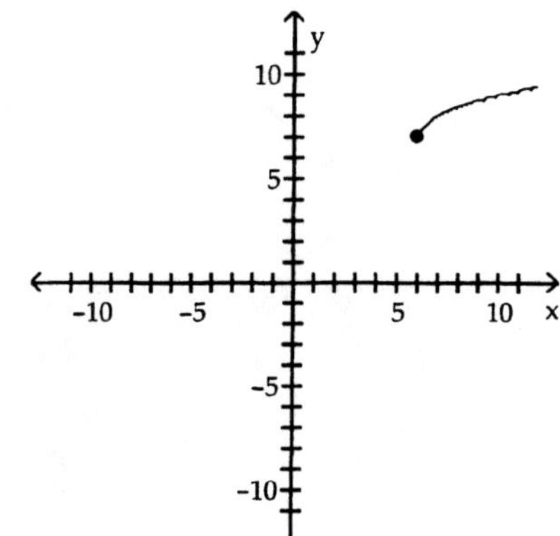

B)

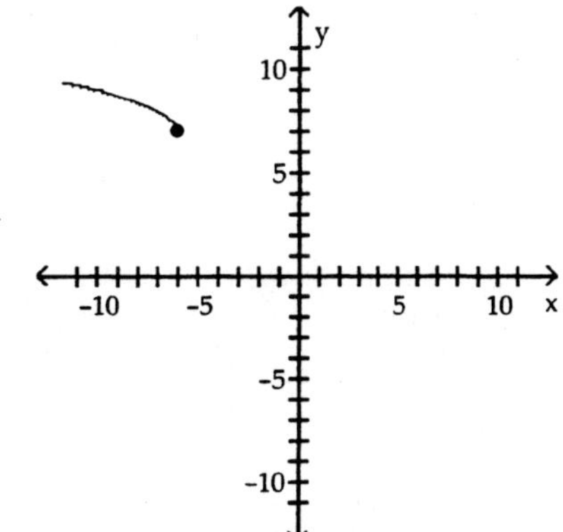

C)

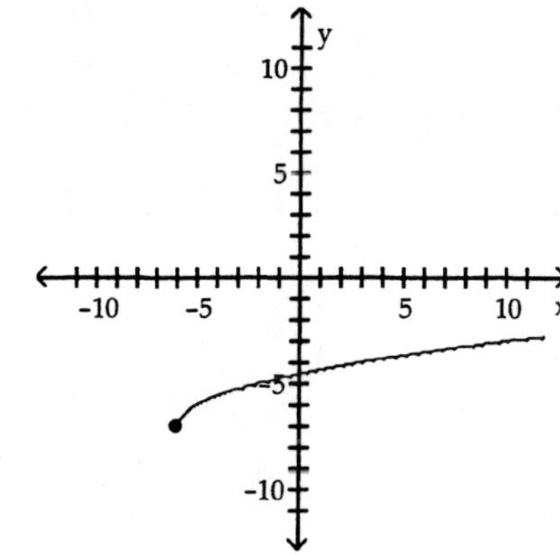

D)

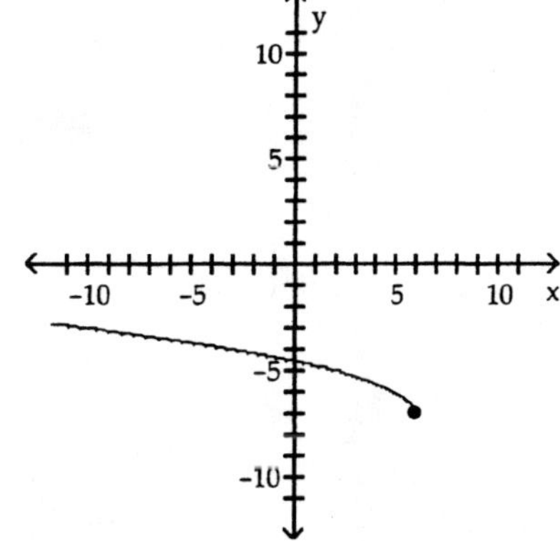

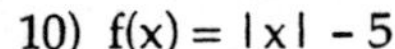
10) $f(x) = |x| - 5$

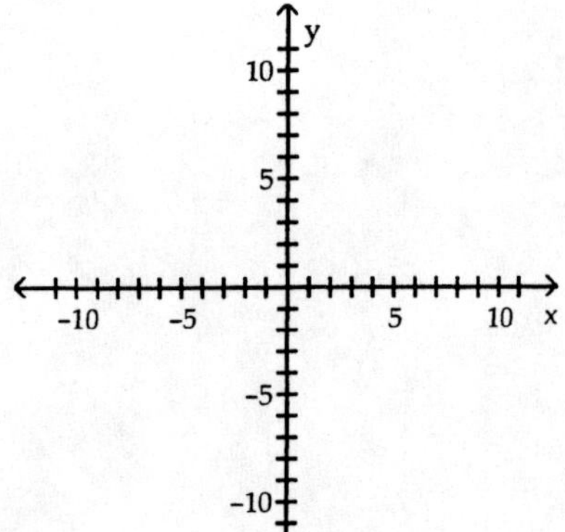

A)

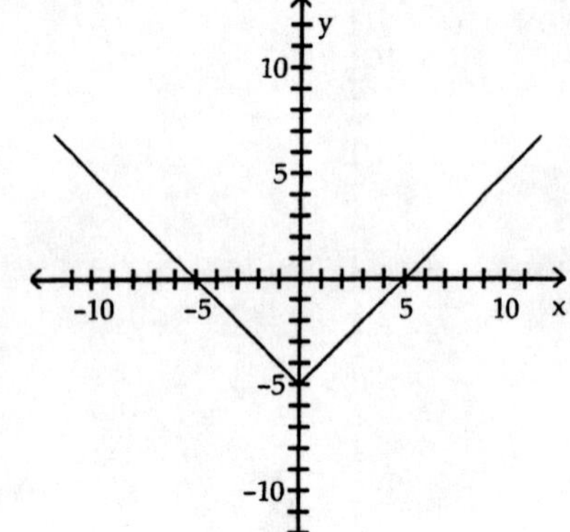

B)

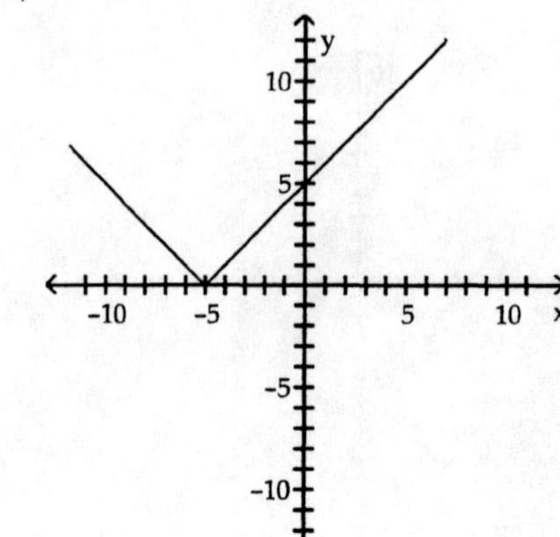

C)

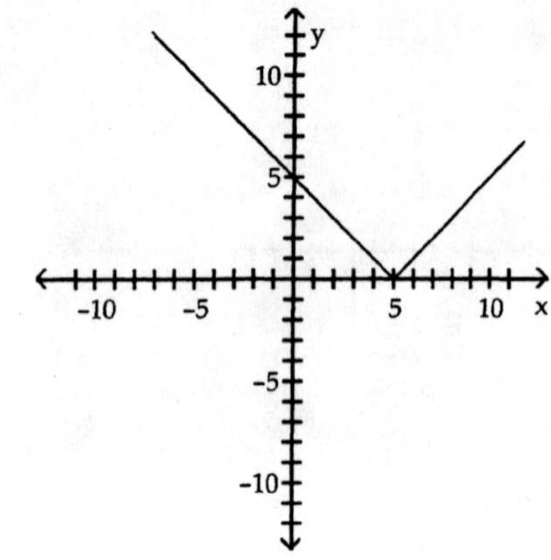

D)

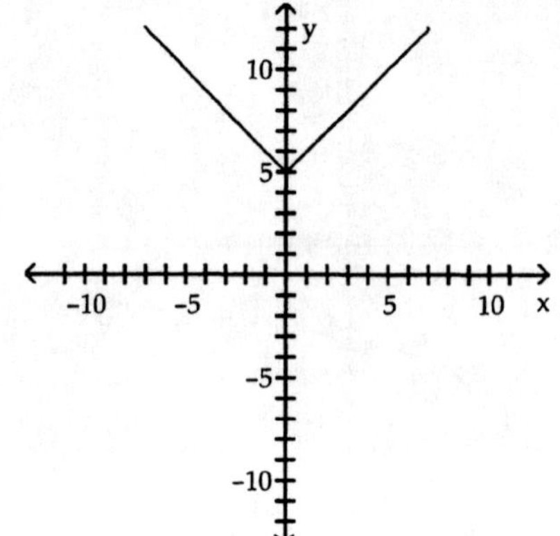

11) $f(x) = |x + 1|$

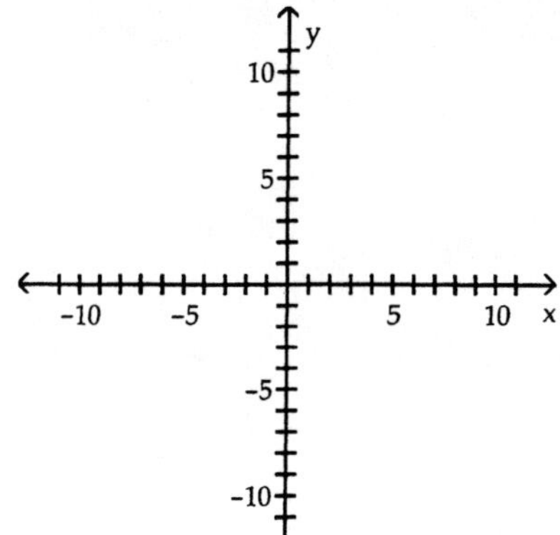

A)

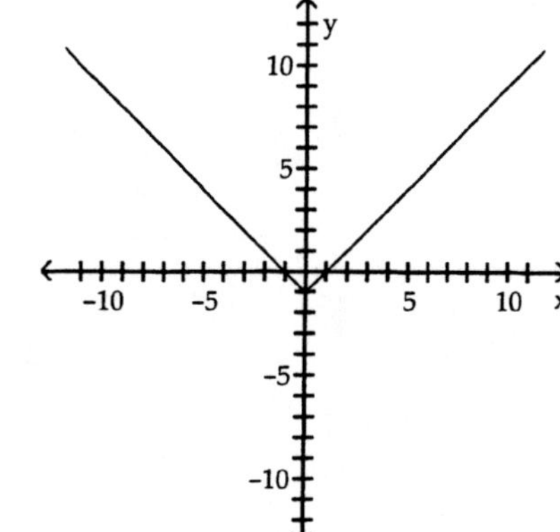

B)

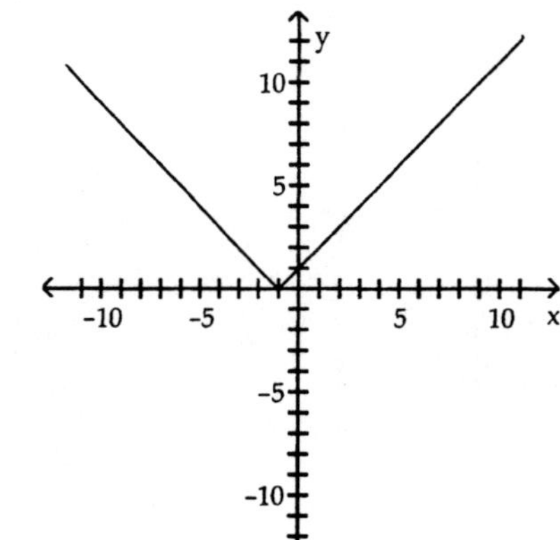

C)

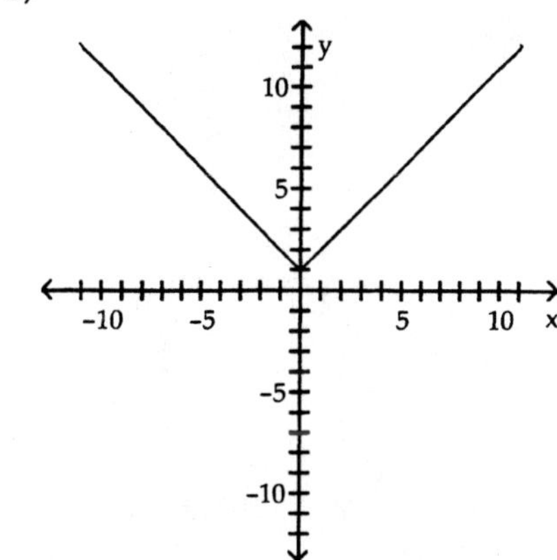

D)

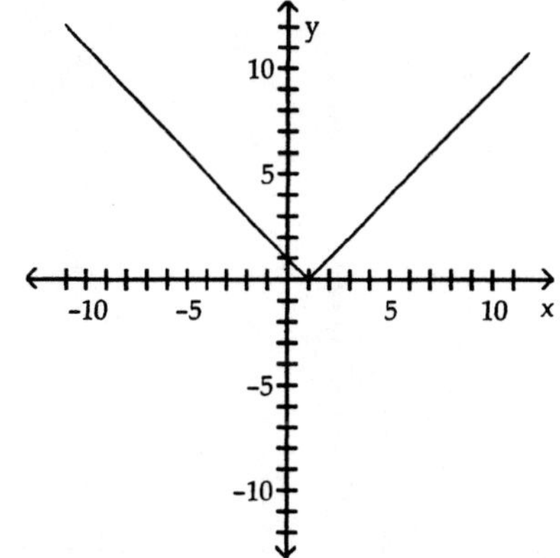

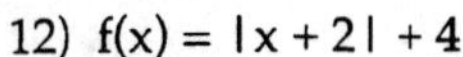

12) $f(x) = |x + 2| + 4$

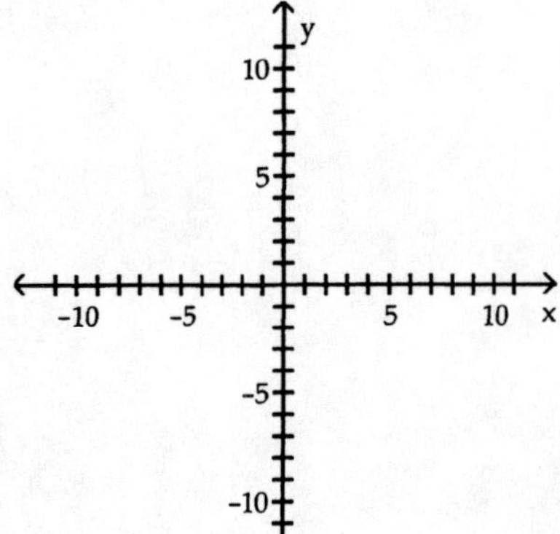

A)

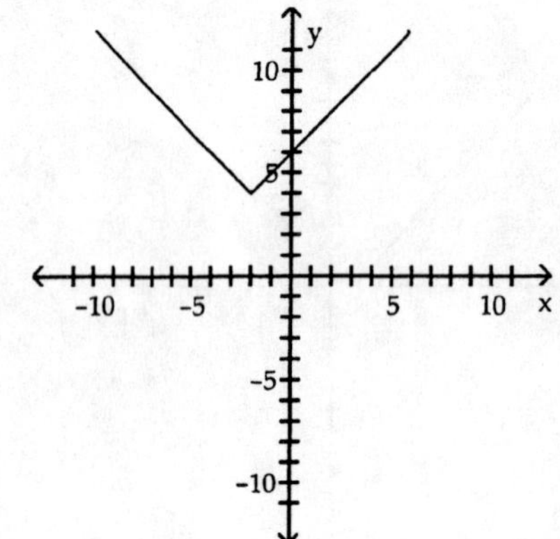

B)

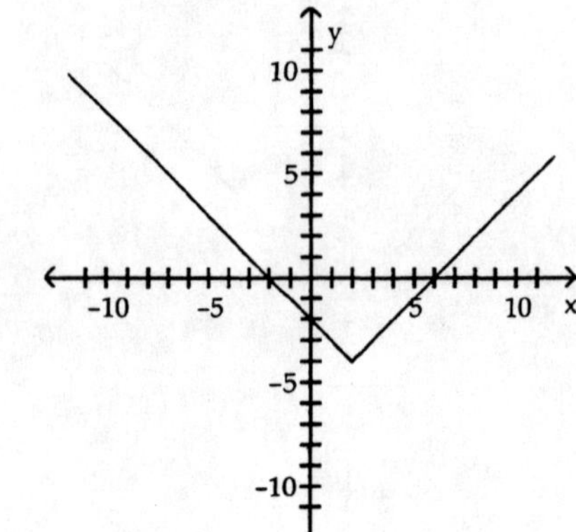

C)

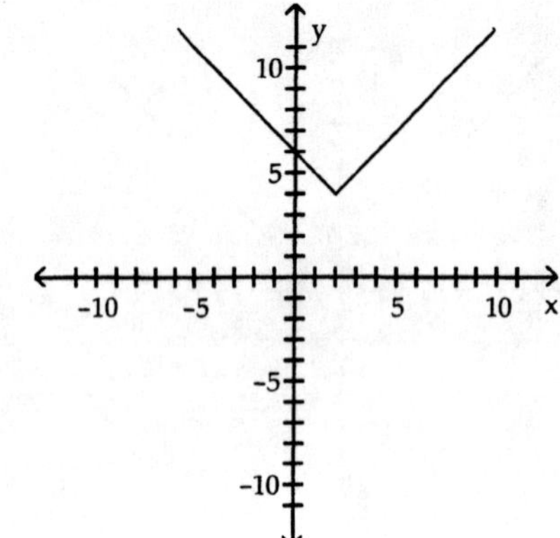

D)

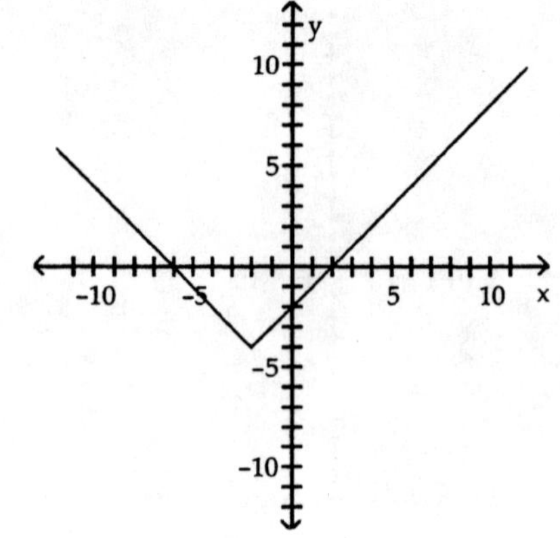

13) $f(x) = \frac{1}{x} - 4$

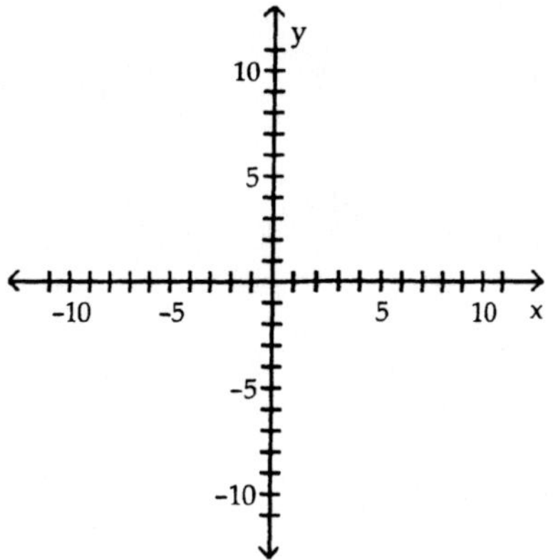

A)

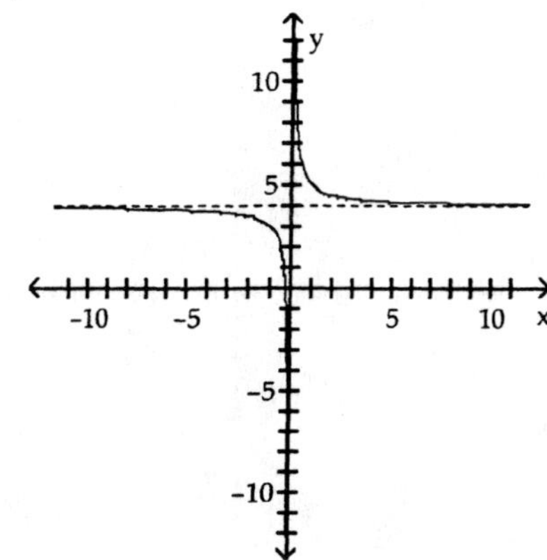

B)

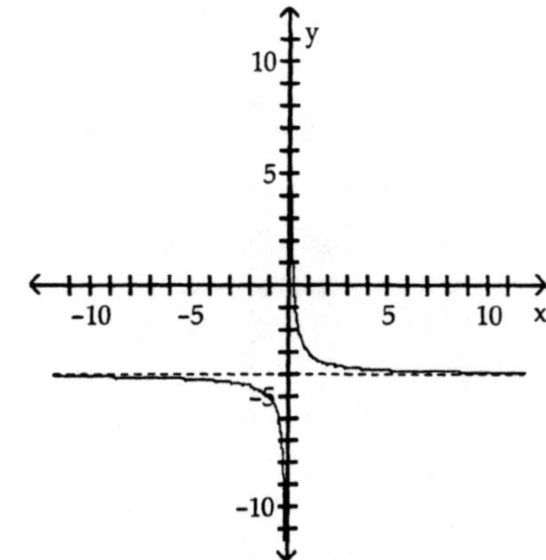

C)

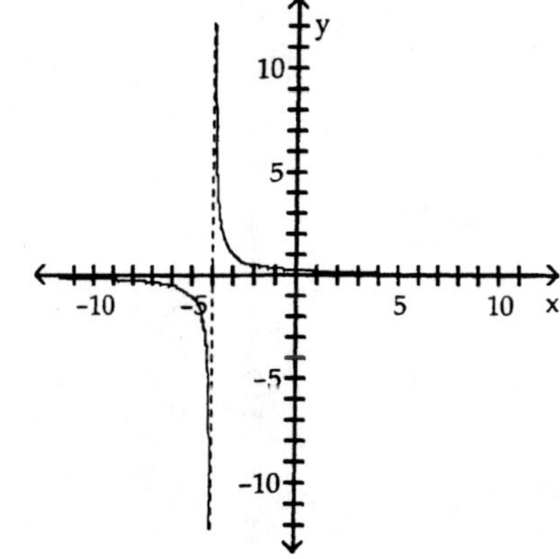

D)

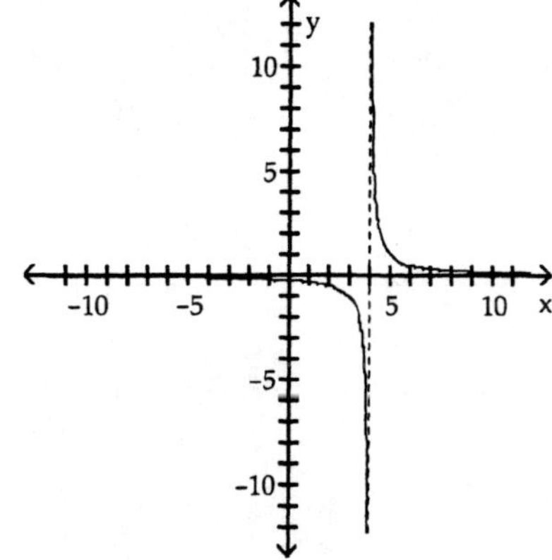

14) $f(x) = \frac{1}{x - 5}$

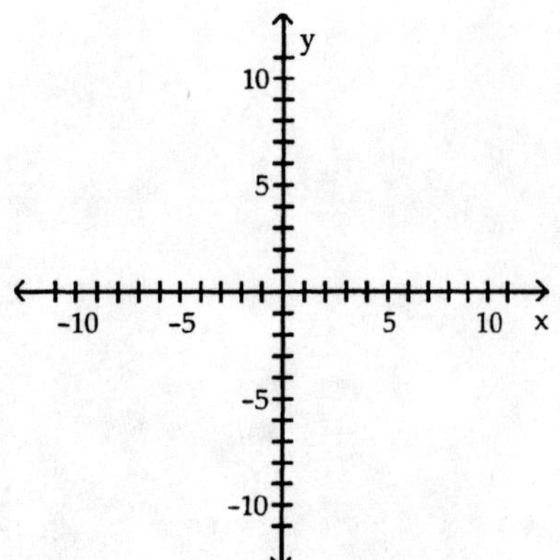

A)

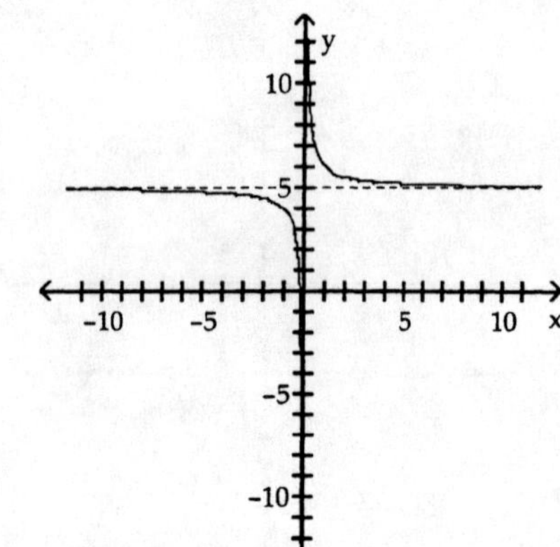

B)

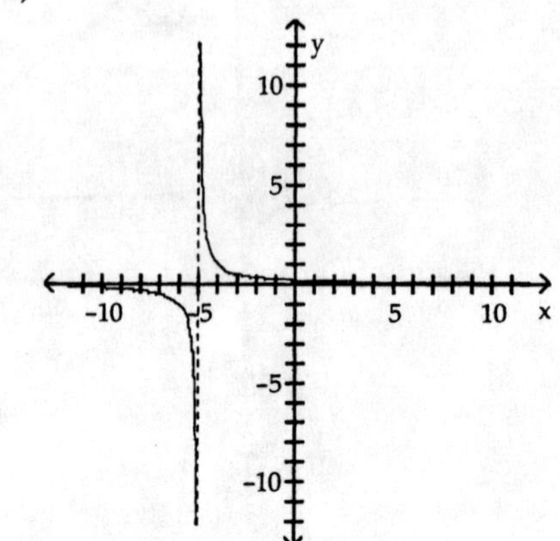

C)

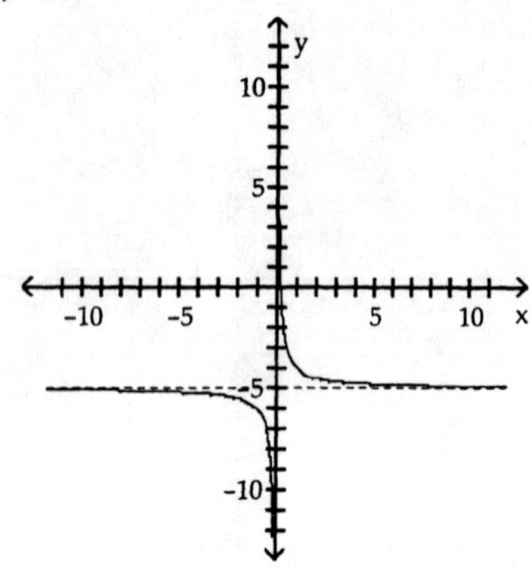

D)

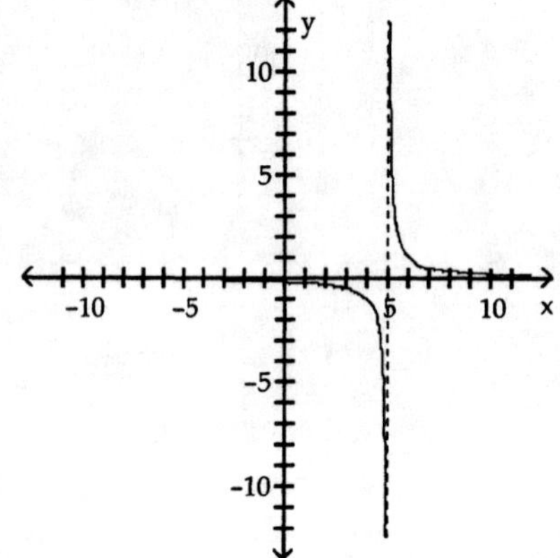

15) $f(x) = \frac{1}{x + 6} - 6$

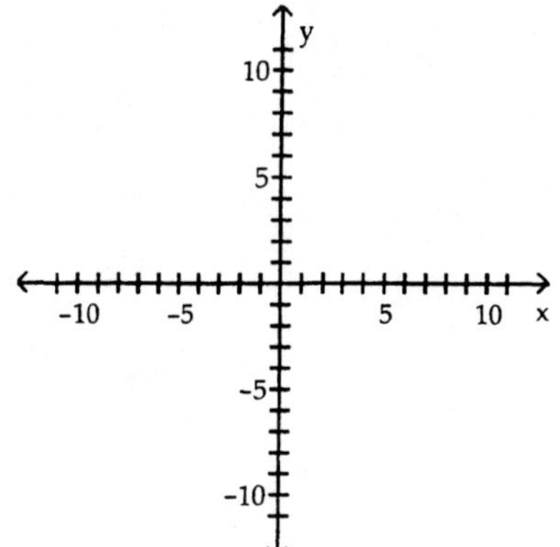

A)

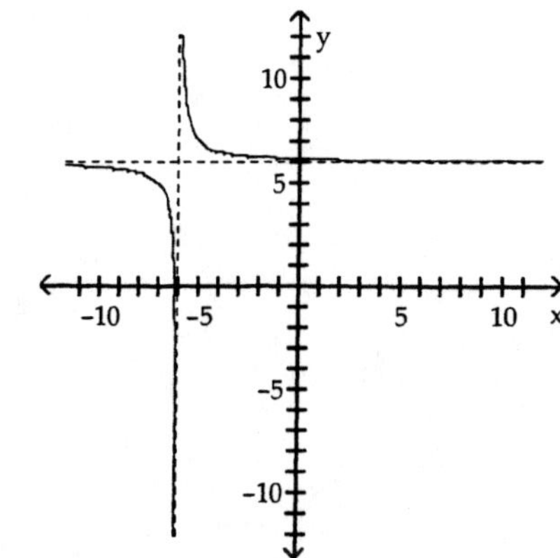

B)

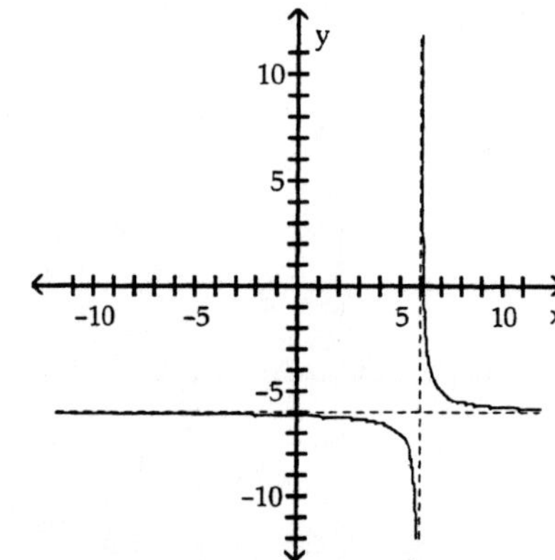

C)

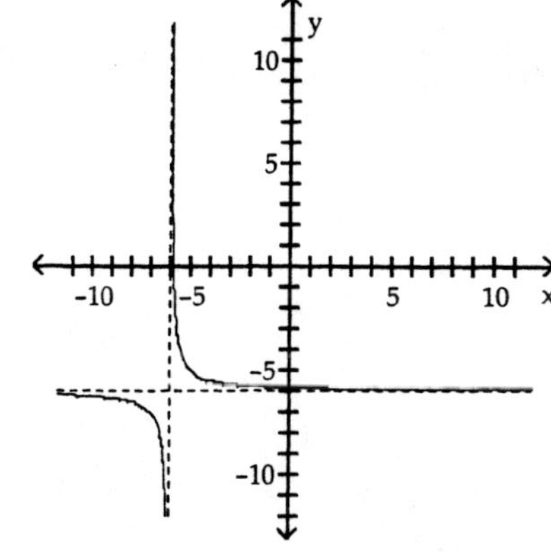

D)

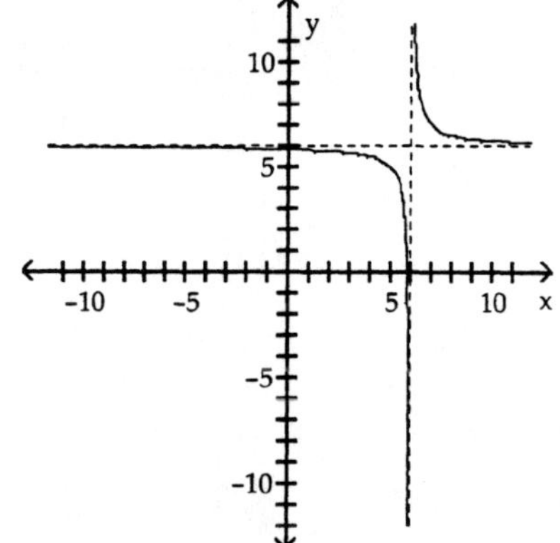

Match the correct function to the graph.

16)

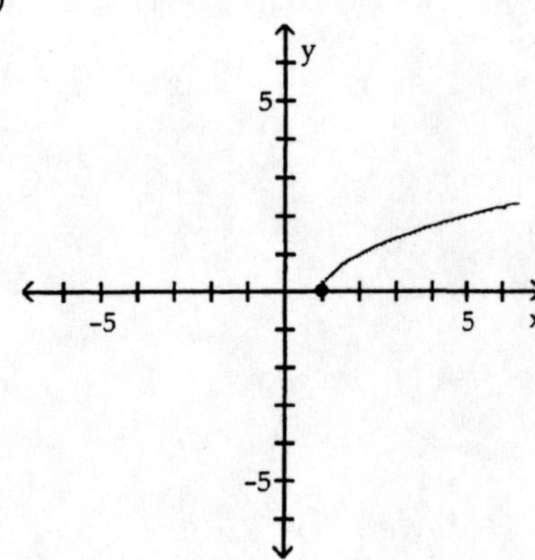

A) $y = \sqrt{x + 1}$ B) $y = \sqrt{x}$ C) $y = x - 1$ D) $y = \sqrt{x - 1}$

17)

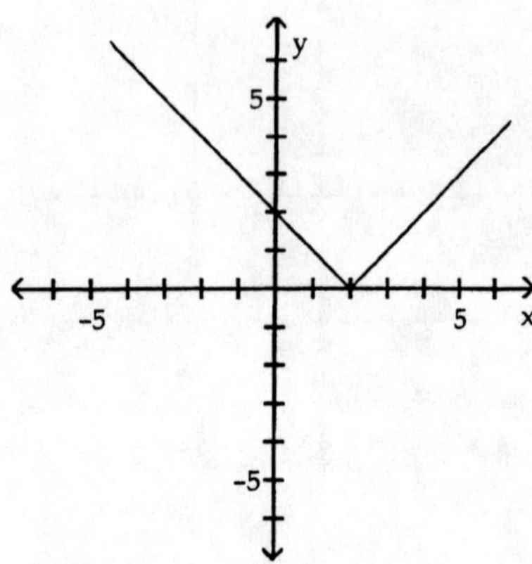

A) $y = x - 2$ B) $y = |2 - x|$ C) $y = |x + 2|$ D) $y = |1 - x|$

2 Graph Functions Using Compressions and Stretches

Graph the function by starting with the graph of the basic function and then using the techniques of shifting, compressing, stretching, and/or reflecting.

1) $f(x) = 4x^2$

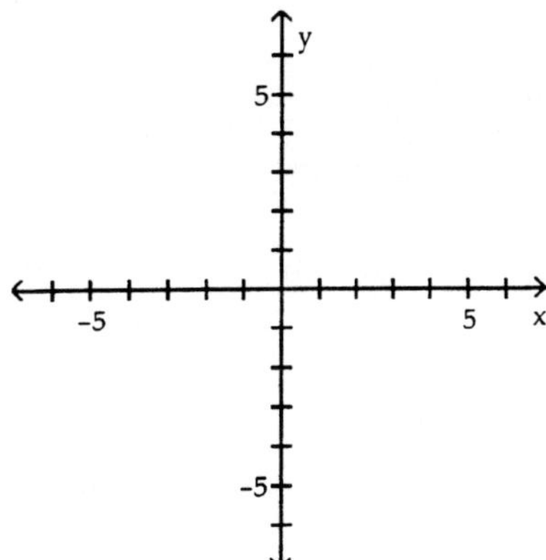

A)

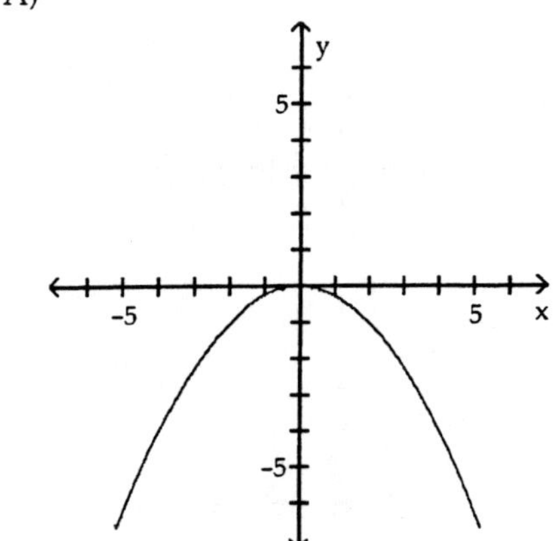

B)

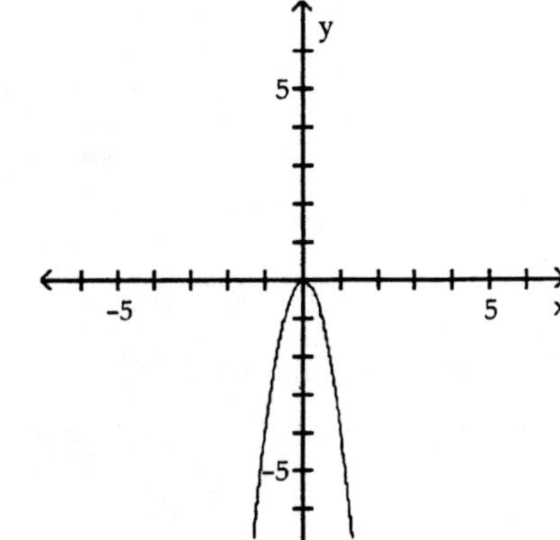

C)

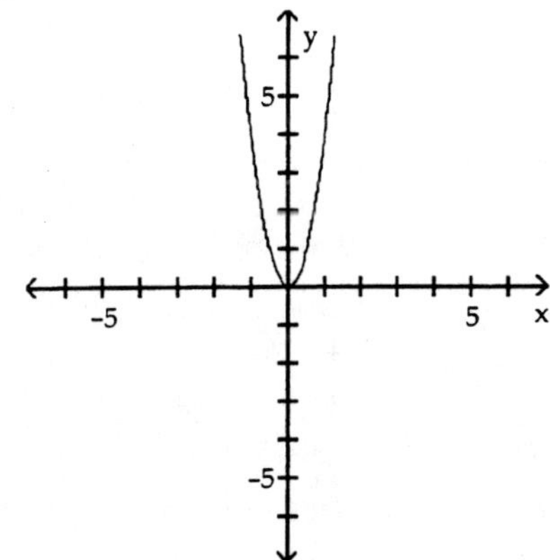

D)

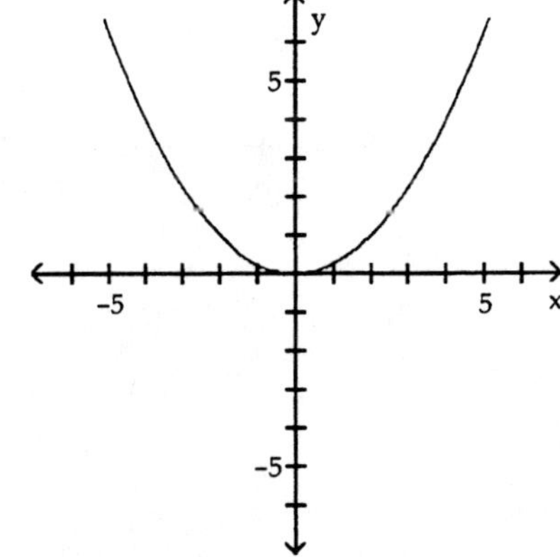

2) $f(x) = \frac{1}{5}x^2$

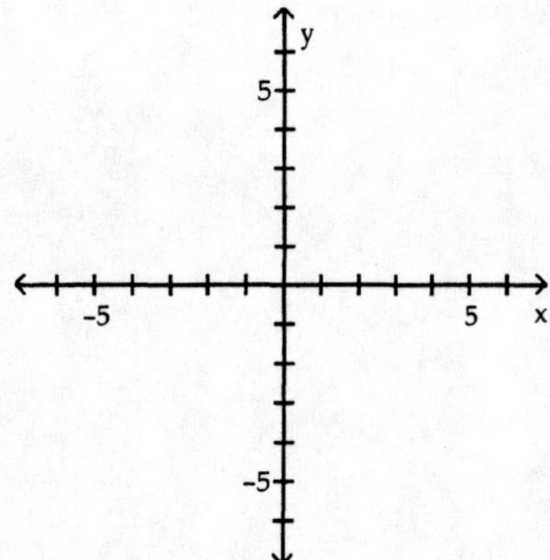

A)

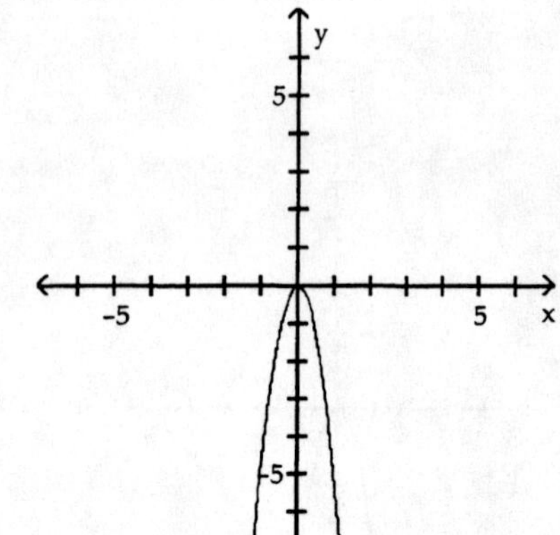

B)

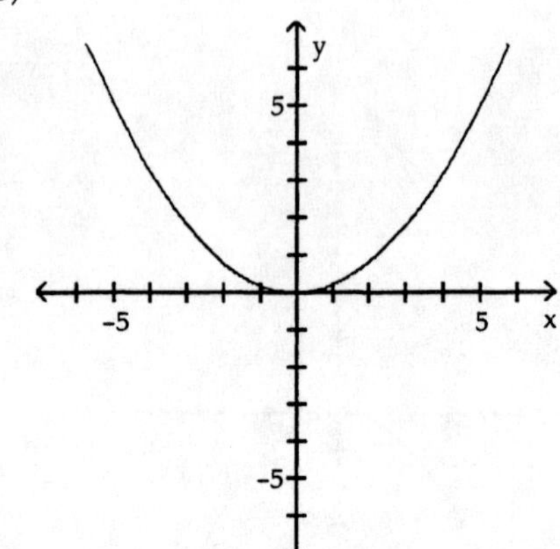

C)

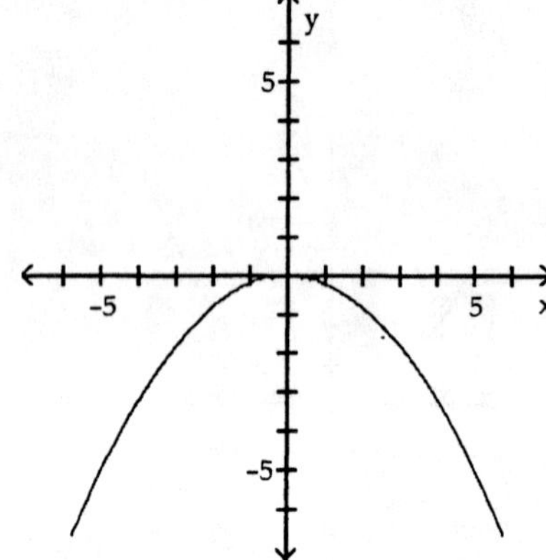

D)

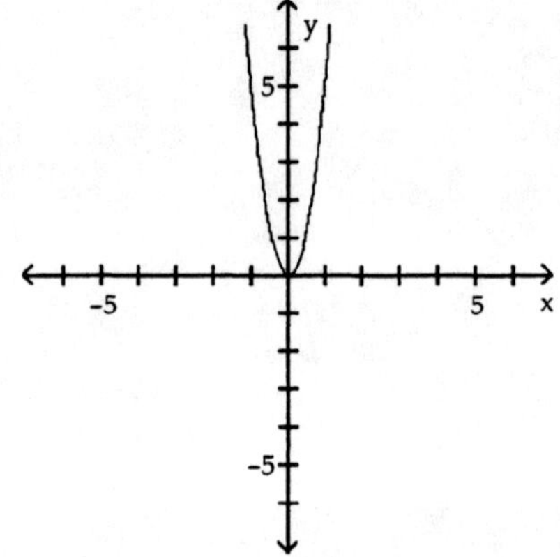

3) $f(x) = 7x^3$

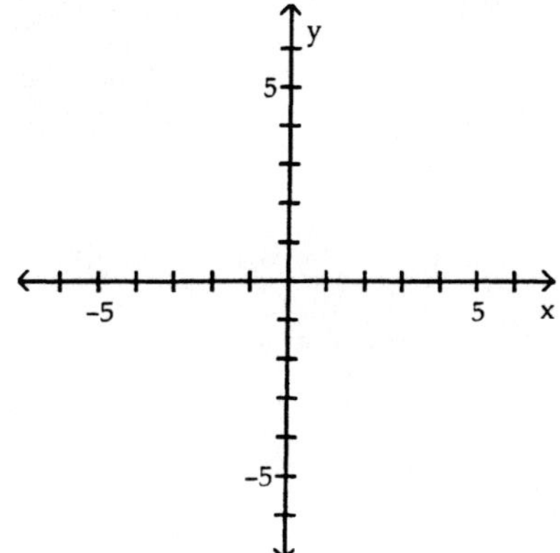

A)

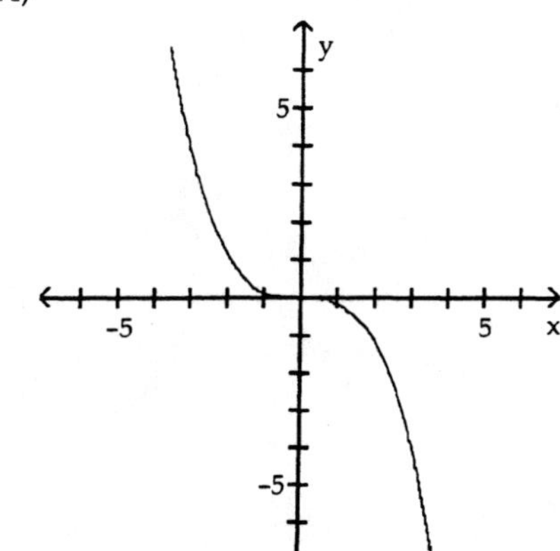

B)

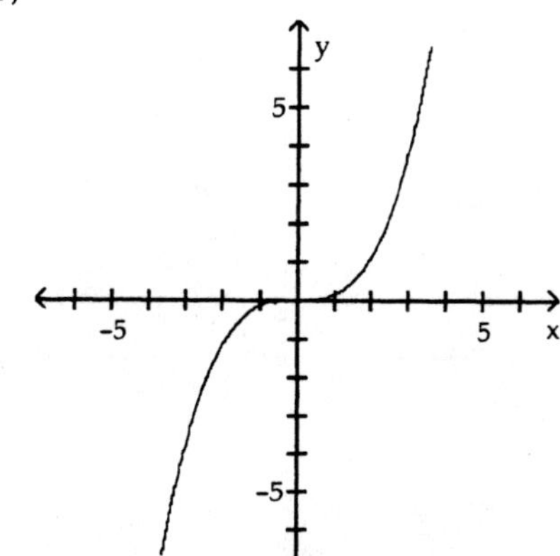

C)

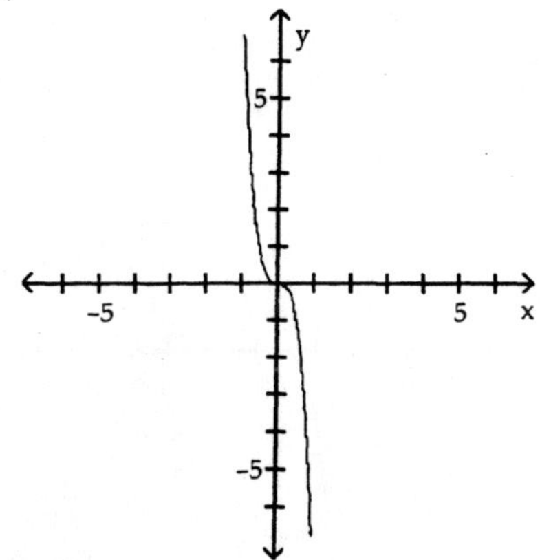

D)

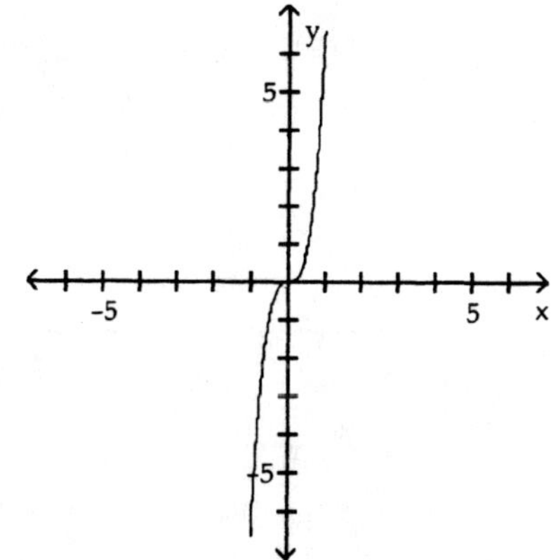

4) $f(x) = \frac{1}{2}x^3$

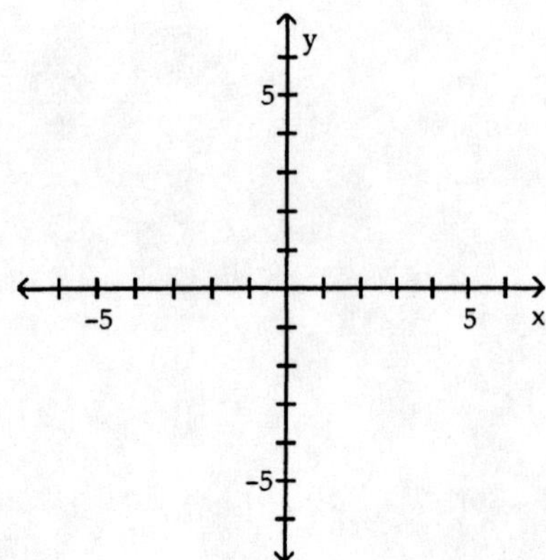

A)

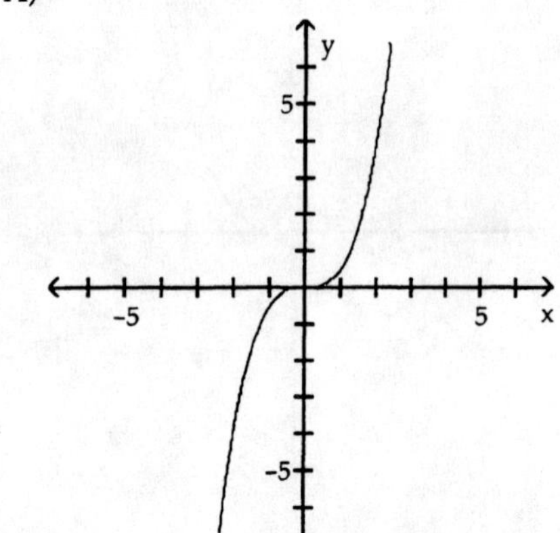

B)

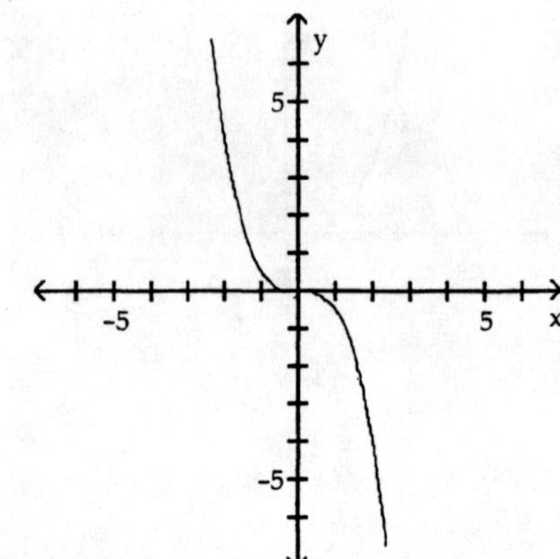

C)

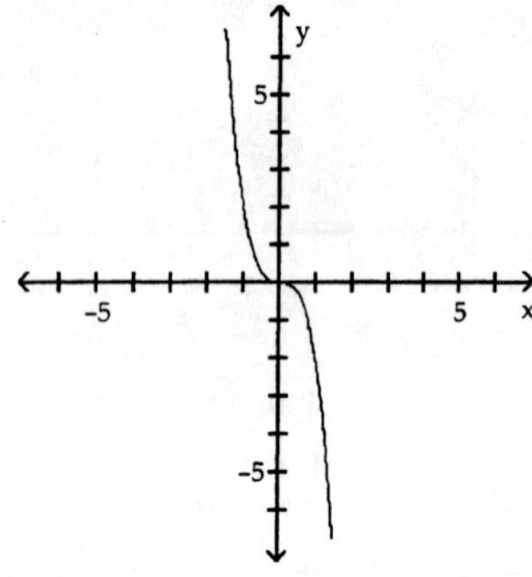

D)

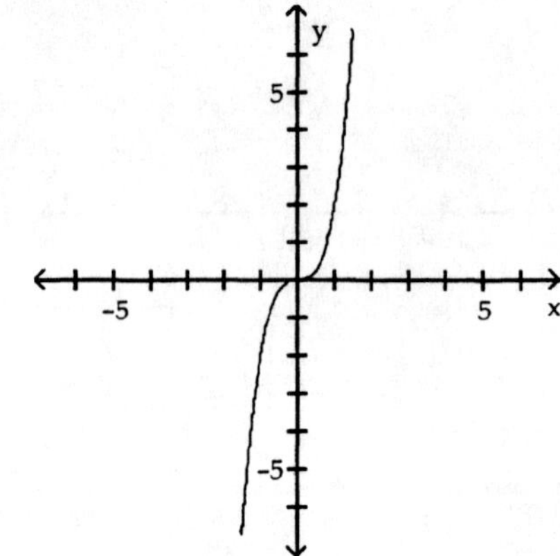

5) $f(x) = 4\sqrt{x}$

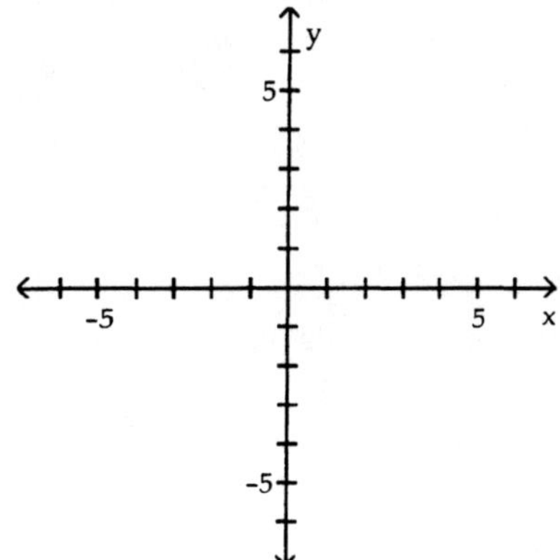

A)

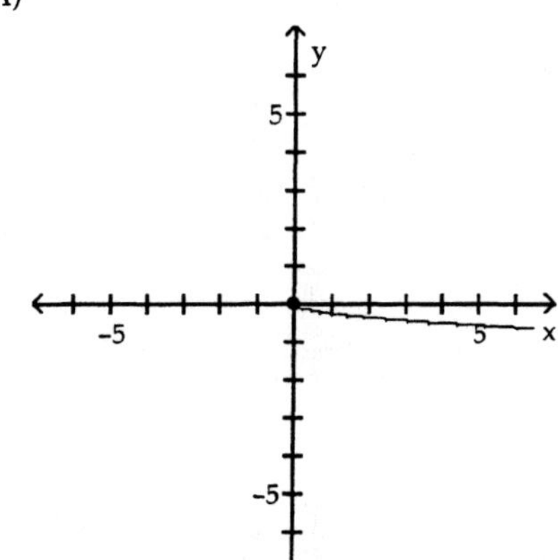

B)

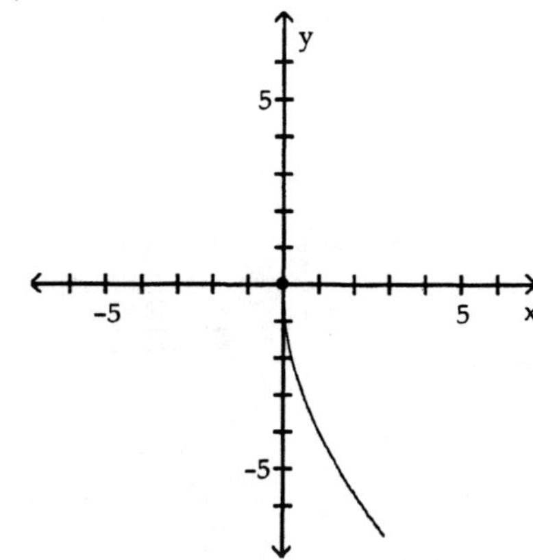

C)

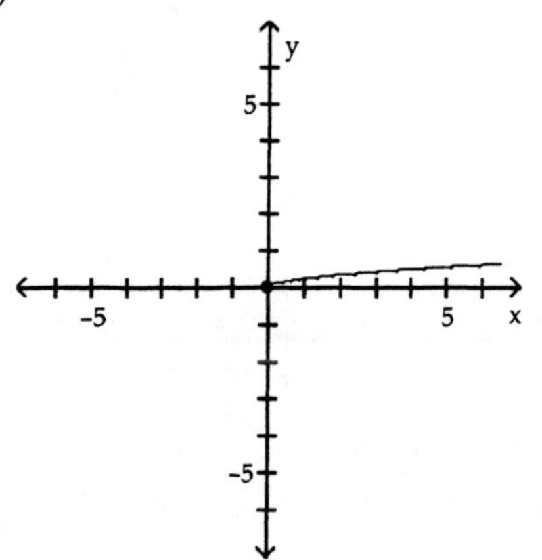

D)

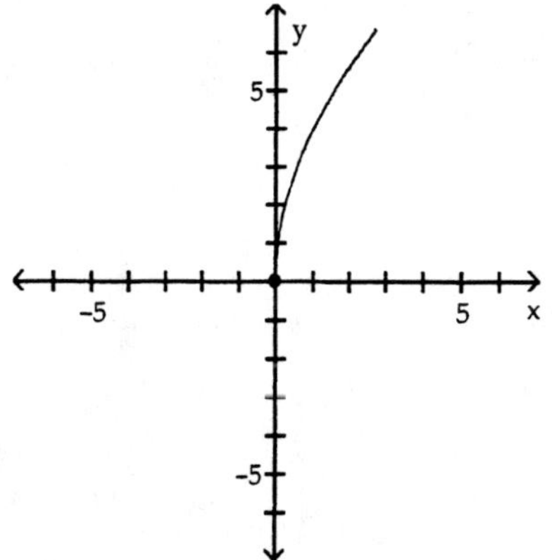

6) $f(x) = \frac{1}{4}\sqrt{x}$

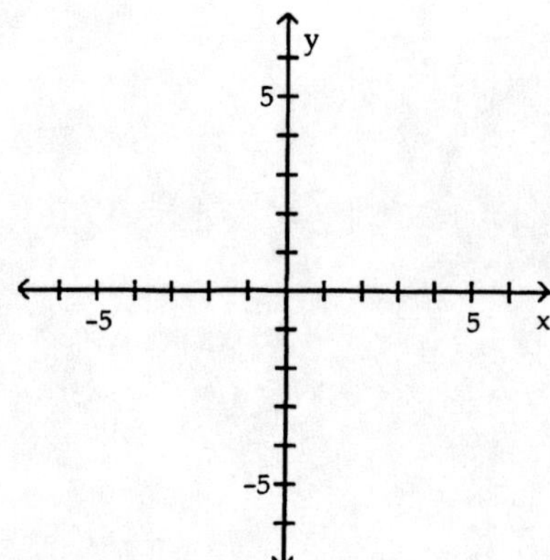

A)

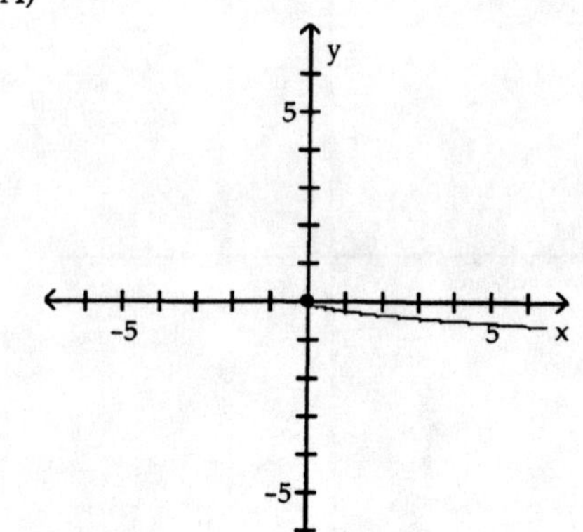

B)

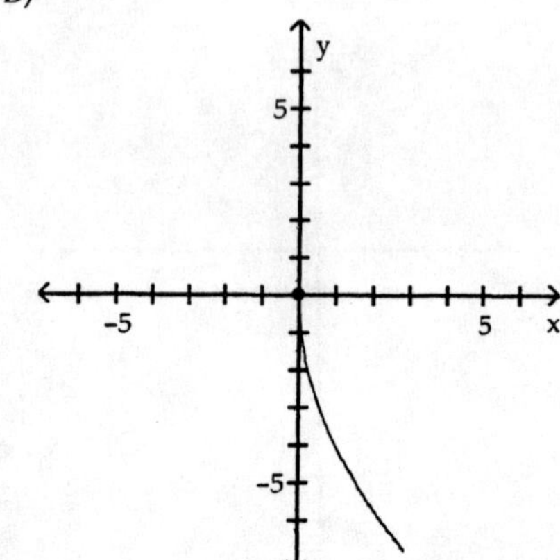

C)

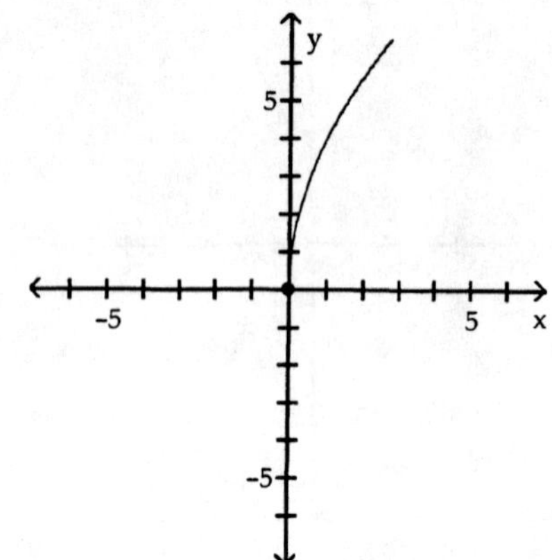

D)

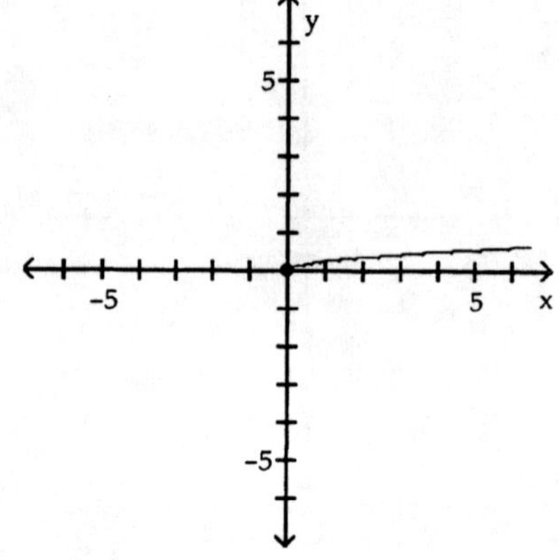

7) $f(x) = 3|x|$

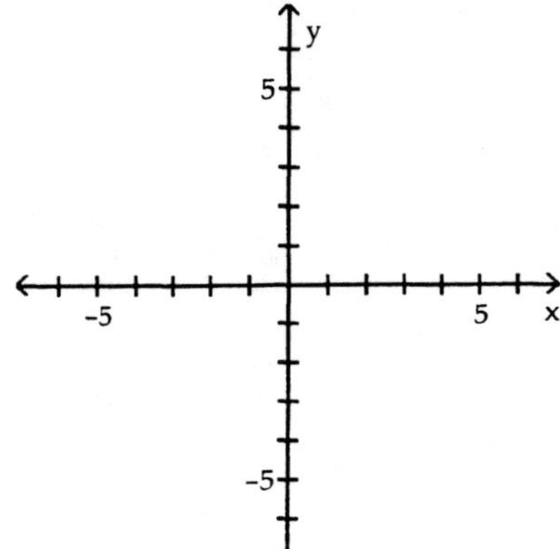

A)

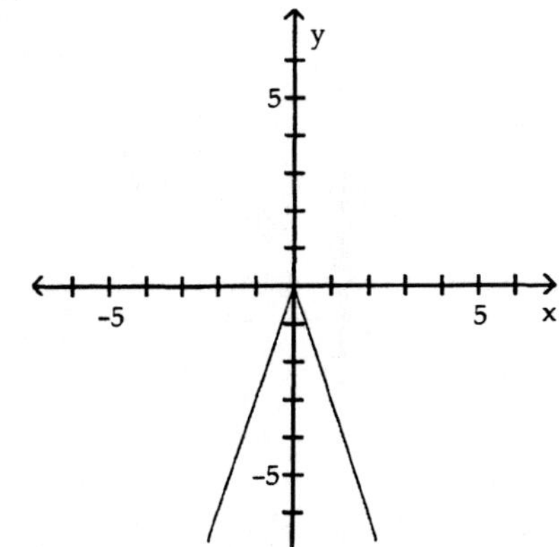

B)

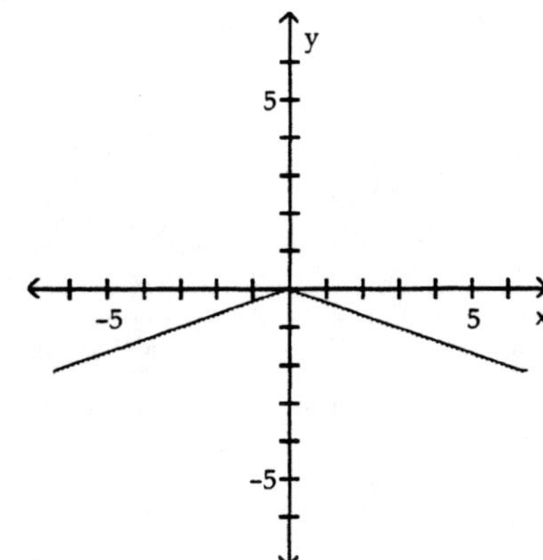

C)

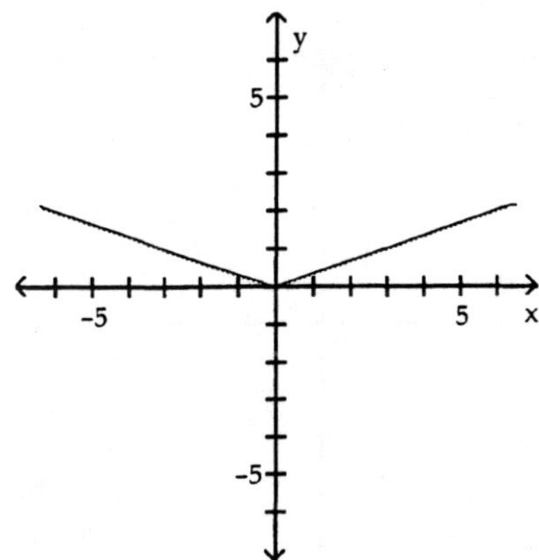

D)

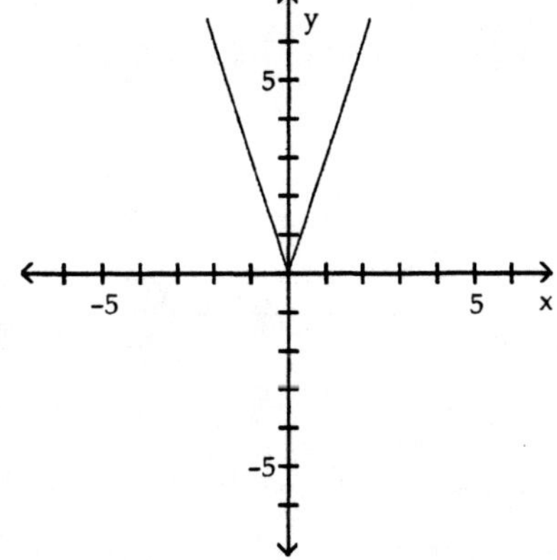

8) $f(x) = \frac{1}{6}|x|$

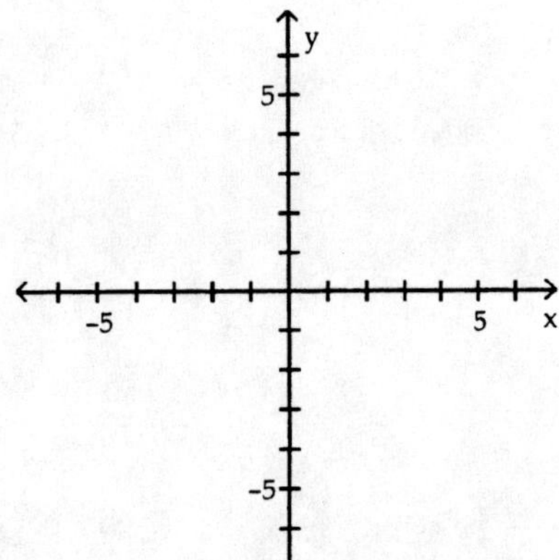

A)

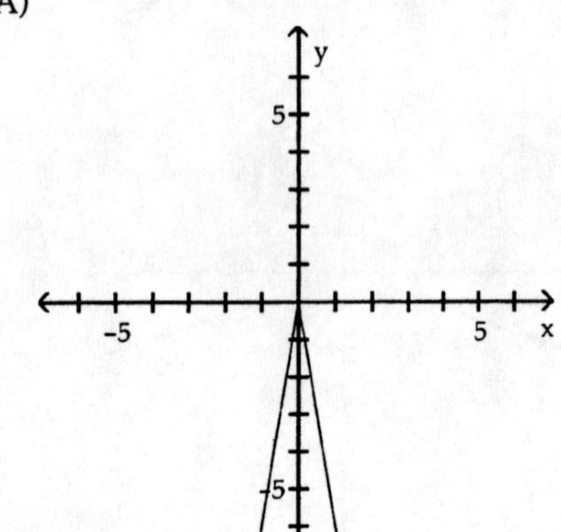

B)

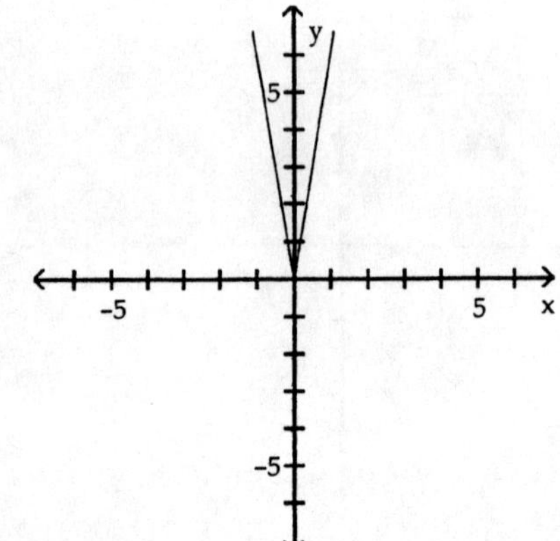

C)

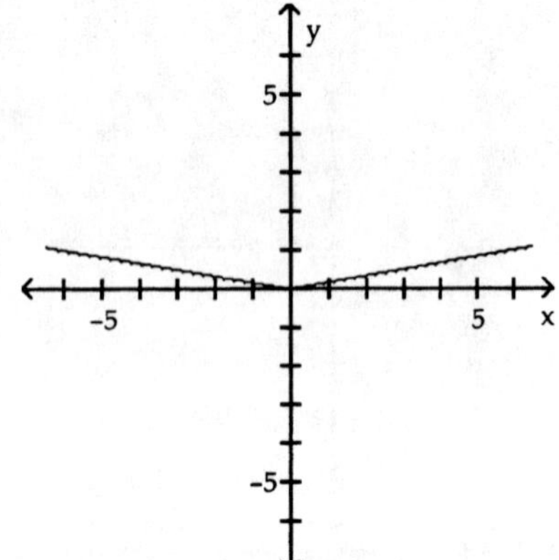

D)

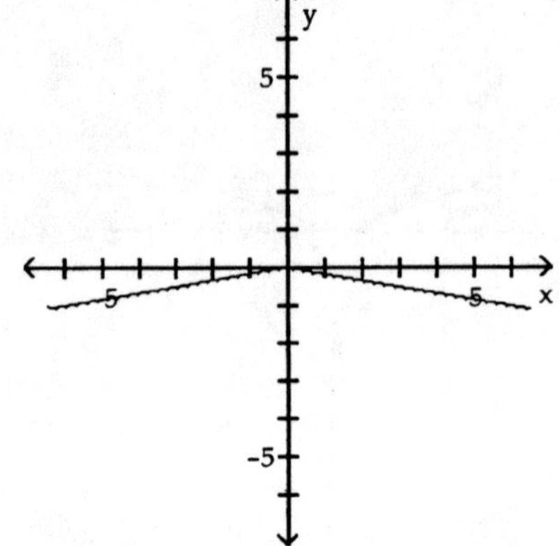

9) $f(x) = \frac{2}{x}$

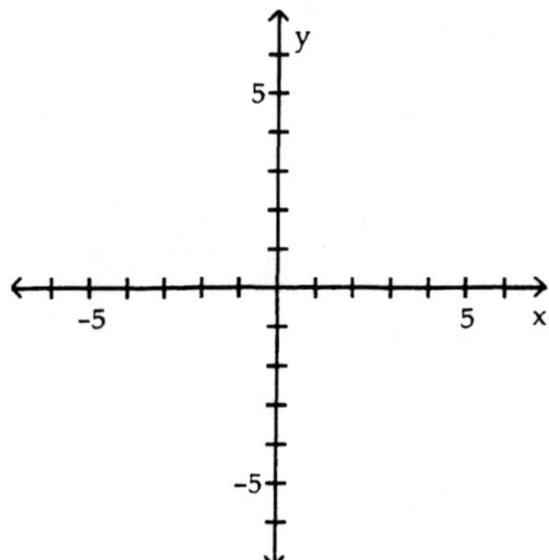

A)

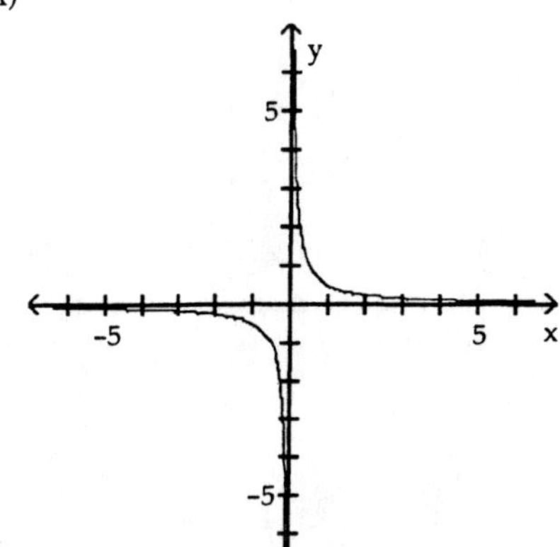

B)

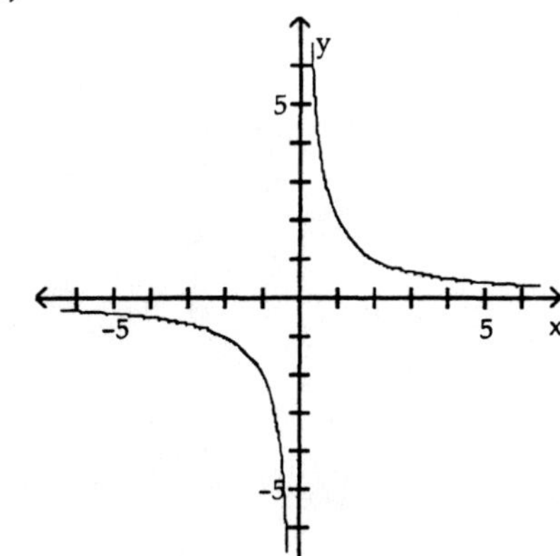

C)

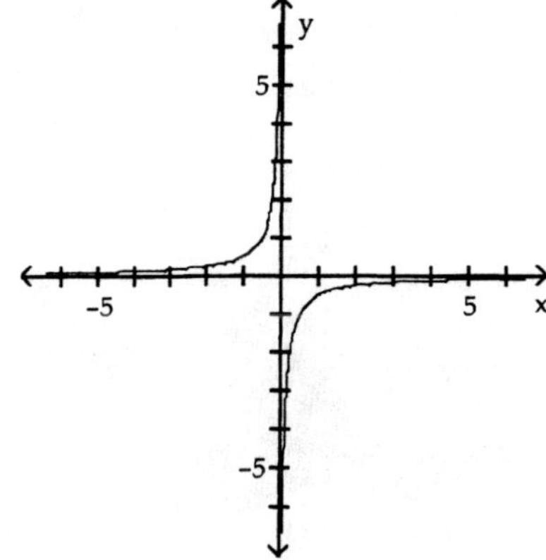

D)

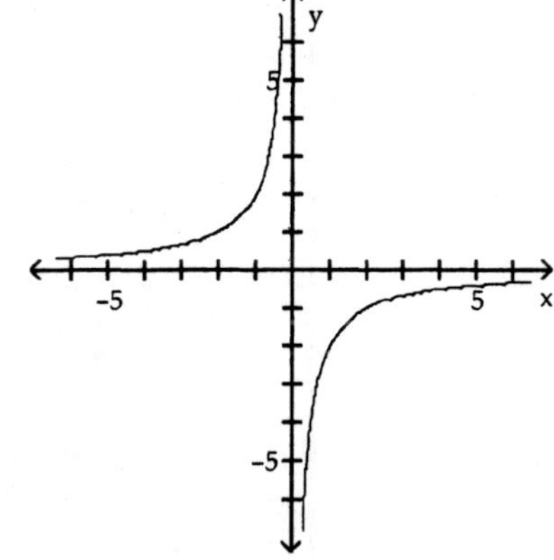

10) $f(x) = \frac{1}{5x}$

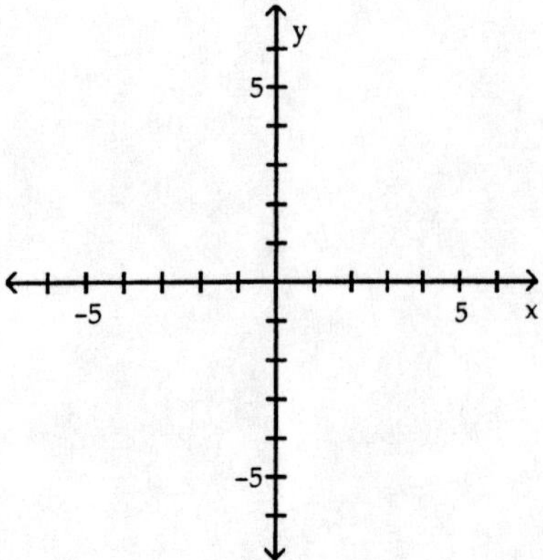

A)

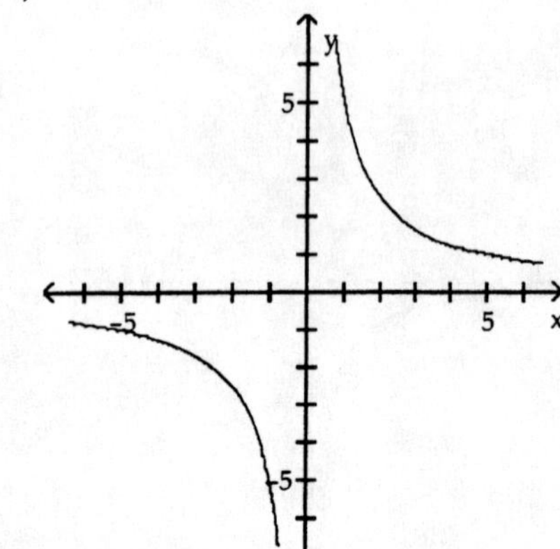

B)

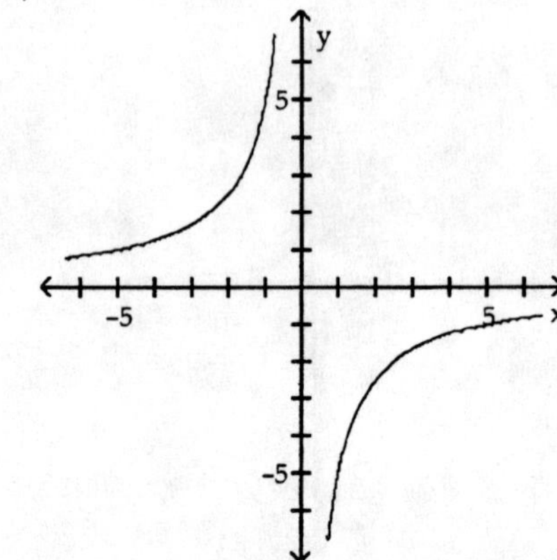

C)

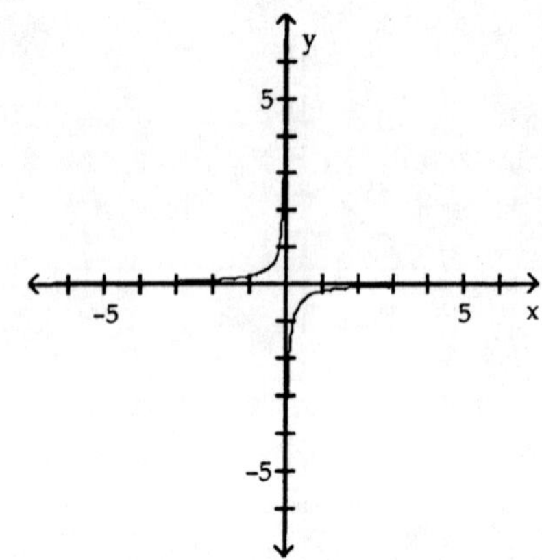

D)

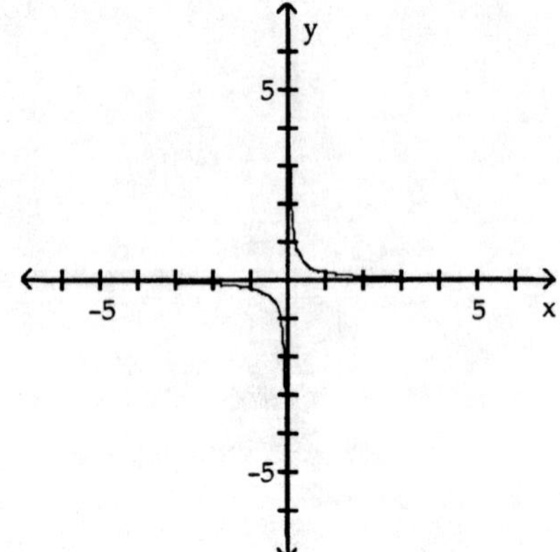

3 Graph Functions Using Reflections about the x–Axis or y–Axis

Graph the function by starting with the graph of the basic function and then using the techniques of shifting, compressing, stretching, and/or reflecting.

1) $f(x) = -x^2$

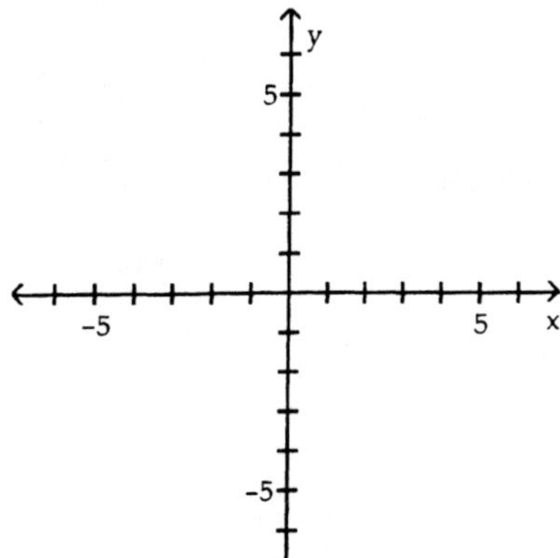

A)

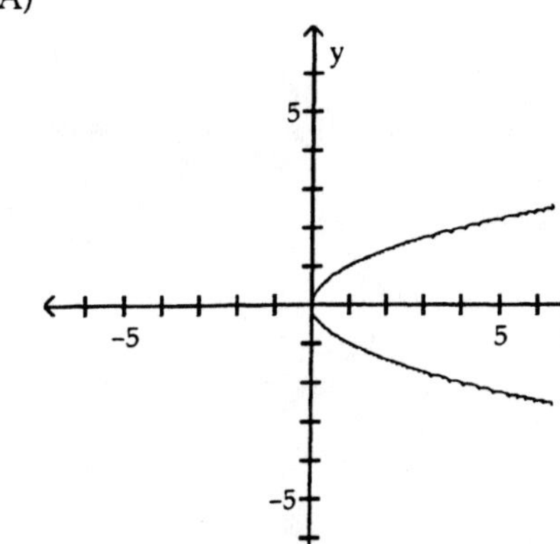

B)

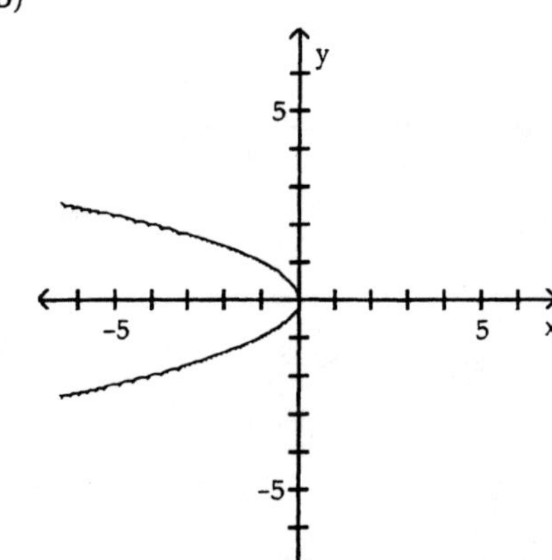

C)

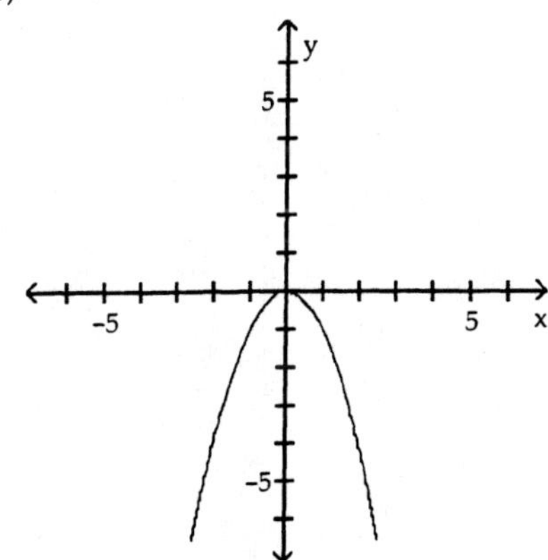

D)

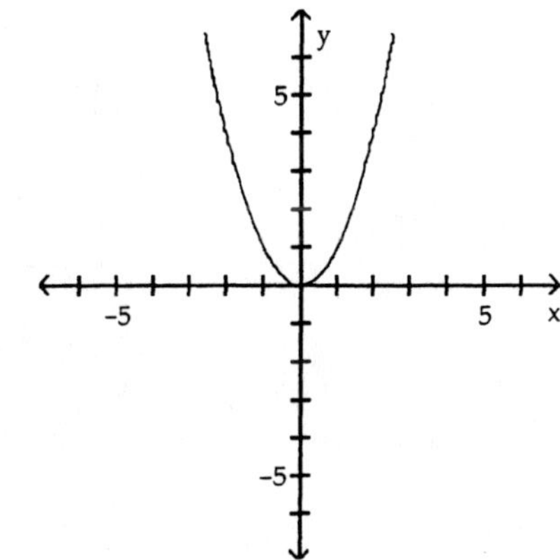

2) $f(x) = (-x)^2$

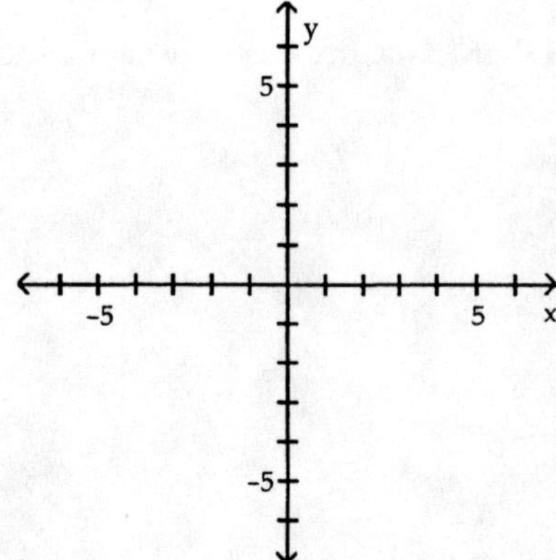

A)

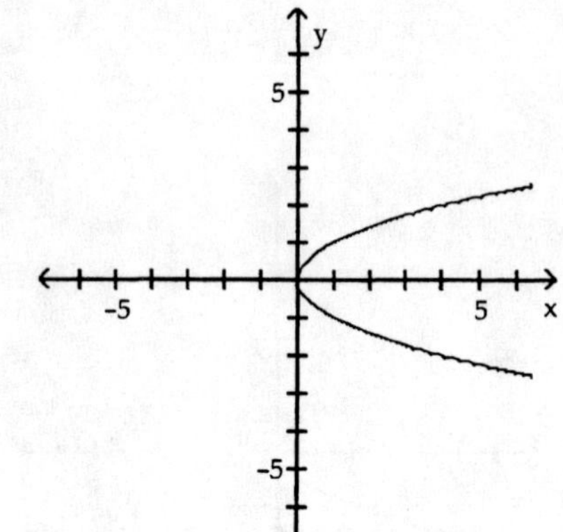

B)

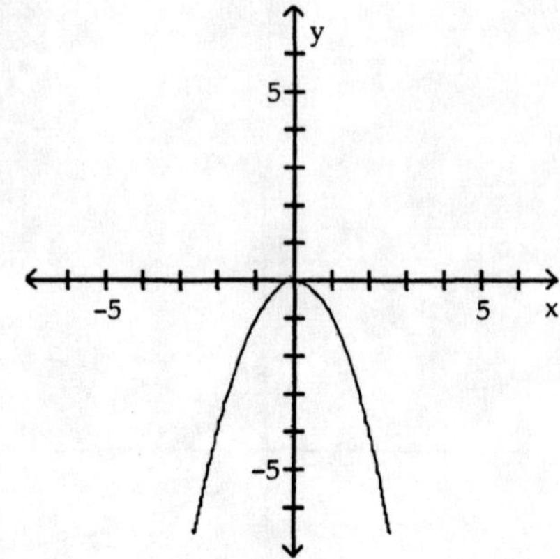

C)

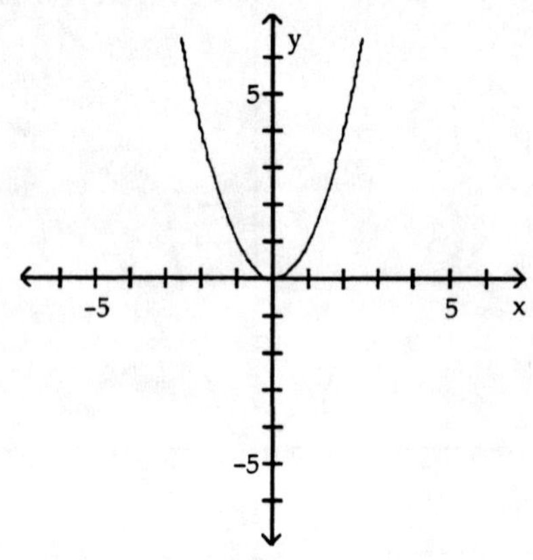

D)

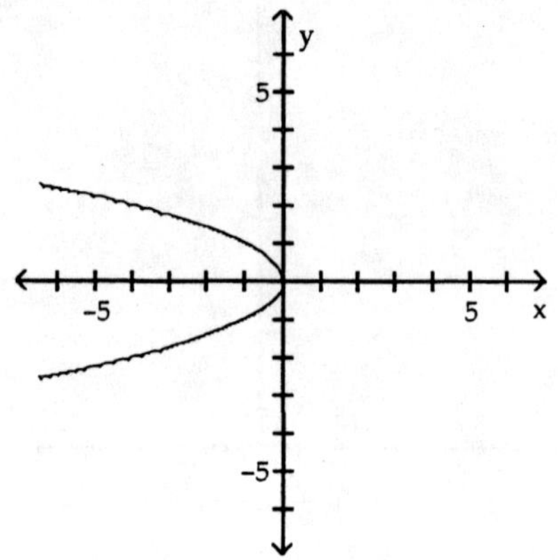

3) $f(x) = -x^3$

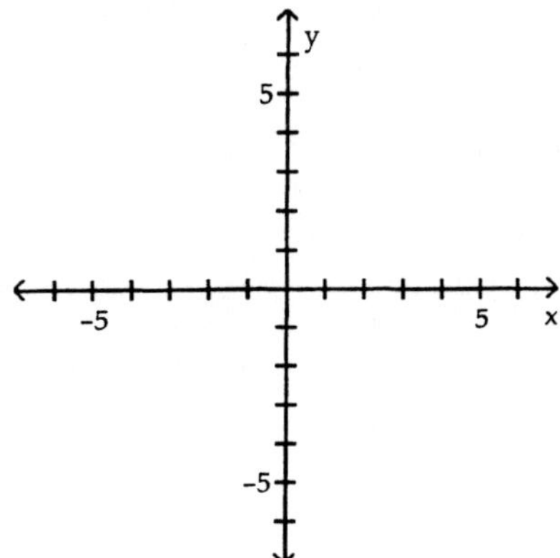

A)

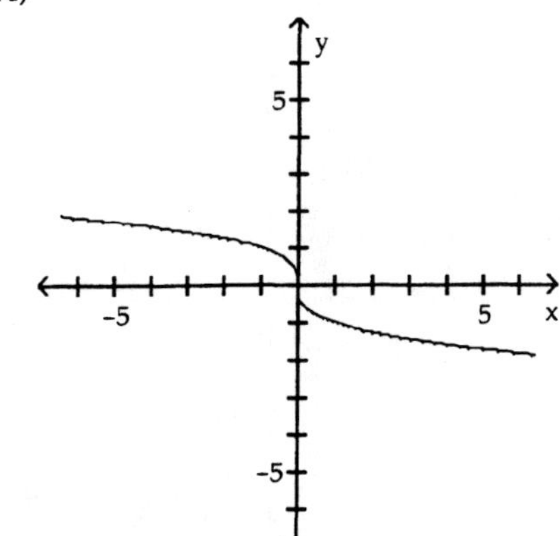

B)

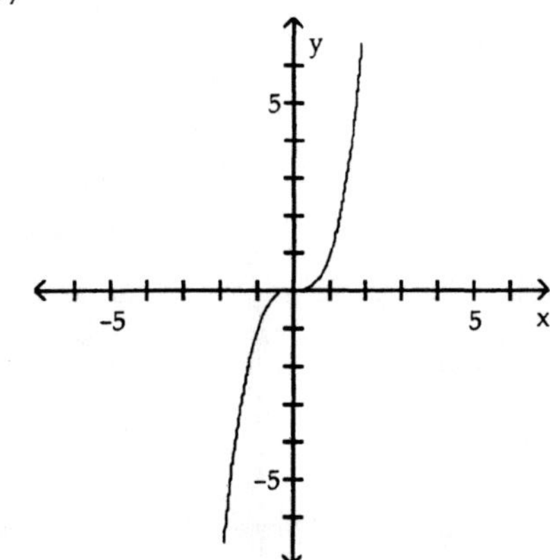

C)

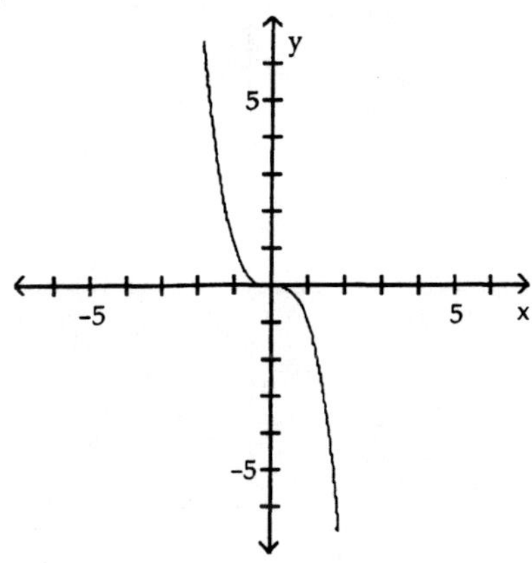

D)

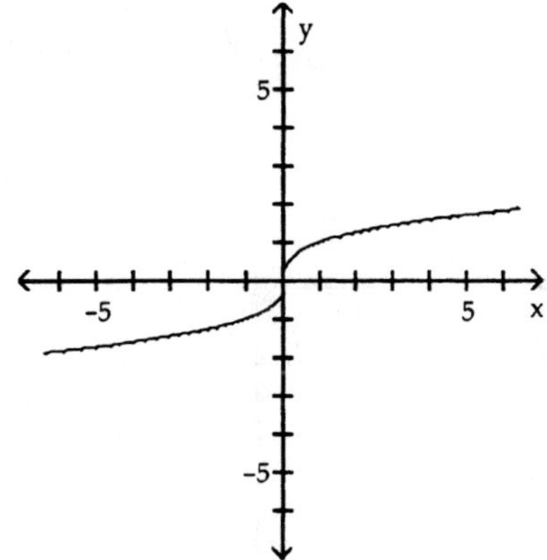

4) $f(x) = (-x)^3$

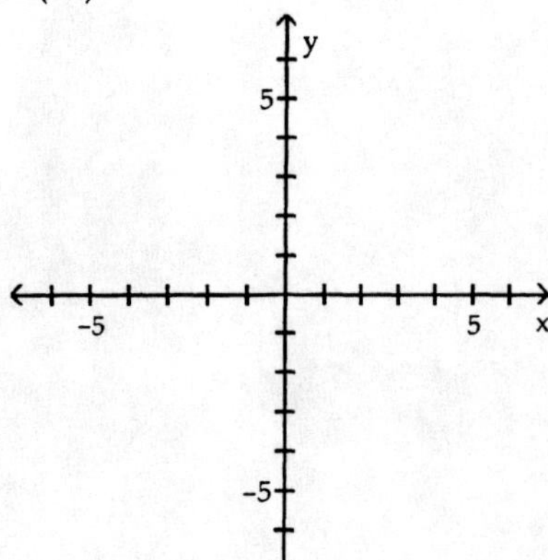

A)

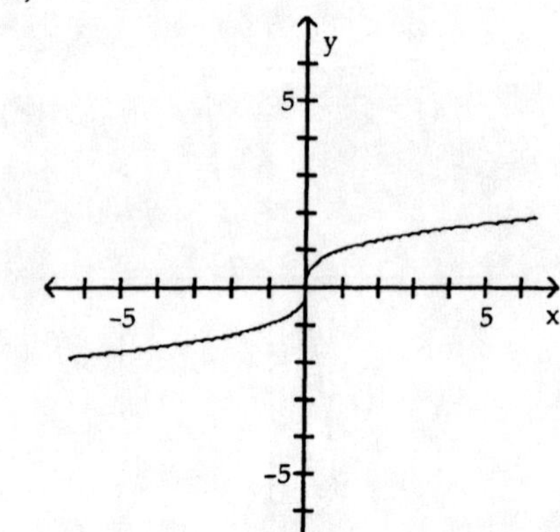

B)

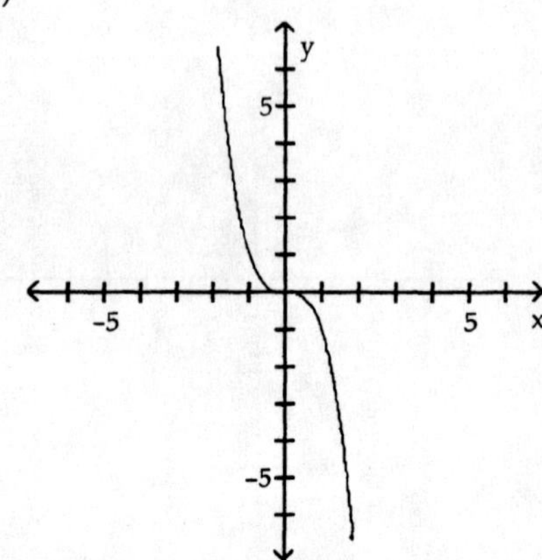

C)

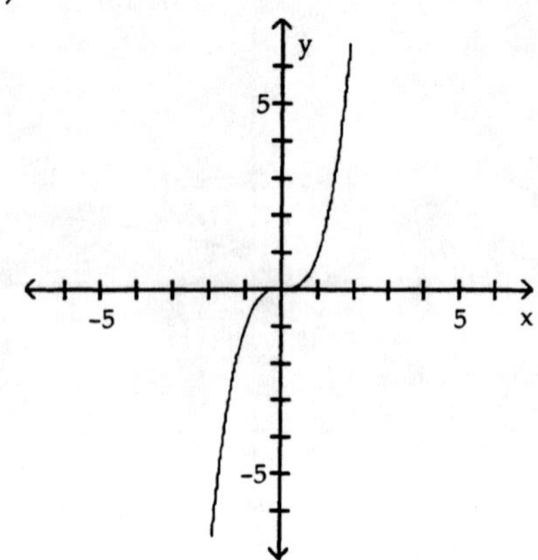

D)

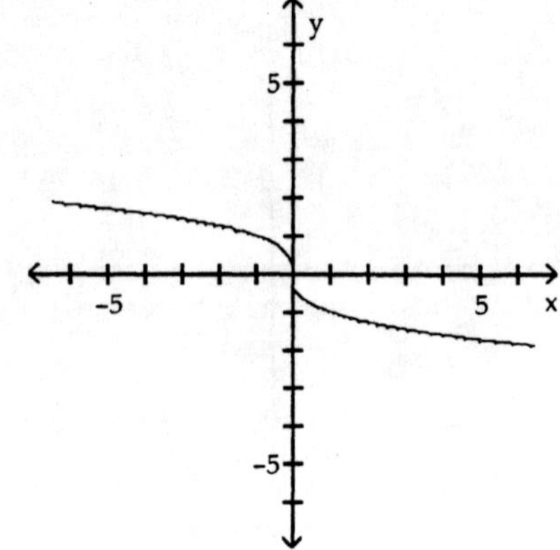

5) $f(x) = -\sqrt{x}$

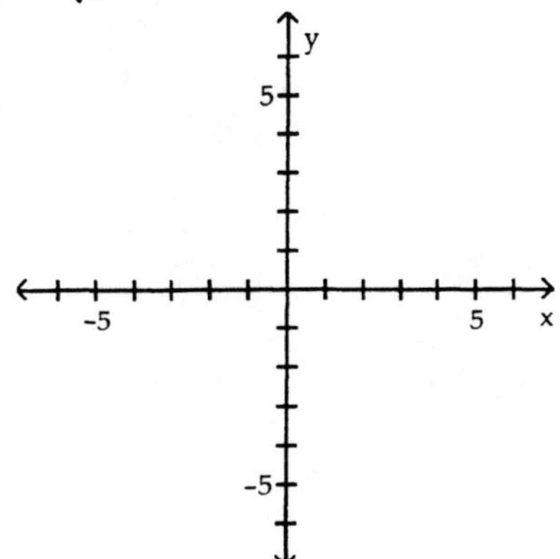

A)

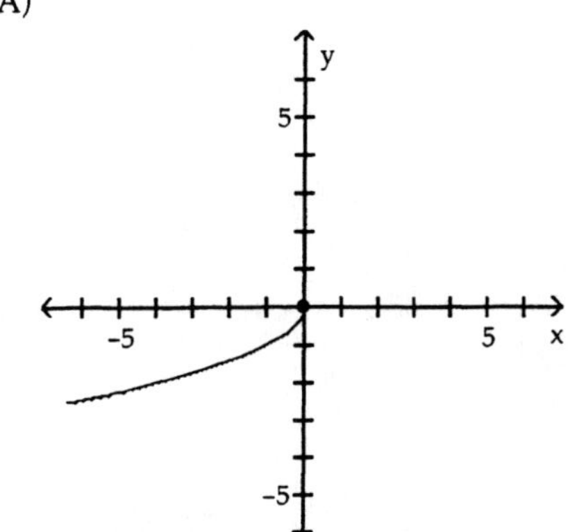

B)

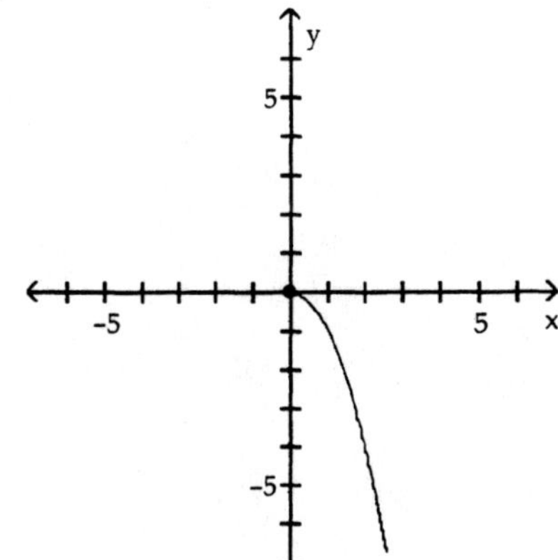

C)

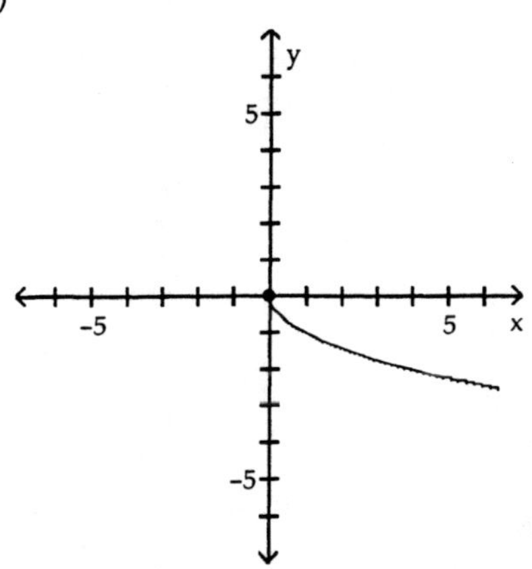

D)

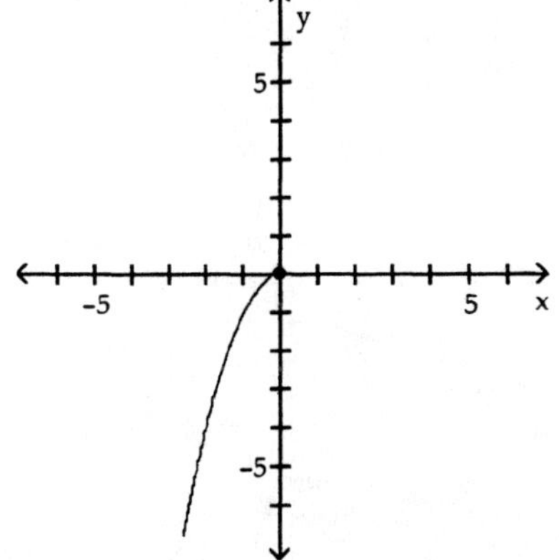

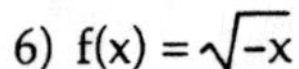
6) $f(x) = \sqrt{-x}$

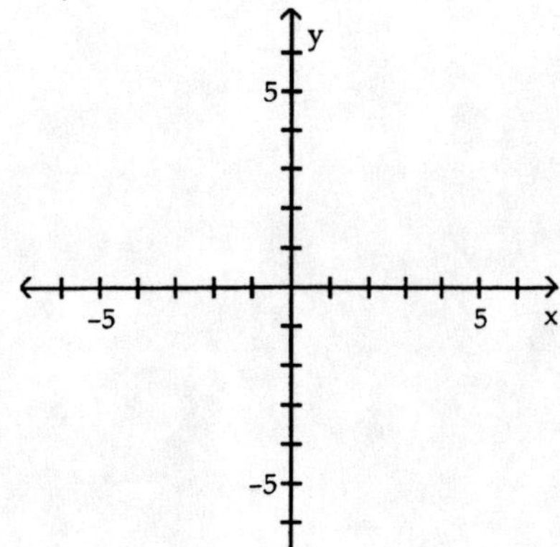

A)

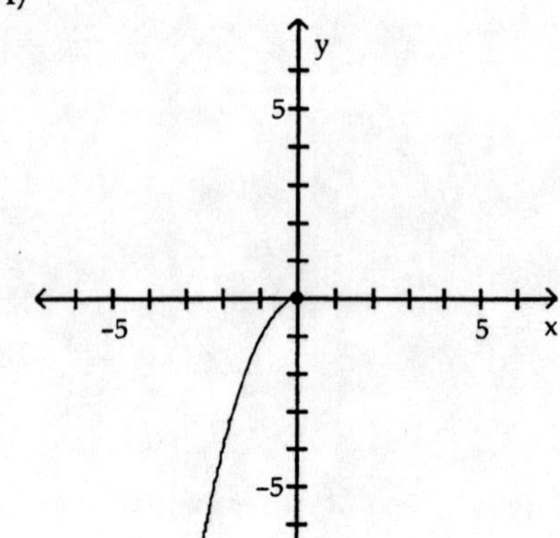

B)

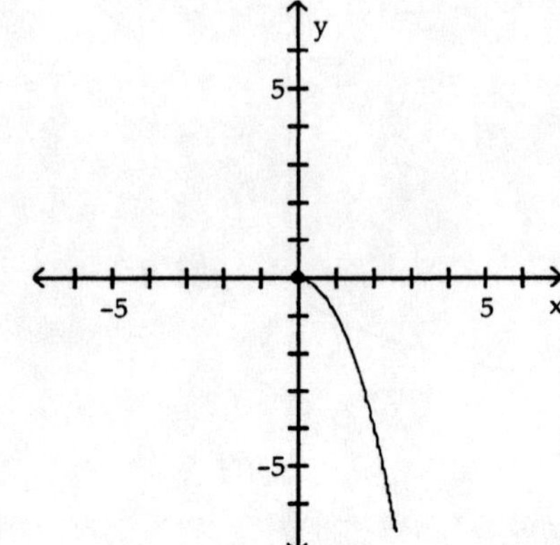

C)

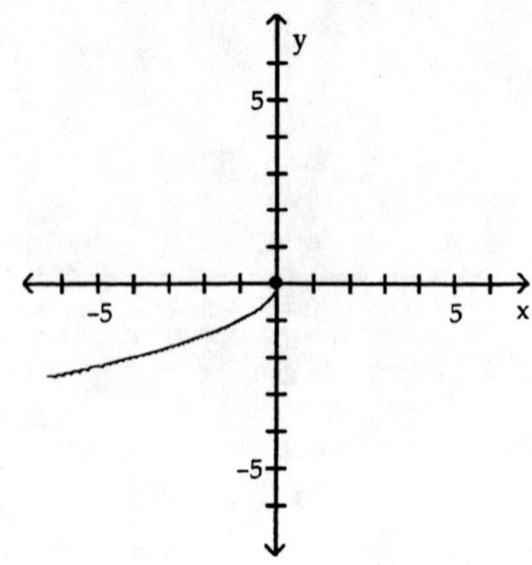

D)

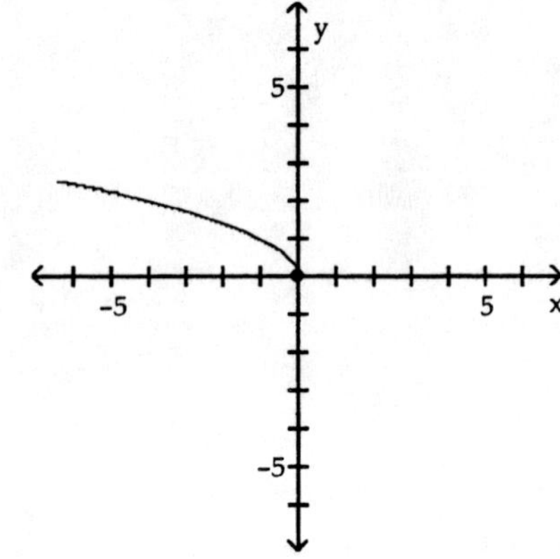

7) $f(x) = -|x|$

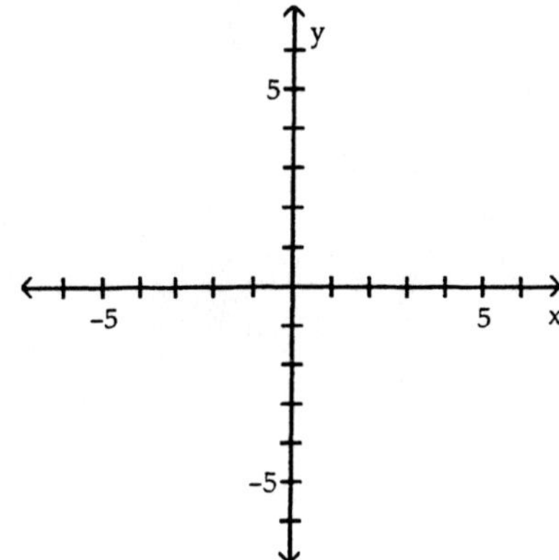

A)

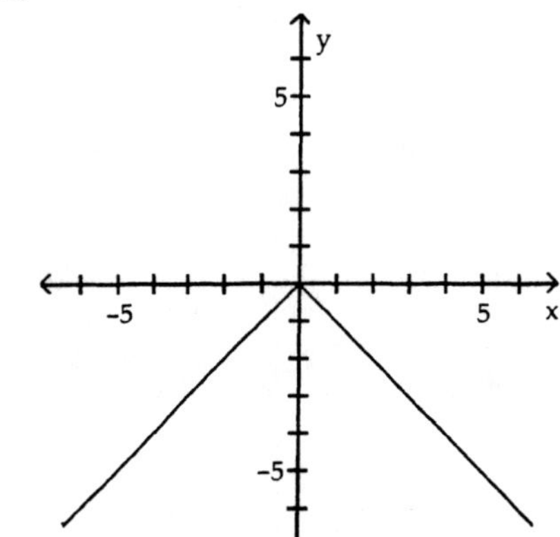

B)

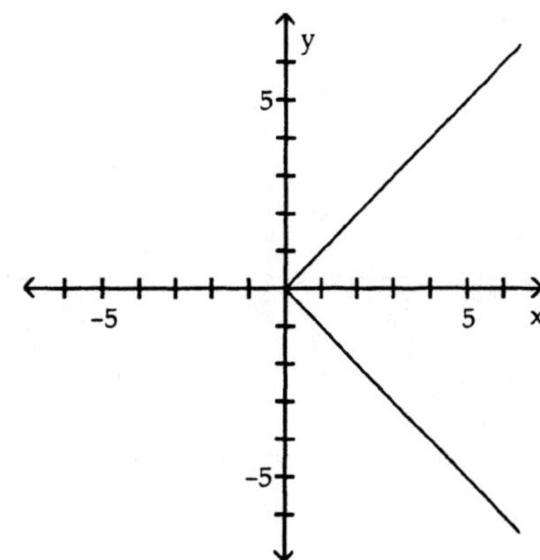

C)

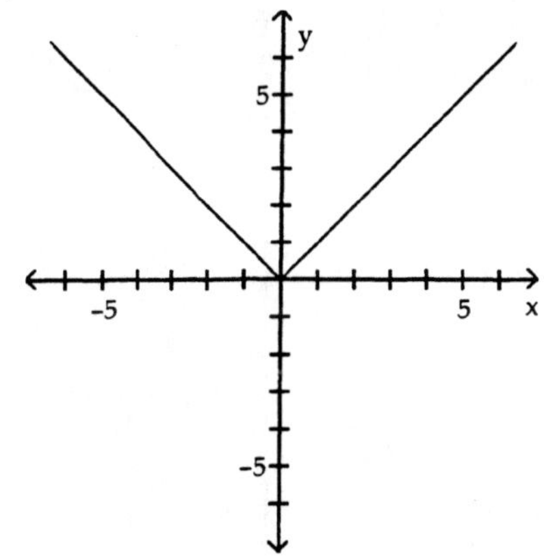

D)

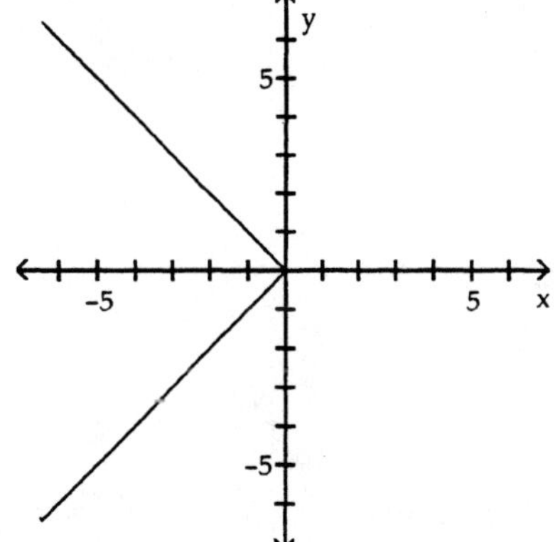

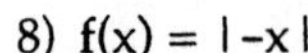

8) $f(x) = |-x|$

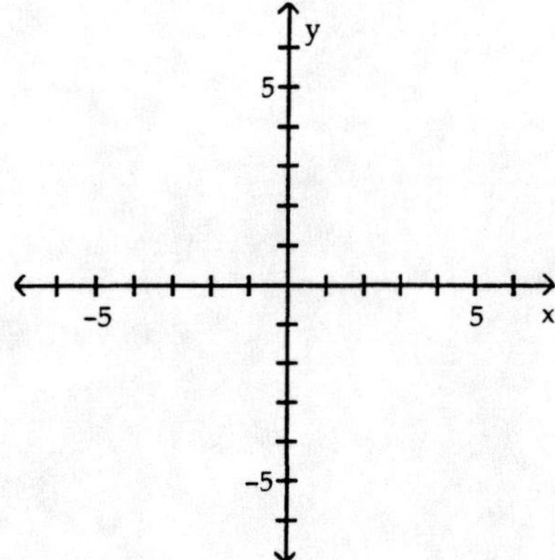

A)

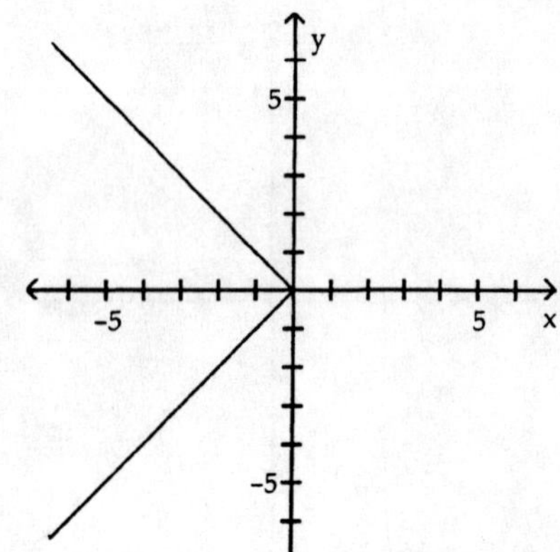

B)

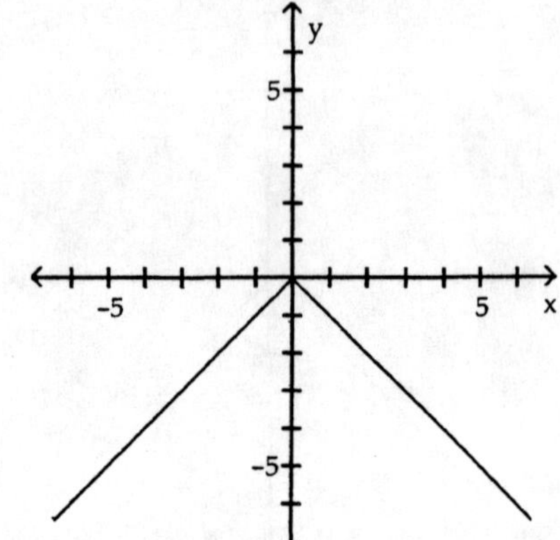

C)

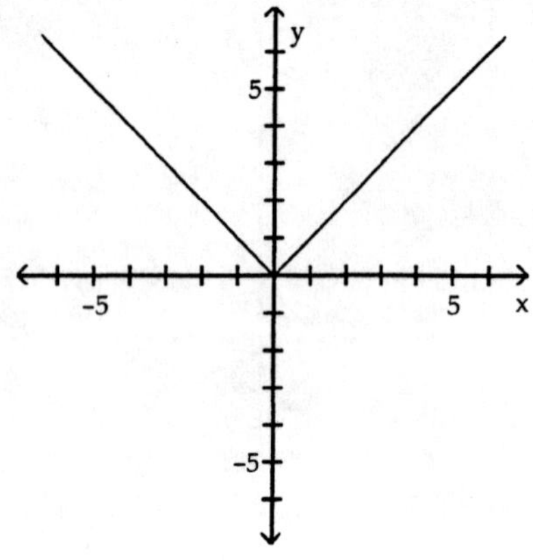

D)

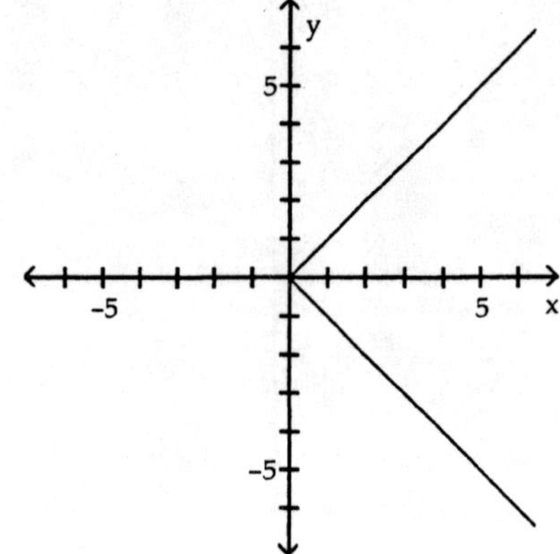

9) $f(x) = -\frac{1}{x}$

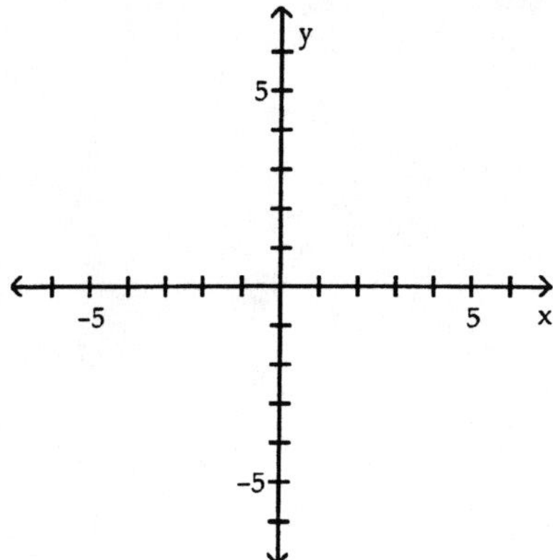

A)

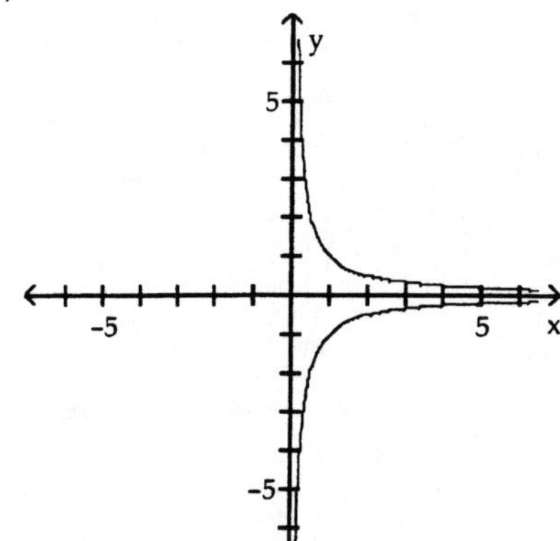

B)

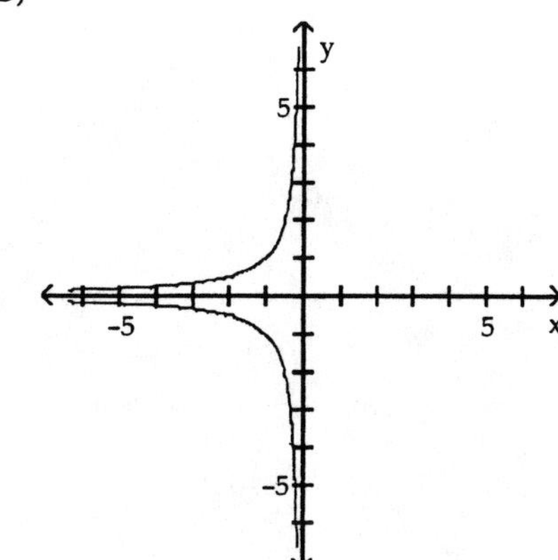

C)

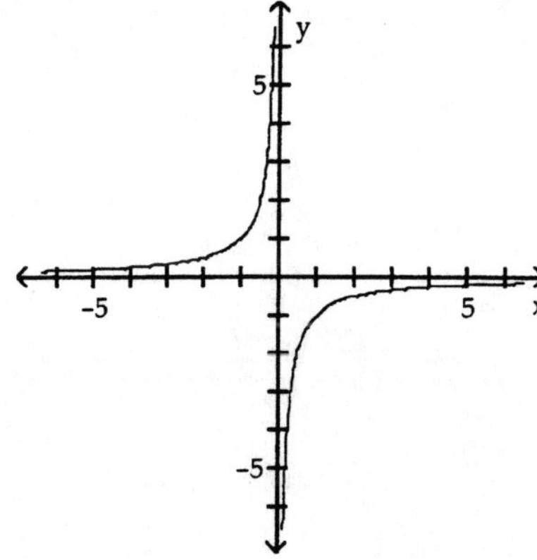

D)

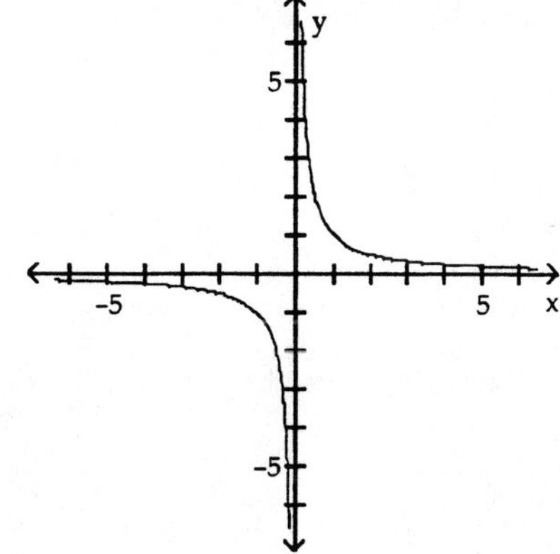

Graph the function by starting with the graph of the basic function and then using the techniques of shifting, compressing, stretching, and/or reflecting.

1) $f(x) = x^2 - 1$

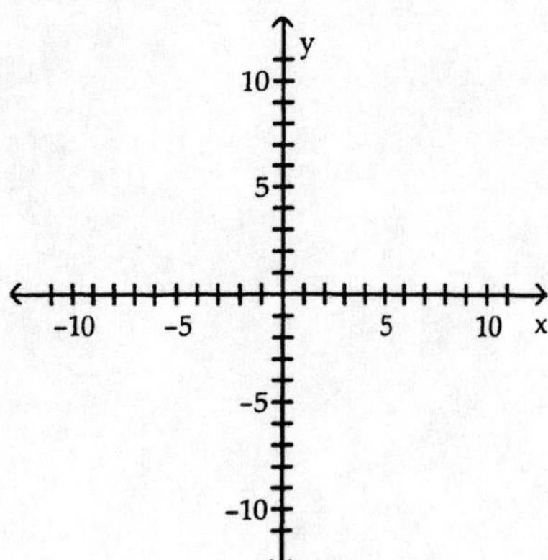

A)

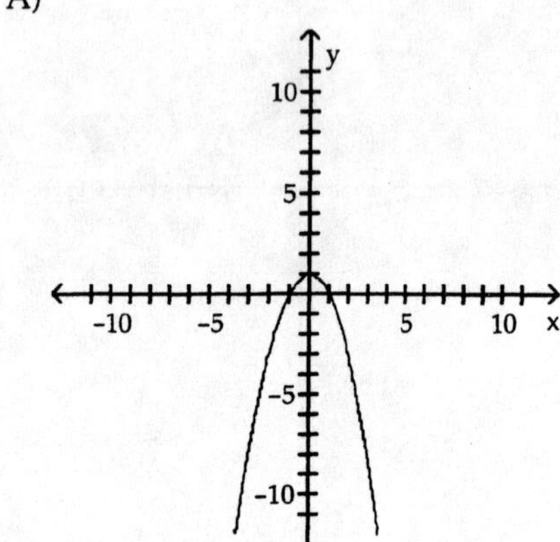

B)

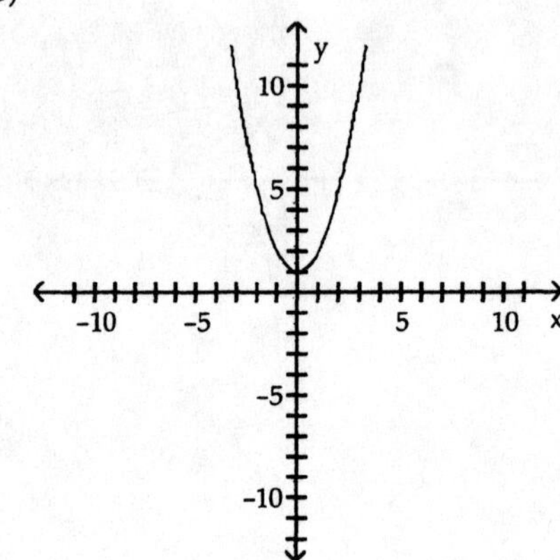

C)

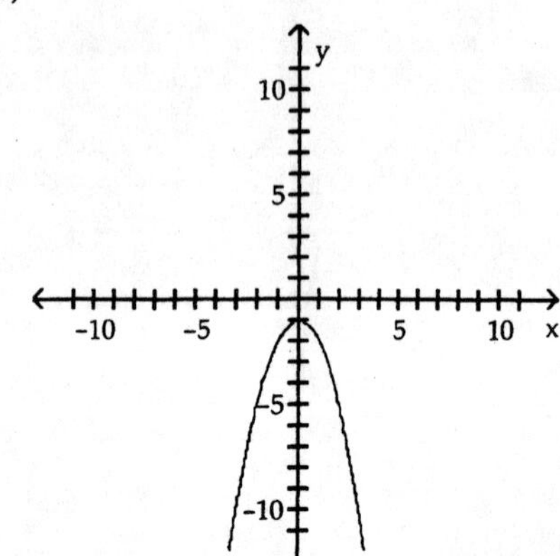

D)

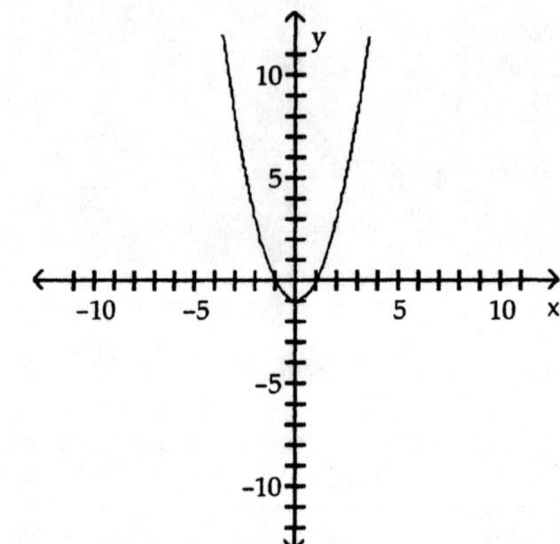

2) $f(x) = 3(x + 1)^2 - 2$

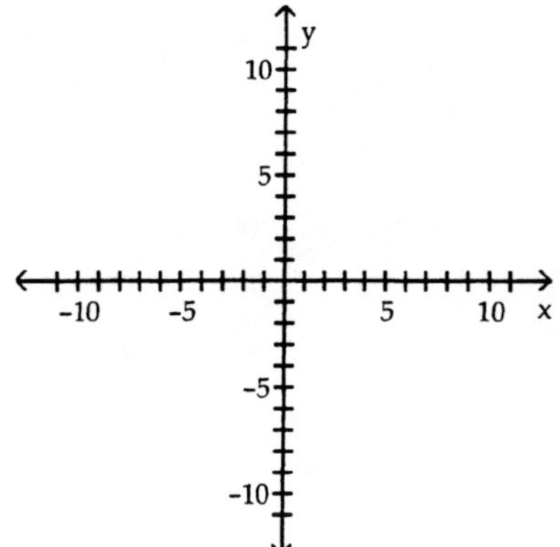

A)

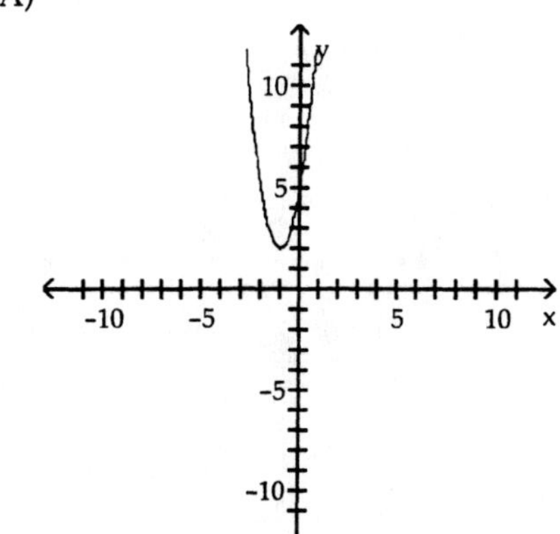

B)

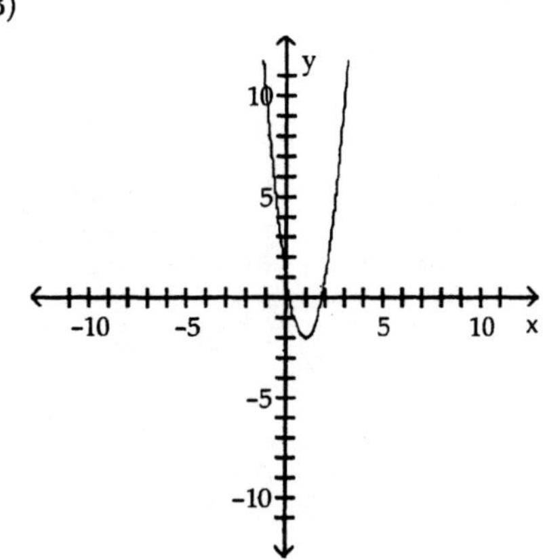

C)

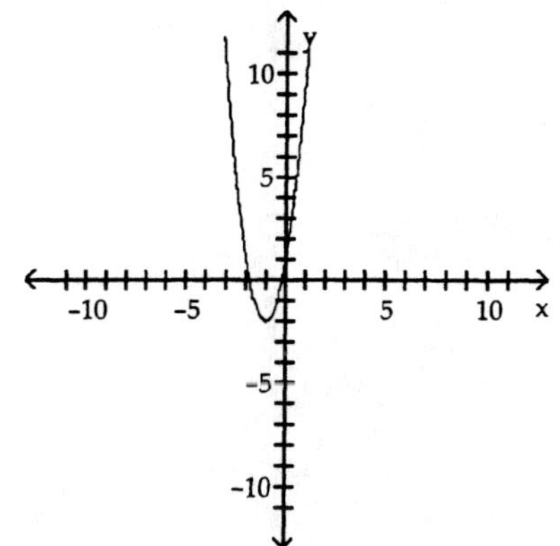

D)

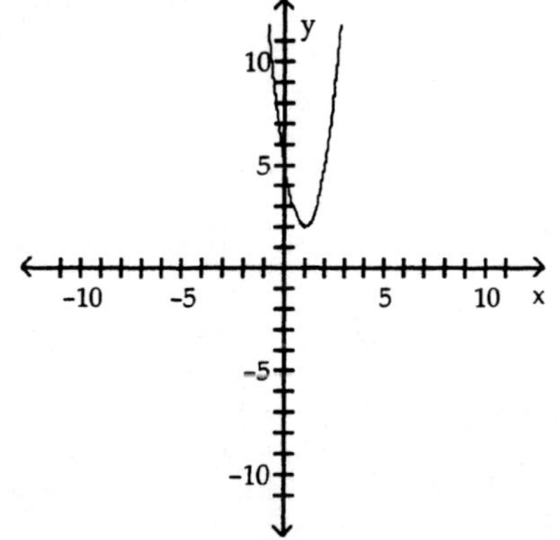

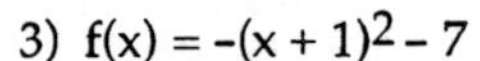

3) $f(x) = -(x + 1)^2 - 7$

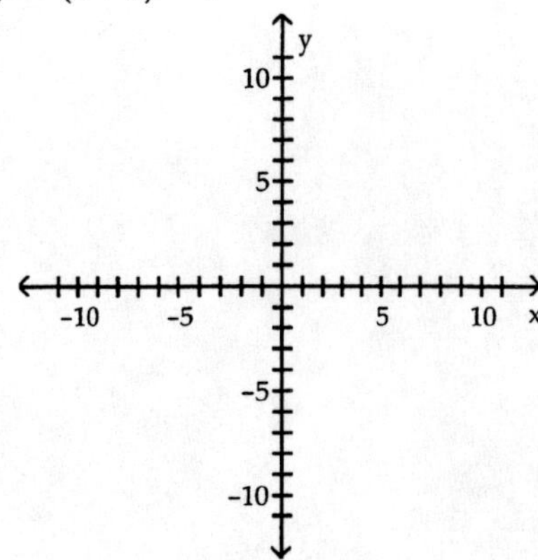

A)

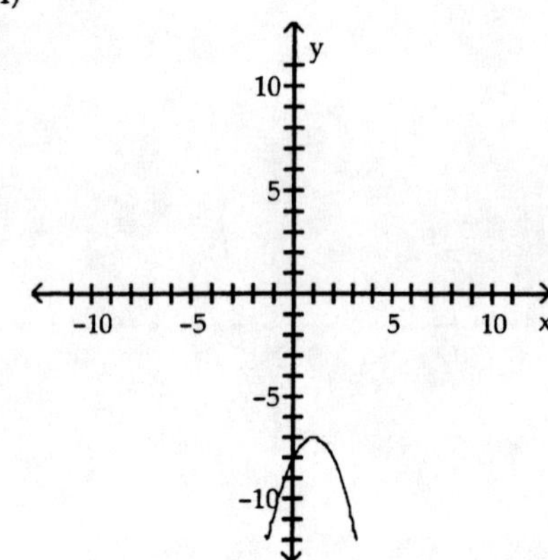

B)

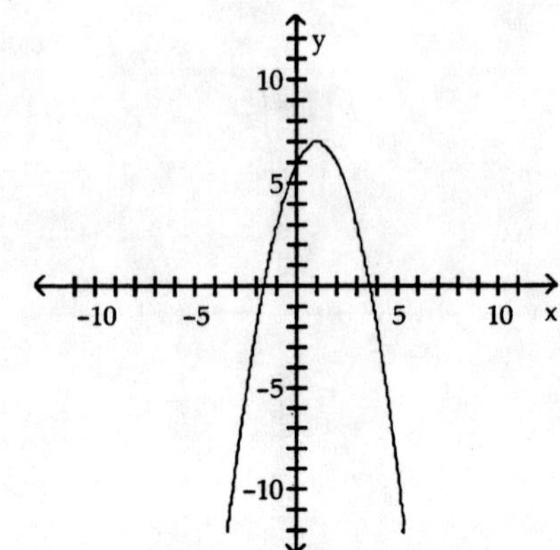

C)

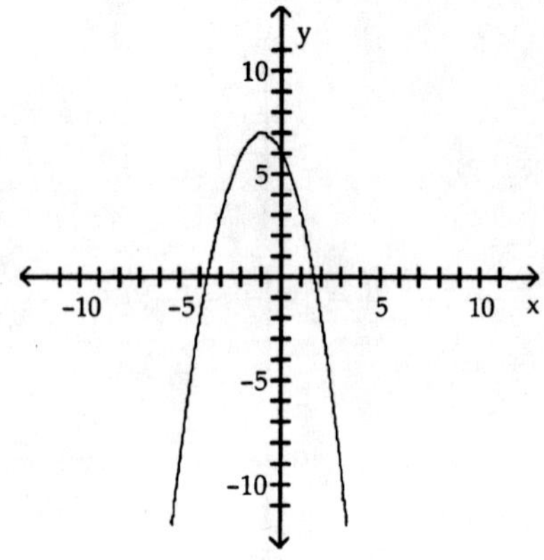

D)

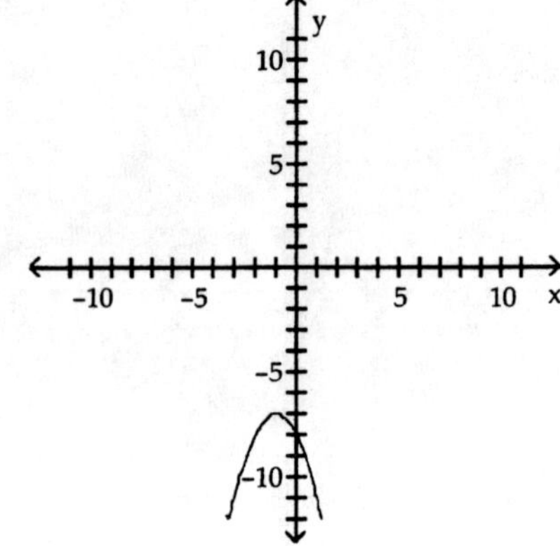

4) $f(x) = -3(x + 1)^2 + 2$

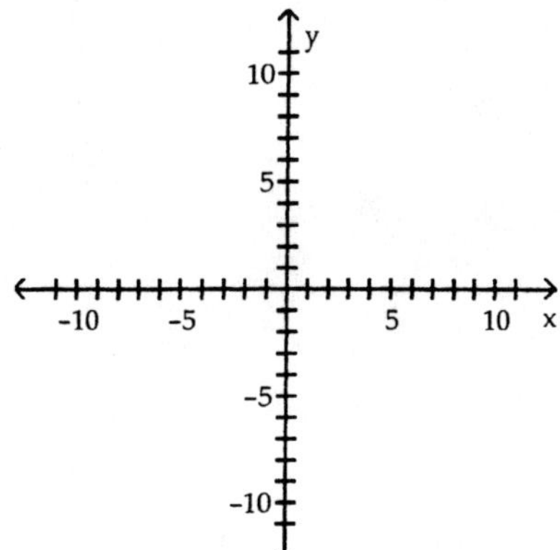

A)

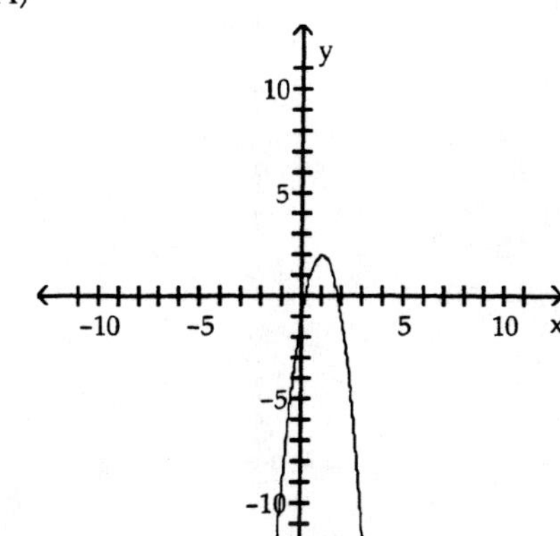

B)

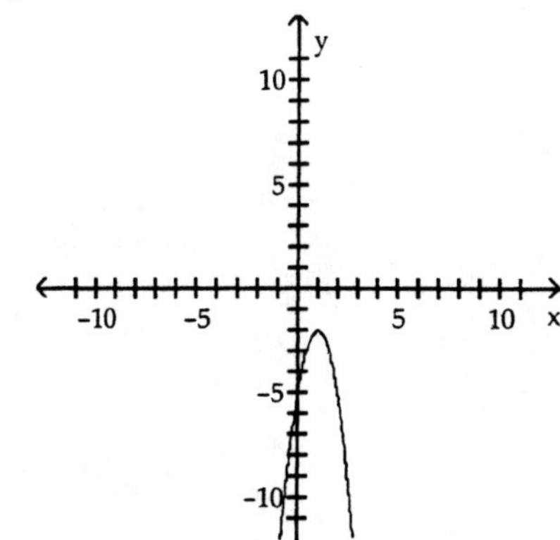

C)

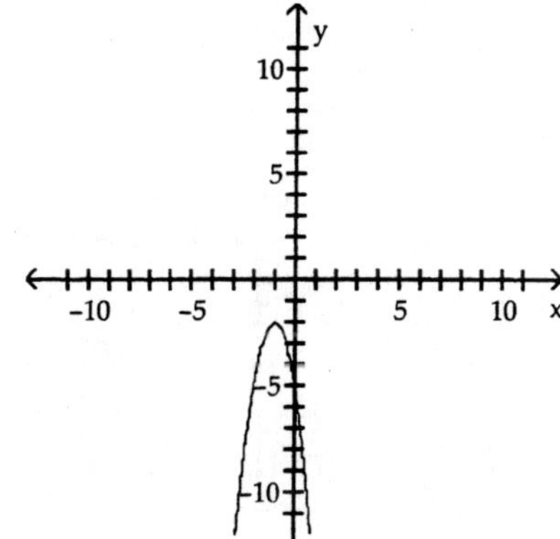

D)

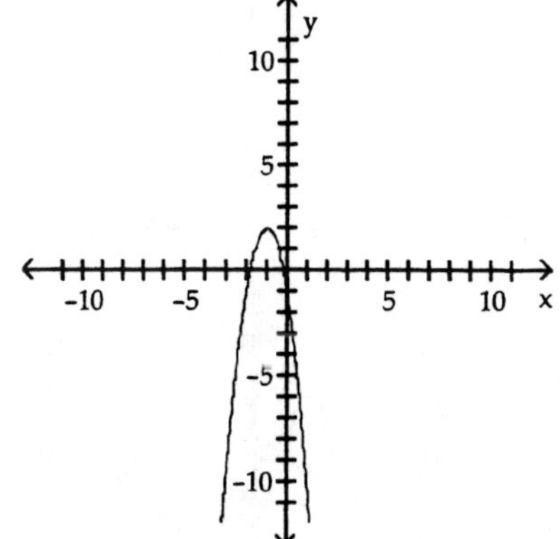

5) $f(x) = \sqrt{x + 5} - 4$

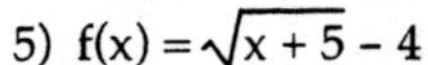

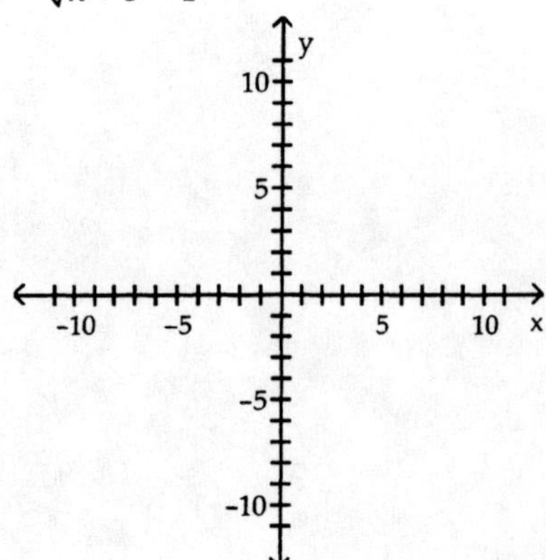

A)

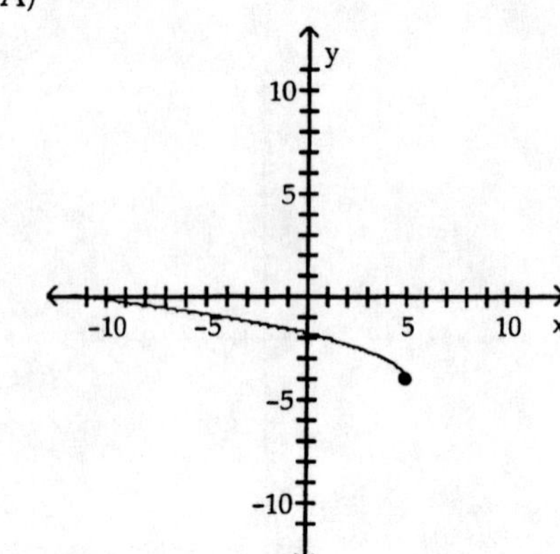

B)

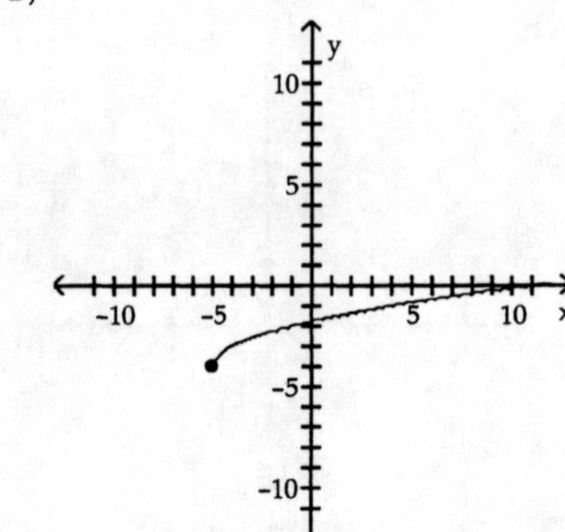

C)

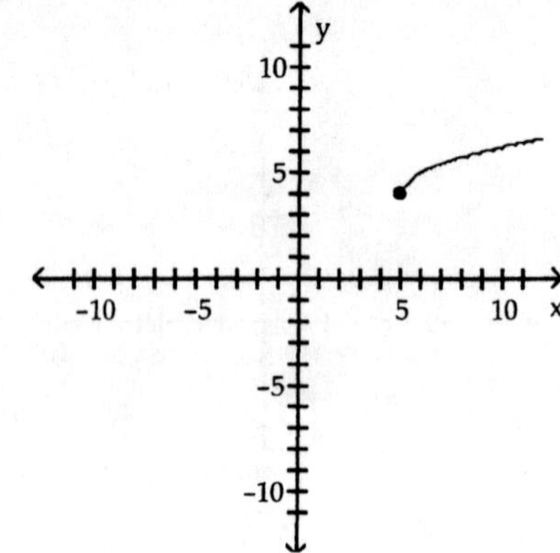

D)

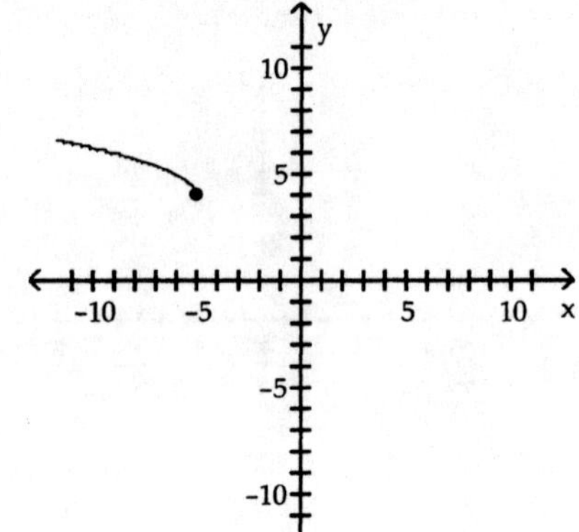

Using transformations, sketch the graph of the requested function.

6) The graph of a function f is illustrated. Use the graph of f as the first step toward graphing the function F(x), where F(x) = f(x + 2) − 1.

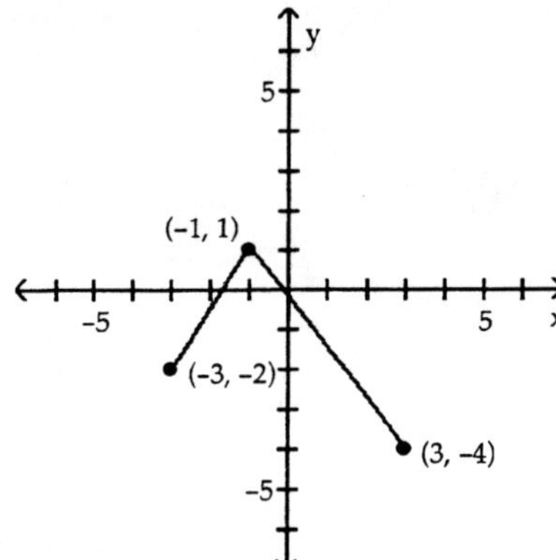

A)

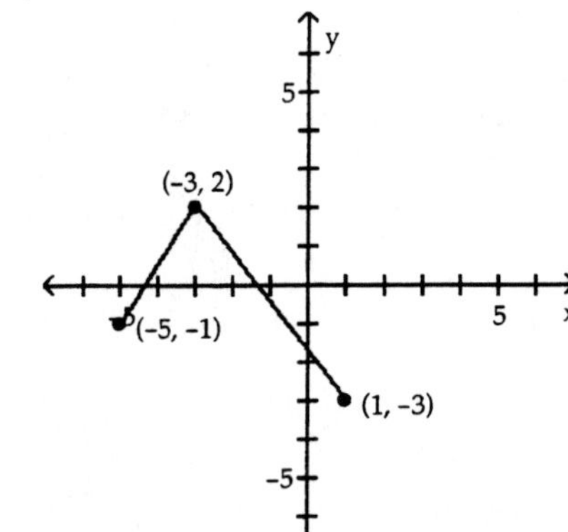

B)

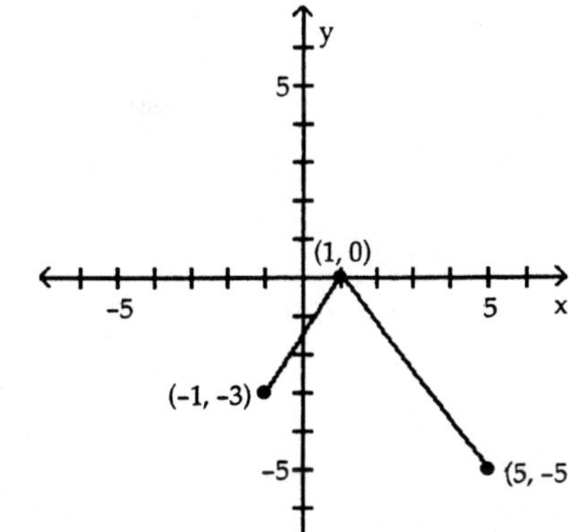

C)

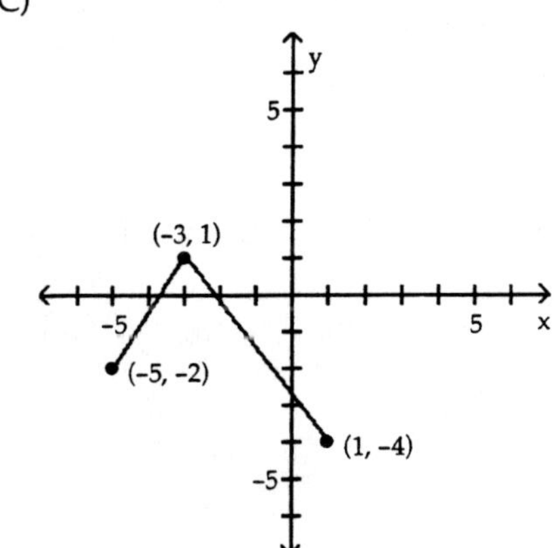

D)

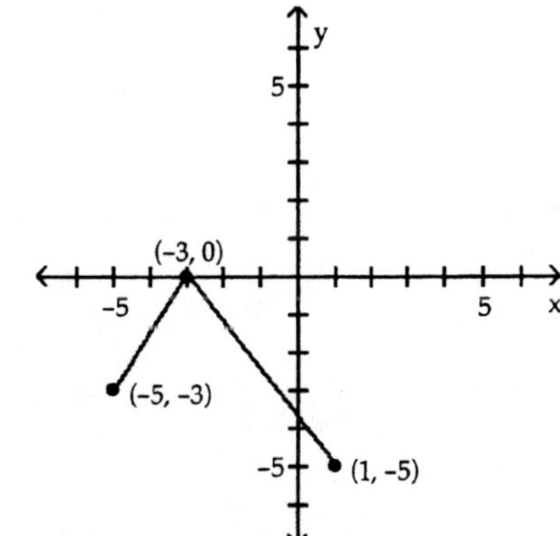

Match the correct function to the graph.

7)

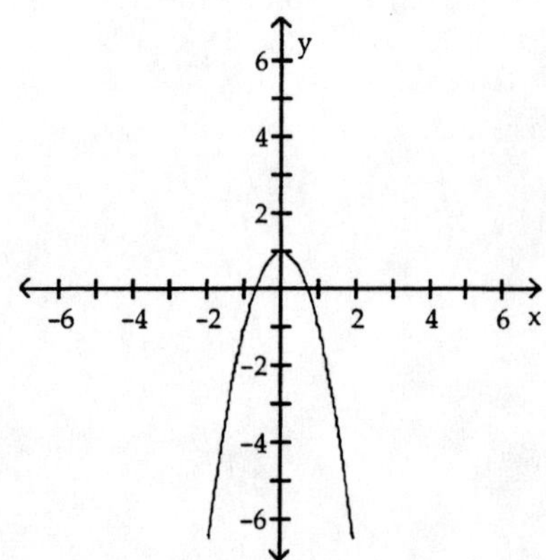

A) $y = -2x^2 - 1$ B) $y = -2x^2 + 1$ C) $y = 1 - x^2$ D) $y = -2x^2$

Find the function.

8) Find the function that is finally graphed after the following transformations are applied to the graph of $y = |x|$. The graph is shifted right 3 units, stretched by a factor of 3, shifted vertically down 2 units, and finally reflected across the x-axis.

A) $y = -3|x - 3| - 2$ B) $y = -(3|x + 3| - 2)$ C) $y = -(3|x - 3| - 2)$ D) $y = 3|-x - 3| - 2$

Find the function that is finally graphed after the following transformations are applied to the graph of $y = \sqrt{x}$.

9) i) Shift down 8 units
ii) Reflect about the y-axis
iii) Shift left 5 units

A) $y = -\sqrt{x + 5} - 8$ B) $y = \sqrt{-x + 5} + 8$ C) $y = \sqrt{-x - 5} - 8$ D) $y = \sqrt{-x - 5} + 8$

3.6 Mathematical Models: Constructing Functions

1 Construct and Analyze Functions

Solve the problem.

1) The volume V of a square-based pyramid with base sides s and height h is $V = \frac{1}{3}s^2h$. If the height is half of the length of a base side, express the volume V as a function of s.

2) The price p and the quantity x sold of a certain product obey the demand equation:

$p = -\frac{1}{5}x + 400, \{x \mid 0 \le x \le 500\}$

What is the revenue to the nearest dollar when 400 units are sold?

A) \$210,000 B) \$128,000 C) \$10,000 D) \$192,000

3) A farmer has 1000 yards of fencing to enclose a rectangular garden. Express the area A of the rectangle as a function of the width x of the rectangle. What is the domain of A?

A) $A(x) = -x^2 + 500x;\ \{x \mid 0 < x < 1000\}$ B) $A(x) = -x^2 + 500x;\ \{x \mid 0 < x < 500\}$

C) $A(x) = x^2 + 500x;\ \{x \mid 0 < x < 500\}$ D) $A(x) = -x^2 + 1000x;\ \{x \mid 0 < x < 1000\}$

4) A wire of length 8x is bent into the shape of a square. Express the area A of the square as a function of x.

A) $A(x) = \frac{1}{16}x^2$ B) $A(x) = 2x^2$ C) $A(x) = 4x^2$ D) $A(x) = 8x^2$

5) Let P = (x, y) be a point on the graph of $y = \sqrt{x}$. Express the distance d from P to the point (1, 0) as a function of x.

A) $d(x) = x^2 + 2x + 2$ B) $d(x) = x^2 - x + 1$ C) $d(x) = \sqrt{x^2 - x + 1}$ D) $d(x) = \sqrt{x^2 + 2x + 2}$

6) The price p and x, the quantity of a certain product sold, obey the demand equation

$p = -\frac{1}{10}x + 100, \{x \mid 0 \le x \le 1000\}$

a) Express the revenue R as a function of x.
b) What is the revenue if 450 units are sold?
c) Graph the revenue function using a graphing utility.
d) What quantity x maximizes revenue? What is the maximum revenue?
e) What price should the company charge to maximize revenue?

7) A right triangle has one vertex on the graph of $y = x^2$ at (x, y), another at the origin, and the third on the (positive) y-axis at (0, y). Express the area A of the triangle as a function of x.

8) A wire 20 feet long is to be cut into two pieces. One piece will be shaped as a square and the other piece will be shaped as an equilateral triangle. Express the total area A enclosed by the pieces of wire as a function of the length x of a side of the equilateral triangle. What is the domain of A?

9) Two boats leave a dock at the same time. One boat is headed directly east at a constant speed of 35 knots (nautical miles per hour), and the other is headed directly south at a constant speed of 22 knots. Express the distance d between the boats as a function of the time t.

Ch. 3 Functions and Their Graphs
Answer Key

3.1 Functions

1 Determine Whether a Relation Represents a Function

1) A
2) C
3) A
4) A
5) C
6) B
7) B
8) A
9) A
10) A
11) B
12) B
13) B
14) B
15) A
16) A
17) B
18) A
19) A

2 Find the Value of a Function

1) B
2) B
3) C
4) B
5) D
6) B
7) D
8) A
9) B
10) C
11) B
12) C
13) C
14) B
15) B
16) B
17) D
18) A
19) B

3 Find the Domain of a Function

1) D
2) D
3) A
4) C
5) A
6) C
7) B

4 Form the Sum, Difference, Product and Quotient of Two Functions

1) A
2) C

3) C
4) A
5) A
6) C
7) B
8) D
9) B
10) B
11) D
12) D

5 Additional Applications

1) B
2) C
3) C
4) B
5) B
6) A

3.2 The Graph of a Function

1 Identify the Graph of a Function

1) D
2) C
3) A
4) B
5) D
6) B
7) D

2 Obtain Information from or about the Graph of a Function

1) C
2) A
3) B
4) B
5) D
6) C
7) B
8) B
9) D
10) D
11) C
12) B
13) A
14) B
15) A
16) A
17) D
18) B
19) A
20) B
21) A
22) A
23) D
24) B
25) D
26) B
27) D

28)

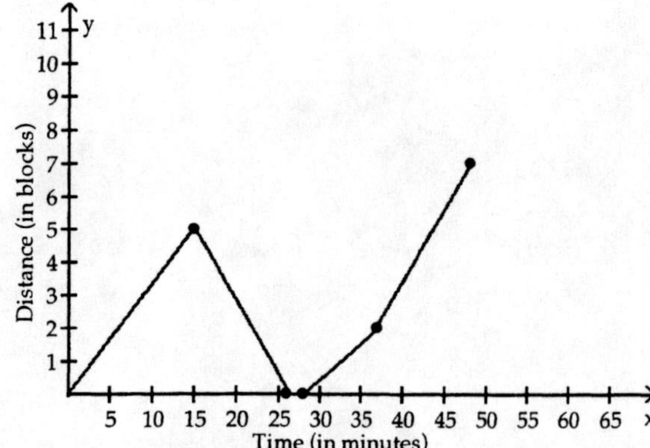

3.3 Properties of Functions

1 Determine Even and Odd Functions from a Graph

1) A
2) A
3) C
4) C
5) B
6) B
7) B
8) A

2 Identify Even and Odd Functions from the Equation

1) B
2) A
3) A
4) C
5) B
6) C
7) A
8) A
9) B
10) B
11) B

3 Use a Graph to Determine Where a Function Is Increasing, Is Decreasing, or Is Constant

1) C
2) A
3) A
4) B
5) C
6) C
7) C
8) B
9) C
10) C
11) A
12) B

4 Use a Graph to Locate Local Maxima and Minima

1) C
2) B
3) C

4) B
5) A
6) C

5 Use a Graphing Utility to Approximate Local Maxima and Minima and to Determine Where a Function Is Increasing or Decreasing

1) B
2) local maximum at (0, 6)
local minimum at (2.67, -3.48)
increasing on (−1, 0) and (2.67, 4)
decreasing on (0, 2.67)
3) local maximum at (0, 0)
local minimum at (0.74, −0.33)
increasing on (−2, 0) and (0.74, 2)
decreasing on (0, 0.74)
4) local maximum at (2.34, 1.61)
local minimum at (−1.9, −9.82)
increasing on (−1.9, 2.34)
decreasing on (−4, −1.9) and (2.34, 5)
5) local maximum at (0, 5)
local minima at (−2.55, 1.17) and (1.05, 4.65)
increasing on (−2.55, 0) and (1.05, 2)
decreasing on (−4, −2.55) and (0, 1.05)
6) D
7) C
8) B
9) C
10) C

6 Find the Average Rate of Change of a Function

1) C
2) C
3) A
4) A
5) C
6) A
7) A
8) D
9) D
10) D
11) D
12) D
13) B
14) C
15) B
16) D
17) B

3.4 Library of Functions; Piecewise-defined Functions

1 Graph the Functions Listed in the Library of Functions

1) B
2) C
3) A
4) C
5) C
6) A
7) C
8) C

9) C
10) D
11) C
12) D
13) D
14) D
15) C
16) B

2 Graph Piecewise-defined Functions

1) C
2) B
3) B
4) A
5) B
6) D
7) D
8) B
9) B
10) A
11) C
12) B
13) A
14) D
15) B

3 Additional Applications

1) A
2) \$25.52
\$42.69
$$C(x) = \begin{cases} 8.8 + 0.6686x & \text{if } 0 \le x \le 25 \\ 4.0475 + 0.8587x & \text{if } x > 25 \end{cases}$$
3) \$39.70
\$49.69
$$C(x) = \begin{cases} 4.93 + 0.11589x & \text{if } 0 \le x \le 300 \\ -0.266 + 0.13321x & \text{if } x > 300 \end{cases}$$
4) \$18.00
\$24.25
\$65.50
5) 6.0°C
6) \$27.50
\$32.50;
$$C(x) = \begin{cases} 20 & \text{if } 0 \le x \le 100 \\ 12.5 + 0.075x & \text{if } 100 < x \le 200 \\ 7.5 + 0.1x & \text{if } x > 200 \end{cases}$$

3.5 Graphing Techniques: Transformations

1 Graph Functions Using Horizontal and Vertical Shifts

1) B
2) B
3) A
4) C
5) C
6) D
7) A
8) B
9) C
10) A

11) B
12) A
13) B
14) D
15) C
16) D
17) B

2 Graph Functions Using Compressions and Stretches

1) C
2) B
3) D
4) A
5) D
6) D
7) D
8) C
9) B
10) D

3 Graph Functions Using Reflections about the x–Axis or y–Axis

1) C
2) C
3) C
4) B
5) C
6) D
7) A
8) C
9) C

4 Additional Applications

1) D
2) C
3) D
4) D
5) B
6) D
7) B
8) C
9) C

3.6 Mathematical Models: Constructing Functions

1 Construct and Analyze Functions

1) $V(s) = \frac{1}{6}s^3$

2) B
3) B
4) C
5) C

6) a. $R(x) = -\frac{1}{10}x^2 + 100x$

b. $R(450) = \$24{,}750.00$

c.

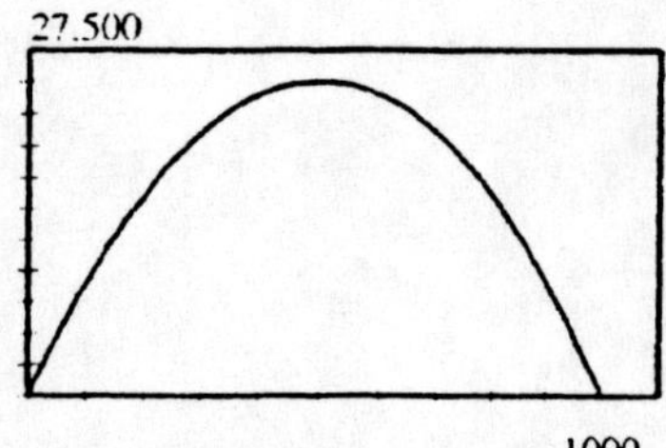

d. 500; \$25,000.00

e. \$50.00

7) $A(x) = \frac{1}{2}x^3$

8) $A(x) = \frac{4\sqrt{3} + 9}{16}x^2 - \frac{15}{2}x + 25$; $\{x \mid 0 \le x \le \frac{20}{3}\}$

9) $d(t) = \sqrt{1709}t$

Ch. 4 Polynomial and Rational Functions

4.1 Quadratic Functions and Models

1 Graph a Quadratic Function Using Transformations

Match the graph to one of the listed functions without using a graphing utility.

1)

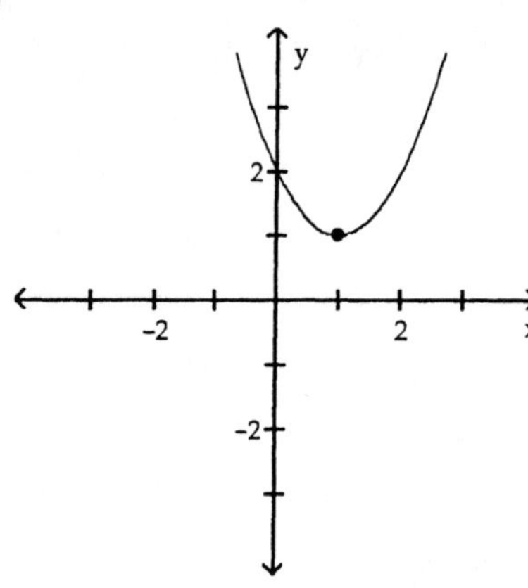

A) $x^2 - 2x + 1$ B) $x^2 + 2x + 2$ C) $x^2 + 2x + 1$ D) $x^2 - 2x + 2$

Graph the function f by starting with the graph of $y = x^2$ and using transformations (shifting, compressing, stretching, and/or reflection).

2) $f(x) = -x^2 + 2$

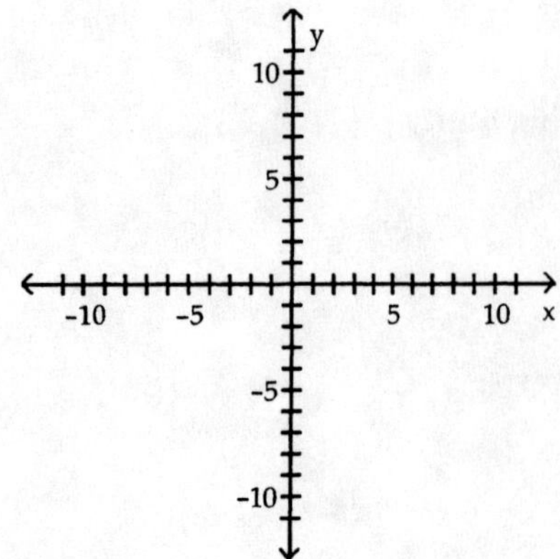

A)

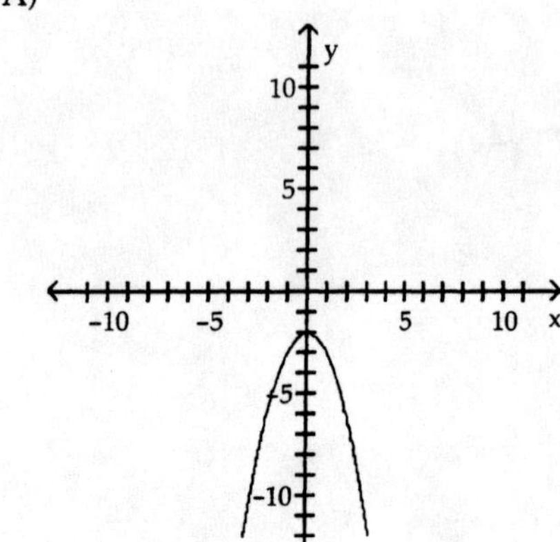

B)

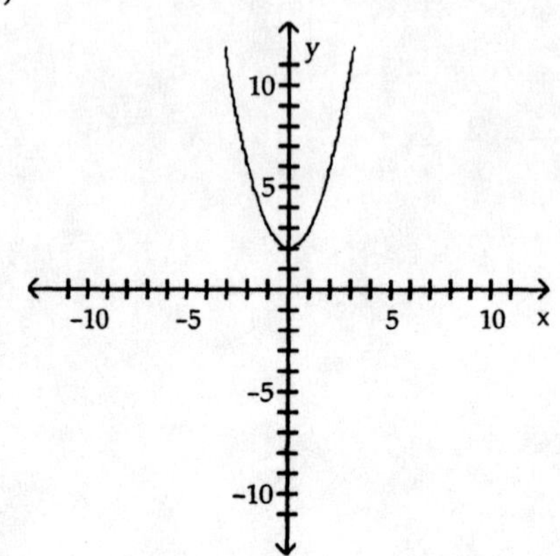

C)

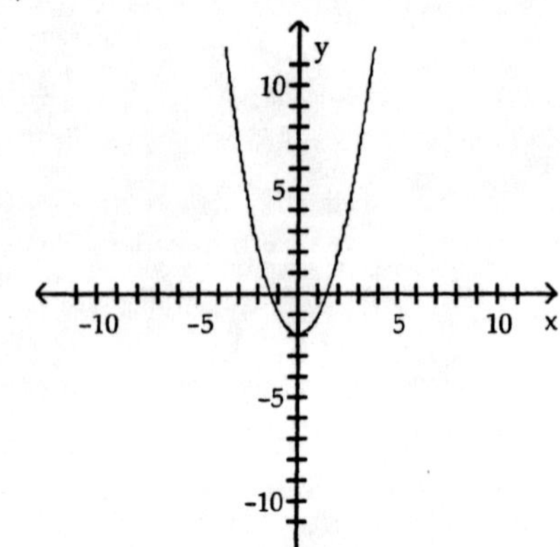

D)

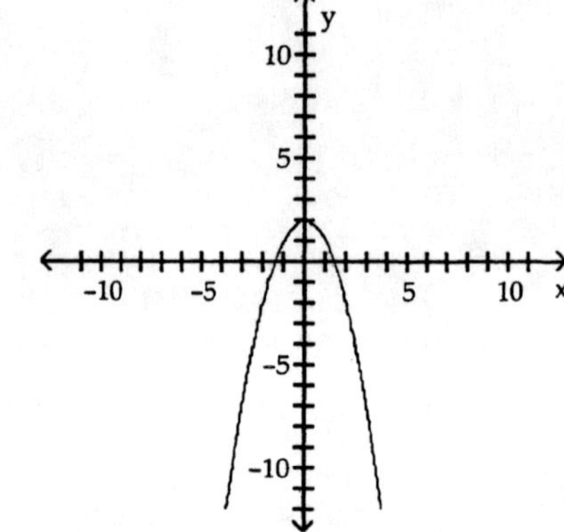

3) $f(x) = (x + 3)^2$

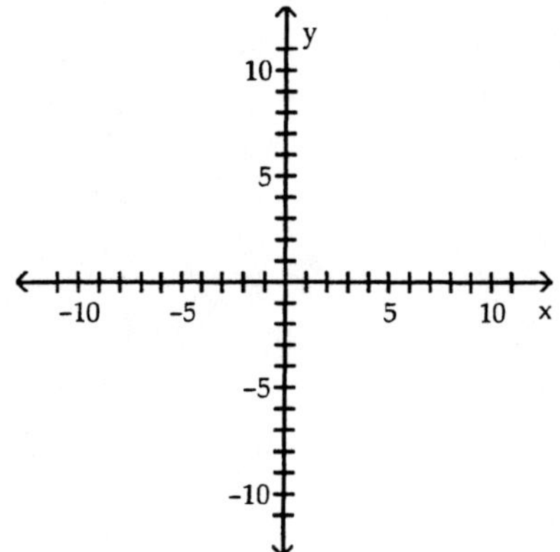

A)

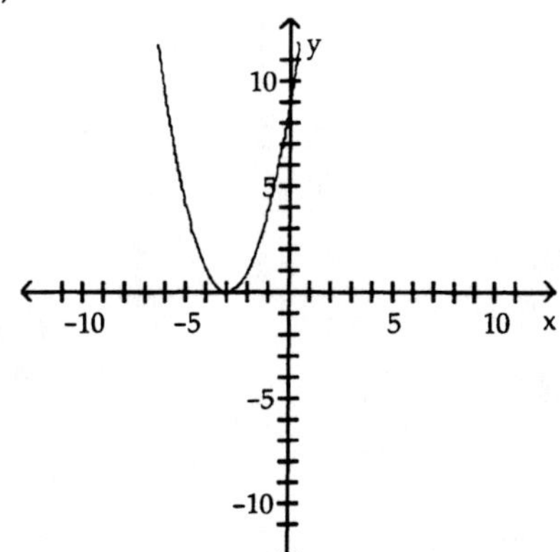

B)

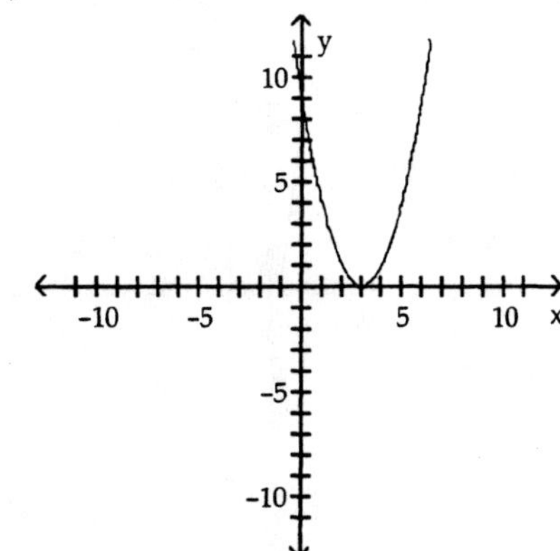

C)

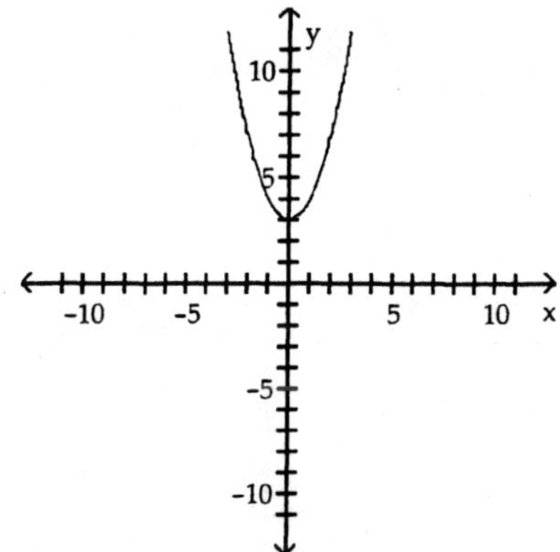

D)

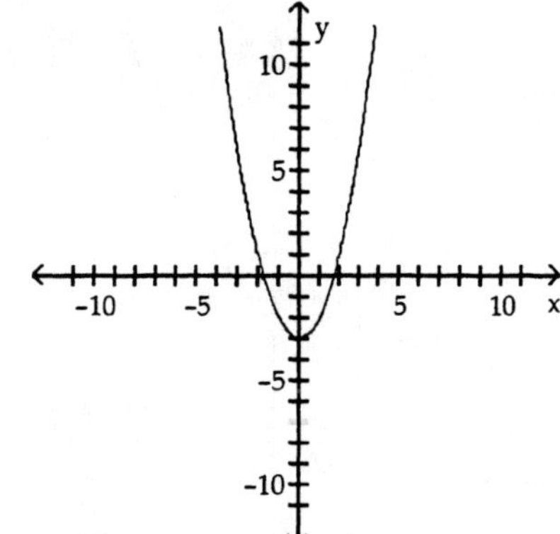

4) $f(x) = 4x^2$

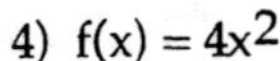

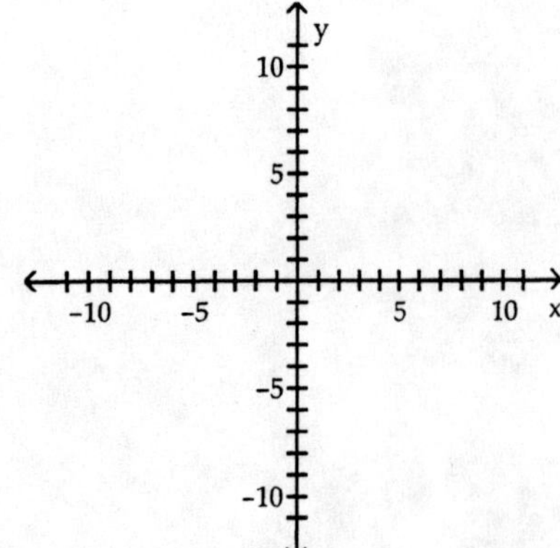

A)

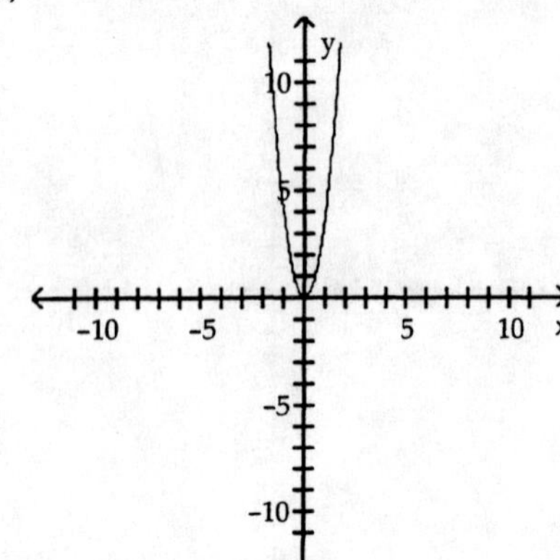

B)

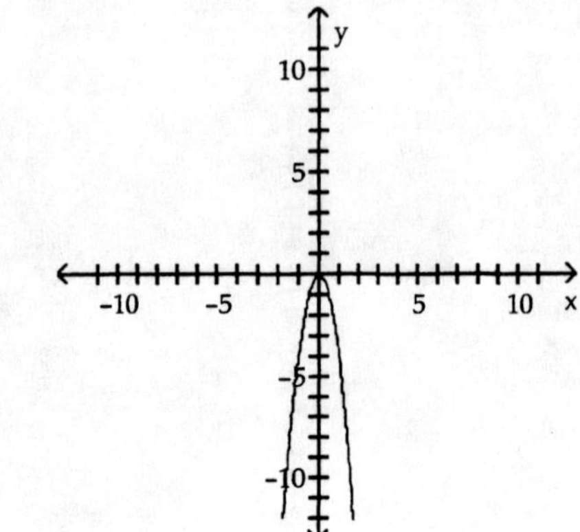

C)

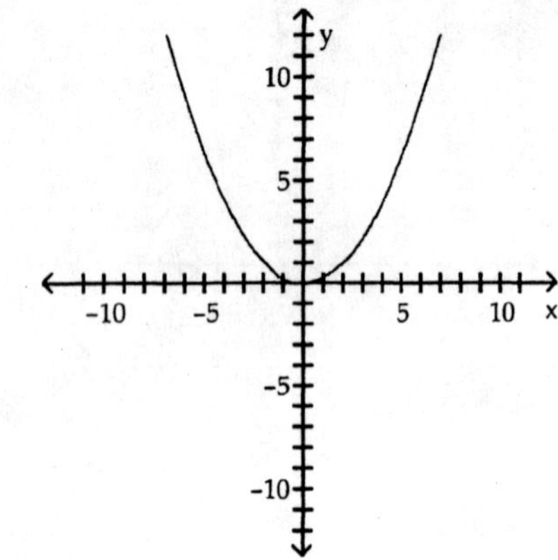

D)

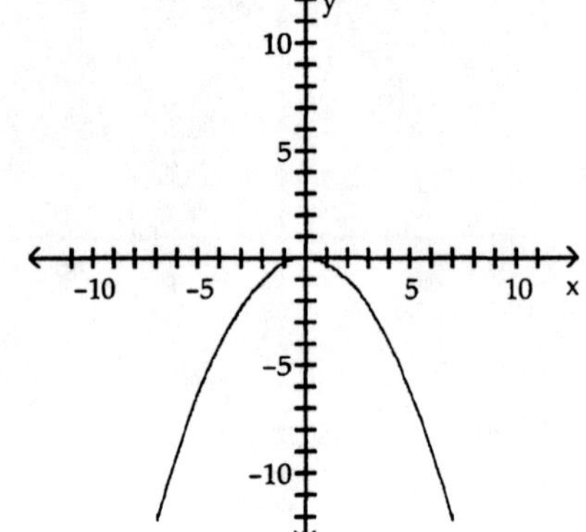

5) $f(x) = -\frac{1}{3}x^2$

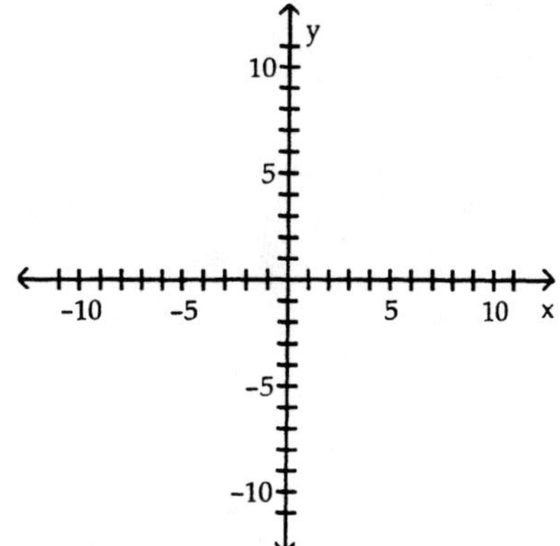

A)

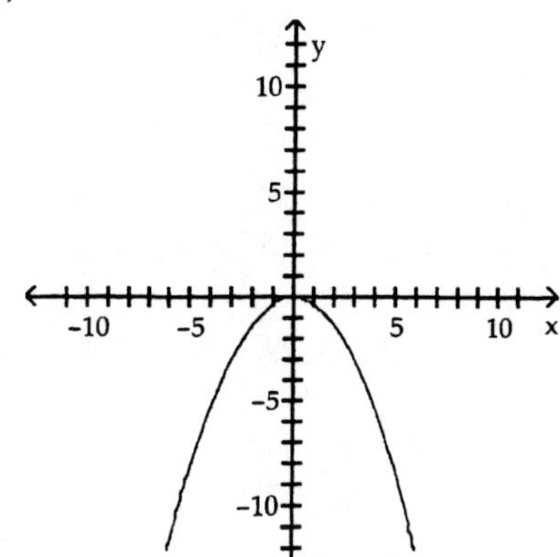

B)

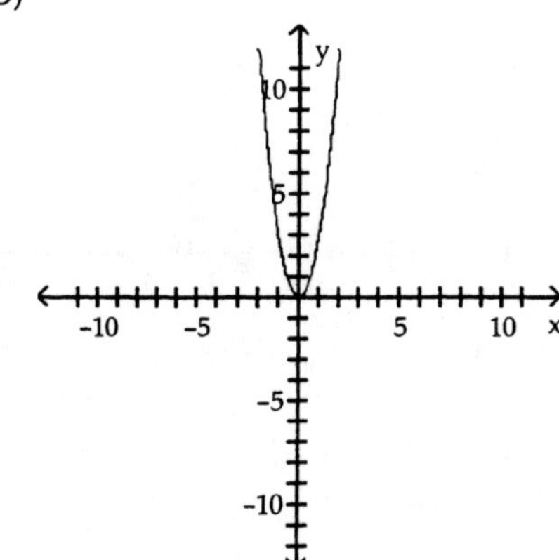

C)

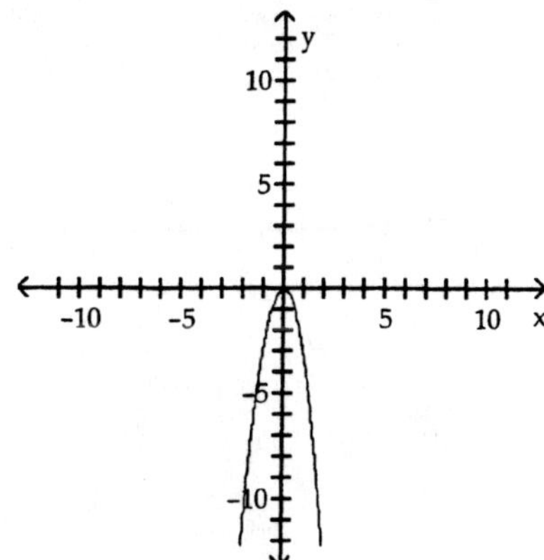

D)

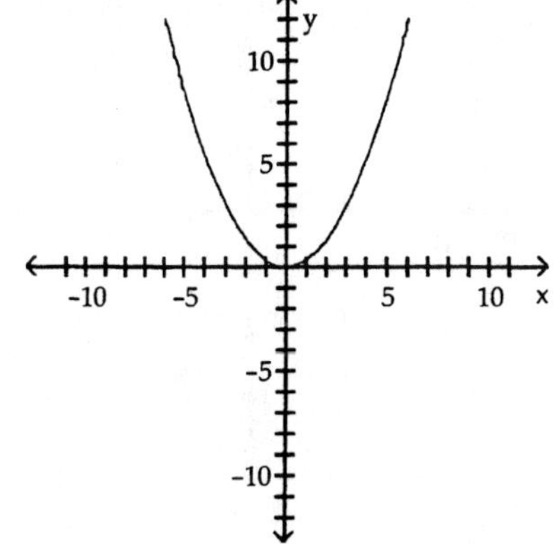

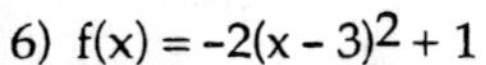

6) $f(x) = -2(x - 3)^2 + 1$

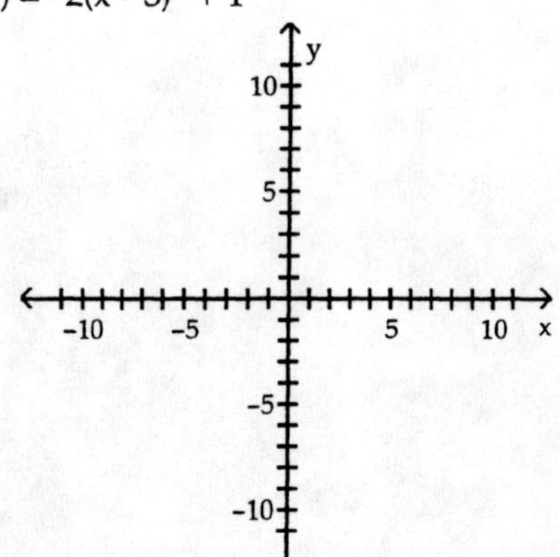

A)

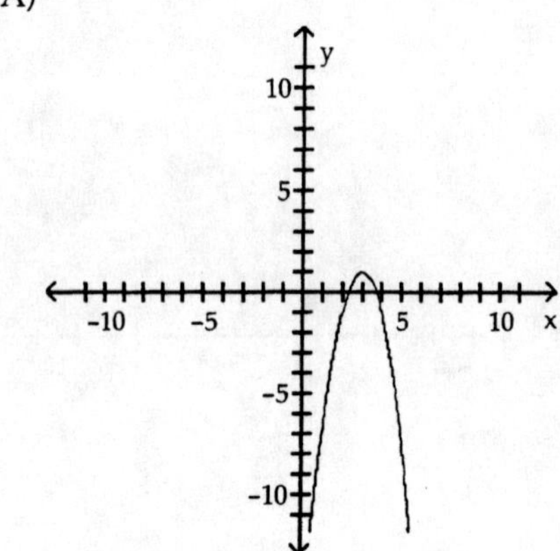

B)

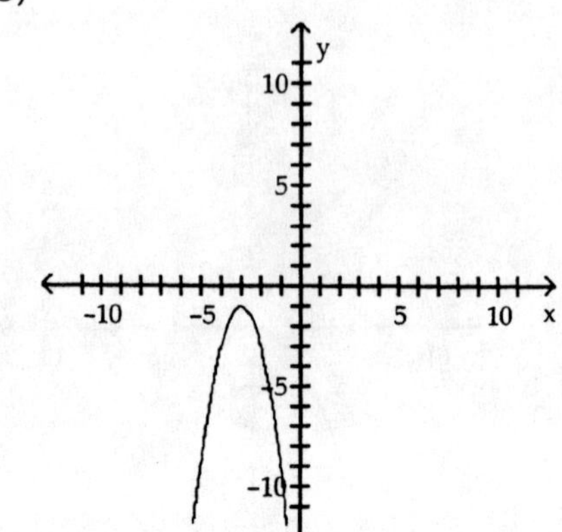

C)

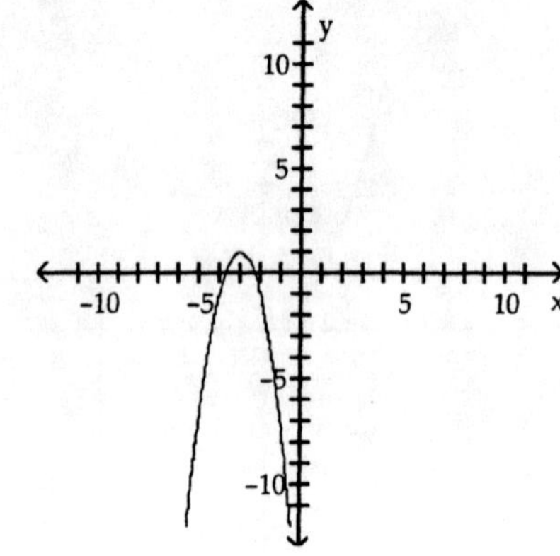

D)

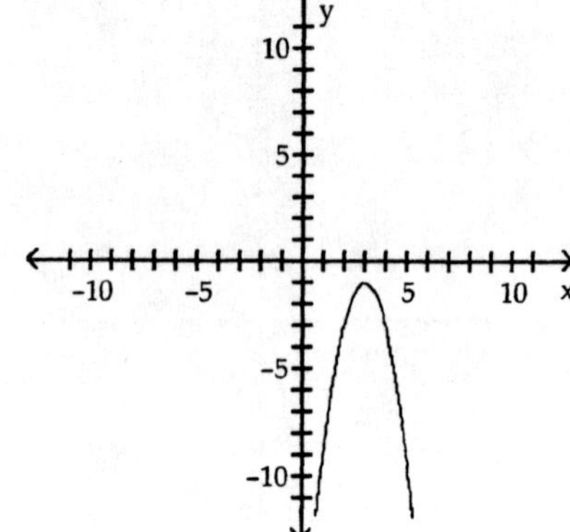

7) $f(x) = 4x^2 + 8x - 1$

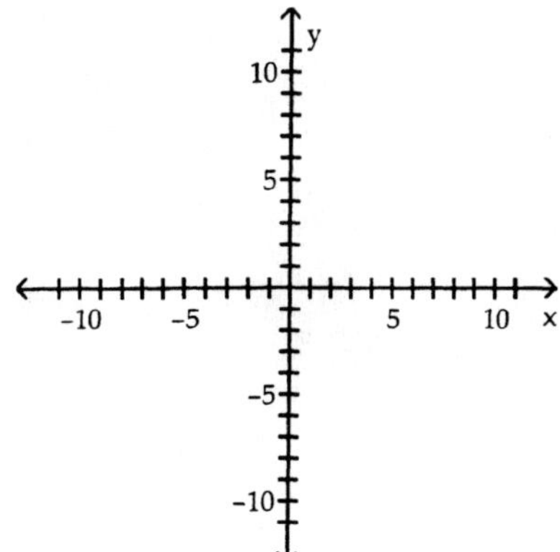

A)

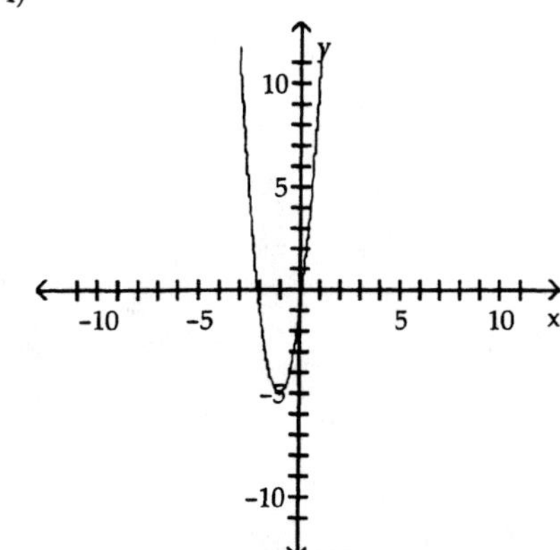

B)

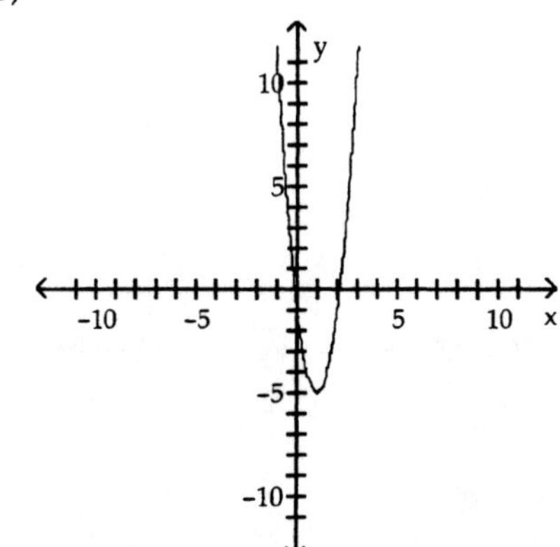

C)

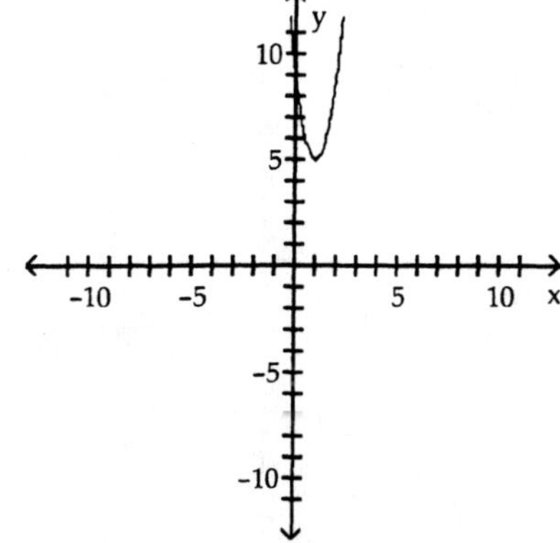

D)

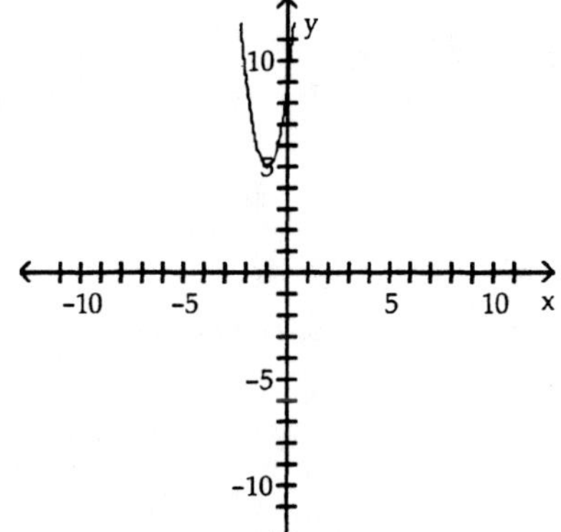

8) $f(x) = \frac{2}{3}x^2 + \frac{4}{3}x - 1$

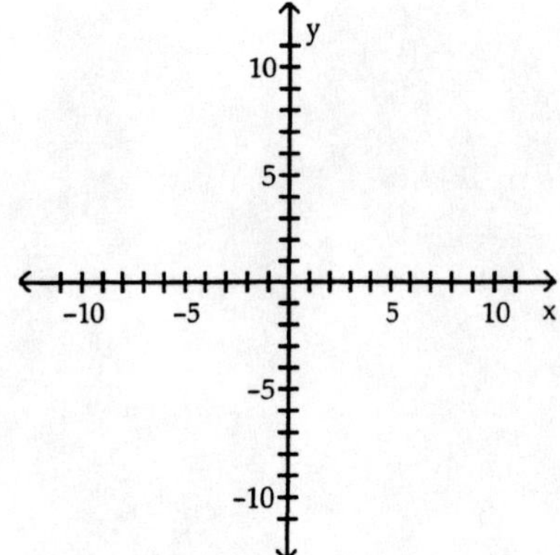

A)

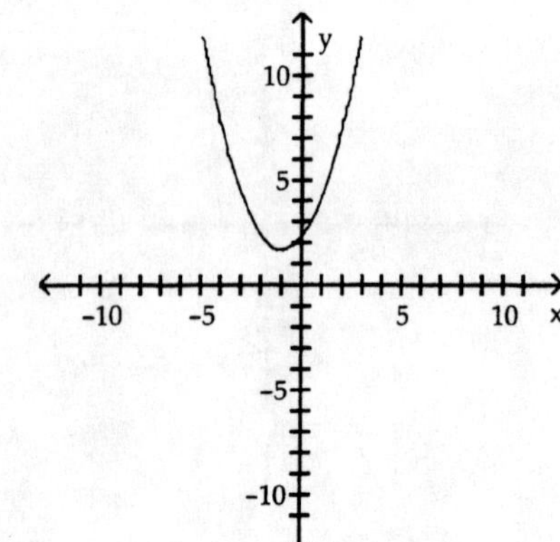

B)

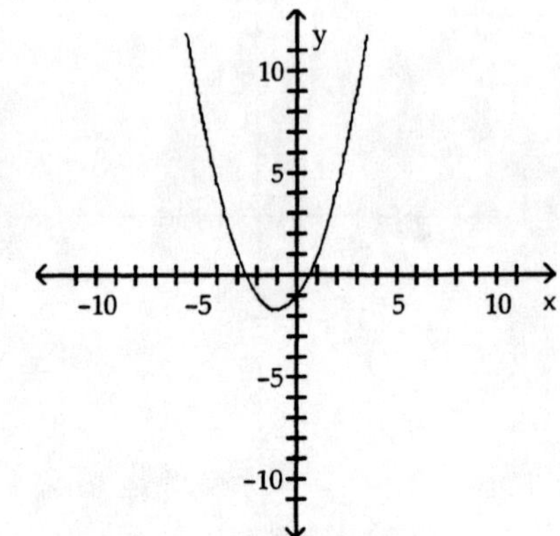

C)

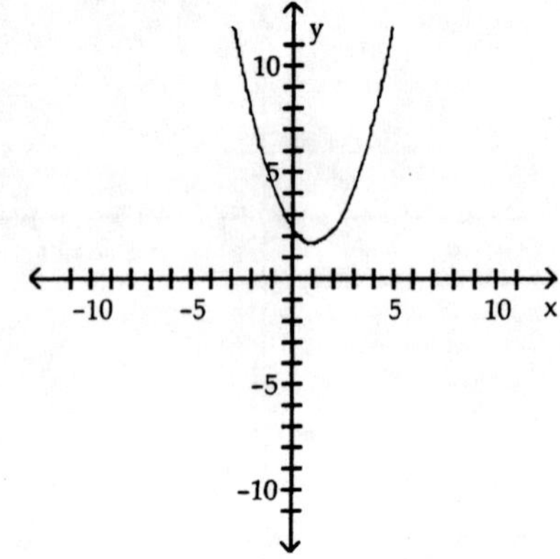

D)

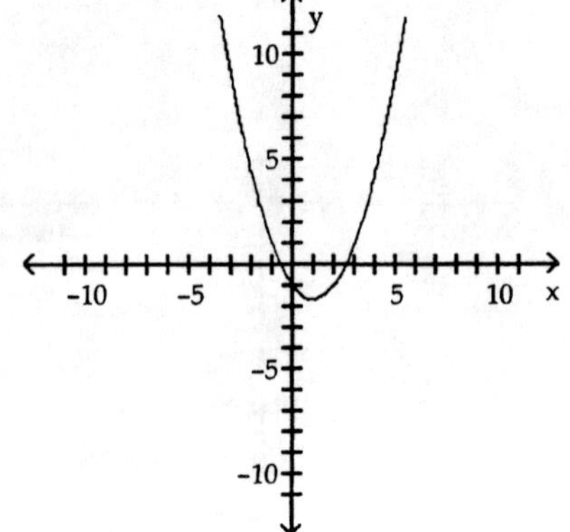

9) $f(x) = -x^2 + 6x - 3$

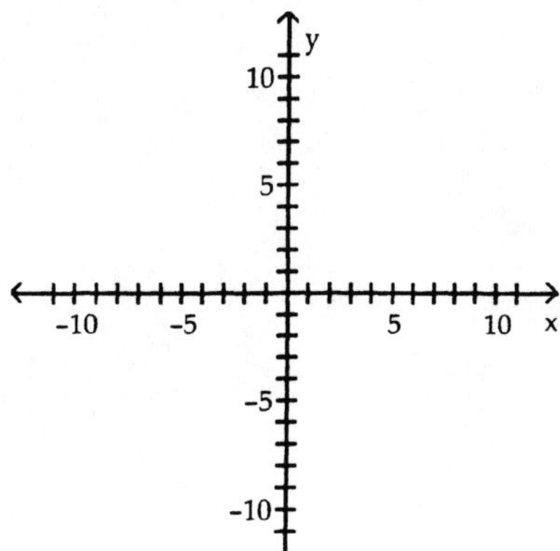

A)

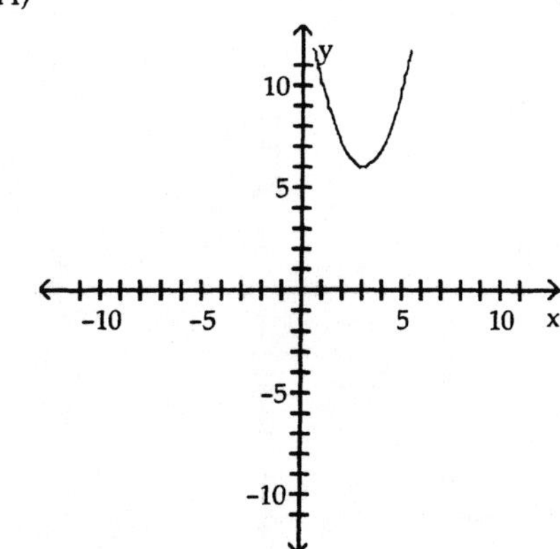

B)

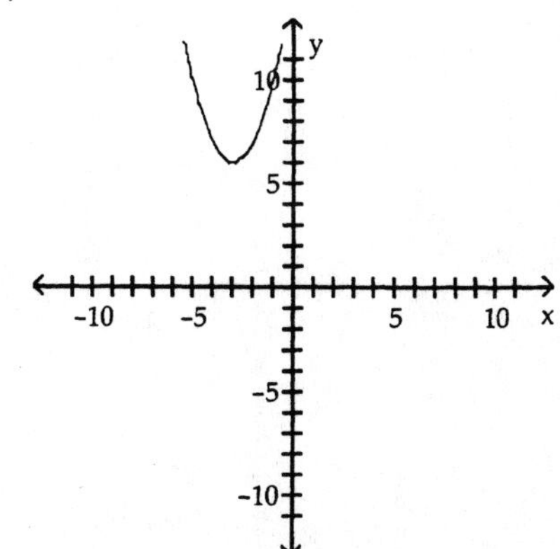

C)

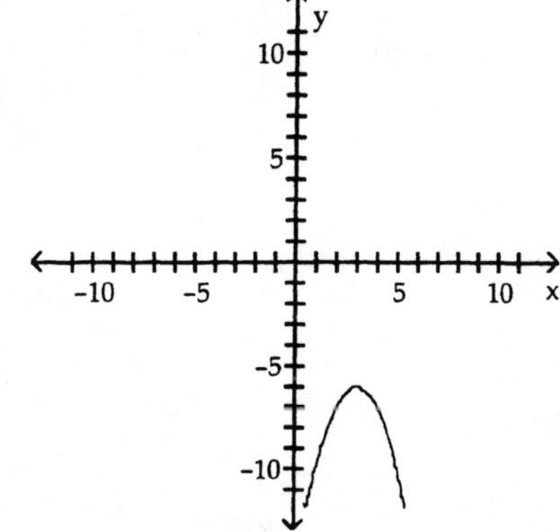

D)

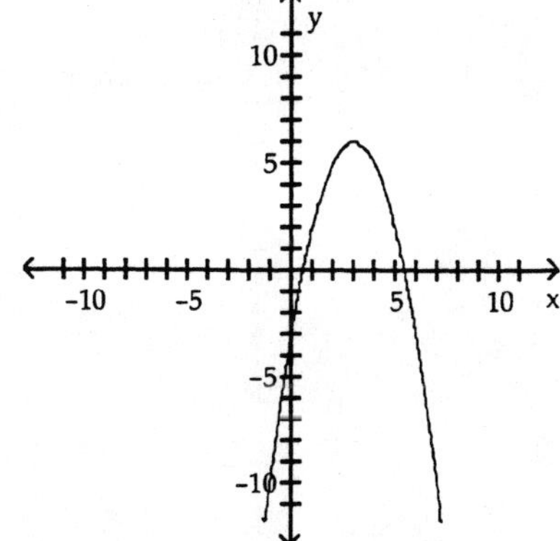

10) $f(x) = \frac{1}{5}x^2 + 4$

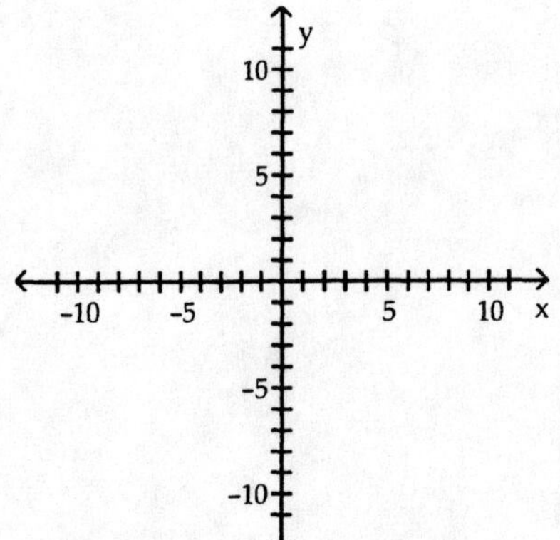

A)

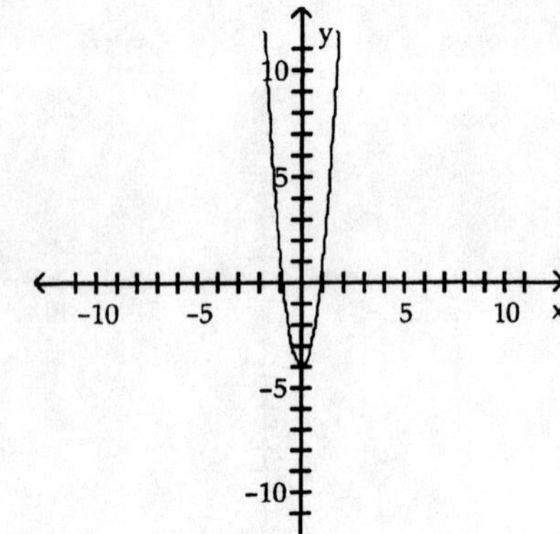

B)

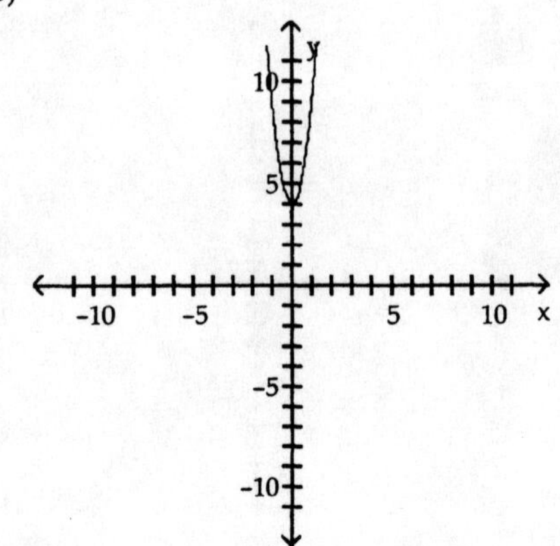

C)

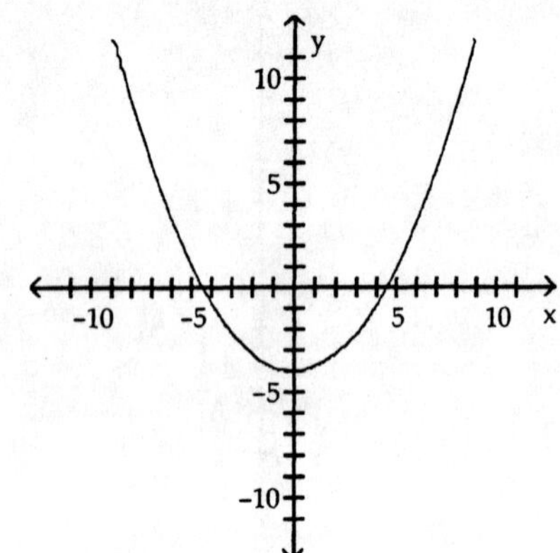

D)

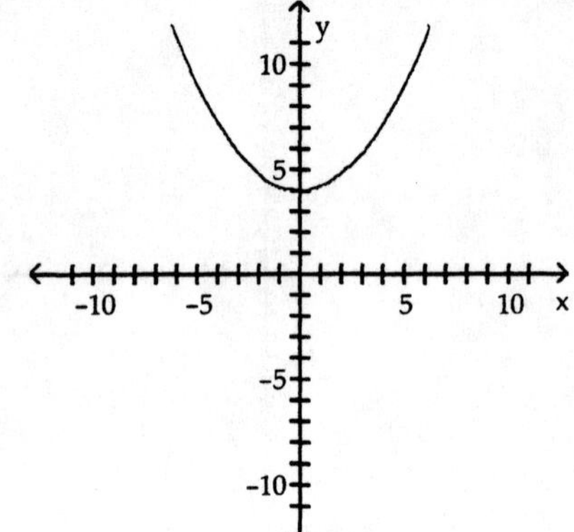

11) $f(x) = -4x^2 - 3$

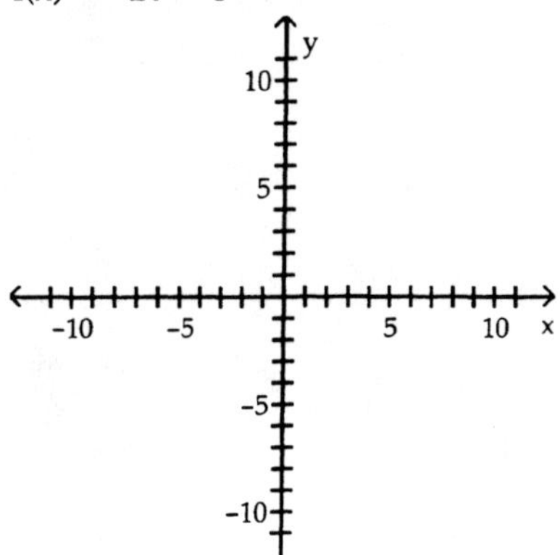

A)

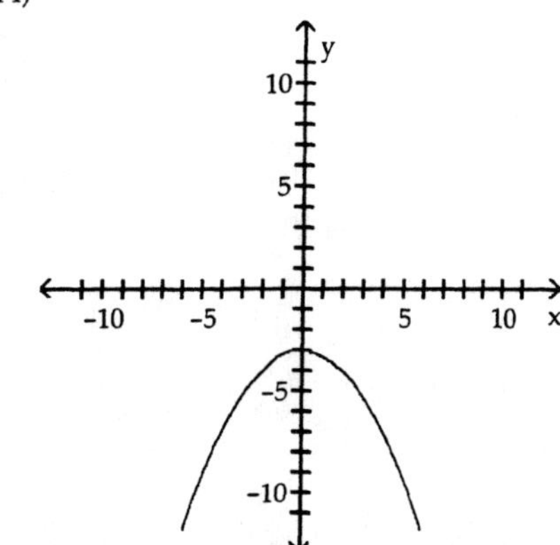

B)

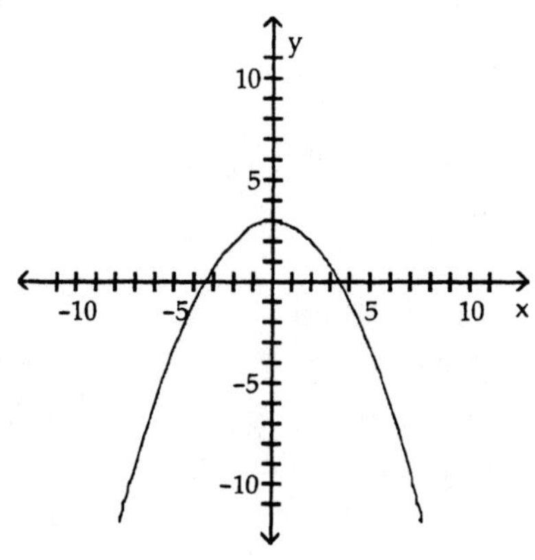

C)

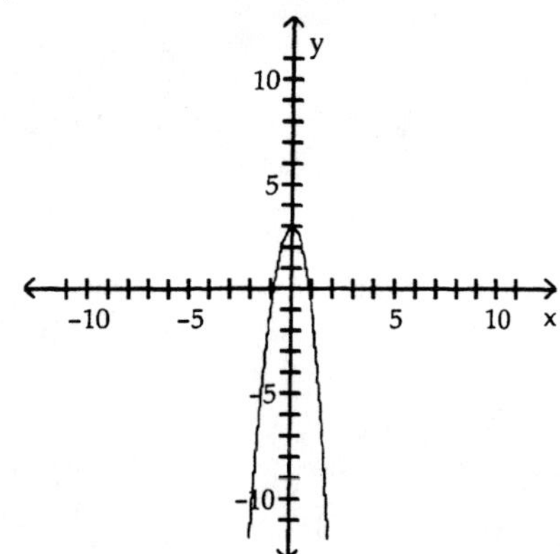

D)

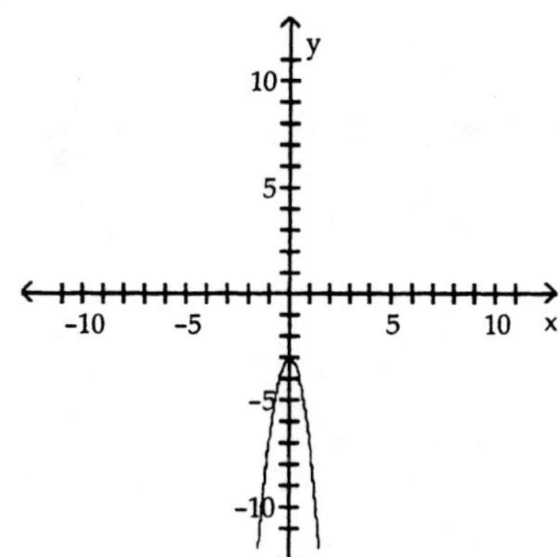

2 Identify the Vertex and Axis of Symmetry of a Quadratic Function

Find the vertex and axis of symmetry of the graph of the function.

1) $f(x) = x^2 + 4x$

A) (-2, -4); x = -2 B) (-4, 2); x = -4 C) (4, -2); x = 4 D) (2, -4); x = 2

2) $f(x) = x^2 - 4x$

A) (-2, 4); x = -2 B) (-4, 2); x = -4 C) (4, -2); x = 4 D) (2, -4); x = 2

3) $f(x) = -x^2 + 6x$

A) (9, -3); x = 9 B) (3, 9); x = 3 C) (-3, -9); x = -3 D) (-9, 3); x = -9

4) $f(x) = -x^2 - 4x$

A) $(4, -2)$; $x = 4$ B) $(2, -4)$; $x = 2$ C) $(-4, 2)$; $x = -4$ D) $(-2, 4)$; $x = -2$

5) $f(x) = 4x^2 + 24x$

A) $(-3, -36)$; $x = -3$ B) $(-3, 0)$; $x = -3$ C) $(3, 0)$; $x = 3$ D) $(3, -36)$; $x = 3$

6) $f(x) = x^2 + 8x + 7$

A) $(4, 9)$; $x = 4$ B) $(-4, -9)$; $x = -4$ C) $(-4, 9)$; $x = -4$ D) $(4, -9)$; $x = 4$

7) $f(x) = x^2 + 10x + 30$

A) $(5, 0)$; $x = 5$ B) $(0, 5)$; $x = 0$ C) $(-5, 5)$; $x = -5$ D) $(-5, -5)$; $x = -5$

8) $f(x) = x^2 + 1$

A) $(1, 0)$; $x = 1$ B) $(0, -1)$; $x = 0$ C) $(0, 1)$; $x = 0$ D) $(-1, 0)$; $x = -1$

9) $f(x) = x^2 + 6x + 13$

A) $(4, -9)$; $x = 4$ B) $(4, -3)$; $x = 4$ C) $(-4, 3)$; $x = -4$ D) $(-3, 4)$; $x = -3$

10) $f(x) = 4x^2 + 24x - 1$

A) $(3, 35)$; $x = 3$ B) $(-3, 107)$; $x = -3$ C) $(-3, -37)$; $x = -3$ D) $(3, 107)$; $x = 3$

11) $f(x) = -2x^2 + 4x + 1$

A) $(1, 3)$; $x = 1$ B) $(-1, -5)$; $x = -1$ C) $(1, 7)$; $x = 1$ D) $(-1, 3)$; $x = -1$

12) $f(x) = \frac{1}{4}x^2 - x + 5$

A) $(2, 4)$; $x = 2$ B) $(-2, 8)$; $x = -2$ C) $(-2, 8)$; $x = 8$ D) $(2, 4)$; $x = 4$

13) $f(x) = 3x^2 - 2x + 5$

A) $(\frac{1}{3}, \frac{14}{3})$; $x = \frac{14}{3}$ B) $(-\frac{1}{3}, 6)$; $x = -\frac{1}{3}$ C) $(-\frac{1}{3}, 6)$; $x = 6$ D) $(\frac{1}{3}, \frac{14}{3})$; $x = \frac{1}{3}$

14) $f(x) = 3x^2 - 2x + 6$

A) $(-\frac{1}{3}, 7)$; $x = 7$ B) $(-\frac{1}{3}, 7)$; $x = -\frac{1}{3}$ C) $(\frac{1}{3}, \frac{17}{3})$; $x = \frac{1}{3}$ D) $(\frac{1}{3}, \frac{17}{3})$; $x = \frac{17}{3}$

3 Graph a Quadratic Function Using Its Vertex, Axis and Intercepts

Graph the function using its vertex, axis of symmetry, and intercepts.

1) $f(x) = x^2 + 8x$

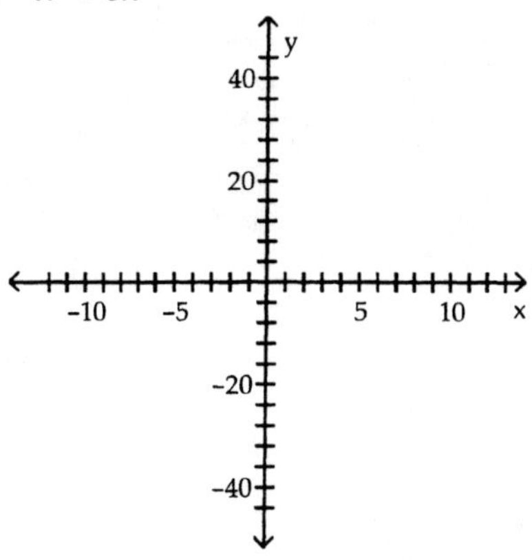

A) vertex (4, 16)
intercept (0, 32)

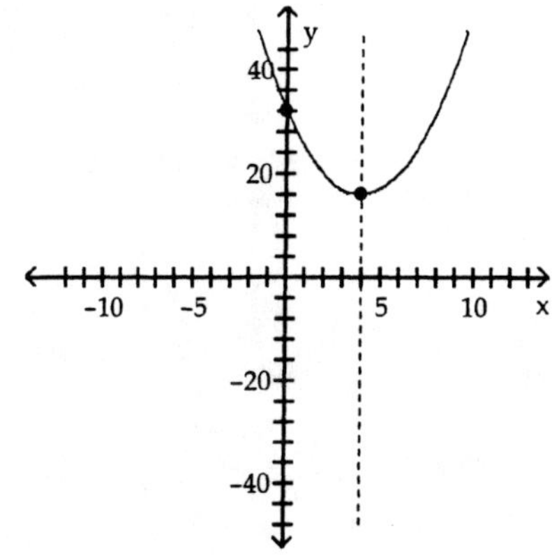

B) vertex (−4, 16)
intercept (0, 32)

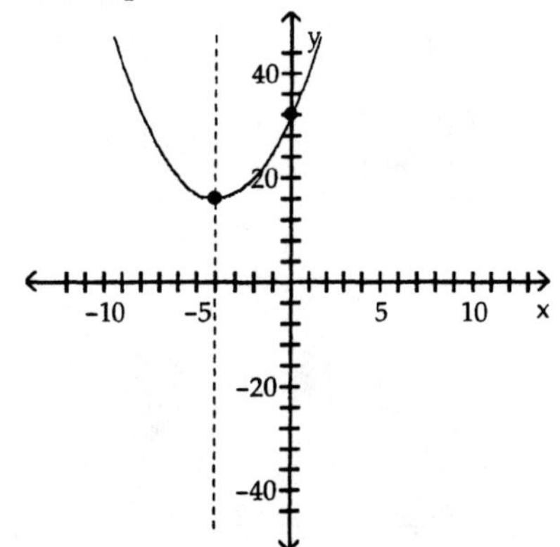

C) vertex (−4, −16)
intercepts (0, 0), (− 8, 0)

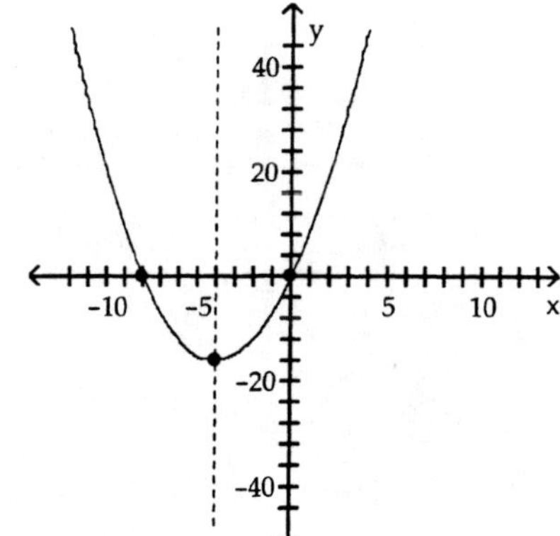

D) vertex (4, −16)
intercepts (0, 0), (8, 0)

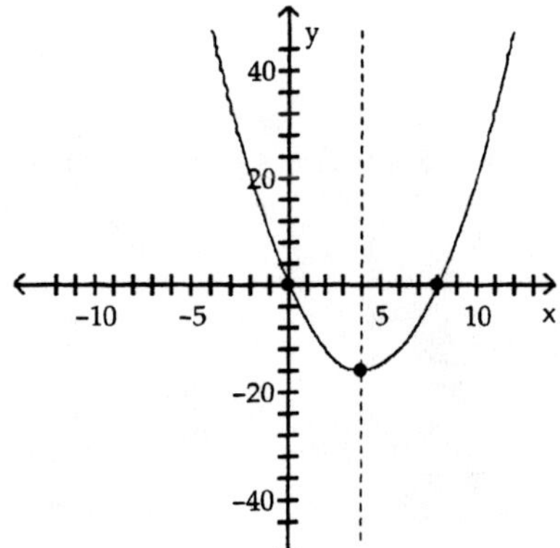

2) $f(x) = -x^2 - 12x$

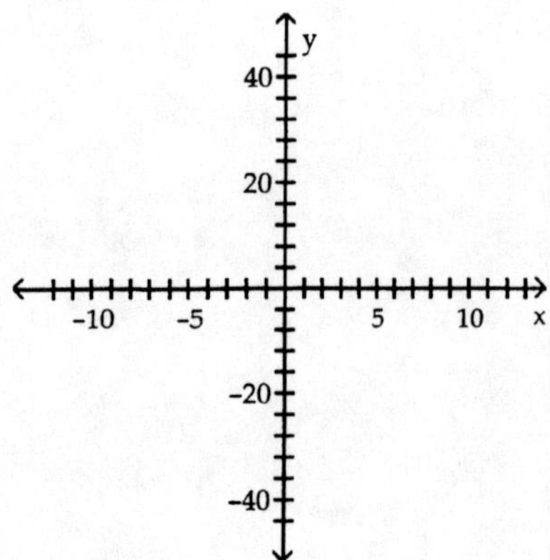

A) vertex (6, -36)
intercept (0, -72)

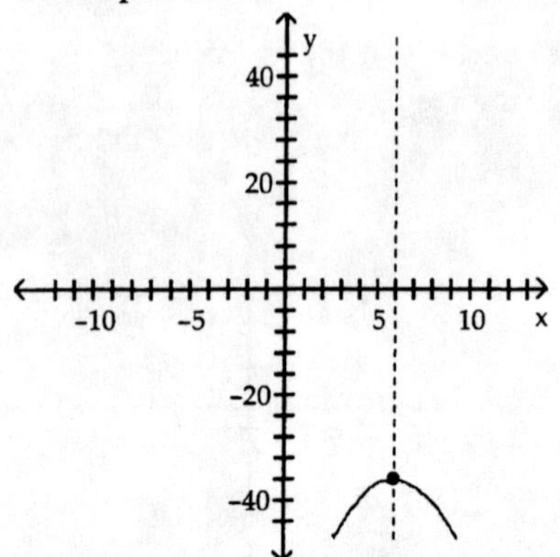

B) vertex (6, 36)
intercepts (0, 0), (12, 0)

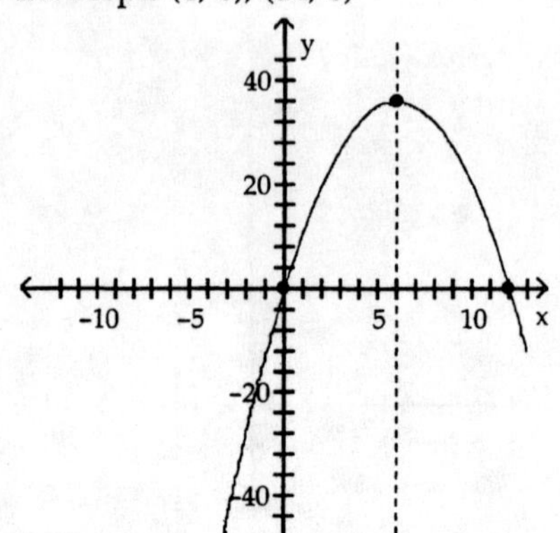

C) vertex (-6, 36)
intercepts (0, 0), (-12, 0)

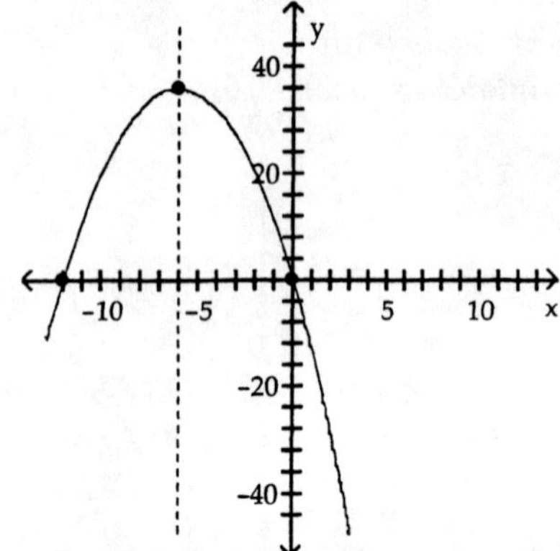

D) vertex (-6, -36)
intercept (0, -72)

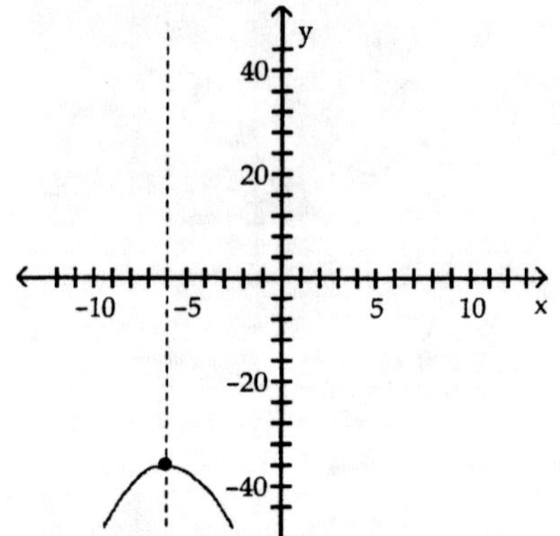

3) $f(x) = x^2 + 12x + 36$

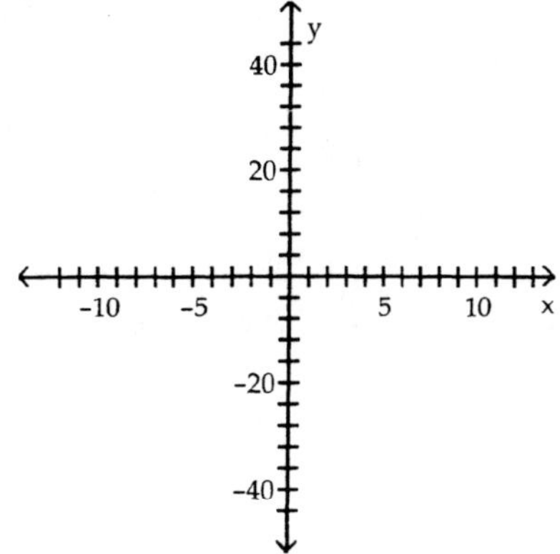

A) vertex (6, 36)
intercept (0, 72)

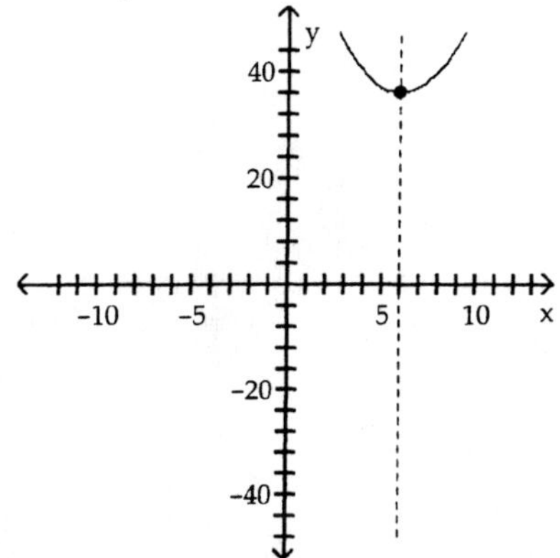

B) vertex (6, 0)
intercepts (0, 36), (6, 0)

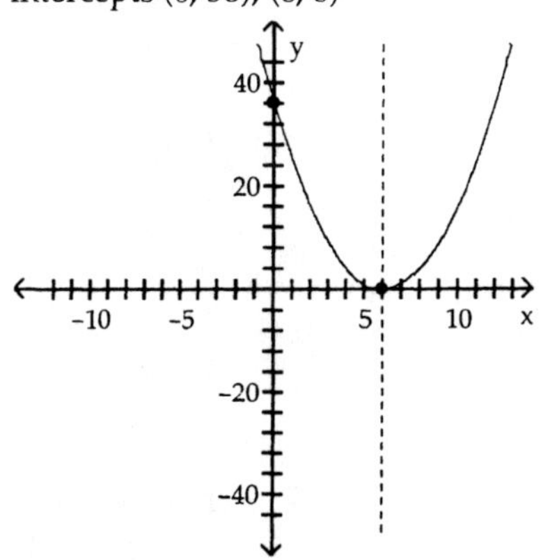

C) vertex (−6, 36)
intercept (0, 72)

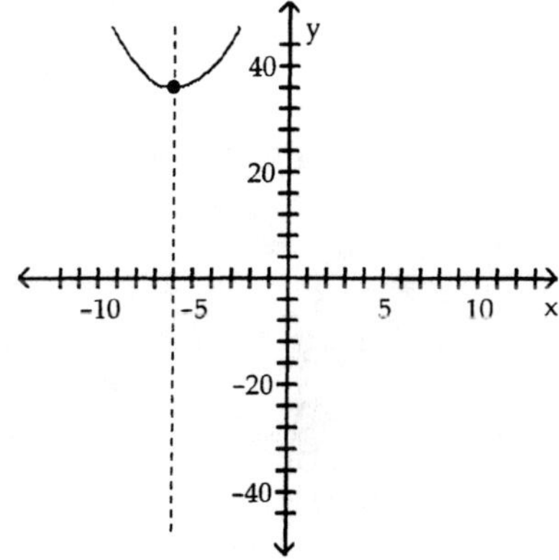

D) vertex (−6, 0)
intercepts (0, 36), (−6, 0)

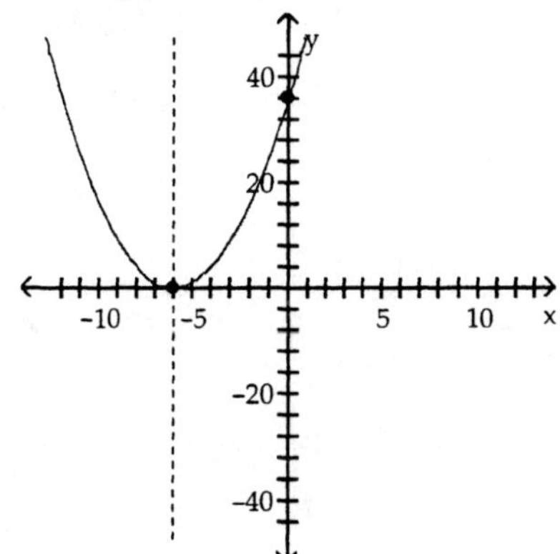

4) $f(x) = x^2 + 6x + 5$

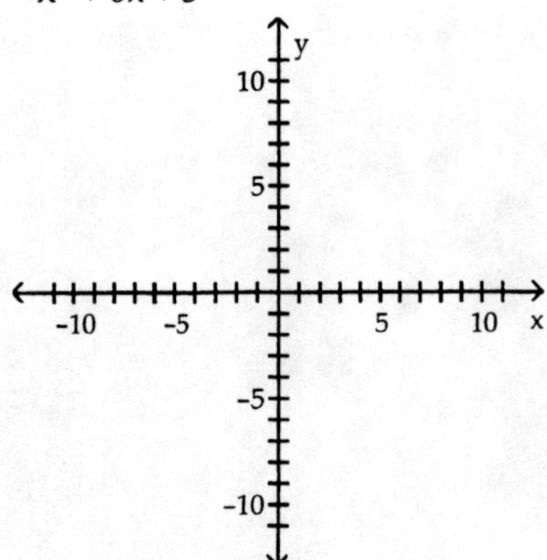

A) vertex (3, 4)
intercepts (1, 0), (5, 0), (0, -5)

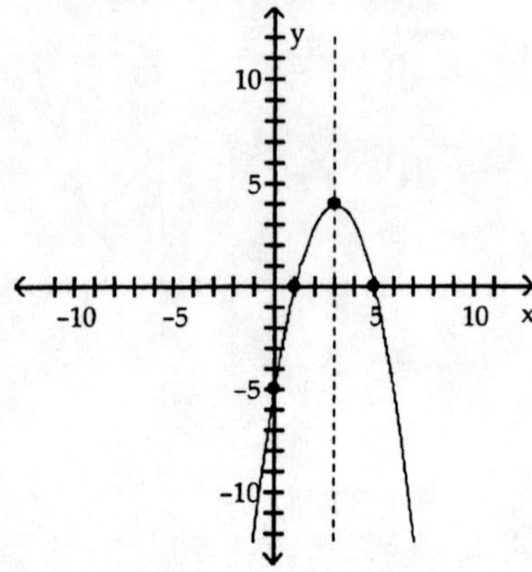

B) vertex (3, -4)
intercepts (1, 0), (5, 0), (0, 5)

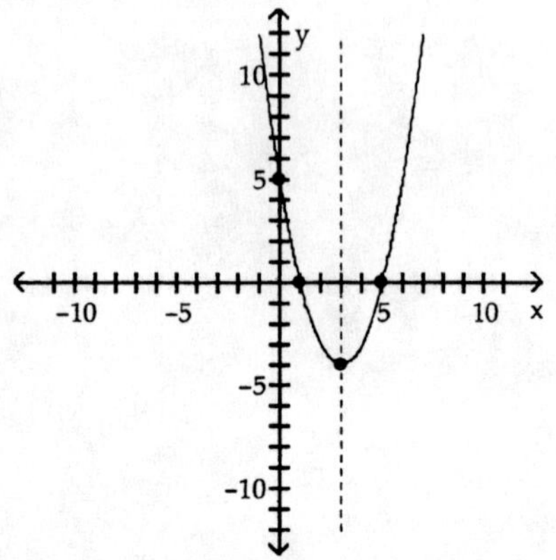

C) vertex (-3, -4)
intercepts (-1, 0), (- 5, 0), (0, 5)

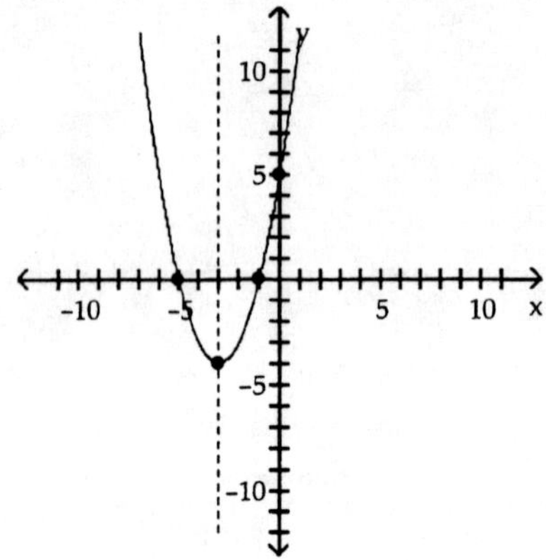

D) vertex (-3, 4)
intercepts (-1, 0), (- 5, 0), (0, -5)

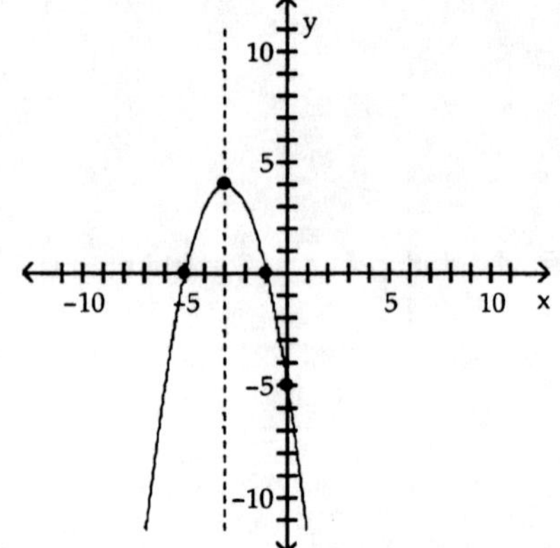

5) $f(x) = -x^2 - 6x - 8$

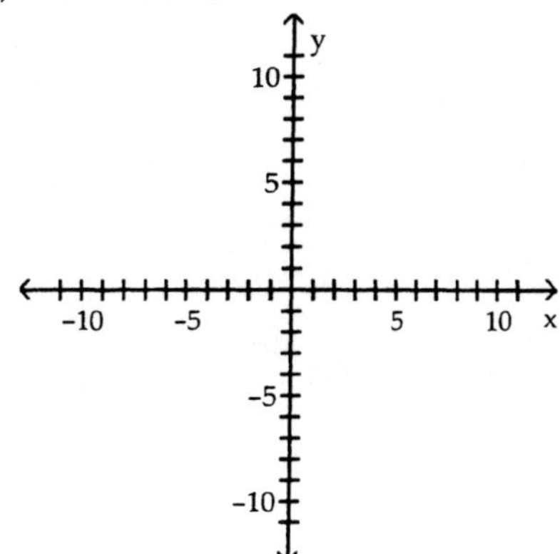

A) vertex (3, −1)
intercepts (2, 0), (4, 0), (0, 8)

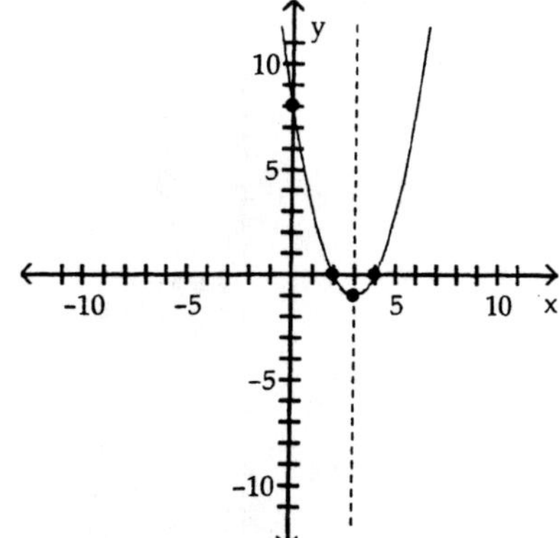

B) vertex (−3, −1)
intercepts (−2, 0), (− 4, 0), (0, 8)

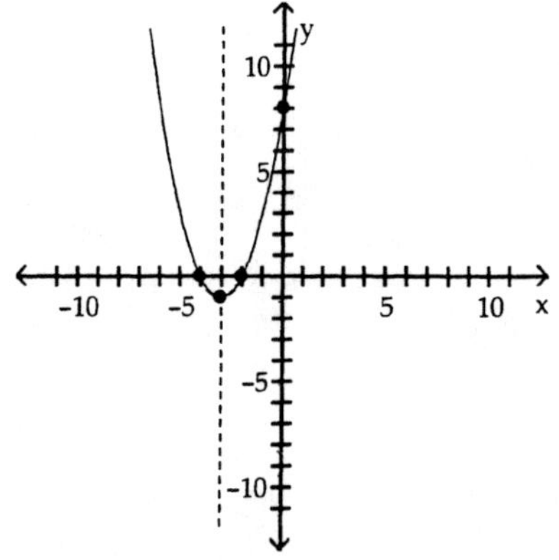

C) vertex (3, 1)
intercepts (2, 0), (4, 0), (0, −8)

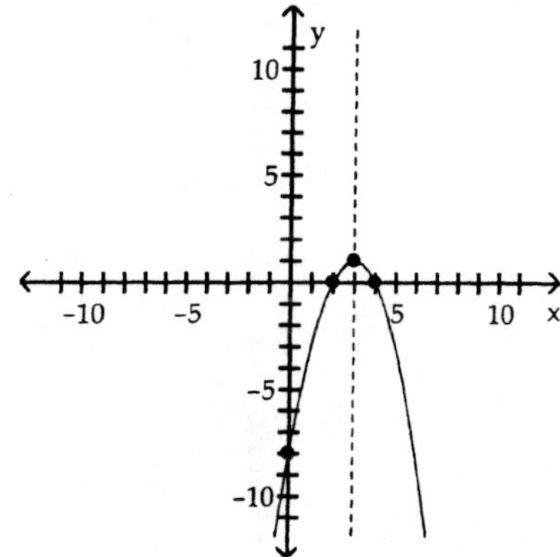

D) vertex (−3, 1)
intercepts (−2, 0), (− 4, 0), (0, −8)

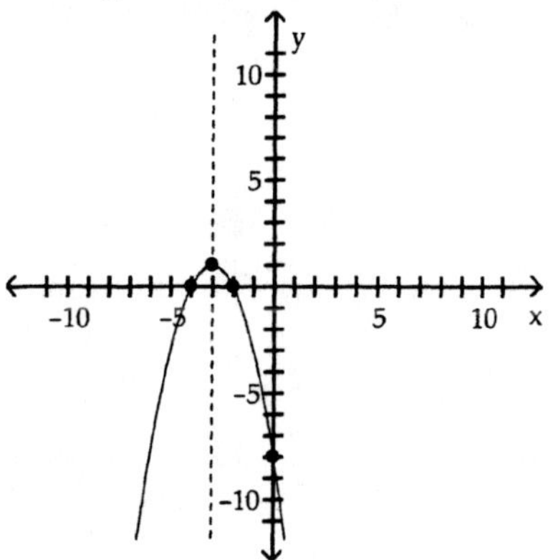

6) $f(x) = x^2 - 2x - 3$

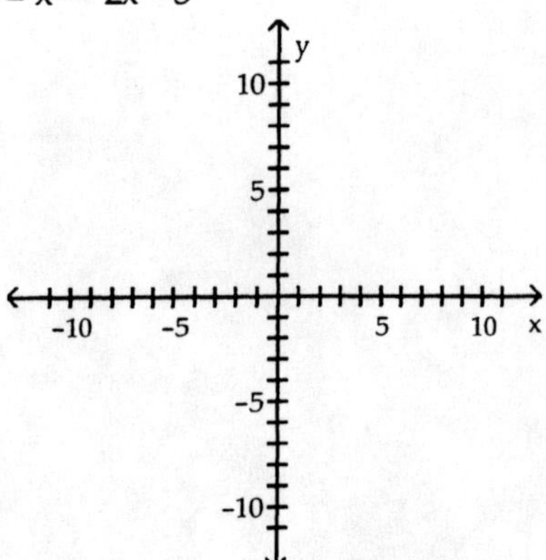

A) vertex (−1, −4)
intercepts (−3, 0), (1, 0), (0, −3)

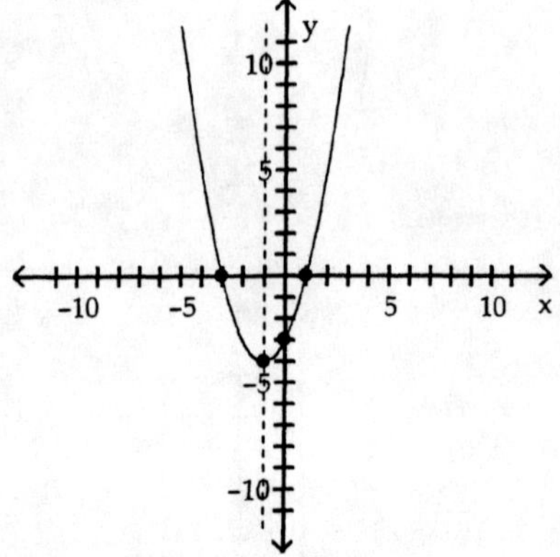

B) vertex (1, −4)
intercepts (3, 0), (− 1, 0), (0, −3)

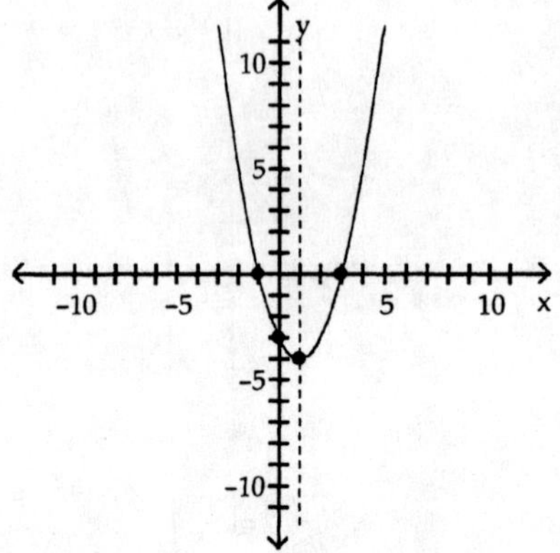

C) vertex (−1, 4)
intercepts (−3, 0), (1, 0), (0, 3)

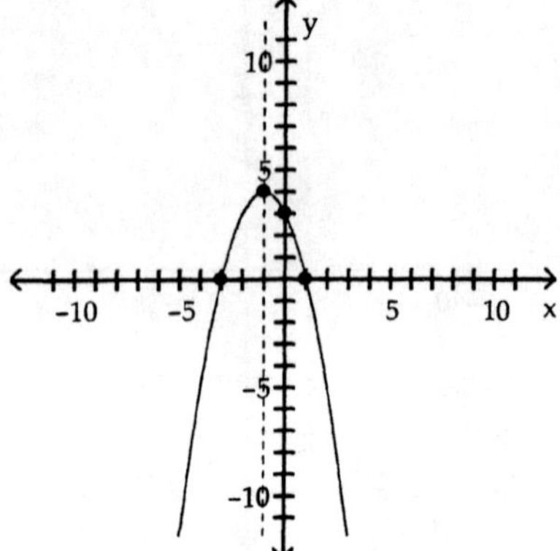

D) vertex (1, 4)
intercepts (3, 0), (− 1, 0), (0, 3)

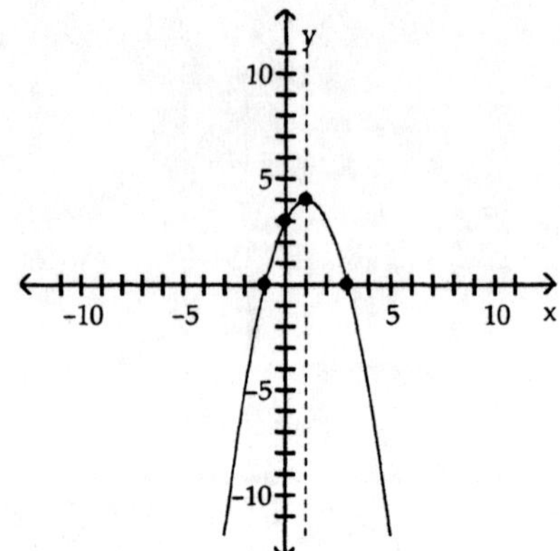

7) $f(x) = -x^2 + 4x + 5$

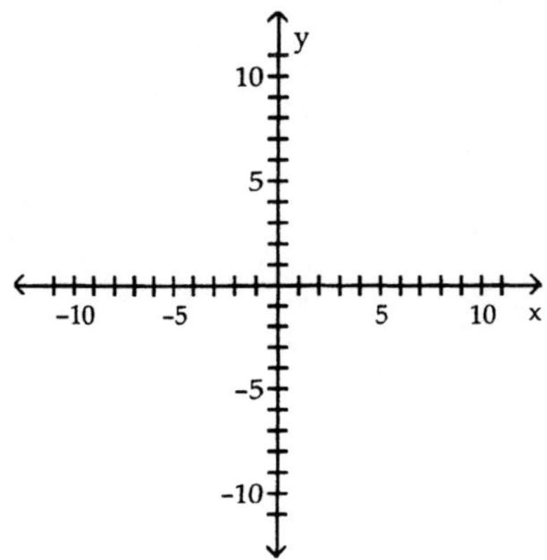

A) vertex (−2, 9)
intercepts (−5, 0), (1, 0), (0, 5)

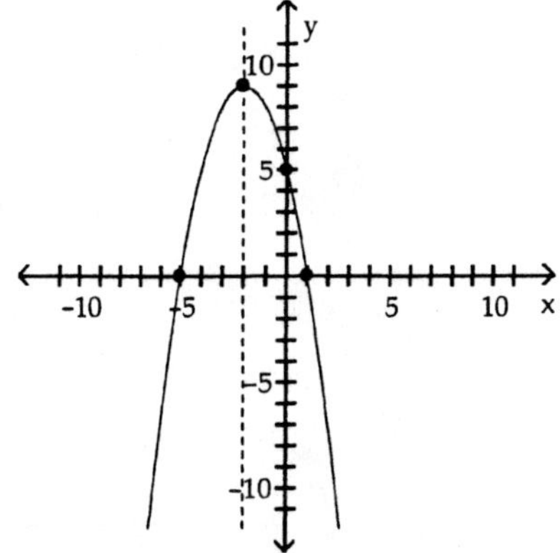

B) vertex (2, 9)
intercepts (5, 0), (− 1, 0), (0, 5)

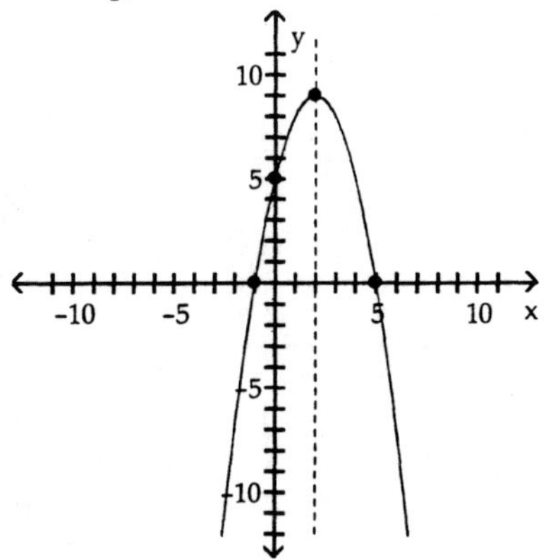

C) vertex (2, −9)
intercepts (5, 0), (− 1, 0), (0, −5)

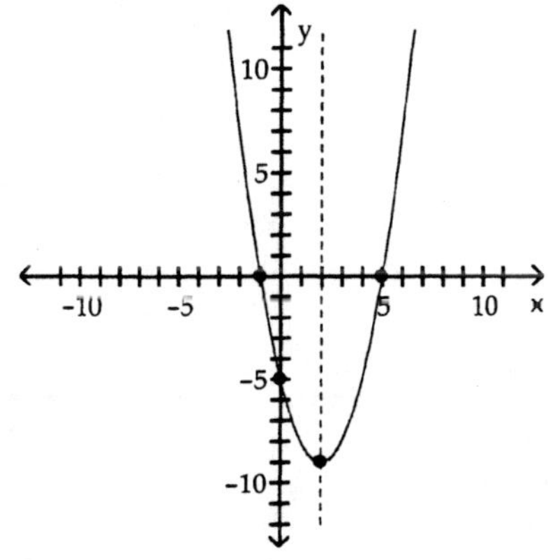

D) vertex (−2, −9)
intercepts (−5, 0), (1, 0), (0, −5)

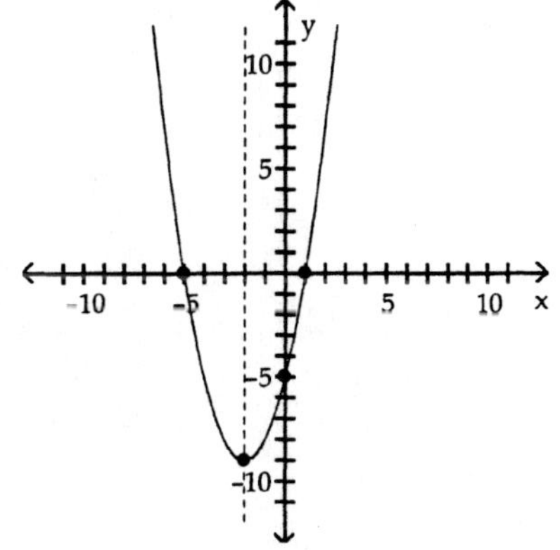

8) $f(x) = 2x^2 + 12x + 19$

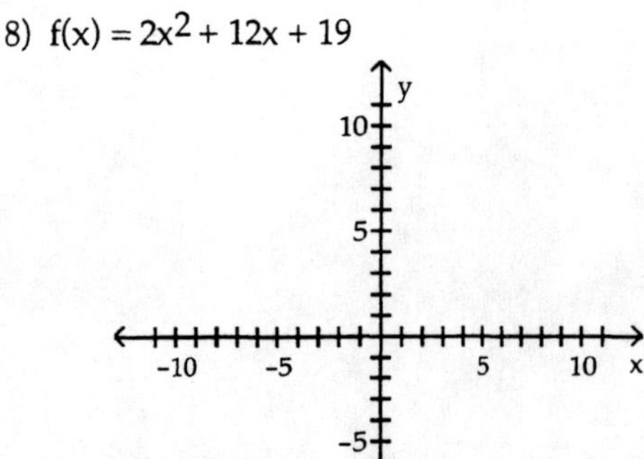

A) vertex (3, 1)

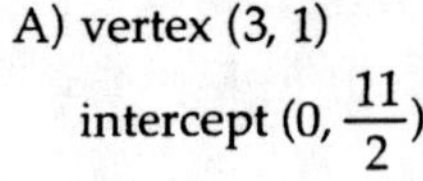
intercept $(0, \frac{11}{2})$

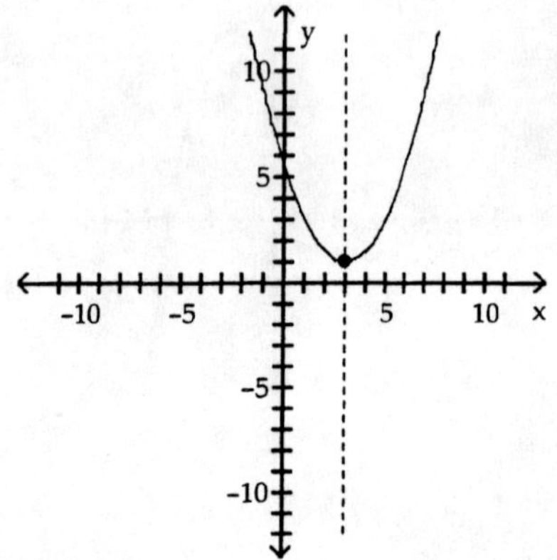

B) vertex (−3, 1)
intercept (0, 19)

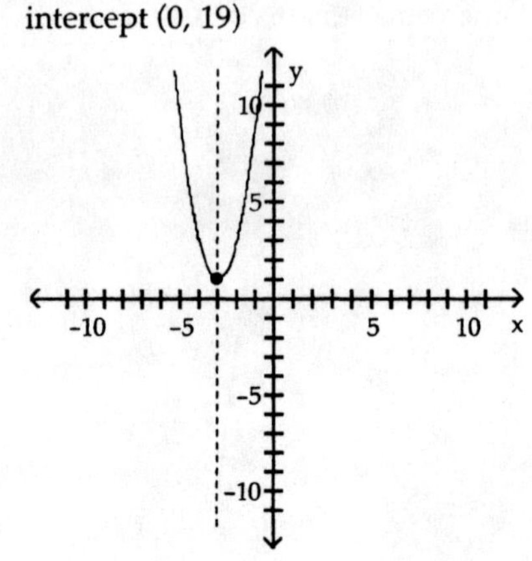

C) vertex (3, 1)
intercept (0, 19)

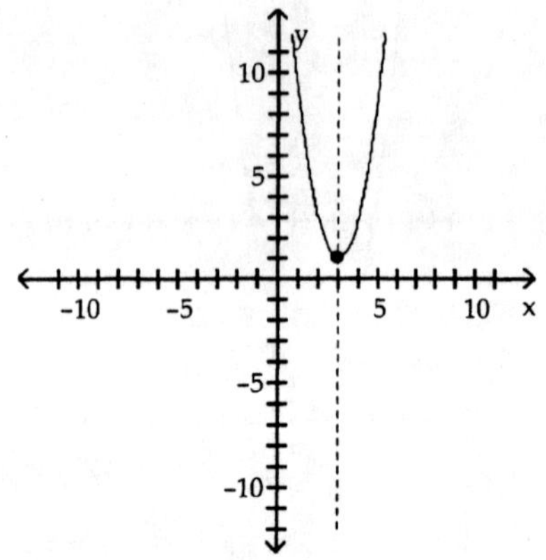

D) vertex (−3, 1)

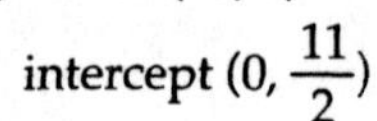
intercept $(0, \frac{11}{2})$

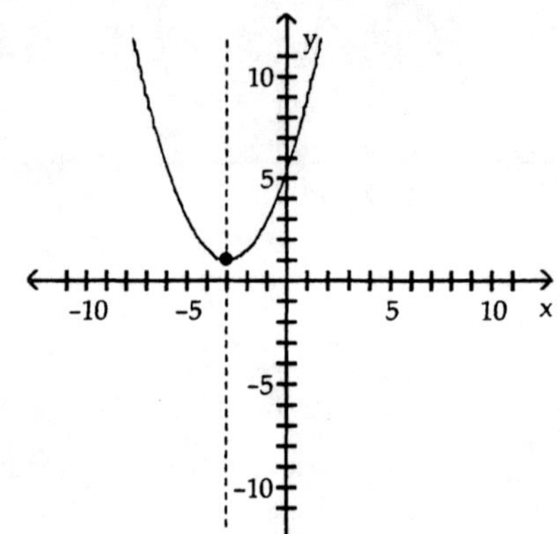

9) $f(x) = 2x^2 + 3x - 1$

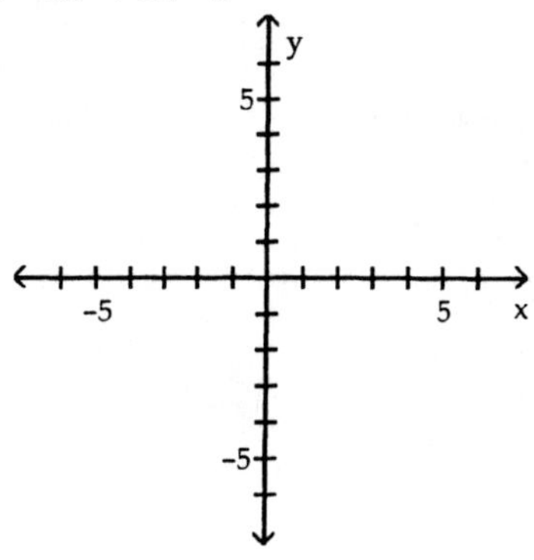

A) vertex $(-\frac{3}{2}, -\frac{13}{4})$

intercepts $(0, -1)$, $(\frac{-3 - \sqrt{13}}{2}, 0)$, $(\frac{-3 + \sqrt{13}}{2}, 0)$

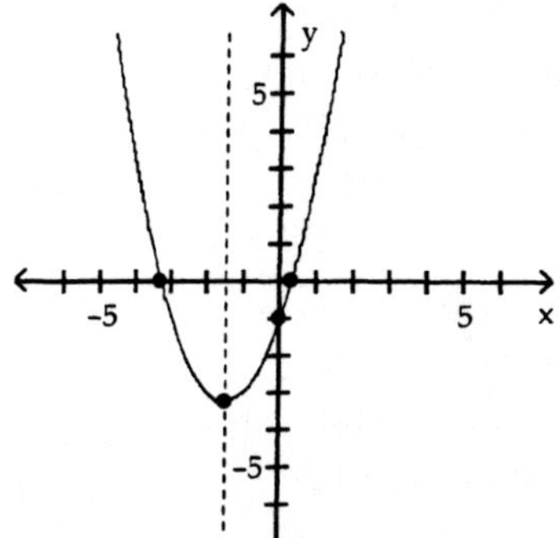

B) vertex $(-\frac{3}{4}, 0)$

intercepts $(0, 1)$ $(-\frac{3}{4}, 0)$

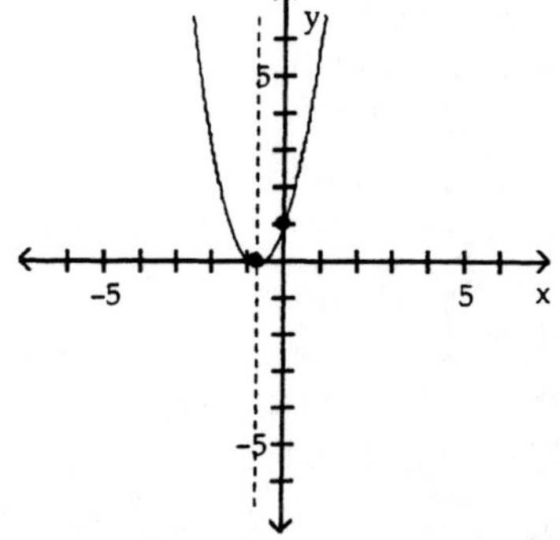

C) vertex $(-\frac{3}{4}, -\frac{17}{8})$

intercepts $(0, -1)$, $(\frac{-3 - \sqrt{17}}{4}, 0)$, $(\frac{-3 + \sqrt{17}}{4}, 0)$

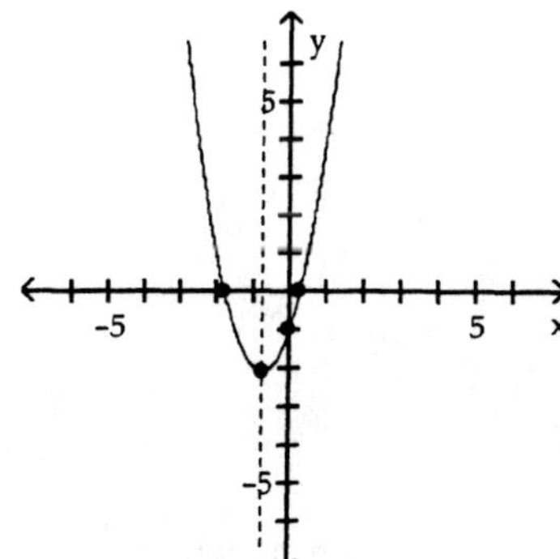

D) vertex $(\frac{3}{4}, 0)$

intercept $(0, 1)$, $(\frac{3}{4}, 0)$

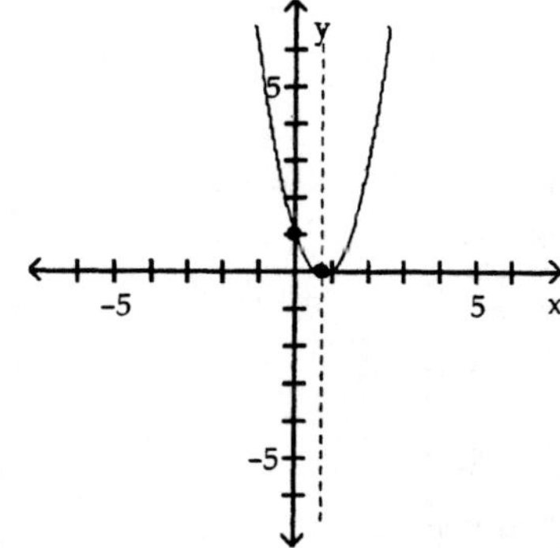

10) $f(x) = 4x^2 - x - 3$

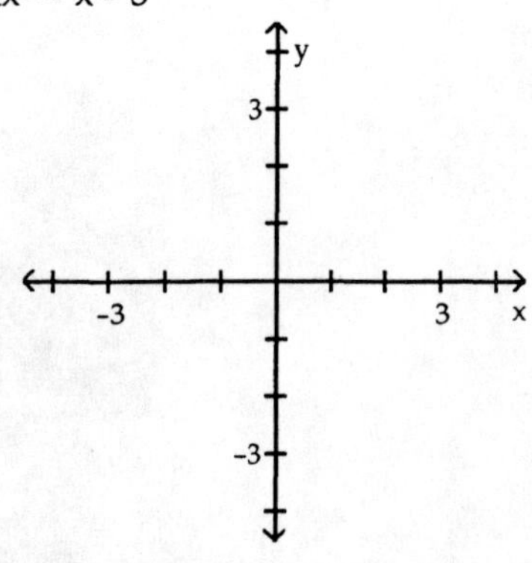

Determine the domain and the range of the function.

11) $f(x) = x^2 - 12x$

A) domain: all real numbers
range: $\{y \mid y \geq -36\}$

B) domain: $\{x \mid x \geq -6\}$
range: $\{y \mid y \geq 36\}$

C) domain: $\{x \mid x \geq 6\}$
range: $\{y \mid y \geq -36\}$

D) domain: all real numbers
range: $\{y \mid y \geq 36\}$

12) $f(x) = -x^2 + 2x$

A) domain: $\{x \mid x \leq 1\}$
range: $\{y \mid y \leq 1\}$

B) domain: $\{x \mid x \leq -1\}$
range: $\{y \mid y \leq 1\}$

C) domain: all real numbers
range: $\{y \mid y \leq -1\}$

D) domain: all real numbers
range: $\{y \mid y \leq 1\}$

13) $f(x) = x^2 + 2x + 1$

A) domain: $\{x \mid x \geq 1\}$
range: $\{y \mid y \geq 0\}$

B) domain: all real numbers
range: $\{y \mid y \geq 1\}$

C) domain: $\{x \mid x \geq -1\}$
range: $\{y \mid y \geq 0\}$

D) domain: all real numbers
range: $\{y \mid y \geq 0\}$

14) $f(x) = x^2 + 4x + 3$

A) domain: range: $\{x \mid x \geq 2\}$
range: $\{y \mid y \geq 1\}$

B) domain: all real numbers
range: $\{y \mid y \geq -1\}$

C) domain: all real numbers
range: $\{y \mid y \geq 1\}$

D) domain: range: $\{x \mid x \geq 2\}$
range: $\{y \mid y \geq -1\}$

15) $f(x) = -x^2 - 4x + 5$

A) domain: all real numbers
range: $\{y \mid y \leq 9\}$

B) domain: all real numbers
range: $\{y \mid y \leq -9\}$

C) domain: $\{x \mid x \leq -2\}$
range: $\{y \mid y \leq -9\}$

D) domain: $\{x \mid x \leq -2\}$
range: $\{y \mid y \leq 9\}$

16) $f(x) = x^2 - 8x + 7$

A) domain: all real numbers
range: $\{y \mid y \leq 9\}$

B) domain: $\{x \mid x \geq -4\}$
range: $\{y \mid y \geq -9\}$

C) domain: all real numbers
range: $\{y \mid y \geq -9\}$

D) domain: all real numbers
range: all real numbers

17) $f(x) = -x^2 + 6x - 5$

A) domain: $\{x \mid x \le -3\}$
range: $\{y \mid y \le 4\}$

B) domain: all real numbers
range: $\{y \mid y \le -4\}$

C) domain: all real numbers
range: all real numbers

D) domain: all real numbers
range: $\{y \mid y \le 4\}$

18) $f(x) = 2x^2 + 3x - 1$

A) domain: all real numbers
range: $\{y \mid y \ge -\frac{17}{8}\}$

B) domain: all real numbers
range: all real numbers

C) domain: $\{x \mid x \ge -\frac{3}{4}\}$
range: all real numbers

D) domain: $\{x \mid x \ge -\frac{3}{4}\}$
range: $\{y \mid y \ge -\frac{17}{8}\}$

19) $f(x) = 4x^2 - x - 3$

Determine where the function is increasing and where it is decreasing.

20) $f(x) = x^2 + 10x$

A) increasing on $(-\infty, -5)$
decreasing on $(-5, \infty)$

B) increasing on $(-\infty, 5)$
decreasing on $(5, \infty)$

C) increasing on $(-5, \infty)$
decreasing on $(-\infty, -5)$

D) increasing on $(5, \infty)$
decreasing on $(-\infty, 5)$

21) $f(x) = -x^2 - 6x$

A) increasing on $(-3, \infty)$
decreasing on $(-\infty, -3)$

B) increasing on $(-\infty, 3)$
decreasing on $(3, \infty)$

C) increasing on $(-\infty, -3)$
decreasing on $(-3, \infty)$

D) increasing on $(3, \infty)$
decreasing on $(-\infty, 3)$

22) $f(x) = x^2 - 8x + 16$

A) increasing on $(-\infty, 4)$
decreasing on $(4, \infty)$

B) increasing on $(-4, \infty)$
decreasing on $(-\infty, -4)$

C) increasing on $(4, \infty)$
decreasing on $(-\infty, 4)$

D) increasing on $(-\infty, -4)$
decreasing on $(-4, \infty)$

23) $f(x) = x^2 + 6x + 5$

A) increasing on $(-\infty, -3)$
decreasing on $(-3, \infty)$

B) increasing on $(-\infty, -4)$
decreasing on $(-4, \infty)$

C) increasing on $(-3, \infty)$
decreasing on $(-\infty, -3)$

D) increasing on $(-4, \infty)$
decreasing on $(-\infty, -4)$

24) $f(x) = -x^2 - 2x + 3$

A) increasing on $(-\infty, -1)$
decreasing on $(-1, \infty)$

B) increasing on $(4, \infty)$
decreasing on $(-\infty, 4)$

C) increasing on $(-\infty, 4)$
decreasing on $(4, \infty)$

D) increasing on $(-1, \infty)$
decreasing on $(-\infty, -1)$

25) $f(x) = x^2 - 8x + 7$

A) increasing on $(-\infty, 4)$
decreasing on $(4, \infty)$

B) increasing on $(4, \infty)$
decreasing on $(-\infty, 4)$

C) increasing on $(-9, \infty)$
decreasing on $(-\infty, -9)$

D) increasing on $(-\infty, -9)$
decreasing on $(-9, \infty)$

26) $f(x) = -x^2 + 4x - 3$

A) increasing on $(-\infty, 2)$
decreasing on $(2, \infty)$

B) increasing on $(-\infty, 1)$
decreasing on $(1, \infty)$

C) increasing on $(1, \infty)$
decreasing on $(-\infty, 1)$

D) increasing on $(2, \infty)$
decreasing on $(-\infty, 2)$

27) $f(x) = 2x^2 + 3x - 1$

A) increasing on $(-\infty, -\frac{3}{4})$
decreasing on $(-\frac{3}{4}, \infty)$

B) increasing on $-\infty, -\frac{17}{8})$
decreasing on $(-\frac{17}{8}, \infty)$

C) increasing on $(-\frac{17}{8}, \infty)$
decreasing on $-\infty, -\frac{17}{8})$

D) increasing on $(-\frac{3}{4}, \infty)$
decreasing on $(-\infty, -\frac{3}{4})$

28) $f(x) = 4x^2 - x - 3$

A) increasing on $(-\frac{49}{16}, \infty)$
decreasing on $(-\infty, -\frac{49}{16})$

B) increasing on $(-\infty, -\frac{49}{16})$
decreasing on $(-\frac{49}{16}, \infty)$

C) increasing on $(\frac{1}{8}, \infty)$
decreasing on $(-\infty, \frac{1}{8})$

D) increasing on $(-\infty, \frac{1}{8})$
decreasing on $(\frac{1}{8}, \infty)$

4 Use the Maximum or Minimum Value of a Quadratic Function to Solve Applied Problems

Determine, without graphing, whether the given quadratic function has a maximum value or a minimum value and then find that value.

1) $f(x) = x^2 + 6$

A) maximum; 0 B) minimum; 6 C) minimum; 0 D) maximum; 6

2) $f(x) = x^2 - 9$

A) minimum; 0 B) minimum; -9 C) maximum; 0 D) maximum; -9

3) $f(x) = x^2 - 2x - 2$

A) minimum; 1 B) minimum; - 3 C) maximum; - 3 D) maximum; 1

4) $f(x) = -x^2 - 2x - 9$

A) maximum; - 1 B) minimum; - 8 C) minimum; - 1 D) maximum; - 8

5) $f(x) = 2x^2 + 2x + 2$

A) maximum; $\frac{3}{2}$
B) minimum; $-\frac{1}{2}$
C) maximum; $-\frac{1}{2}$
D) minimum; $\frac{3}{2}$

6) $f(x) = 2x^2 - 2x$

A) maximum; $\frac{1}{2}$
B) minimum; $-\frac{1}{2}$
C) maximum; $-\frac{1}{2}$
D) minimum; $\frac{1}{2}$

7) $f(x) = -2x^2 - 2x$

A) maximum; $\frac{1}{2}$
B) maximum; $-\frac{1}{2}$
C) minimum; $-\frac{1}{2}$
D) minimum; $\frac{1}{2}$

Solve the problem.

8) The manufacturer of a CD player has found that the revenue R (in dollars) is $R(p) = -5p^2 + 1840p$, when the unit price is p dollars. If the manufacturer sets the price p to maximize revenue, what is the maximum revenue to the nearest whole dollar?

A) \$338,560
B) \$169,280
C) \$1,354,240
D) \$677,120

9) The owner of a video store has determined that the cost C, in dollars, of operating the store is approximately given by $C(x) = 2x^2 - 28x + 730$, where x is the number of videos rented daily. Find the lowest cost to the nearest dollar.

A) \$632
B) \$828
C) \$534
D) \$338

10) A projectile is fired from a cliff 100 feet above the water at an inclination of 45° to the horizontal, with a muzzle velocity of 130 feet per second. The height h of the projectile above the water is given by $h(x) = \frac{-32x^2}{(130)^2} + x + 100$, where x is the horizontal distance of the projectile from the base of the cliff. Find the maximum height of the projectile.

A) 496.09 ft
B) 264.06 ft
C) 132.03 ft
D) 232.03 ft

11) Consider the quadratic model $h(t) = -16t^2 + 40t + 50$ for the height (in feet), h, of an object t seconds after the object has been projected straight up into the air. Find the maximum height attained by the object. How much time does it take to fall back to the ground? Assume that it takes the same time for going up and coming down.

A) maximum height = 50 ft; time to reach ground = 2.5 seconds
B) maximum height = 75 ft; time to reach ground = 1.25 seconds
C) maximum height = 50 ft; time to reach ground = 1.25 seconds
D) maximum height = 75 ft; time to reach ground = 2.5 seconds

12) A developer wants to enclose a rectangular grassy lot that borders a city street for parking. If the developer has 352 feet of fencing and does not fence the side along the street, what is the largest area that can be enclosed?

A) 30,976 ft^2
B) 7744 ft^2
C) 15,488 ft^2
D) 23,232 ft^2

13) Alan is building a garden shaped like a rectangle with a semicircle attached to one short side. If he has 60 feet of fencing to go around it, what dimensions will give him the maximum area in the garden?

A) width $= \frac{120}{\pi + 4} \approx 16.8$, length = 8.4
B) width $= \frac{120}{\pi + 4} \approx 16.8$, length = 21.6
C) width $= \frac{120}{\pi + 8} \approx 10.8$, length = 16.1
D) width $= \frac{60}{\pi + 4} \approx 8.4$, length = 16.8

14) The price p dollars and the quantity x sold of a certain product obey the demand equation

$x = -15p + 450;\ 0 \le p \le 30.$

a) Express the revenue R as a function of x.

b) What quantity x maximizes the revenue?

c) What price should the company charge to maximize revenue?

15) A suspension bridge has twin towers that are 1300 feet apart. Each tower extends 180 feet above the road surface. The cables are parabolic in shape and are suspended from the tops of the towers. The cables touch the road surface at the center of the bridge. Find the height of the cable at a point 200 feet from the center of the bridge.

16) The price p and x, the quantity of a certain product sold, obey the demand equation

$p = -\frac{1}{10}x + 100,\ 0 \le x \le 1000.$

a) Express the revenue R as a function of x.

b) What is the revenue if 450 units are sold?

c) Graph the revenue function using a graphing utility.

d) What quantity x maximizes revenue? What is the maximum revenue?

e) What price should the company charge to maximize revenue?

5 Use a Graphing Utility to Find the Quadratic Function of Best Fit to Data

Use a graphing calculator to plot the data and find the quadratic function of best fit.

1) Southern Granite and Marble sells granite and marble by the square yard. One of its granite patterns is price sensitive. If the price is too low, customers perceive that it has less quality. If the price is too high, customers perceive that it is overpriced. The company conducted a pricing test with potential customers. The following data was collected. Use a graphing calculator to plot the data. What is the quadratic function of best fit?

Price, x	Buyers, B
$20	30
$30	50
$40	65
$60	75
$80	72
$100	50
$110	25

A) $B(x) = 0.0243x^2 - 3.115x - 22.13$

B) $B(x) = -0.0243x^2 + 3.115x - 22.13$

C) $B(x) = -0.0243x^2 + 3.115x + 22.13$

D) $B(x) = -0.243x^2 + 3.115x - 22.13$

2) An engineer collects data showing the speed s of a given car model and its average miles per gallon M. Use a graphing calculator to plot the scatter diagram. What is the quadratic function of best fit?

Speed, s	mph, M
20	18
30	20
40	23
50	25
60	28
70	24
80	22

A) $M(s) = -0.631x^2 + 0.720x + 5.142$

B) $M(s) = 0.063x^2 + 0.720x + 5.142$

C) $M(s) = -6.309x^2 + 0.720x + 5.142$

D) $M(s) = -0.0063x^2 + 0.720x + 5.142$

3) The number of housing starts in one beachside community remained fairly level until 1992 and then began to increase. The following data shows the number of housing starts since 1992 ($x = 1$). Use a graphing calculator to plot a scatter diagram. What is the quadratic function of best fit?

Year, x	Housing Starts, H
1	200
2	205
3	210
4	240
5	245
6	230
7	220
8	210

A) $H(x) = -2.679x^2 - 26.607x + 168.571$

B) $H(x) = -2.679x^2 + 26.607x - 168.571$

C) $H(x) = 2.679x^2 + 26.607x + 168.571$

D) $H(x) = -2.679x^2 + 26.607x + 168.571$

4) The number of housing starts in one beachside community remained fairly level until 1992 and then began to increase. The following data shows the number of housing starts since 1992 ($x = 1$). Use a graphing calculator to plot a scatter diagram. What is the quadratic function of best fit?

Year, x	Housing Starts, H
1	200
2	210
3	230
4	240
5	250
6	230
7	215
8	208

A) $H(x) = 3.268x^2 + 30.494x + 168.982$

B) $H(x) = -3.268x^2 - 30.494x + 168.982$

C) $H(x) = -3.268x^2 + 30.494x + 168.982$

D) $H(x) = -3.268x^2 + 30.494x - 168.982$

5) A small manufacturing firm collected the following data on advertising expenditures (in thousands of dollars) and total revenue (in thousands of dollars).

Advertising, x	Total Revenue, R
25	6430
28	6432
31	6434
32	6434
34	6434
39	6431
40	6432
45	6420

Find the quadratic function of best fit.

A) $R(x) = -0.31x^2 + 2.63x + 6128$
B) $R(x) = -0.015x^2 + 4.53x + 6123$
C) $R(x) = -0.091x^2 + 5.95x + 6337$
D) $R(x) = -0.024x^2 + 7.13x + 6209$

6) The following data represents the total revenue, R (in dollars), received from selling x bicycles at Tunney's Bicycle Shop. Using a graphing utility, find the quadratic function of best fit using coefficients rounded to the nearest hundredth.

Number of Bicycles, x	Total Revenue, R (in dollars)
0	0
22	27,000
70	46,000
96	55,200
149	61,300
200	64,000
230	64,500
250	67,000

7) The following table shows the median number of hours of leisure time that Americans had each week in various years.

Year	1973	1980	1987	1993	1997
Median # of Leisure hrs per Week	26.2	19.2	16.6	18.8	19.5

Use $x = 0$ to represent the year 1973. Using a graphing utility, determine the quadratic regression equation for the data given. What year corresponds to the time when Americans had the least time to spend on leisure?

6 Additional Applications

Determine the quadratic function whose graph is given.

1)

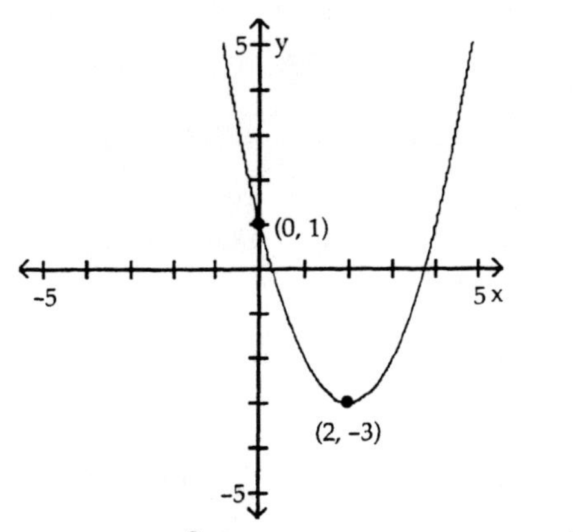

A) $f(x) = -x^2 + 4x - 1$ B) $f(x) = x^2 - 4x + 1$ C) $f(x) = x^2 + 8x + 1$ D) $f(x) = -x^2 - 4x + 1$

4.2 Polynomial Functions

1 Identify Polynomials and Their Degree

State whether the function is a polynomial function or not. If it is, give its degree. If it is not, tell why not.

1) $f(x) = 3x + 6x^5$

A) Yes; degree 6 B) Yes; degree 1 C) Yes; degree 3 D) Yes; degree 5

2) $f(x) = 20x^3 - 2x^2 - 4$

A) No; the last term has no variable
B) Yes; degree 5
C) Yes; degree 3
D) Yes; degree 6

3) $f(x) = \dfrac{9 - x^2}{4}$

A) No; x is a negative term
B) No; it is a ratio
C) Yes; degree 2
D) Yes; degree 1

4) $f(x) = \dfrac{1}{2} - \dfrac{1}{4}x$

A) Yes; degree 0
B) Yes; degree 4
C) No; x has a fractional coefficient
D) Yes; degree 1

5) $f(x) = 6$

A) Yes; degree 0
B) Yes; degree 1
C) No; it is a constant
D) No; it contains no variables

6) $f(x) = 1 + \dfrac{5}{x}$

A) Yes; degree 0
B) No; x is raised to a negative power
C) Yes; degree 5
D) Yes; degree 1

7) $f(x) = x(x - 14)$

A) Yes; degree 2 B) No; it is a product C) Yes; degree 1 D) Yes; degree 0

8) $f(x) = 8 - \frac{2}{x^2}$

A) Yes; degree $\frac{1}{2}$

B) Yes; degree 2

C) Yes; degree -2

D) No; x is raised to the negative 2 power

9) $f(x) = \frac{x^3 - 7}{x^2}$

A) Yes; degree -2

B) No; it is a ratio of polynomials

C) Yes; degree 2

D) Yes; degree 3

10) $f(x) = x^{5/4} - x^2 + 2$

A) Yes; degree 5

B) Yes; degree 2

C) Yes; degree 5/4

D) No; x is raised to non-integer 5/4 power

11) $9(x - 1)^{11}(x + 1)^6$

A) Yes; degree 99 B) Yes; degree 11 C) Yes; degree 17 D) Yes; degree 9

12) $f(x) = \sqrt{x}(\sqrt{x} - 10)$

A) No; x is raised to non-integer power

B) No; it is a product

C) Yes; degree 1

D) Yes; degree 2

13) $f(x) = -12x^3 + \pi x^2 + 1$

A) Yes; degree 3

B) No; x^2 has a non-integer coefficient

C) Yes; degree 5

D) Yes; degree 6

2 Graph Polynomial Functions Using Transformations

Use transformations of the graph of $y = x^4$ or $y = x^5$ to graph the function.

1) $f(x) = (x - 4)^4$

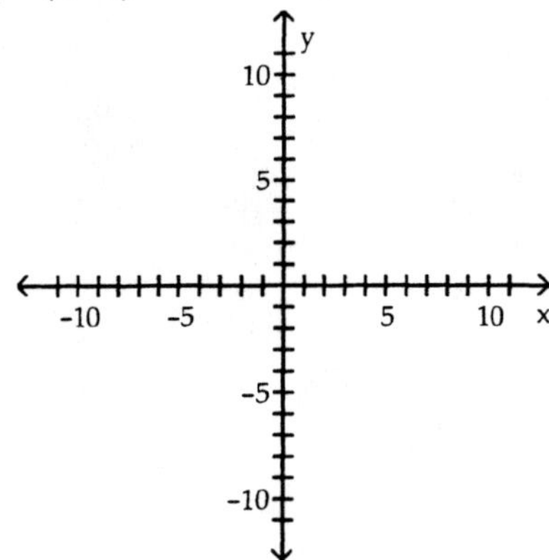

A)

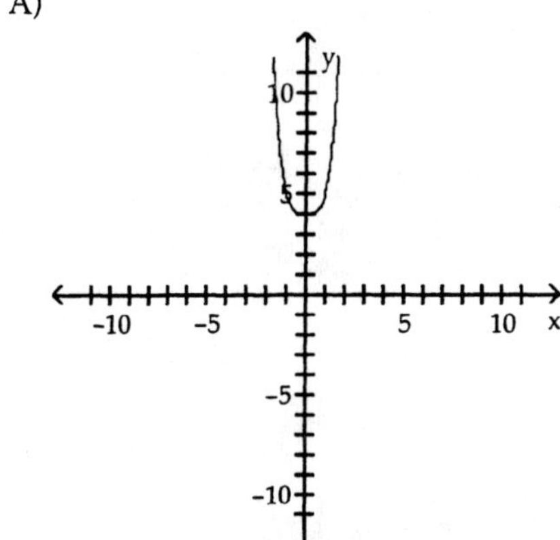

B)

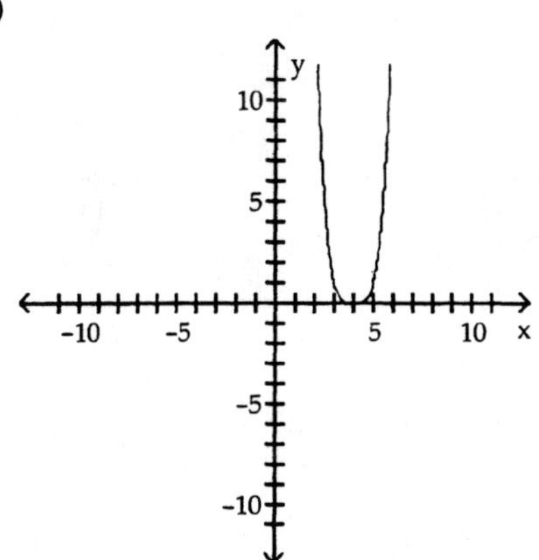

C)

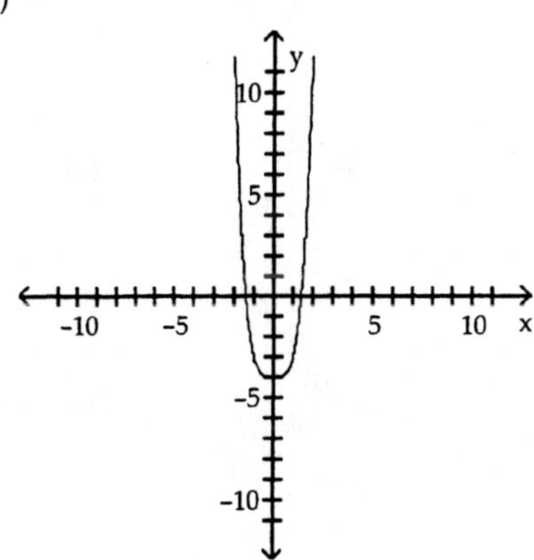

D)

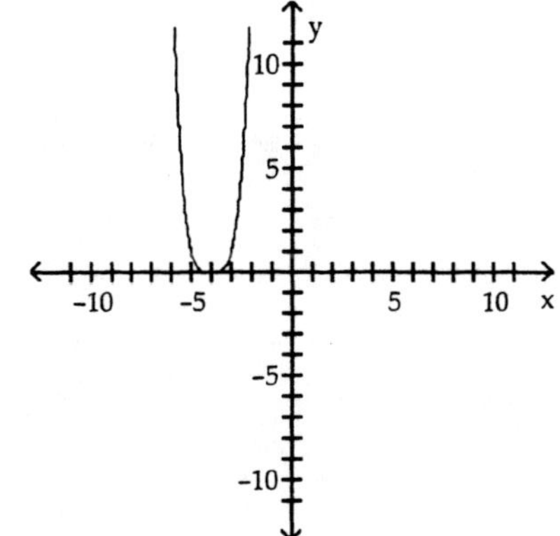

2) $f(x) = x^4 - 4$

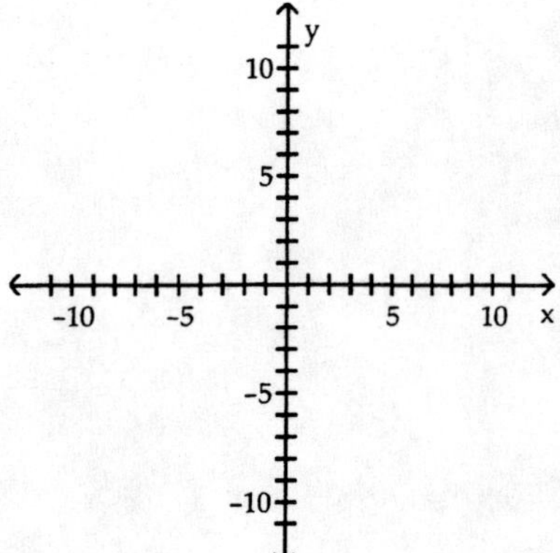

A)

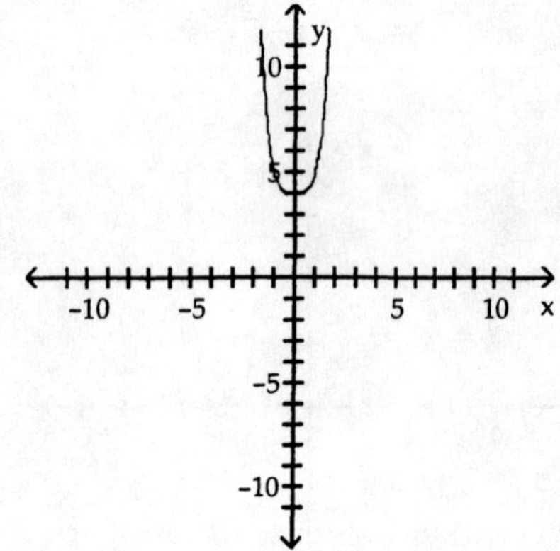

B)

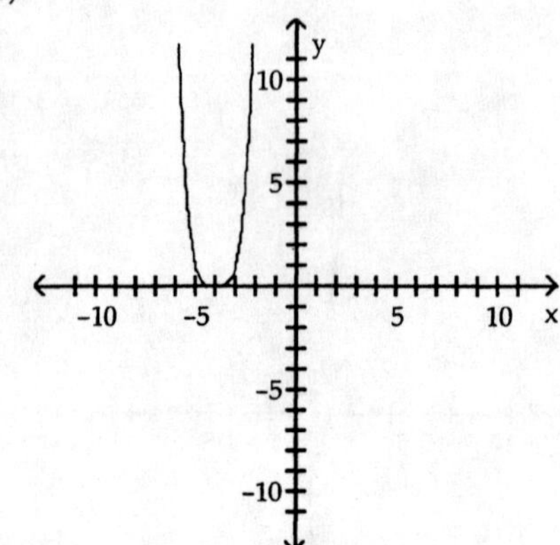

C)

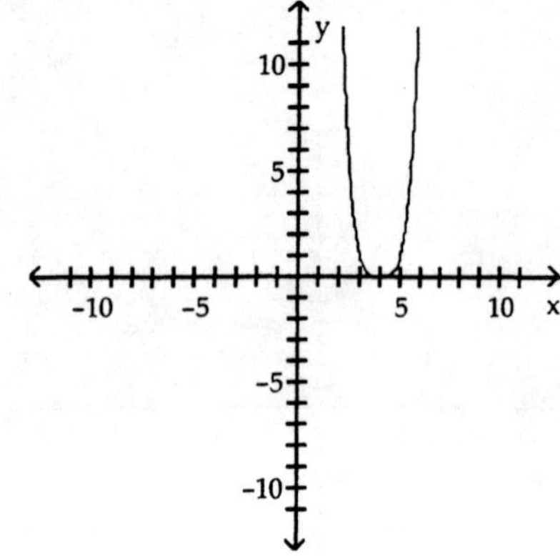

D)

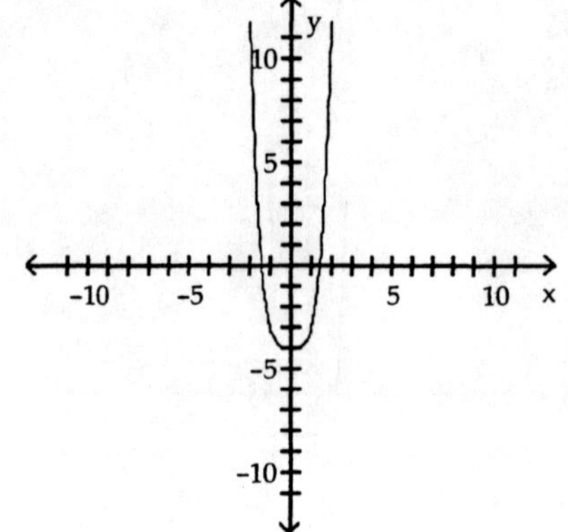

3) $f(x) = \frac{1}{3}x^4$

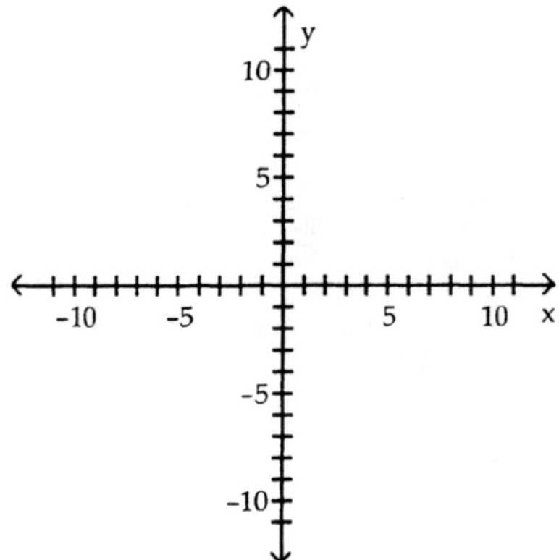

A)

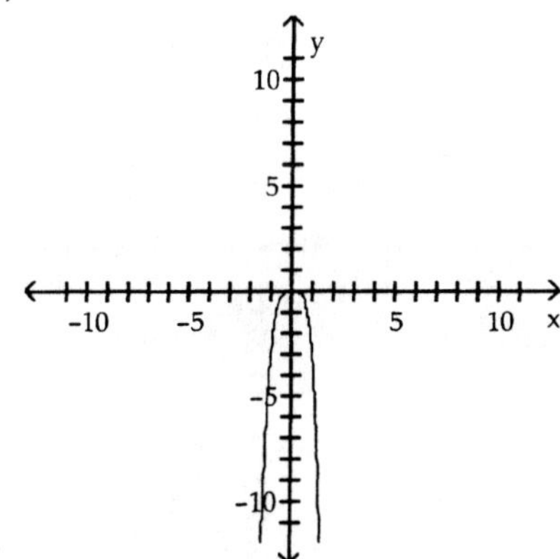

B)

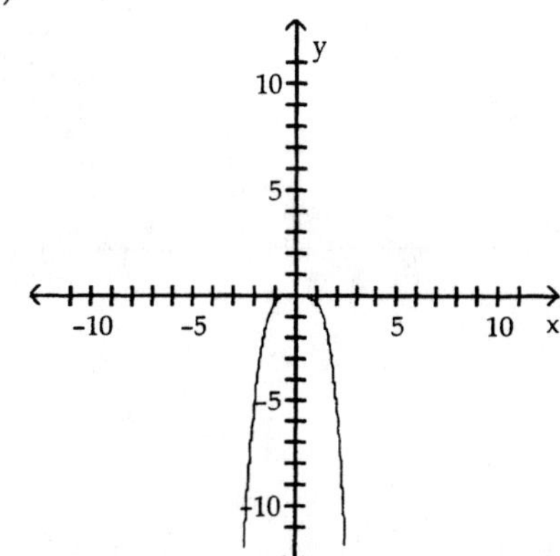

C)

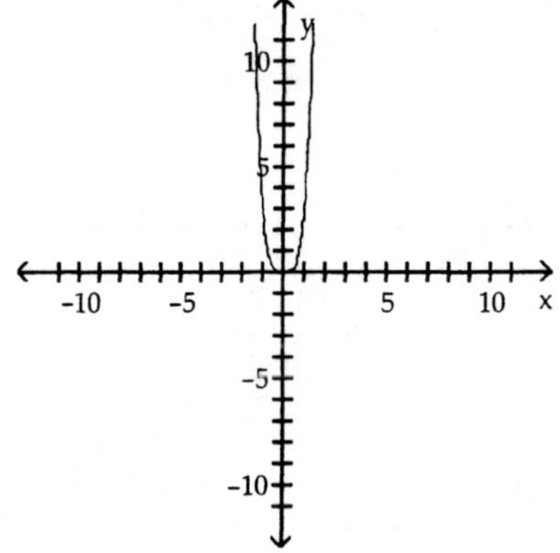

D)

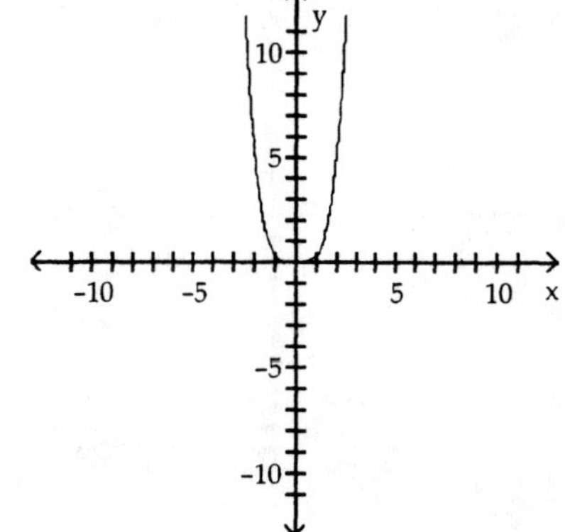

4) $f(x) = 2x^4$

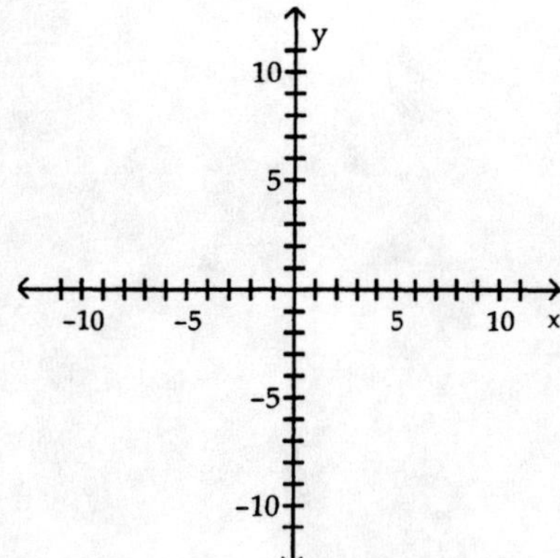

A)

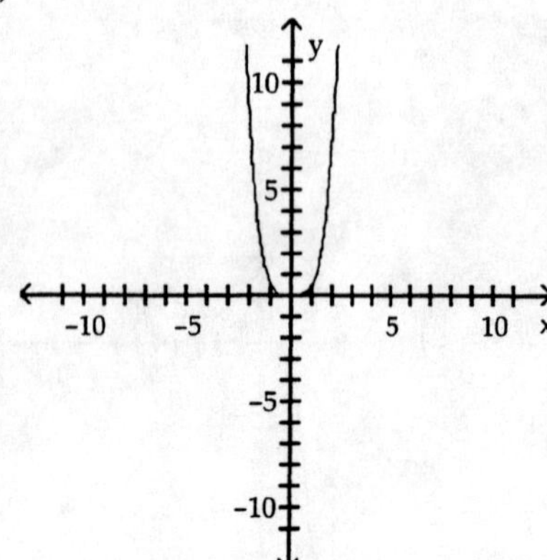

B)

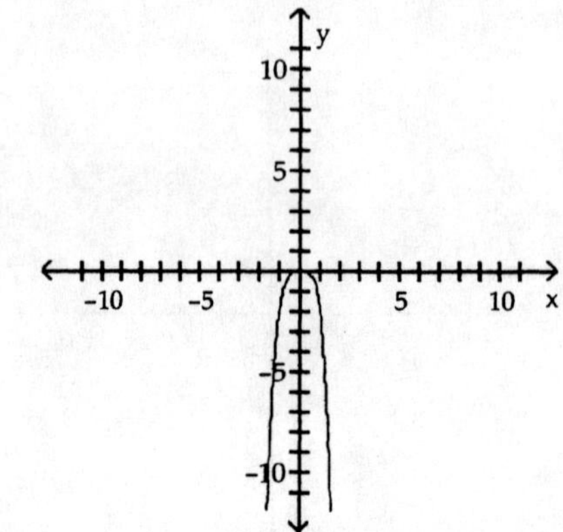

C)

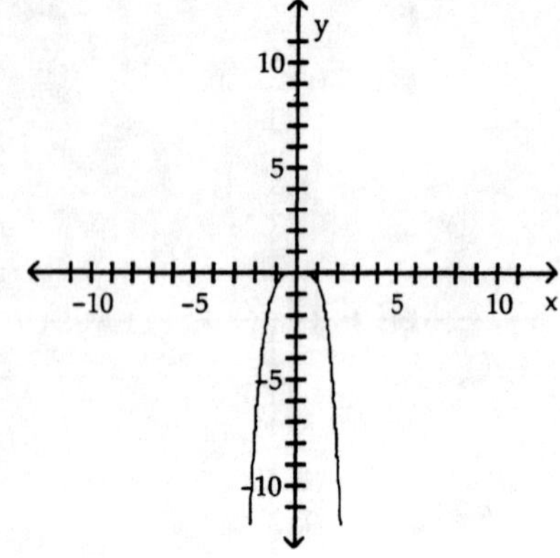

D)

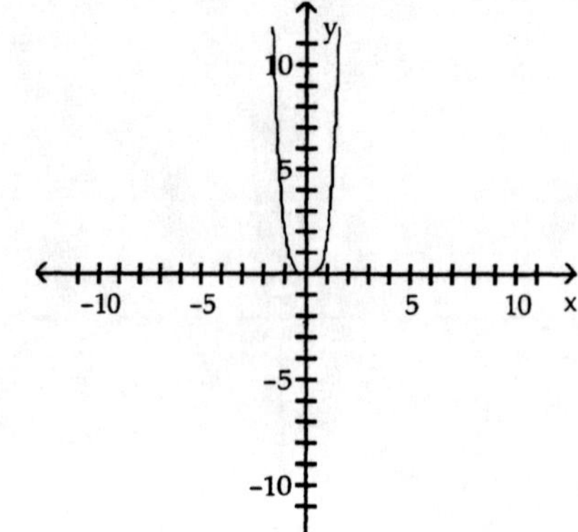

5) $f(x) = (x - 2)^4 + 4$

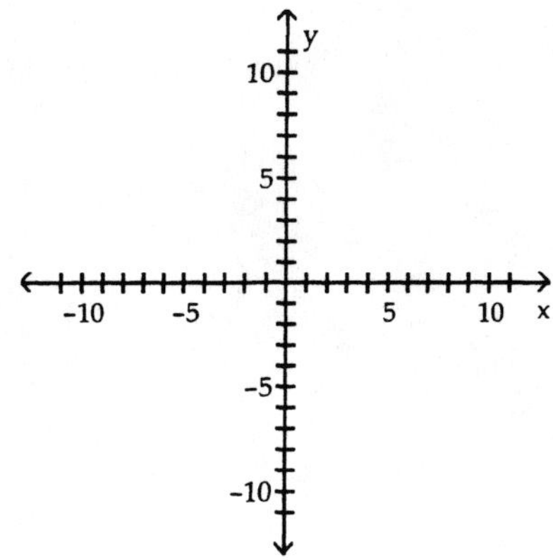

A)

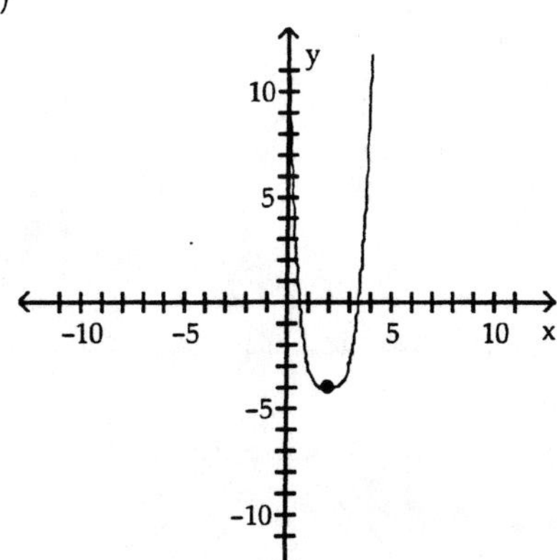

B)

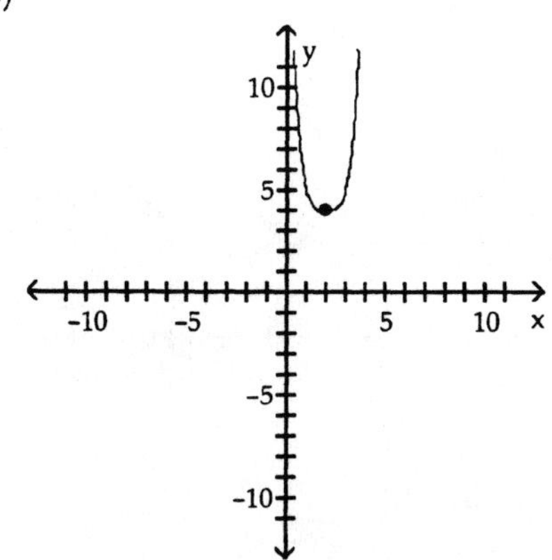

C)

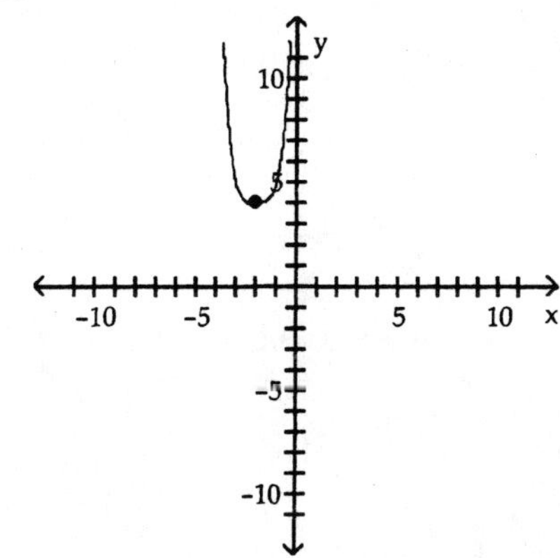

D)

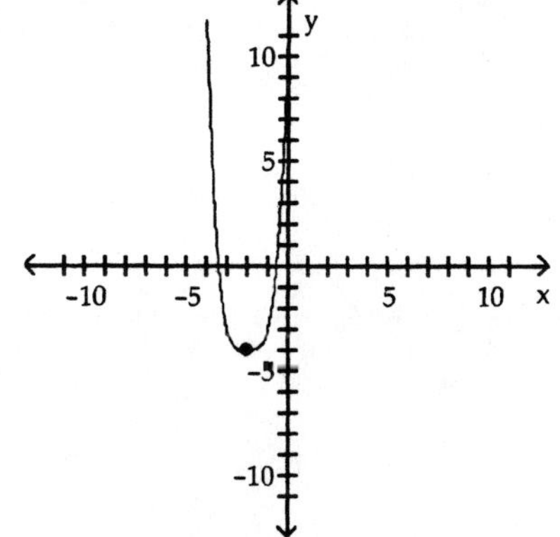

6) $f(x) = \frac{1}{2}(x + 5)^4 + 4$

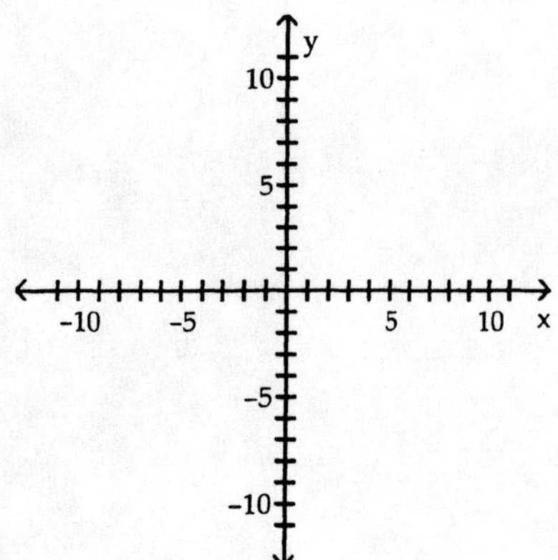

A)

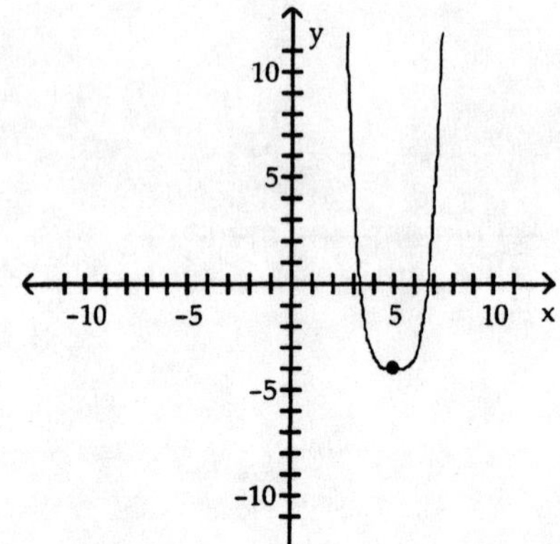

B)

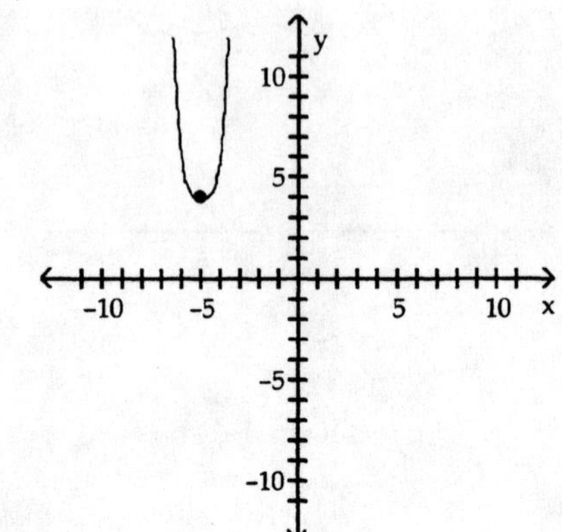

C)

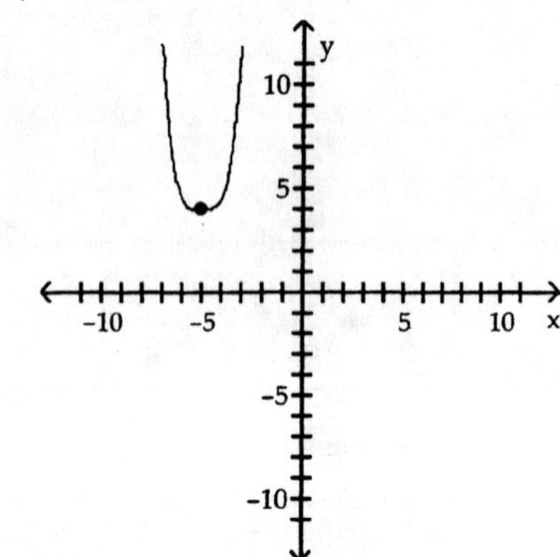

D)

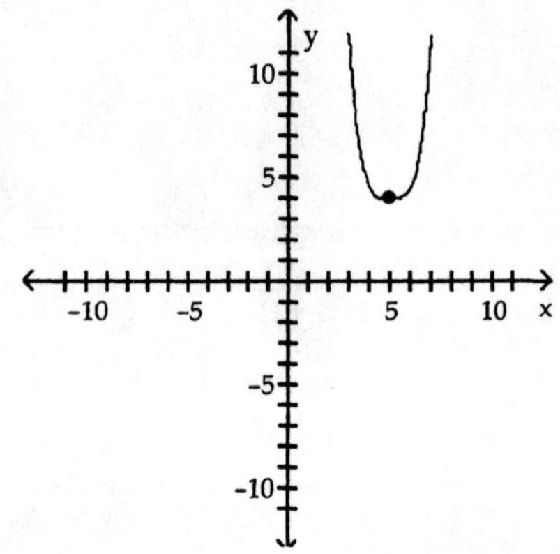

7) $f(x) = -2(x - 5)^4 + 4$

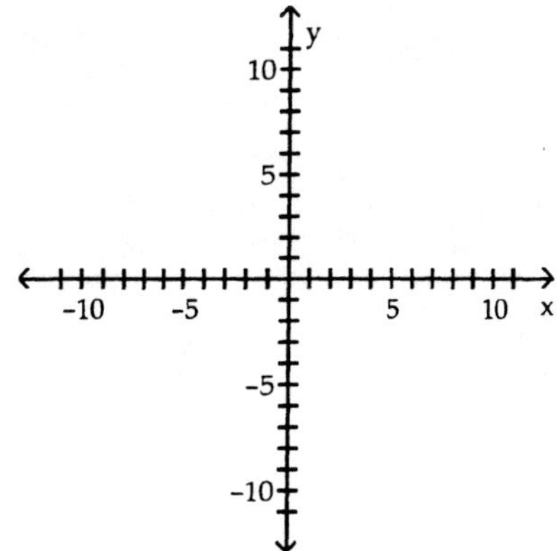

A)

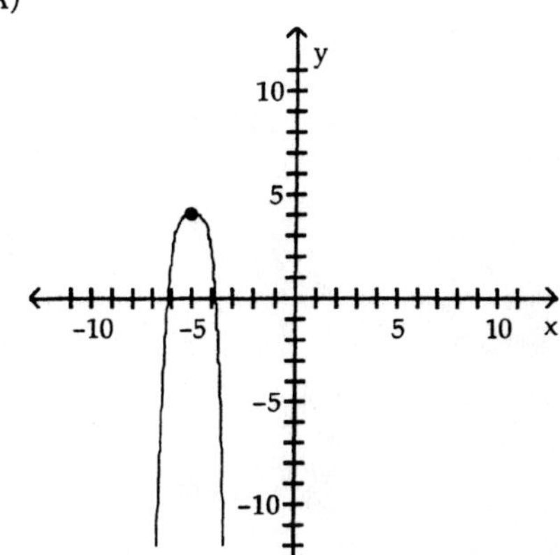

B)

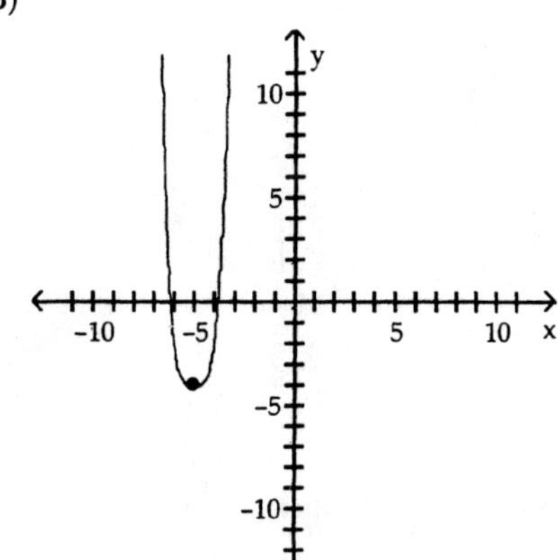

C)

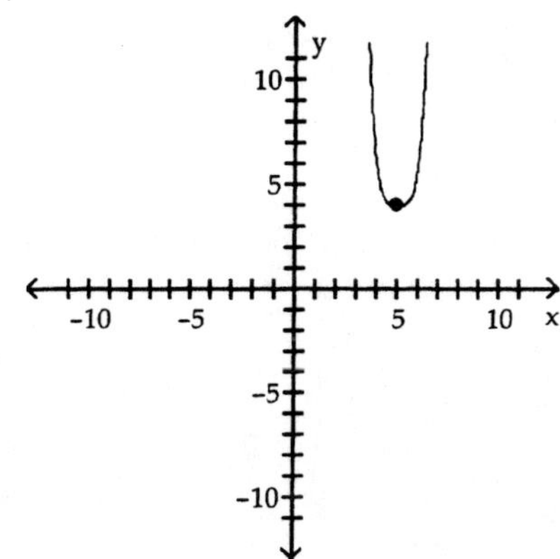

D)

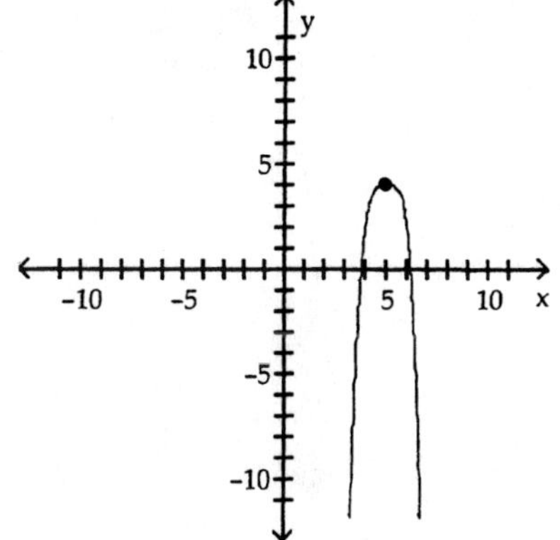

8) $f(x) = 2 - (x - 3)^4$

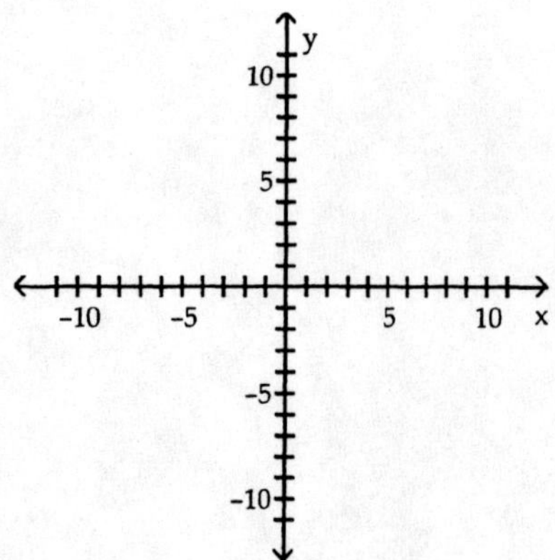

A)

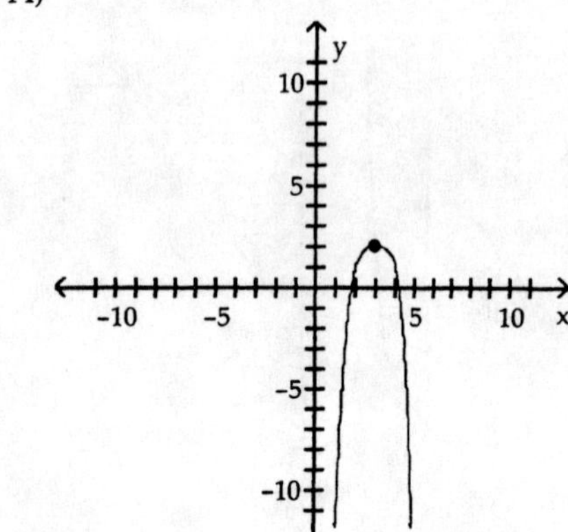

B)

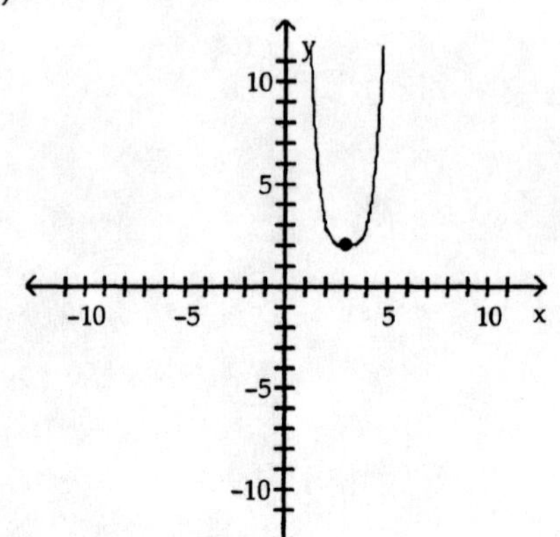

C)

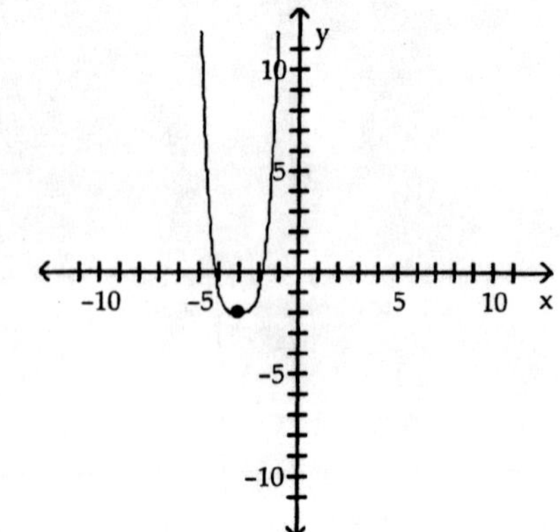

D)

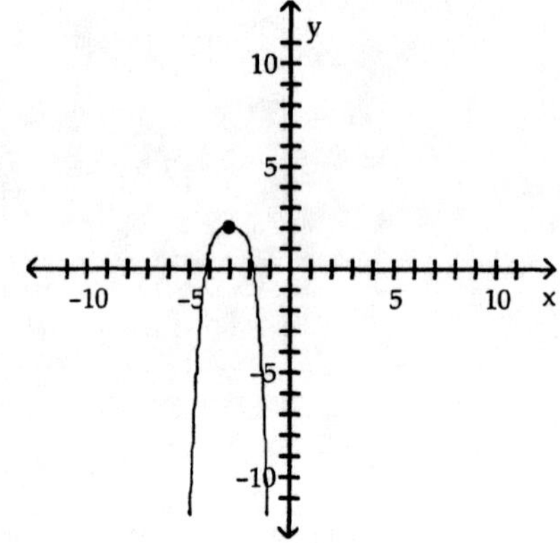

9) $f(x) = (x + 4)^5$

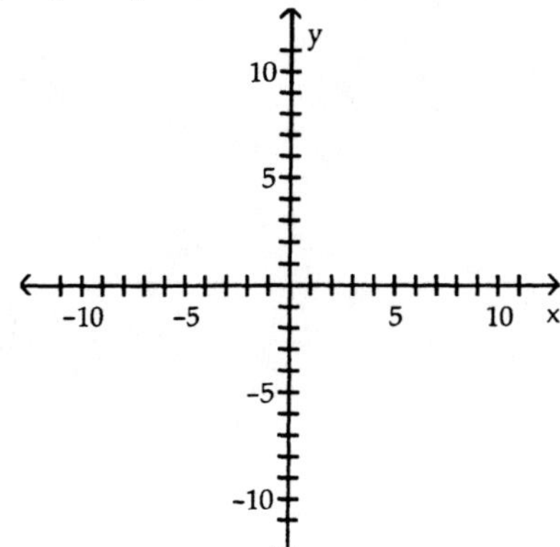

A)

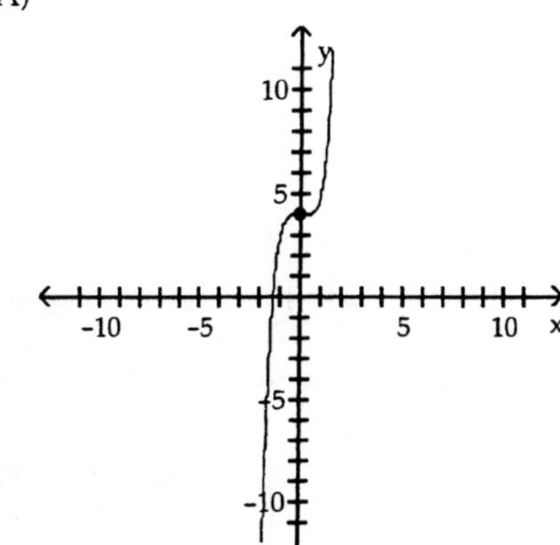

B)

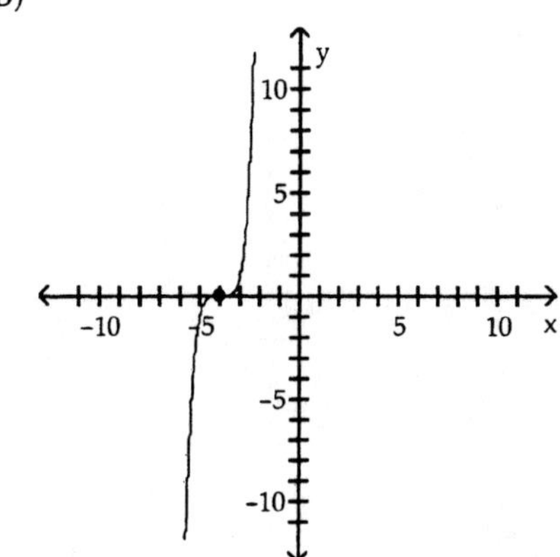

C)

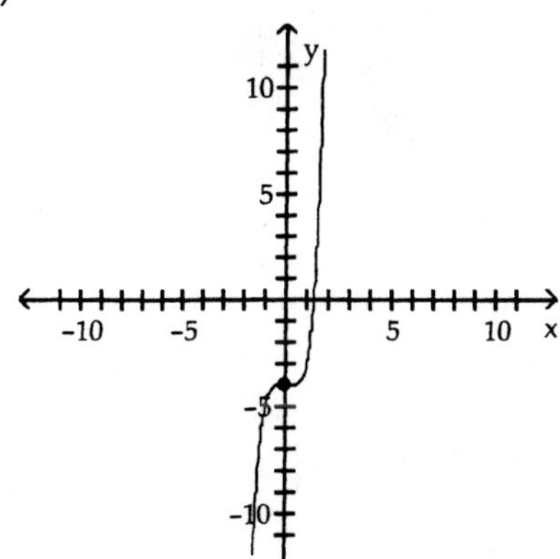

D)

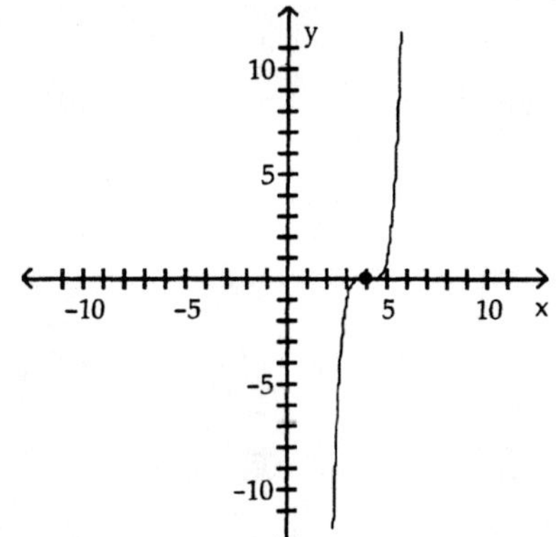

10) $f(x) = x^5 + 4$

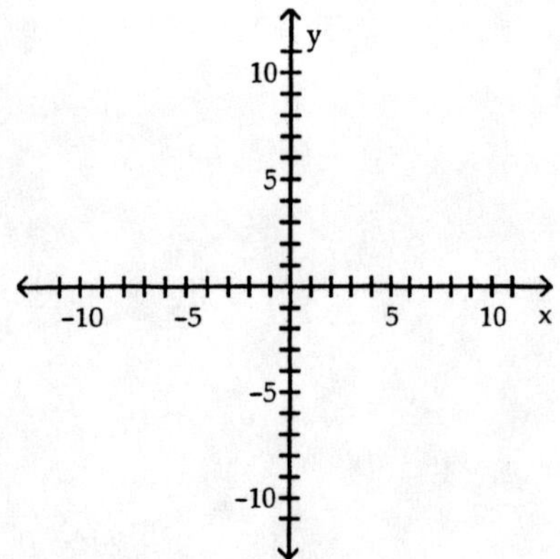

A)

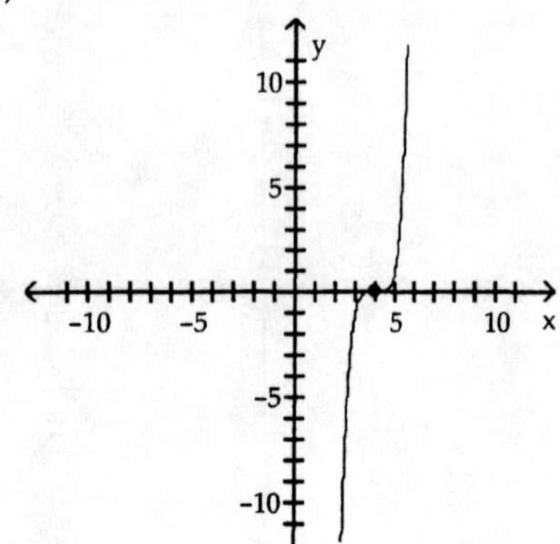

B)

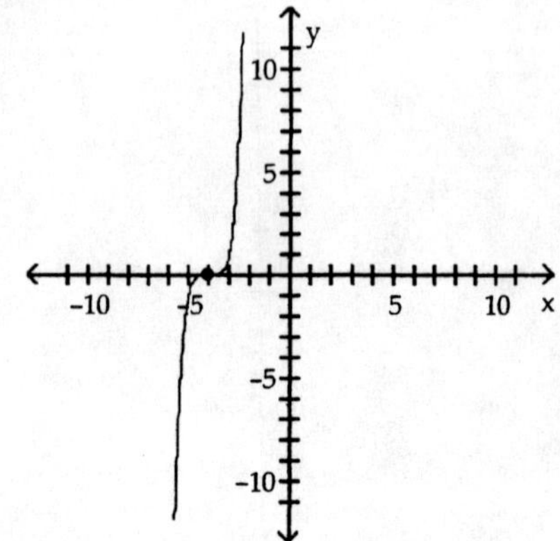

C)

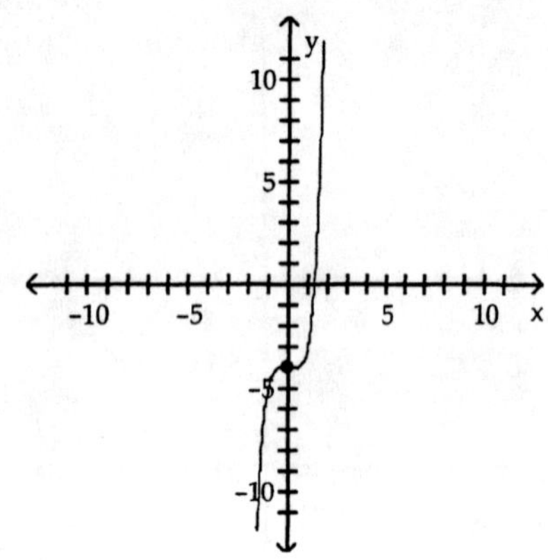

D)

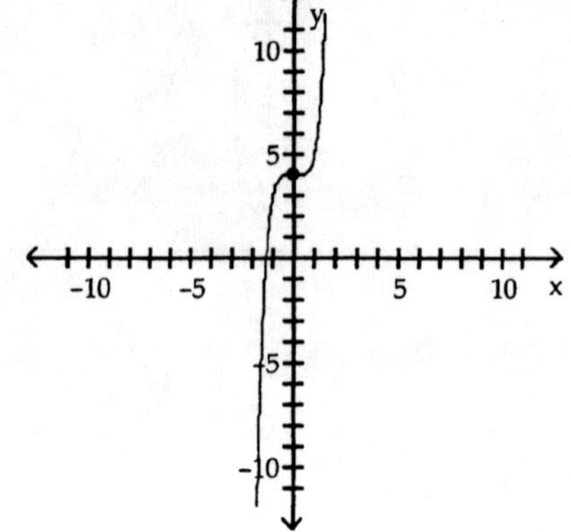

11) $f(x) = -\frac{1}{4}x^5$

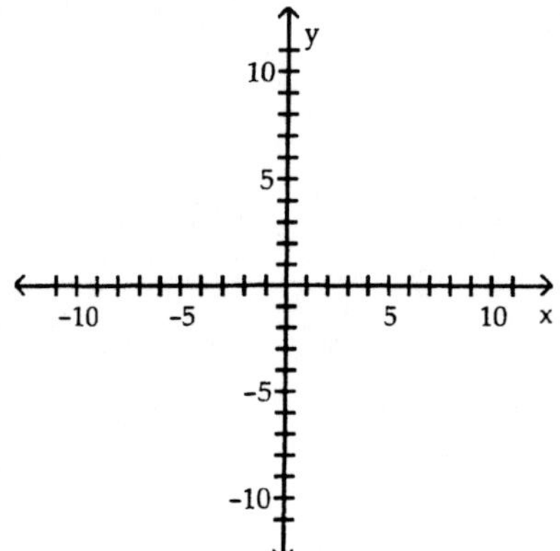

A)

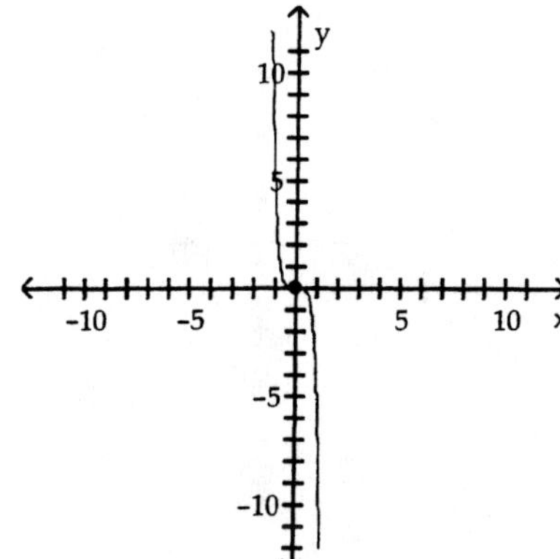

B)

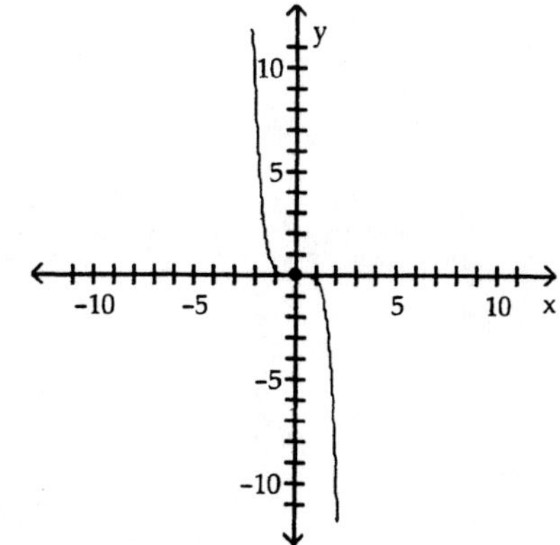

C)

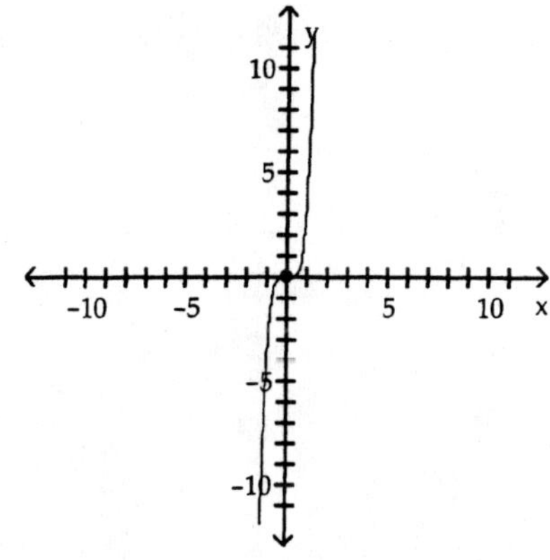

D)

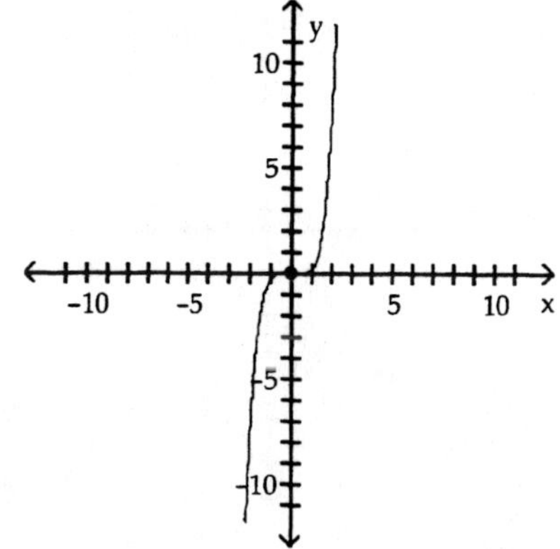

12) 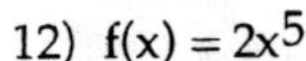 $f(x) = 2x^5$

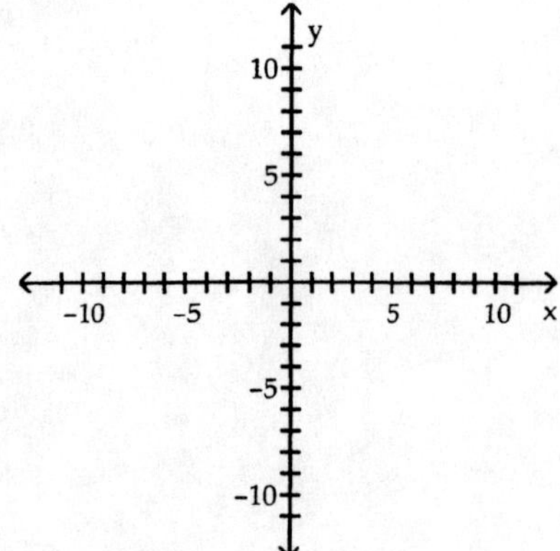

A)

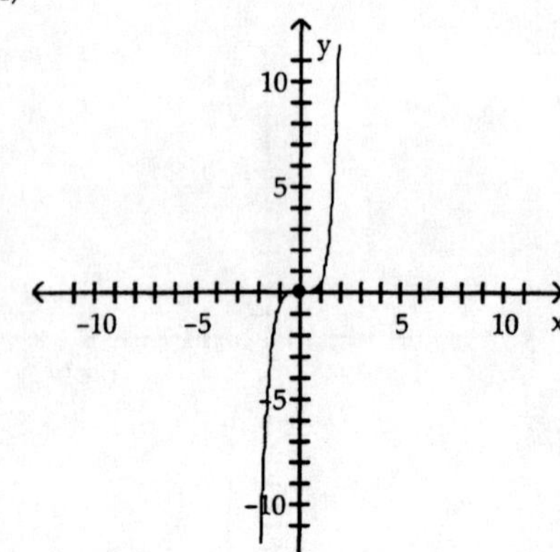

B)

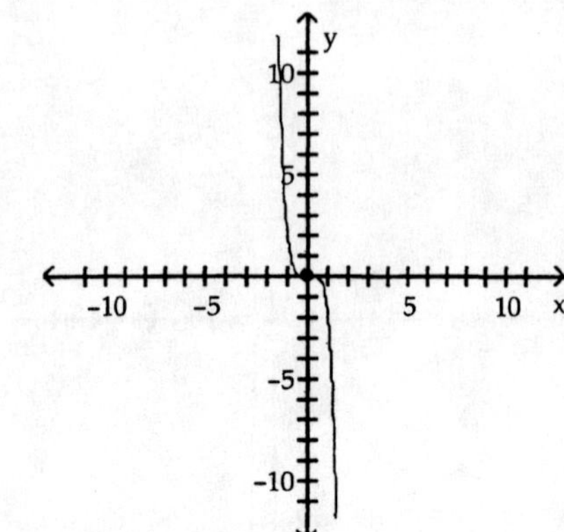

C)

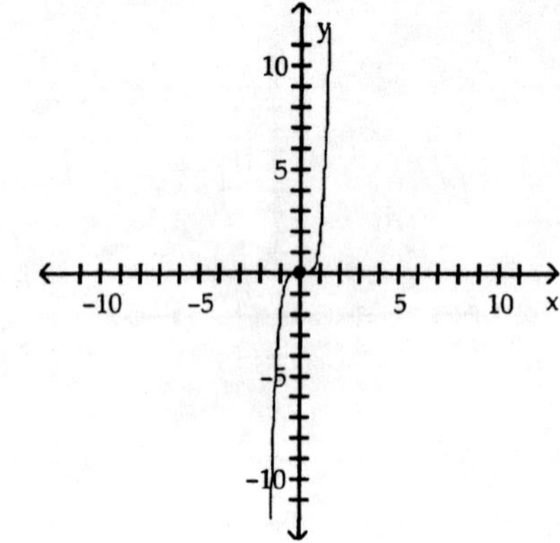

D)

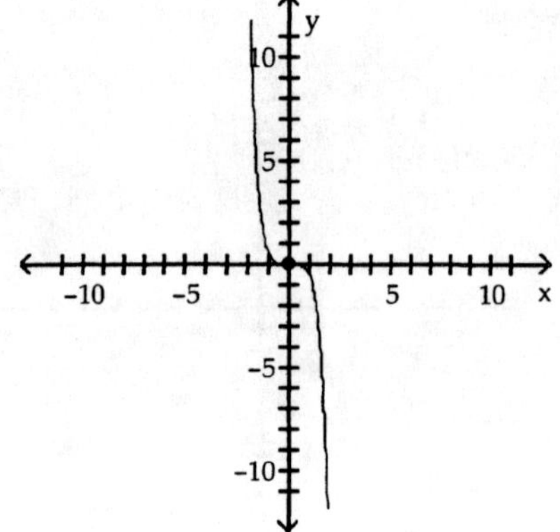

13) $f(x) = (x - 4)^5 + 4$

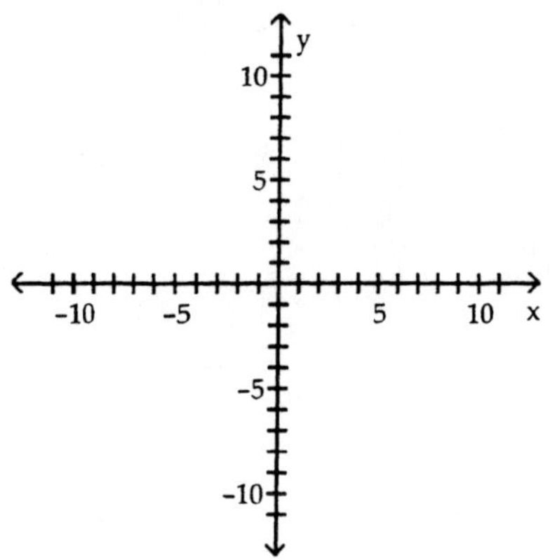

A)

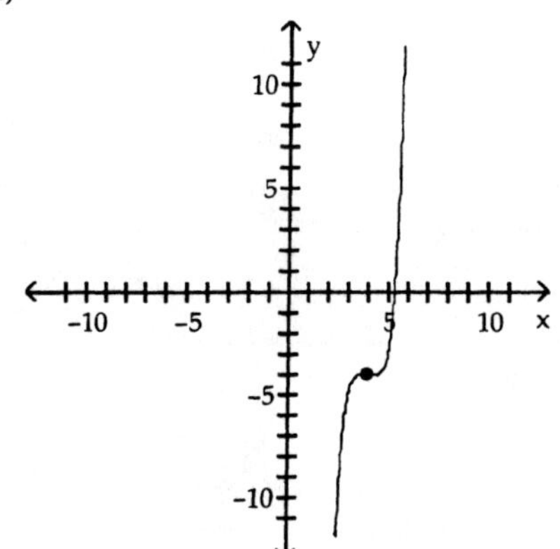

B)

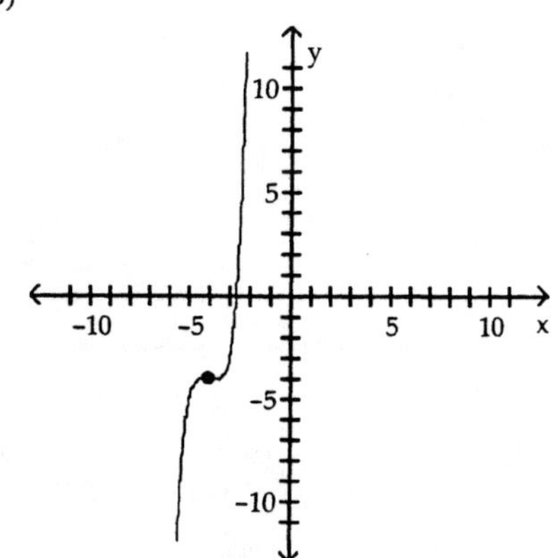

C)

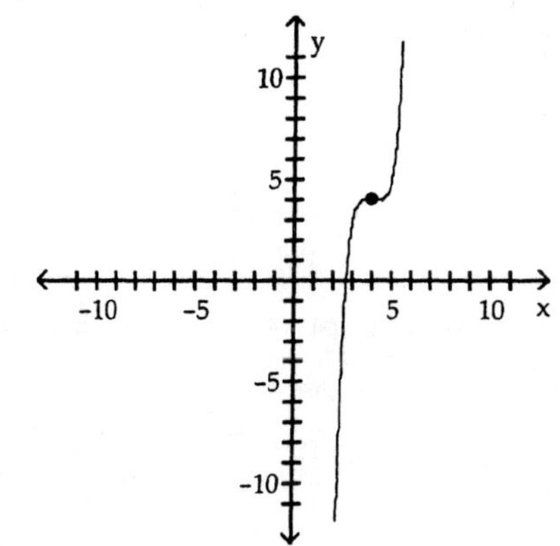

D)

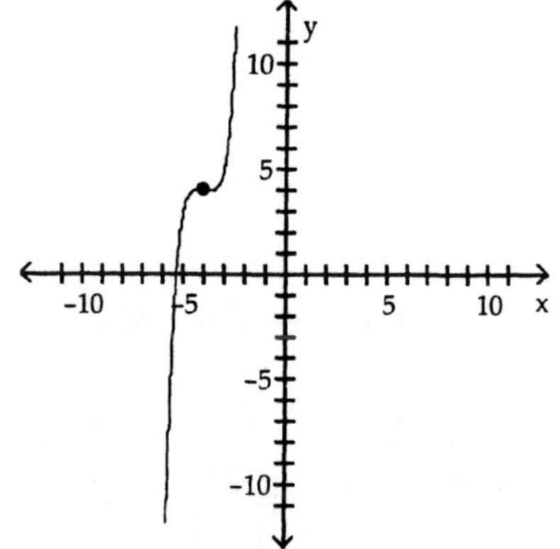

14) $f(x) = \frac{1}{2}(x - 4)^5 + 3$

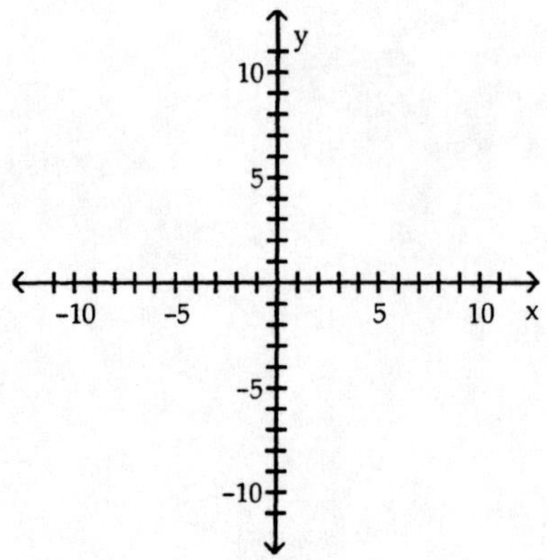

A)

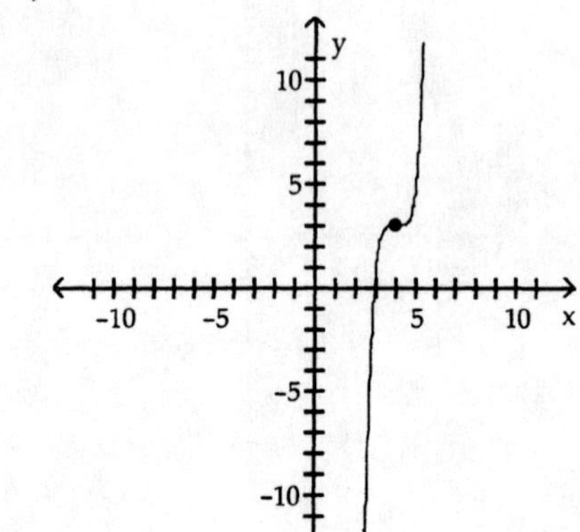

B)

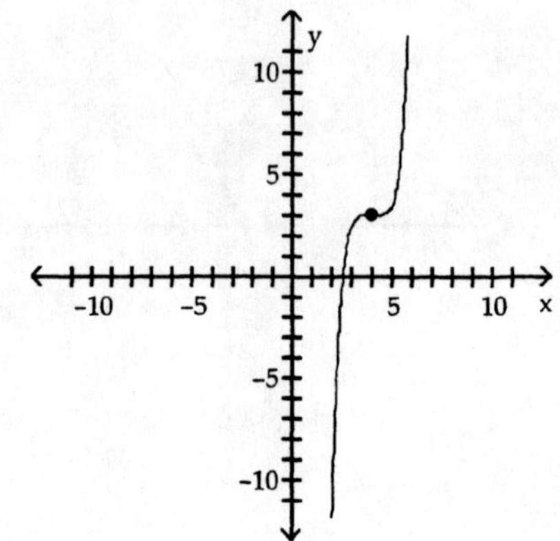

C)

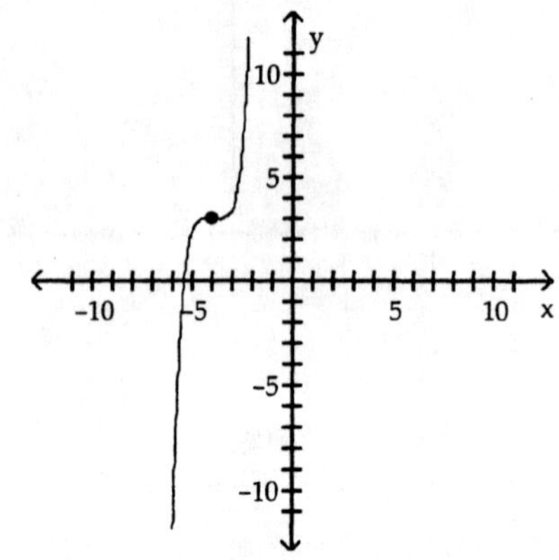

D)

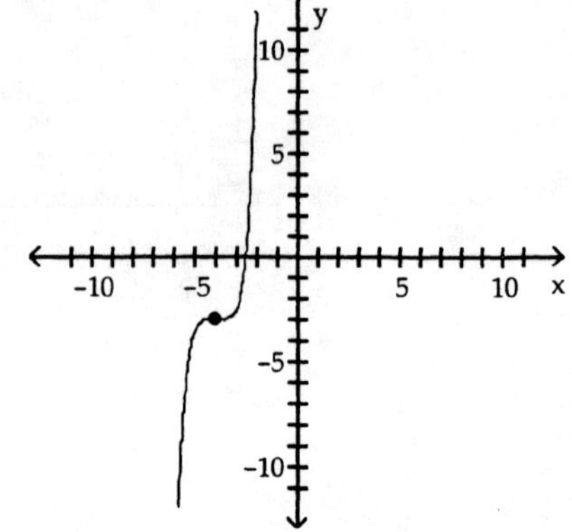

15) $f(x) = -2(x + 4)^5 + 4$

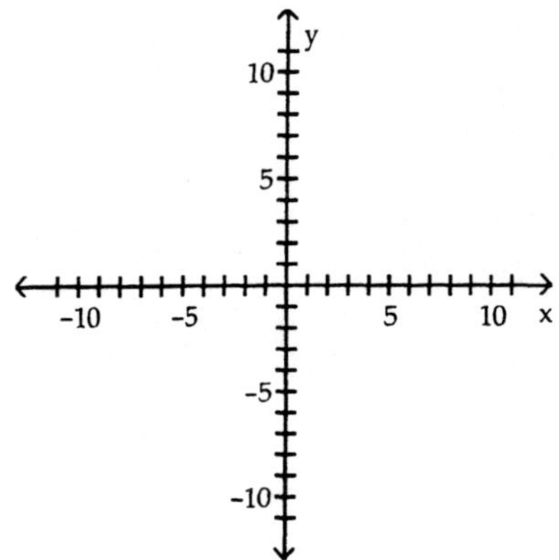

A)

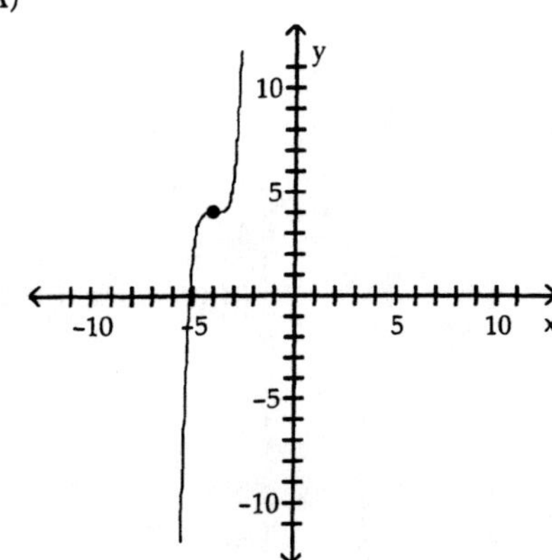

B)

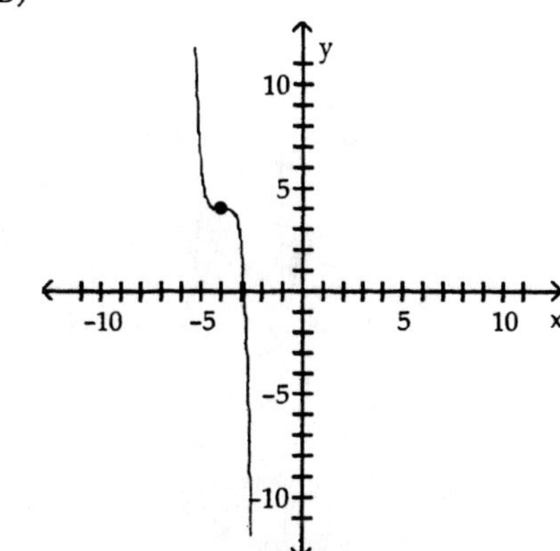

C)

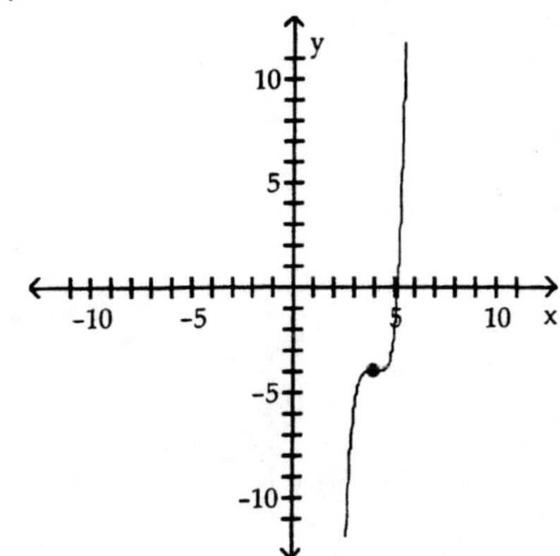

D)

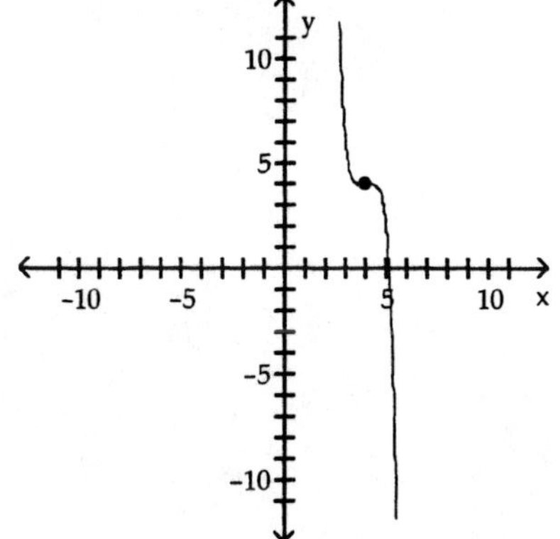

16) $f(x) = 4 - (x + 3)^5$

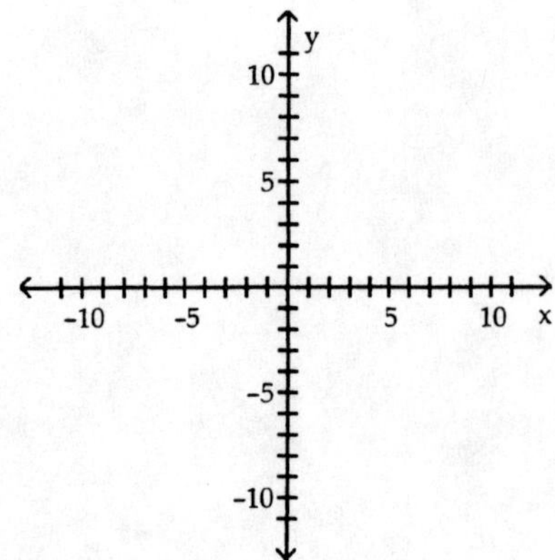

A)

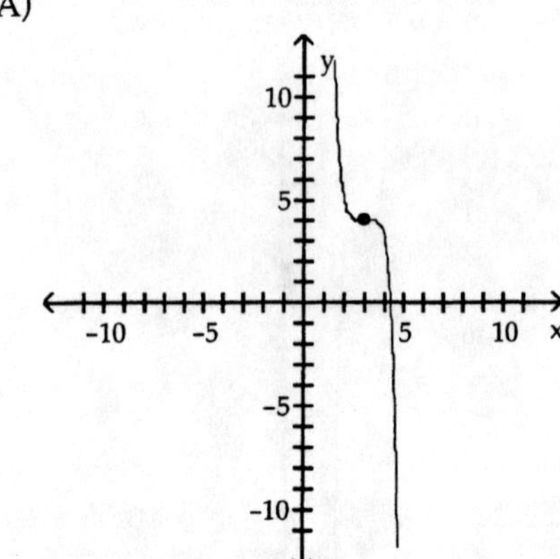

B)

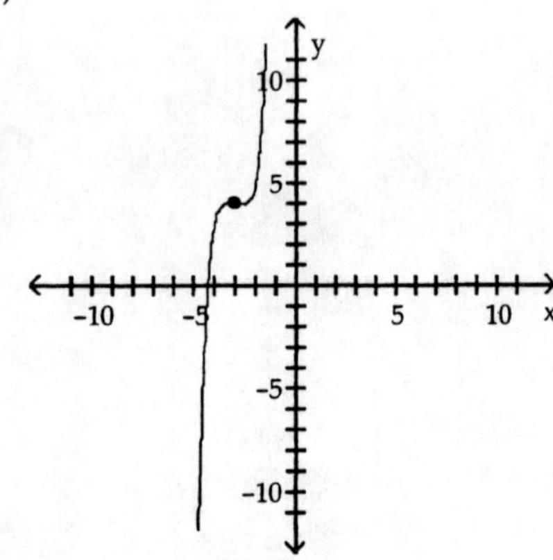

C)

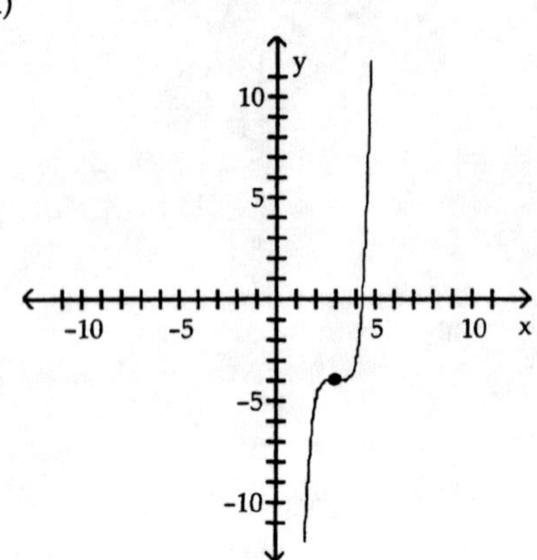

D)

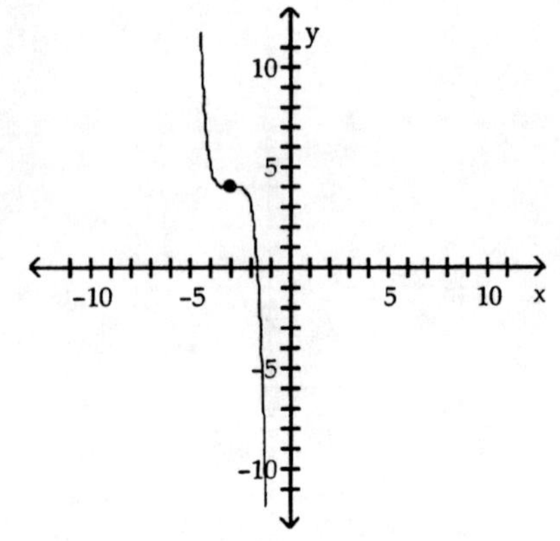

3 Identify the Zeros of a Polynomial Function and Their Multiplicity

Form a polynomial whose zeros and degree are given.

1) Zeros: –3, –2, 3; degree 3

A) $f(x) = x^3 - 2x^2 + 9x - 18$ for $a = 1$

B) $f(x) = x^3 + 2x^2 - 9x - 18$ for $a = 1$

C) $f(x) = x^3 - 2x^2 - 9x + 18$ for $a = 1$

D) $f(x) = x^3 + 2x^2 + 9x + 18$ for $a = 1$

2) Zeros: 0, – 7, 6; degree 3

A) $f(x) = x^3 + x^2 + x - 42$ for $a = 1$

B) $f(x) = x^3 + x^2 + x + 42$ for $a = 1$

C) $f(x) = x^3 + x^2 + 42x$ for $a = 1$

D) $f(x) = x^3 + x^2 - 42x$ for $a = 1$

3) Zeros: -1, 1, -7; degree 3

A) $f(x) = x^3 - 7x^2 + x - 7$ for $a = 1$
B) $f(x) = x^3 - 7x^2 - x + 7$ for $a = 1$
C) $f(x) = x^3 + 7x^2 + x + 7$ for $a = 1$
D) $f(x) = x^3 + 7x^2 - x - 7$ for $a = 1$

4) Zeros: -6, -5, 5; degree 3

A) $f(x) = x^3 + 25x + 6x^2 + 150$ for $a = 1$
B) $f(x) = x^3 - 25x - 6x^2 + 150$ for $a = 1$
C) $f(x) = x^3 - 25x + 6x^2 - 150$ for $a = 1$
D) $f(x) = x^3 + 25x - 6x^2 - 150$ for $a = 1$

5) Zeros: 3, multiplicity 2; -3, multiplicity 2; degree 4

A) $f(x) = x^4 + 6x^3 - 18x^2 + 27x - 81$
B) $f(x) = x^4 - 6x^3 + 18x^2 - 27x + 81$
C) $f(x) = x^4 + 18x^2 + 81$
D) $f(x) = x^4 - 18x^2 + 81$

For the polynomial, list each real zero and its multiplicity. Determine whether the graph crosses or touches the x-axis at each x -intercept.

6) $f(x) = 3(x - 7)(x - 1)^4$

A) 7, multiplicity 1, crosses x-axis; 1, multiplicity 4, touches x-axis
B) -7, multiplicity 1, touches x-axis; -1, multiplicity 4, crosses x-axis
C) 7, multiplicity 1, touches x-axis; 1, multiplicity 4, crosses x-axis
D) -7, multiplicity 1, crosses x-axis; -1, multiplicity 4, touches x-axis

7) $f(x) = 4(x^2 + 4)(x - 2)^2$

A) -4, multiplicity 1, crosses x-axis; 2, multiplicity 2, touches x-axis
B) -4, multiplicity 1, touches x-axis; 2, multiplicity 2, crosses x-axis
C) 2, multiplicity 2, touches x-axis
D) 2, multiplicity 2, crosses x-axis

8) $f(x) = (x + \frac{1}{2})^4 (x - 5)^5$

A) $-\frac{1}{2}$, multiplicity 4, crosses x-axis; 5, multiplicity 5, touches x-axis

B) $-\frac{1}{2}$, multiplicity 4, touches x-axis; 5, multiplicity 5, crosses x-axis

C) $\frac{1}{2}$, multiplicity 4, crosses x-axis; -5, multiplicity 5, touches x-axis

D) $\frac{1}{2}$, multiplicity 4, touches x-axis; -5, multiplicity 5, crosses x-axis

9) $f(x) = (x + \frac{1}{5})^4 (x^2 + 2)^3$

A) $-\frac{1}{5}$, multiplicity 4, touches x-axis

B) $\frac{1}{5}$, multiplicity 4, touches x-axis; 2, multiplicity 3, crosses x-axis

C) $-\frac{1}{5}$, multiplicity 4, crosses x-axis

D) $-\frac{1}{5}$, multiplicity 4, touches x-axis; -2, multiplicity 3, crosses x-axis

10) $f(x) = 3(x^2 + 5)(x^2 + 1)^2$

A) -5, multiplicity 1, touches x-axis; -1, multiplicity 2, crosses x-axis

B) -5, multiplicity 1, crosses x-axis; -1, multiplicity 2, touches x-axis

C) No real zeros

D) $\sqrt{5}$, multiplicity 1, crosses x-axis; $-\sqrt{5}$, multiplicity 1, crosses x-axis; 1, multiplicity 2, touches x-axis; -1, multiplicity 2, touches x-axis

11) $f(x) = \frac{1}{3}x^2(x^2 - 3)(x - 3)$

A) 0, multiplicity 2, touches x-axis; 3, multiplicity 1, crosses x-axis

B) 0, multiplicity 2, touches x-axis; 3, multiplicity 1, crosses x-axis; $\sqrt{3}$, multiplicity 1, crosses x-axis; $-\sqrt{3}$, multiplicity 1, crosses x-axis

C) 0, multiplicity 2, cross eses x-axis; 3, multiplicity 1, touches x-axis; $\sqrt{3}$, multiplicity 1, touches x-axis; $-\sqrt{3}$, multiplicity 1, touches x-axis

D) 0, multiplicity 2, crosses x-axis; 3, multiplicity 1, touches x-axis

4 Analyze the Graph of a Polynomial Function

Find the x- and y-intercepts of f.

1) $f(x) = (x + 11)^2$

A) x-intercept: -11; y-intercept: 121

B) x-intercept: -11; y-intercept: 0

C) x-intercept: 11; y-intercept: 0

D) x-intercept: 11; y-intercept: 121

2) $f(x) = (x + 2)^3$

A) x-intercept: -2; y-intercept: -8

B) x-intercept: -2; y-intercept: 6

C) x-intercept: -2; y-intercept: -6

D) x-intercept: -2; y-intercept: 8

3) $f(x) = 2x^2(x - 2)^5$

A) x-intercepts: 0, 2; y-intercept: 0

B) x-intercepts: 0, -2; y-intercept: 2

C) x-intercepts: 0, -2; y-intercept: 0

D) x-intercepts: 0, 2; y-intercept: 2

4) $f(x) = (x + 2)(x - 6)(x + 6)$

A) x-intercepts: -2, -6, 6; y-intercept: -72

B) x-intercepts: -6, 6, 2; y-intercept: 72

C) x-intercepts: -6, 6, 2; y-intercept: -72

D) x-intercepts: -2, -6, 6; y-intercept: 72

5) $f(x) = 5x - x^3$

A) x-intercepts: 0, -5; y-intercept: 0

B) x-intercepts: 0, -5; y-intercept: 5

C) x-intercepts: 0, $\sqrt{5}$, $-\sqrt{5}$; y-intercept: 5

D) x-intercepts: 0, $\sqrt{5}$, $-\sqrt{5}$; y-intercept: 0

6) $f(x) = (x + 1)(x - 7)(x - 1)^2$

A) x-intercepts: -1, 1, 7; y-intercept: -7

B) x-intercepts: -1, 1, 7; y-intercept: 7

C) x-intercepts: -1, 1, -7; y-intercept: 7

D) x-intercepts: -1, 1, -7; y-intercept: -7

7) $f(x) = (x - 1)(x - 5)(x + 1)^2$

A) x-intercepts: −1, 1, 5; y-intercept: −5
B) x-intercepts: −1, 1, 5; y-intercept: 5
C) x-intercepts: −1, 1, −5; y-intercept: −5
D) x-intercepts: −1, 1, −5; y-intercept: 5

8) $f(x) = -x^2(x + 7)(x^2 - 1)$

A) x-intercepts: −1, 0, 1, 7; y-intercept: 0
B) x-intercepts: −7, 0, 1; y-intercept: −7
C) x-intercepts: −7, −1, 0, 1; y-intercept: 0
D) x-intercepts: −7, −1, 0, 1; y-intercept: −7

9) $f(x) = -x^2(x + 9)(x^2 + 1)$

A) x-intercepts: −9, −1, 0; y-intercept: −9
B) x-intercepts: −9, −1, 0, 1; y-intercept: 0
C) x-intercepts: −9, 0; y-intercept: 0
D) x-intercepts: −9, −1, 0; y-intercept: 9

10) $f(x) = (x - 1)(x - 2)$

A) x-intercepts: −1, −2; y-intercept: −3
B) x-intercepts: −1, −2; y-intercept: 2
C) x-intercepts: 1, 2; y-intercept: −3
D) x-intercepts: 1, 2; y-intercept: 2

11) $f(x) = x^2(x - 6)(x - 2)$

A) x-intercepts: 0, −6, −2; y-intercept: 12
B) x-intercepts: 0, 6, 2; y-intercept: 0
C) x-intercepts: 0, 6, 2; y-intercept: 12
D) x-intercepts: 0, −6, −2; y-intercept: 0

12) $f(x) = (x + 2)(x - 7)$

A) x-intercepts: −2, 7; y-intercept: −14
B) x-intercepts: −2, 7; y-intercept: −5
C) x-intercepts: −7, 2; y-intercept: −14
D) x-intercepts: −7, 2; y-intercept: −5

13) $f(x) = -x^2(x + 2)(x - 7)$

A) x-intercepts: −7, 2; y-intercept: −14
B) x-intercepts: −2, 0, 7; y-intercept: −14
C) x-intercepts: −7, 0, 2; y-intercept: 0
D) x-intercepts: −2, 0, 7; y-intercept: 0

14) $f(x) = (x - 2)^2(x + 5)^2$

A) x-intercepts: −5, 2; y-intercept: 100
B) x-intercepts: −5, 2; y-intercept: −10
C) x-intercepts: −2, 5; y-intercept: 100
D) x-intercepts: −2, 5; y-intercept: −10

15) $f(x) = (x - 3)^2(x^2 - 25)$

A) x-intercepts: 3, 25; y-intercept: 75
B) x-intercepts: −5, 3, 5; y-intercept: −225
C) x-intercepts: −3, −25; y-intercept: 75
D) x-intercepts: −5, 3, 5; y-intercept: 225

Find the power function that the graph of f resembles for large values of $|x|$.

16) $f(x) = (x + 6)^2$

A) $y = x^{12}$
B) $y = x^6$
C) $y = x^{36}$
D) $y = x^2$

17) $f(x) = (x - 4)^3$

A) $y = x^{-64}$
B) $y = x^3$
C) $y = x^{12}$
D) $y = x^{-4}$

18) $f(x) = (x + 2)^2(x + 1)^4$

A) $y = x^2$
B) $y = x^8$
C) $y = x^6$
D) $y = x^4$

19) $f(x) = -x^2(x + 3)^3(x^2 - 1)$

A) $y = x^3$ B) $y = x^5$ C) $y = x^7$ D) $y = x^2$

20) $f(x) = 9x - x^3$

A) $y = x^4$ B) $y = x$ C) $y = x^2$ D) $y = x^3$

Determine the maximum number of turning points of f.

21) $f(x) = -x^2(x + 6)^3(x^2 - 1)$

A) 6 B) 7 C) 2 D) 5

22) $f(x) = 5x - x^3$

A) 4 B) 1 C) 2 D) 3

23) $f(x) = (x - 2)^2(x + 4)^2$

A) 1 B) 3 C) 4 D) 2

Use the x-intercepts to find the intervals on which the graph of f is above and below the x-axis.

24) $f(x) = (x + 16)^2$

A) above the x-axis: $(-\infty, -16)$
below the x-axis: $(-16, \infty)$

B) above the x-axis: $(-16, \infty)$
below the x-axis: $(-\infty, -16)$

C) above the x-axis: $(-\infty, -16)$, $(-16, \infty)$
below the x-axis: no intervals

D) above the x-axis: no intervals
below the x-axis: $(-\infty, -16)$, $(-16, \infty)$

25) $f(x) = (x + 3)^3$

A) above the x-axis: $(-\infty, -3)$, $(-3, \infty)$
below the x-axis: no intervals

B) above the x-axis: $(-\infty, -3)$
below the x-axis: $(-3, \infty)$

C) above the x-axis: no intervals
below the x-axis: $(-\infty, -3)$, $(-3, \infty)$

D) above the x-axis: $(-3, \infty)$
below the x-axis: $(-\infty, -3)$

26) $f(x) = (x - \frac{1}{9})^4(x - 2)^5$

A) above the x-axis: $(-\infty, \frac{1}{9})$, $(2, \infty)$
below the x-axis: $(\frac{1}{9}, 2)$

B) above the x-axis: $(2, \infty)$
below the x-axis: $(-\infty, \frac{1}{9})$, $(\frac{1}{9}, 2)$

C) above the x-axis: $(\frac{1}{9}, 2)$
below the x-axis: $(-\infty, \frac{1}{9})$, $(2, \infty)$

D) above the x-axis: $(-\infty, \frac{1}{9})$, $(\frac{1}{9}, 2)$
below the x-axis: $(2, \infty)$

27) $f(x) = (x - 2)^2(x + 3)^2$

A) above the x-axis: $(-\infty, -3)$, $(-3, 2)$, $(2, \infty)$
below the x-axis: no intervals

B) above the x-axis: no intervals
below the x-axis: $(-\infty, -3)$, $(-3, 2)$, $(2, \infty)$

C) above the x-axis: $(-3, 2)$
below the x-axis: $(-\infty, -3)$, $(2, \infty)$

D) above the x-axis: $(-\infty, -3)$, $(2, \infty)$
below the x-axis: $(-3, 2)$

Solve the problem.

28) For the polynomial function $f(x) = (x - 2)^3(x - 3)^2(x - 4)$

a) Find the x- and y-intercepts of the graph of f.

b) Determine whether the graph crosses or touches the x-axis at each x-intercept.

c) End behavior: find the power function that the graph of f resembles for large values of $|x|$.

d) Use a graphing utility to graph the function. Approximate the local maxima rounded to two decimal places, if necessary. Approximate the local minima rounded to two decimal places, if necessary.

e) Determine the number of turning points on the graph.

f) Put all the information together, and connect the points with a smooth, continuous curve to obtain the graph of f.

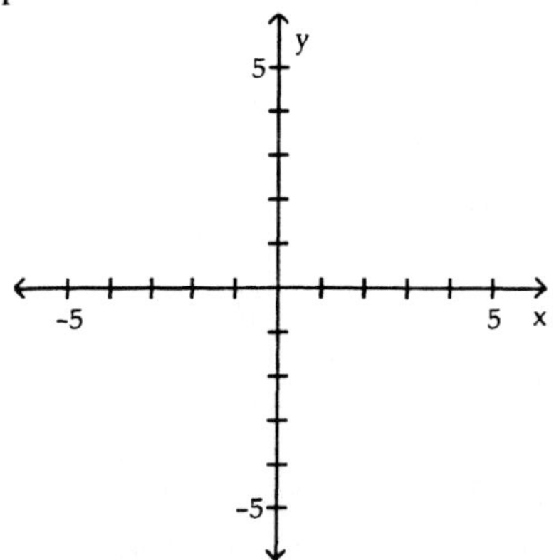

29) For the polynomial function $f(x) = x^3 - 4x + 2$.

a) Find the x- and y-intercepts of the graph of f. Round to two decimal places, if necessary.

b) Determine whether the graph crosses or touches the x-axis at each x-intercept.

c) End behavior: find the power function that the graph of f resembles for large values of $|x|$.

d) Use a graphing utility to graph the function. Approximate the local maxima rounded to two decimal places, if necessary. Approximate the local minima rounded to two decimal places, if necessary.

e) Determine the number of turning points on the graph.

f) Put all the information together, and connect the points with a smooth, continuous curve to obtain the graph of f.

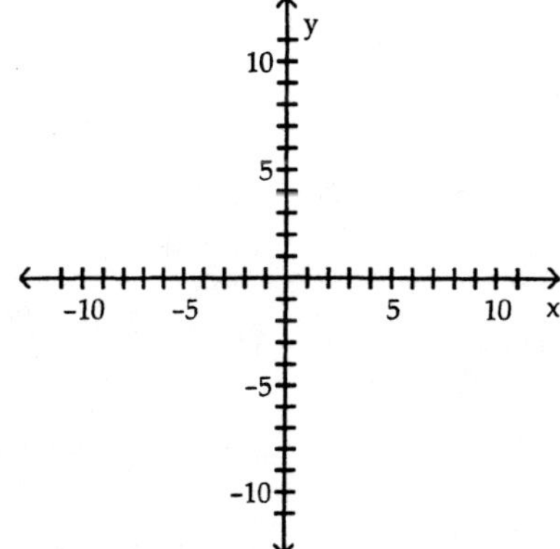

30) For the polynomial function $f(x) = 2x^4 - 7x^3 + 11x - 4$

a) Find the x- and y-intercepts of the graph of f. Round to two decimal places, if necessary.

b) Determine whether the graph crosses or touches the x-axis at each x-intercept.

c) End behavior: find the power function that the graph of f resembles for large values of $|x|$.

d) Use a graphing utility to graph the function.Approximate the local maxima rounded to two decimal places, if necessary. Approximate the local minima rounded to two decimal places, if necessary.

e) Determine the number of turning points on the graph.

f) Put all the information together, and connect the points with a smooth, continuous curve to obtain the graph of f.

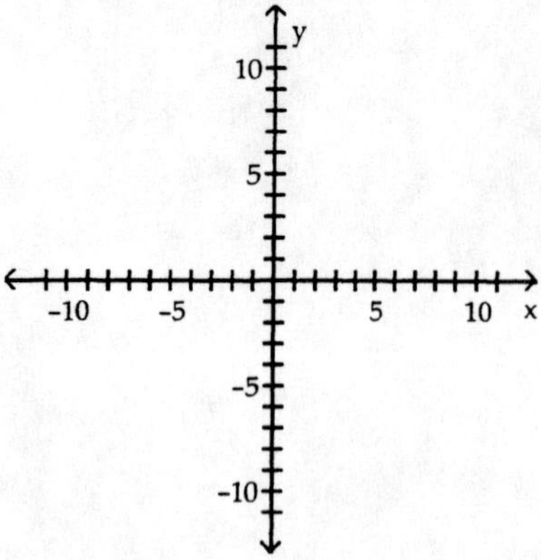

31) Which of the following polynomial functions might have the graph shown in the illustration below?

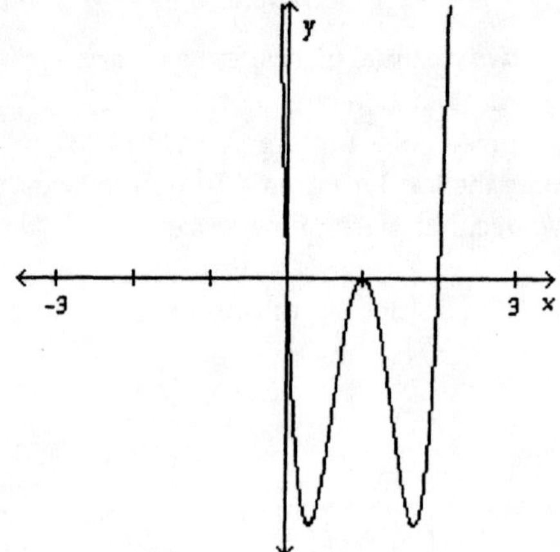

A) $f(x) = x(x - 2)^2(x - 1)$

B) $f(x) = x^2(x - 2)^2(x - 1)^2$

C) $f(x) = x(x - 2)(x - 1)^2$

D) $f(x) = x^2(x - 2)(x - 1)$

32) Determine the number of turning points on the graph of $f(x) = 5x^4 + 4x^3 - 4x^2 - 2x - 3$. Approximate the turning points, if any, rounded to two decimal places.

A) 3; (−1.39, 0), (0, −3), (1, 0)

B) 2; (−0.92, −4.08), (0.52, −4.19)

C) 3; (−0.92, −4.08), (−0.21, −2.78), and (0.52, −4.19)

D) 1; (−0.21, −2.78)

33) Construct a polynomial P(x) whose graph crosses the x-axis at -2 and 3, touches the x-axis at 5, crosses the y-axis at -5 and is below the x-axis between -2 and 3.

A) $P(x) = (x - 2)(x + 3)(x + 5)^2$

B) $P(x) = (x - 2)(x + 3)(x + 5)^2 - 145$

C) $P(x) = \frac{1}{30}(x + 2)(x - 3)(x - 5)^2$

D) $P(x) = (x + 2)(x - 3)(x - 5)^2$

34) Form a polynomial function P(x) with zeros: -3, multiplicity 2; 1, multiplicity 1; 5, multiplicity 3; and degree = 6.

A) $P(x) = (x + 3)^2(x - 1)(x - 5)^3$

B) $P(x) = (x - 2)^3(x + 1)(x + 3)^5$

C) $P(x) = (x + 2)^3(x - 1)(x - 3)^5$

D) $P(x) = (x - 3)^2(x + 1)(x + 5)^3$

5 Additional Applications

Solve the problem.

1) The profits (in millions) for a company for 8 years were as follows:

Year, x	Profits, P
1993, 1	1.1
1994, 2	1.7
1995, 3	2.0
1996, 4	1.4
1997, 5	1.3
1998, 6	1.5
1999, 7	1.8
2000, 8	2.1

Which of the following polynomials is the best model for this data?

A) $P(x) = 0.03x^3 - 0.3x^2 + 1.3x + 0.17$

B) $P(x) = -0.08x^3 + 7x^2 + 1.3x - 0.18$

C) $P(x) = -0.03x^4 - 0.3x^2 + 1.3x + 0.17$

D) $P(x) = 0.05x^2 - 0.8x + 6$

2) The width of a box is 2 units less than its length and 5 units more than its height. Write a polynomial function in x, where x equals the width, that represents the volume of the box. (Note: x may need to be defined in the problem.)

4.3 Rational Functions I

1 Find the Domain of a Rational Function

Find the domain of the rational function.

1) $g(x) = \frac{5x}{x - 3}$

A) $\{x \mid x \neq 3\}$

B) all real numbers

C) $\{x \mid x \neq -3\}$

D) $\{x \mid x \neq 0\}$

2) $g(x) = \frac{5x^2}{(x + 8)(x - 2)}$

A) all real numbers

B) $\{x \mid x \neq -8, x \neq 2, x \neq -5\}$

C) $\{x \mid x \neq -8, x \neq 2\}$

D) $\{x \mid x \neq 8, x \neq -2\}$

3) $f(x) = \frac{x + 4}{x^2 - 64}$

A) $\{x \mid x \neq 0, x \neq 64\}$
B) $\{x \mid x \neq -8, x \neq 8\}$
C) all real numbers
D) $\{x \mid x \neq -8, x \neq 8, x \neq -4\}$

4) $f(x) = \frac{x + 9}{x^2 + 25}$

A) $\{x \mid x \neq 0, x \neq -25\}$
B) $\{x \mid x \neq -5, x \neq 5\}$
C) $\{x \mid x \neq -5, x \neq 5, x \neq -9\}$
D) all real numbers

5) $g(x) = \frac{x + 3}{x^2 - 25x}$

A) $\{x \mid x \neq 0, x \neq 25\}$
B) all real numbers
C) $\{x \mid x \neq -5, x \neq 5\}$
D) $\{x \mid x \neq -5, x \neq 5, x \neq -3\}$

6) $R(x) = \frac{x^2 + x - 20}{x^2 - 14x + 48}$.

A) $\{x \mid x \neq -5, x \neq 4\}$
B) $\{x \mid x \neq 6, x \neq 8\}$
C) $\{x \mid x \neq -6, x \neq -8\}$
D) $\{x \mid x \neq 5, x \neq -4\}$

7) $f(x) = \frac{2x^2 - 4}{3x^2 + 6x - 45}$.

A) $\{x \mid x \neq 3, x \neq -3, x \neq -5\}$
B) $\{x \mid x \neq 3, x \neq -5\}$
C) all real numbers
D) $\{x \mid x \neq -3, x \neq 5\}$

8) $R(x) = \frac{3x^2 - 5x - 2}{x^2 - 4}$

A) all real numbers
B) $\{x \mid x \neq -2\}$
C) $\{x \mid x \neq 2, x \neq -2\}$
D) $\{x \mid x \neq 2\}$

Use the graph to determine the domain and range of the function.

9)

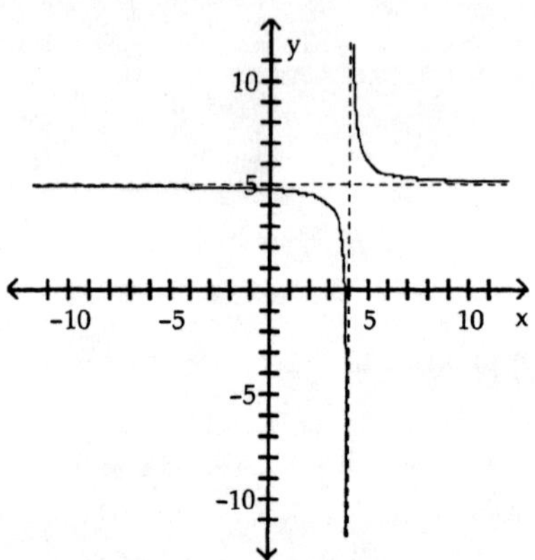

A) domain: $\{x \mid x \neq 5\}$
range: $\{y \mid y \neq 4\}$

B) domain: $\{x \mid x \neq -4\}$
range: $\{y \mid y \neq 5\}$

C) domain: $\{x \mid x \neq 5\}$
range: $\{y \mid y \neq -4\}$

D) domain: $\{x \mid x \neq 4\}$
range: $\{y \mid y \neq 5\}$

10)

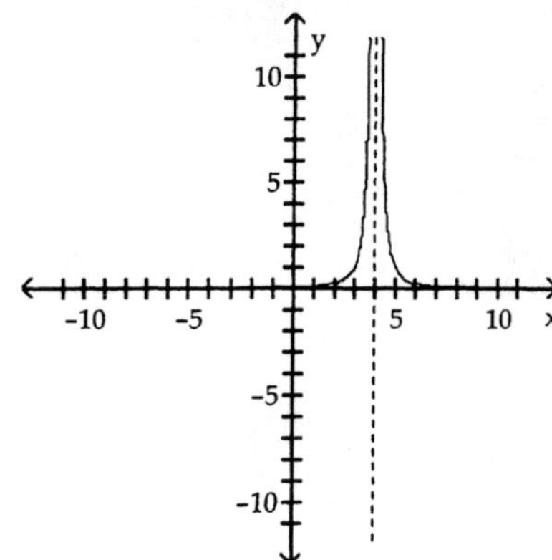

A) domain: $\{x \mid x > 0\}$
range: $\{y \mid y \neq 4\}$

B) domain: $\{x \mid x \neq 4\}$
range: $\{y \mid y \geq 0\}$

C) domain: $\{x \mid x \neq 4\}$
range: $\{y \mid y > 0\}$

D) domain: $\{x \mid x \geq 0\}$
range: $\{y \mid y \neq 4\}$

11)

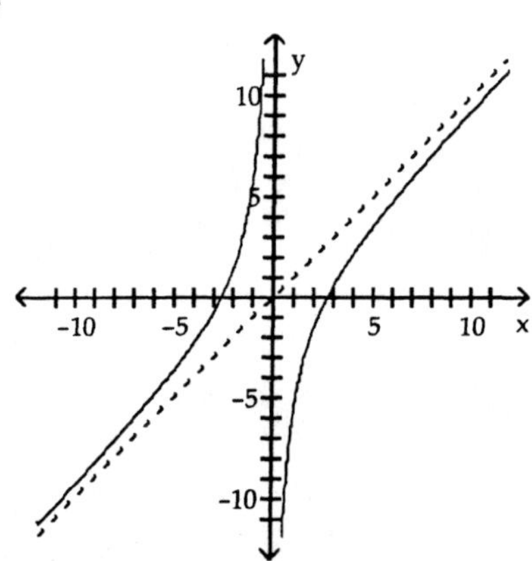

A) domain: $\{x \mid x \neq 0\}$
range: all real numbers

B) domain: all real numbers
range: $\{y \mid y \neq 0\}$

C) domain: $\{x \mid x \neq 0\}$
range: $\{y \mid y \neq 0\}$

D) domain: all real numbers
range: all real numbers

12)

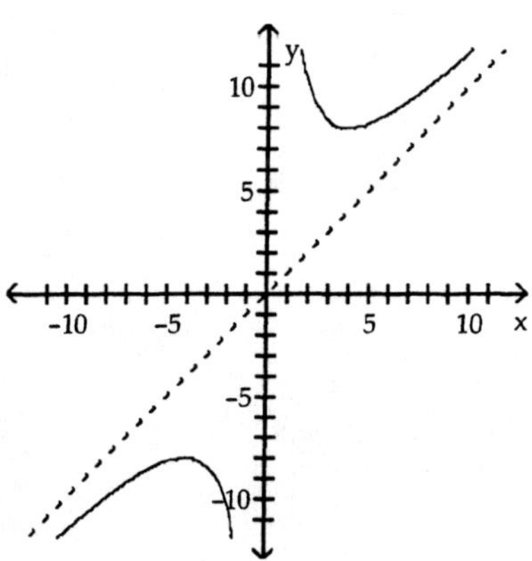

A) domain: $\{x \mid x \neq 0\}$
range: all real numbers

B) domain: $\{x \mid x \neq 0\}$
range: $\{y \mid y \leq -8 \text{ or } y \geq 8\}$

C) domain: $\{x \mid x \leq -8 \text{ or } x \geq 8\}$
range: $\{y \mid y \neq 0\}$

D) domain: all real numbers
range: $\{y \mid y \leq -8 \text{ or } y \geq 8\}$

13)

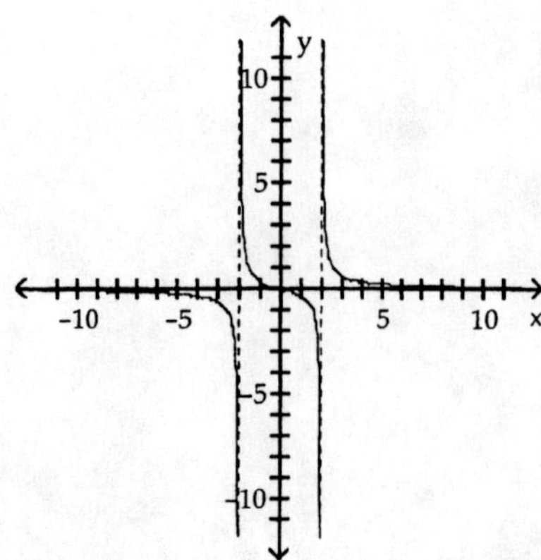

A) domain: {x | x ≠ -2, x ≠ 2}
range: all real numbers

B) domain: all real numbers
range: all real numbers

C) domain: all real numbers
range: {y | y ≠ -2, y ≠ 2}

D) domain: {x | x ≠ -2, x ≠ 2}
range: {y | y ≠ 0}

14)

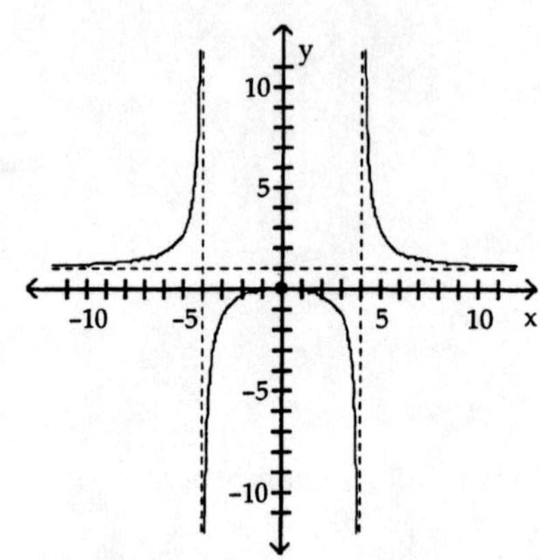

A) domain: {x | x ≠ -4, x ≠ 4}
range: {y | y ≤ 0 or y > 1}

B) domain: all real numbers
range: all real numbers

C) domain: {x | x ≠ -4, x ≠ 4}
range: {y | y ≤ 0 or y ≥ 1}

D) domain: {x | x ≤ 0 or x > 1}
range: {y | y ≠ -4, y ≠ 4}

2 Determine the Vertical Asymptotes of a Rational Function

Give the equation of the specified asymptote(s).

1) Vertical asymptote(s): $f(x) = \dfrac{x - 9}{x^2 + 2}$

A) x = 3, x = -3

B) x = -2

C) x = 2

D) no vertical asymptotes

2) Vertical asymptote(s): $f(x) = \dfrac{x - 1}{x^2 - 36}$

A) x = 1

B) x = -6

C) x = 6

D) x = 6, x = -6

3) Vertical asymptote(s): $f(x) = \dfrac{3x + 10}{x^2 + 12x + 35}$

A) $x = -7, x = -5$ B) $x = -\frac{10}{3}, x = 3$ C) $x = -\frac{10}{3}, x = -7$ D) $x = 3$

4) Vertical asymptote(s): $f(x) = \dfrac{3x - 7}{x^2 - 5x - 14}$

A) $x = -7, x = 2$ B) $x = \frac{7}{3}, x = 7, x = -2$

C) $x = 7, x = -2$ D) no vertical asymptotes

5) Vertical asymptote(s): $f(x) = \dfrac{x - x^3}{x^2 - 4x + 3}$

A) $x = 0, x = 1, x = 3$ B) $x = 3, x = 1$

C) $x = 3$ D) no vertical asymptotes

3 Determine the Horizontal or Oblique Asymptotes of a Rational Function

Give the equation of the specified asymptote(s).

1) Horizontal asymptote: $g(x) = \dfrac{x^2 + 2x - 8}{x - 8}$

A) $y = 7$ B) $y = 8$

C) $y = -2$ D) no horizontal asymptotes

2) Horizontal asymptote: $f(x) = \dfrac{8x^2 + 9}{8x^2 - 9}$

A) $y = 9$ B) $y = -9$

C) $y = 1$ D) no horizontal asymptotes

3) Horizontal asymptote: $h(x) = \dfrac{2x^2 - 8x - 7}{4x^2 - 4x + 5}$

A) $y = 0$ B) $y = \frac{1}{2}$

C) $y = 2$ D) no horizontal asymptotes

4) Horizontal asymptote: $f(x) = \dfrac{x^4 - 1}{3x^5 + 3}$.

A) $y = 3$ B) $y = \frac{1}{3}$

C) $y = 0$ D) no horizontal asymptotes

5) Oblique asymptote: $f(x) = \frac{x^2 + 9x - 6}{x - 4}$

A) $x = y + 13$
B) $y = x + 5$
C) $y = x + 13$
D) no oblique asymptotes

6) Oblique asymptote: $f(x) = \frac{3x^2}{5x^2 + 4}$

A) $x = y + 3$
B) $y = x + \frac{3}{5}$
C) $y = x + 3$
D) no oblique asymptotes

7) Oblique asymptote: $f(x) = \frac{x^2 - 3x + 4}{x + 7}$

A) $y = x + 7$
B) $x = y + 3$
C) $y = x - 10$
D) no oblique asymptotes

8) Oblique asymptote: $f(x) = \frac{x^2 + 2x + 5}{x + 7}$

A) $y = x - 9$
B) $y = x - 5$
C) $x = y - 5$
D) no oblique asymptotes

9) Oblique asymptote: $f(x) = \frac{2x^3 + 11x^2 + 5x - 1}{x^2 + 6x + 5}$.

A) $y = 2x$
B) $y = 2x - 1$
C) $y = 2x + 1$
D) $y = 0$

10) Horizontal and/or oblique asymptote(s): $R(x) = \frac{x^3 - 5x^2 + 4}{x^2 + 2}$

A) R(x) has an oblique asymptote, $y = x - 3$
B) R(x) has a horizontal asymptote, $y = 0$
C) R(x) has a horizontal asymptote, $y = 1$
D) R(x) has an oblique asymptote, $y = x - 5$

Graph the function using transformations.

11) $f(x) = \dfrac{4}{(6 + x)^2}$

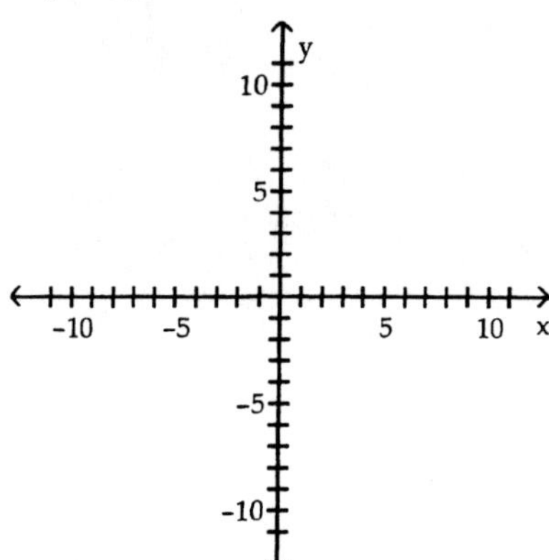

A)

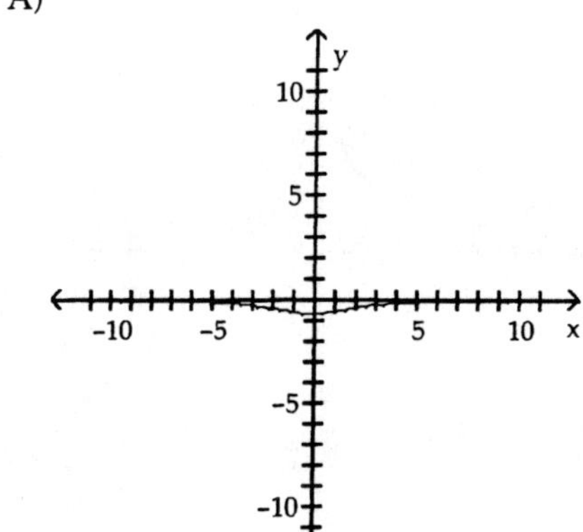

B)

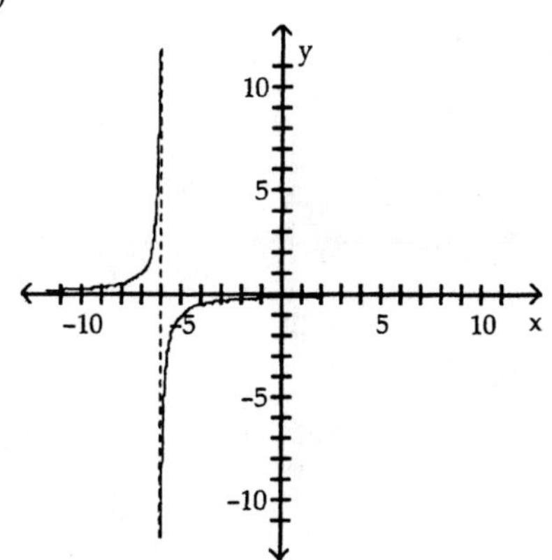

C)

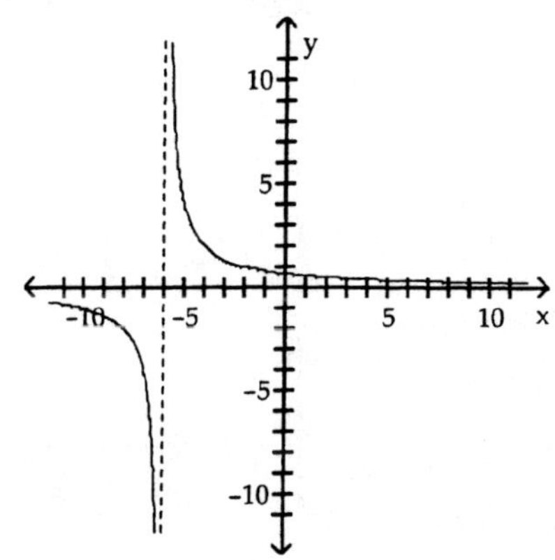

D)

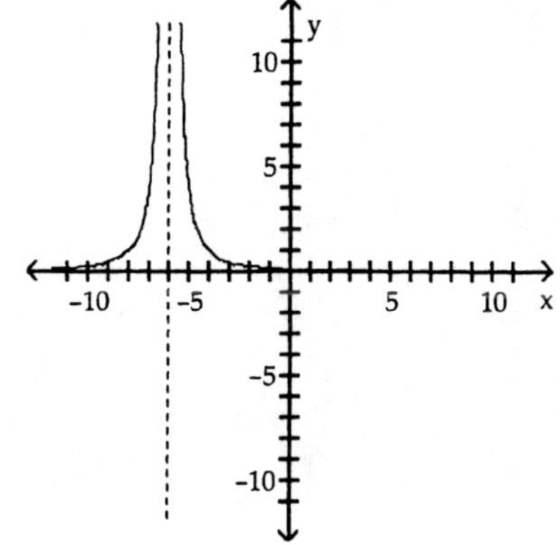

12) $f(x) = \frac{1}{x} + 1$

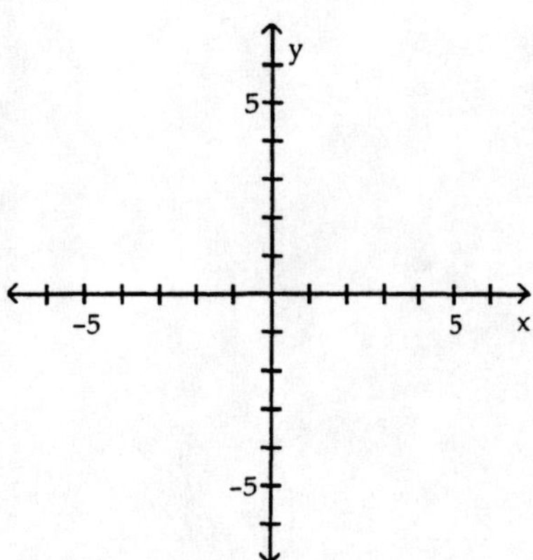

A)

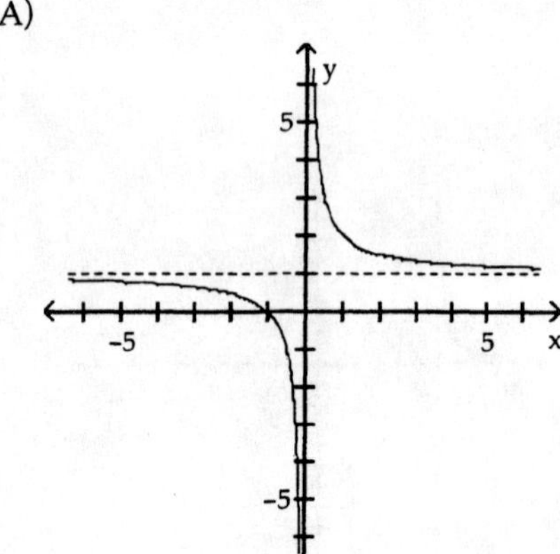

B)

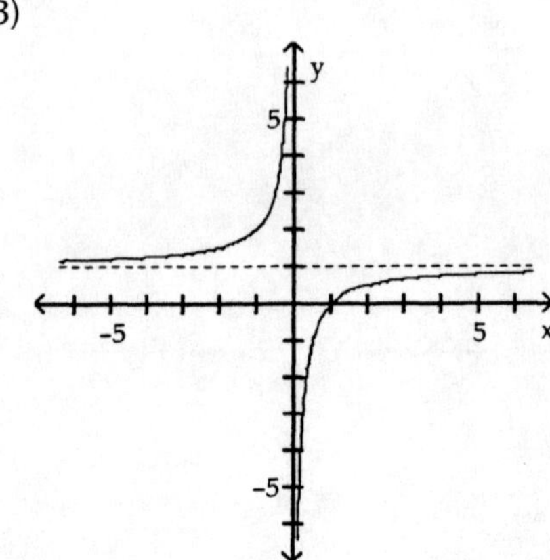

C)

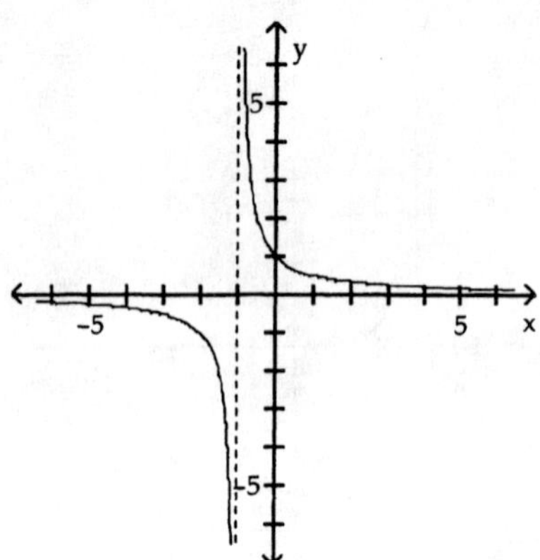

D)

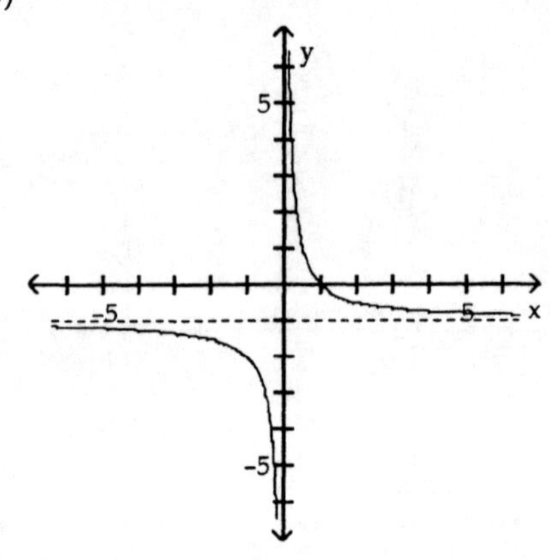

Solve the problem.

13) The acceleration due to gravity g (in meters per second per second) at a height h meters above sea level is given by $g(h) = \frac{3.99 \times 10^{14}}{(6.374 \times 10^{6} + h)^2}$ where 6.374×10^6 is the radius of Earth in meters. Death Valley in California is 86 m below sea level.

a) Find the value of g(h) at Death Valley to four decimal places.

b) Compare the value in (a) to the value of g(h) at sea level.

14) The distance formula states that $d = rt$. If a car drives 50 miles, the function $r = 50/t$ is a rational function. Find the asymptotes of this function.

15) A lens can be used to create an image of an object on the opposite side of the lens, such as the image created on a movie screen. Every lens has a measurement called its focal length, f. The distance s_1 of the object to the lens is related to the distance s_2 of the lens to the image by the function

$$s_1 = \frac{fs_2}{s_2 - f}.$$

For a lens with f = 0.3 m, what are the asymptotes of this function?

16) When two lenses are placed next to each other, their combined focal length (a measurement that can be negative or positive) is described by the equation

$$f = \frac{f_1 f_2}{f_1 + f_2}.$$

If $f_1 = 0.001$, what are the asymptotes of this function?

4.4 Rational Functions II: Analyzing Graphs

1 Analyze the Graph of a Rational Function

Find the indicated intercept(s) of the graph of the function.

1) x-intercepts of $f(x) = \frac{x - 6}{x^2 + 7x - 2}$

A) (6, 0) B) (7, 0) C) (3, 0) D) none

2) x-intercepts of $f(x) = \frac{x^2 + 8}{x^2 + 9x + 2}$

A) $(\sqrt{8}, 0), (-\sqrt{8}, 0)$ B) (2, 0) C) (8, 0) D) none

3) x-intercepts of $f(x) = \frac{x + 2}{x^2 + 4x - 5}$

A) $(\frac{2}{5}, 0)$ B) (2, 0) C) (-2, 0) D) none

4) x-intercepts of $f(x) = \frac{x^2 + 3x}{x^2 + 3x - 8}$

A) (-3, 0) B) (3, 0) C) (0, 0) and (-3, 0) D) (0, 0) and (3, 0)

5) x-intercepts of $f(x) = \frac{(x - 8)(2x + 9)}{x^2 + 6x - 2}$

A) (8, 0) and (-9, 0) B) (-8, 0) and $(\frac{9}{2}, 0)$ C) (8, 0) and $(-\frac{9}{2}, 0)$ D) none

6) y-intercept of $f(x) = \frac{x - 8}{x^2 + 9x - 14}$

A) $(0, \frac{4}{7})$ B) $(0, -\frac{7}{4})$ C) (0, 8) D) none

7) y-intercept of $f(x) = \frac{x^2 - 2x}{x^2 + 15x - 13}$

A) $(0, -\frac{13}{2})$ B) (0, 2) C) (0, 0) D) $(0, \frac{2}{13})$

8) y-intercept of $f(x) = \frac{x^2 - 9}{x^2 + 2x - 12}$

A) (0, 9) B) $(0, \frac{3}{4})$ C) $(0, -\frac{4}{3})$ D) none

9) y-intercept of $f(x) = \frac{x^2 - 8x + 2}{9x}$

A) $(0, \frac{2}{9})$ B) $(0, -\frac{9}{2})$ C) (0, 2) D) none

10) y-intercept of $f(x) = \frac{x^2 - 6x - 9}{x^2 + 15x + 3}$

A) (0, -9) B) (0, 15) C) (0, -3) D) none

Solve the problem.

11) Is there y-axis symmetry for the rational function $f(x) = \frac{-7x^2}{9x^4 - 10}$?

A) Yes B) No

12) Is there y-axis symmetry for the rational function $f(x) = \frac{-2x^2}{9x^3 - 3}$?

A) Yes B) No

13) Is there y-axis symmetry for the rational function $f(x) = \frac{-3x^2 - 3x - 14}{7x + 16}$?

A) Yes B) No

14) Is there origin symmetry for the rational function $f(x) = \frac{-3x}{2x^2 + 8}$?

A) Yes B) No

15) Is there origin symmetry for the rational function $f(x) = \frac{5x^2 + 5}{-4x}$?

A) Yes B) No

16) Is there origin symmetry for the rational function $f(x) = \frac{-6x^2 - 11}{-4x^2 + 8}$?

A) Yes B) No

17) Find the vertical asymptote(s) and/or hole(s) for $R(x) = \frac{x^2 + x - 20}{x^2 - x - 12}$.

A) vertical asymptote: $x = -3$; hole at $(4, \frac{9}{7})$

B) vertical asymptotes: $x = -4$, $x = 3$

C) vertical asymptotes: $x = 4$, $x = -3$

D) vertical asymptote: $x = -3$

18) Find the vertical asymptote(s) and/or hole(s) for $R(x) = \frac{2}{(x^2 - 25)(x + 4)}$.

A) vertical asymptotes: $x = 25$, $x = -4$

B) vertical asymptotes: $x = -5$, $x = 5$, $x = -4$

C) vertical asymptote: $x = -4$

D) vertical asymptotes: $x = -5$, $x = 5$; hole at $(-4, -\frac{2}{9})$

Graph the function.

19) $f(x) = x - \frac{4}{x}$

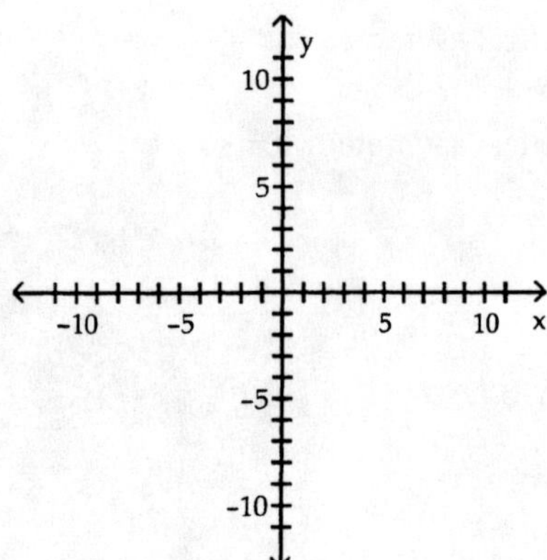

A)

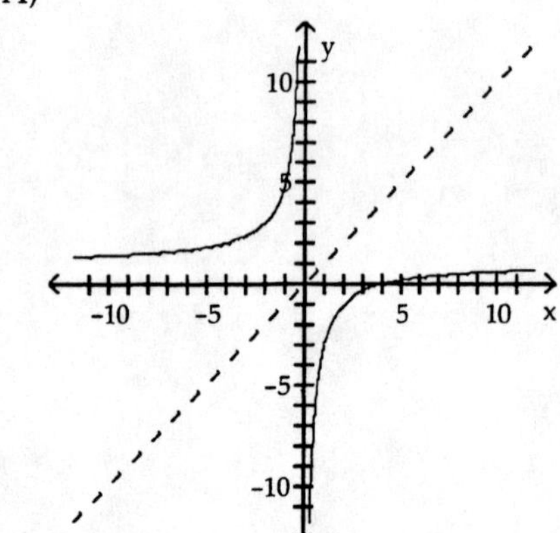

B)

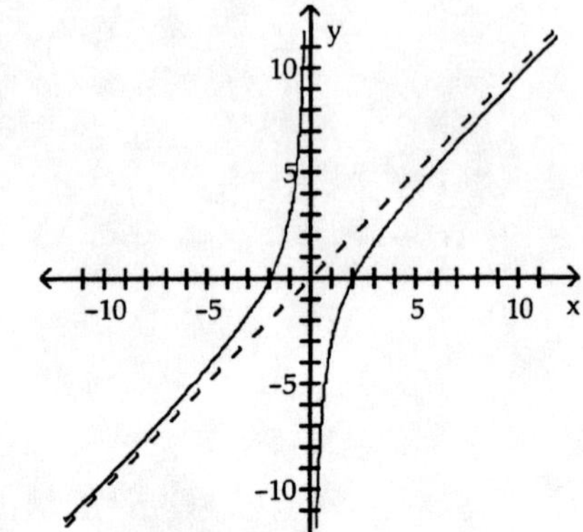

C)

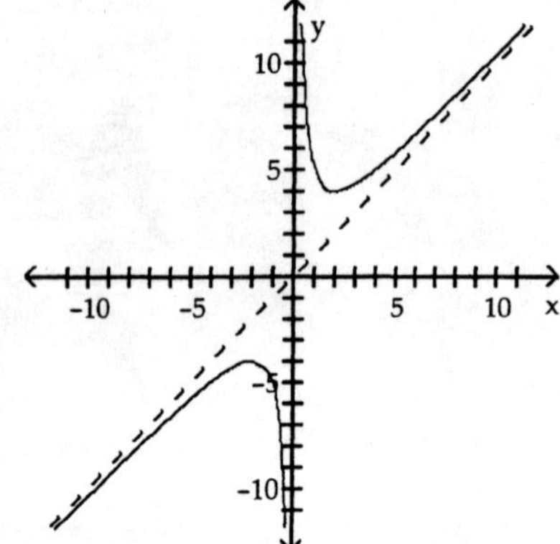

D)

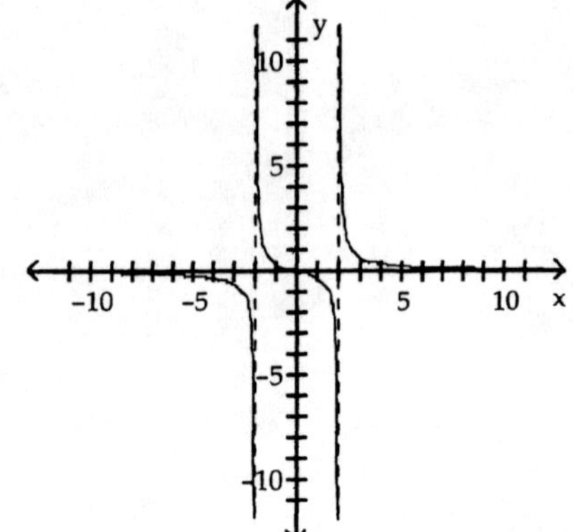

20) $f(x) = \dfrac{x}{x^2 - 1}$

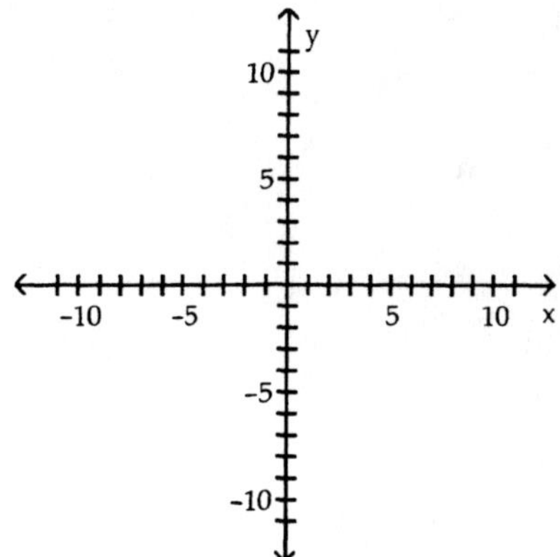

A)

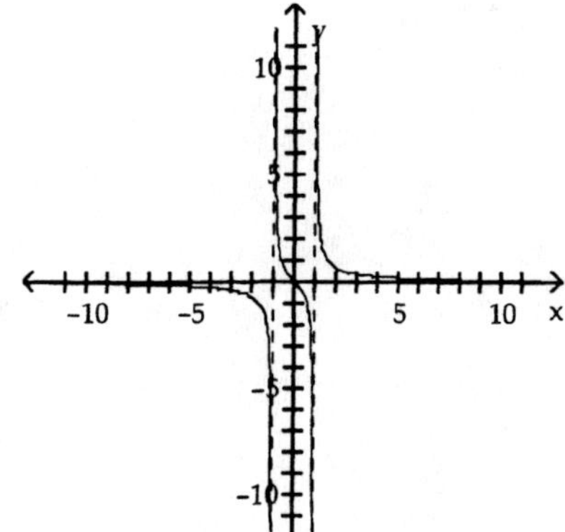

B)

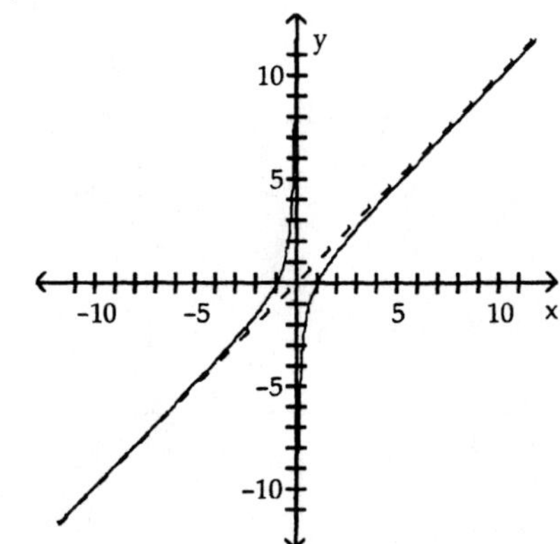

C)

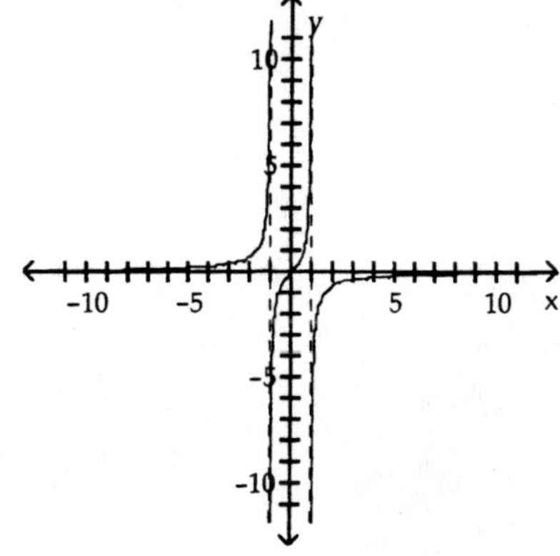

D)

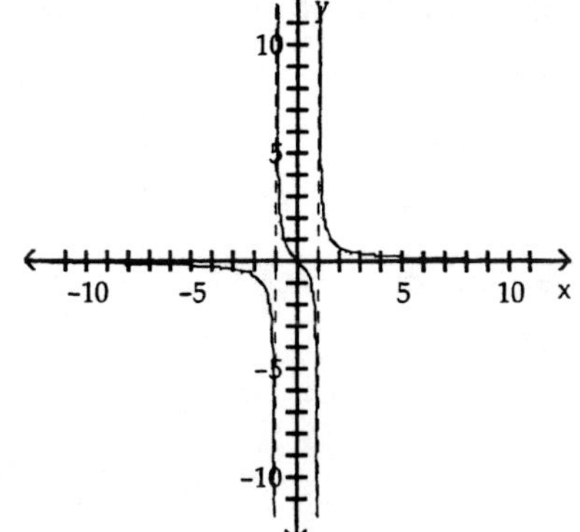

Solve the problem.

21) Decide which of the rational functions might have the given graph.

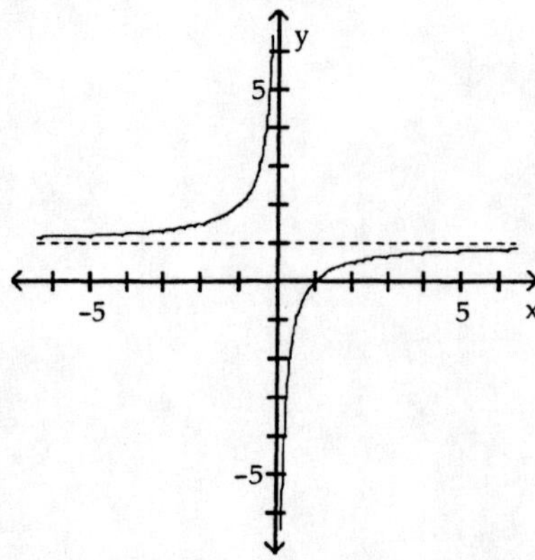

A) $f(x) = 1 - x$ B) $f(x) = 1 + \frac{1}{x}$ C) $f(x) = 1 - \frac{1}{x}$ D) $f(x) = \frac{1}{x} - 1$

22) Decide which of the rational functions might have the given graph.

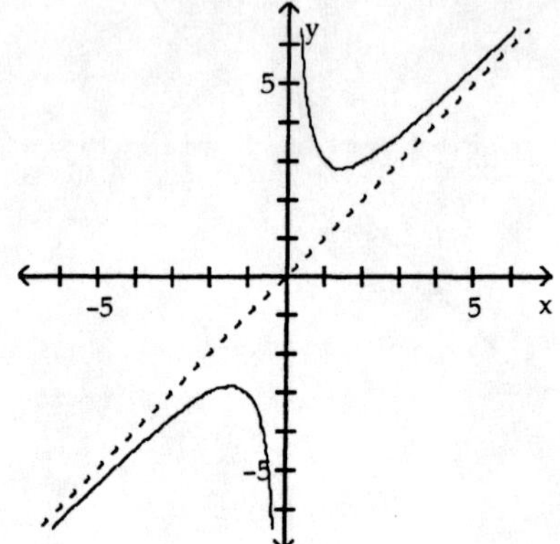

A) $f(x) = x + \frac{1}{x}$ B) $f(x) = x + 2$ C) $f(x) = 2x + \frac{1}{x}$ D) $f(x) = x + \frac{2}{x}$

23) Decide which of the rational functions might have the given graph.

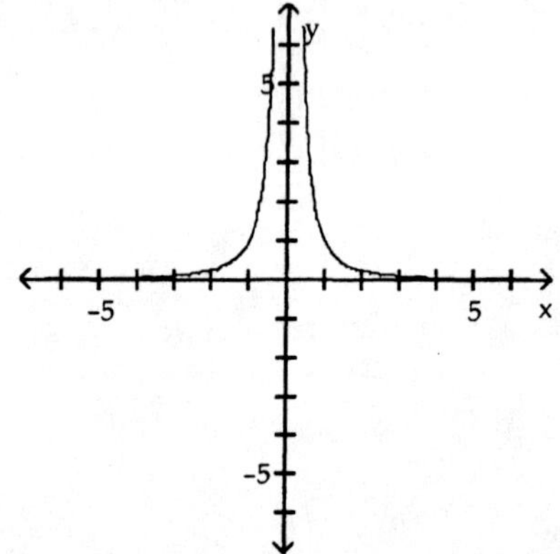

A) $f(x) = \frac{1}{x^2}$ B) $f(x) = x^2$ C) $f(x) = \frac{1}{x}$ D) $f(x) = \frac{1}{2x}$

24) Decide which of the rational functions might have the given graph.

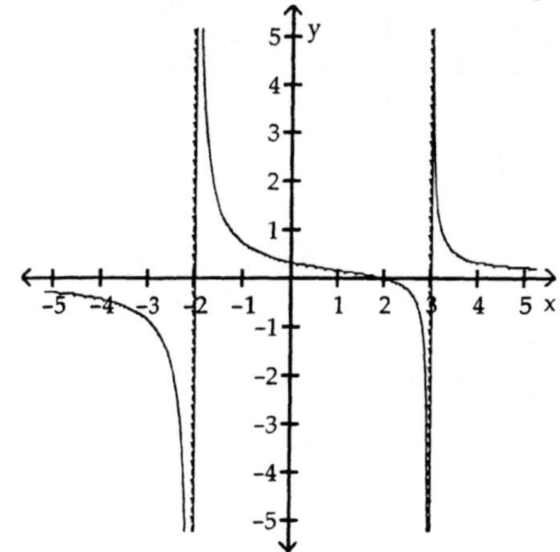

A) $R(x) = \frac{2 - x}{(x + 2)(x - 3)}$

B) $R(x) = \frac{x - 2}{(x + 2)^2(x - 3)^2}$

C) $R(x) = \frac{x + 2}{(x - 2)(x + 3)}$

D) $R(x) = \frac{x - 2}{(x + 2)(x - 3)}$

25) Determine which rational function R(x) has a graph that crosses the x-axis at -1, touches the x-axis at -4, has vertical asymptotes at x = -2 and x = 3, and has one horizontal asymptote at y = -2.

A) $R(x) = \frac{-2(x + 1)(x + 4)^2}{(x + 2)^2(x - 3)}, x \neq -2, 3$

B) $R(x) = \frac{-2(x + 1)(x + 4)}{(x + 2)(x - 3)}, x \neq -2, 3$

C) $R(x) = \frac{-(x + 1)(x + 4)^2}{2(x - 2)^2(x + 3)}, x \neq 2, -3$

D) $R(x) = \frac{-2(x - 3)(x + 2)^2}{(x + 4)^2(x + 1)}, x \neq -4, -1$

2 Solve Applied Problems Involving Rational Functions

Solve the problem.

1) A rare species of insect was discovered in the rain forest of Costa Rica. Environmentalists transplant the insect into a protected area. The population of the insect t months after being transplanted is

$$P(t) = \frac{45(1 + 0.6t)}{(3 + 0.02t)}.$$

a) What was the population when t = 0?
b) What will the population be after 10 years?
c) What is the largest value the population could reach?

2) The concentration C of a certain drug in a patient's bloodstream is given by

$$\frac{30t}{t^2 + 49}.$$

a) Find the horizontal asymptote of C(t).
b) Using a graphing utility, determine the time at which the concentration is highest.

3) A can in the shape of a right circular cylinder is required to have a volume of 700 cubic centimeters. The top and bottom are made up of a material that costs 8¢ per square centimeter, while the sides are made of material that costs 5¢ per square centimeter. Which function below describes the total cost of the material as a function of the radius r of the cylinder?

A) $C(r) = 0.16\pi r^2 + \frac{70}{r}$
B) $C(r) = 0.08\pi r^2 + \frac{70}{r}$

C) $C(r) = 0.08\pi r^2 + \frac{140}{r}$
D) $C(r) = 0.16\pi r^2 + \frac{140}{r}$

4) The concentration of a drug in the bloodstream, measured in milligrams per liter, can be modeled by the function, $C(t) = \frac{12t + 4}{3t^2 + 2}$, where t is the number of minutes after injection of the drug. When will the drug be at its highest concentration? Approximate your answer rounded to two decimal places.

A) t = 3.65 minutes after the injection is given
B) t = 0.55 minutes after the injection is given
C) t = 4 minutes after the injection is given
D) at the time of injection

5) A closed box with a square base has to have a volume of 13,000 cubic inches. Find a function for the surface area of the box.

A) $S(x) = x^2 + \frac{52{,}000}{x}$
B) $S(x) = 2x^2 + \frac{52{,}000}{x}$
C) $S(x) = 2x^2 + \frac{13{,}000}{x}$
D) $S(x) = 2x^2 + \frac{78{,}000}{x}$

6) Economists use what is called a Leffer curve to predict the government revenue for tax rates from 0% to 100%. Economists agree that the end points of the curve generate 0 revenue, but disagree on the tax rate that produces the maximum revenue. Suppose an economist produces this rational function

$R(x) = \frac{10x(100 - x)}{50 + x}$, where R is revenue in millions at a tax rate of x percent. Use a graphing calculator to graph the function. What tax rate produces the maximum revenue? What is the maximum revenue?

A) 35.8%; $276 million
B) 34.0%; $271 million
C) 36.6%; $268 million
D) 41.2%; $264 million

7) Economists use what is called a Leffer curve to predict the government revenue for tax rates from 0% to 100%. Economists agree that the end points of the curve generate 0 revenue, but disagree on the tax rate that produces the maximum revenue. Suppose an economist produces this rational function

$R(x) = \frac{10x(100 - x)}{25 + x}$, where R is revenue in millions at a tax rate of x percent. Use a graphing calculator to graph the function. What tax rate produces the maximum revenue? What is the maximum revenue?

A) 38.4%; $383 million
B) 30.9%; $382 million
C) 27.0%; $379 million
D) 28.8%; $272 million

8) Economists use what is called a Leffer curve to predict the government revenue for tax rates from 0% to 100%. Economists agree that the end points of the curve generate 0 revenue, but disagree on the tax rate that produces the maximum revenue. Suppose an economist produces this rational function

$R(x) = \frac{10x(100 - x)}{75 + x}$, where R is revenue in millions at a tax rate of x percent. Use a graphing calculator to graph the function. What tax rate produces the maximum revenue? What is the maximum revenue?

A) 39.6%; $209 million
B) 35.8%; $209 million
C) 37.5%; $210 million
D) 34.9%; $207 million

9) Economists use what is called a Leffer curve to predict the government revenue for tax rates from 0% to 100%. Economists agree that the end points of the curve generate 0 revenue, but disagree on the tax rate that produces the maximum revenue. Suppose an economist produces this rational function

$R(x) = \dfrac{10x(100 - x)}{15 + x}$, where R is revenue in millions at a tax rate of x percent. Use a graphing calculator to graph the function. What tax rate produces the maximum revenue? What is the maximum revenue?

A) 28.1%; $470 million B) 29.7%; $467 million C) 26.5%; $469 million D) 31.4%; $464 million

10) A box has a base whose length is twice its width. The volume of the box is 7000 cubic inches.

a) Find a function for the surface area of the box.

b) What are the dimensions of the box that minimizes surface area?

11) The formula $y = \dfrac{fx}{x - f}$ models the relationships needed to focus an image where y is the distance between the film and projector lens, x is the distance between the move screen and the projector lens, and f is the focal length.

a) Sketch the graph of this rational function for f = 5 centimeters.

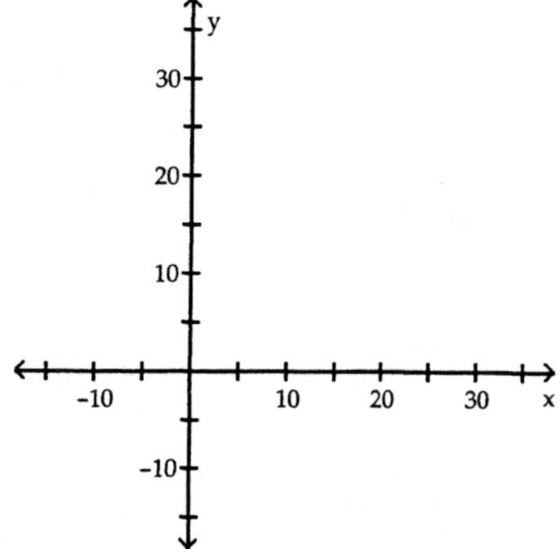

b) Bob and Carol are showing home movies. Use the graph to describe the desired distance between the film and the projector lens as Carol moves the projector further from the screen.

12) Which of the following functions could have this graph?

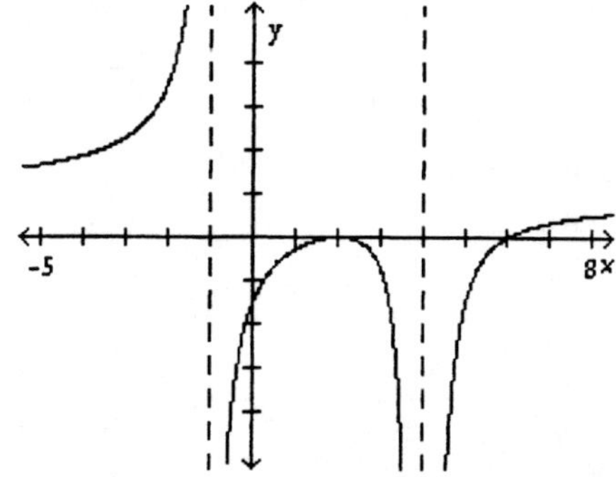

A) $y = \dfrac{2(x - 2)^2(x - 6)}{(x + 1)(x - 4)^2}$ B) $y = \dfrac{(x + 1)(x - 4)^2}{(x - 2)2(x - 6)}$ C) $y = \dfrac{(x - 2)^2(x - 6)}{(x + 1)(x - 4)^2}$ D) $y = \dfrac{(x - 2)(x - 6)^2}{(x + 1)2(x - 4)}$

4.5 Polynomial and Rational Inequalities

1 Solve Polynomial Inequalities

Solve the inequality.

1) $(x + 7)(x - 2) > 0$

A) $(-\infty, -7)$ B) $(-\infty, -7)$ or $(2, \infty)$ C) $(2, \infty)$ D) $(-7, 2)$

2) $(x + 1)(x - 4) \leq 0$

A) $(-\infty, -1]$ B) $[-1, 4]$ C) $[4, \infty)$ D) $(-\infty, -1]$ or $[4, \infty)$

3) $x^2 - 5x \geq 0$

A) $[0, 5]$ B) $(-\infty, -5]$ or $[0, \infty)$ C) $(-\infty, 0]$ or $[5, \infty)$ D) $[-5, 0]$

4) $x^2 + 3x \geq 0$

A) $(-\infty, 0]$ or $[3, \infty)$ B) $[0, 3]$ C) $[-3, 0]$ D) $(-\infty, -3]$ or $[0, \infty)$

5) $x^2 + 2x \leq 0$

A) $(-\infty, 0]$ or $[2, \infty)$ B) $[-2, 0]$ C) $(-\infty, -2]$ or $[0, \infty)$ D) $[0, 2]$

6) $x^2 - 7x \leq 0$

A) $[-7, 0]$ B) $(-\infty, 0]$ or $[7, \infty)$ C) $[0, 7]$ D) $(-\infty, -7]$ or $[0, \infty)$

7) $x^2 - 25 > 0$

A) $(-\infty, -25)$ or $(25, \infty)$ B) $(-5, 5)$ C) $(-25, 25)$ D) $(-\infty, -5)$ or $(5, \infty)$

8) $x^2 - 81 \leq 0$

A) $[-9, 9]$ B) $(-\infty, -9]$ or $[9, \infty)$ C) $[-81, 81]$ D) $(-\infty, -81]$ or $[81, \infty)$

9) $56(x^2 - 1) > 15x$

A) $(-\infty, -\frac{7}{8})$ or $(\frac{8}{7}, \infty)$ B) $(-\frac{8}{7}, \frac{7}{8})$ C) $(-\frac{7}{8}, \frac{8}{7})$ D) $(-\infty, -\frac{8}{7})$ or $(\frac{7}{8}, \infty)$

10) $x^2 + 3x \geq 4$

A) $[1, \infty)$ B) $(-\infty, -4]$ C) $(-\infty, -4]$ or $[1, \infty)$ D) $[-4, 1]$

11) $(a + 3)(a - 2)(a - 6) > 0$

A) $(6, \infty)$ B) $(-3, 2)$ or $(6, \infty)$ C) $(-\infty, 2)$ D) $(-\infty, -3)$ or $(2, 6)$

12) $(b + 7)(b - 5)(b - 7) < 0$

A) $(7, \infty)$ B) $(-\infty, 5)$ C) $(-7, 5)$ or $(7, \infty)$ D) $(-\infty, -7)$ or $(5, 7)$

13) $x^2 + 9x + 14 > 0$

A) $(-\infty, -7)$ or $(-2, \infty)$ B) $(-\infty, -7)$ C) $(-7, -2)$ D) $(-2, \infty)$

14) $x^2 - 6x - 7 \leq 0$

A) $(-\infty, -1]$ B) $(-\infty, -1]$ or $[7, \infty)$ C) $[7, \infty)$ D) $[-1, 7]$

15) $x^3 + 3x^2 - 18x > 0$

A) (-3, 0) or (6, ∞) B) (-6, ∞) C) (-∞,-6) or (0, 3) D) (-6, 0) or (3, ∞)

16) $2x^2 - 7x > 9$

A) $(\frac{9}{2}, \infty)$ B) $(-\infty, -1)$ or $(\frac{9}{2}, \infty)$ C) $(-1, \frac{9}{2})$ D) $(-\infty, -\frac{9}{2})$ or $(1, \infty)$

17) $x(x + 3)(5 - x) \geq 0$

A) (-3, 5) B) [-3, 5] C) (-∞, -3) or (0, 5) D) (-∞, -3] or [0, 5)

18) $x^4 < 25x^2$

A) (-5, 0) or (0, 5) B) (-∞, -5) or (0, 5) C) (-∞, -5) or (5, ∞) D) (-5, 0) or (5, ∞)

Solve the problem.

19) The revenue achieved by selling x graphing calculators is figured to be x(26 - 0.2x) dollars. The cost of each calculator is $14. How many graphing calculators must be sold to make a profit (revenue - cost) of at least $170.20?

A) $\{x|24 < x < 22\}$ B) $\{x|23 < x < 37\}$ C) $\{x|8 < x < 22\}$ D) $\{x|25 < x < 35\}$

2 Solve Rational Inequalities

Solve the inequality.

1) $\frac{x - 7}{x + 3} < 0$

A) (-3, 7) B) (-∞, -3) or (7, ∞) C) (-∞, -3) D) (7, ∞)

2) $\frac{x - 6}{x + 1} > 0$

A) (6, ∞) B) (-1, 6) C) (-∞, -1) or (6, ∞) D) (-∞, -1)

3) $\frac{x + 20}{x + 8} < 7$

A) (-∞, - 6) or (8, ∞) B) (-∞, -8) or (- 6, ∞) C) (-8, - 6) D) (-∞, -8) or (8, ∞)

4) $x + \frac{56}{x} < 15$

A) (-∞, 0) or (7, 8) B) (0, 7) or (8, ∞) C) (-∞, 0) or (8, ∞) D) (0, 7) or (7, 8)

5) $\frac{(x - 4)(x + 4)}{x} \leq 0$

A) [-4, 0) or (0, 4] B) (-∞, -4] or [4, ∞) C) (-∞, -4] or (0, 4] D) [-4, 0) or [4, ∞)

6) $\frac{(x + 11)(x - 9)}{x - 1} \geq 0$

A) (-∞, -11] or (1, 9] B) [-11, 1) or (1, 9] C) (-∞, -11] or [9, ∞) D) [-11, 1) or [9, ∞)

7) $\frac{(x - 2)^2}{x^2 - 9} > 0$

A) (-3, 2) or (2, 3) B) (-∞, -3) or (2, 3) C) (-∞, -3) or (3, ∞) D) (-3, 2) or (3, ∞)

8) $\frac{(x-1)(3-x)}{(x-2)^2} \le 0$

A) $(-\infty, 1)$ or $(3, \infty)$
B) $(-\infty, -3)$ or $(-1, \infty)$
C) $(-\infty, -3]$ or $(-2, -1)$ or $[1, \infty)$
D) $(-\infty, 1]$ or $[3, \infty)$

9) $\frac{4x}{6-x} < x$

A) $(6, \infty)$
B) $(0, 2)$ or $(6, \infty)$
C) $(2, 6)$
D) $(-\infty, 2)$ or $(6, \infty)$

10) $\frac{4x}{6-x} \ge 4x$

A) $(-\infty, 5]$ or $[6, \infty)$
B) $(-\infty, 0]$ or $[5, 6)$
C) $[0, 5]$ or $[6, \infty)$
D) $[6, \infty)$

11) $\frac{12}{x-5} > \frac{10}{x+1}$

A) $(-31, -1)$ or $(5, \infty)$
B) $(-31, -1)$ or $(-1, 5)$
C) $(-\infty, -31)$ or $(-1, 5)$
D) $(-\infty, -31)$ or $(5, \infty)$

12) $\frac{x^2(x-11)(x+1)}{(x-5)(x+8)} \ge 0$

A) $(-8, -1]$ or $(5, 11]$
B) $(-\infty, -8)$ or $[11, \infty)$
C) $(-\infty, -8)$ or $[-1, 0)$ or $(0, 5)$ or $[11, \infty)$
D) $(-\infty, -8)$ or $[-1, 5)$ or $[11, \infty)$

13) $\frac{x-2}{x} + \frac{2}{x-1} \le 2$

A) $(-\infty, 0]$ or $[1, \infty)$
B) $(-\infty, \infty)$
C) $[-3, 0)$ or $(1, \infty)$
D) $(-\infty, -1]$ or $(0, 1)$ or $[2, \infty)$

Solve the problem.

14) A ball is thrown vertically upward with an initial velocity of 160 feet per second. The distance in feet of the ball from the ground after t seconds is $s = 160t - 16t^2$. For what interval of time is the ball more than 384 above the ground?

A) $\{x \mid 9 \text{ sec} < x < 11 \text{ sec}\}$
B) $\{x \mid 4.5 \text{ sec} < x < 5.5 \text{ sec}\}$
C) $\{x \mid 3.5 \text{ sec} < x < 6.5 \text{ sec}\}$
D) $\{x \mid 4 \text{ sec} < x < 6 \text{ sec}\}$

15) A ball is thrown vertically upward with an initial velocity of 160 feet per second. The distance in feet of the ball from the ground after t seconds is $s = 160t - 16t^2$. For what intervals of time is the ball less than 336 above the ground (after it is tossed until it returns to the ground)?

A) $\{x \mid 0 \text{ sec} < x < 4.5 \text{ sec and } 5.5 \text{ sec} > x > 10 \text{ sec}\}$
B) $\{x \mid 3 \text{ sec} < x < 7 \text{ sec}\}$
C) $\{x \mid 0 \text{ sec} < x < 3 \text{ sec and } 7 \text{ sec} > x > 10 \text{ sec}\}$
D) $\{x \mid 0 \text{ sec} < x < 2.5 \text{ sec and } 7.5 \text{ sec} > x > 10 \text{ sec}\}$

16) The revenue achieved by selling x graphing calculators is figured to be $x(31 - 0.2x)$ dollars. The cost of each calculator is \$19. How many graphing calculators must be sold to make a profit (revenue - cost) of at least \$160.00?

A) $\{x \mid 22 < x < 38\}$
B) $\{x \mid 5 < x < 25\}$
C) $\{x \mid 20 < x < 40\}$
D) $\{x \mid 21 < x < 19\}$

17) The revenue achieved by selling x graphing calculators is figured to be $x(35 - 0.5x)$ dollars. The cost of each calculator is $19. How many graphing calculators must be sold to make a profit (revenue - cost) of at least $115.50?

A) $\{x \mid 12 < x < 20\}$ B) $\{x \mid 15 < x < 25\}$ C) $\{x \mid 11 < x < 21\}$ D) $\{x \mid 13 < x < 19\}$

18) A ball is thrown vertically upward with an initial velocity of 90 feet per second. The distance s (in feet) of the ball from the ground after t seconds is $s = 90t - 16t^2$. When is the ball more than 90 feet above the ground?

19) The monthly revenue achieved by selling x calendars is calculated to be $x(10 - 0.3x)$ dollars. If each calender is produced at a cost of 15 cents, how many calendars must be sold each month to achieve a profit of at least $80?

4.6 The Real Zeros of a Polynomial Function

1 Use the Remainder and Factor Theorems

Use the Factor Theorem to determine whether x - c is a factor of f(x).

1) $f(x) = x^3 + 3x^2 - 8x + 10;\ x + 5$

A) Yes B) No

2) $f(x) = x^3 + 7x^2 - 16x + 18;\ x - 9$

A) Yes B) No

3) $f(x) = x^4 - 45x^2 - 196;\ x - 7$

A) Yes B) No

4) $f(x) = x^4 - 12x^2 - 64;\ x - 8$

A) Yes B) No

5) $f(x) = x^4 + 6x^3 + 3x^2 + 10x - 48;\ x + 6$

A) Yes B) No

6) $f(x) = x^4 + 8x^3 + 5x^2 + 31x - 72;\ x - 8$

A) Yes B) No

7) $f(x) = 18x^3 + 95x^2 - 42x - 110;\ x + \frac{11}{2}$

A) Yes B) No

8) $f(x) = 12x^3 + 38x^2 - 28x - 22;\ x - \frac{11}{3}$

A) Yes B) No

9) $7x^4 + 20x^3 - 3x^2 + x - 3;\ x + 3$

A) Yes B) No

10) $7x^4 + 13x^3 - 2x^2 + x + 2;\ x + 2$

A) Yes B) No

11) $6x^3 + 14x^2 - 11x + 3$; $x + 3$

A) Yes

B) No

12) $8x^3 + 21x^2 - 8x - 3$; $x + 3$

A) Yes

B) No

Solve the problem.

13) Use the Remainder Theorem to find the remainder when f(x) is divided by g(x).

$f(x) = x^4 + 8x^3 + 12x^2$; $g(x) = x + 1$

A) -21

B) 5

C) -5

D) 21

14) Use the Remainder Theorem to find the remainder when f(x) is divided by g(x).

$f(x) = 5x^6 - 3x^3 + 8$; $g(x) = x + 1$

A) 6

B) 16

C) 8

D) 10

2 Use Descartes' Rule of Signs to Determine the Number of Positive and the Number of Negative Real Zeros of a Polynomial Function

Give the maximum number of zeros the polynomial function may have. Use Descarte's Rule of Signs to determine how many positive and how many negative zeros it may have.

1) $f(x) = -6x^7 + x^5 - x^2 + 9$

A) 7; 2 or 0 positive zeros; 2 or 0 negative zeros

B) 7; 3 or 1 positive zeros; 2 or 0 negative zeros

C) 7; 3 or 1 positive zeros; 3 or 1 negative zeros

D) 7; 2 or 0 positive zeros; 3 or 1 negative zeros

2) $f(x) = 5x^7 - 6x^2 + x + 3$

A) 7; 2 or 0 positive zeros; 2 or 0 negative zeros

B) 7; 2 or 0 positive zeros; 1 or 0 negative zeros

C) 7; 2 or 0 positive zeros; 1 negative zero

D) 7; 3 or 1 positive zeros; 3 or 1 negative zeros

3) $f(x) = x^7 + x^6 + x^2 + x + 5$

A) 7; 0 positive zeros; 2 or 0 negative zeros

B) 7; 0 positive zeros; 1 negative zero

C) 7; 0 positive zeros; 3 or 1 negative zeros

D) 7; 0 positive zeros; 0 negative zeros

4) $f(x) = x^5 - 2.4x^4 - 11.51x^3 + 26.855x^2 + 45.81x - 8.533$

A) 5; 3 or 1 positive zeros; 3 or 1 negative zeros

B) 5; 2 or 0 positive zeros; 2 or 0 negative zeros

C) 5; 2 or 0 positive zeros; 3 or 1 negative zeros

D) 5; 3 or 1 positive zeros; 2 or 0 negative zeros

5) $f(x) = x^8 - 13$

A) 8; 1 positive zero; 0 negative zeros

B) 8; 1 positive zero; 1 negative zero

C) 8; 0 positive zeros; 1 negative zero

D) 8; 0 positive zeros; 0 negative zeros

6) $f(x) = x^6 + 4x^5 + 4x^4 + 3x^3 - x^2 - 5x + 4$

A) 6; 2 or 0 positive zeros; 4, 2, or 0 negative zeros

B) 6; 3 or 1 positive zeros; 4, 2, or 0 negative zeros

C) 6; 2 or 0 positive zeros; 5, 3, or 1 negative zeros

D) 6; 4, 2, or 0 positive zeros; 2, or 0 negative zeros

7) $f(x) = x^6 - x^5 - x^4 - 5x^3 + 2x^2 + 2x + 3$

A) 6; 3 or 1 positive zeros; 3 or 1 negative zeros

B) 6; 2 or 0 positive zeros; 4, 2, or 0 negative zeros

C) 6; 2 or 0 positive zeros; 2, or 0 negative zeros

D) 6; 4, 2, or 0 positive zeros; 2 or 0 negative zeros

3 Use the Rational Zeros Theorem to List the Potential Rational Zeros of a Polynomial Function

List the potential rational zeros of the polynomial function. Do not find the zeros.

1) $f(x) = 7x^3 - x^2 + 5$

A) $\pm\frac{1}{7}, \pm\frac{1}{5}, \pm 1, \pm 5, \pm 7$

B) $\pm\frac{1}{7}, \pm\frac{5}{7}, \pm 1, \pm 5, \pm 7$

C) $\pm\frac{1}{5}, \pm\frac{7}{5}, \pm 1, \pm 7$

D) $\pm\frac{1}{7}, \pm\frac{5}{7}, \pm 1, \pm 5$

2) $f(x) = 6x^4 + 2x^3 - 3x^2 + 2$

A) $\pm\frac{1}{6}, \pm\frac{1}{3}, \pm\frac{1}{2}, \pm\frac{2}{3}, \pm 1, \pm 2$

B) $\pm\frac{1}{6}, \pm\frac{1}{3}, \pm\frac{1}{2}, \pm 1, \pm 2$

C) $\pm\frac{1}{2}, \pm\frac{3}{2}, \pm 1, \pm 2, \pm 3, \pm 6$

D) $\pm\frac{1}{6}, \pm\frac{1}{3}, \pm\frac{1}{2}, \pm\frac{2}{3}, \pm 1, \pm 2, \pm 3$

3) $f(x) = x^5 - 4x^2 + 4x + 3$

A) $\pm\frac{1}{4}, \pm\frac{3}{4}, \pm 3$

B) $\pm 3, \pm\frac{1}{3}$

C) $\pm 1, \pm 3$

D) $\pm 1, \pm\frac{1}{3}$

4) $f(x) = x^5 - 5x^2 + 2x + 10$

A) $\pm 1, \pm 5, \pm 2, \pm 10$

B) $\pm 1, \pm 5, \pm 2$

C) $\pm 1, \pm\frac{1}{5}, \pm\frac{1}{2}, \pm\frac{1}{10}$

D) $\pm 1, \pm\frac{1}{5}, \pm\frac{1}{2}, \pm\frac{1}{10}, \pm 5, \pm 2, \pm 10$

4 Find the Real Zeros of a Polynomial Function

Find all of the real zeros of the polynomial function, then use the real zeros to factor f over the real numbers.

1) $f(x) = x^3 + 2x^2 - 9x - 18$

A) $-3, 2, 3;\ f(x) = (x + 3)(x - 2)(x - 3)$

B) $-3;\ f(x) = (x + 3)(x^2 - x - 6)$

C) $-2;\ f(x) = (x + 2)(x^2 + x - 9)$

D) $-3, -2, 3;\ f(x) = (x + 3)(x + 2)(x - 3)$

2) $f(x) = x^4 + x^2 - 20$

A) $3, -2, 2;\ (x + 2)(x - 2)(x - 3)$

B) $-3, 3;\ (x + 3)(x - 3)(x^2 + 4)$

C) $-2, 2;\ (x + 2)(x - 2)(x^2 + 5)$

D) $-5, -2, 2;\ (x + 2)(x - 2)(x + 5)$

3) $f(x) = 2x^3 + 3x^2 - 9x - 10$

A) $-2, -\frac{5}{2}, 1;\ f(x) = (x + 2)(2x + 5)(x - 1)$

B) $-2, \frac{5}{2}, 1;\ f(x) = (x + 2)(2x - 5)(x + 2)$

C) $-\frac{5}{2}, -1, 2;\ f(x) = (2x + 5)(x + 1)(x - 2)$

D) $2;\ f(x) = (x - 2)(2x^2 + 7x + 5)$

The equation has a solution r in the interval indicated. Approximate this solution correct to two decimal places.

4) $x^3 - 8x - 3 = 0;\ -3 \le r \le -2$

5) $x^3 - 8x - 3 = 0;\ -1 \le r \le 0$

6) $x^4 - x^3 - 7x^2 + 5x + 10 = 0;\ -3 \le r \le -2$

7) $x^4 - x^3 - 7x^2 + 5x + 10 = 0$; $2 < r \le 3$

5 Solve Polynomial Equations

Solve the equation in the real number system.

1) $3x^3 - x^2 - 21x + 7 = 0$

A) $\{\frac{1}{3}, \sqrt{7}, -\sqrt{7}\}$ B) $\{-3, \sqrt{7}, -\sqrt{7}\}$ C) $\{-\frac{1}{3}, \sqrt{7}, -\sqrt{7}\}$ D) $\{3, \sqrt{7}, -\sqrt{7}\}$

2) $x^3 + 3x^2 - 4x - 12 = 0$

A) $\{-2, 2, 3\}$ B) $\{-3, -2\}$ C) $\{-3, -2, 2\}$ D) $\{2, 3\}$

3) $3x^3 - x^2 + 3x - 1 = 0$

A) $\{-3, \frac{1}{3}, -1\}$ B) $\{\frac{1}{3}, -1\}$ C) $\{\frac{1}{3}\}$ D) $\{-3, -\frac{1}{3}, -1\}$

4) $x^4 - 3x^3 + 5x^2 - x - 10 = 0$

A) $\{-1, 2\}$ B) $\{-2, 1\}$ C) $\{1, 2\}$ D) $\{-1, -2\}$

5) $x^4 - 12x^2 - 64 = 0$

A) $\{-8, 8\}$ B) $\{-4, -2, 2, 4\}$ C) $\{-4, 4\}$ D) $\{-2, 2\}$

6) $x^3 + 2x^2 - 6x + 8 = 0$

A) $\{4\}$ B) $\{-4\}$ C) $\{1\}$ D) $\{-4, 4\}$

7) $x^4 - 8x^3 + 16x^2 + 8x - 17 = 0$

A) $\{-4, 4\}$ B) $\{-1, 1\}$ C) $\{-4, 1\}$ D) $\{-1, 4\}$

8) $2x^4 - 2x^3 + x^2 - 5x - 10 = 0$

A) $\{-\frac{\sqrt{10}}{2}, \frac{\sqrt{10}}{2}\}$ B) $\{-\frac{5}{2}, \frac{5}{2}\}$ C) $\{1, -2\}$ D) $\{-1, 2\}$

Solve the problem.

9) One solution of $x^3 - 5x^2 + 5x - 1 = 0$ is 1. Find the other two solutions.

A) $\{2 + \sqrt{3}, 2 - \sqrt{3}\}$ B) $\{2 + 2\sqrt{3}, 2 - 2\sqrt{3}\}$ C) $\{4 + 2\sqrt{3}, 4 - 2\sqrt{3}\}$ D) $\{4 + \sqrt{3}, 4 - \sqrt{3}\}$

10) Find k such that $f(x) = x^4 + kx^3 + 2$ has the factor $x + 1$.

A) 3 B) -3 C) -2 D) 2

6 Use the Theorem for Bounds on Zeros

Use the Theorem for bounds on zeros to find a bound on the real zeros of the polynomial function.

1) $f(x) = x^4 + 2x^2 - 3$

A) -6 and 6 B) -4 and 4 C) -3 and 3 D) -5 and 5

2) $f(x) = 3x^4 + 9x^3 - 6x^2 - 3x - 12$

A) -10 and 10 B) -33 and 33 C) - 12 and 12 D) -13 and 13

7 Use the Intermediate Value Theorem

Use the intermediate value theorem to determine whether the polynomial function has a zero in the given interval.

1) $f(x) = 2x^3 - 8x^2 - 4x + 1$; [-1, 0]

A) f(-1) = 5 and f(0) = -1; yes
B) f(-1) = 5 and f(0) = 1; no
C) f(-1) = -5 and f(0) = 1; yes
D) f(-1) = -5 and f(0) = -1; no

2) $f(x) = 10x^5 - 3x^3 - 8x^2 + 6$; [-1, 0]

A) f(-1) = -9 and f(0) = 6; yes
B) f(-1) = -9 and f(0) = -6; no
C) f(-1) = 9 and f(0) = -6; yes
D) f(-1) = 9 and f(0) = 6; no

3) $f(x) = -10x^4 - 2x^2 + 8$; [-1, 0]

A) f(-1) = -4 and f(0) = 8; yes
B) f(-1) = -4 and f(0) = -8; no
C) f(-1) = 4 and f(0) = 9; no
D) f(-1) = 4 and f(0) = -8; yes

4) $f(x) = 3x^4 - 8x^3 - 10x - 10$; [-1, 0]

A) f(-1) = 11 and f(0) = 10; no
B) f(-1) = -11 and f(0) = -10; no
C) f(-1) = 11 and f(0) = -10; yes
D) f(-1) = -11 and f(0) = 10; yes

5) $f(x) = 10x^3 + 5x + 6$; [-1, 0]

A) f(-1) = 9 and f(0) = 6; no
B) f(-1) = 9 and f(0) = -6; yes
C) f(-1) = -9 and f(0) = 6; yes
D) f(-1) = -9 and f(0) = -6; no

4.7 Complex Zeros; Fundamental Theorem of Algebra

1 Use the Conjugate Pairs Theorem to Find the Complex Zeros of a Polynomial

Information is given about a polynomial f(x) whose coefficients are real numbers. Find the remaining zeros of f.

1) Degree 3; zeros: 1, 3 - i

A) 3 + i
B) -3 + i
C) -1
D) no other zeros

2) Degree 4; zeros: i, 3 + i

A) -3 + i, 3 - i
B) -i, 3 - i
C) -i, -3 + i
D) 3 - i

3) Degree 5; zeros: -1, i, 2i

A) 1, -i, -2i
B) 1, -i
C) 1, -2i
D) -i, -2i

4) Degree 6; zeros: 3, 2 + i, -3 - i, 0

A) -3, 2 - i, -3 + i
B) -2 + i, 3 - i
C) -2 - i, 3 + i
D) 2 - i, -3 + i

2 Find a Polynomial Function with Specified Zeros

Form a polynomial f(x) with real coefficients having the given degree and zeros.

1) Degree: 3; zeros: 1 and 3 + i.

A) $f(x) = x^3 - 7x^2 + 4x - 10$
B) $f(x) = x^3 - 7x^2 + 16x - 10$
C) $f(x) = x^3 + 7x^2 + 16x + 10$
D) $f(x) = x^3 - 5x^2 + 4x + 10$

2) Degree: 3; zeros: -2 and 3 + i.

A) $f(x) = x^3 - 4x^2 - 2x + 20$

B) $f(x) = x^3 - 4x^2 - 10x + 20$

C) $f(x) = x^3 - 8x^2 + 2x + 20$

D) $f(x) = x^3 - 6x^2 - 10x + 20$

3) Degree: 4; zeros: -1, 2, and 1 - 2i.

A) $f(x) = x^4 - 3x^3 + 5x^2 - x - 10$

B) $f(x) = x^4 - x^3 + x^2 + 9x - 10$

C) $f(x) = x^4 - 3x^3 - 3x^2 + 7x + 6$

D) $f(x) = x^4 - x^3 + 3x^2 - 5x - 10$

3 Find the Complex Zeros of a Polynomial

Use the given zero to find the remaining zeros of the function.

1) $f(x) = x^4 - 12x^2 - 64$; zero: $-2i$

A) 2i, 4i, -4i

B) 2i, 8, -8

C) 2i, 4, -4

D) 2i, 8i, -8i

2) $f(x) = x^3 + 8x^2 - 18x + 20$; zero: $1 + i$

A) -10, 10

B) 1 - i, 10

C) 1 - i, 10i

D) 1 - i, -10

3) $f(x) = x^3 - 2x^2 - 11x + 52$; zero: -4

A) 3 + 2i, 3 - 2i

B) $1 + 2\sqrt{13}i, 1 - 2\sqrt{13}i$

C) 1 + 2i, 1 - 2i

D) 6 + 4i, 6 - 4i

Find the complex zeros of the polynomial function.

4) $f(x) = x^4 - 8x^3 + 16x^2 + 8x - 17$

A) 1, -1, 4 - i, 4 + i

B) 1, -1, 4 - 2i, 4 + 2i

C) 4, -4, 1 - i, 1 + i

D) 4, -4, 1 - 2i, 1 + 2i

5) $f(x) = 2x^4 - 2x^3 + x^2 - 5x - 10$

A) $1, -2, \frac{5}{2}i, -\frac{5}{2}i$

B) $-1, 2, \frac{\sqrt{10}}{2}i, -\frac{\sqrt{10}}{2}i$

C) $-1, 2, \frac{5}{2}i, -\frac{5}{2}i$

D) $1, -2, \frac{\sqrt{10}}{2}i, -\frac{\sqrt{10}}{2}i$

6) $f(x) = x^4 - 4x^3 - 44x^2 + 196x - 245$

4 Additional Applications

Solve the problem.

1) If $f(x) = x^4 - 5x^2 - 36 = (x + 2i)(x - r_2)(x - r_3)(x - 3)$, what are the values of r_2 and r_3?

2) A polynomial f(x) of degree 7 with real coefficients has six complex, non-real zeros. Is the seventh zero real or not real?

3) A polynomial g(x) of degree 5 whose coefficients are real numbers has zeros 2, 3, -3, and 7 - i. Find the fifth zero and then determine g(x).

Ch. 4 Polynomial and Rational Functions
Answer Key

4.1 Quadratic Functions and Models

1 Graph a Quadratic Function Using Transformations

1) D
2) D
3) B
4) B
5) B
6) A
7) A
8) B
9) D
10) D
11) D

2 Identify the Vertex and Axis of Symmetry of a Quadratic Function

1) A
2) D
3) B
4) D
5) A
6) B
7) C
8) C
9) D
10) C
11) A
12) A
13) D
14) C

3 Graph a Quadratic Function Using Its Vertex, Axis and Intercepts

1) C
2) C
3) D
4) C
5) D
6) B
7) B
8) B
9) A

10) vertex $(\frac{1}{8}, -\frac{49}{16})$

intercepts (0, -3), $(-\frac{3}{4}, 0)$, (1, 0)

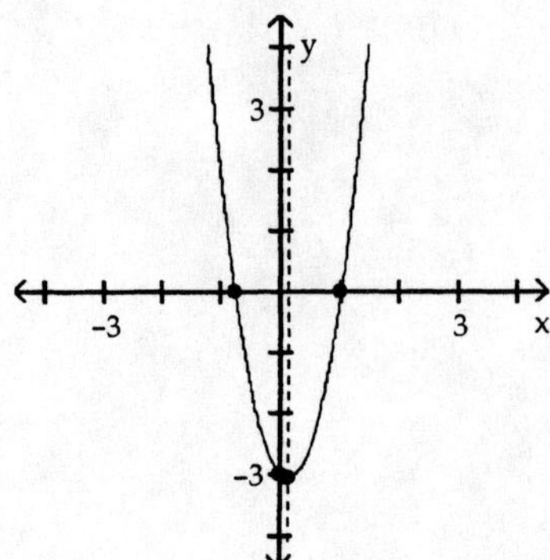

11) A
12) D
13) D
14) B
15) A
16) C
17) D
18) A
19) domain: all real numbers

range: $\{y \mid y \geq -\frac{49}{16}\}$

20) C
21) C
22) C
23) C
24) A
25) B
26) A
27) D
28) C

4 Use the Maximum or Minimum Value of a Quadratic Function to Solve Applied Problems

1) B
2) B
3) B
4) D
5) D
6) B
7) A
8) B
9) A
10) D
11) D
12) C
13) A
14) a) $R(x) = \frac{-x^2}{15} + 30x$ dollars, $0 \leq x \leq 450$

b) $x = 225$ units

c) $p = \$15$

15) approximately 17 ft.

16) a) $R(x) = -\frac{1}{10}x^2 + 100x$

b) $R(450) = \$24{,}750.00$

c)

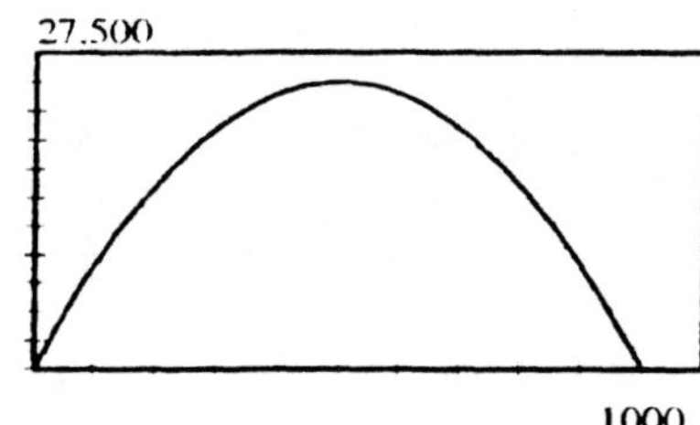

d) 500; $25,000.00

e) 50.00

5 Use a Graphing Utility to Find the Quadratic Function of Best Fit to Data

1) B
2) D
3) D
4) C
5) C
6) $R(x) = -1.65x^2 + 634.42x + 7089.93$
7) $M(x) = 0.04x^2 - 1.21x + 26.03$; 1988

6 Additional Applications

1) B

4.2 Polynomial Functions

1 Identify Polynomials and Their Degree

1) D
2) C
3) C
4) D
5) A
6) B
7) A
8) D
9) B
10) D
11) C
12) A
13) A

2 Graph Polynomial Functions Using Transformations

1) B
2) D
3) D
4) D
5) B
6) C
7) D
8) A
9) B
10) D
11) B
12) C
13) C
14) B
15) B

16) D

3 Identify the Zeros of a Polynomial Function and Their Multiplicity

1) B
2) D
3) D
4) C
5) D
6) A
7) C
8) B
9) A
10) C
11) B

4 Analyze the Graph of a Polynomial Function

1) A
2) D
3) A
4) A
5) D
6) A
7) B
8) C
9) C
10) D
11) B
12) A
13) D
14) A
15) B
16) D
17) B
18) C
19) C
20) D
21) A
22) C
23) B
24) C
25) D
26) B
27) A

28) a) The x–intercepts are 2, 3, and 4. The y–intercept is 288.
b) The graph crosses the x–axis at 2 and 4, and touches it at 3.
c) The graph resembles $f(x) = x^6$ for large values of $|x|$.
d) Maximum at (3, 0); minima at (2.57, –0.05) and (3.77, –0.76)
e) The graph has 3 turning points.
f)

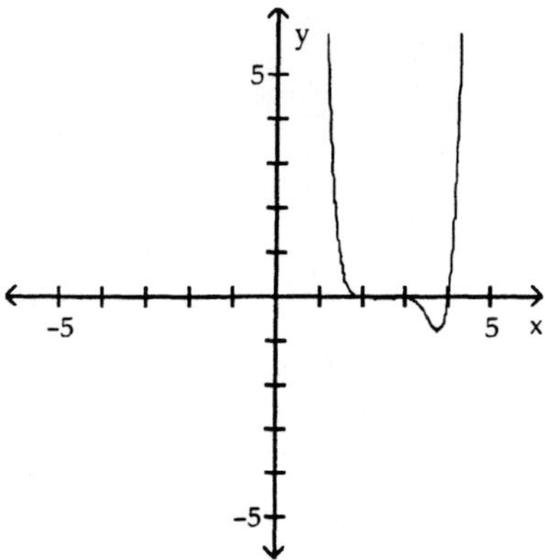

29) a) The x–intercepts are –2.21, 0.54, and 1.68. The y–intercept is 2.
b) The graph crosses the x–axis at each x–intercept.
c) The graph resembles $f(x) = x^3$ for large values of $|x|$.
d) Maximum at (–1.15, 5.08); minimum at (1.15, –1.08)
e) The graph has 2 turning points.
f)

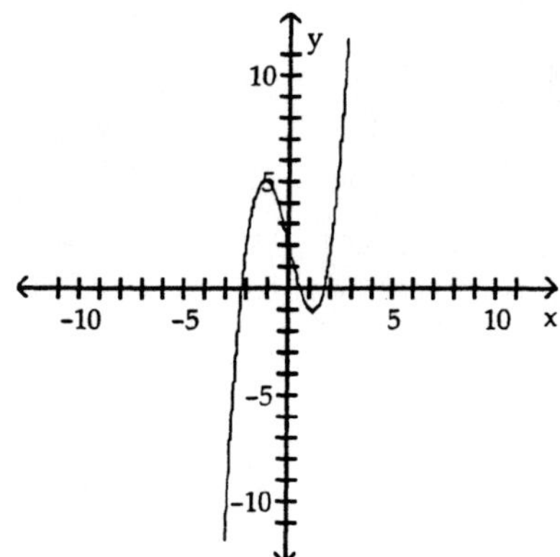

30) a) The x–intercepts are –1.23, 0.40, 1.38, and 2.94. The y–intercept is –4.
b) The graph crosses the x–axis at each x–intercept.
c) The graph resembles $f(x) = x^4$ for large values of $|x|$.
d) Maximum at (0.89, 2.11); minima at (–0.65, –8.87) and (2.38, –8.02)
e) The graph has 3 turning points.
f)

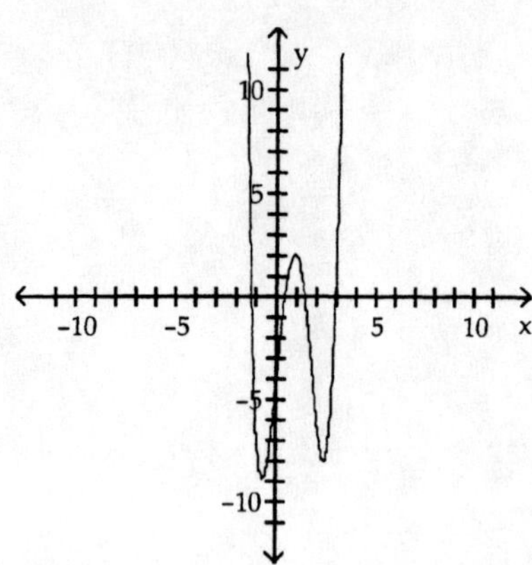

31) C
32) C
33) C
34) A

5 Additional Applications

1) A
2) Volume equals $V = x(x + 2)(x - 5) = x^3 - 3x^2 - 10x$, where x is the width of the box.

4.3 Rational Functions I

1 Find the Domain of a Rational Function

1) A
2) C
3) B
4) D
5) A
6) B
7) B
8) C
9) D
10) C
11) A
12) B
13) A
14) A

2 Determine the Vertical Asymptotes of a Rational Function

1) D
2) D
3) A
4) C
5) C

3 Determine the Horizontal or Oblique Asymptotes of a Rational Function

1) D
2) C
3) B
4) C
5) C

6) D
7) C
8) B
9) B
10) D
11) D
12) A
13) a) $g(-86) \approx 9.8211 \text{ m/sec}^2$
b) $g(0) \approx 9.8208 \text{ m/sec}^2$
14) horizontal asymptote: $r = 0$; vertical asymptote: $t = 0$
15) vertical asymptote: $s_2 = 0.3$; horizontal asymptote: $s_1 = 0.3$
16) vertical asymptote: $f_2 = -0.001$; horizontal asymptote: $f = 0.001$

4.4 Rational Functions II: Analyzing Graphs

1 Analyze the Graph of a Rational Function

1) A
2) D
3) C
4) C
5) C
6) A
7) C
8) B
9) D
10) C
11) A
12) B
13) B
14) A
15) A
16) B
17) A
18) B
19) B
20) A
21) C
22) D
23) A
24) D
25) A

2 Solve Applied Problems Involving Rational Functions

1) a) $P(0) = 15$ insects
b) $P(120) \approx 608$ insects
c) 1,350
2) a) $y = 0$
b) $t = 7$
3) A
4) B
5) B
6) C
7) B
8) A
9) C

10) a) Surface area equals $A(x) = 4x^2 + \frac{21,000}{x}$, $x > 0$, where x equals the width of the box.

b) width ≈ 13.8 in, length ≈ 27.6 in, height ≈ 18.4 in

11) a)

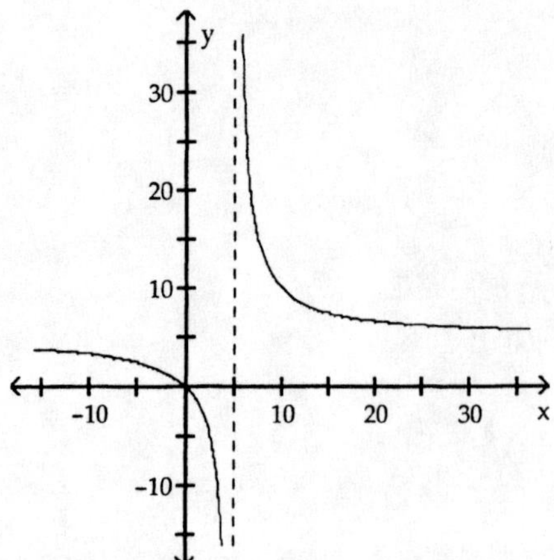

b) To remain focused, the desired distance between the film and the projector lens decreases and approaches 5 centimeters as the distance between the movie screen and the projector lens increases.

12) C

4.5 Polynomial and Rational Inequalities

1 Solve Polynomial Inequalities

1) B
2) B
3) C
4) D
5) B
6) C
7) D
8) A
9) A
10) C
11) B
12) D
13) A
14) D
15) D
16) B
17) D
18) A
19) B

2 Solve Rational Inequalities

1) A
2) C
3) B
4) A
5) C
6) D
7) C
8) D
9) B
10) B
11) A
12) D

13) D
14) D
15) C
16) C
17) C
18) when $1.30 < t < 4.32$
19) between 15 and 18 calendars

4.6 The Real Zeros of a Polynomial Function

1 Use the Remainder and Factor Theorems

1) A
2) B
3) A
4) B
5) A
6) B
7) A
8) B
9) B
10) A
11) A
12) B
13) B
14) B

2 Use Descartes' Rule of Signs to Determine the Number of Positive and the Number of Negative Real Zeros of a Polynomial Function

1) B
2) C
3) C
4) D
5) B
6) A
7) B

3 Use the Rational Zeros Theorem to List the Potential Rational Zeros of a Polynomial Function

1) D
2) A
3) C
4) A

4 Find the Real Zeros of a Polynomial Function

1) D
2) C
3) C
4) -2.62
5) -0.38
6) -2.24
7) 2.24

5 Solve Polynomial Equations

1) A
2) C
3) C
4) A
5) C
6) B
7) B
8) D
9) A

10) A

6 Use the Theorem for Bounds on Zeros

1) B

2) D

7 Use the Intermediate Value Theorem

1) C

2) A

3) A

4) C

5) C

4.7 Complex Zeros; Fundamental Theorem of Algebra

1 Use the Conjugate Pairs Theorem to Find the Complex Zeros of a Polynomial

1) A

2) B

3) D

4) D

2 Find a Polynomial Function with Specified Zeros

1) B

2) A

3) A

3 Find the Complex Zeros of a Polynomial

1) C

2) D

3) A

4) A

5) B

6) -7, 7, 2 - i, 2 + i

4 Additional Applications

1) 2i, -3

2) real number

3) $g(x) = x^5 - 16x^4 + 69x^3 + 44x^2 - 702x + 900$, or any non-zero constant multiple of g(x).

Ch. 5 Exponential and Logarithmic Functions

5.1 Composite Functions

1 Form the Composite Function and Find Its Domain

For the given functions f and g, find the requested composite function value.

1) $f(x) = \sqrt{x + 4}$; $g(x) = 3x$; Find $(f \circ g)(1)$.

A) $3\sqrt{5}$ B) $\sqrt{7}$ C) $3\sqrt{15}$ D) $\sqrt{15}$

2) $f(x) = 2x + 2$; $g(x) = 4x^2 + 3$; Find $(f \circ f)(0)$.

A) 8 B) 6 C) 19 D) 39

3) Given $f(x) = \dfrac{x - 6}{x}$ and $g(x) = x^2 + 9$, find $(g \circ f)(-2)$.

A) 25 B) $\dfrac{145}{16}$ C) 13 D) $\dfrac{7}{13}$

For the functions f and g and the number c, compute $(f \circ g)(c)$.

4) $f(x) = 9x^2 - 4x$
$g(x) = 14x - 10$
$c = 8$

A) 85,622 B) 55,488 C) 93,228 D) 7606

5) $f(t) = \sqrt{t^4 + 30t^2 + 225}$
$g(t) = \dfrac{t + 3}{3}$
$c = 15$

A) 81 B) 1440 C) 51 D) 2601

6) $f(x) = x^2 + 2x - 3$
$g(x) = x^2 - 2x + 1$
$c = 4$

A) 96 B) 36 C) 396 D) 456

Find the indicated composite for the pair of functions.

7) $(f \circ g)(x)$: $f(x) = 3x + 15$, $g(x) = 5x - 1$

A) $15x + 14$ B) $15x + 12$ C) $15x + 74$ D) $15x + 18$

8) $(g \circ f)(x)$: $f(x) = -2x + 4$, $g(x) = 5x + 4$

A) $-10x - 16$ B) $-10x + 24$ C) $10x + 24$ D) $-10x + 12$

9) $(f \circ g)(x)$: $f(x) = \dfrac{7}{x - 4}$, $g(x) = \dfrac{4}{5x}$

A) $\dfrac{7x}{4 - 20x}$ B) $\dfrac{4x - 16}{35x}$ C) $\dfrac{35x}{4 + 20x}$ D) $\dfrac{35x}{4 - 20x}$

10) $(g \circ f)(x)$: $f(x) = \frac{x-4}{3}$, $g(x) = 3x + 4$

A) $x - \frac{4}{3}$ B) $3x + 8$ C) x D) $x + 8$

11) $(f \circ g)(x)$: $f(x) = \sqrt{x+7}$, $g(x) = 8x - 11$

A) $8\sqrt{x+7} - 11$ B) $8\sqrt{x-4}$ C) $2\sqrt{2x+1}$ D) $2\sqrt{2x-1}$

12) $(g \circ f)(x)$: $f(x) = 4x^2 + 6x + 4$, $g(x) = 6x - 8$

A) $4x^2 + 36x + 16$ B) $24x^2 + 36x + 16$ C) $4x^2 + 6x - 4$ D) $24x^2 + 36x + 32$

Find the domain of the composite function $f \circ g$.

13) $f(x) = 3x + 9$; $g(x) = x + 7$

A) $\{x \mid x \neq -10\}$ B) $\{x \mid x \text{ is any real number}\}$
C) $\{x \mid x \neq 10\}$ D) $\{x \mid x \neq -7, x \neq -3\}$

14) $f(x) = \frac{6}{x+9}$; $g(x) = x + 1$

A) $\{x \mid x \text{ is any real number}\}$ B) $\{x \mid x \neq -9, x \neq -1\}$
C) $\{x \mid x \neq -9\}$ D) $\{x \mid x \neq -10\}$

15) $f(x) = x + 6$; $g(x) = \frac{7}{x+7}$

A) $\{x \mid x \neq -13\}$ B) $\{x \mid x \neq -7\}$
C) $\{x \mid x \neq -7, x \neq -6\}$ D) $\{x \mid x \text{ is any real number}\}$

16) $f(x) = \frac{5}{x+5}$; $g(x) = \frac{-5}{x}$

A) $\{x \mid x \neq 0, x \neq -5\}$ B) $\{x \mid x \text{ is any real number}\}$
C) $\{x \mid x \neq 0, x \neq -5, x \neq 1\}$ D) $\{x \mid x \neq 0, x \neq 1\}$

17) $f(x) = \frac{-70}{x}$; $g(x) = \frac{8}{x+7}$

A) $\{x \mid x \neq -7, x \neq 0\}$ B) $\{x \mid x \text{ is any real number}\}$
C) $\{x \mid x \neq -7\}$ D) $\{x \mid x \neq 0, x \neq -7, x \neq 10\}$

18) $f(x) = \frac{x}{x+3}$; $g(x) = \frac{6}{x+4}$

A) $\{x \mid x \neq 0, x \neq -4, x \neq -6\}$ B) $\{x \mid x \text{ is any real number}\}$
C) $\{x \mid x \neq -4, x \neq -6\}$ D) $\{x \mid x \neq -4, x \neq -3\}$

19) $f(x) = \sqrt{x}$; $g(x) = 6x + 18$

A) $\{x \mid x \geq 0\}$ B) $\{x \mid x \geq -3\}$
C) $\{x \mid x \text{ is any real number}\}$ D) $\{x \mid x \leq -3 \text{ or } x \geq 0\}$

20) $f(x) = 4x + 20$; $g(x) = \sqrt{x}$

A) $\{x \mid x \le -5 \text{ or } x \ge 0\}$
B) $\{x \mid x \ge -5\}$
C) $\{x \mid x \ge 0\}$
D) $\{x \mid x \text{ is any real number}\}$

21) $f(x) = \sqrt{x - 3}$; $g(x) = \dfrac{3}{x - 7}$

A) $\{x \mid x \ge 3, x \ne 7\}$
B) $\{x \mid x \text{ is any real number}\}$
C) $\{x \mid x \ne 7, x \ne 3\}$
D) $\{x \mid 7 < x \le 8\}$

22) $f(x) = \dfrac{4}{x - 9}$; $g(x) = \sqrt{x - 4}$

A) $\{x \mid x \text{ is any real number}\}$
B) $\{x \mid x \ge 4, x \ne 9\}$
C) $\{x \mid x \ge 4, x \ne 9, x \ne 85\}$
D) $\{x \mid x \ge 4, x \ne 85\}$

Solve the problem.

23) If $f(x) = x^2$ and $g(x) = -1 + 5x$, find $(f \circ g)(x)$ and find the domain of $(f \circ g)(x)$.

Decide whether the composite functions, f ∘ g and g ∘ f, are equal to x.

24) $f(x) = \sqrt[5]{x - 12}$, $g(x) = x^5 + 12$

A) Yes, yes
B) No, yes
C) No, no
D) Yes, no

25) $f(x) = x^2 + 5$, $g(x) = \sqrt{x} - 5$

A) Yes, no
B) No, yes
C) No, no
D) Yes, yes

26) $f(x) = \sqrt{x}$, $g(x) = x^2$

A) Yes, yes
B) No, yes
C) No, no
D) Yes, no

27) $f(x) = \dfrac{1}{x}$, $g(x) = x$

A) No, yes
B) No, no
C) Yes, no
D) Yes, yes

28) $f(x) = \sqrt{x + 1}$, $g(x) = x^2$

A) Yes, no
B) No, yes
C) No, no
D) Yes, yes

29) $f(x) = x^3 + 7$, $g(x) = \sqrt[3]{x - 7}$

A) Yes, no
B) No, yes
C) No, no
D) Yes, yes

Find functions f and g so that the composition of f and g is H.

30) $H(x) = \sqrt[3]{x + 1}$

A) $f(x) = x + 1$; $g(x) = \sqrt[3]{x}$
B) $f(x) = \sqrt[3]{x}$; $g(x) = x + 1$
C) $f(x) = \sqrt{x}$; $g(x) = x + 1$
D) $f(x) = \sqrt[3]{x}$; $g(x) = 1$

31) $H(x) = (5 - 2x^3)^2$

A) $f(x) = (5 - 2x)^2$; $g(x) = x^3$
B) $f(x) = x^3$; $g(x) = (5 - 2x)^2$
C) $f(x) = x^2$; $g(x) = 5 - 2x^3$
D) $f(x) = 5 - 2x^3$; $g(x) = x^2$

32) $H(x) = |4 - 3x^2|$

A) $f(x) = |x|$; $g(x) = 4 - 3x^2$

B) $f(x) = 4 - 3|x|$; $g(x) = x^2$

C) $f(x) = 4 - 3x^2$; $g(x) = |x|$

D) $f(x) = x^2$; $g(x) = 4 - 3|x|$

Solve the problem.

33) The population P of a predator mammal depends upon the number x of a smaller animal that is its primary food source. The population s of the smaller animal depends upon the amount a of a certain plant that is its primary food source. If $P(x) = 2x^2 + 9$ and $s(a) = 3a + 4$, what is the relationship between the predator mammal and the plant food source?

A) $P(s(a)) = 9a^2 + 24a + 25$

B) $P(s(a)) = 18a^2 + 24a + 41$

C) $P(s(a)) = 6a + 13$

D) $P(s(a)) = 18a^2 + 48a + 41$

34) An oil well off the Gulf Coast is leaking, with the leak spreading oil over the surface of the gulf as a circle. At any time t, in minutes, after the beginning of the leak, the radius of the oil slick on the surface is $r(t) = 5t$ ft. Find the area A of the oil slick as a function of time.

A) $A(r(t)) = 25t^2$

B) $A(r(t)) = 5\pi t^2$

C) $A(r(t)) = 25\pi t^2$

D) $A(r(t)) = 25\pi t$

35) An airline charter service charges a fare per person of $400 plus $40 for each unsold seat. The airplane holds 100 passengers. Let x represent the number of unsold seats and write an expression for the total revenue R for a charter flight.

A) $R(x) = x(400 + 40x)$ or $400x + 40x^2$

B) $R(x) = 100(400 + 40x)$ or $40{,}000 + 4000x$

C) $R(x) = (100 - x)(400 + 40x)$ or $40{,}000 + 4000x - 40x^2$

D) $R(x) = (100 - x)(400 + 40x)$ or $40{,}000 + 3600x - 40x^2$

36) The surface area of a balloon is given by $S(r) = 4\pi r^2$, where r is the radius of the balloon. If the radius is increasing with time t, as the balloon is being blown up, according to the formula $r(t) = \frac{4}{5}t^3$, $t \geq 0$, find the surface area S as a function of the time t.

A) $S(r(t)) = \frac{16}{25}\pi t^6$

B) $S(r(t)) = \frac{64}{25}\pi t^9$

C) $S(r(t)) = \frac{64}{25}\pi t^3$

D) $S(r(t)) = \frac{64}{25}\pi t^6$

37) The surface area S (in square inches) of a cylindrical pipe with length 12 inches is given by $S(r) = 2\pi r^2 + 24\pi r$, where r is the radius of the piston (in inches). If the radius is increasing with time t (in minutes) according to the formula $r(t) = \frac{1}{6}t^2$, $t \geq 0$, find the surface area S of the pipe as a function of the time t.

38) The volume V (in cubic inches) of a cylindrical pipe with length 12 inches is given by $V(r) = 12\pi r^2$, where r is the radius of the piston (in inches). If the radius is increasing with time t (in minutes) according to the formula $r(t) = \frac{1}{6}t^2$, $t \geq 0$, find the volume V of the pipe as a function of the time t.

39) The price p of a certain product and the quantity sold x obey the demand equation $p = -\frac{2}{3}x + 200$, $0 \leq x \leq 300$.

Suppose that the cost C of producing x units is $C = \frac{\sqrt{x}}{20} + 800$. Assuming that all items produced are sold, find the cost C as a function of the price p.

40) If $f(x) = \frac{1}{2}x^2 + 4$ and $g(x) = 2x - a$, find a so that the graph of $f \circ g$ crosses the y–axis at 36.

5.2 Inverse Functions

1 Determine the Inverse of a Function

Find the inverse. Determine whether the inverse represents a function.

1) {(–6, –7), (12, –7), (14, 2)}

A) {(–7, –6), (14, 12), (2, –7)}; not a function
B) {(–7, –6), (–7, 12), (2, 14)}; a function
C) {(–7, –6), (–7, 12), (2, 14)}; not a function
D) {(–6, –7), (–7, 12), (2, 14)}; not a function

2) {(6, 5), (3, 6), (1, 7), (–1, 8)}

A) {(5, 6), (6, 3), (7, 1), (8, –1)}; a function
B) {(6, 5), (8, 1), (6, 1), (6, 7)}; a function
C) {(6, 5), (5, 1), (6, 3), (6, 7)}; not a function
D) {(5, 6), (6, 3), (7, 1), (8, –1)}; not a function

3) {(–3, 4), (–1, 5), (0, 2), (2, 4), (5, 7)}

A) {(4, –3), (5, –1), (2, 0), (4, 2), (7, 5)}; not a function
B) {(3, –4), (1, –5), (0, –2), (–2, –4), (–5, –7)}; a function
C) {(3, 4), (1, 5), (0, 2), (–2, 4), (–5, 7)}; a function
D) {(–3, –4), (–1, –5), (0, –2), (2, –4), (5, –7)}; a function

Solve the problem.

4) The relationship between a person and his or her height forms a function, such as the following:

Domain	Range
Randy	68 in.
Moesha	62 in.
Jason	68 in.
Amanda	66 in.

What is the domain of the inverse of the function given above? Is the inverse a function?

5) List 5 ordered pairs that satisfy the equation $x = 5$. Does this equation represent a function?

Decide whether or not the functions are inverses of each other.

6) $f(x) = 7x + 6$; $g(x) = \frac{1}{7}(x - 6)$

A) No
B) Yes

7) $f(x) = 7x + 8$; $g(x) = \frac{x}{7} - 8$

A) Yes
B) No

8) $f(x) = 8 - 4x$; $g(x) = -\frac{x}{4}(x - 8)$

A) Yes
B) No

9) $f(x) = 5 - 3x;\;\; g(x) = \frac{x}{3}(x - 5)$

A) No

B) Yes

10) $f(x) = (x - 4)^2,\ x \geq 4;\;\; g(x) = \sqrt{x} + 4$

A) Yes

B) No

11) $f(x) = (x - 3)^2,\ x \geq 3;\;\; g(x) = \sqrt{x + 3}$

A) Yes

B) No

Solve the problem.

12) Show that f and g are inverse functions or state that they are not.

$$f(x) = \sqrt[3]{-8x - 6};\; g(x) = -\frac{x^3 + 6}{8}$$

The graph of a function f is given. Use the horizontal line test to determine whether f is one-to-one.

13)

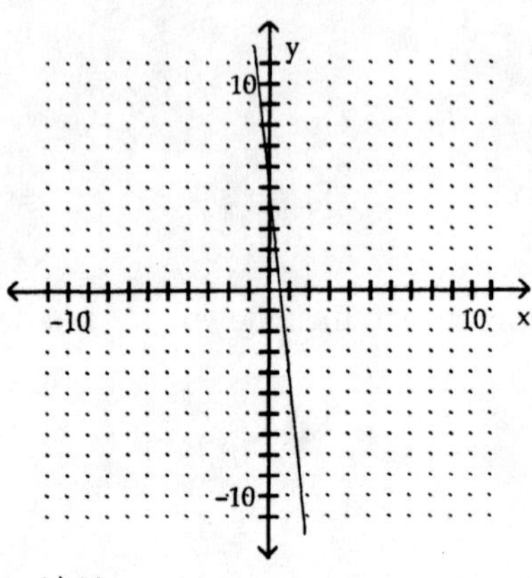

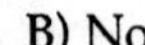

A) Yes

B) No

14)

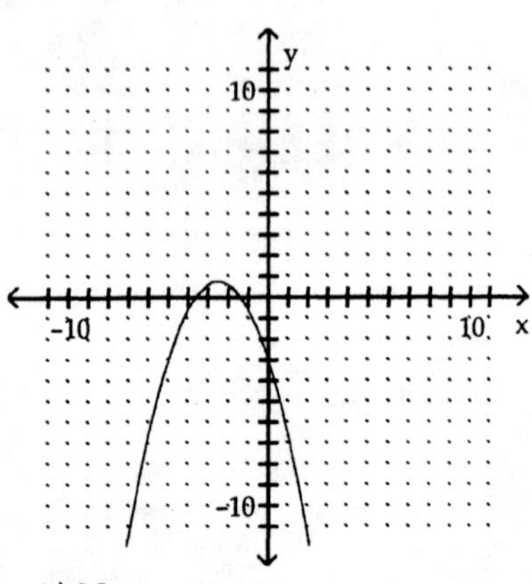

A) No

B) Yes

15)

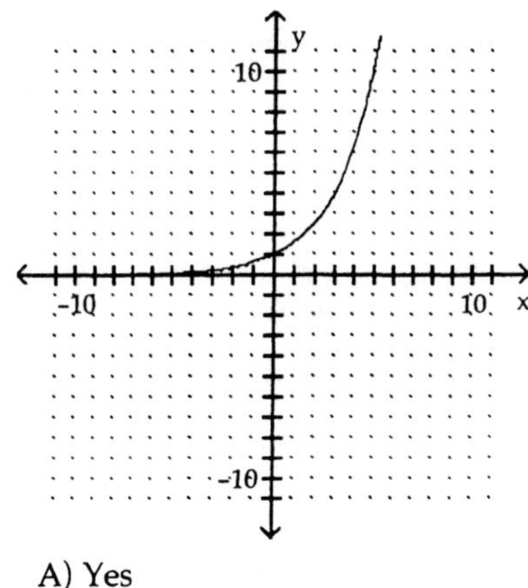

A) Yes B) No

16)

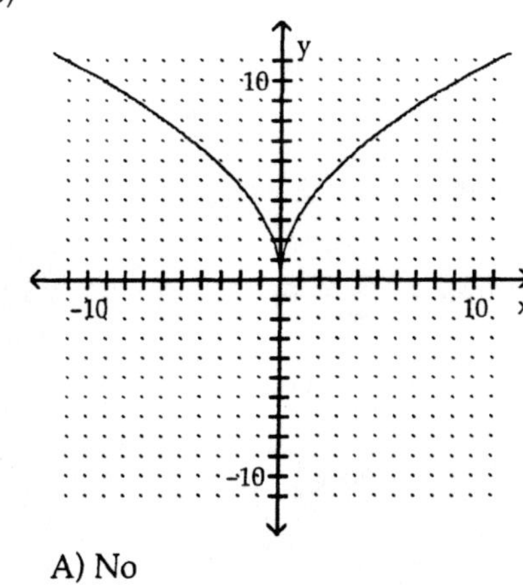

A) No B) Yes

Solve the problem.

17) In a grocery store, every item is marked with a bar code. The bar code is scanned for the price of the item. What is the domain and range in this situation? Does the situation represent a function? Is the inverse a function?

18) In a wholesale discount store, every item is marked with a bar code. The bar code is scanned for the price of the item. What is the domain and range of the inverse of the function represented in this situation? Does the inverse represent a function?

2 Obtain the Graph of the Inverse Function from the Graph of the Function

Graph the function as a solid line or curve and its inverse as a dashed line or curve on the same axes.

1) $f(x) = 2x$

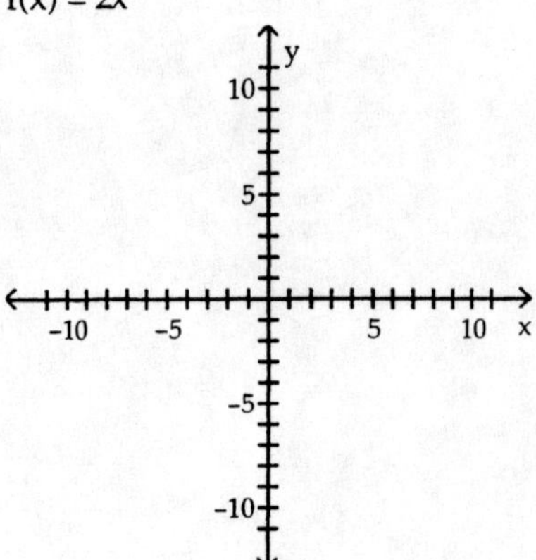

A)

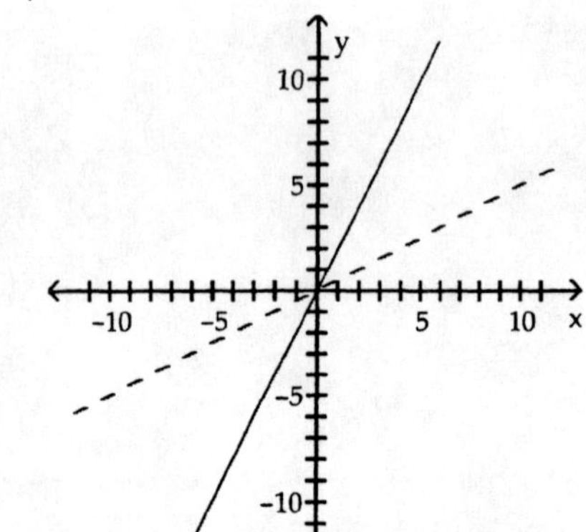

B)

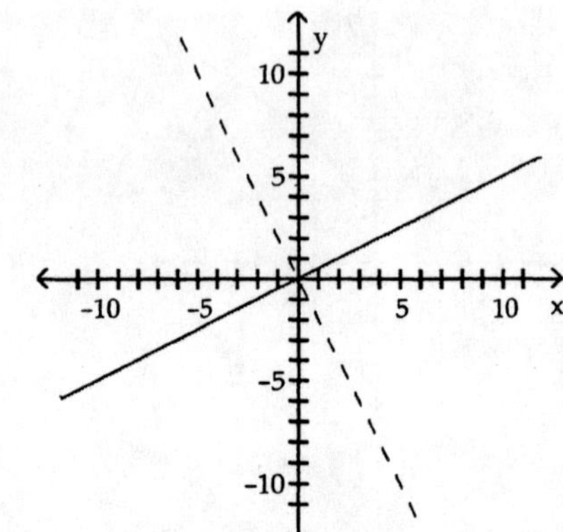

C)

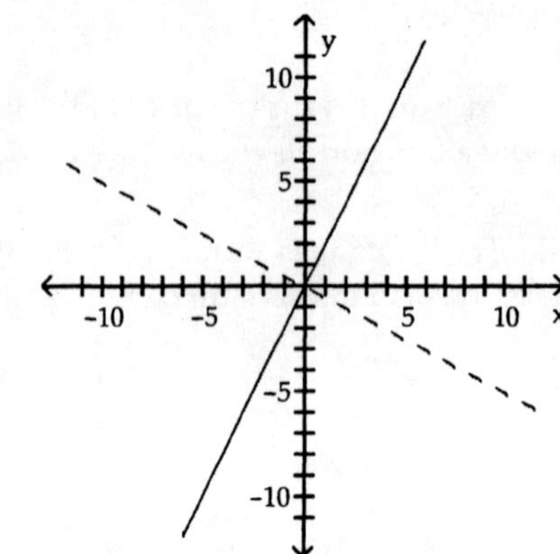

D)

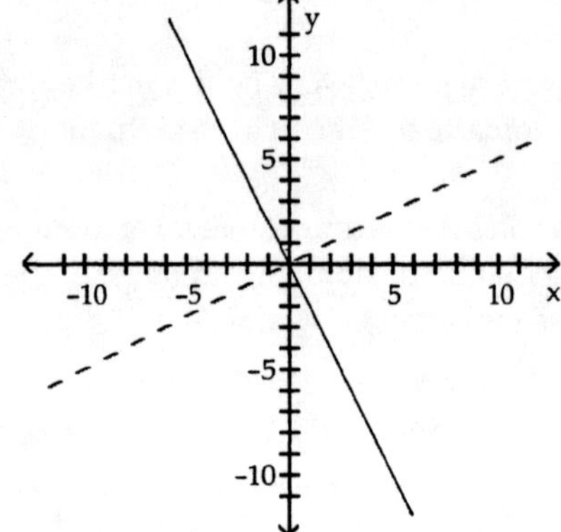

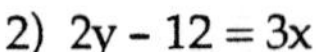
2) $2y - 12 = 3x$

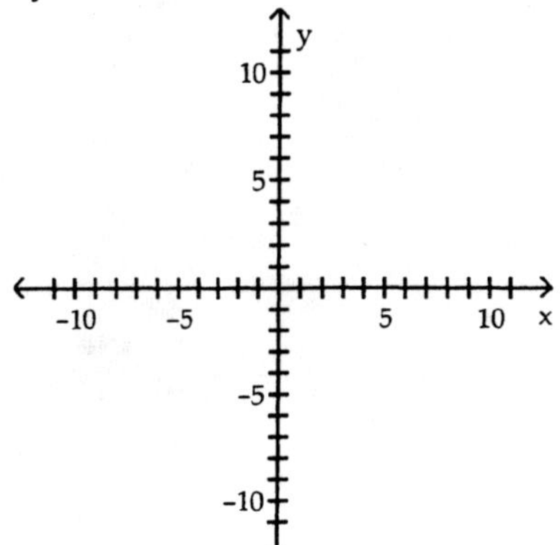

A)

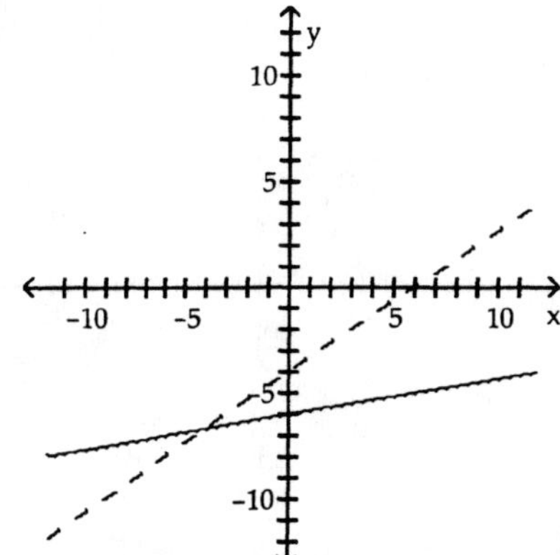

B)

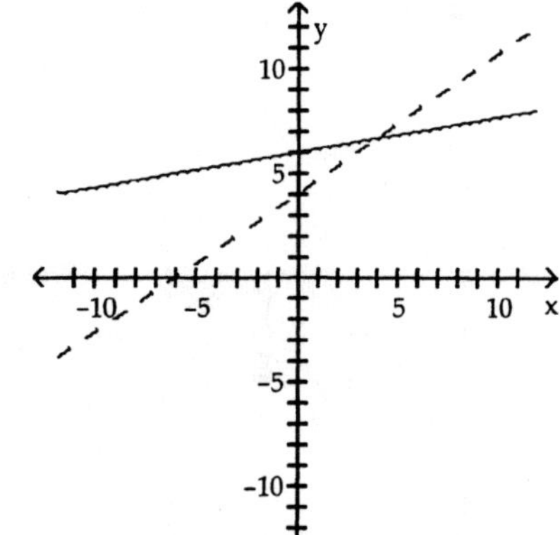

C)

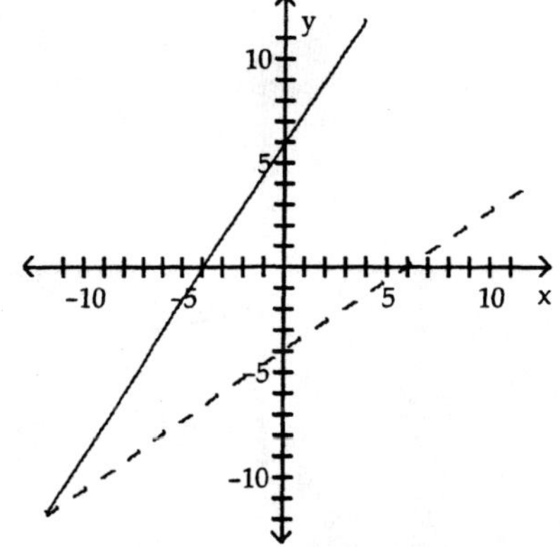

D)

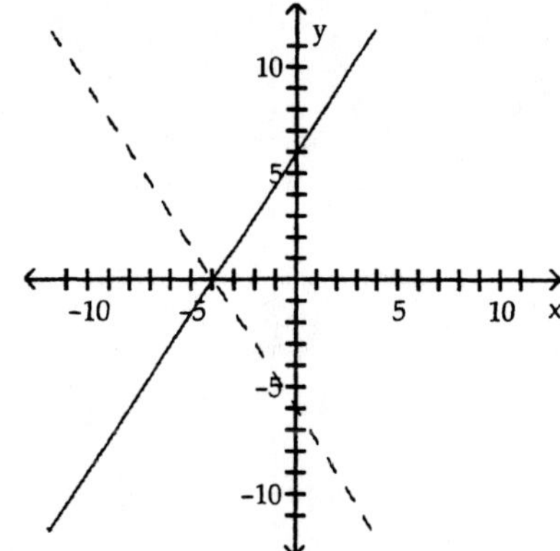

3) $f(x) = \sqrt{x + 2}$

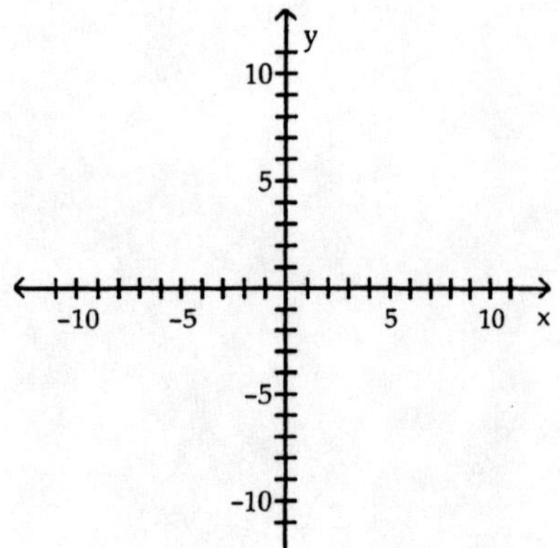

A)

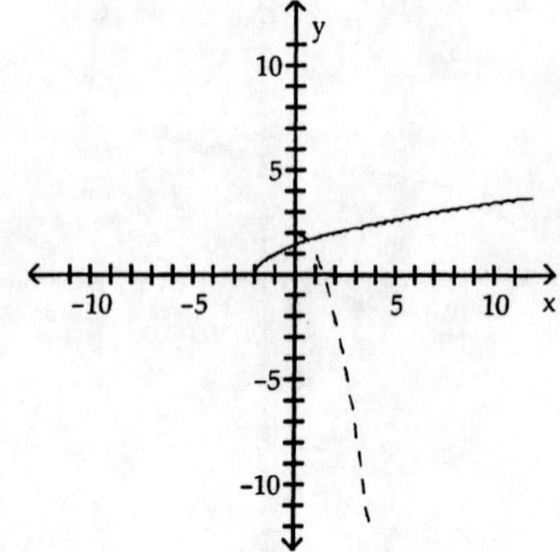

B)

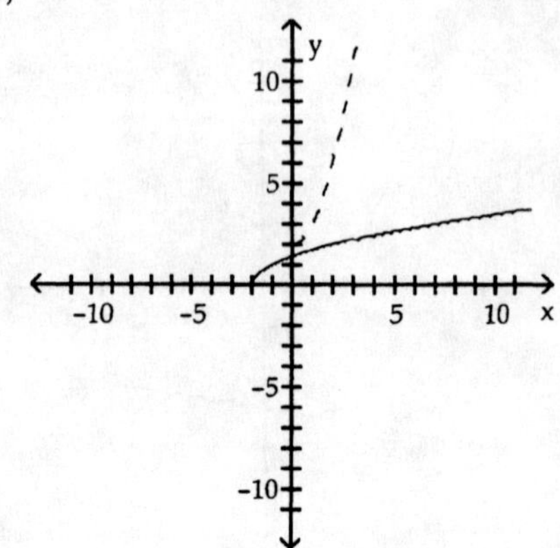

C)

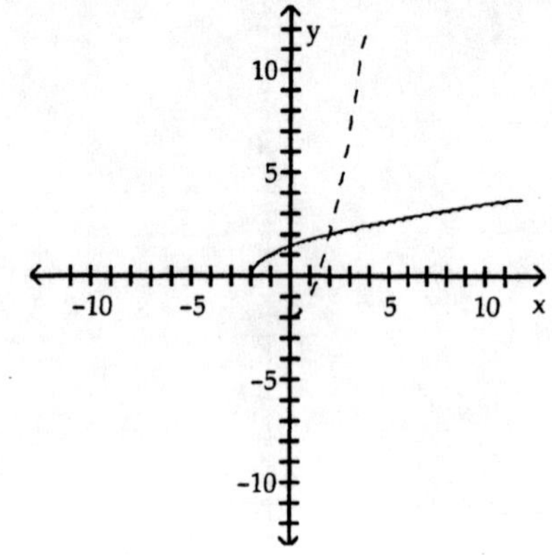

D)

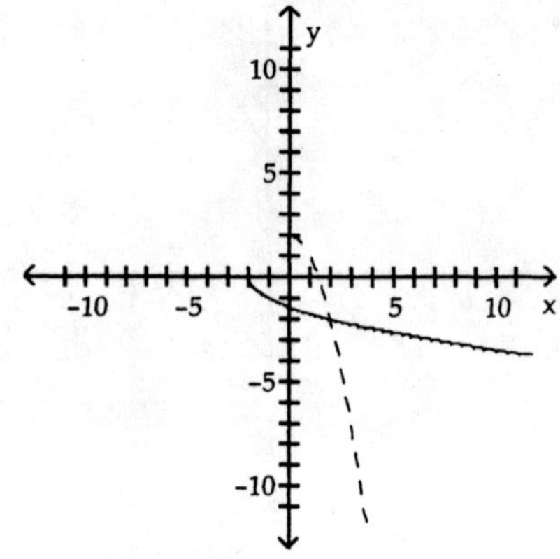

4) $f(x) = x^3 + 1$

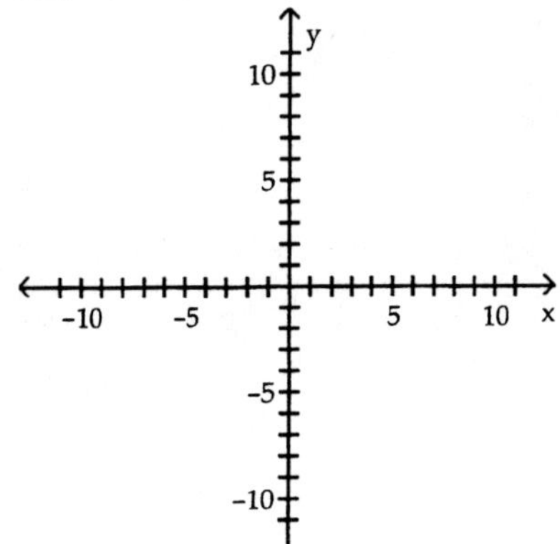

A)

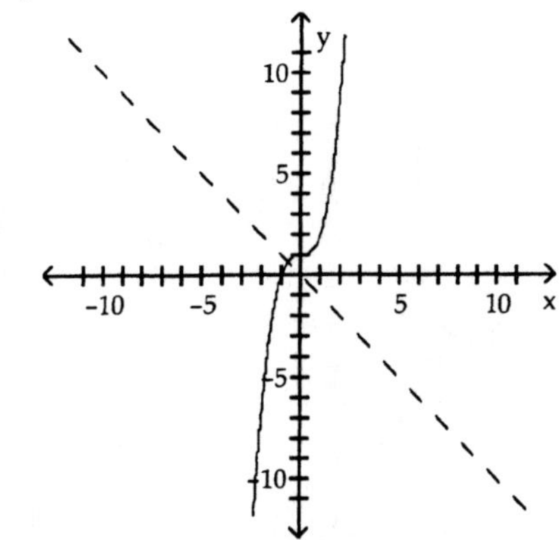

B)

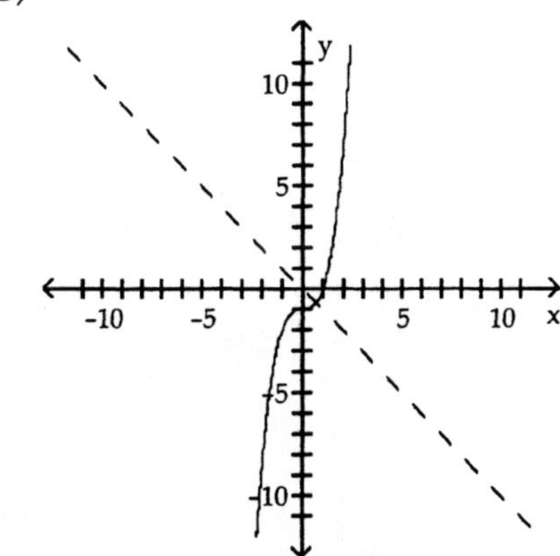

C)

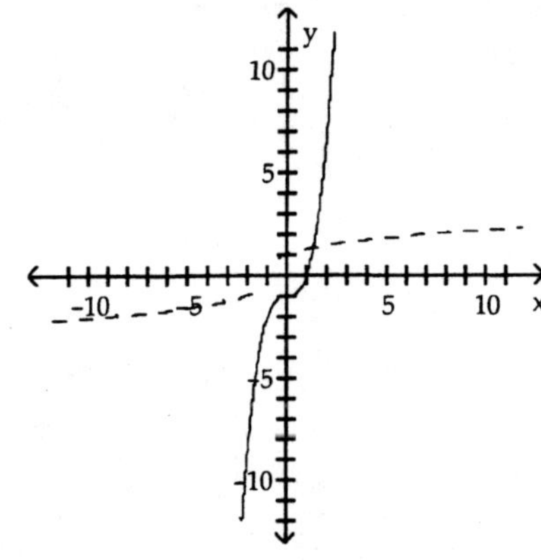

D)

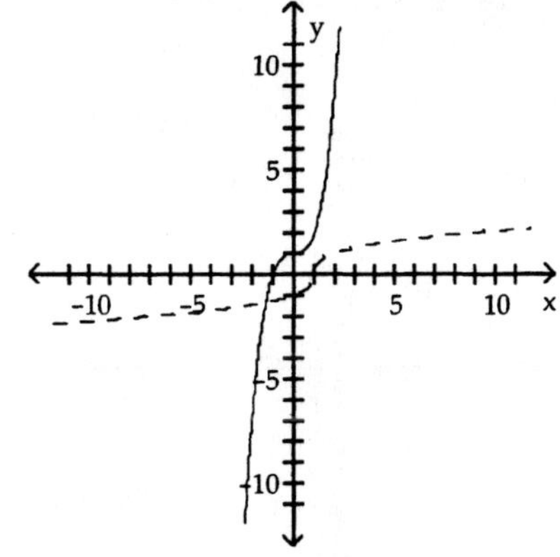

5) $f(x) = \frac{2}{x}$

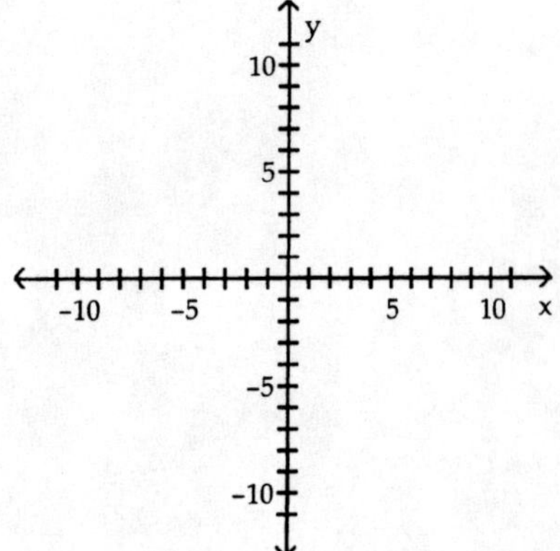

A)

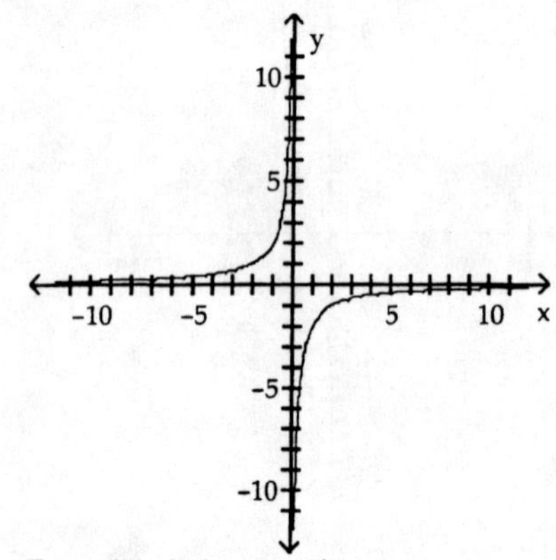

Function is its own inverse

B)

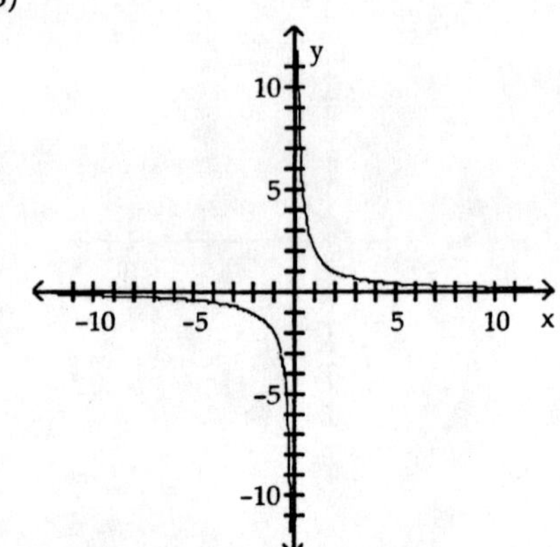

Function is its own inverse

C)

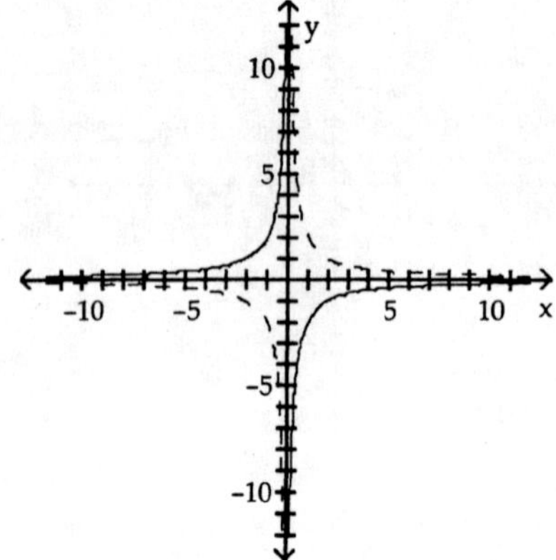

D)

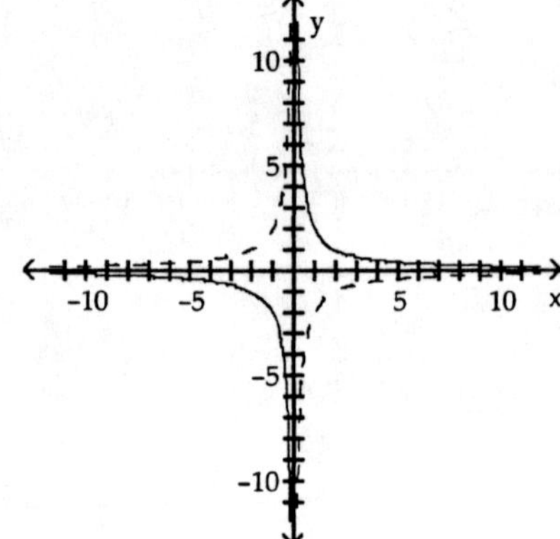

The graph of a one-to-one function is given. Draw the graph of the inverse function f^{-1}. For convenience, the graph of $y = x$ is also give.

6)

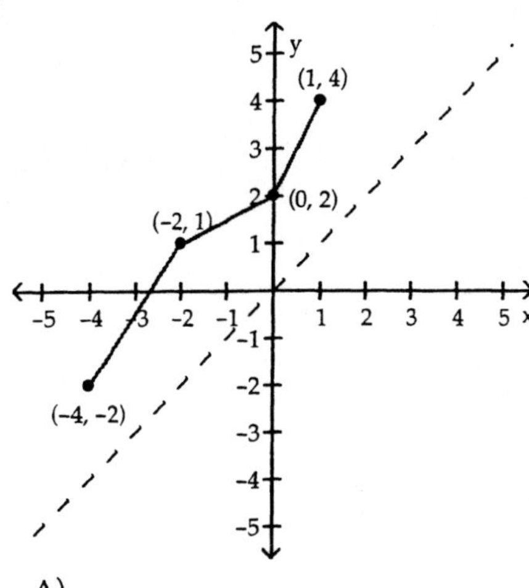

A)

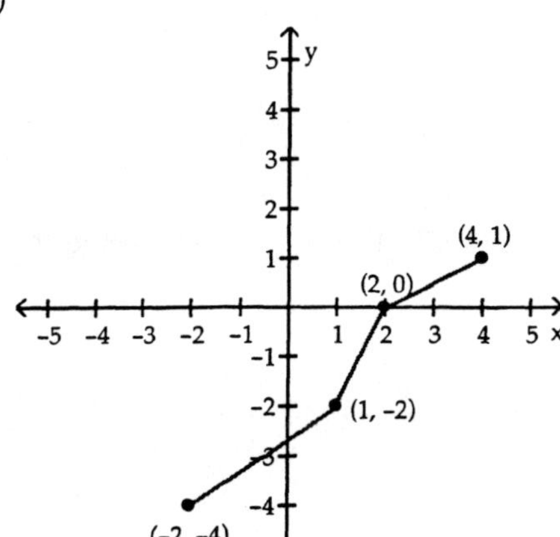

B)

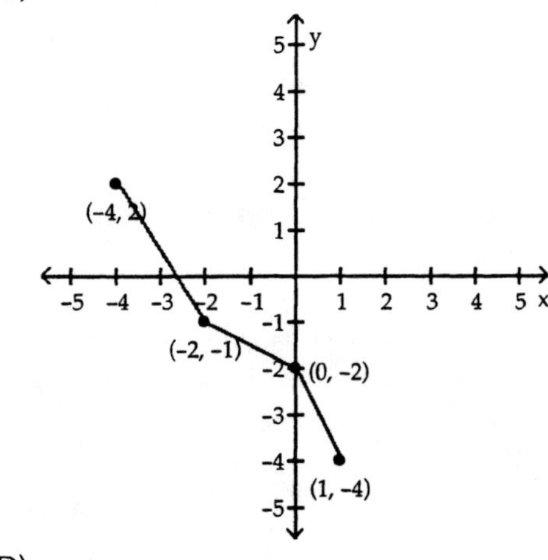

C)

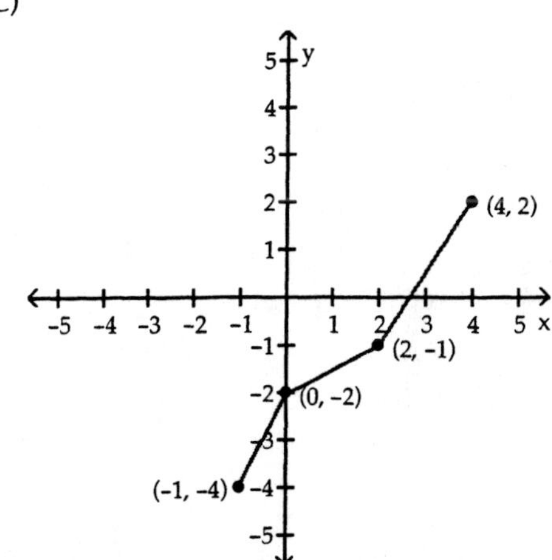

D)

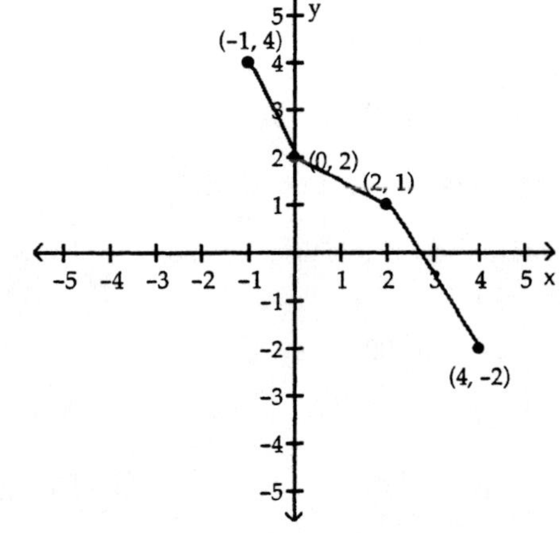

3 Find the Inverse Function f^{-1}

If the following defines a one-to-one function, find the inverse.

1) $f(x) = 5x + 7$

A) $f^{-1}(x) = -\frac{x + 5}{7}$ B) $f^{-1}(x) = -\frac{x - 7}{5}$ C) $f^{-1}(x) = \frac{x + 7}{5}$ D) $f^{-1}(x) = \frac{x - 7}{5}$

2) $f(x) = x^2 - 4$

A) Not a one-to-one function
B) $f^{-1}(x) = x^2 - 4$
C) $f^{-1}(x) = x^2 + 4$
D) $f^{-1}(x) = x + 4$

3) $f(x) = (x + 2)^3 - 8.$

A) $f^{-1}(x) = \sqrt[3]{x + 6}$
B) $f^{-1}(x) = \sqrt[3]{x - 2} + 8$
C) $f^{-1}(x) = \sqrt[3]{x + 8} - 2$
D) $f^{-1}(x) = \sqrt[3]{x + 10}$

Find the inverse function of f. State the domain and range of f.

4) $f(x) = \dfrac{3x - 2}{x + 5}$

A) $f^{-1}(x) = \dfrac{x + 5}{3x - 2}$; domain of f: $\{x \mid x \neq -5\}$; range of f: $\{y \mid y \neq \frac{2}{3}\}$

B) $f^{-1}(x) = \dfrac{3x + 2}{x - 5}$; domain of f: $\{x \mid x \neq -5\}$; range of f: $\{y \mid y \neq 5\}$

C) $f^{-1}(x) = \dfrac{5x + 2}{3 - x}$; domain of f: $\{x \mid x \neq -5\}$; range of f: $\{y \mid y \neq 3\}$

D) $f^{-1}(x) = \dfrac{5x + 2}{3 + x}$; domain of f: $\{x \mid x \neq -5\}$; range of f: $\{y \mid y \neq -3\}$

Solve the problem.

5) The profit P for selling x items is given by the equation $P = 2x - 500$. Express the sales amount x as a function of the profit P.

6) The function $f(x) = |x| - 5$ is not one-to-one.
(a) Find a suitable restriction on the domain of f so that the new function that results is one-to-one.
(b) Find the inverse of f.

7) The weight W of a bird's brain (in ounces) is related to the volume V of the bird's skull (in cubic ounces) through the function $W(V) = 3.38\sqrt[3]{V} + 1.21$.
(a) Express the skull volume V as a function of brain weight W.
(b) Predict the skull volume of a bird whose brain weighs 4 oz.

5.3 Exponential Functions

1 Evaluate Exponential Functions

Approximate the value using a calculator. Express answer rounded to three decimal places.

1) $4^{4.1}$

A) 256.000
B) 294.067
C) 282.576
D) 16.400

2) $3.5^{2.5}$

A) 22.918
B) 80.212
C) 8.750
D) 24.705

3) $5.439^{3.982}$

A) 10,011.247
B) 848.862
C) 21.658
D) 1836.330

4) 2.8^{π}

A) 25.397
B) 36.462
C) 24.661
D) 8.796

5) $4^{\sqrt{5}}$

A) 512.000 B) 8.944 C) 22.195 D) 25.000

6) Find the value of y (rounded to 3 decimal places) when x = 3: $y = 14(2.2)^x$

A) 92.400 B) 120.736 C) 29,218.112 D) 149.072

Determine whether the given function is exponential or not. If it is exponential, identify the value of the base a.

7)

x	H(x)
-1	8
0	11
1	14
2	17
3	20

A) exponential; a = 3 B) exponential; a = 11 C) exponential; a = 8 D) not exponential

8)

x	H(x)
-1	$\frac{8}{5}$
0	1
1	$\frac{5}{8}$
2	$\frac{25}{64}$
3	$\frac{125}{512}$

A) exponential; a = 5 B) exponential; $a = \frac{8}{5}$ C) exponential; $a = \frac{5}{8}$ D) not exponential

2 Graph Exponential Functions

The graph of an exponential function is given. Match the graph to one of the following functions.

1)

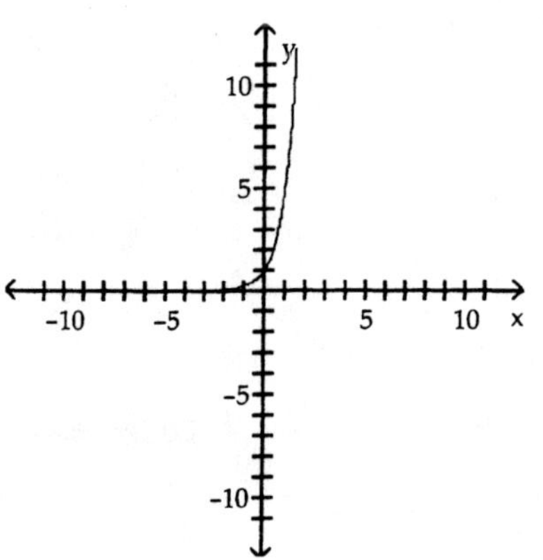

A) $f(x) = 5^x$ B) $f(x) = 5^{x-2}$ C) $f(x) = 5^x + 2$ D) $f(x) = 5^x - 2$

2)

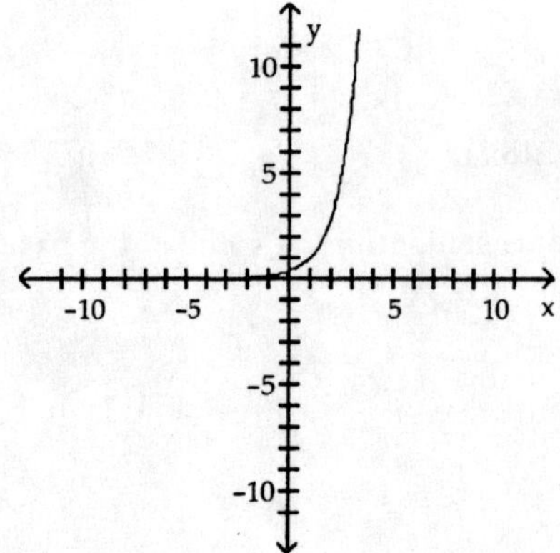

A) $f(x) = 3^x$ B) $f(x) = 3^{x - 1}$ C) $f(x) = 3^x - 1$ D) $f(x) = 3^x + 1$

3)

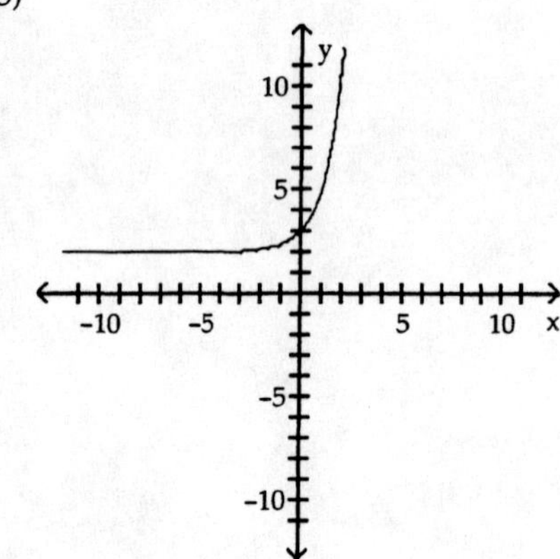

A) $f(x) = 3^x - 2$ B) $f(x) = 3^{x + 2}$ C) $f(x) = 3^x$ D) $f(x) = 3^x + 2$

4)

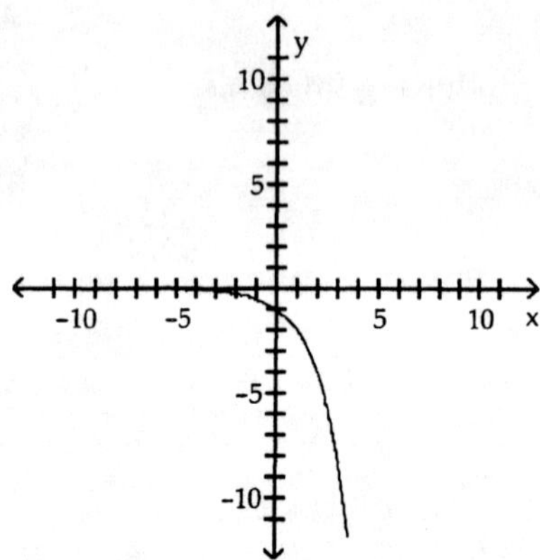

A) $f(x) = 2^{-x}$ B) $f(x) = -2^{-x}$ C) $f(x) = 2^x$ D) $f(x) = -2^x$

5)

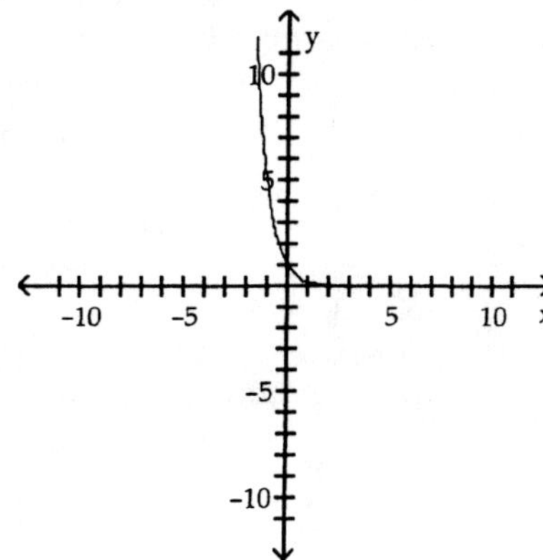

A) $f(x) = -5^x$

B) $f(x) = -5^{-x}$

C) $f(x) = 5^{-x}$

D) $f(x) = 5^x$

6)

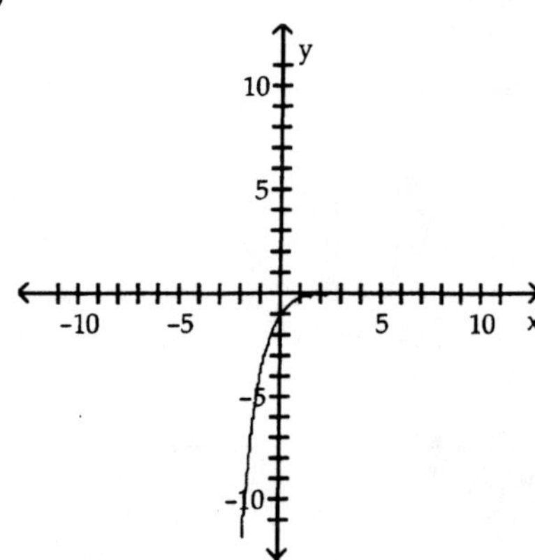

A) $f(x) = 4^x$

B) $f(x) = -4^x$

C) $f(x) = -4^{-x}$

D) $f(x) = 4^{-x}$

Solve the problem.

7)

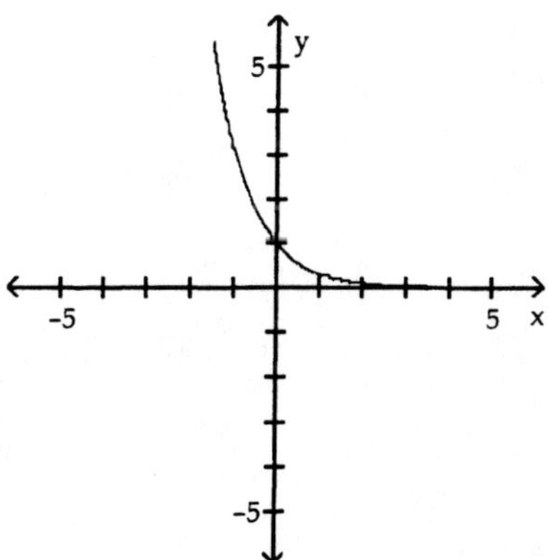

A) $y = 3.5^x$

B) $y = 2.4^x$

C) $y = 0.65^x$

D) $y = 0.32^x$

Use transformations to graph the function. Determine the domain, range, and vertical asymptote of the function.

8) Graph the function $f(x) = -1 + e^x$. Determine the domain, range, and horizontal asymptote.

A)

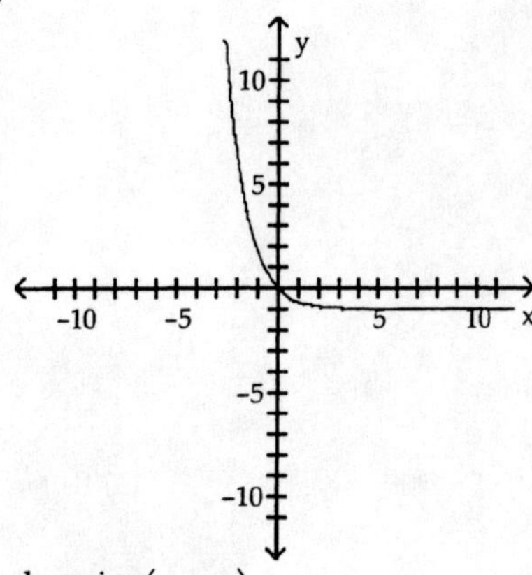

domain: $(-\infty, \infty)$
range: $(-1, \infty)$
horizontal asymptote: $y = -1$

B)

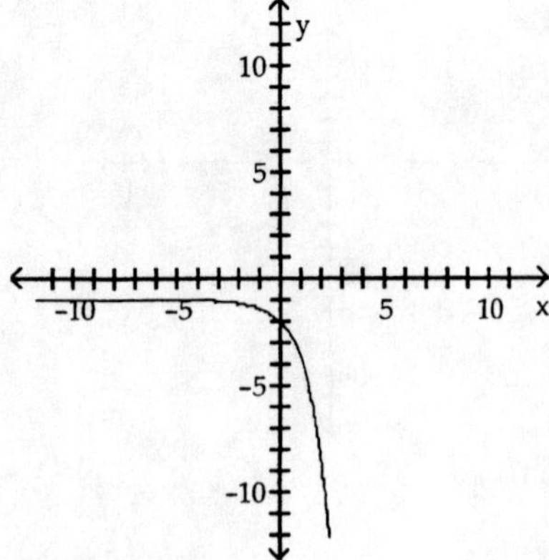

domain: $(-\infty, \infty)$
range: $(-1, \infty)$
horizontal asymptote: $y = -1$

C)

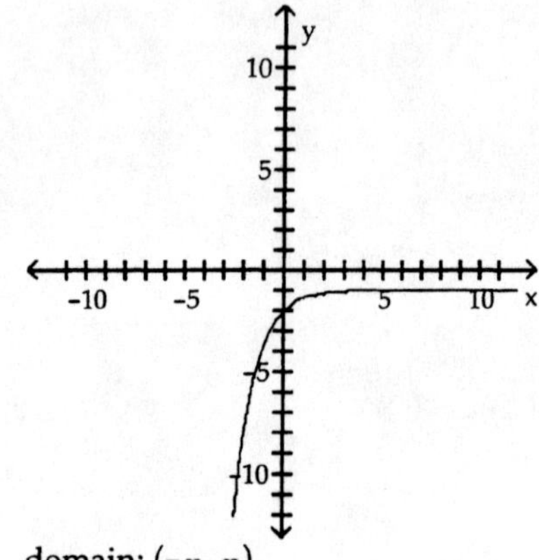

domain: $(-\infty, \infty)$
range: $(-1, \infty)$
horizontal asymptote: $y = -1$

D)

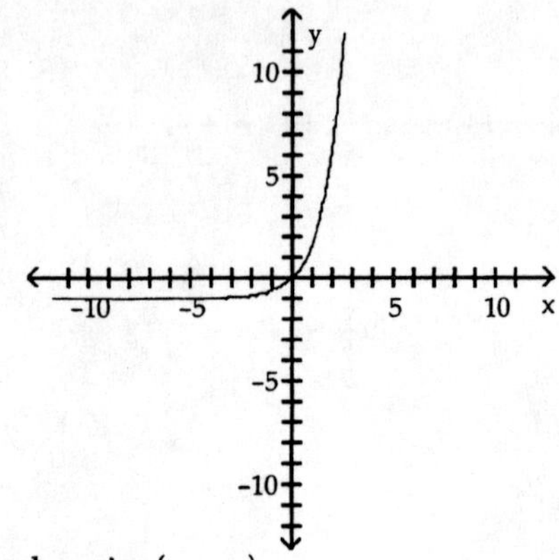

domain: $(-\infty, \infty)$
range: $(-1, \infty)$
horizontal asymptote: $y = -1$

9) $f(x) = -2^{x+3} + 4$

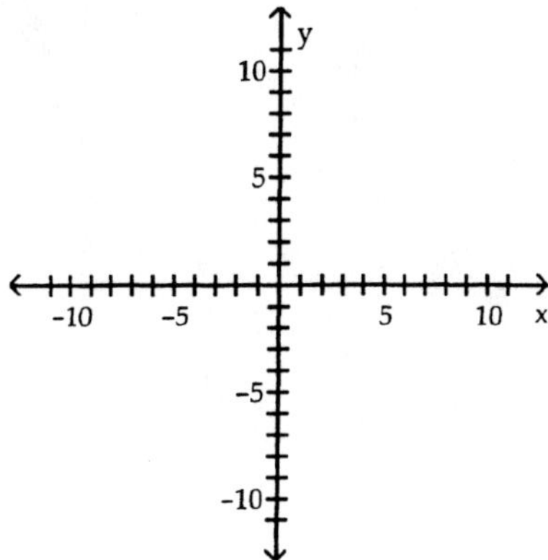

A)

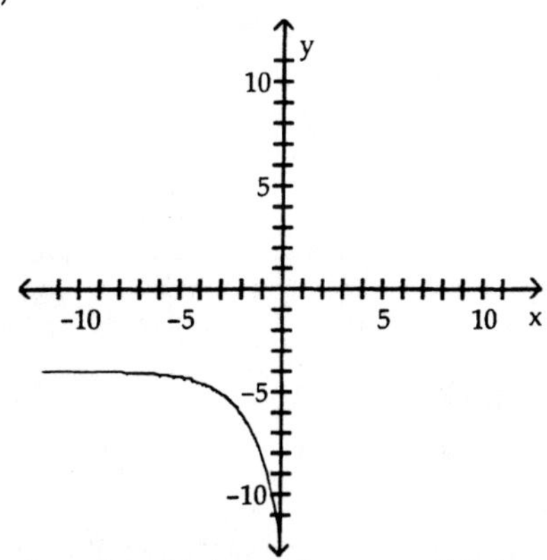

domain of f: $(-\infty, \infty)$; range of f: $(-\infty, -4)$;
horizontal asymptote: $y = -4$

B)

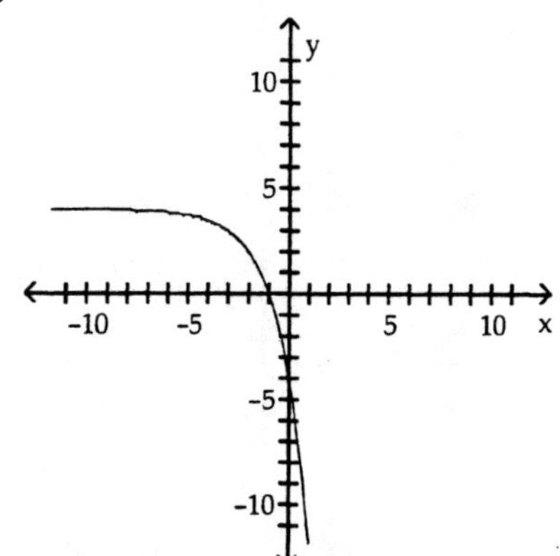

domain of f: $(-\infty, \infty)$; range of f: $(-\infty, 4)$;
horizontal asymptote: $y = 4$

C)

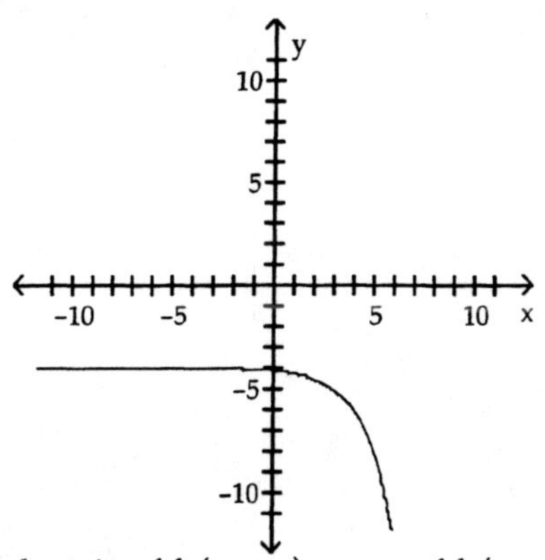

domain of f: $(-\infty, \infty)$; range of f: $(-\infty, -4)$;
horizontal asymptote: $y = -4$

D)

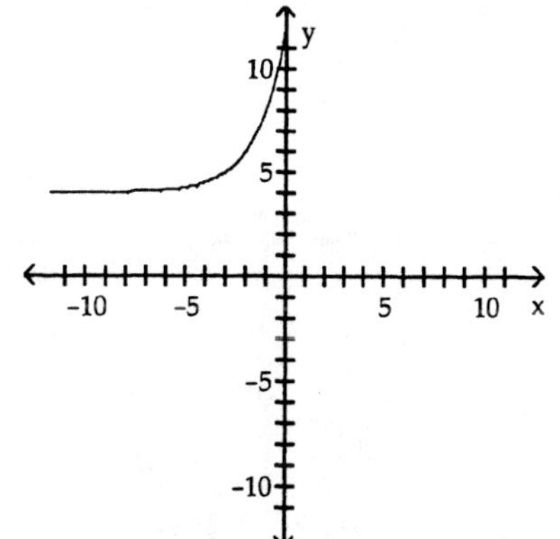

domain of f: $(-\infty, \infty)$; range of f: $(-4, \infty)$;
horizontal asymptote: $y = 4$

10) $f(x) = 3^{(x - 2)}$

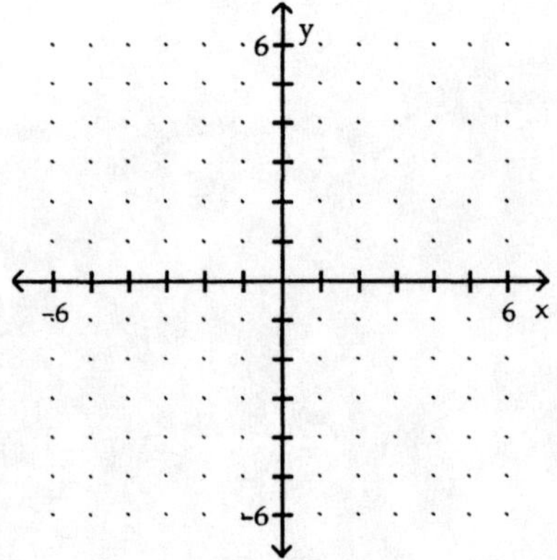

A)

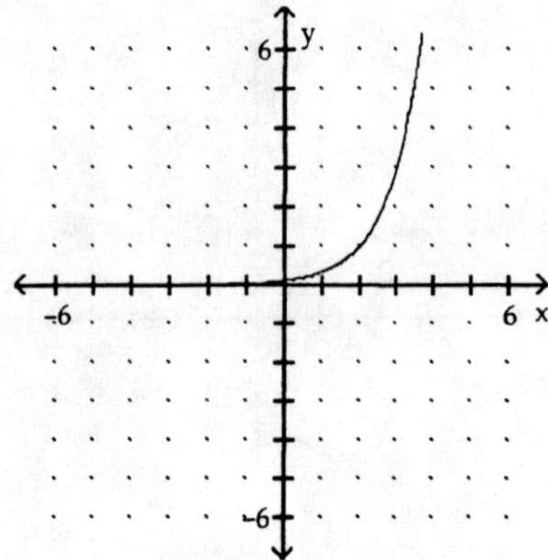

domain of f: $(-\infty, \infty)$; range of f:$(0, \infty)$
horizontal asymptote: $y = 0$

B)

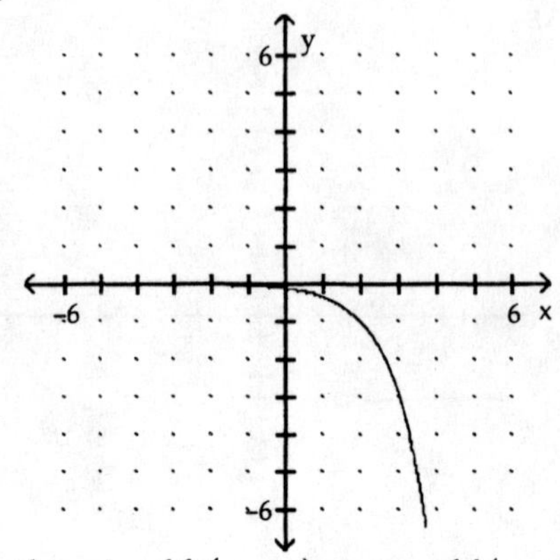

domain of f: $(-\infty, \infty)$; range of f:$(-\infty, 0)$
horizontal asymptote: $y = 0$

C)

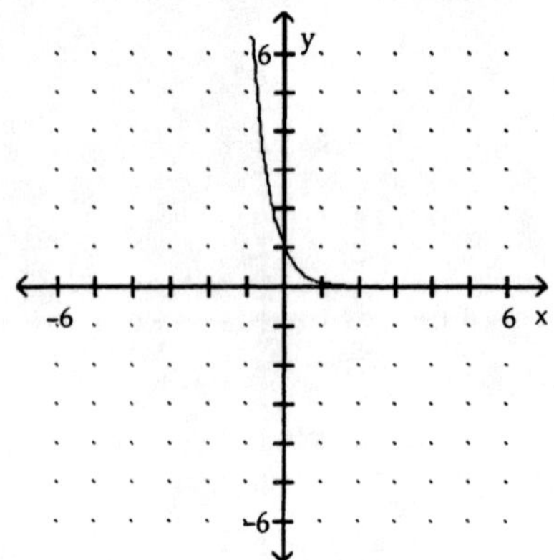

domain of f: $(-\infty, \infty)$; range of f:$(0, \infty)$
horizontal asymptote: $y = 0$

D)

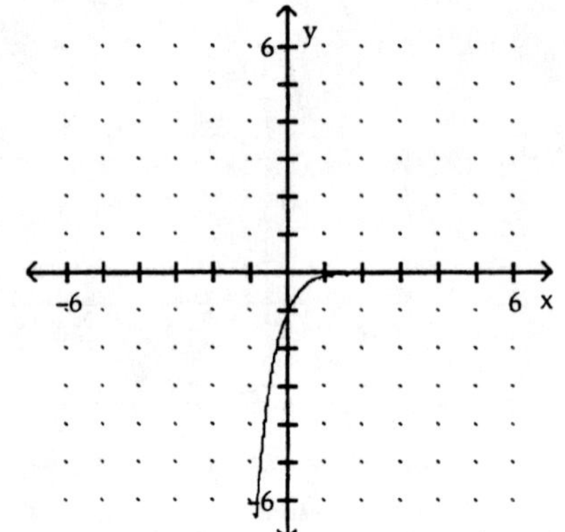

domain of f: $(-\infty, \infty)$; range of f:$(-\infty, 0)$
horizontal asymptote: $y = 0$

11) $f(x) = 3^{-x} + 2$

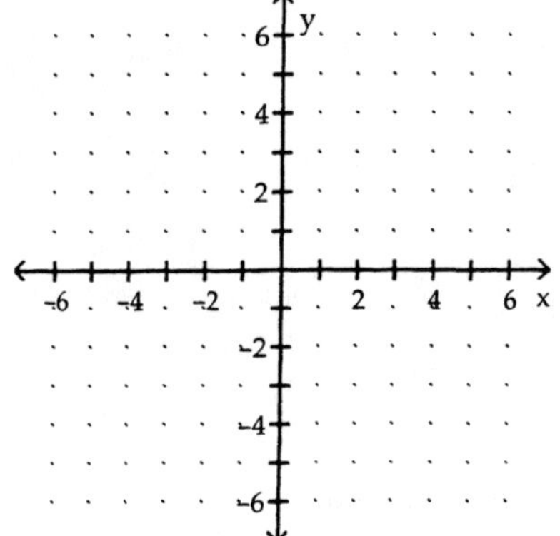

A)

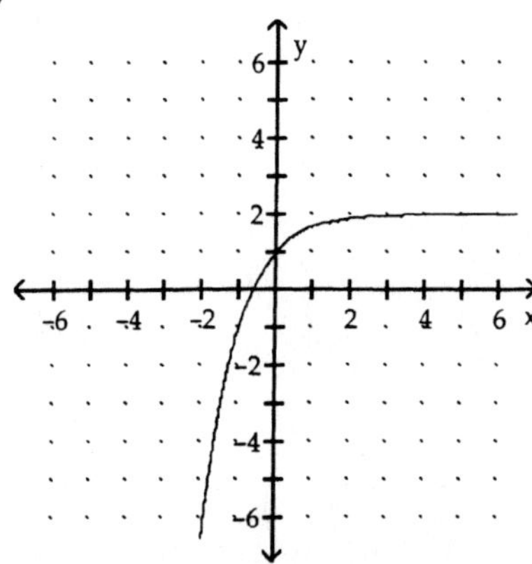

domain of f: $(-\infty, \infty)$; range of f:$(2, \infty)$
horizontal asymptote: $y = 2$

B)

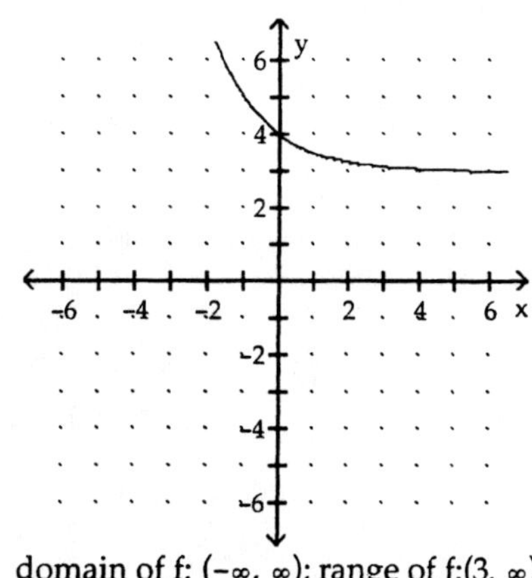

domain of f: $(-\infty, \infty)$; range of f:$(3, \infty)$
horizontal asymptote: $y = 3$

C)

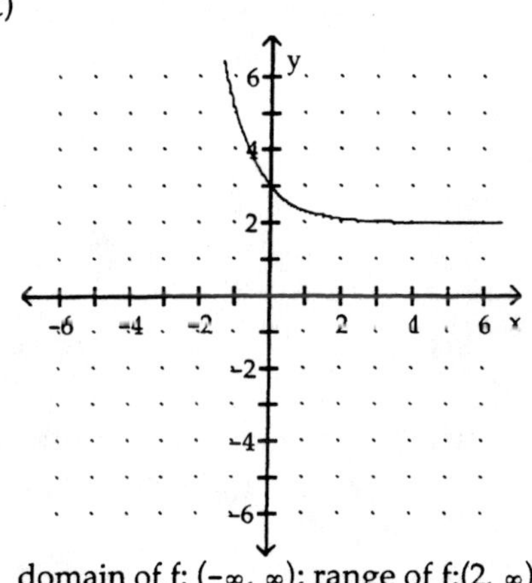

domain of f: $(-\infty, \infty)$; range of f:$(2, \infty)$
horizontal asymptote: $y = 2$

D)

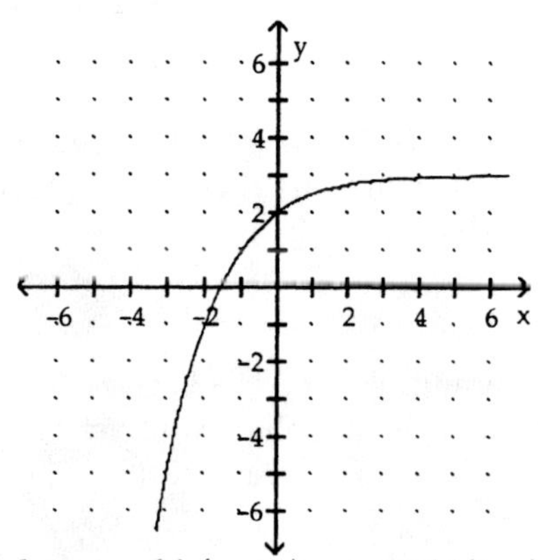

domain of f: $(-\infty, \infty)$; range of f:$(3, \infty)$
horizontal asymptote: $y = 3$

Answer the question.

1) Define the number e.

A) The number defined by $e = \lim_{n \to \infty} \left(1 + \frac{1}{n}\right)^n$ in Calculus.

B) The number approximately equal to 2.72.

C) The number that the expression, $\left(1 + \frac{1}{n}\right)^n$, approaches as $n \to \infty$.

D) All of the above.

Approximate the value using a calculator. Express answer rounded to three decimal places.

2) $e^{3.9}$

A) 49.402 B) 40.428 C) 10.601 D) 49.702

3) $e^{3.33}$

A) 9.052 B) 15.154 C) 26.312 D) 27.938

4) $e^{-2.2}$

A) 0.411 B) 0.111 C) -0.111 D) -5.980

5) 2^e

A) 5.437 B) 4.718 C) 7.389 D) 6.581

6) π^e

A) 5.860 B) 23.141 C) 22.459 D) 8.540

7) e^π

A) 5.860 B) 23.141 C) 22.459 D) 8.540

Solve the problem.

8) The function $D(h) = 5e^{-0.4h}$ can be used to determine the milligrams D of a certain drug in a patient's bloodstream h hours after the drug has been given. How many milligrams (to two decimals) will be present after 6 hours?

A) 0.81 mg B) 3.78 mg C) 0.45 mg D) 55.12 mg

9) The formula $P = 14.7e^{-0.21x}$ gives the average atmospheric pressure, P, in pounds per square inch, at an altitude x, in miles above sea level. Find the average atmospheric pressure for an altitude of 2.3 miles. Round your answer to the nearest tenth.

A) 8.4 lb/in^2 B) 9.1 lb/in^2 C) 11.0 lb/in^2 D) 7.8 lb/in^2

10) Instruments on a satellite measure the amount of power generated by the satellite's power supply. The time t and the power P can be modeled by the function $P = 50e^{-t/300}$, where t is in days and P is in watts. How much power will be available after 378 days? Round to the nearest hundredth.

11) A cancer patient undergoing chemotherapy is injected with a particular drug. The function $D(h) = 4e^{-0.35h}$ gives the number of milligrams D of this drug that is in the patient's bloodstream h hours after the drug has been administered. How many milligrams of the drug were injected? To the nearest milligram, how much of the drug will be present after 2 hours?

12) A grocery store normally sells 5 jars of caviar per week. Use the Poisson Distribution $P(x) = \dfrac{5^x e^{-5}}{x!}$ to find the probability (to three decimals) of selling 3 jars in a week. ($x! = x \cdot (x - 1) \cdot (x - 2) \cdot \ldots \cdot (3)(2)(1)$).

A) 0.421 B) 0.281 C) 0.094 D) 0.14

4 Solve Exponential Equations

Solve the equation.

1) $2^{1 + 2x} = 32$

A) 16 B) -2 C) 4 D) 2

2) $2^{-x} = \dfrac{1}{8}$

A) $\dfrac{1}{4}$ B) -3 C) 3 D) $\dfrac{1}{3}$

3) $3^{6 - 3x} = \dfrac{1}{27}$

A) $\dfrac{1}{9}$ B) 9 C) -3 D) 3

4) $4^x = \dfrac{1}{256}$

A) -4 B) $\dfrac{1}{4}$ C) $\dfrac{1}{64}$ D) 4

5) $2^{x^2 - 3} = 64$

A) 3 B) 6 C) 3, -3 D) $\sqrt{35}, -\sqrt{35}$

6) $9^{2x} \cdot 27^{(3 - x)} = \dfrac{1}{9}$

A) $x = -8$

B) $x = \dfrac{9 + \sqrt{87}}{6}$ and $x = \dfrac{9 - \sqrt{87}}{6}$

C) $x = -11$

D) $x = 10$

Solve the problem.

7) A rumor is spread at an elementary school with 1200 students according to the model $N = 1200(1 - e^{-0.16d})$ where N is the number of students who have heard the rumor and d is the number of days that have elapsed since the rumor began. How many students will have heard the rumor after 5 days?

8) The bacteria in a 12-liter container double every 3 minutes. After 48 minutes the container is full. How long did it take to fill a quarter of the container?

A) 24 min B) 12 min C) 42 min D) 36 min

5.4 Logarithmic Functions

1 Change Exponential Expressions to Logarithmic Expressions

Change the exponential expression to an equivalent expression involving a logarithm.

1) $5^2 = 25$

A) $\log_5 2 = 25$ B) $\log_2 25 = 5$ C) $\log_5 25 = 2$ D) $\log_{25} 5 = 2$

2) $5^{-3} = \frac{1}{125}$

A) $\log_{1/125} 5 = -3$ B) $\log_5 \frac{1}{125} = -3$ C) $\log_{-3} \frac{1}{125} = 5$ D) $\log_5 -3 = \frac{1}{125}$

3) $5^3 = x$

A) $\log_5 x = 3$ B) $\log_5 3 = x$ C) $\log_x 5 = 3$ D) $\log_3 x = 5$

4) $4^{5/2} = 32$

A) $\log_5 4 = \frac{5}{2}$ B) $\frac{\log_2 32}{\log_5 4} = 4$ C) $\log_4 32 = \frac{5}{2}$ D) $\log_{32} 4 = \frac{5}{2}$

5) $4^x = 16$

A) $\log_{16} 4 = x$ B) $\log_x 16 = 4$ C) $\log_{16} x = 4$ D) $\log_4 16 = x$

6) $e^x = 6$

A) $\ln 6 = x$ B) $\ln x = 6$ C) $\log_x e = 6$ D) $\log_6 x = e$

7) $2.3^{x+5} = 15$

A) $\log_{2.3} x = 10$ B) $\log_{15} (x + 5) = 2.3$ C) $\log_{2.3} 15 = x + 5$ D) $\log_{(x+5)} 15 = 2.3$

8) $x^{\sqrt{5}} = \pi$

A) $x = \log_{\sqrt{5}} \pi$ B) $\sqrt{5} = \log_x \pi$ C) $\sqrt{5} = \log_\pi x$ D) $x = \log_\pi \sqrt{5}$

Solve the problem.

9) Write the equation $6^5 = 7776$ in logarithmic form.

A) $\log_6 7776 = 5$ B) $\log_5 7776 = 6$ C) $\log_{7776} 6 = 5$ D) $\log_{1/5} 7776 = 6$

10) Write the equation $16^{3/2} = 64$ in logarithmic form.

2 Change Logarithmic Expressions to Exponential Expressions

Change the logarithmic expression to an equivalent expression involving an exponent.

1) $\log_{1/3} 81 = -4$

A) $\left(\frac{1}{3}\right)^4 = 81$ B) $81^{1/3} = 4$ C) $\left(\frac{1}{3}\right)^{-4} = 81$ D) $(-4)^{1/3} = 81$

2) $\log_2 \frac{1}{4} = -2$

A) $2^4 = 2$ B) $2^2 = \frac{1}{4}$ C) $(\frac{1}{4})^2 = 2$ D) $2^{-2} = \frac{1}{4}$

3) $\log_3 9 = 2$

A) $9^2 = 3$ B) $3^2 = 9$ C) $3^9 = 2$ D) $2^3 = 9$

4) $\log_4 x = 2$

A) $2^4 = x$ B) $x^2 = 4$ C) $4^x = 2$ D) $4^2 = x$

5) $\log_b 36 = 2$

A) $b^2 = 36$ B) $36^2 = b$ C) $2^b = 36$ D) $36^b = 2$

6) $\log_6 36 = x$

A) $36^x = 6$ B) $36^6 = x$ C) $x^6 = 36$ D) $6^x = 36$

7) $\log_b 49 = \frac{2}{3}$

A) $b^{2/3} = 49$ B) $b^{3/2} = 49$ C) $49^{2/3} = b$ D) $(\frac{2}{3})^b = 49$

8) $\log_{\sqrt{3}} 64 = \frac{7}{11}$

A) $\sqrt{3}^{11/7} = 64$ B) $64^{7/11} = \sqrt{3}$ C) $\sqrt{3}^{7/11} = 64$ D) $(\frac{7}{11})^{\sqrt{3}} = 64$

9) $\log_\pi 37 = x$

A) $x^\pi = 37$ B) $37^x = \pi$ C) $\pi^x = 37$ D) $\pi^x = \frac{1}{37}$

10) Write the equation $\log_{1024} 256 = \frac{4}{5}$ in exponential form.

A) $256^{4/5} = 1024$ B) $256^{5/4} = 1024$ C) $1024^{4/5} = 256$ D) $\left(\frac{4}{5}\right)^{1024} = 256$

11) Write in exponential form: $y = \log_{61} x$

3 Evaluate Logarithmic Functions

Find the exact value of the logarithmic expression.

1) $\log_8 512$

A) 24 B) 512 C) 3 D) 8

2) $\log_9 \frac{1}{81}$

A) 9 B) -9 C) 2 D) -2

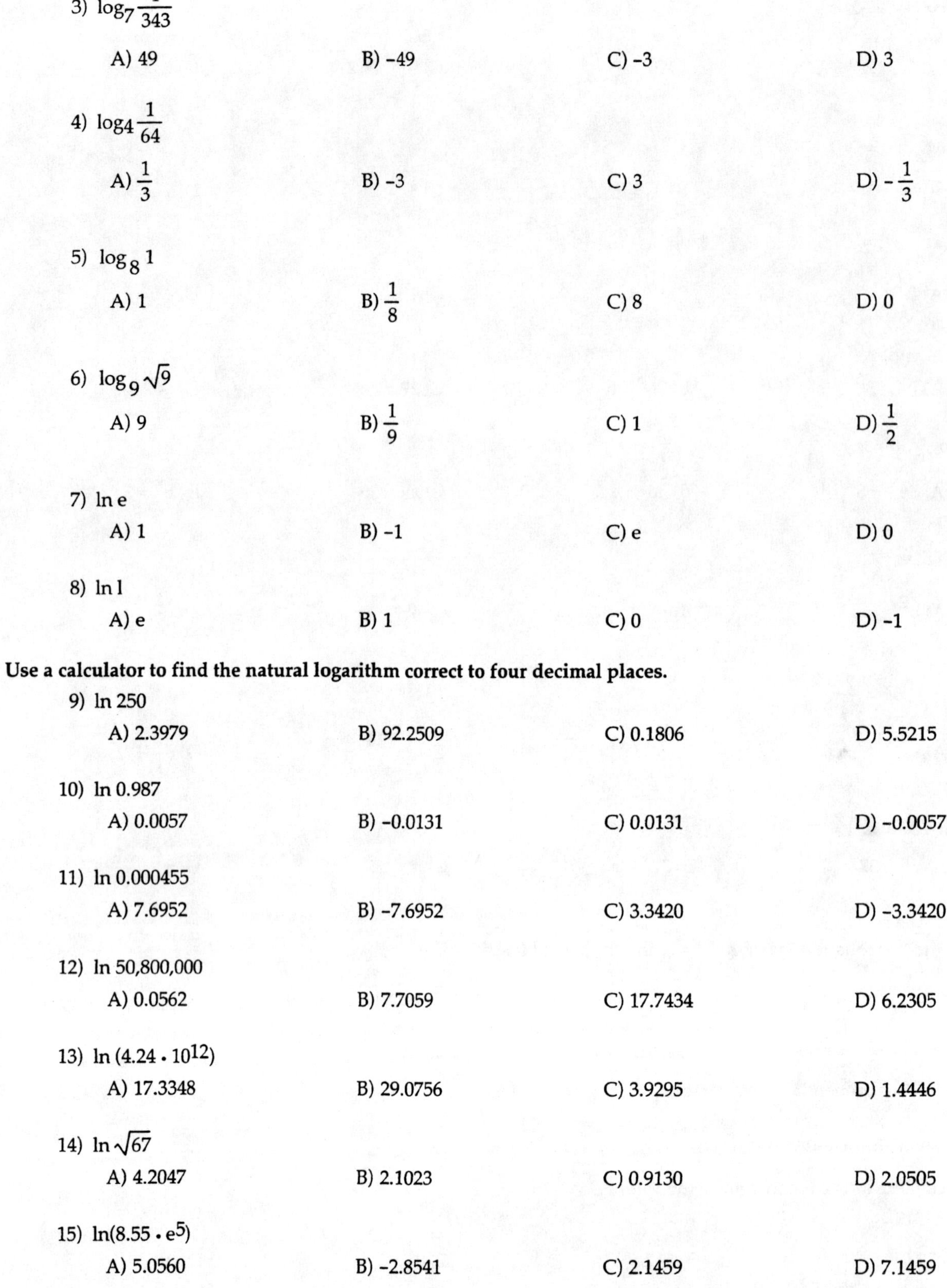

3) $\log_7 \frac{1}{343}$

A) 49 B) -49 C) -3 D) 3

4) $\log_4 \frac{1}{64}$

A) $\frac{1}{3}$ B) -3 C) 3 D) $-\frac{1}{3}$

5) $\log_8 1$

A) 1 B) $\frac{1}{8}$ C) 8 D) 0

6) $\log_9 \sqrt{9}$

A) 9 B) $\frac{1}{9}$ C) 1 D) $\frac{1}{2}$

7) ln e

A) 1 B) -1 C) e D) 0

8) ln 1

A) e B) 1 C) 0 D) -1

Use a calculator to find the natural logarithm correct to four decimal places.

9) ln 250

A) 2.3979 B) 92.2509 C) 0.1806 D) 5.5215

10) ln 0.987

A) 0.0057 B) -0.0131 C) 0.0131 D) -0.0057

11) ln 0.000455

A) 7.6952 B) -7.6952 C) 3.3420 D) -3.3420

12) ln 50,800,000

A) 0.0562 B) 7.7059 C) 17.7434 D) 6.2305

13) $\ln(4.24 \cdot 10^{12})$

A) 17.3348 B) 29.0756 C) 3.9295 D) 1.4446

14) $\ln\sqrt{67}$

A) 4.2047 B) 2.1023 C) 0.9130 D) 2.0505

15) $\ln(8.55 \cdot e^5)$

A) 5.0560 B) -2.8541 C) 2.1459 D) 7.1459

Solve the problem.

16) The pH of a chemical solution is given by the formula

$$pH = -\log_{10}[H^+]$$

where $[H^+]$ is the concentration of hydrogen ions in moles per liter.
Find the $[H^+]$ if the pH = 8.4.

A) 2.51×10^{-9} B) 2.51×10^{-8} C) 3.98×10^{-9} D) 3.98×10^{-8}

17) The pH of a chemical solution is given by the formula

$$pH = -\log_{10}[H^+]$$

where $[H^+]$ is the concentration of hydrogen ions in moles per liter.
Find the pH if the $[H^+] = 8 \times 10^{-6}$.

A) 6.9 B) 6.1 C) 5.9 D) 5.1

18) The long jump record, in feet, at a particular school can be modeled by $f(x) = 20 + 2.4 \ln(x + 1)$ where x is the number of years since records began to be kept at the school. What is the record for the long jump 18 years after record started being kept? Round your answer to the nearest tenth.

A) 26.8 feet B) 22.4 feet C) 27.1 feet D) 26.9 feet

19) The function $f(x) = 1 + 1.3 \ln(x+1)$ models the average number of free-throws a basketball player can make consecutively during practice as a function of time, where x is the number of consecutive days the basketball player has practiced for two hours. After 100 days of practice, what is the average number of consecutive free throws the basketball player makes?

A) 7 consecutive free throws B) 11 consecutive free throws
C) 8 consecutive free throws D) 10 consecutive free throws

20) Find a so that the graph of $f(x) = \log_a x$ contains the point (5, 12).

A) $\sqrt[12]{5}$ B) $\frac{12}{5}$ C) $\sqrt[5]{12}$ D) $\frac{5}{12}$

4 Determine the Domain of a Logarithmic Function

Find the domain of the function.

1) $f(x) = \log(x - 5)$

A) $x > 5$ B) $x > 1$ C) $x > -5$ D) $x > 0$

2) $f(x) = \ln(7 - x)$

A) $x > 7$ B) $x > -7$ C) $x < -7$ D) $x < 7$

3) $f(x) = \log_4(16 - x^2)$

A) $-4 \le x \le 4$ B) $-4 < x < 4$ C) $-16 < x < 16$ D) $x < -4$ and $x > 4$

4) $f(x) = \ln(7x - x^2)$

A) $0 < x < 7$ B) $-7 \le x < 0$ C) $x \le 7$ D) $-7 < x < 7$

5) $f(x) = \log_{10}(x^2 - 10x + 16)$

A) $(-\infty, 2) \cup (8, \infty)$ B) $(8, \infty)$ C) $(-\infty, -2)$ D) $(-2, 8)$

6) $f(x) = \log_{10} (\frac{x + 3}{x - 5})$

A) $(-\infty, -3)$ B) $(-\infty, -3) \cup (5, \infty)$ C) $(-3, 5)$ D) $(5, \infty)$

7) $f(x) = \ln (\frac{1}{x + 8})$

A) $x > 1$ B) $x > 0$ C) $x > -8$ D) $x > 8$

8) $f(x) = 4 - \ln (7x)$

A) $(0, \infty)$ B) $(-4, 7)$ C) $(-\infty, 4) \cup (7, \infty)$ D) $(7, \infty)$

Solve the problem.

9) Determine the domain of the function $f(x) = \log_{1/2}(x + 4)$.

A) $(4, \infty)$ B) $(-4, \infty)$ C) $(-\infty, -4)$ D) $(-\infty, 4)$

10) Determine the domain of the function $f(x) = \log_5(x + 2)$.

11) Determine the domain of the function $f(x) = \ln \sqrt{x}$.

A) $(1, \infty)$ B) $(-\infty, 0)$ C) $(0, \infty)$ D) $(-\infty, 1)$

5 Graph Logarithmic Functions

Graph the function.

1) $y = \log_4 x$

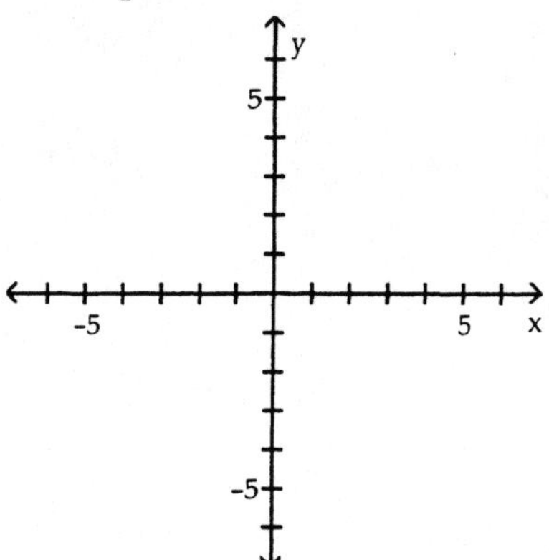

A)

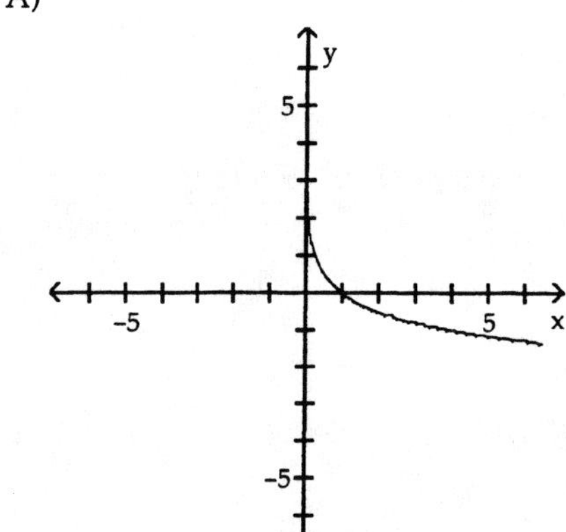

B)

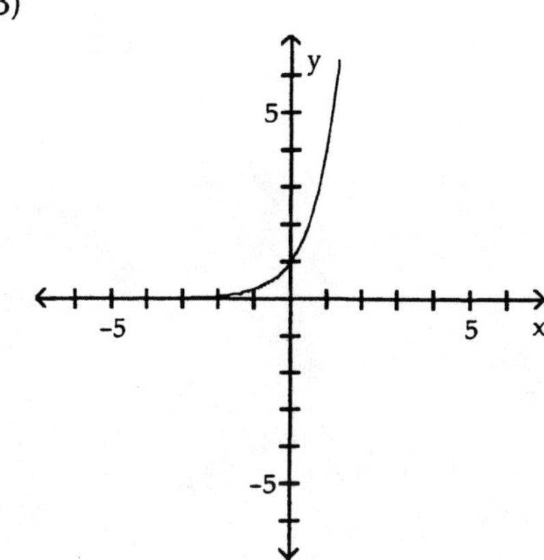

C)

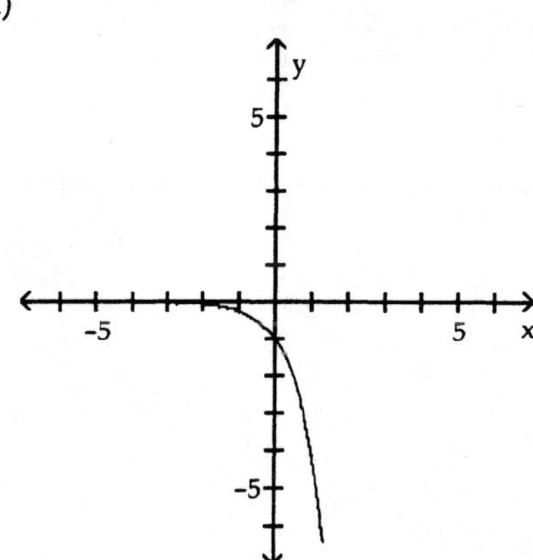

D)

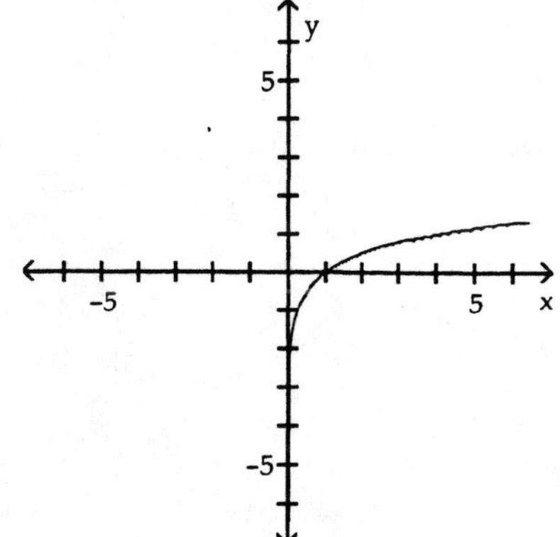

2) $y = \log_{1/6} x$

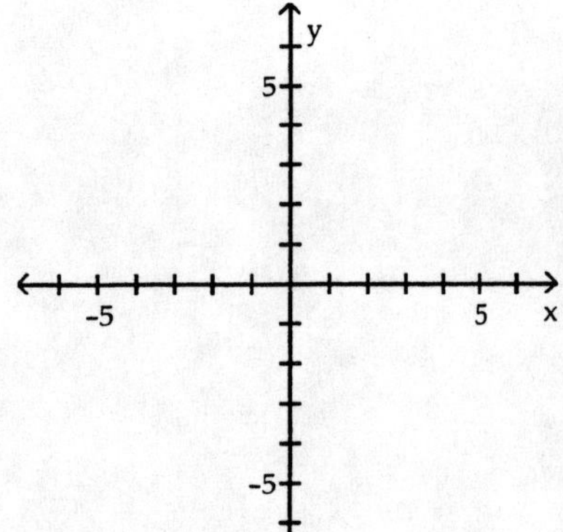

A)

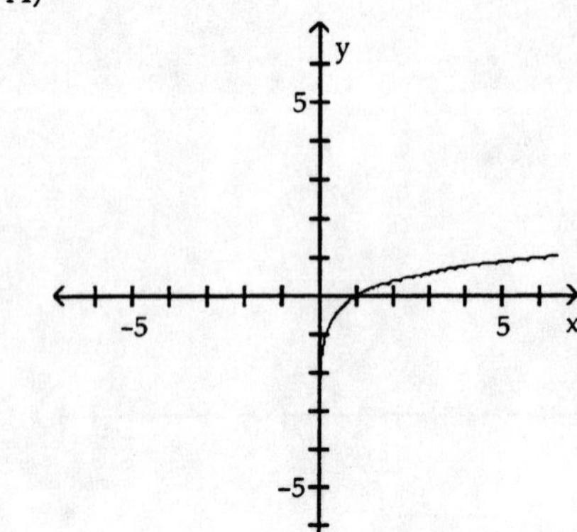

B)

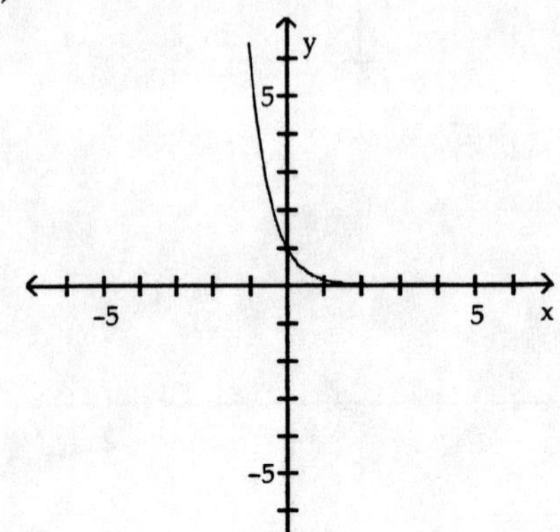

C)

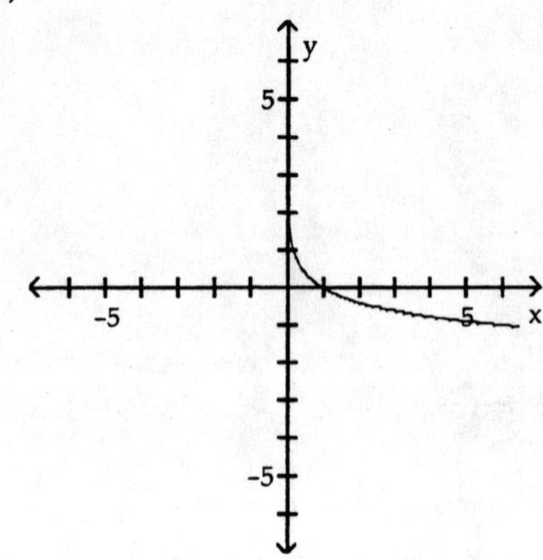

D)

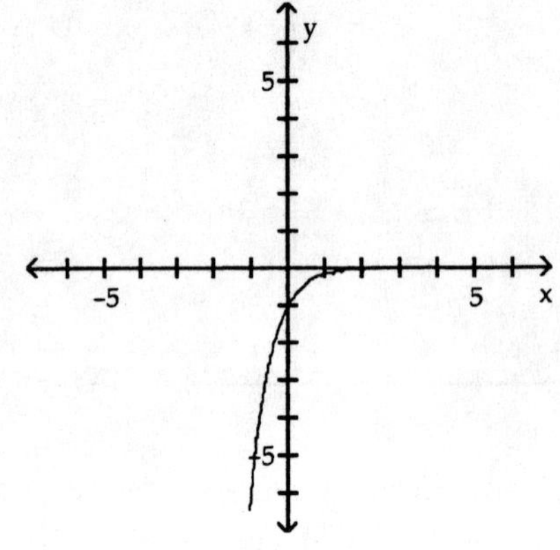

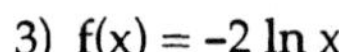
3) $f(x) = -2 \ln x$

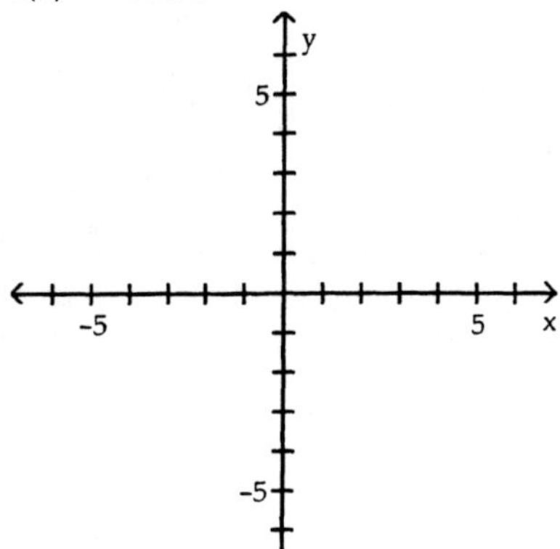

A)

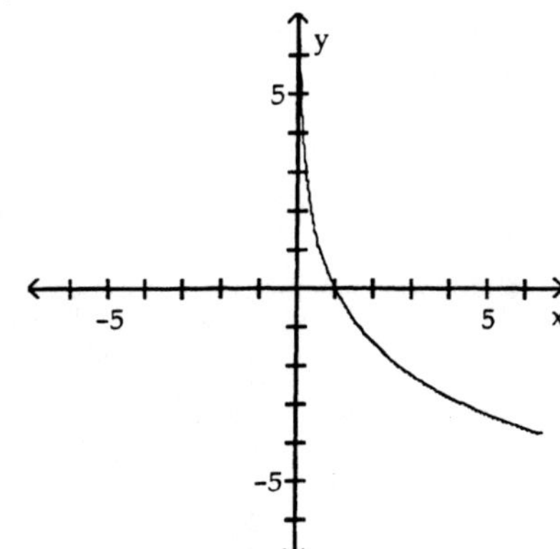

B)

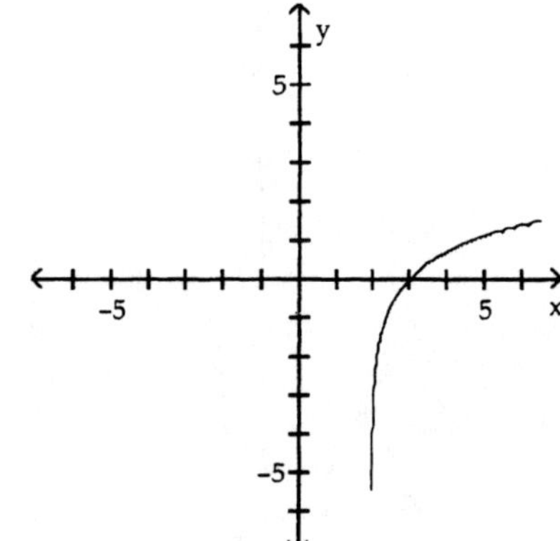

C)

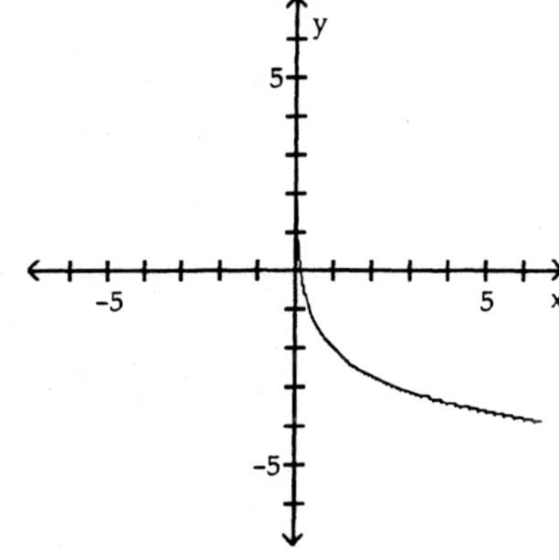

D)

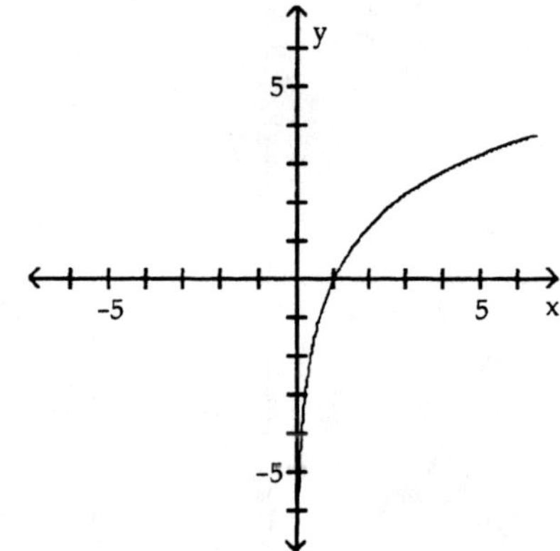

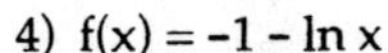
4) $f(x) = -1 - \ln x$

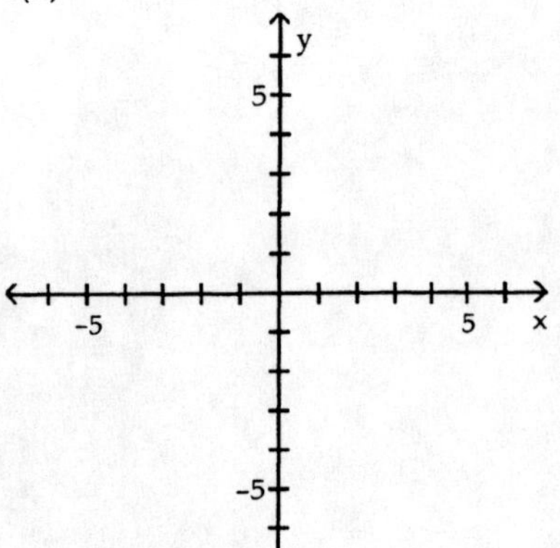

A)

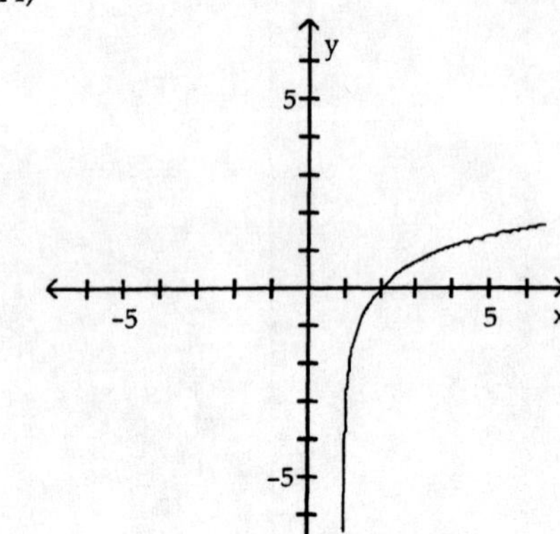

B)

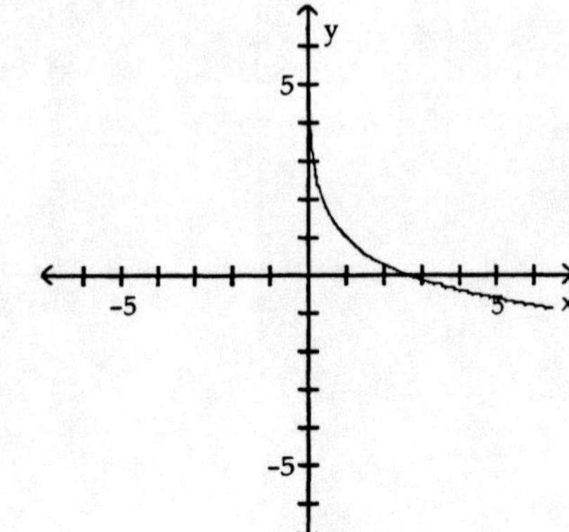

C)

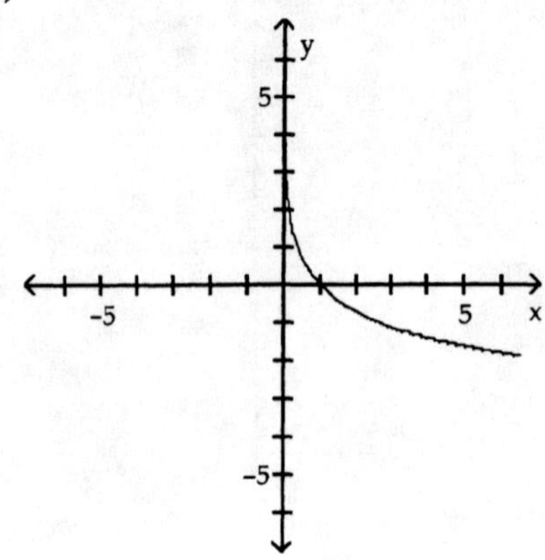

D)

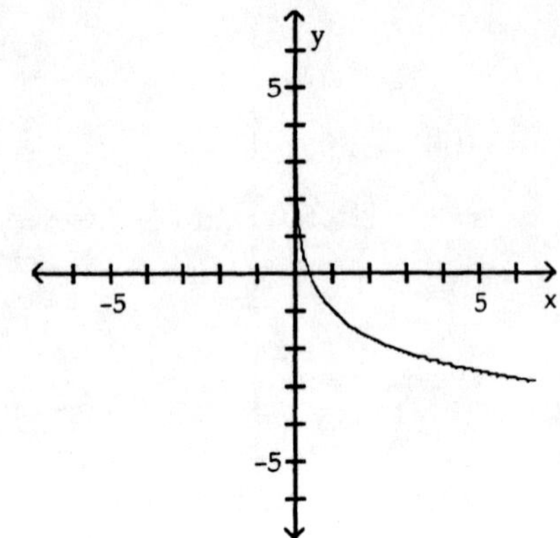

The graph of a logarithmic function is shown. Select the function which matches the graph.

5)

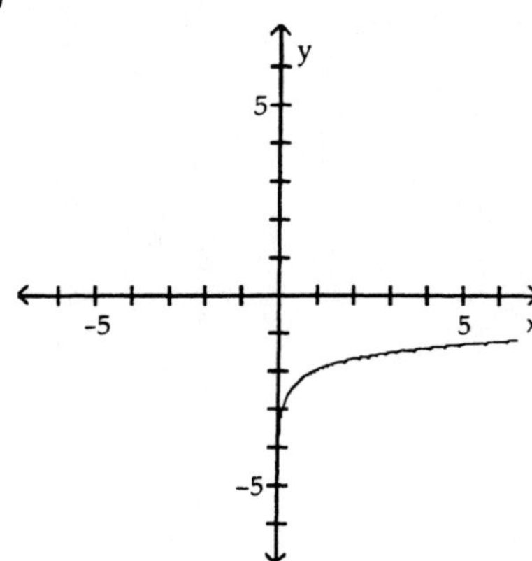

A) $y = \log(x - 2)$ B) $y = \log(2 - x)$ C) $y = 2 - \log(x)$ D) $y = \log(x) - 2$

6)

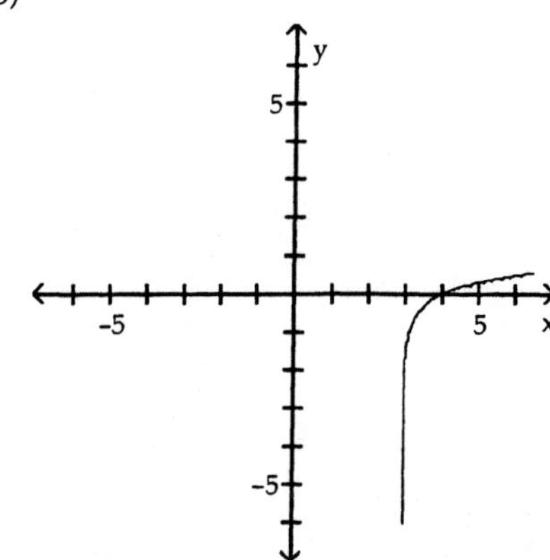

A) $y = 3 - \log(x)$ B) $y = \log(x - 3)$ C) $y = \log(x) - 3$ D) $y = \log(3 - x)$

7)

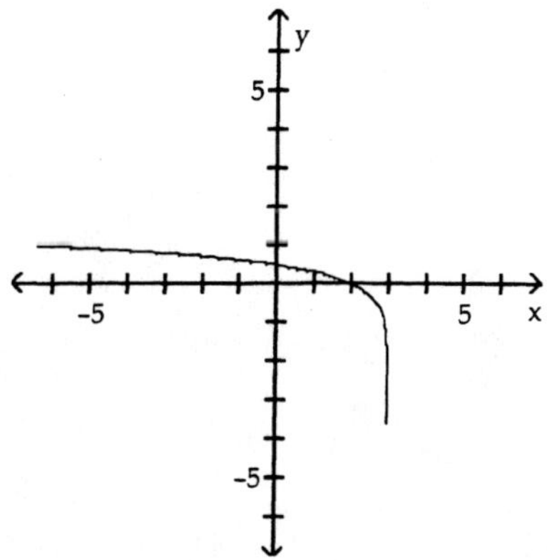

A) $y = 3 - \log(x)$ B) $y = \log(x) - 3$ C) $y = \log(x - 3)$ D) $y = \log(3 - x)$

8)

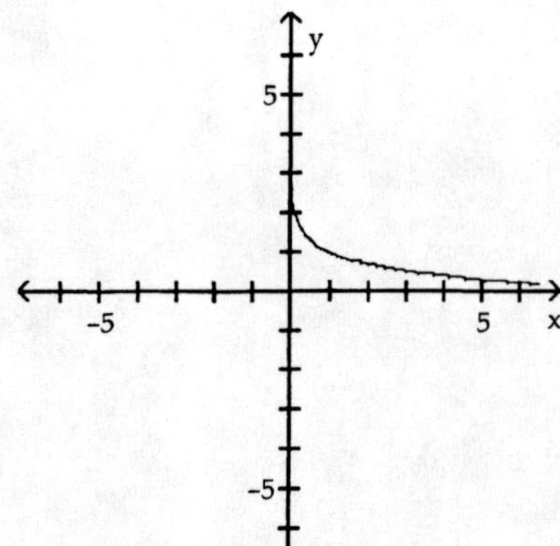

A) $y = 1 - \log(x)$ B) $y = \log(1 - x)$ C) $y = \log(x) - 1$ D) $y = \log(x - 1)$

9)

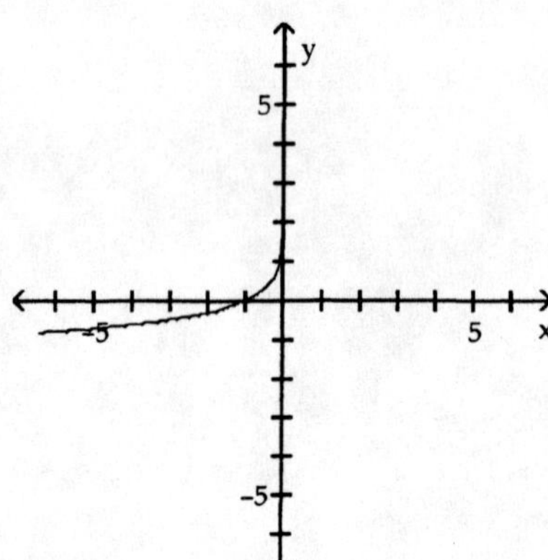

A) $y = \log(x)$ B) $y = -\log(x)$ C) $y = -\log(-x)$ D) $y = \log(-x)$

10)

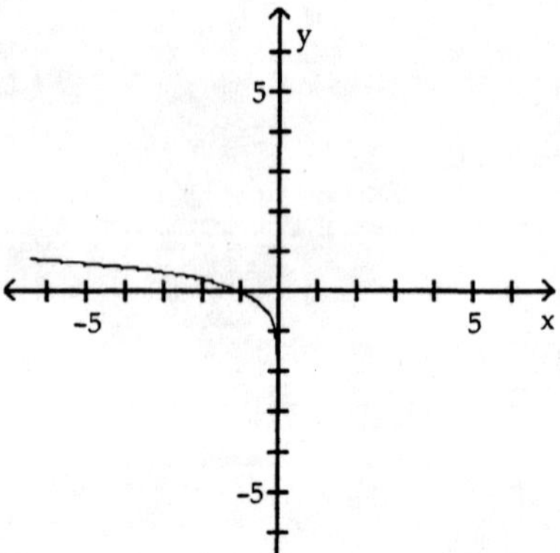

A) $y = \log(-x)$ B) $y = \log(x)$ C) $y = -\log(x)$ D) $y = -\log(-x)$

11)

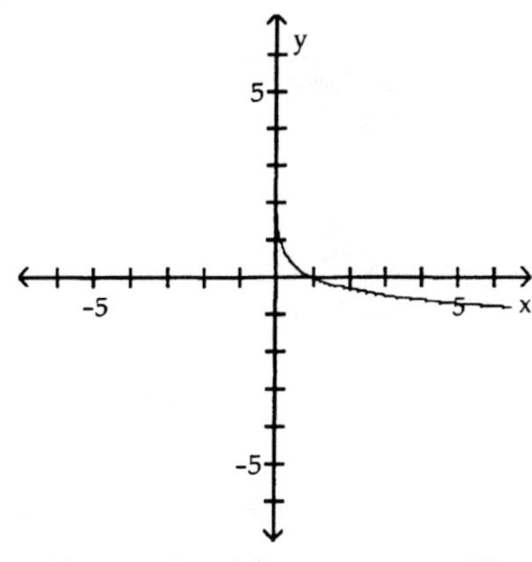

A) $y = -\log(x)$ B) $y = \log(x)$ C) $y = \log(-x)$ D) $y = -\log(-x)$

Use transformations to graph the function. Determine the domain, range, and vertical asymptote of the function.

12) $h(x) = -2 + \log(x + 0)$

6 Solve Logarithmic Equations

Solve the equation.

1) $\log_5 x = 3$

A) 8 B) 125 C) 15 D) 243

2) $\log_x \left(\frac{216}{125}\right) = 3$

A) 216 B) 3 C) $\frac{6}{5}$ D) $\frac{5}{6}$

3) $\log_5 25 = x$

A) 125 B) 30 C) 2 D) 5

Solve the problem.

4) The formula $D = 8e^{-0.04h}$ can be used to find the number of milligrams D of a certain drug in a patient's bloodstream h hours after the drug has been given. When the number of milligrams reaches 4, the drug is to be given again. What is the time between injections?

A) 17.33 hrs B) 51.99 hrs C) 34.66 hrs D) 18.71 hrs

The Richter scale converts seismographic readings into numbers for measuring the magnitude of an earthquake according to this function

$M(x) = \log\left(\frac{x}{x_0}\right)$, where $x_0 = 10^{-3}$.

5) What is the magnitude of an earthquake whose seismographic reading is 6.6 millimeters at a distance of 100 kilometers from its epicenter? Round the answer to the nearest tenth.

A) 3.8 B) 819.5 C) 2.5 D) 2.8

6) What is the magnitude of an earthquake whose seismographic reading is 7.6 millimeters at a distance of 100 kilometers from its epicenter?

A) 2.0281 B) 0.20281 C) 3.8808 D) 0.38808

7) What is the magnitude of an earthquake whose seismographic reading is 0.94 millimeters at a distance of 100 kilometers from its epicenter?

8) Find the magnitude (to one decimal place) of an earthquake whose seismographic reading is 2000 millimeters at a distance of 100 kilometers from its epicenter.

9) Two earthquakes differ by 0.1 when measured on the Richter scale. How would the seismographic readings differ at a distance of 100 kilometers from the epicenter?

The loudness of a sound of intensity x, measured in watts per square meter, is defined as $L(x) = \log\left(\frac{x}{x_0}\right)$, where $x_0 = 10^{-3}$.

10) A company with loud machinery needs to cut its sound intensity to 26% of its original level. By how many decibels should the loudness be reduced?

A) 6.237 decibels B) 6.880 decibels C) 5.760 decibels D) 5.850 decibels

11) A company with load machinery needs to cut its sound intensity to 72% of its original level. By how many decibels should the loudness be reduced?

12) A particular Boeing 747 jetliner produces noise at a loudness level of 113 decibels. Find the intensity level (round to the nearest hundredth) in watt per square meter for this noise.

13) At a recent Phish rock concert, sound intensity reached a level of 0.50 watt per square meter. To the nearest whole number, calculate the loudness of this sound in decibels.

14) At a rock concert by The Who, the music registered a loudness level of 120 decibels. The human threshold of pain due to sound averages 130 decibels. Compute the ratio of the intensities associated with these two loudness level to determine by how much the intensity of a sound that crosses the human threshold of pain exceeds that of this particular rock concert.

15) You have two friends, Jim and Amy. Jim always yells when he speaks, and Amy always whispers. The loudness of Jim's voice is 120 decibels, and the loudness of Amy's voice is 20 decibels. Determine how many times as intense Jim's voice is as compared at Amy's.

Solve the problem.

16) Between 7:00 AM and 8:00 AM, trains arrive at a subway station at a rate of 5 trains per hour (0.08 trains per minute). The following formula from statistics can be used to determine the probability that a train will arrive within t minutes of 7:00 AM.

$$F(t) = 1 - e^{-0.08t}$$

Determine how many minutes are needed for the probability to reach 30%.

A) 4.46 min B) 15.70 min C) 11.92 min D) 6.37 min

17) The concentration of alcohol in a person's blood is measurable. Suppose that the risk R (given as a percent) of having an accident while driving a car can be modeled by the equation

$$R = 6e^{kx}$$

where x is the variable concentration of alcohol in the blood and k is a constant.

(a) Suppose that a concentration of alcohol in the blood of 0.07 results in a 14% risk (R = 14) of an accident. Find the constant k in the equation.
(b) Using this value of k, what is the risk if the concentration is 0.014?
(c) Using the same value of k, what concentration of alcohol corresponds to a risk of 100%?
(d) If the law asserts that anyone with a risk of having an accident of 11 or more should not have driving privileges, at what concentration of alcohol in the blood should a driver be arrested and charged with a DUI?

5.5 Properties of Logarithms

1 Work with the Properties of Logarithms

Use the properties of logarithms to find the exact value of the expression. Do not use a calculator.

1) $\log_5 5^{-2}$

A) 1 B) -10 C) -2 D) 5

2) $\ln e^{\sqrt{7}}$

A) e B) 49 C) $\sqrt{7}$ D) 7

3) $\log_{30} 3 + \log_{30} 10$

A) 10 B) 1 C) 3 D) 30

4) $\log_4 36 - \log_4 9$

A) 4 B) 1 C) 9 D) 36

5) $\log_4 24 - \log_4 6$

A) 4 B) 24 C) 6 D) 1

6) $\log_4 18 \cdot \log_{18} 64$

A) 64 B) 4 C) 3 D) 18

7) $2 \ln e^{4.2}$

A) 8.4 B) $e^{8.4}$ C) 2.1 D) 4.2

Find the value of the expression.

8) Let $\log_b A = 5$ and $\log_b B = -2$. Find $\log_b AB$.

A) -10 B) 3 C) 7 D) 10

9) Let $\log_b A = 3$ and $\log_b B = -6$. Find $\log_b \frac{A}{B}$.

A) 9 B) $-\frac{1}{2}$ C) -3 D) $\frac{1}{2}$

10) Let $\log_b A = 3$ and $\log_b B = -5$. Find $\log_b B^2$.

A) −10 B) −25 C) 6 D) 25

11) Let $\log_b A = 2$ and $\log_b B = -5$. Find $\log_b \sqrt[2]{AB}$.

A) −3.162 B) $2\sqrt{-10}$ C) 3.162 D) -1.500

12) Let $\log_b A = 2.432$ and $\log_b B = 0.306$. Find $\log_b AB$.

A) 2.738 B) 2.126 C) 0.744 D) 7.953

13) Let $\log_b A = 2.262$ and $\log_b B = 0.275$. Find $\log_b \frac{A}{B}$.

A) 0.623 B) 2.537 C) 2.262 D) 1.987

Express y as a function of x. The constant C is a positive number.

14) $\ln y = \ln 4x + \ln C$

A) $y = 4x + C$ B) $y = x + 4C$ C) $y = 4Cx$ D) $y = (4x)^C$

15) $3\ln y = \frac{1}{3}\ln x - \ln \frac{x^2 - 1}{x^{4/3}} + \ln C$

A) $y = \sqrt[3]{\frac{Cx}{x^2 - 1}}$ B) $y = \sqrt[3]{\frac{C}{x^{1/3}(x^2 - 1)}}$ C) $y = \sqrt[3]{\frac{Cx^{5/3}}{x^2 - 1}}$ D) $y = \sqrt[3]{\frac{Cx^{1/3}}{x^2 - 1}}$

Using the properties of logarithms, find the exact value of the expression. Do not use a calculator.

16) $\log_2 2^{52.2}$

A) 524 B) 52.2 C) 2 D) 0.019

17) $3^{\log_3 4.86}$

A) 0.206 B) 4.86 C) 489 D) 3

2 Write a Logarithmic Expression as a Sum or Difference of Logarithms

Write as the sum and/or difference of logs. Express powers as factors.

1) $\log_{11} \frac{19}{12}$

A) $\log_{11} 19 \div \log_{11} 12$ B) $\log_{11} 19 - \log_{11} 12$

C) $\log_{11} 12 - \log_{11} 19$ D) $\log_{11} 19 + \log_{11} 19$

2) $\log_9 \frac{13\sqrt{m}}{n}$

A) $\log_9 (13\sqrt{m}) - \log_9 n$ B) $\log_9 13 + \frac{1}{2} \log_9 m - \log_9 n$

C) $\log_9 13 \cdot \frac{1}{2} \log_9 m \div \log_9 n$ D) $\log_9 n - \log_9 13 - \frac{1}{2} \log_9 m$

3) $\log_{17} \frac{\sqrt[9]{20}}{y^2 x}$

A) $\frac{1}{9} \log_{17} 20 - 2 \log_{17} y - \log_{17} x$

B) $\log_{17} 20 - \log_{17} y - \log_{17} x$

C) $\frac{1}{9} \log_{17} 20 - 2 \log_{17} y - 2 \log_{17} x$

D) $9 \log_{17} 20 - 2 \log_{17} y - \log_{17} 9$

4) $\log_{18} \sqrt{\frac{mn}{19}}$

A) $\frac{1}{2}\log_{18} m \cdot \frac{1}{2}\log_{18} n \div \frac{1}{2}\log_{18} 19$

B) $\frac{1}{2}\log_{18} m + \frac{1}{2}\log_{18} n - \frac{1}{2}\log_{18} 19$

C) $\frac{1}{2}\log_{18} m + \frac{1}{2}\log_{18} n - \log_{18} 19$

D) $\frac{1}{2}\log_{18} mn - \frac{1}{2}\log_{18} 19$

5) $\log_5 \frac{\sqrt[3]{x}\sqrt[4]{y}}{z^2}$

A) $\frac{1}{3}\log_5 x \cdot \frac{1}{4}\log_5 y \div 2\log_5 z$

B) $\frac{1}{3}\log_5 x + \frac{1}{4}\log_5 y - 2\log_5 z$

C) $3\log_5 x + 4\log_5 y - 2\log_5 z$

D) $\frac{3}{5}\log_5 x + \frac{4}{5}\log_5 y - \frac{2}{5}\log_5 z$

6) $\log_b \sqrt[3]{\frac{x^5 y^8}{z^2}}$

Solve the problem.

7) Which of the following is equivalent to $\log\left(1 - \frac{1}{x^3}\right)$?

A) $\log(x - 1) + \log(x^2 + x + 1) - 3 \log x$

B) $\log x^3 - \log 1 - 3 \log x$

C) $\log 1 - 3 \log 1 - 3 \log x$

D) $\log(x - 1) + \log(x^2 + 1) - 3 \log x$

Write as the sum and/or difference of logs. Express powers as factors.

8) $\ln \left(\frac{(x+8)(x-9)}{(x-5)^3}\right)^{3/2}$, $x > 9$

A) $\frac{3}{2}\ln(x+8) + \frac{3}{2}\ln(x-9) - \frac{9}{2}\ln(x-5)$

B) $\ln(x+8) + \ln(x-9) + \ln 3 - 9\ln(x-5) - \ln 2$

C) $3\ln(x+8) - 2\ln(x-9) - \frac{9}{2}\ln(x-5)$

D) $\frac{3}{2}\ln(x^2 + 17x - 72) - \frac{9}{2}\ln(x-5)$

9) $\ln \frac{(3x)\sqrt[9]{1+2x}}{(x-7)^7}$, $x > 7$

A) $\ln 3 + \ln x + \frac{1}{9}\ln(1+2x) - \ln 7 - \ln(x-7)$

B) $3\ln x + \frac{2}{9}\ln(1+2x) - 7\ln(x-7)$

C) $\ln 3 + \ln x - 9\ln(1+2x) - 7\ln(x-7)$

D) $\ln 3 + \ln x + \frac{1}{9}\ln(1+2x) - 7\ln(x-7)$

3 Write a Logarithmic Expression as a Single Logarithm

Express as a single logarithm.

1) $\log_c m + \log_c n$

A) $\log_c (mn)$ B) $\log_c mn$ C) $\log_c m \cdot \log_c n$ D) $\log_c \frac{m}{n}$

2) $3\log_b q - \log_b r$

A) $\log_b q^3 \div \log_b r$ B) $\log_b (q^3 - r)$ C) $\log_b \frac{3q}{r}$ D) $\log_b \frac{q^3}{r}$

3) $(\log_a x - \log_a y) + 2\log_a z$

A) $\log_a \frac{x}{z^2 y}$ B) $\log_a \frac{xz^2}{y}$ C) $\log_a \frac{2xz}{y}$ D) $\log_a xz^2 y$

4) $9\log_c 6 + 10\log_c 10$

A) $\log_c 6^9 \cdot \log_c 10^{10}$ B) $\log_c \frac{6^9}{10^{10}}$ C) $\log_c (54 + 100)$ D) $\log_c 6^9 10^{10}$

5) $6\log_a t - \frac{6}{5}\log_a s + \frac{1}{2}\log_a v - 5\log_a u$

A) $\log_a \frac{t^6\ s^{6/5}}{v^{1/2}\ u^5}$ B) $\log_a \frac{t^6 u^5}{v^{1/2}\ s^{6/5}}$

C) $\log_a (6t - \frac{6}{5}s + \frac{1}{2}v - 5u)$ D) $\log_a \frac{t^6\ v^{1/2}}{s^{6/5}\ u^5}$

6) $3\log_6 x + 5\log_6 (x - 6)$

A) $15 \log_6 x(x - 6)$ B) $\log_6 x(x - 6)^{15}$ C) $\log_6 x^3(x - 6)^5$ D) $\log_6 x(x - 6)$

7) $3 \log_a (2x + 1) - 2 \log_a (2x - 1) + 2$

A) $\log_a \frac{a^2(2x + 1)^3}{(2x - 1)^2}$ B) $\log_a 2(x + 1)$ C) $\log_a (2x + 3)$ D) $\log_a (2x + 1) + 2$

8) $\log_a 6x + 3(\log_a x - \log_a y)$

9) $\ln \frac{x^2 - 3x - 28}{x - 1} - \ln \frac{x^2 + 3x - 4}{x + 3} + \ln (x^2 - 14x + 49)$, $a > 0$

A) $\ln \frac{3(x - 7)(x + 3)}{2(x - 1)}$ B) $\ln \frac{(x - 7)^3(x + 3)}{(x - 1)^2}$ C) $\ln \frac{(x - 7)^3}{(x - 1)^2(x + 3)}$ D) $\ln \frac{3(x - 7)}{2(x - 1)(x + 3)}$

10) $12 \log_4 \sqrt[4]{x} + \log_4 (12x^6) - \log_4 12$

A) $\log_4 x^9$ B) $\log_4 x^{7/6}$ C) $\log_4 x^{10/3}$ D) $\log_4 x^{9/4}$

4 Evaluate Logarithms Whose Base Is Neither 10 nor e

Use the Change-of-Base Formula and a calculator to evaluate the logarithm. Round your answer to three decimal places.

1) $\log_3 77.30$

A) 25.767 B) 1.888 C) 3.957 D) 0.253

2) $\log_2 0.159$

A) -2.653 B) -0.799 C) 12.579 D) -0.377

3) $\log_{4.6} 248$

A) 53.913 B) 3.613 C) 2.394 D) 0.277

4) $\log_{7.8} 4.3$

A) 0.633 B) 0.710 C) 1.408 D) 0.551

5) $\log_{\sqrt{5}} 273.1$

A) 0.349 B) 0.143 C) 6.971 D) 3.486

Use natural logs to evaluate the logarithm to two decimal places.

6) $\log_{6.8} 94$

A) 1.97 B) 13.82 C) 0.42 D) 2.37

7) $\log_{5.4} 3.5$

A) 0.54 B) 1.35 C) 0.74 D) 0.65

8) $\log_{\sqrt{2}} 288.5$

A) 0.06 B) 0.15 C) 8.17 D) 16.34

9) Evaluate $\log_3 25$. Round your answer to the nearest hundredth.

A) 1.10 B) 0.34 C) 2.93 D) 3.22

10) Evaluate $\log_{(2/3)} 19$. Round your answer to the nearest hundredth.

Solve the problem.

11) Find the value of $\log_3 4 \cdot \log_4 5 \cdot \log_5 6 \cdot \log_6 7 \cdot \log_7 8 \cdot \log_8 9$

Graph the function using a graphing utility and the Change-of-Base Formula.

12) $\log_{x-5}(x+5)$

A)

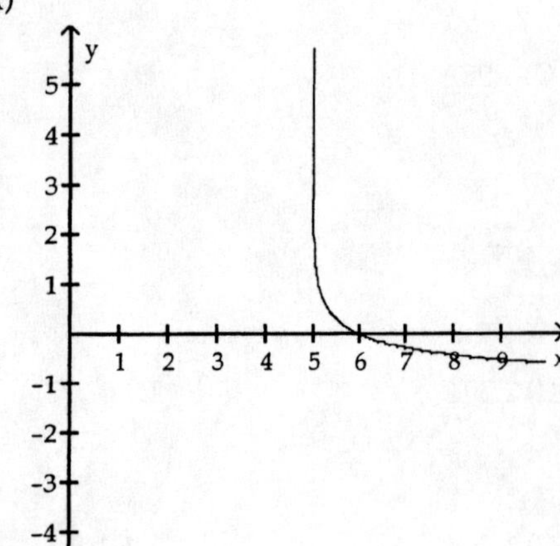

B)

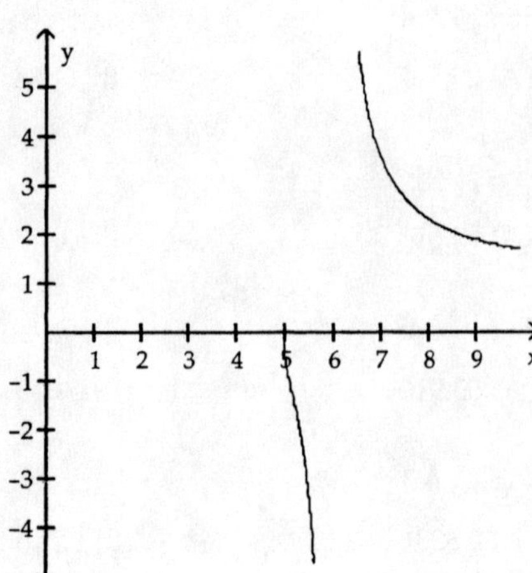

C)

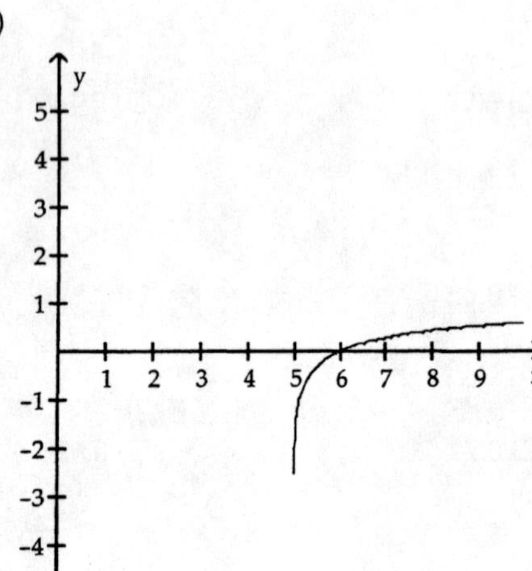

D)

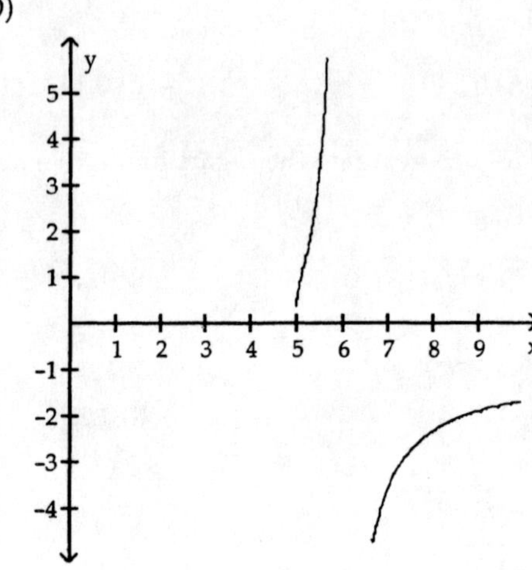

5.6 Logarithmic and Exponential Equations

1 Solve Logarithmic Equations Using the Properties of Logarithms

Solve the equation.

1) $\log(x+2) = \log(5x-3)$

A) $\frac{5}{3}$ B) $\frac{1}{4}$ C) -1.25 D) 1.25

2) $\log_3 x = 4$

A) 1.26 B) 81 C) 12 D) 64

3) $\log_y 13 = 2$

A) $\frac{13}{2}$ B) $2^{1/13}$ C) 13^2 D) $13^{1/2}$

4) $\log(3+x) - \log(x-2) = \log 2$

A) -7 B) $\varnothing$ C) $\frac{1}{2}$ D) 7

5) $\log_6 (2x + 8) = \log_6 (2x + 3)$

A) 5 B) $\frac{11}{5}$ C) $\varnothing$ D) 0

6) $\frac{1}{3} \log_2 (x + 6) = \log_8 3x$

A) 9 B) 3 C) 3, 0 D) $\varnothing$

7) Find all real solutions of the following equation.
$\log_3 x + \log_3(x - 24) = 4$

A) x = 27 B) x = 53 C) $x = -3, 27$ D) no real solutions

Solve the problem.

8) Solve: $\log_a(x - 7) - \log_a(x - 2) = \log_a(x - 4) - \log_a(x + 12)$

9) Solve for x correct to two decimal places: $\log x + \log 6 = 0.3031$

Solve the equation. Express irrational answers in exact form and as a decimal rounded to 3 decimal places.

10) $\ln x + \ln (x + 6) = 5$

A) $-6 + \sqrt{36 + 4e^5} \approx 19.093$ B) $\frac{-6 + 2\sqrt{36 + e^5}}{2} \approx 10.580$

C) $\frac{-6 + \sqrt{36 + 4e^5}}{2} \approx 9.546$ D) $\frac{-6 - \sqrt{36 + 4e^5}}{2} \approx -15.546$

2 Solve Exponential Equations

Solve the equation.

1) $3^x = 81$

A) 3 B) 5 C) 27 D) 4

2) $3^{(1 + 2x)} = 27$

A) 3 B) -1 C) 9 D) 1

3) $4^{(5 - 3x)} = \frac{1}{256}$

A) 128 B) -3 C) 3 D) $\frac{1}{64}$

4) $3^{(6 + 3x)} = \frac{1}{27}$

A) 9 B) 3 C) $\frac{1}{9}$ D) -3

5) $(\frac{1}{4})^x = 10$

A) -2.50 B) 1.66 C) -0.60 D) -1.66

6) $3^{(x - 1)} = 19$

A) 2.81 B) 1.68 C) 3.68 D) 7.33

7) $2^{(3x - 1)} = 18$

A) 0.23 B) 1.72 C) 1.06 D) 3.33

8) $3^{2x} + 3^x - 6 = 0$

Solve the problem.

9) Solve the equation $4^{x+4} = 5^{2x+5}$ and express the answer in terms of natural logarithms.

A) $x = 7 \ln 5 - 5 \ln 4$ B) $x = \ln\left[\frac{5^5}{4^4} - \frac{4}{5^2}\right]$ C) $x = \frac{5 \ln 5 - 4 \ln 4}{\ln 4 - 2 \ln 5}$ D) $x = \ln 5 - \ln 4$

10) Solve: $9^{7x+3} = 27$

11) Solve for x correct to four decimal places: $7^x = 6^{x+7}$

A) 40.6820 B) 81.3640 C) 42.0000 D) 3.3556

12) Solve $\pi^{x+1} = e^{2x}$ and express the answer interms of natural logarithms.

13) Solve $\frac{e^x - e^{-x}}{2} = -1$ and express the answer interms of natural logarithms.

Solve the equation. Express irrational answers in exact form and as a decimal rounded to 3 decimal places.

14) $\left(\frac{9}{2}\right)^x = 3^{1 - x}$

A) $\frac{\ln 3}{\ln\left(\frac{9}{2}\right) + \ln 3} \approx 0.422$ B) $\frac{\ln\left(\frac{9}{2}\right) + \ln 3}{\ln 3} \approx 2.369$

C) $\ln\left(\frac{9}{2}\right) - \ln 3 \approx 0.405$ D) $\frac{\ln 27}{\ln 6} \approx 1.839$

3 Solve Logarithmic and Exponential Equations Using a Graphing Utility

Use a graphing calculator to solve the equation. Round your answer to two decimal places.

1) $\log_5 x + \log_2 x = 3$

A) 0.63 B) 4.28 C) 1.62 D) 5.58

2) $\log_4(x + 2) - \log_5(x - 1) = 1$

A) -0.69 B) 1.75 C) 1.98 D) 2.05

3) $e^x = -x$

A) -1.05 B) -0.57 C) 1.05 D) 0.57

4) $e^x - \ln x = 4$

A) 0.57 B) 2.17 C) 1.27 D) 1.48

5) $e^x = x^3$

A) 2.54 B) 1.86 C) -0.79 D) -0.71

6) $e^x = x^2 - 1$

A) -1.15 B) 2.54 C) 0 D) -0.71

5.7 Compound Interest

1 Determine the Future Value of a Lump Sum of Money

Compute the amount in m years if a principal P is invested at a nominal annual interest rate of r compounded as given. Round to the nearest cent.

1) P = $1,000, m = 4, r = 6% compounded annually

A) $1191.02 B) $1262.48 C) $1338.23 D) $262.48

2) P = $1,000, m = 3, r = 14% compounded semiannually

A) $1402.55 B) $1500.73 C) $500.73 D) $1481.54

3) P = $14,000, m = 5, r = 12% compounded semiannually

A) $23,652.71 B) $25,071.87 C) $24,672.78 D) $11,071.87

4) P = $480, m = 2, r = 9% compounded quarterly

A) $560.90 B) $573.52 C) $93.52 D) $570.29

5) P = $12,000, m = 9, r = 12% compounded quarterly

A) $34,779.34 B) $33,766.35 C) $22,779.34 D) $33,276.95

Solve the problem.

6) Find the amount owed at the end of 8 years if $5000 is loaned at a rate of 5% compounded monthly.

A) $7452.93 B) $12,911.25 C) $9093.60 D) $8060.16

7) If $5,000 is invested for 6 years at 5%, compounded continuously, find the future value.

8) John Forgetsalot deposited $100 at a 3% annual interest rate in a savings account fifty years ago, and then he promptly forgot he had done it. Recently, he was cleaning out his home office and discovered the forgotten bank book. How much money is in the account?

9) Meike earned $1565 in tips while working a summer job at a coffee shop. She wants to use this money to take a trip to Europe next summer. If she places the money in an account which pays 6.5% compounded continuously, how much money will she have in nine months?

10) Carla has just inherited a building that is worth $250,000. The building is in a high demand area, and the value of the building is projected to increase at a rate of 25% per year for the next 4 years. How much more money will she make if she waits four years to sell the building instead of selling now?

2 Calculate Effective Rates of Return

Find the effective rate of interest.

1) 8% compounded continuously

A) 8.451% B) 8.089% C) 8.374% D) 8.329%

2) 7.2% compounded continuously

A) 7.289% B) 7.574% C) 7.466% D) 7.651%

3) 50.12% compounded daily

A) 65.013% B) 50.325% C) 50.243% D) 51.025%

4) 11.75% compounded quarterly

A) 12.655% B) 11.955% C) 11.873% D) 12.278%

5) 4.75% compounded monthly

A) 4.955% B) 4.855% C) 4.873% D) 5.655%

Solve the problem.

6) Find the effective rate of interest for $6\frac{1}{4}\%$ compounded monthly.

A) 6.39% B) 6.29% C) 6.43% D) 6.25%

7) Find the effective rate of interest for $4\frac{3}{4}\%$ compounded quarterly.

3 Determine the Present Value of a Lump Sum of Money

Find the present value required to achieve the amount A when compounded at a rate of r for t years. Round to the nearest cent.

1) A = $5,600, t = 5, r = 6% compounded annually

A) $1415.35 B) $4435.72 C) $7472.58 D) $4184.65

2) A = $10,500, t = 15, r = 5% compounded annually

A) $5050.68 B) $4810.17 C) $5303.21 D) $5449.32

3) A = $2,000, t = 5, r = 5% compounded semiannually

A) $437.6 B) $1601.46 C) $1567.05 D) $1562.40

4) A = $25,000, t = 4, r = 10% compounded semiannually

A) $8079.02 B) $17,767.03 C) $16,920.98 D) $17,075.34

5) A = $6,500, t = 8, r = 4% compounded quarterly

A) $4749.49 B) $4727.48 C) $1772.52 D) $4774.75

6) A = $10,000, t = 2, r = 6% compounded monthly

A) $10,616.78 B) $5000.00 C) $9419.05 D) $8871.86

Solve the problem.

7) What principal invested at 8% compounded continuously for 4 years will yield \$1190? Round the answer to two decimal places.

A) \$627.48 B) \$1188.62 C) \$864.12 D) \$1638.78

8) What principal invested at 6%, compounded continuously for 3 years, will yield \$1500? Round the answer to two decimal places.

4 Determine the Time Required to Double or Triple a Lump Sum of Money

Solve the problem. Round your answer to three decimals.

1) What annual rate of interest is required to double an investment in 5 years?

A) 14.87% B) 24.573% C) 13.863% D) 7.435%

2) What annual rate of interest is required to triple an investment in 8 years?

A) 7.36% B) 9.051% C) 14.72% D) 13.733%

3) How long will it take for an investment to double in value if it earns 12.25% compounded continuously?

A) 5.658 years B) 2.829 years C) 5.821 years D) 8.968 years

4) How long will it take for an investment to triple in value if it earns 5.75% compounded continuously?

A) 19.106 years B) 12.055 years C) 9.553 years D) 21.054 years

Solve the problem.

5) How long does it take \$1125 to triple if it is invested at 7% interest, compounded quarterly? Round your answer to the nearest tenth.

A) 18.1 years B) 15.8 years C) 18.1 months D) 15.8 months

6) How long does it take \$1700 to double if it is invested at 5% interest, compounded monthly? Round your answer to the nearest tenth.

7) Gillian has \$10,000 to invest in a mutual fund. The average annual rate of return for the past five years was 12.25%. Assuming this rate, determine how long it will take for her investment to double.

5 Additional Applications

Solve the problem.

1) If Emery has \$2600 to invest at 11% per year compounded monthly, how long will it be before he has \$3500? If the compounding is continuous, how long will it be? (Round your answers to three decimal places.)

A) 55.286 yrs, 2.884 yrs B) 3.391 yrs, 0.27 yrs
C) 0.237 yrs, 0.225 yrs D) 2.715 yrs, 2.702 yrs

2) Cindy will require \$16,000 in 3 years to return to college to get an MBA degree. How much money should she ask her parents for now so that, if she invests it at 9% compounded continuously, she will have enough for school? (Round your answer to the nearest dollar.)

A) \$12,355 B) \$12,214 C) \$20,959 D) \$9324

3) Tracey bought a diamond ring appraised at \$1400 at an antique store. If diamonds have appreciated in value at an annual rate of 10%, what was the value of the ring 9 years ago? (Round your answer to the nearest dollar.)

A) \$3301 B) \$231 C) \$594 D) \$569

4) The Feldmans bought their first house for $13,000. Over the years they moved three times into bigger and bigger houses. Now, 42 years later, they are ready to retire and want a smaller house like the first one they bought. If inflation in property values has averaged 3.7% per year during that time, how much will such a house cost them now? (Round your answer to the nearest dollar.)

A) $2748 B) $61,495 C) $59,793 D) $2826

5) Larry has $1400 to invest and needs $2000 in 11 years. What annual rate of return will he need to get in order to accomplish his goal? (Round your answer to two decimals.)

A) 2.57% B) 3.57% C) 3.24% D) 1.57%

6) A venture capital firm invested $2,000,000 in a new company in 1995. In 1999, they sold their stake in the company for $10,500,000. What was the average annual rate of return on their investment?

7) Julio figures that he can save $5000 per year. If, at the end of each year, he invests the money in a certificate of deposit (CD) which pays 7% interest annually, how much money will he have saved in six years?

5.8 Exponential Growth and Decay; Newton's Law; Logistic Models

1 Find Equations of Populations That Obey the Law of Uninhibited Growth

Solve the problem.

1) The size P of a small herbivore population at time t (in years) obeys the function $P(t) = 900e^{0.18t}$ if they have enough food and the predator population stays constant. After how many years will the population reach 1800?

A) 3.85 yrs B) 9.41 yrs C) 37.79 yrs D) 12.21 yrs

2) Conservationists tagged 50 black-nosed rabbits in a national forest in 1990. In 1993, they tagged 100 black-nosed rabbits in the same range. If the rabbit population follows the exponential law, how many rabbits will be in the range 8 years from 1990?

A) 635 B) 130 C) 317 D) 65

3) The revenue for a dot.com company is projected to double each year for the first 5 years. If the revenue for the first year is $2 million, write a function showing the revenue R after x years. What is the revenue for the fourth year?

4) In a networking marketing plan for a company, each distributor is expected to recruit 3 new distributors. Jack was the first distributor hired by the company, so he is considered a level 1 distributor. The 3 people he recruits are considered level 2 distributors. The people recruited by the level 2 distributors are considered level 3 distributors, and so on. Write a function that models the number of distributors D at each level L. How many distributors would there be at the fifth level?

5) The bacteria in a container quadruples every day. If there are initially 100 bacteria, write an equation that models the number of bacteria A after d days. How many bacteria will there be after 1 week?

6) The concentration of alcohol in a person's blood is measurable. Suppose that the risk R (given as a percent) of having an accident while driving a car can be modeled by the equation

$$R = 5e^{kx}$$

where x is the variable concentration of alcohol in the blood and k is a constant.

Suppose that a concentration of alcohol in the blood of 0.07 results in a 10% risk (R = 10) of an accident. Find the constant k in the equation.

Using this value of k, what is the risk if the concentration is 0.11?

7) During 1991, 200,000 people visited Rave Amusement Park. During 1997, the number had grown to 834,000. If the number of visitors to the park obeys the law of uninhibited growth, find the exponential growth function that models this data.

A) $f(t) = 634{,}000e^{0.248t}$ B) $f(t) = 200{,}000e^{0.238t}$ C) $f(t) = 634{,}000e^{0.238t}$ D) $f(t) = 200{,}000e^{0.248t}$

8) A culture of bacteria obeys the law of uninhibited growth. If 140,000 bacteria are present initially and there are 609,000 after 6 hours, how long will it take for the population to reach one million?

9) The size P of a certain insect population at time t (in days) obeys the function $P = 700e^{0.03t}$. After how many days will the population reach 1500?

10) Bob, the incredible shrinking man, loses half of his height each day after he was exposed to a mysterious form of cosmic radiation. How many days before he is literally "knee-high to a grasshopper"? Assume that a grasshopper's knee is 4 millimeters high and that Bob is 2 meters tall. Round your answer to the nearest whole day. (1000 millimeters = 1 meter)

2 Find Equations of Populations That Obey the Law of Decay

Solve the problem.

1) The half-life of silicon-32 is 710 years. If 70 grams is present now, how much will be present in 400 years? (Round your answer to three decimal places.)

A) 0 B) 1.41 C) 47.37 D) 67.319

2) The half-life of plutonium-234 is 9 hours. If 20 milligrams is present now, how much will be present in 3 days? (Round your answer to three decimal places.)

A) 11.487 B) 1.984 C) 15.874 D) 0.078

3) A fossilized leaf contains 16% of its normal amount of carbon 14. How old is the fossil (to the nearest year)? Use 5600 years as the half-life of carbon 14.

A) 1406 B) 22,360 C) 35,732 D) 14,779

4) If a single pane of glass obliterates 15% of the light passing through it, then the percent P of light that passes through n successive panes can be approximated by the equation

$$P = 100e^{-0.15n}$$

How many panes are necessary to block at least 50% of the light?

5) The half-life of a radioactive element is 130 days, but your sample will not be useful to you after 80% of the radioactive nuclei originally present have disintegrated. About how many days can you use the sample?

A) 302 B) 287 C) 297 D) 312

6) The half-life of carbon-14 is 5700 years. Find the age of a sample in which 8% of the radioactive nuclei originally present have decayed.

7) The half-life of radium is 1690 years. If 150 grams is present now, how long (to the nearest year) till only 100 grams are present?

8) Assume that the half-life of Carbon-14 is 5700 years. Find the age (to the nearest year) of a wooden axe in which the amount of Carbon-14 is 30% of what it originally had.

9) The formula

$$D = 8e^{-0.6h}$$

can be used to find the number of milligrams D of a certain drug that is in a patient's bloodstream h hours after the drug has been administered. The drug is to be administered again when the amount in the bloodstream reaches 4 milligrams. What is the time between injections?

10) Between 8:30 a.m. and 9:30 a.m., cars drive through the Cappuccino Express at a rate of 12 cars per hour (0.2 per minute). The following formula from probability can be used to determine the probability that a car will arrive within t minutes of 8:30a.m.

$$F(t) = 1 - e^{-0.2t}$$

Determine how many minutes are needed for the probability to reach 0.6.

11) A rumor is spread at an elementary school with 1200 students according to the model $N = 1200(1 - e^{-0.16d})$ where N is the number of students who have heard the rumor and d is the number of days that have elapsed since the rumor began. How many days must elapse for 500 to have heard the rumor?

3 Use Newton's Law of Cooling

Solve the problem.

1) Sandy manages a ceramics shop and uses a 650°F kiln to fire ceramic greenware. After turning off her kiln, she must wait until its temperature gauge reaches 190°F before opening it and removing the ceramic pieces. If room temperature is 75°F and the gauge reads 500°F in 9 minutes, how long must she wait before opening the kiln? Assume the kiln cools according to Newton's Law of Cooling:

$$U = T + (U_0 - T)e^{kt}.$$

(Round your answer to the nearest whole minute.)

A) 104 minutes B) 71 minutes C) 48 minutes D) 37 minutes

2) A thermometer reading 84°F is placed inside a cold storage room with a constant temperature of 38°F. If the thermometer reads 75°F in 11 minutes, how long before it reaches 59°F? Assume the cooling follows Newton's Law of Cooling:

$$U = T + (U_0 - T)e^{kt}.$$

(Round your answer to the nearest whole minute.)

A) 6 minutes B) -18 minutes C) 19 minutes D) 40 minutes

3) A thermometer reading 8°C is brought into a room with a constant temperature of 25°C. If the thermometer reads 17°C after 4 minutes, what will it read after being in the room for 8 minutes? Assume the cooling follows Newton's Law of Cooling:

$$U = T + (U_0 - T)e^{kt}.$$

(Round your answer to two decimal places.)

A) 21.24°C B) -2.54°C C) 24.96°C D) 28.76°C

4) A thermometer reading 34°F is brought into a room with a constant temperature of 70°F. If the thermometer reads 43°F after 3 minutes, what will it read after being in the room for 10 minutes? Assume the cooling follows Newton's Law of Cooling:

$$U = T + (U_0 - T)e^{kt}.$$

(Round your answer to two decimal places.)

A) 83.8°F B) 67.97°F C) 22.53°F D) 56.2°F

5) A cup of coffee is heated to 194°and is then allowed to cool in a room whose air temperature is 72°. After 11 minutes, the temperature of the cup of coffee is 140°. Find the time needed for the coffee to cool to a temperature of 102°. Assume the cooling follows Newton's Law of Cooling:

$$U = T + (U_0 - T)e^{kt}.$$

(Round your answer to one decimal place.)

A) 26.4 minutes B) 41.1 minutes C) 15.1 minutes D) 29.7 minutes

6) A thermometer is taken from a room at 71°F to the outdoors where the temperature is 14°F. Determine what the reading on the thermometer will be after 5 minutes, if the reading drops to 45°F after 1 minute. Assume the cooling follows Newton's Law of Cooling:

$$U = T + (U_0 - T)e^{kt}.$$

(Round your answer to two decimal places.)

7) The temperature (in degrees Fahrenheit) of a dead body that has been cooling in a room set at 70° is measured as 88°. One hour later, the body temperature is 87.5°. How long (to the nearest hour) before the first measurement was the time of death, assuming that the body temperature of the deceased at the time of death was 98.6°. Assume the cooling follows Newton's Law of Cooling:

$$U = T + (U_0 - T)e^{kt}.$$

8) A fully cooked turkey is taken out of an oven set at 200°C (Celsius) and placed in a sink of chilled water of temperature 4°C. After 3 minutes, the temperature of the turkey is measured to be 50°C. How long (to the nearest minute) will it take for the temperature of the turkey to reach 15°C? Assume the cooling follows Newton's Law of Cooling:

$$U = T + (U_0 - T)e^{kt}.$$

(Round your answer to the nearest minute.)

4 Use Logistic Models

Solve the problem.

1) The logistic growth model $P(t) = \dfrac{1450}{1 + 31.22e^{-0.317t}}$ represents the population of a bacterium in a culture tube after t hours. What was the initial amount of bacteria in the population?

A) 50 B) 45 C) 46 D) 44

2) The logistic growth model $P(t) = \dfrac{940}{1 + 30.33e^{-0.344t}}$ represents the population of a bacterium in a culture tube after t hours. When will the amount of bacteria be 680?

A) 8.98 hours B) 6.9 hours C) 3.16 hours D) 12.71 hours

3) The logistic growth model $P(t) = \dfrac{180}{1 + 44e^{-0.158t}}$ represents the population of a species introduced into a new territory after t years. When will the population be 70?

A) 5.31 years B) 17.97 years C) 8.43 years D) 21.09 years

4) The logistic growth model $P(t) = \dfrac{400}{1 + 39e^{-0.153t}}$ represents the population of a species introduced into a new territory after t years. What will the population be in 30 years?

A) 400 B) 299 C) 1010 D) 287

5) The logistic growth model $P(t) = \dfrac{1}{1 + 19e^{-0.848t}}$ represents the proportion of the total market of a new product as it penetrates the market t years after introduction. When will the product have 80% of the market?

A) 3.21 years B) 5.11 years C) 4.21 years D) 6.11 years

6) In 1990, the population of a country was estimated at 4 million. For any subsequent year the population, P(t) (in millions), can be modeled by the equation $P(t) = \dfrac{240}{5 + 54.99e^{-0.0208t}}$, where t is the number of years since 1990. Estimate the year when the population will be 21 million.

A) approximately the year 2093 B) approximately the year 2016
C) approximately the year 2088 D) approximately the year 2041

7) In 1992, the population of a country was estimated at 5 million. For any subsequent year, the population, P(t) (in millions), can be modeled using the equation $P(t) = \dfrac{250}{5 + 44.99e^{-0.0208t}}$, where t is the number of years since 1992. Determine the year when the population will be 39 million.

5.9 Fitting Data to Exponential, Logarithmic, and Logistic Functions

1 Use a Graphing Utility to Fit an Exponential Function to Data

Solve the problem.

1) The population (in hundred thousands) for the Colonial United States in ten-year increments for the years 1700–1780 is given in the table. (Source: 1998 Information Please Almanac)

Decade	Population	Decade	Population
0	251	5	1171
1	332	6	1594
2	466	7	2148
3	629	8	2780
4	906		

State whether the data can be more accurately modeled using an exponential function or a logarithmic function. Using a graphing utility, find a model for population (in hundred thousands) as a function of decades since 1700.

2) A biologist has a bacteria sample. She records the amount of bacteria every week for 8 weeks and finds that the exponential function of best fit to the data is $A = 150 \cdot 1.79^{t}$. Express the function of best fit in the form $A = A_0e^{kt}$.

3) A music store manager collected data regarding price and quantity demanded of cassette tapes every week for 10 weeks, and found that the exponential function of best fit to the data was $p = 25 \cdot 0.89^{q}$. Express the function of best fit in the form $p = p_0e^{kq}$, and use this expression to predict the quantity demanded if the price is $8.50.

4) A life insurance company uses the following rate table for annual premiums for women for term life insurance. Use a graphing utility to fit an exponential function to the data. Predict the annual premium for a woman aged 70 years.

Age	35	40	45	50	55	60	65
Premium	$103	$133	$190	$255	$360	$503	$818

A) $y = 0.0000398x^{4.06}$, $1233
B) $y = -9306.4 + 2516.3 \ln (x)$, $1723
C) $y = 8.94e^{0.068x}$, $1044
D) $y = 6.367e^{0.068x}$, $743

5) Money magazine reports that the percentage of trading days in which the Dasdaq loses or gains 2% or more has been increasing since 1995 indicating more volatility in the Dasdaq. Use a graphing utility to fit an exponential function to the data. Predict the percentage of trading days in 2001 having such swings in value.

Year	% of trading days
1995, 0	2
1996, 1	5
1997, 2	8
1998, 3	18
1999, 4	23
2000, 5	49

A) $y = 2.353e^{0.610x}$, 91 days
B) $y = 1.669x^{1.706}$, 46 days
C) $y = 1.849e^{0.531x}$, 76 days
D) $y = 1.278e^{0.611x}$, 92 days

6) A nuclear scientist has a sample of 100 mg of a radioactive material which has a half-life in hours. She monitors the amount of radioactive material over a period of a day and obtains the following data. Use a graphing utility to fit an exponential function to the data. Predict the amount of material remaining at 40 hours.

Hours	0	5	10	15	20	25	30
mg	100	68.3	45.2	31.3	21.5	14.6	9.8

A) $y = 92e^{-0.0686x}$, 5.9 mg
B) $y = 100e^{-0.077x}$, 6.7 mg
C) $y = 100e^{-0.077x}$, 4.6 mg
D) $y = 86e^{-0.071x}$, 5.0 mg

2 Use a Graphing Utility to Fit a Logarithmic Function to Data

Solve the problem.

1) Data representing the price and quantity demanded for hand-held electronic organizers were analyzed every day for 15 days. The logarithmic function of best fit to the data was found to be $p = 398 - 73 \ln(q)$. Use this to predict the number of hand-held electronic organizers that would be demanded if the price were $275.

2) The rates of death (in number of deaths per 100,000 population) for 1–4 year olds in the United States between 1980–1995 are given below. (Source: NCHS Data Warehouse)

Year	Rate of Death
1980	91.4
1985	74.5
1990	69.3
1995	61.3

A logarithmic equation that models this data is $y = 822.99 - 167.55 \ln x$ where x represents the number of years since 1900. Use this equation to predict the rate of death for 1–4 year olds in 2005.

3) The rates of death (in number of deaths per 100,000 population) for 20–24 year olds in the United States between 1985–1993 are given below. (Source: NCHS Data Warehouse)

Year	Rate of Death
1985	134.9
1987	154.7
1989	162.9
1991	174.5
1993	182.2

A logarithmic equation that models this data is $y = 57.76 + 48.56 \ln x$ where x represents the number of years since 1980 and y represents the rate of death in that year. Use this equation to predict the year in which the rate of death for 20–24 year olds first exceeds 200.

4) After introducing an inhibitor into a culture of luminescent bacteria, a scientist monitors the luminosity produced by the culture. Use a graphing utility to fit a logarithmic function to the data. Predict the luminosity after 20 hours.

Time, hrs	2	3	4	5	8	10	15
Luminosity	77.4	60.8	54.5	45.8	30.0	24.3	10.5

A) $y = 100.5 - 32.7 \ln(x)$, 2.54
B) $y = 112.97 - 45.97 \ln(x)$, −24.74
C) $y = 107.55 - 41 \ln(x)$, −15.27
D) $y = 98.75 - 32.66 \ln(x)$, 0.91

5) In a Psychology class, the students were tested at the end of the course on a final exam. Then they were retested with an equivalent test at subsequent time intervals. Their average scores after t months are given in the table.

Time, t (in months)	1	2	3	4	5
Score, y (in percentage)	86.2	85.7	85.4	85.2	85.0

Using a graphing utility, fit a logarithmic function $y = a + b \ln x$ to the data. Using the function you found, estimate how long will it take for the test scores to fall below 84%. Express your answer to the nearest month.

A) 8 months
B) 10 months
C) 20 months
D) 12 months

Solve the problem.

1) A mechanic is testing the cooling system of a boat engine. He measures the engine's temperature over time. Use a graphing utility to fit a logistic function to the data. What is the carrying capacity of the cooling system?

time, min	5	10	15	20	25
temperature, °F	100	180	270	300	305

A) $y = \dfrac{314.79}{1 + 7.86e^{-0.246x}}$, 315°F

B) $y = \dfrac{311.63}{1 + 8.1e^{-0.253x}}$, 312°F

C) $y = \dfrac{314.79}{1 + 7.86e^{-1.22x}}$, 315°F

D) $y = \dfrac{306.53}{1 + 7.92e^{-0.254x}}$, 307°F

Ch. 5 Exponential and Logarithmic Functions
Answer Key

5.1 Composite Functions

1 Form the Composite Function and Find Its Domain

1) B
2) B
3) A
4) C
5) C
6) A
7) B
8) B
9) D
10) C
11) D
12) B
13) B
14) D
15) B
16) D
17) C
18) C
19) B
20) C
21) D
22) D
23) $(f \circ g)(x) = (-1 + 5x)^2$; Domain: all real numbers
24) A
25) C
26) A
27) B
28) C
29) D
30) B
31) C
32) A
33) D
34) C
35) D
36) D
37) $S(t) = \frac{1}{18}\pi t^4 + 4\pi t^2$
38) $V(t) = \frac{1}{3}\pi t^4$
39) $C = \frac{\sqrt{1200 - 6p}}{40} + 800$
40) 8 or -8

5.2 Inverse Functions

1 Determine the Inverse of a Function

1) C
2) A
3) A

4) {62, 66, 68}; no
5) ordered pairs may vary, but each should have x–coordinate 5; no
6) B
7) B
8) B
9) A
10) A
11) B
12) Inverses. Check to see that $f(g(x)) = x$ and that $g(f(x)) = x$.
13) A
14) A
15) A
16) A
17) Domain: {all bar codes}, Range: {all item prices}; yes; no
18) Domain: {all item prices}, Range: {all bar codes}; no

2 Obtain the Graph of the Inverse Function from the Graph of the Function

1) A
2) C
3) C
4) D
5) B
6) A

3 Find the Inverse Function f^{-1}

1) D
2) A
3) C
4) C
5) $x = \frac{1}{2}P + 250$
6) (a) $x \ge 0$ is one correct answer; another equally correct answer is $x \le 0$.
   (b) For the case $x \ge 0$, the inverse is $f^{-1}(x) = x + 5$.
   For the case $x \le 0$, the inverse is $f^{-1}(x) = -x - 5$.
7) (a) $V(W) = (\frac{W - 1.21}{3.38})^3$
   (b) 0.56

5.3 Exponential Functions

1 Evaluate Exponential Functions

1) B
2) A
3) B
4) A
5) C
6) D
7) D
8) C

2 Graph Exponential Functions

1) A
2) B
3) D
4) D
5) C
6) C
7) D
8) D

9) B
10) A
11) C

3 Define the Number e

1) D
2) A
3) D
4) B
5) D
6) C
7) B
8) C
9) B
10) 14.18 watts
11) 4 mg; 2 mg
12) D

4 Solve Exponential Equations

1) D
2) C
3) D
4) A
5) C
6) C
7) 661 students
8) C

5.4 Logarithmic Functions

1 Change Exponential Expressions to Logarithmic Expressions

1) C
2) B
3) A
4) C
5) D
6) A
7) C
8) B
9) A
10) $\log_{16} 64 = \frac{3}{2}$

2 Change Logarithmic Expressions to Exponential Expressions

1) C
2) D
3) B
4) D
5) A
6) D
7) A
8) C
9) C
10) C
11) $61^y = x$

3 Evaluate Logarithmic Functions

1) C
2) D
3) C

4) B
5) D
6) D
7) A
8) C
9) D
10) B
11) B
12) C
13) B
14) B
15) D
16) C
17) D
18) C
19) A
20) A

4 Determine the Domain of a Logarithmic Function

1) A
2) D
3) B
4) A
5) A
6) B
7) C
8) A
9) B
10) $(-2, \infty)$
11) C

5 Graph Logarithmic Functions

1) D
2) C
3) A
4) D
5) D
6) B
7) D
8) A
9) C
10) A
11) A

12)

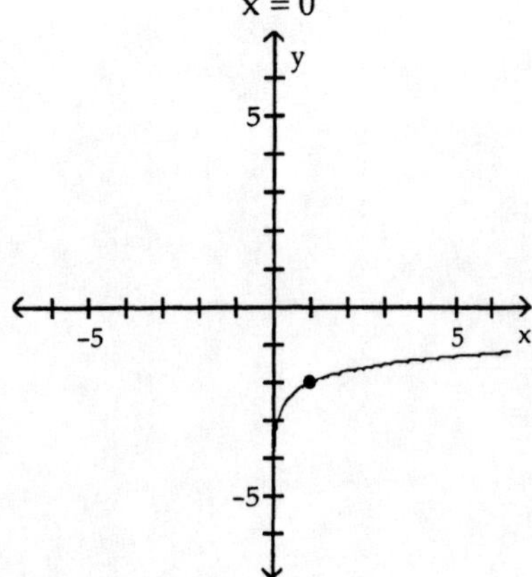

domain $(0, \infty)$
range $(-\infty, \infty)$
vertical asymptote x = 0

6 Solve Logarithmic Equations

1) B
2) C
3) C
4) A
5) A
6) C
7) 2.9731
8) The magnitude of the earthquake measures 6.3 on the Richter scale.
9) The earthquake of greater magnitude has a seismographic reading that is $10^{0.1} \approx 1.26$ times that of the lesser earthquake.
10) D
11) 1.427 decibels
12) 0.20 watt per square meter
13) 117 decibels
14) The intensity of a sound that crosses the human threshold of pain is 10 times as intense as this rock concert.
15) Jim's voice is 10^{10} times as intense as Amy's.
16) A
17) (a) k = 12.10
(b) 7%
(c) 0.23
(d) 0.05

5.5 Properties of Logarithms

1 Work with the Properties of Logarithms

1) C
2) C
3) B
4) B
5) D
6) C
7) A
8) B
9) A
10) A
11) D
12) A
13) D

14) C
15) C
16) B
17) B

2 Write a Logarithmic Expression as a Sum or Difference of Logarithms

1) B
2) B
3) A
4) B
5) B
6) $\frac{5}{3}\log_b x + \frac{8}{3}\log_b y - \frac{2}{3}\log_b z$
7) A
8) A
9) D

3 Write a Logarithmic Expression as a Single Logarithm

1) B
2) D
3) B
4) D
5) D
6) C
7) A
8) $\log_a \frac{6x^4}{y^3}$
9) B
10) A

4 Evaluate Logarithms Whose Base Is Neither 10 nor e

1) C
2) A
3) B
4) B
5) C
6) D
7) C
8) D
9) C
10) -7.26
11) 2
12) B

5.6 Logarithmic and Exponential Equations

1 Solve Logarithmic Equations Using the Properties of Logarithms

1) D
2) B
3) D
4) D
5) C
6) B
7) A
8) 8.36
9) 0.33
10) C

2 Solve Exponential Equations

1) D

2) D
3) C
4) D
5) D
6) C
7) B
8) $x = \frac{\ln 2}{\ln 3}$

9) C
10) $-\frac{3}{14}$

11) B
12) $x = \frac{\ln \pi}{2 - \ln \pi}$

13) $x = \ln(-1 + \sqrt{2})$
14) A

3 Solve Logarithmic and Exponential Equations Using a Graphing Utility

1) B
2) C
3) B
4) D
5) B
6) A

5.7 Compound Interest

1 Determine the Future Value of a Lump Sum of Money

1) B
2) B
3) B
4) B
5) A
6) A
7) \$6749
8) \$438.39
9) \$1643.18
10) \$360,351.56

2 Calculate Effective Rates of Return

1) D
2) C
3) A
4) D
5) B
6) C
7) 4.84%

3 Determine the Present Value of a Lump Sum of Money

1) D
2) A
3) D
4) C
5) B
6) D
7) C
8) \$1252.91

4 Determine the Time Required to Double or Triple a Lump Sum of Money

1) A

2) D
3) A
4) A
5) B
6) 13.9 years
7) 6 years

5 Additional Applications

1) D
2) B
3) C
4) C
5) C
6) 51.37%
7) $35,766.45

5.8 Exponential Growth and Decay; Newton's Law; Logistic Models

1 Find Equations of Populations That Obey the Law of Uninhibited Growth

1) A
2) C
3) $R(x) = 2000000 \cdot 2^{x-1}$; $16,000,000
4) $D = 3^{L-1}$; 81 distributors
5) $A = 100 \cdot 4^{d}$; 1,638,400 bacteria
6) $k = \frac{\ln 2}{0.07} \approx 9.90$; about 14.9%
7) B
8) 8.024 hours
9) 26 days
10) 9 days

2 Find Equations of Populations That Obey the Law of Decay

1) C
2) D
3) D
4) 5 panes
5) A
6) 686 years
7) 989 years
8) 9901 years
9) about 1.2 hours
10) about 4.6 minutes
11) about 3.4 days

3 Use Newton's Law of Cooling

1) C
2) D
3) A
4) D
5) A
6) $16.71°F$
7) The deceased died approximately 16 hours before the first measurement.
8) 6 minutes

4 Use Logistic Models

1) B
2) D
3) D
4) D
5) B

6) A
7) in about 166.47 years or approximately the year 2158

5.9 Fitting Data to Exponential, Logarithmic, and Logistic Functions

1 Use a Graphing Utility to Fit an Exponential Function to Data

1) exponential; $y = 252.68 \cdot 1.36^x$
2) $A = 150e^{0.58t}$
3) $p = 25e^{-0.12q}$; 9 cassettes
4) C
5) A
6) C

2 Use a Graphing Utility to Fit a Logarithmic Function to Data

1) 5 electronic organizers
2) 43.2 deaths per 100,000
3) 1999
4) D
5) C

3 Use a Graphing Utility to Fit a Logistic Function to Data

1) A

Ch. 6 Conics

6.1 Preliminaries

1 Know the Names of the Conics

Name the conic.

1)

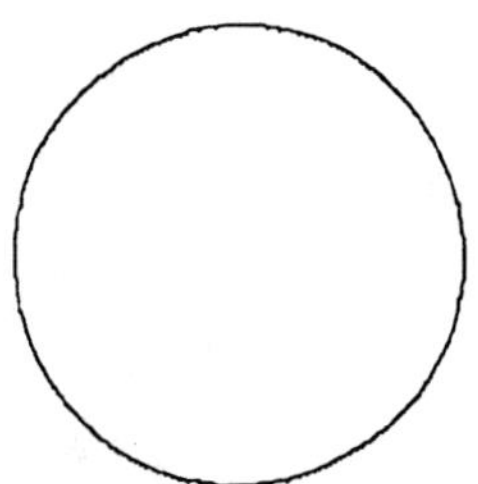

A) parabola B) ellipse C) circle D) hyperbola

2)

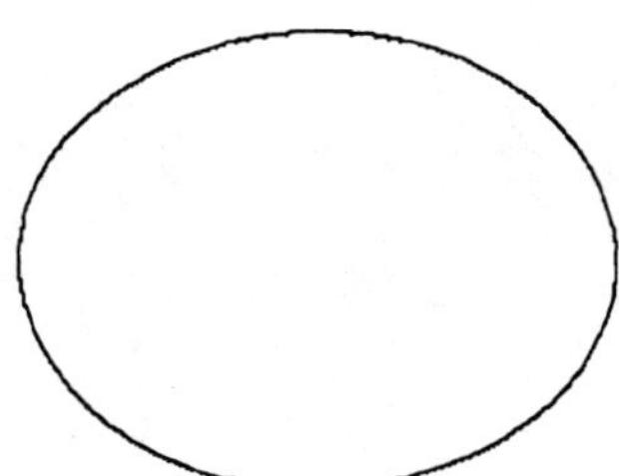

A) circle B) ellipse C) parabola D) hyperbola

3)

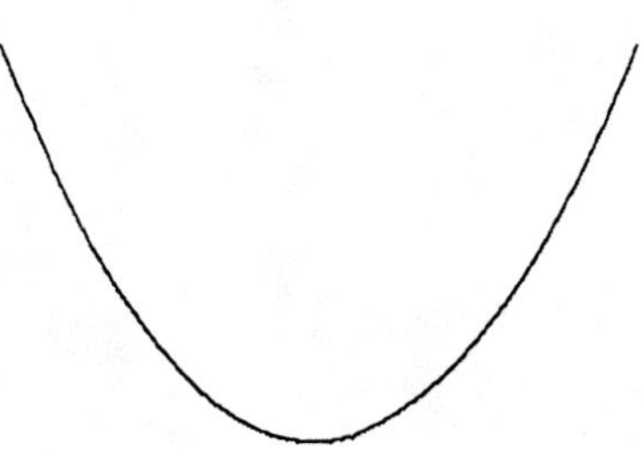

A) parabola B) circle C) hyperbola D) ellipse

4)

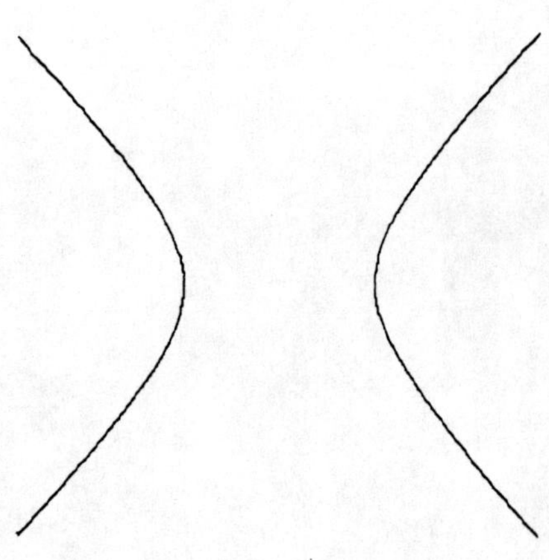

A) ellipse B) parabola C) hyperbola D) circle

6.2 The Parabola

1 Find the Equation of a Parabola

Write an equation for the parabola.

1)

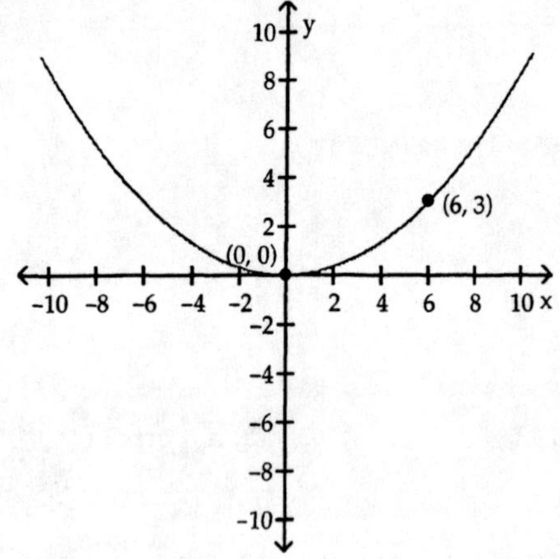

A) $x^2 = -12y$ B) $y^2 = 12x$ C) $y^2 = -12x$ D) $x^2 = 12y$

2)

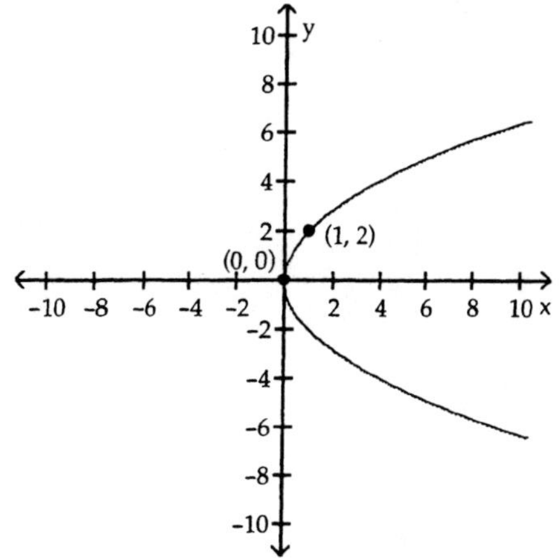

A) $x^2 = 4y$
B) $y^2 = -4x$
C) $y^2 = 4x$
D) $x^2 = -4y$

3) With vertex at (7, 3) and focus at (4, 3)

A) $(y - 3)^2 = 12(x - 7)$
B) $(x - 3)^2 = -4(y - 3)$
C) $(x - 3)^2 = 4(y - 3)$
D) $(y - 3)^2 = -12(x - 7)$

4) With vertex at (2, 8) and focus at (2, 6)

A) $(y - 8)^2 = 16(x - 2)$
B) $(x - 2)^2 = 8(y - 8)$
C) $(y - 8)^2 = -16(x - 2)$
D) $(x - 2)^2 = -8(y - 8)$

Find the equation of the parabola described.

5) Focus at (3, 0); vertex at (0, 0)

A) $x^2 = 3y$
B) $x^2 = 12y$
C) $y^2 = 3x$
D) $y^2 = 12x$

6) Directrix the line y = 3; vertex at (0, 0)

A) $y = -12x^2$
B) $y = -\frac{1}{12}x^2$
C) $x = -\frac{1}{12}y^2$
D) $x = 3y^2$

7) Focus at (5, 0); vertex at (0, 0)

A) $y^2 = 20x$
B) $x = 20y^2$
C) $y = 20x^2$
D) $x^2 = 20y$

8) Vertex at (4, -4); focus at (4, 1)

A) $(x - 4)^2 = 20(y + 4)$
B) $(x + 4)^2 = 5(y - 4)$
C) $(x + 4)^2 = 20(y - 4)$
D) $(x - 4)^2 = 5(y + 4)$

9) Vertex at (3, 4); focus at (3, 2)

A) $(x - 3)^2 = 8(y - 4)$
B) $(x - 3)^2 = -8(y - 4)$
C) $(y - 4)^2 = 8(x - 4)$
D) $(y - 4)^2 = -8(x - 4)$

10) Vertex at (0, 0); axis of symmetry the x-axis; containing the point (8, 1)

A) $y^2 = \frac{1}{32}x$
B) $x^2 = \frac{1}{8}y$
C) $y^2 = \frac{1}{8}x$
D) $x^2 = \frac{1}{32}y$

11) Vertex at (6, -3); focus at (6, -7)

A) $(x - 6)^2 = -16(y + 3)$
B) $(x - 6)^2 = 16(y + 3)$
C) $(y - 3)^2 = -4(x + 6)$
D) $(y - 3)^2 = 4(x + 6)$

12) Vertex at (6, -8); focus at (1, -8)

A) $(x + 6)^2 = 28(y - 8)$
B) $(x + 6)^2 = -28(y - 8)$
C) $(y + 8)^2 = -20(x - 6)$
D) $(y + 8)^2 = 20(x - 6)$

Find the vertex, focus, and directrix of the parabola. Graph the equation.

1) $x^2 = -16y$

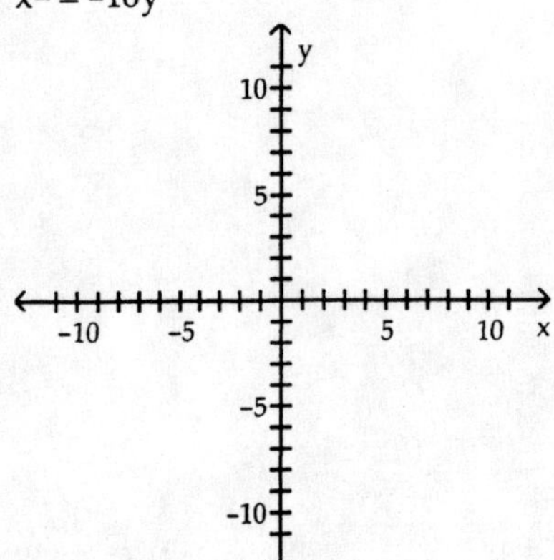

A)

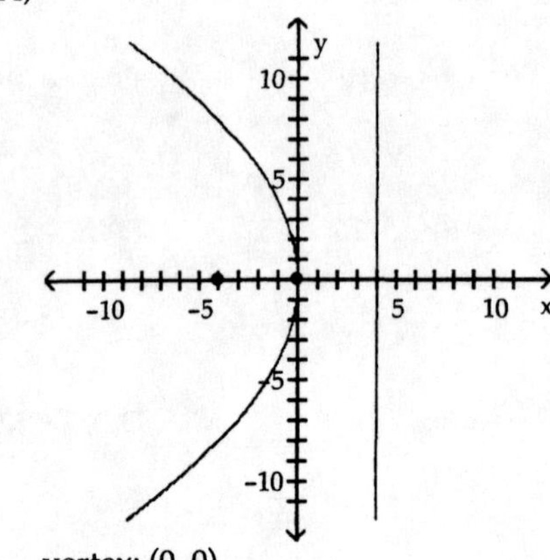

vertex: (0, 0)
focus: (-4, 0)
directrix: x = 4

B)

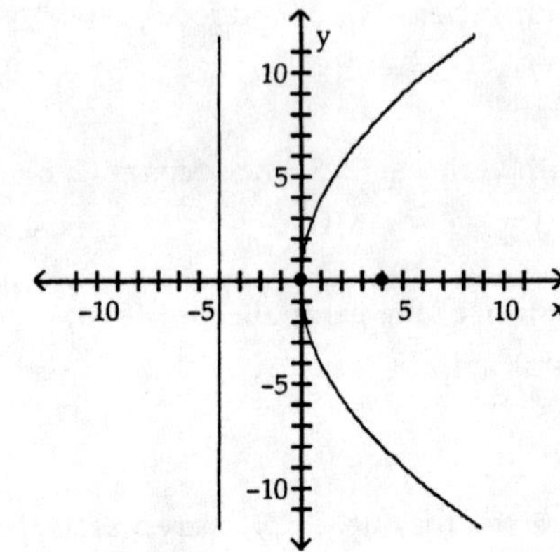

vertex: (0, 0)
focus: (4, 0)
directrix: x = -4

C)

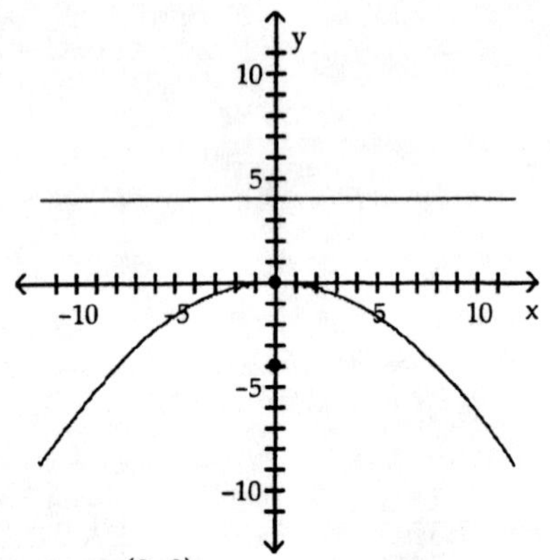

vertex: (0, 0)
focus: (0, -4)
directrix: y = 4

D)

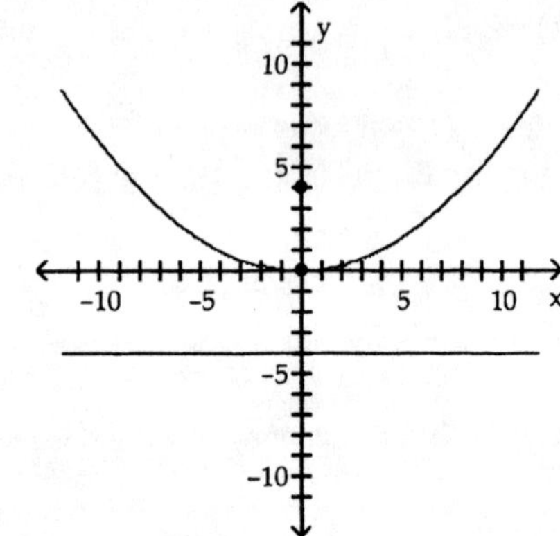

vertex: (0, 0)
focus: (0, 4)
directrix: y = -4

2) $y^2 = 8x$

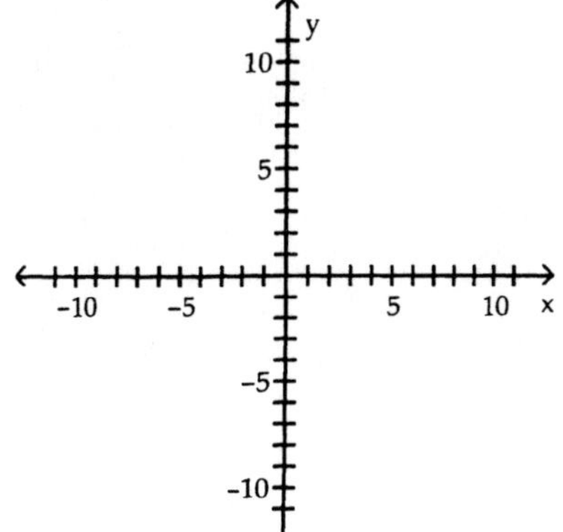

A)

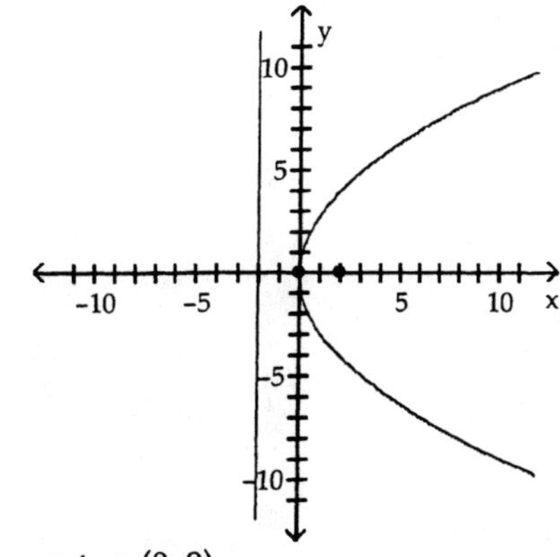

vertex: (0, 0)
focus: (2, 0)
directrix: $x = -2$

B)

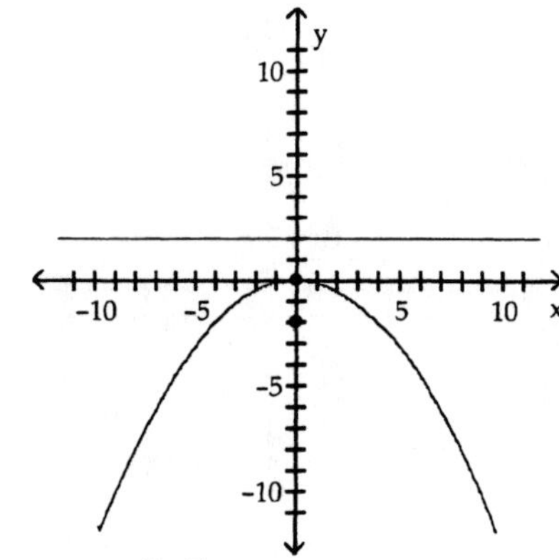

vertex: (0, 0)
focus: (0, 2)
directrix: $y = -2$

C)

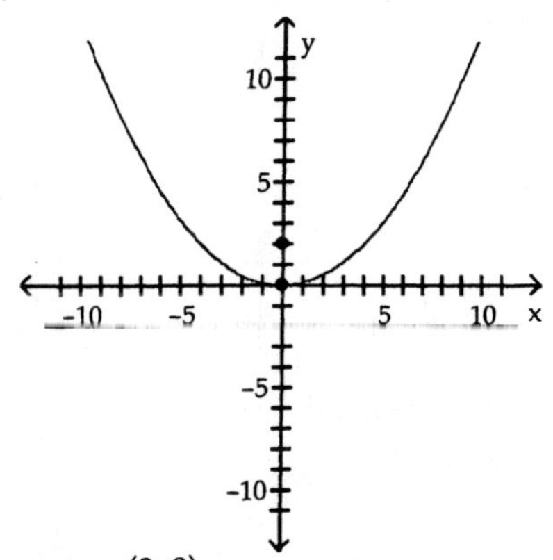

vertex: (0, 0)
focus: (0, 2)
directrix: $y = -2$

D)

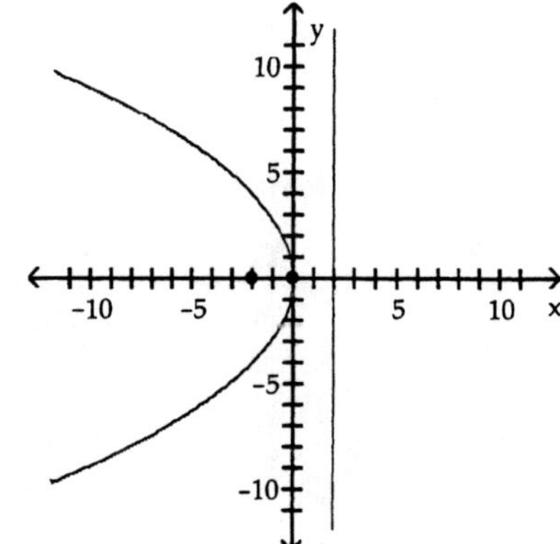

vertex: (0, 0)
focus: (-2, 0)
directrix: $x = 2$

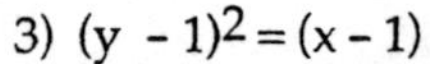

3) $(y - 1)^2 = (x - 1)$

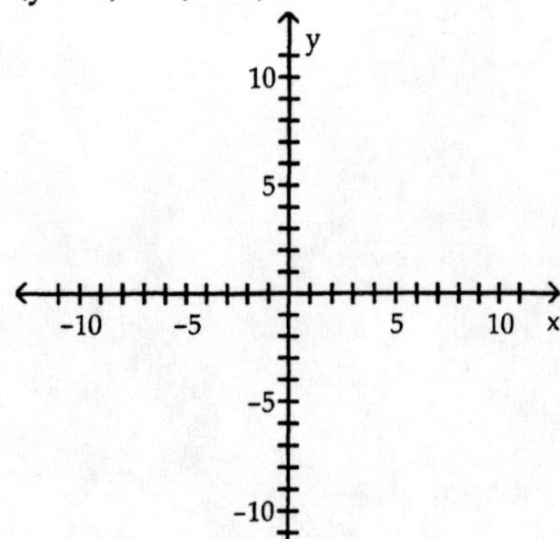

A)

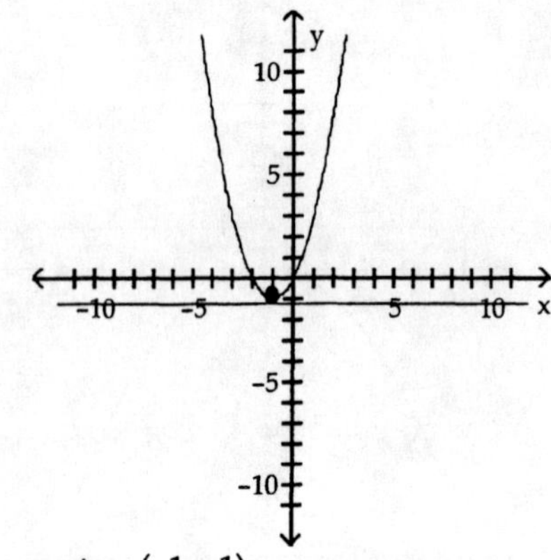

vertex: (−1, −1)
focus: (−1, −0.75)
directrix: y = −1.25

B)

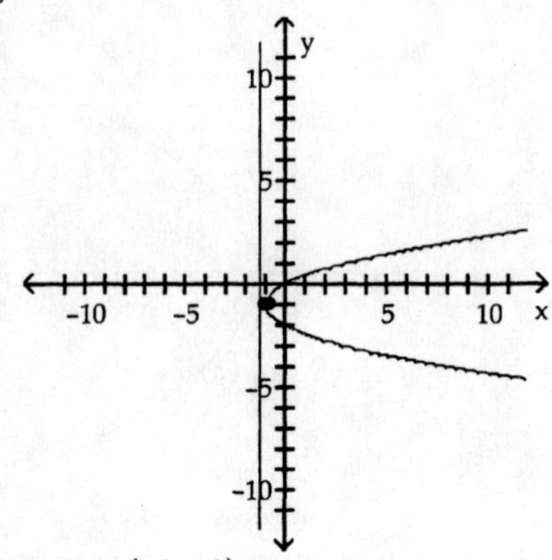

vertex: (−1, −1)
focus: (−0.75, −1)
directrix: x = −1.25

C)

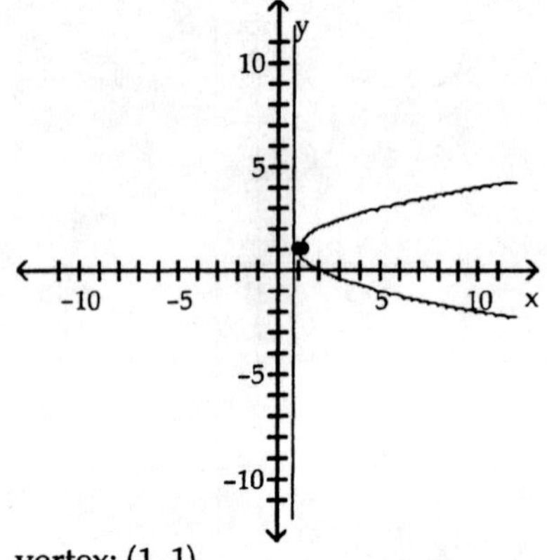

vertex: (1, 1)
focus: (1.25, 1)
directrix: x = 0.75

D)

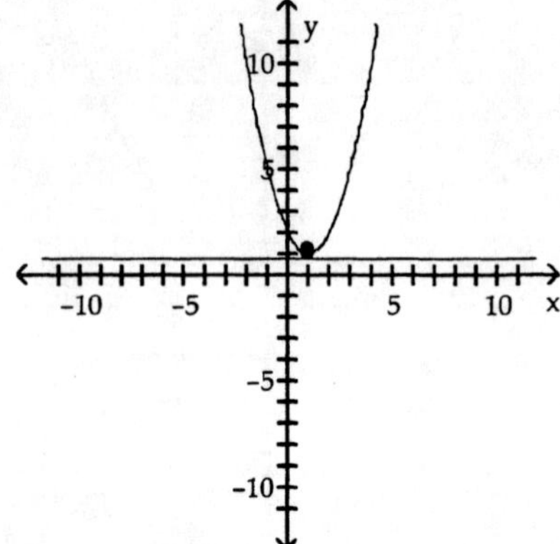

vertex: (1, 1)
focus: (1, 1.25)
directrix: y = 0.75

4) $(x - 1)^2 = 8(y - 1)$

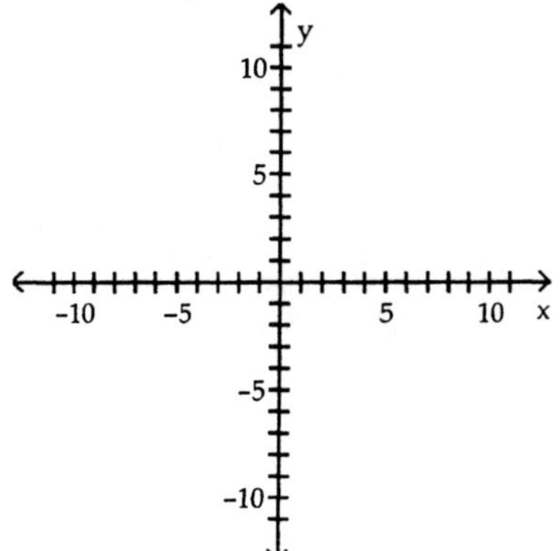

A)

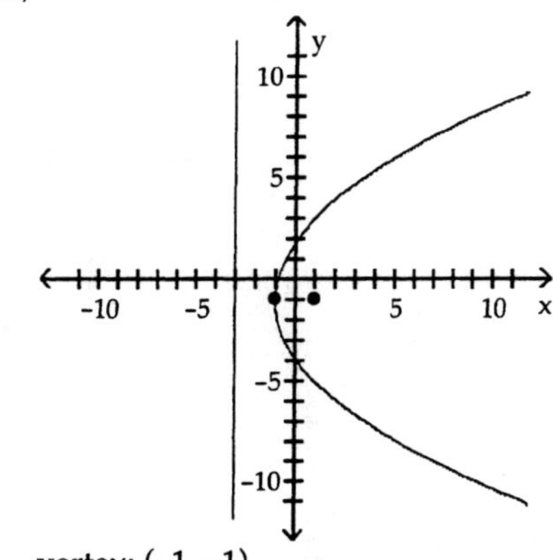

vertex: (-1, -1)
focus: (1, -1)
directrix: x = -3

B)

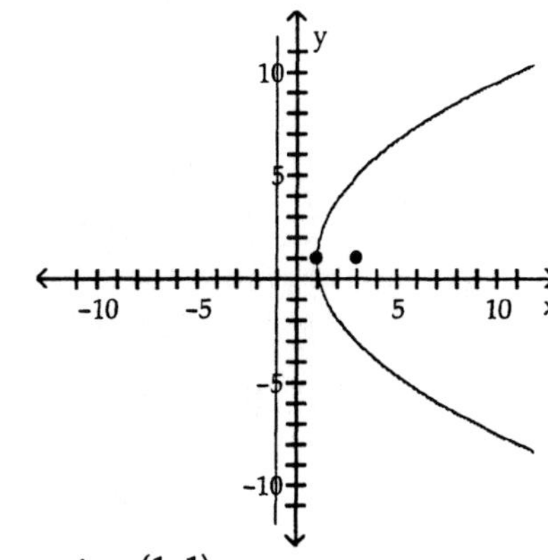

vertex: (1, 1)
focus: (3, 1)
directrix: x = -1

C)

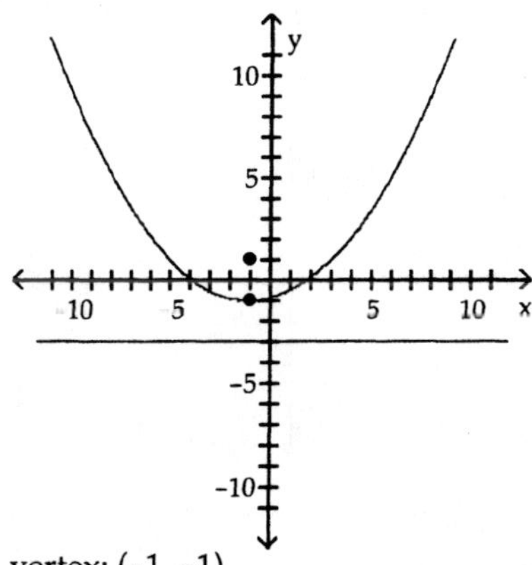

vertex: (-1, -1)
focus: (-1, 1)
directrix: y = -3

D)

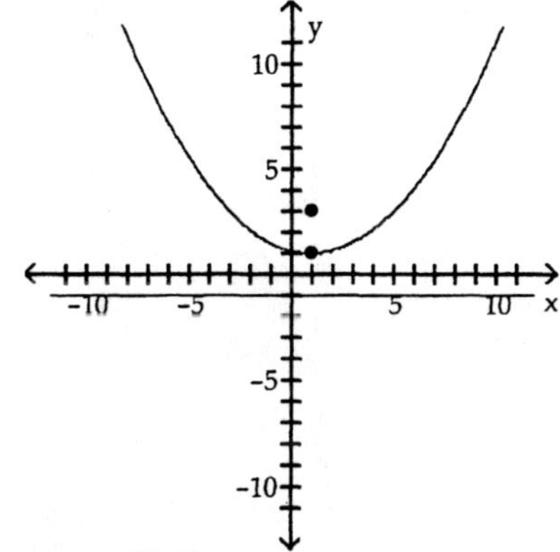

vertex: (1, 1)
focus: (1, 3)
directrix: y = -1

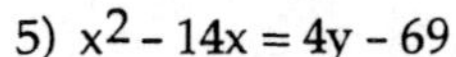

5) $x^2 - 14x = 4y - 69$

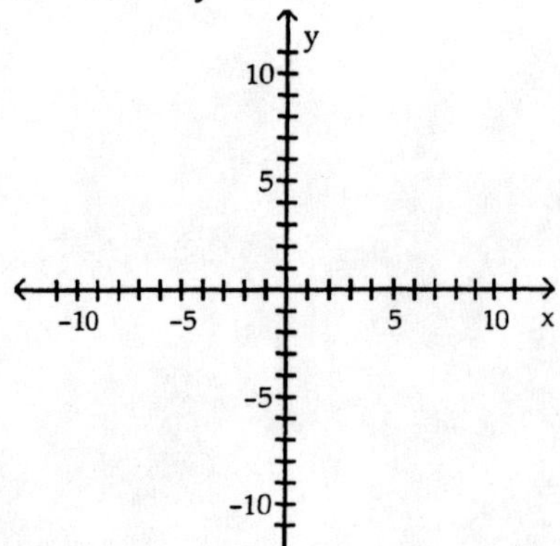

A)

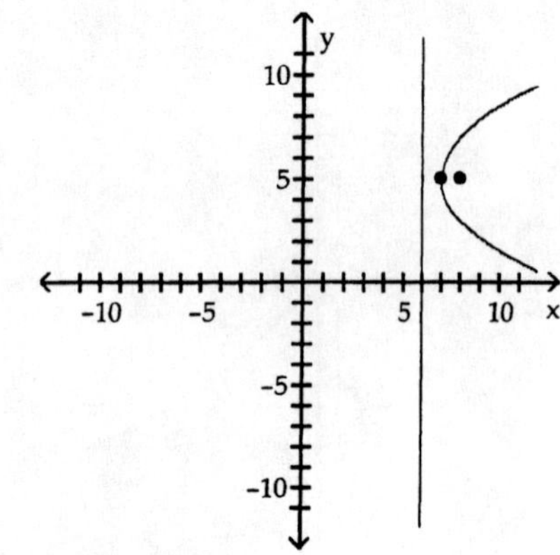

vertex: (7, 5)
focus: (8, 5)
directrix: $x = 6$

B)

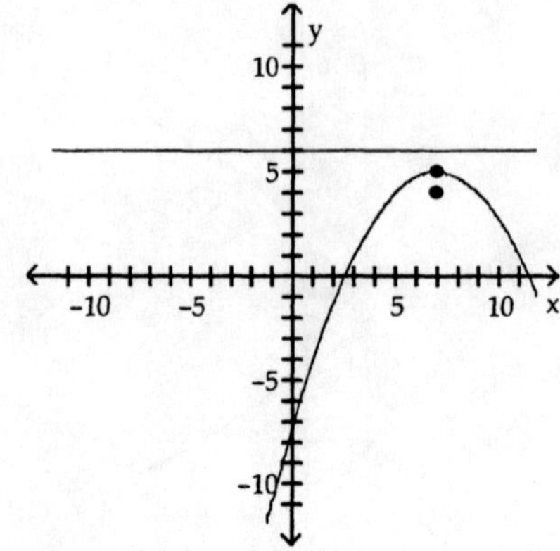

vertex: (7, 5)
focus: (7, 4)
directrix: $y = 6$

C)

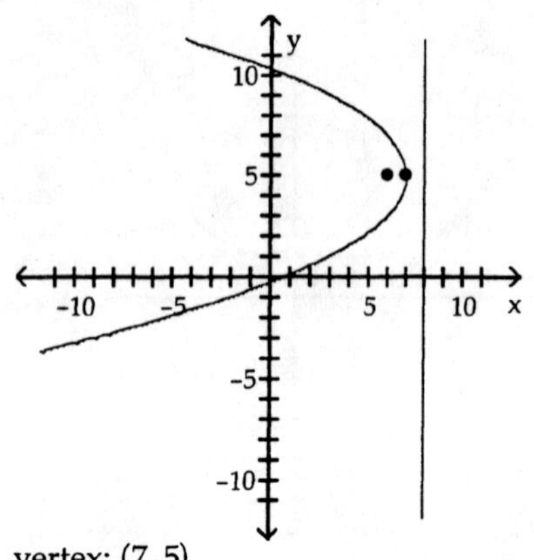

vertex: (7, 5)
focus: (6, 5)
directrix: $x = 8$

D)

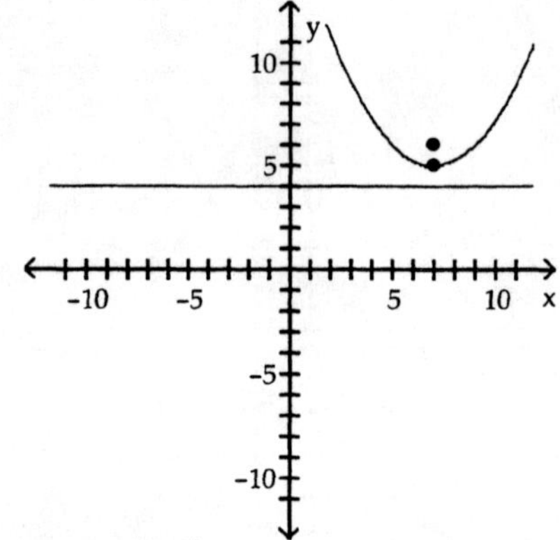

vertex: (7, 5)
focus: (7, 6)
directrix: $y = 4$

6) $y^2 + 14y = 8x - 9$

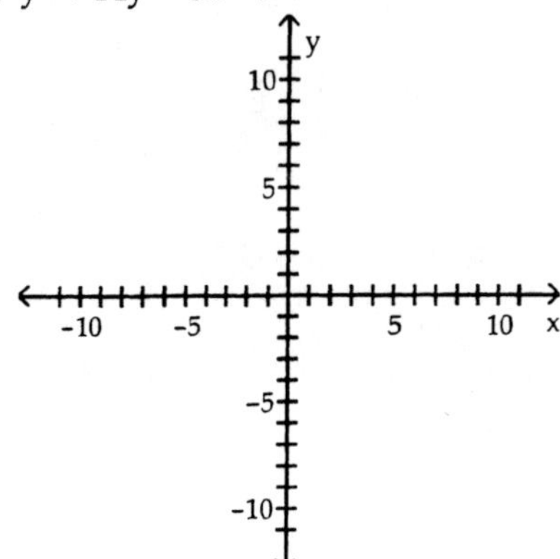

A)

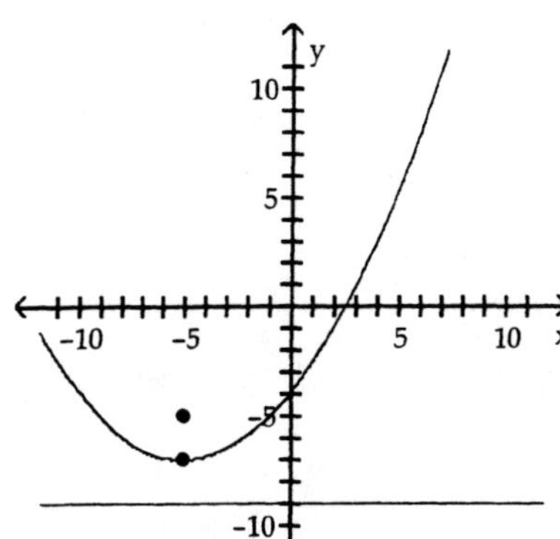

vertex: (-5, -7)
focus: (-5, -5)
directrix: y = -9

B)

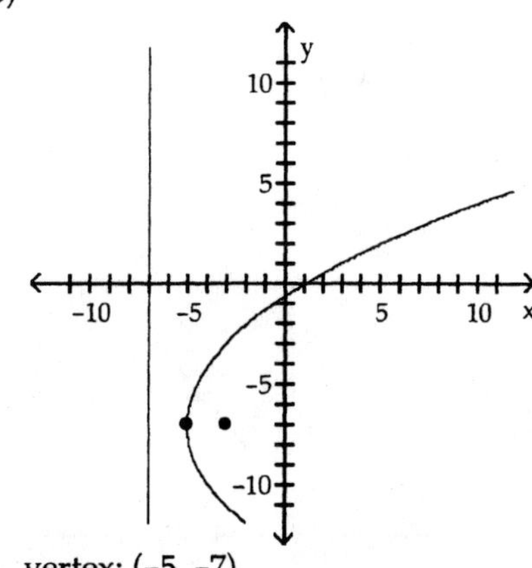

vertex: (-5, -7)
focus: (-3, -7)
directrix: x = -7

C)

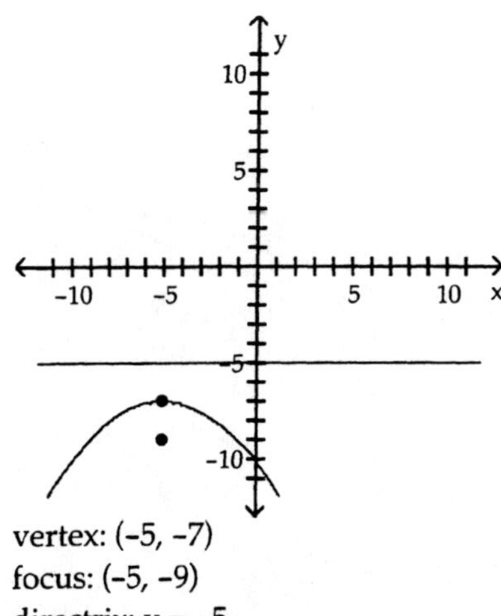

vertex: (-5, -7)
focus: (-5, -9)
directrix: y = -5

D)

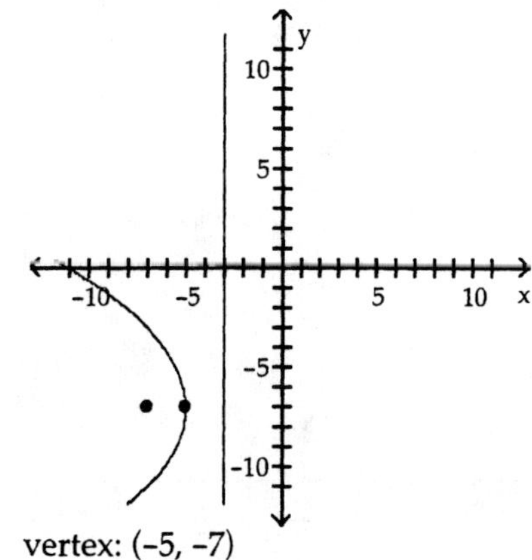

vertex: (-5, -7)
focus: (-7, -7)
directrix: x = -3

3 Discuss the Equation of a Parabola

1) The exercises for this objective are incorporated into other objectives within the section.

4 Work with Parabolas with Vertex at (h, k)

Write an equation for the parabola.

1)

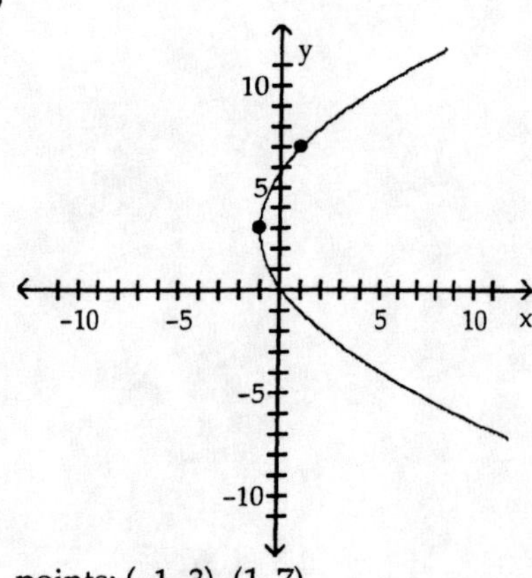

points: (–1, 3), (1, 7)

A) $(x - 3)^2 = 8(y + 1)$ B) $(x + 3)^2 = 8(y - 1)$ C) $(y - 1)^2 = 8(x + 3)$ D) $(y - 3)^2 = 8(x + 1)$

2)

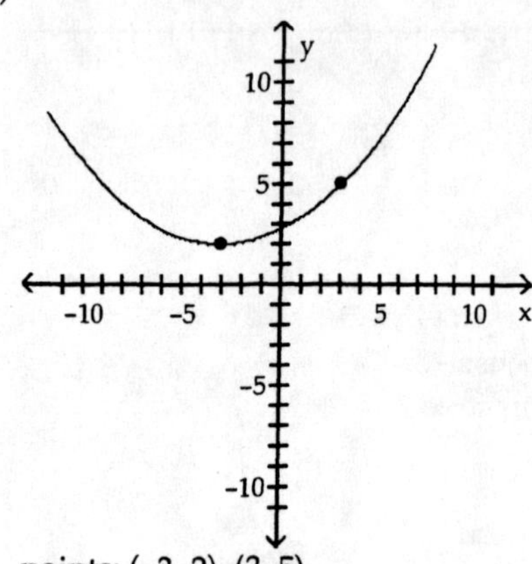

points: (–3, 2), (3, 5)

A) $(x + 2)^2 = 12(y - 3)$ B) $(y + 3)^2 = 12(x - 2)$ C) $(x + 3)^2 = 12(y - 2)$ D) $(x - 2)^2 = 12(y + 3)$

3)

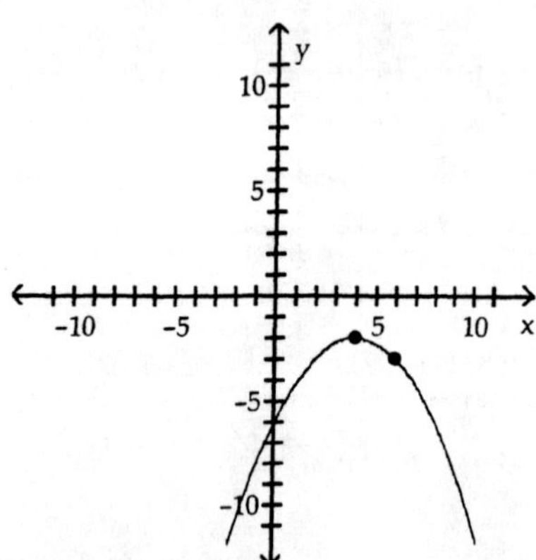

points: (4, –2), (6, –3)

A) $(x + 2)^2 = 4(y - 4)$ B) $(x - 2)^2 = -4(y + 4)$ C) $(x - 4)^2 = -4(y + 2)$ D) $(y - 4)^2 = -4(x + 2)$

5 Solve Applied Problems Involving Parabolas

Solve.

1) A reflecting telescope contains a mirror shaped like a paraboloid of revolution. If the mirror is 22 inches across at its opening and is 3 feet deep, where will the light be concentrated?

A) 0.2 inches from the vertex
B) 10.1 inches from the vertex
C) 0.4 inches from the vertex
D) 0.8 inches from the vertex

2) A searchlight is shaped like a paraboloid of revolution. If the light source is located 2 feet from the base along the axis of symmetry and the opening is 8 feet across, how deep should the searchlight be?

A) 4 feet
B) 2 feet
C) 0.3 feet
D) 8 feet

3) A bridge is built in the shape of a parabolic arch. The bridge arch has a span of 182 feet and a maximum height of 25 feet. Find the height of the arch at 20 feet from its center.

A) 4.8 feet
B) 0.3 feet
C) 58.2 feet
D) 23.8 feet

6.3 The Ellipse

1 Find the Equation of an Ellipse

Write an equation for the ellipse.

1)

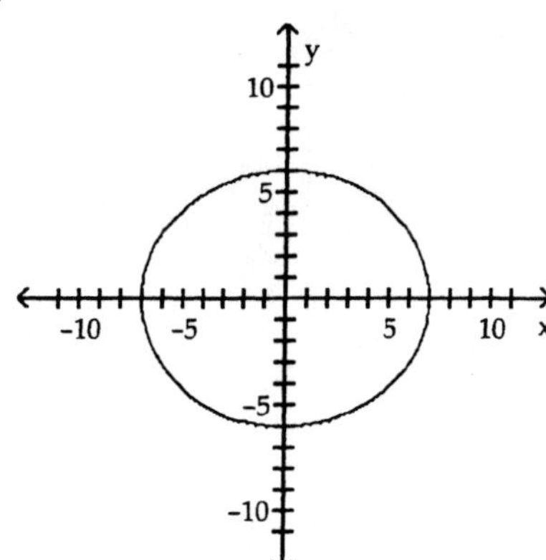

A) $\frac{y^2}{36} + \frac{x^2}{49} = 1$
B) $\frac{x^2}{36} + \frac{y^2}{49} = 1$
C) $\frac{x^2}{49} - \frac{y^2}{36} = 1$
D) $\frac{y^2}{36} - \frac{x^2}{49} = 1$

2)

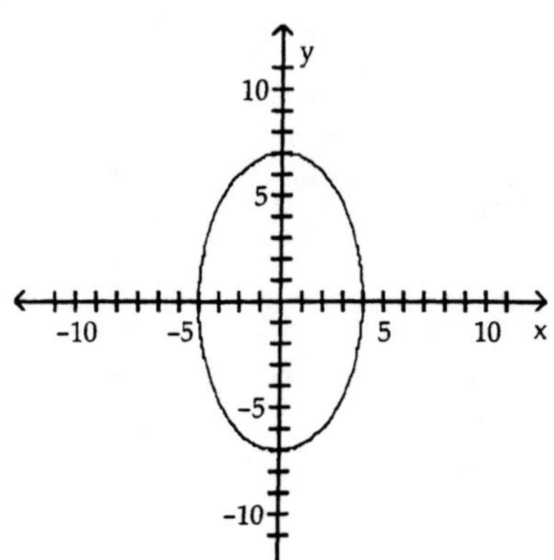

A) $\frac{y^2}{49} + \frac{x^2}{16} = 1$
B) $-\frac{y^2}{49} + \frac{x^2}{16} = 1$
C) $\frac{x^2}{49} + \frac{y^2}{16} = 1$
D) $\frac{y^2}{49} - \frac{x^2}{16} = 1$

Find an equation for the ellipse.

3) Foci $(\pm 6, 0)$; y-intercepts are ± 3

A) $\frac{x^2}{9} + \frac{y^2}{45} = 6$ B) $\frac{x^2}{45} + \frac{y^2}{9} = 1$ C) $\frac{x^2}{9} + \frac{y^2}{45} = 1$ D) $\frac{x^2}{45} + \frac{y^2}{9} = 6$

4) Center at (0, 0); foci at $(0, \pm 4)$; vertices at (0, 7) and (0, −7); major axis along the y-axis.

A) $\frac{x^2}{65} + \frac{y^2}{49} = 1$ B) $\frac{x^2}{49} + \frac{y^2}{65} = 1$ C) $\frac{x^2}{49} + \frac{y^2}{33} = 1$ D) $\frac{x^2}{33} + \frac{y^2}{49} = 1$

Find an equation for the ellipse. Graph the equation.

5) Foci at $(0, \pm 3)$; length of major axis is 12

A) $\frac{x^2}{27} + \frac{y^2}{36}$

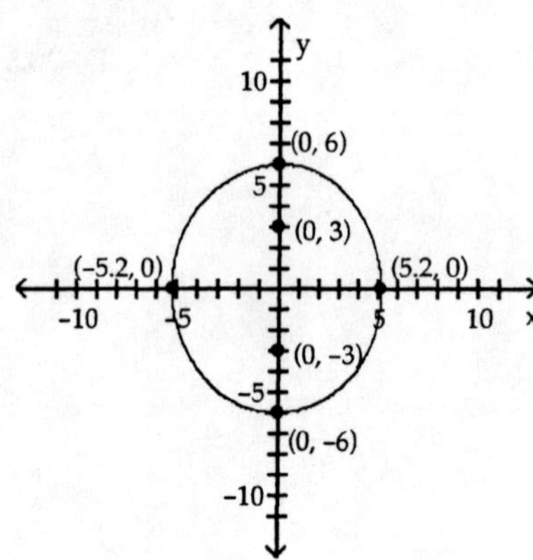

B) $\frac{x^2}{36} + \frac{y^2}{27}$

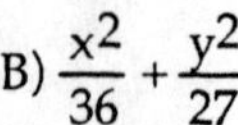

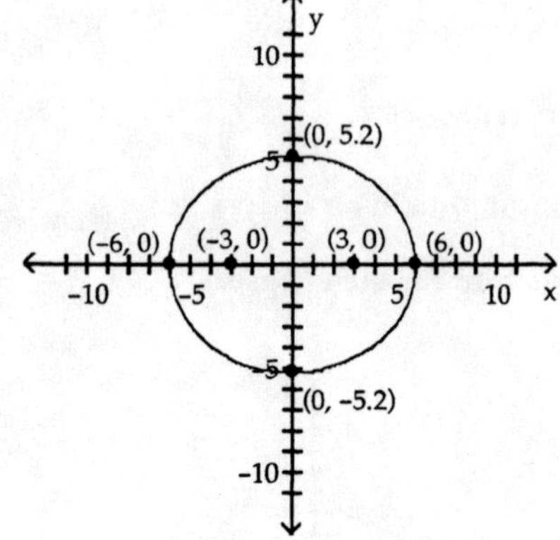

6) Focus at (−3, 0); vertices at $(\pm 4, 0)$

A) $\frac{x^2}{7} + \frac{y^2}{16}$

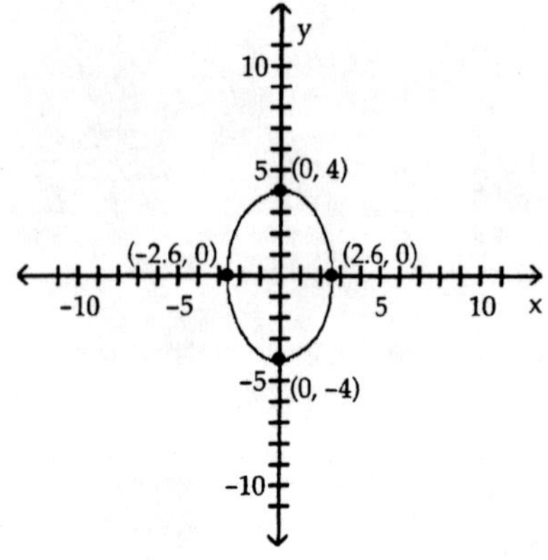

B) $\frac{x^2}{16} + \frac{y^2}{7}$

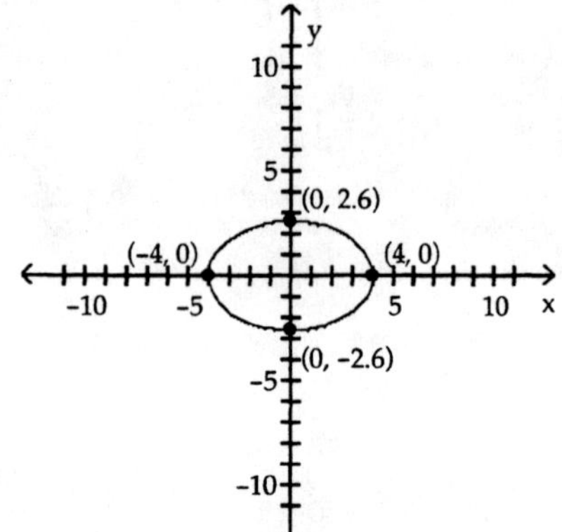

Graph the equation.

1) $\frac{y^2}{25} + \frac{x^2}{49} = 1$

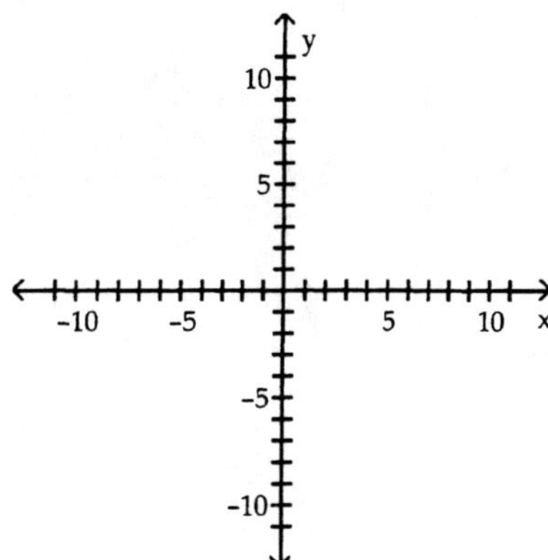

A)

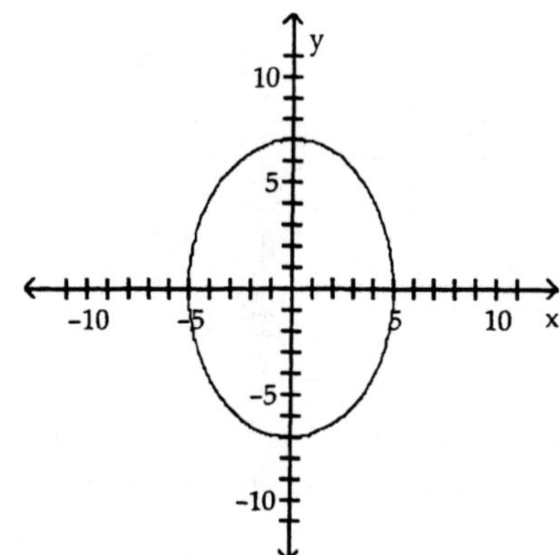

B)

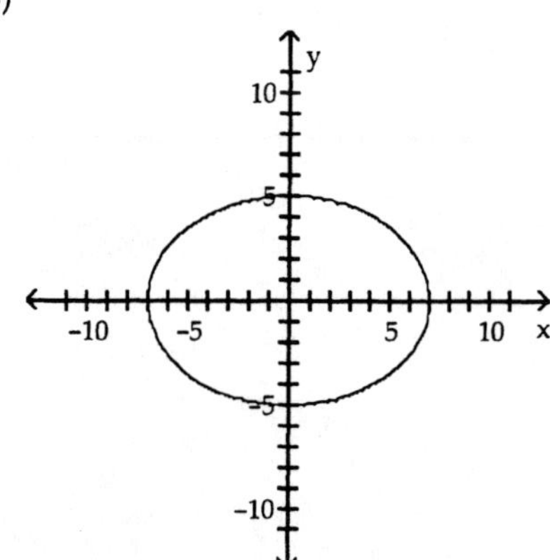

2) $4x^2 + 9y^2 = 36$

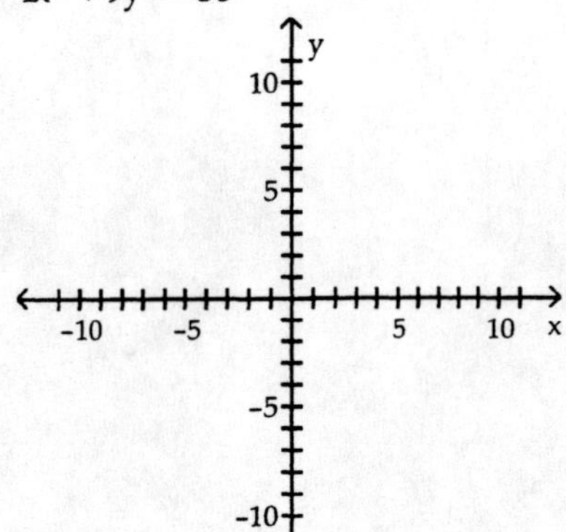

A)

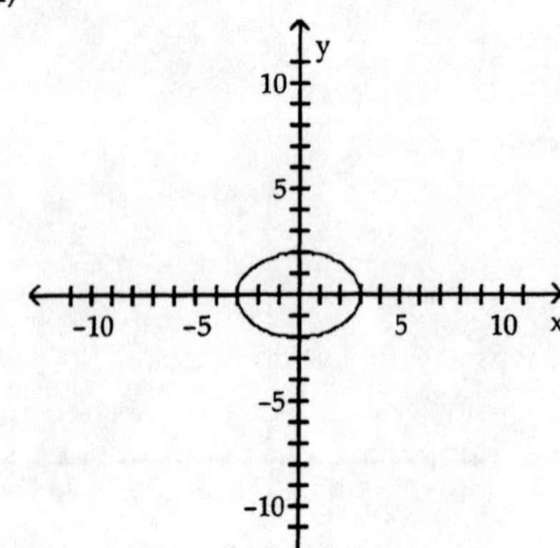

B)

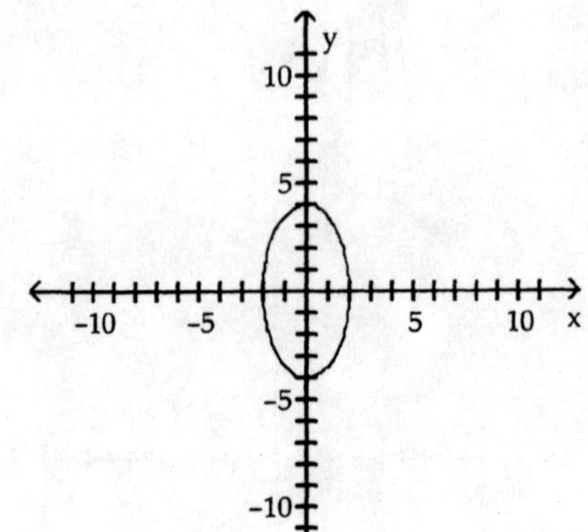

C)

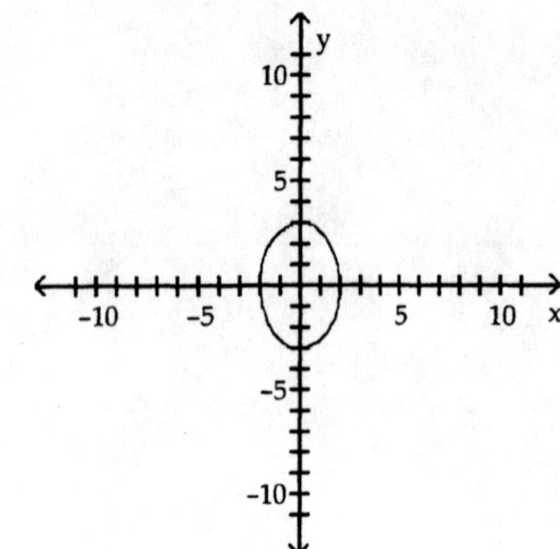

D)

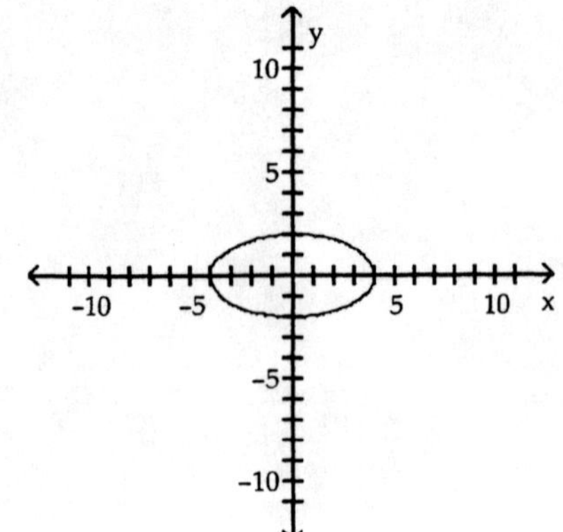

3) $\frac{(x+1)^2}{16} + \frac{(y+2)^2}{9} = 1$

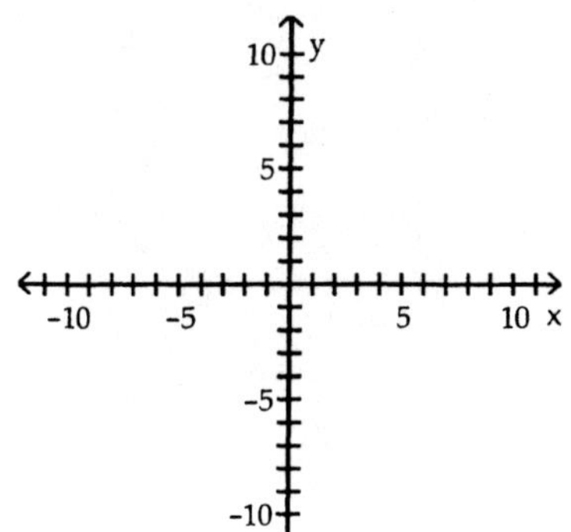

A)

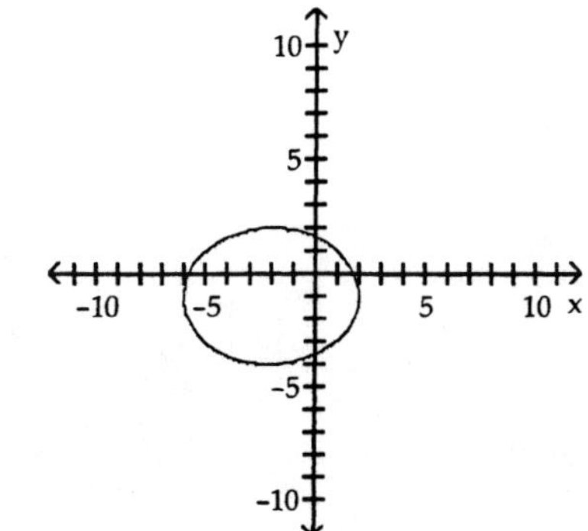

B)

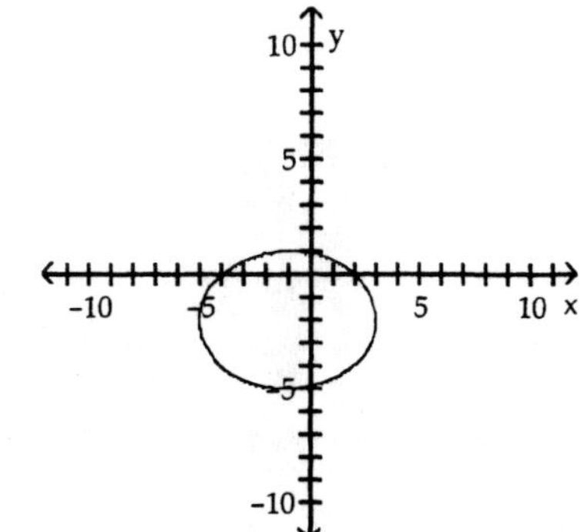

C)

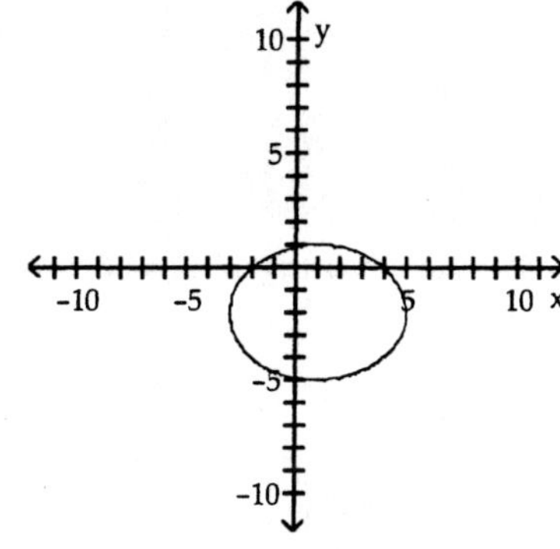

D)

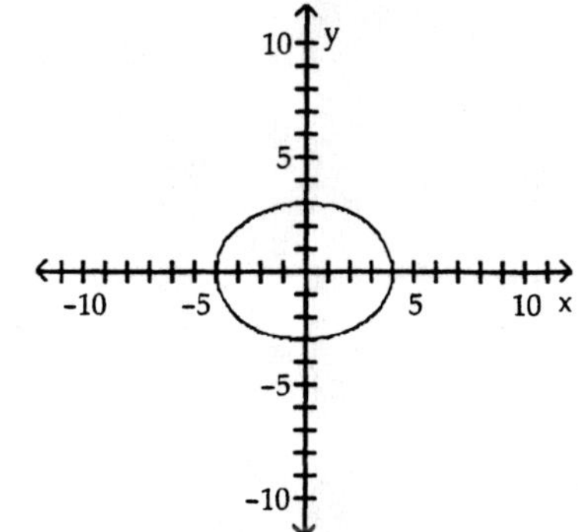

4) $\frac{(x + 1)^2}{9} + \frac{(y + 2)^2}{16} = 1$

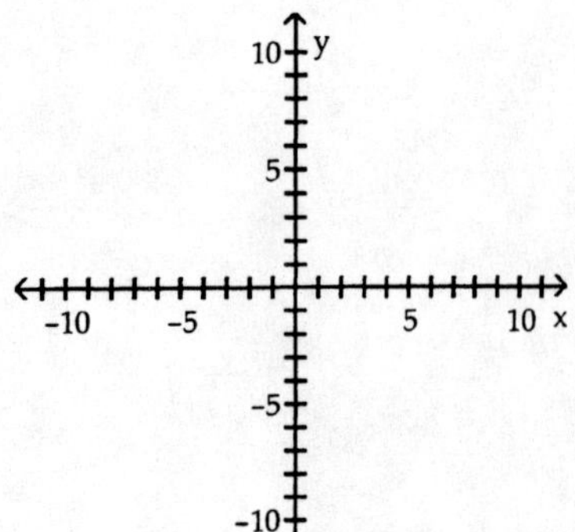

A)

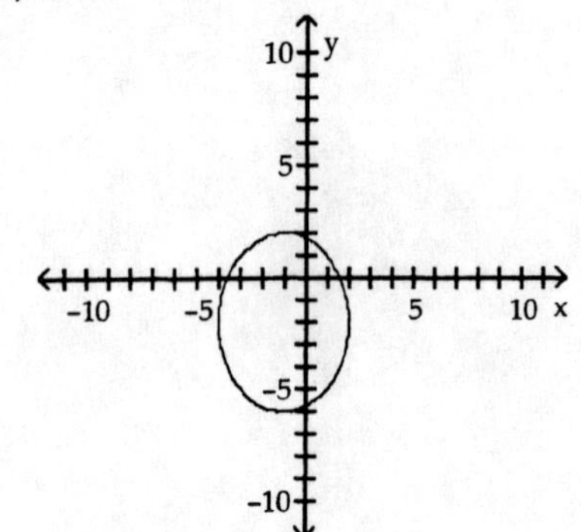

B)

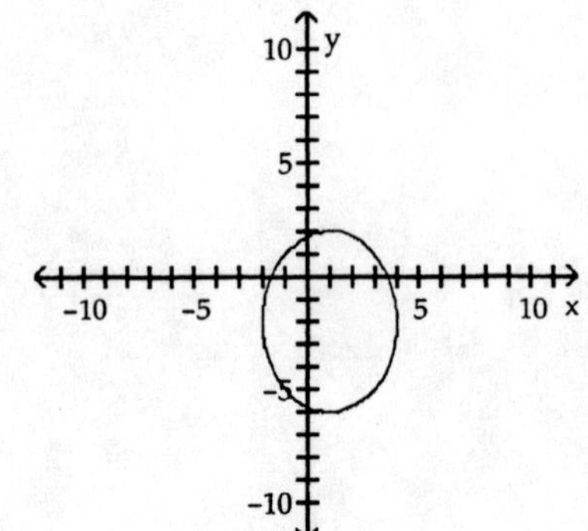

C)

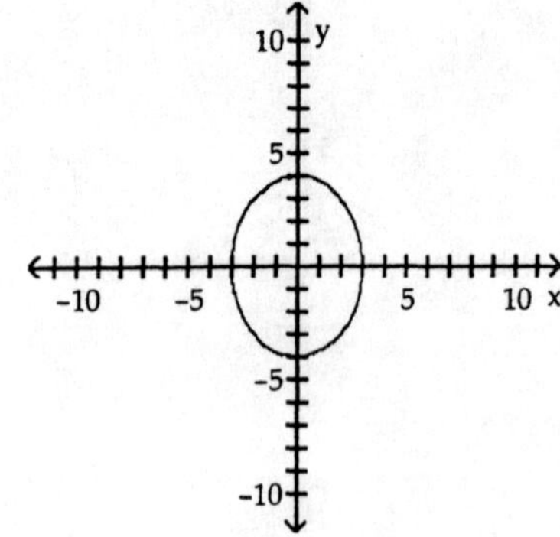

D)

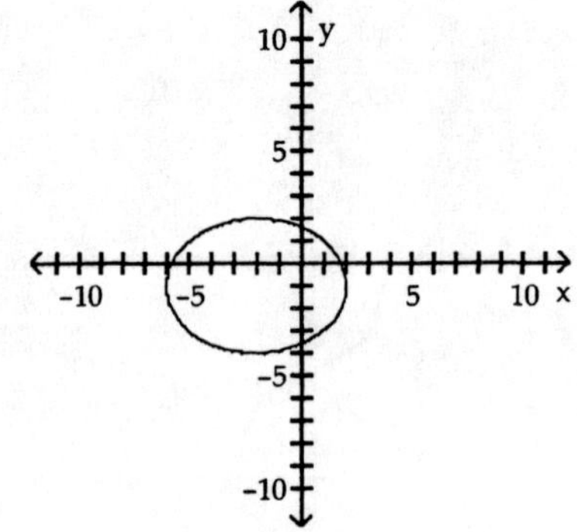

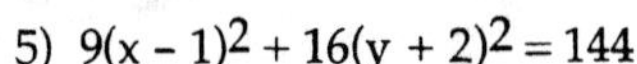

5) $9(x - 1)^2 + 16(y + 2)^2 = 144$

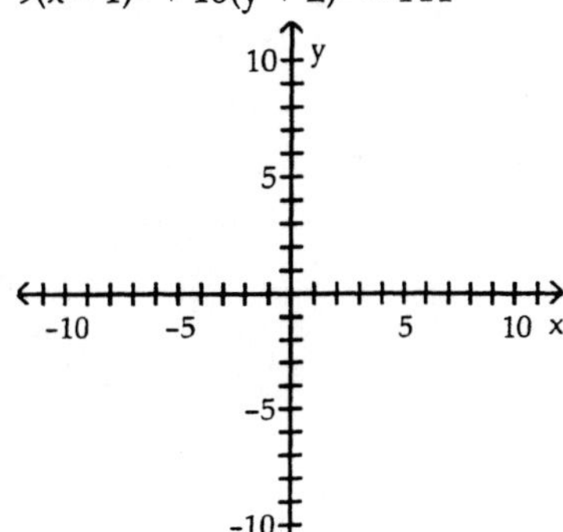

A)

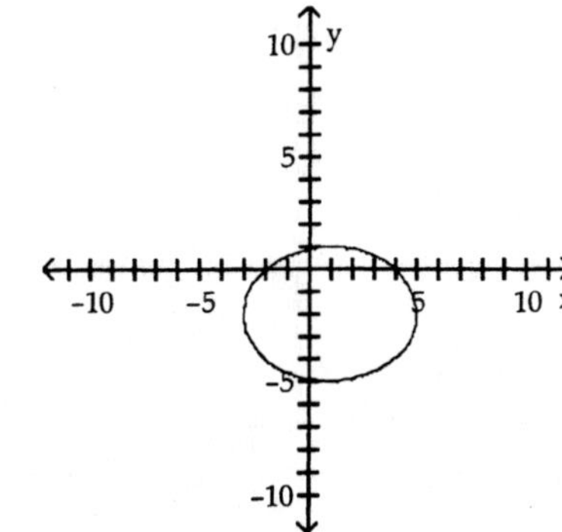

B)

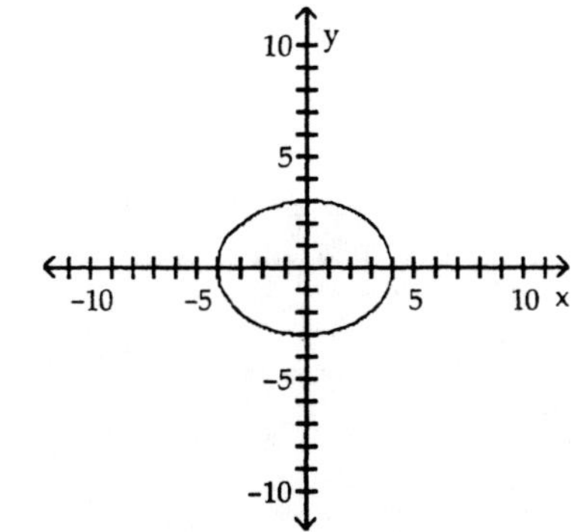

C)

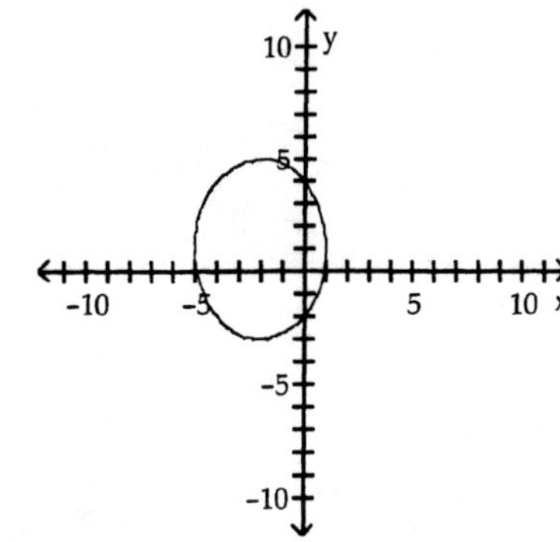

D)

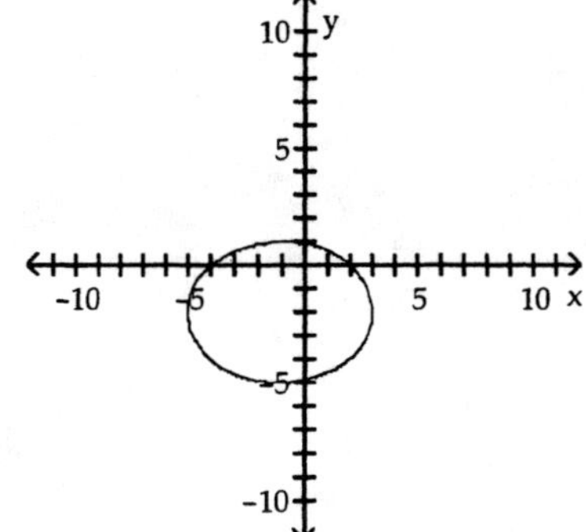

6) $16(x-2)^2 + 4(y+1)^2 = 64$

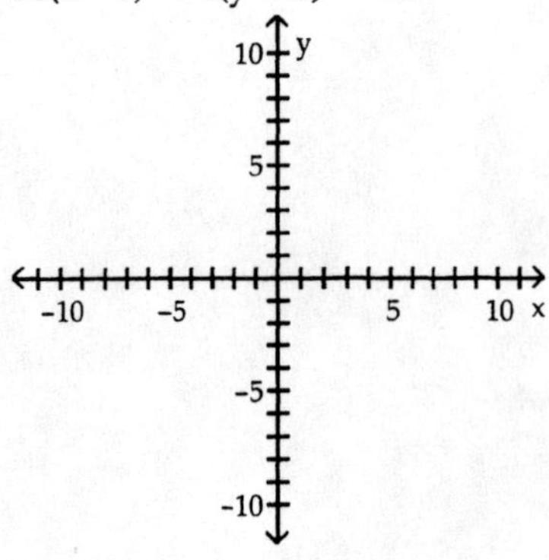

A)

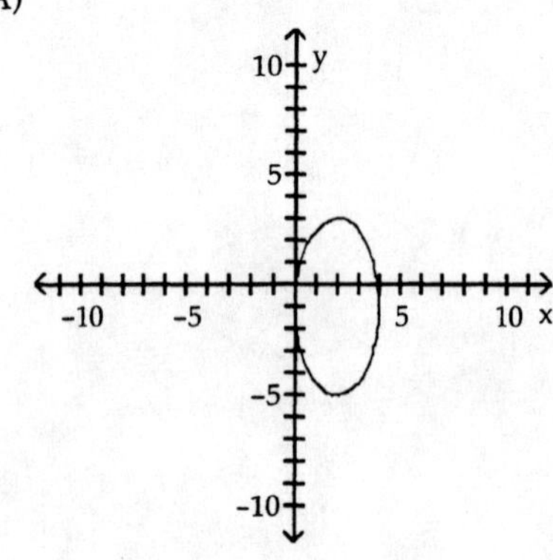

B)

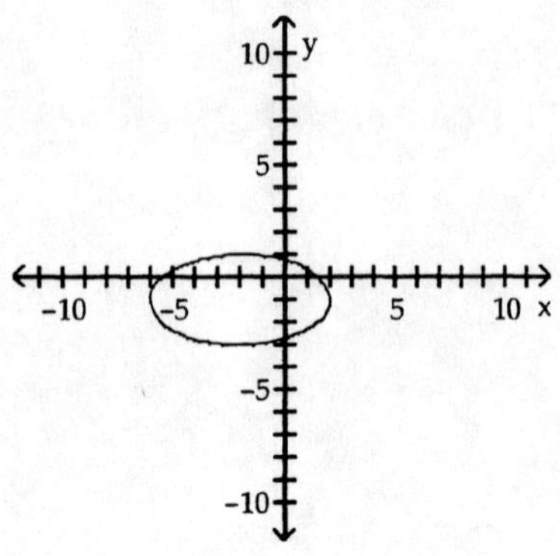

C)

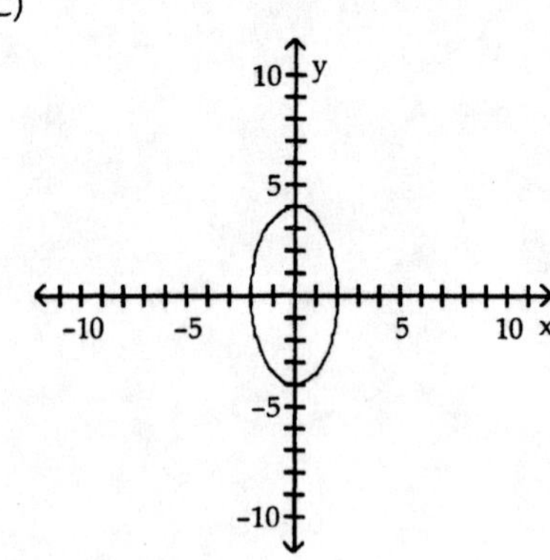

D)

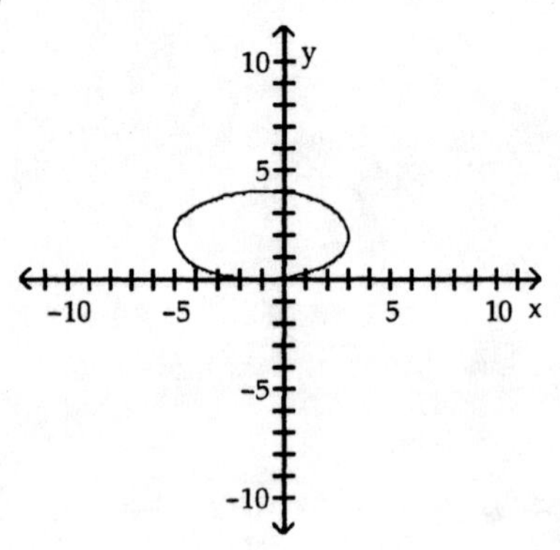

3 Discuss the Equation of an Ellipse

Discuss the equation; that is, find the center, foci, and vertices of the ellipse.

1) $\frac{x^2}{9} + \frac{y^2}{4} = 1$

A) center at (0, 0)
foci at (0, -2) and (0, 2)
vertices at (0, -4), (0, 4)

B) center at (0, 0)
foci at $(0, -\sqrt{5})$ and $(0, \sqrt{5})$
vertices at (0, -3), (0, 3)

C) center at (0, 0)
foci at (-3, 0) and (3, 0)
vertices at (-9, 0), (9, 0)

D) center at (0, 0)
foci at $(-\sqrt{5}, 0)$ and $(\sqrt{5}, 0)$
vertices at (-3, 0), (3, 0)

2) $\frac{x^2}{9} + \frac{y^2}{36} = 1$

A) center at (0, 0)
foci at $(-3\sqrt{3}, 0)$ and $(3\sqrt{3}, 0)$
vertices at (–6, 0), (6, 0)

B) center at (0, 0)
foci at $(0, -3\sqrt{3})$ and $(0, 3\sqrt{3})$
vertices at (0, –6), (0, 6)

C) center at (0, 0)
foci at (0, –6) and (0, 6)
vertices at (0, –36), (0, 36)

D) center at (0, 0)
foci at (0, 6) and (3, 0)
vertices at (0, 36), (9, 0)

3) $25x^2 + 64y^2 = 1600$

A) center at (0, 0)
foci at $(0, -\sqrt{39})$ and $(0, \sqrt{39})$
vertices at (0, –8), (0, 8)

B) center at (0, 0)
foci at (0, –5) and (0, 5)
vertices at (0, –25), (0, 25)

C) center at (0, 0)
foci at $(-\sqrt{39}, 0)$ and $(\sqrt{39}, 0)$
vertices at (–8, 0), (8, 0)

D) center at (0, 0)
foci at (–8, 0) and (8, 0)
vertices at (–64, 0), (64, 0)

4) $25x^2 + 4y^2 = 100$

A) center at (0, 0)
foci at (0, 5) and (2, 0)
vertices at (0, 25) and (4, 0)

B) center at (0, 0)
foci at $(-\sqrt{21}, 0)$ and $(\sqrt{21}, 0)$
vertices at (–5, 0), (5, 0)

C) center at (0, 0)
foci at (0, –5) and (0, 5)
vertices at (0, –25), (0, 25)

D) center at (0, 0)
foci at $(0, -\sqrt{21})$ and $(0, \sqrt{21})$
vertices at (0, –5), (0, 5)

5) $\frac{(x+3)^2}{36} + \frac{(y-3)^2}{25} = 1$

A) center at (–3, 3)
foci at $(-\sqrt{11}, 3)$, $(\sqrt{11}, 3)$
vertices at (6, 3), (–6, 3)

B) center at (–3, 3)
foci at $(-3 + \sqrt{11}, -3)$, $(-3 - \sqrt{11}, -3)$
vertices at (6, 3), (–6, 3)

C) center at (–3, 3)
foci at $(-3 + \sqrt{11}, 3)$, $(-3 - \sqrt{11}, 3)$
vertices at (–9, 3), (3, 3)

D) center at (3, –3)
foci at $(3 + \sqrt{11}, -3)$, $(3 - \sqrt{11}, -3)$
vertices at (–9, 3), (3, 3)

6) $36(x + 1)^2 + 9(y + 3)^2 = 324$

A) center at (0, –3)
foci at $(0, -3 - 3\sqrt{3})$, $(0, -3 + 3\sqrt{3})$
vertices at (0, 3), (0, –9)

B) center at (1, –3)
foci at $(1, -3 - 3\sqrt{3})$, $(1, -3 + 3\sqrt{3})$
vertices at (1, 3), (1, –9)

C) center at (–3, –1)
foci at $(-3, -1 - 3\sqrt{3})$, $(-3, -1 + 3\sqrt{3})$
vertices at (–3, 3), (–3, –9)

D) center at (–1, –3)
foci at $(-1, -3 - 3\sqrt{3})$, $(-1, -3 + 3\sqrt{3})$
vertices at (–1, 3), (–1, –9)

7) $3x^2 + 4y^2 - 36x + 40y + 196 = 0$

A) $\frac{(x-6)^2}{4} + \frac{(y+5)^2}{3} = 1$

center: (–6, 5); foci: (–5, 5), (–7, 5); vertices: (–8, 5), (–4, 5)

B) $\frac{(x-6)^2}{3} + \frac{(y+5)^2}{4} = 1$

center: (–6, 5); foci: (–5, 5), (–7, 5); vertices: (–8, 5), (–4, 5)

C) $\frac{(x-6)^2}{4} + \frac{(y+5)^2}{3} = 1$

center: (6, –5); foci: (7, –5), (5, –5); vertices: (8, –5), (4, –5)

D) $\frac{(x-6)^2}{3} + \frac{(y+5)^2}{4} = 1$

center: (6, –5); foci: (7, –5), (5, –5); vertices: (8, –5), (4, –5)

8) $64x^2 + y^2 - 640x + 1536 = 0$

A) $(x-5)^2 + \frac{y^2}{64} = 1$

center: (5, 0); foci: $(5, 3\sqrt{7})$, $(5, -3\sqrt{7})$; vertices:(5, 8), (5, –8)

B) $\frac{x^2}{25} + (y-8)^2 = 1$

center: (8, 0); foci: $(8, 2\sqrt{6})$, $(8, -3\sqrt{7})$; vertices:(8, 5),
(8, –5)

C) $\frac{x^2}{64} + (y-5)^2 = 1$

center: (5, 0); foci: $(5, 3\sqrt{7})$, $(5, -3\sqrt{7})$; vertices:(5, 8), (5, –8)

D) $(x-8)^2 + \frac{y^2}{25} = 1$

center: (8, 0); foci: $(8, 2\sqrt{6})$, $(8, -2\sqrt{6})$; vertices:(8, 5), (8, –5)

4 Work with Ellipses with Center at (h, k)

Write an equation for the ellipse.

1)

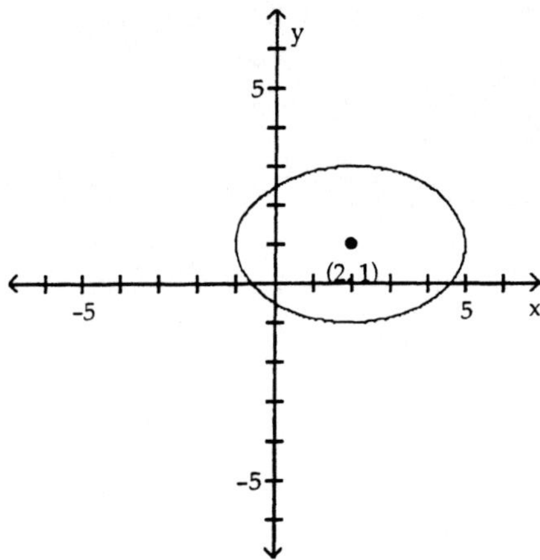

A) $\frac{(x+2)^2}{9}+\frac{(y+1)^2}{4}=1$

B) $\frac{(x-2)^2}{9}+\frac{(y-1)^2}{4}=1$

C) $\frac{(x-2)^2}{4}+\frac{(y-1)^2}{9}=1$

D) $\frac{(x-1)^2}{9}+\frac{(y-2)^2}{4}=1$

Find an equation for the ellipse. Graph the equation.

2) Foci at (-3, 1) and (-3, -5); length of major axis is 10

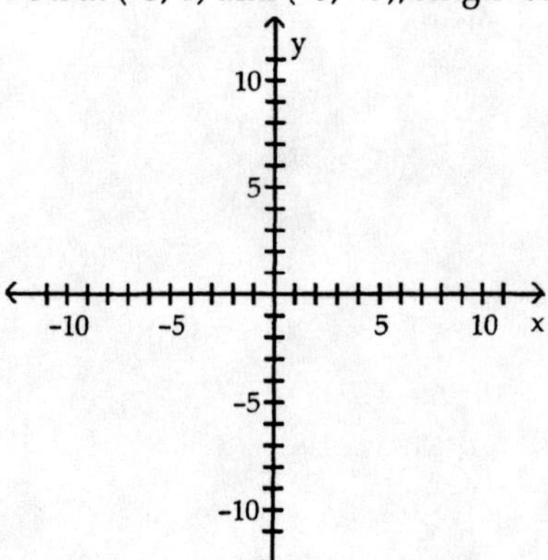

A)

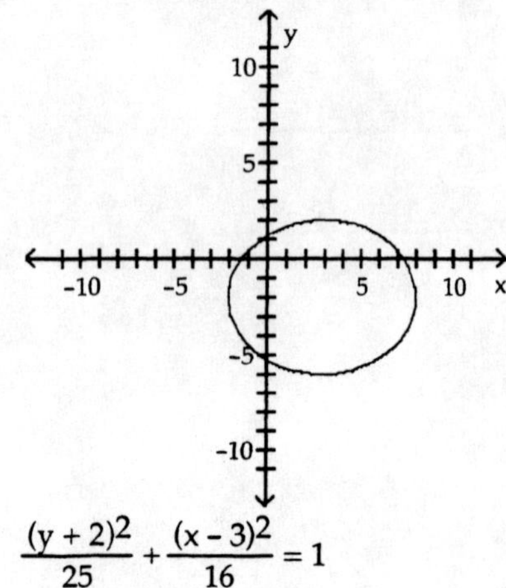

$$\frac{(y+2)^2}{25} + \frac{(x-3)^2}{16} = 1$$

B)

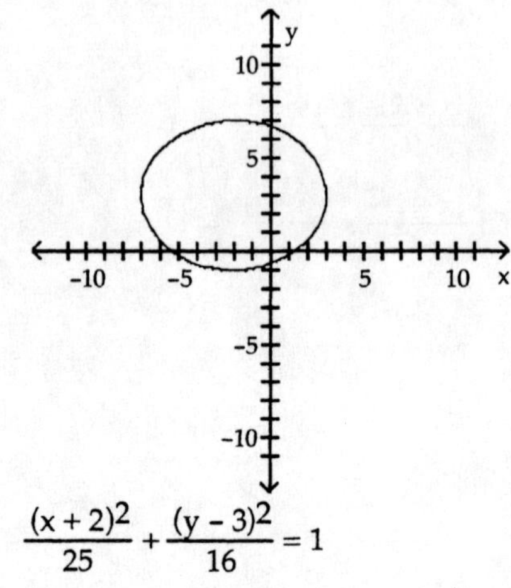

$$\frac{(x+2)^2}{25} + \frac{(y-3)^2}{16} = 1$$

C)

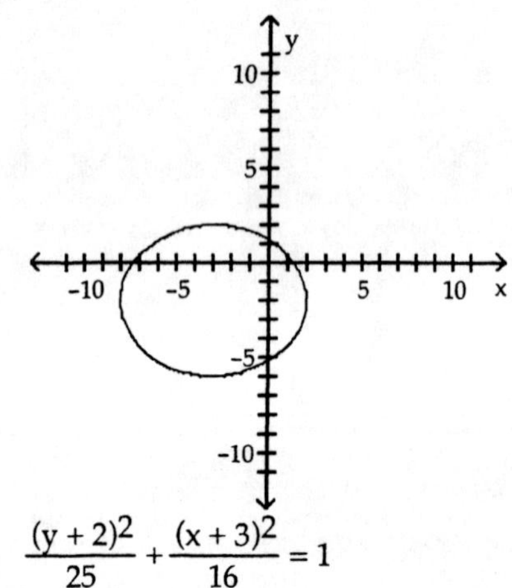

$$\frac{(y+2)^2}{25} + \frac{(x+3)^2}{16} = 1$$

D)

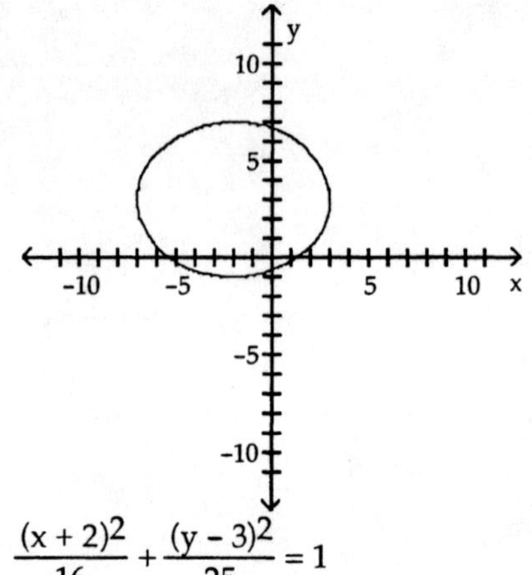

$$\frac{(x+2)^2}{16} + \frac{(y-3)^2}{25} = 1$$

Write an equation for the ellipse satisfying the given conditions.

3) Foci at (−4, −1) and (2, −1); length of major axis is 10

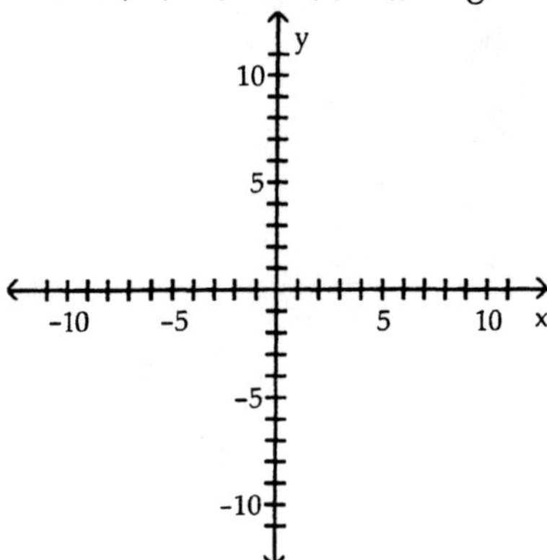

A)

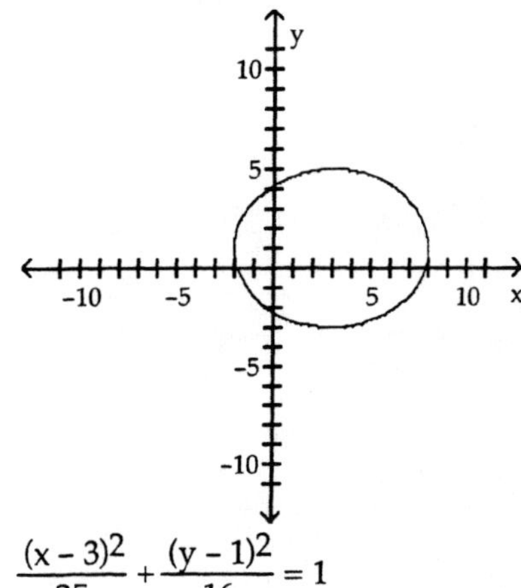

$$\frac{(x-3)^2}{25}+\frac{(y-1)^2}{16}=1$$

B)

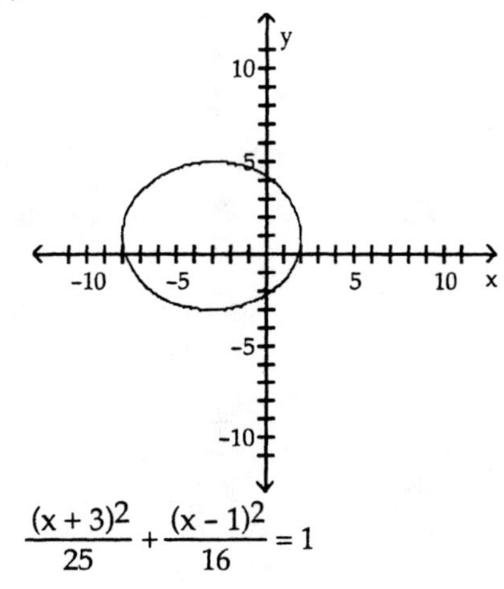

$$\frac{(x+3)^2}{25}+\frac{(x-1)^2}{16}=1$$

C)

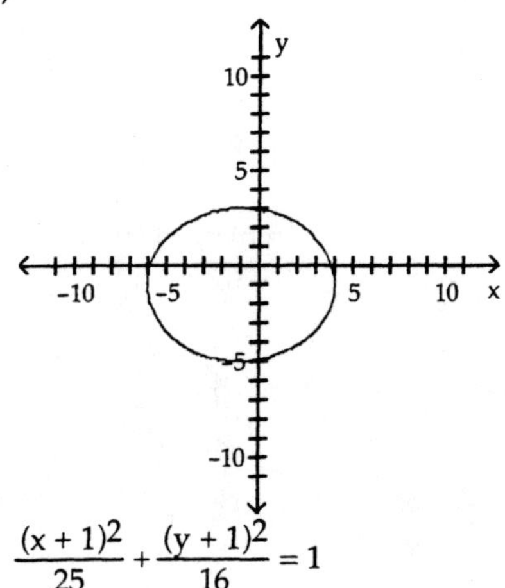

$$\frac{(x+1)^2}{25}+\frac{(y+1)^2}{16}=1$$

D)

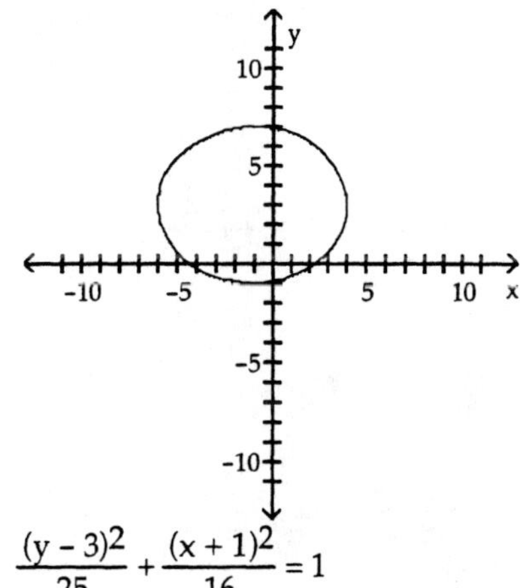

$$\frac{(y-3)^2}{25}+\frac{(x+1)^2}{16}=1$$

4) Vertices at (5, −4) and (5, 8); length of minor axis is 6

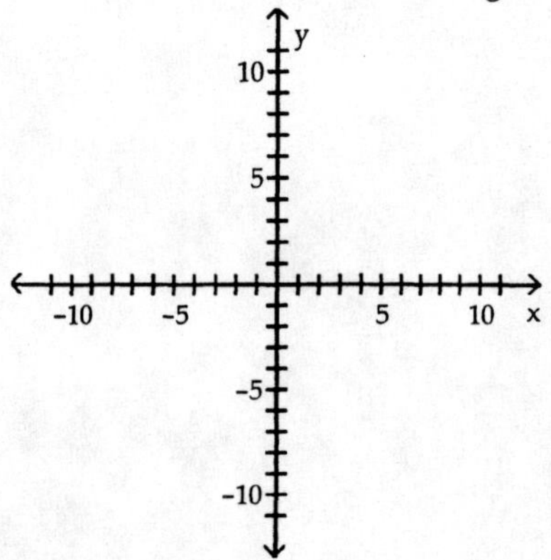

A)

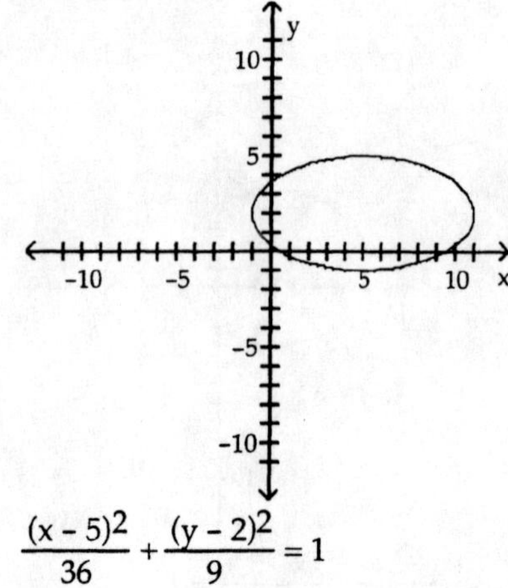

$\frac{(x-5)^2}{36}+\frac{(y-2)^2}{9}=1$

B)

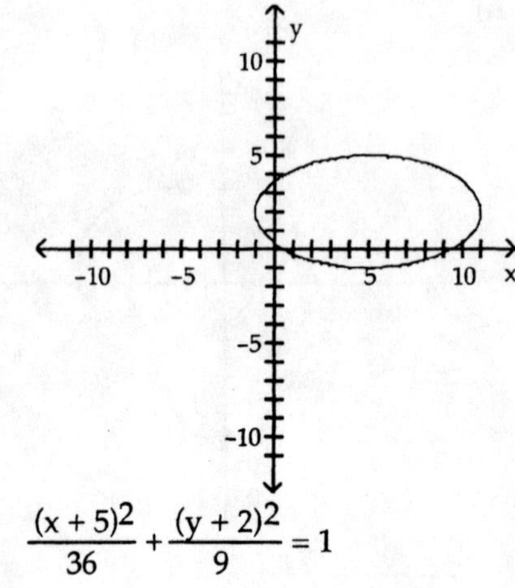

$\frac{(x+5)^2}{36}+\frac{(y+2)^2}{9}=1$

C)

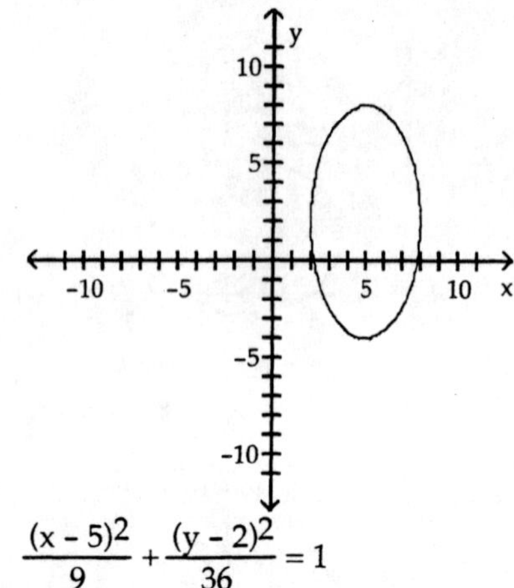

$\frac{(x-5)^2}{9}+\frac{(y-2)^2}{36}=1$

D)

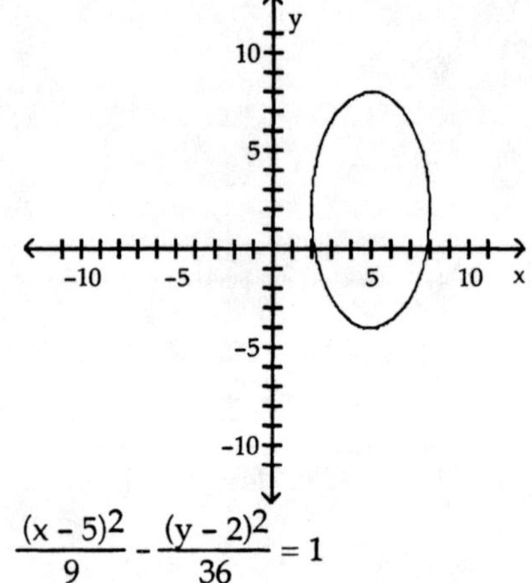

$\frac{(x-5)^2}{9}-\frac{(y-2)^2}{36}=1$

Find an equation for the ellipse. Graph the equation.

5) Foci at (-1, 2), and (-5, 2); vertex at (-8, 2)

A) $\frac{(x-2)^2}{21} + \frac{(y+3)^2}{25} = 1$

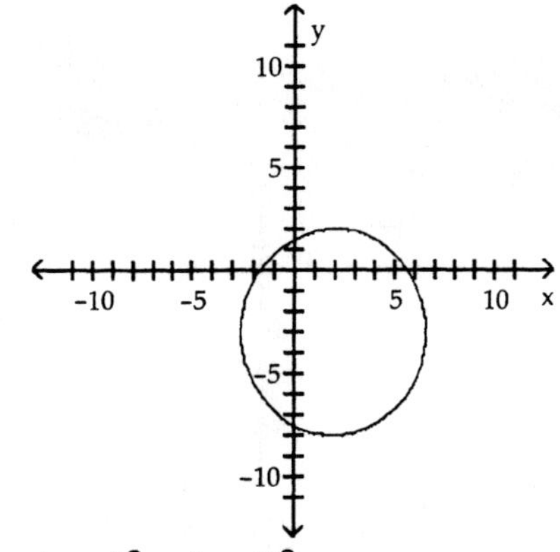

B) $\frac{(x+3)^2}{21} + \frac{(y-2)^2}{25} = 1$

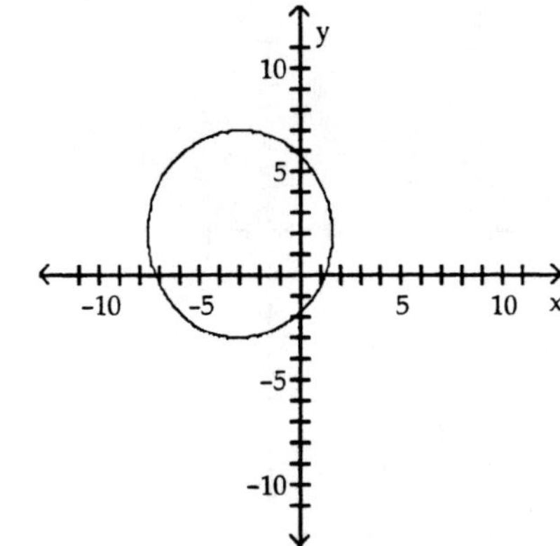

C) $\frac{(x+3)^2}{25} + \frac{(y-2)^2}{21} = 1$

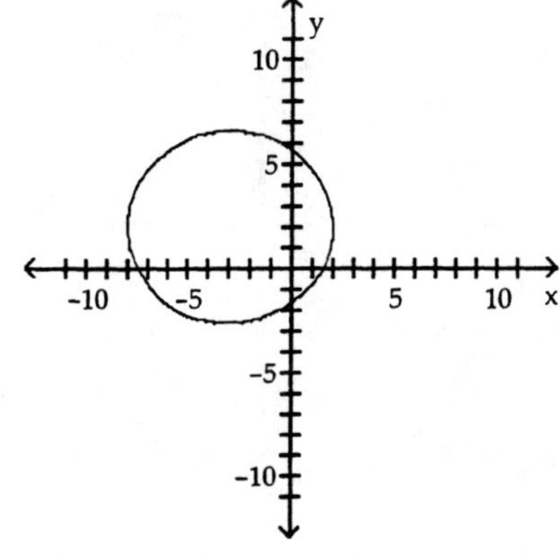

D) $\frac{(x-2)^2}{25} + \frac{(y+3)^2}{21} = 1$

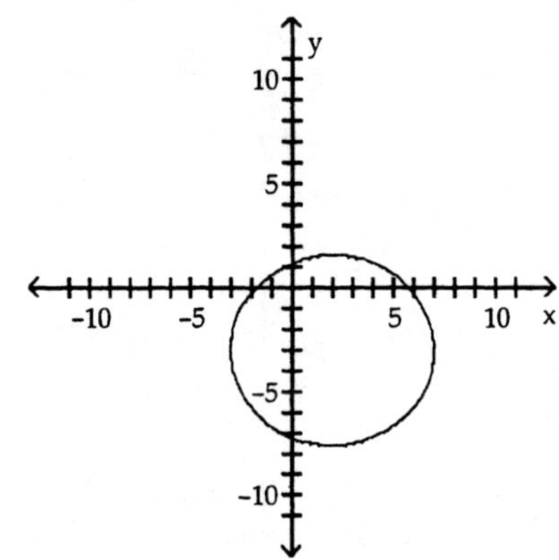

6) center at (5, 2); focus at (8, 2); contains the point (7.6, 2)

A) $\frac{(x+2)^2}{7} + \frac{(y-5)^2}{16} = 1$

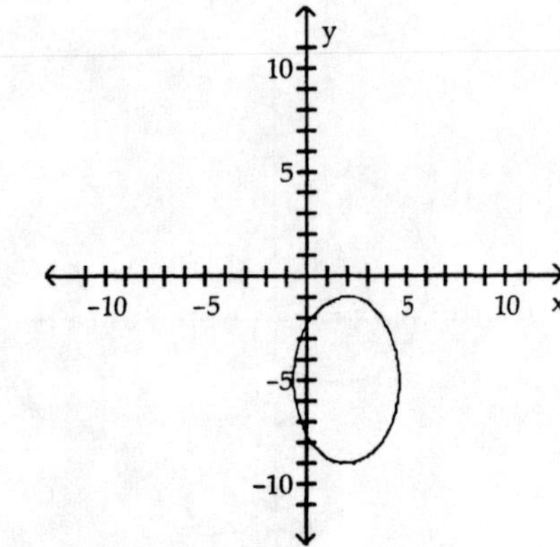

B) $\frac{(x+5)^2}{7} + \frac{(y-2)^2}{16} = 1$

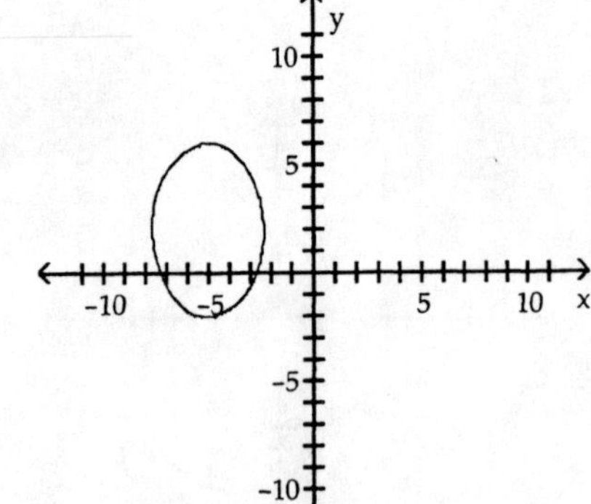

C) $\frac{(x+2)^2}{16} + \frac{(y-5)^2}{7} = 1$

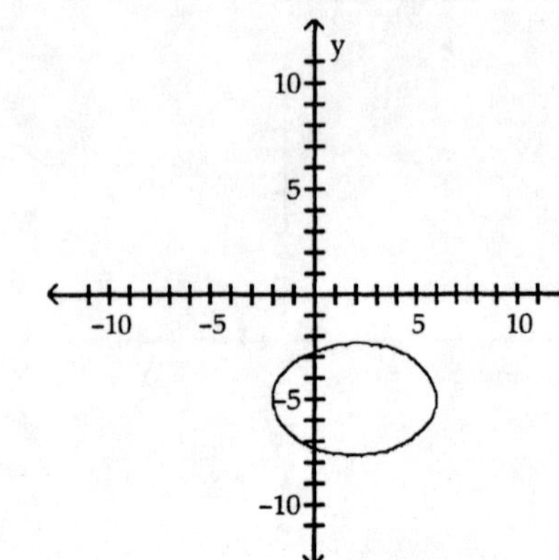

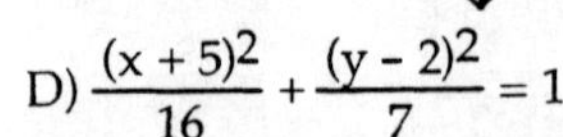

D) $\frac{(x+5)^2}{16} + \frac{(y-2)^2}{7} = 1$

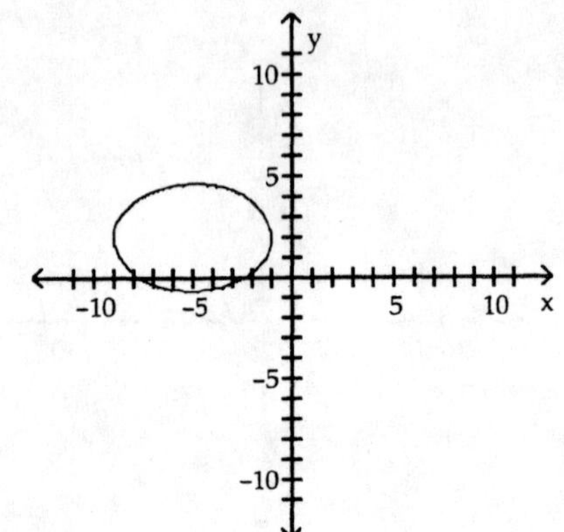

5 Solve Applied Problems Involving Ellipses

Solve.

1) A bridge is built in the shape of a semielliptical arch. It has a span of 90 feet. The height of the arch 29 feet from the center is to be 7 feet. Find the height of the arch at its center.

A) 29.36 feet B) 10.86 feet C) 7.39 feet D) 9.15 feet

2) An arch for a bridge over a highway is in the form of a semiellipse. The top of the arch is 30 feet above ground (the major axis). What should the span of the bridge be (the length of its minor axis) if the height 26 feet from the center is to be 16 feet above ground?

A) 97.5 feet B) 30.74 feet C) 61.47 feet D) 64.14 feet

3) The orbit of a planet around a sun is an ellipse with the sun at one focus. The aphelion of a planet is its greatest distance from the sun, its perihelion is its shortest distance, and its mean distance is the length of the semimajor axis of the elliptical orbit. If a planet has a perihelion of 471.7 million miles and a mean distance of 474 million miles, write an equation for the orbit of the planet around the sun.

A) $\frac{x^2}{474.006^2} + \frac{y^2}{474^2} = 1$

B) $\frac{x^2}{474^2} + \frac{y^2}{2.3^2} = 1$

C) $\frac{x^2}{474^2} + \frac{y^2}{471.7^2} = 1$

D) $\frac{x^2}{474^2} + \frac{y^2}{473.994^2} = 1$

4) An arch in the form of a semiellipse is 52 ft wide at the base and has a height of 20 ft. How wide is the arch at a height of 12 ft above the base?

A) 35.5 feet B) 17.7 feet C) 20.8 feet D) 41.6 feet

6.4 The Hyperbola

1 Find the Equation of a Hyperbola

Write an equation for the hyperbola.

1)

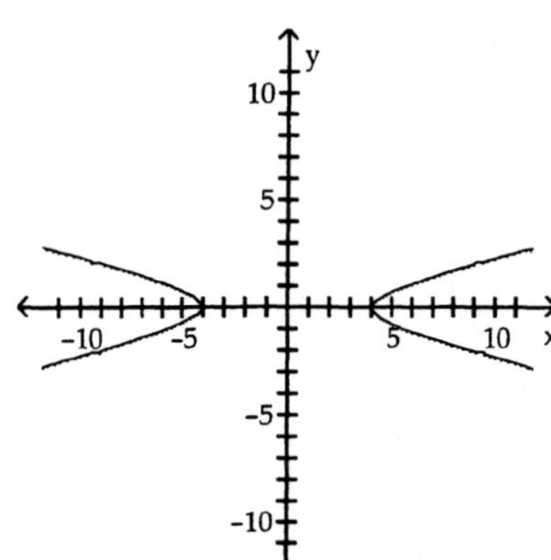

A) $\frac{y^2}{16} - x^2 = 1$ B) $x^2 - \frac{y^2}{16} = 1$ C) $y^2 - \frac{x^2}{16} = 1$ D) $\frac{x^2}{16} - y^2 = 1$

2)

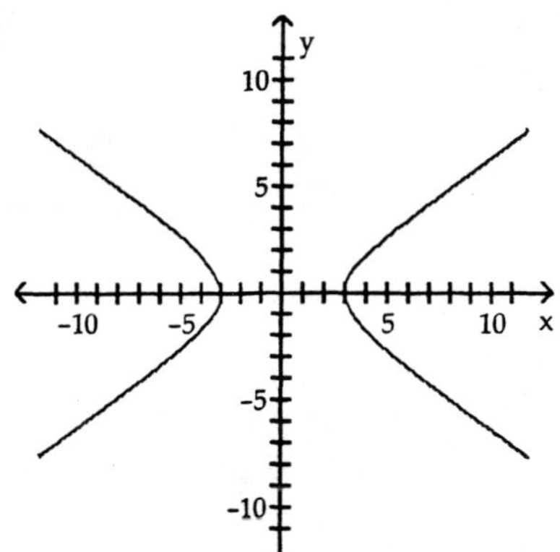

A) $\frac{y^2}{9} - \frac{x^2}{4} = 1$ B) $\frac{y^2}{4} - \frac{x^2}{9} = 1$ C) $\frac{x^2}{9} - \frac{y^2}{4} = 1$ D) $\frac{x^2}{4} - \frac{y^2}{9} = 1$

3)

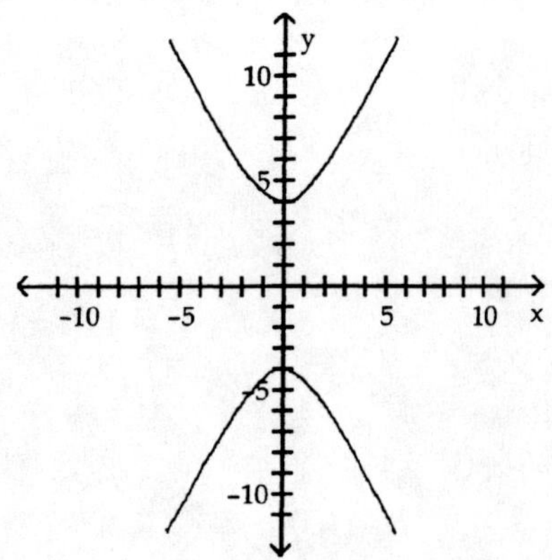

A) $\frac{x^2}{16} - \frac{y^2}{4} = 1$ B) $\frac{x^2}{4} - \frac{y^2}{16} = 1$ C) $\frac{y^2}{4} - \frac{x^2}{16} = 1$ D) $\frac{y^2}{16} - \frac{x^2}{4} = 1$

Find an equation for the hyperbola described.

4) Vertices at (0, ±10); asymptotes at $y = \pm \frac{5}{2}x$

A) $\frac{y^2}{100} - \frac{x^2}{4} = 1$ B) $\frac{y^2}{16} - \frac{x^2}{100} = 1$ C) $\frac{y^2}{4} - \frac{x^2}{25} = 1$ D) $\frac{y^2}{100} - \frac{x^2}{16} = 1$

5) Vertices at (±2, 0); foci at (±9, 0)

A) $\frac{x^2}{81} - \frac{y^2}{4} = 1$ B) $\frac{x^2}{4} - \frac{y^2}{81} = 1$ C) $\frac{x^2}{4} - \frac{y^2}{77} = 1$ D) $\frac{x^2}{77} - \frac{y^2}{4} = 1$

2 Graph Hyperbolas

Graph the equation.

1) $\frac{x^2}{4} - \frac{y^2}{16} = 1$

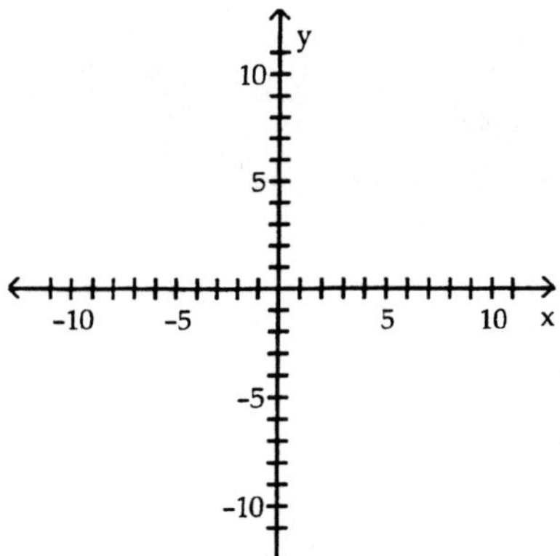

A)

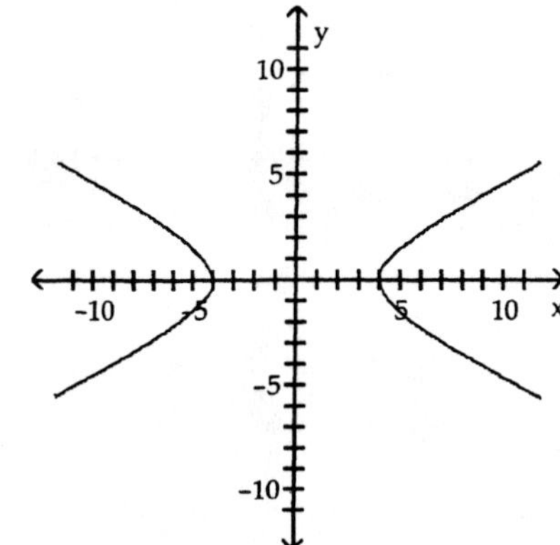

B)

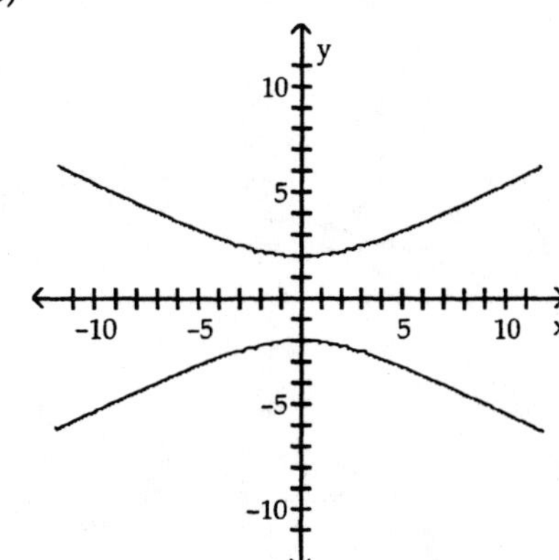

C)

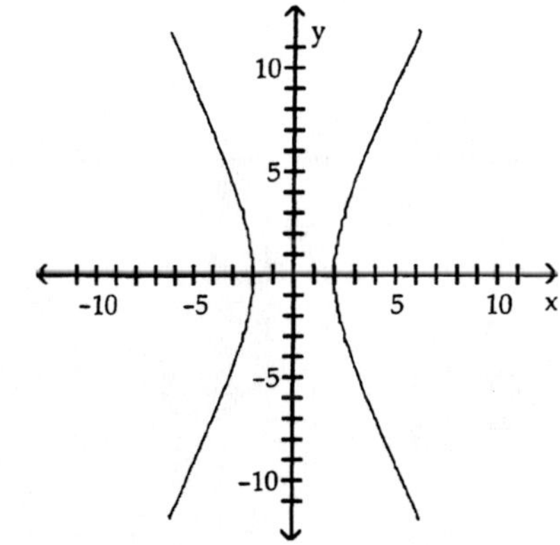

D)

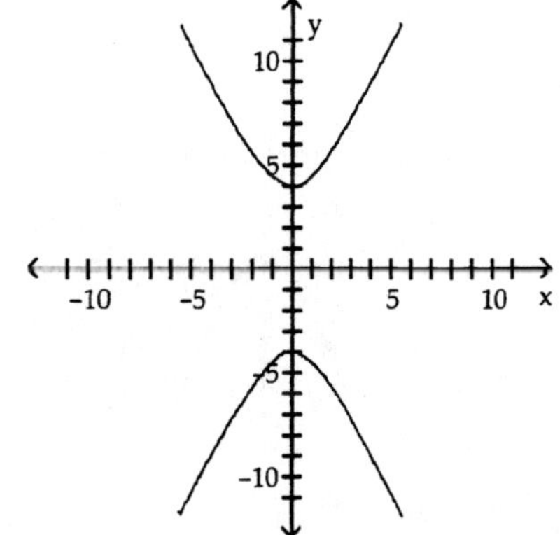

2) $\frac{y^2}{4} - \frac{x^2}{25} = 1$

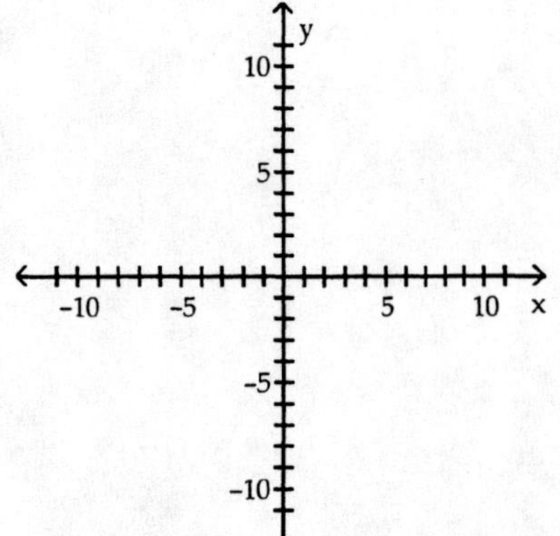

A)

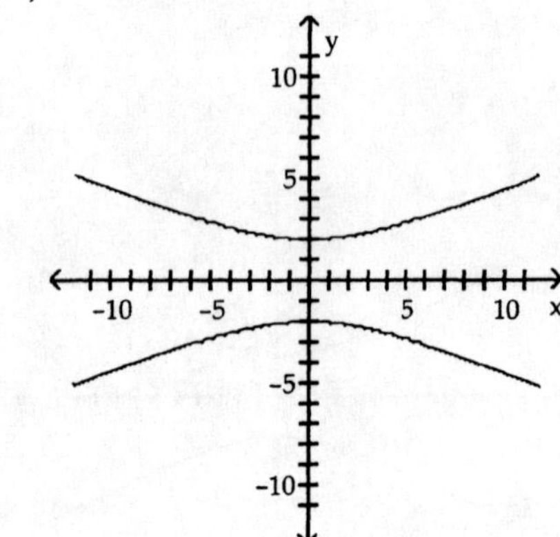

B)

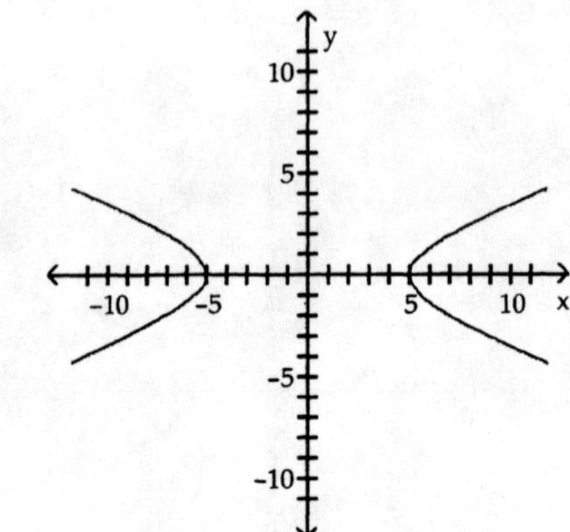

C)

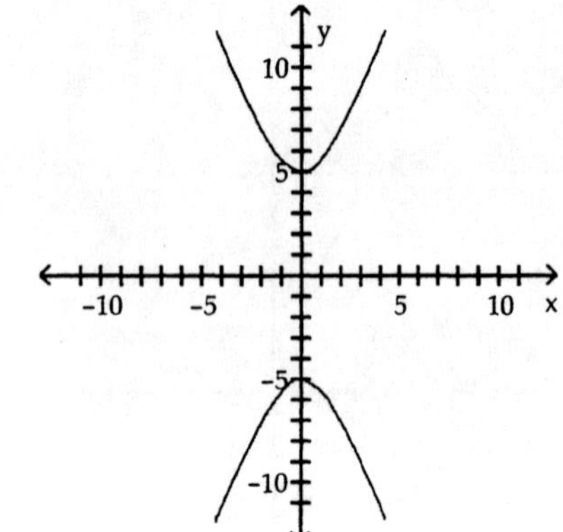

D)

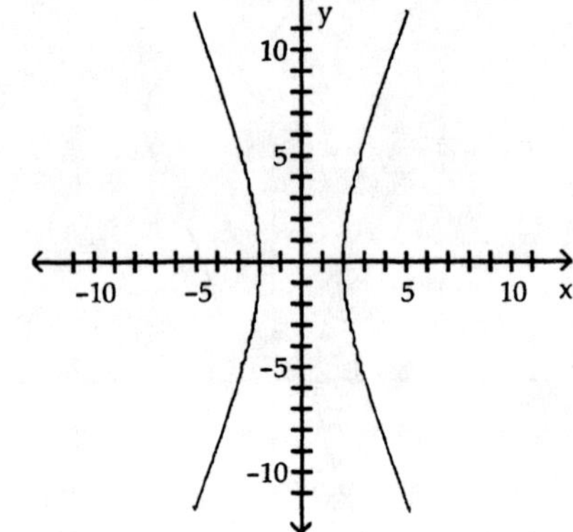

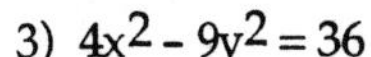

3) $4x^2 - 9y^2 = 36$

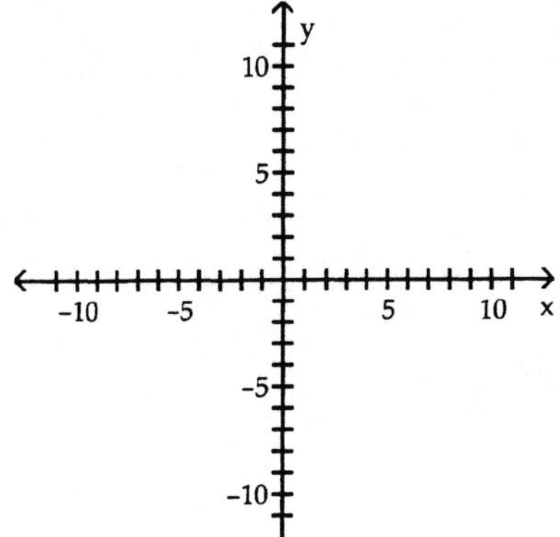

A)

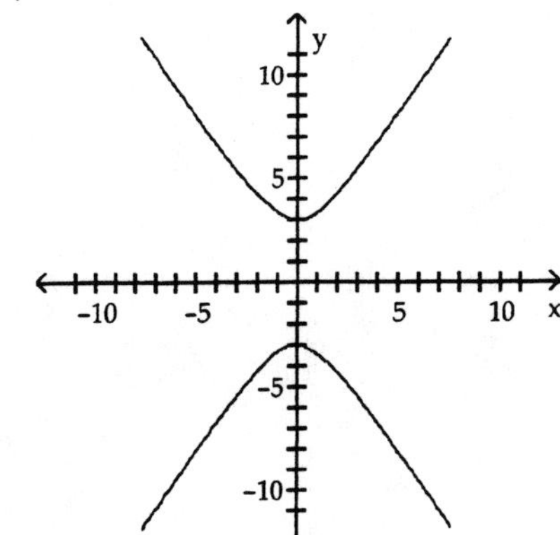

B)

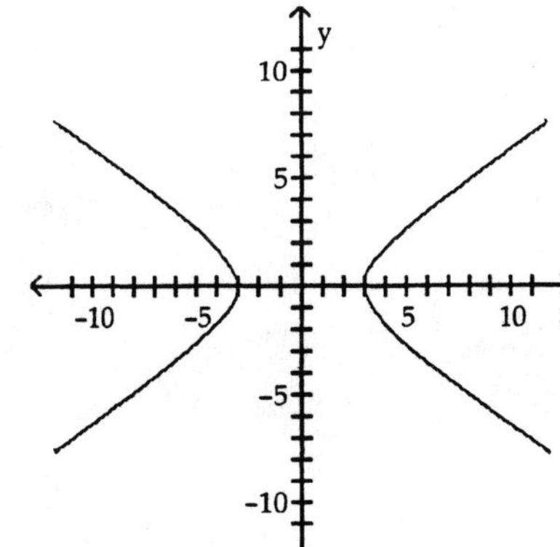

C)

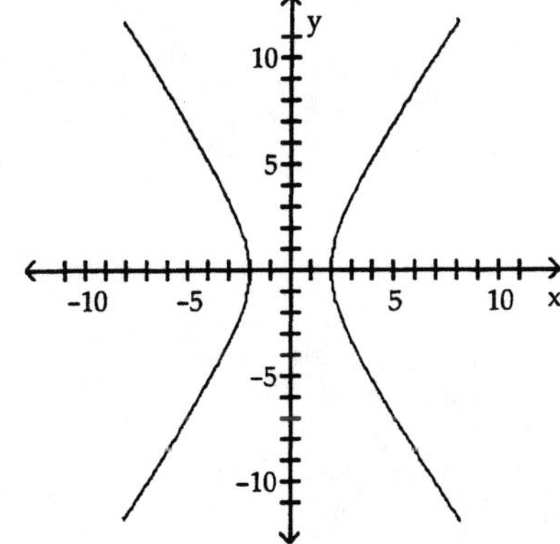

D)

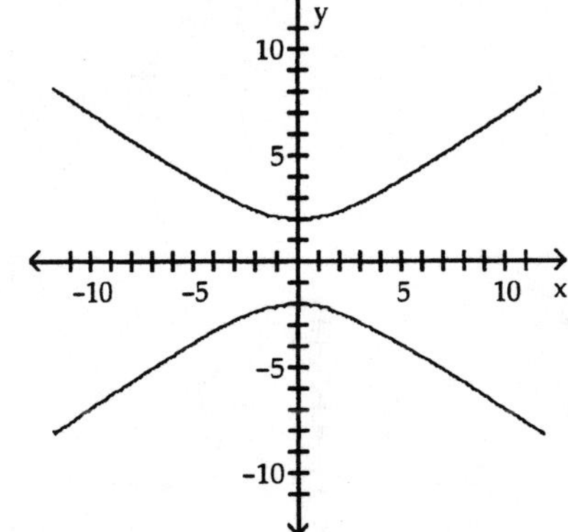

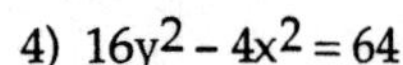

4) $16y^2 - 4x^2 = 64$

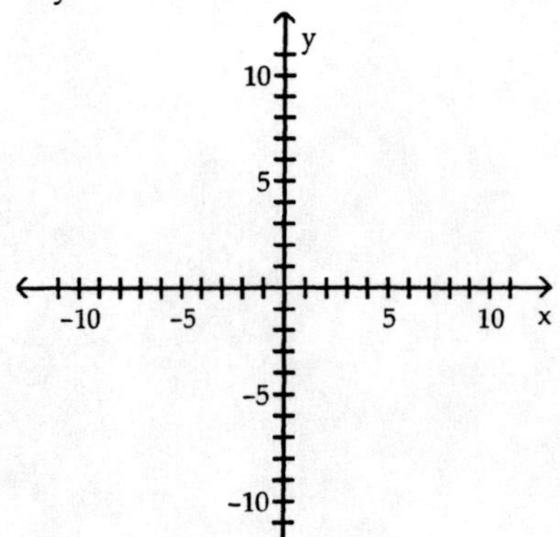

A)

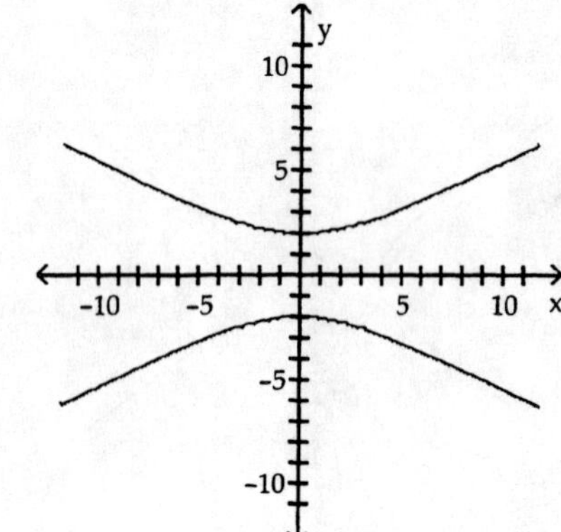

B)

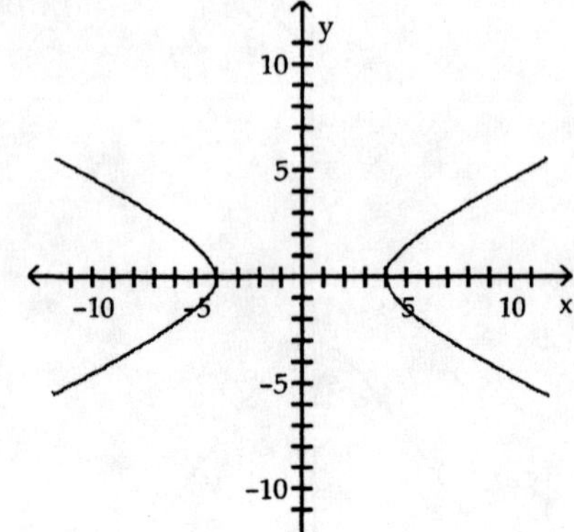

C)

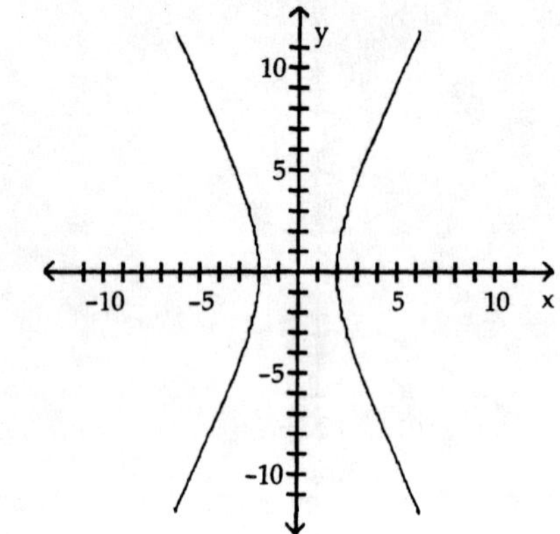

D)

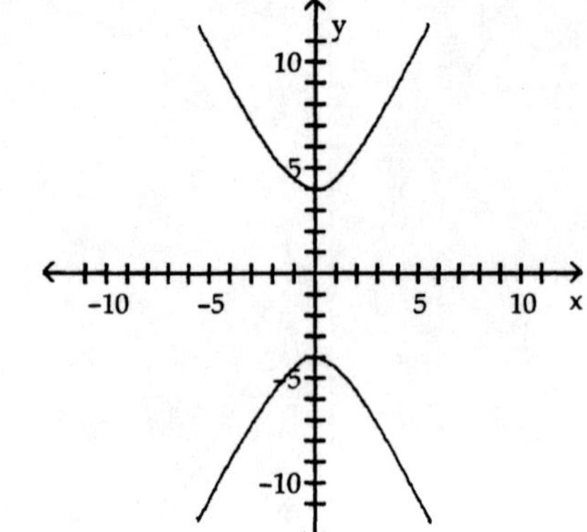

Find an equation for the hyperbola described. Graph the equation.

5) Center at (0, 0); focus at (5, 0); vertex at (3, 0)

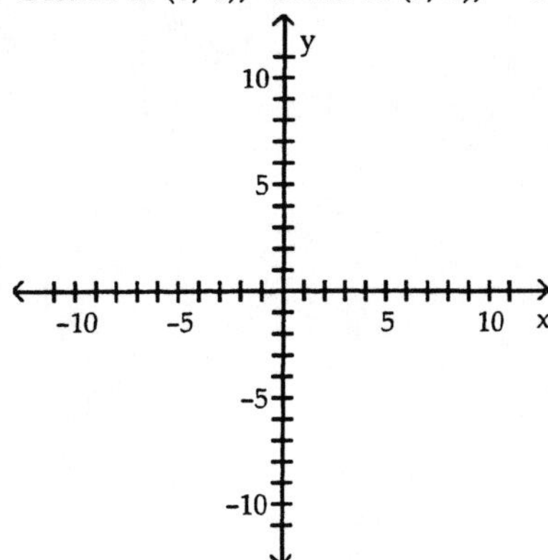

A)

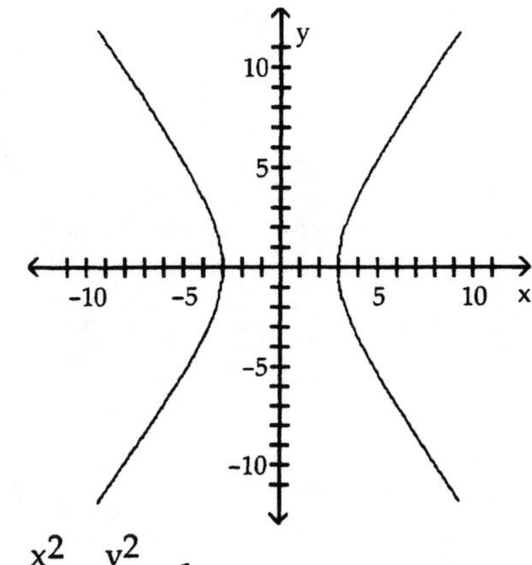

$\frac{x^2}{9} - \frac{y^2}{16} = 1$

B)

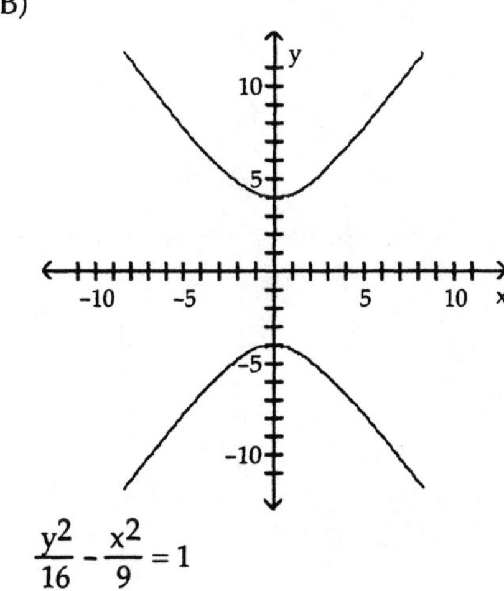

$\frac{y^2}{16} - \frac{x^2}{9} = 1$

C)

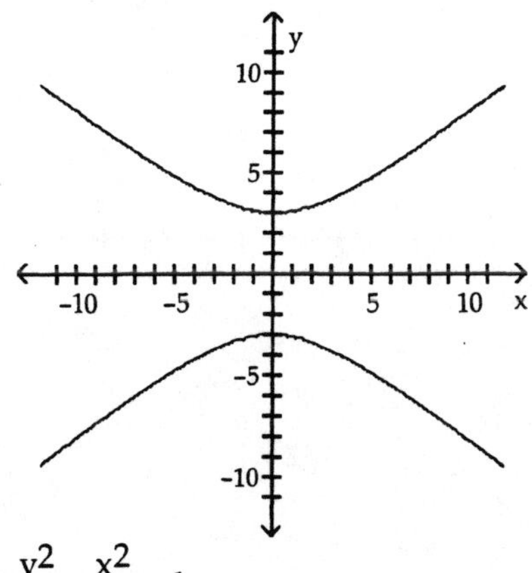

$\frac{y^2}{9} - \frac{x^2}{16} = 1$

D)

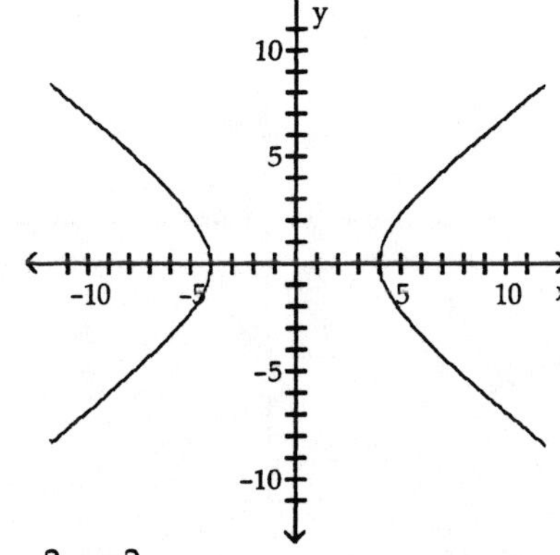

$\frac{x^2}{16} - \frac{y^2}{9} = 1$

6) Center at (0, 0); vertex at (0, 8); focus at $(0, \sqrt{113})$

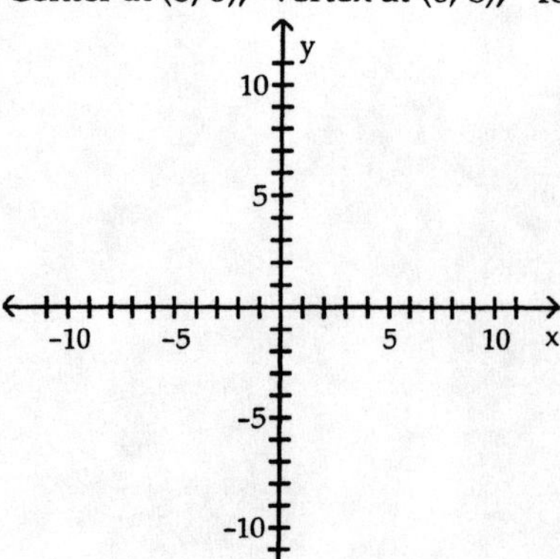

A)

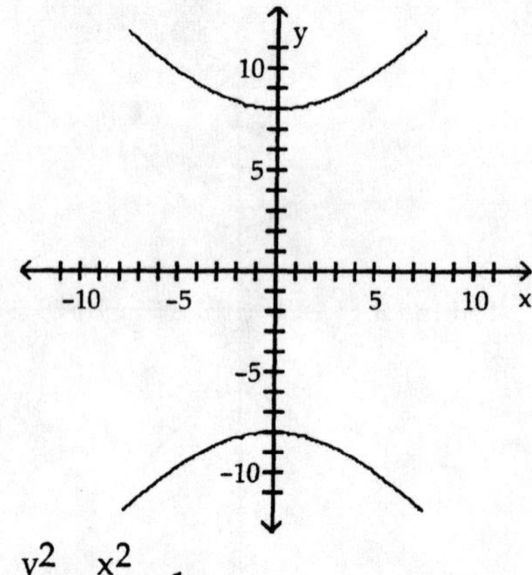

$\frac{y^2}{64} - \frac{x^2}{49} = 1$

B)

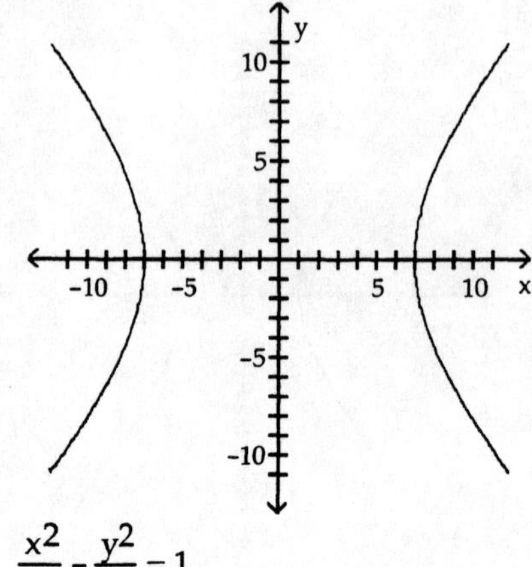

$\frac{x^2}{49} - \frac{y^2}{64} = 1$

C)

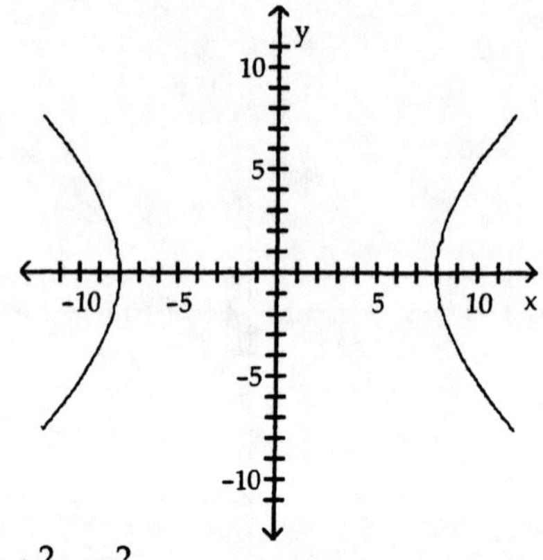

$\frac{x^2}{64} - \frac{y^2}{49} = 1$

D)

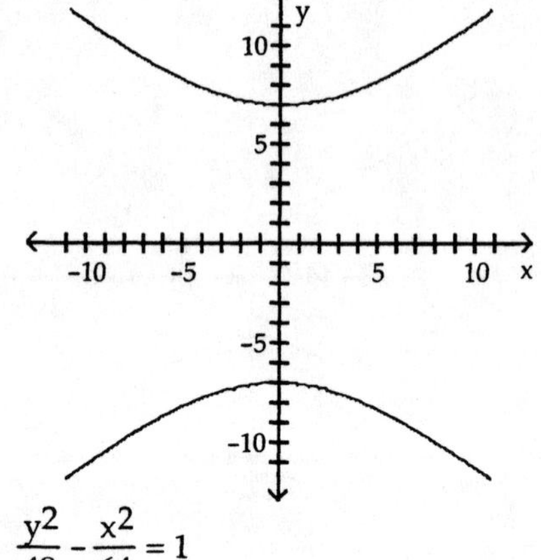

$\frac{y^2}{49} - \frac{x^2}{64} = 1$

Find the center, transverse axis, vertices, foci, and asymptotes of the hyperbola.

1) $\frac{x^2}{144} - \frac{y^2}{64} = 1$

A) center at (0, 0)
transverse axis is x-axis
vertices at (-12, 0) and (12, 0)
foci at $(-4\sqrt{13}, 0)$ and $(4\sqrt{13}, 0)$
asymptotes of $y = -\frac{2}{3}$ and $y = \frac{2}{3}$

B) center at (0, 0)
transverse axis is x-axis
vertices at (-8, 0) and (8, 0)
foci at $(-4\sqrt{13}, 0)$ and $(4\sqrt{13}, 0)$
asymptotes of $y = -\frac{2}{3}$ and $y = \frac{2}{3}$

C) center at (0, 0)
transverse axis is x-axis
vertices at (-12, 0) and (12, 0)
foci at (-8, 0) and (8, 0)
asymptotes of $y = -\frac{2}{3}$ and $y = \frac{2}{3}$

D) center at (0, 0)
transverse axis is y-axis
vertices at (0, -12) and (0, 12)
foci at $(-4\sqrt{13}, 0)$ and $(4\sqrt{13}, 0)$
asymptotes of $y = -\frac{2}{3}$ and $y = \frac{2}{3}$

2) $36y^2 - 25x^2 = 900$

A) center at (0, 0)
transverse axis is x-axis
vertices: (-6, 0), (6, 0)
foci: $(-\sqrt{61}, 0)$, $(\sqrt{61}, 0)$
asymptotes of $y = -\frac{5}{6}$ and $y = \frac{5}{6}$

B) center at (0, 0)
transverse axis is x-axis
vertices: (-5, 0), (5, 0)
foci: (-6, 0), (6, 0)
asymptotes of $y = -\frac{5}{6}$ and $y = \frac{5}{6}$

C) center at (0, 0)
transverse axis is y-axis
vertices: (0, -5), (0, 5)
foci: $(-\sqrt{61}, 0)$, $(\sqrt{61}, 0)$
asymptotes of $y = -\frac{5}{6}$ and $y = \frac{5}{6}$

D) center at (0, 0)
transverse axis is y-axis
vertices at (0, -5) and (0, 5)
foci at $(0, -\sqrt{61})$ and $(0, \sqrt{61})$
asymptotes of $y = -\frac{5}{6}$ and $y = \frac{5}{6}$

3) $\frac{(x+1)^2}{25} - \frac{(y+2)^2}{16} = 1$

A) center at (-1, -2)
transverse axis is parallel to y-axis
vertices at (-1, -7) and (-1, 3)
foci at $(-1, -2 - \sqrt{41})$ and $(-1, -2 + \sqrt{41})$
asymptotes of $y - 2 = -\frac{5}{4}(x - 1)$ and $y - 2 = \frac{5}{4}(x - 1)$

B) center at (-2, -1)
transverse axis is parallel to x-axis
vertices at (-7, -1) and (3, -1)
foci at $(-2 - \sqrt{41}, -1)$ and $(-2 + \sqrt{41}, -1)$
asymptotes of $y + 1 = -\frac{4}{5}(x + 2)$ and $y + 1 = \frac{4}{5}(x + 2)$

C) center at (-1, -2)
transverse axis is parallel to x-axis
vertices at (-5, -2) and (3, -2)
foci at $(-1 - \sqrt{41}, -2)$ and $(-1 + \sqrt{41}, -2)$
asymptotes of $y + 2 = -\frac{5}{4}(x + 1)$ and $y + 2 = \frac{5}{4}(x + 1)$

D) center at (-1, -2)
transverse axis is parallel to x-axis
vertices at (-6, -2) and (4, -2)
foci at $(-1 - \sqrt{41}, -2)$ and $(-1 + \sqrt{41}, -2)$
asymptotes of $y + 2 = -\frac{4}{5}(x + 1)$ and $y + 2 = \frac{4}{5}(x + 1)$

4) $(x + 4)^2 - 25(y - 3)^2 = 25$

A) center at (3, –4)
transverse axis is parallel to x–axis
vertices at (–2, –4) and (8, –4)
foci at $(3 - \sqrt{26}, -4)$ and $(3 + \sqrt{26}, -4)$
asymptotes of $y + 4 = -\frac{1}{5}(x - 3)$ and $y + 4 = \frac{1}{5}(x - 3)$

B) center at (–4, 3)
transverse axis is parallel to x–axis
vertices at (–5, 3) and (–3, 3)
foci at $(-4 - \sqrt{26}, 3)$ and $(-4 + \sqrt{26}, 3)$
asymptotes of $y - 3 = -5(x + 4)$ and $y - 3 = 5(x + 4)$

C) center at (–4, 3)
transverse axis is parallel to x–axis
vertices at (–9, 3) and (1, 3)
foci at $(-4 - \sqrt{26}, 3)$ and $(-4 + \sqrt{26}, 3)$
asymptotes of $y - 3 = -\frac{1}{5}(x + 4)$ and $y - 3 = \frac{1}{5}(x + 4)$

D) center at (–4, 3)
transverse axis is parallel to y–axis
vertices at (–4, –2) and (–4, 8),
foci at $(-4, 3 - \sqrt{26})$ and $(-4, 3 + \sqrt{26})$,
asymptotes of $y + 3 = -5(x - 4)$ and $y + 3 = 5(x - 4)$

5) $x^2 - 9y^2 - 4x + 18y - 14 = 0$

A) center at (2, 1)
transverse axis is parallel to x–axis
vertices at (–1, 1) and (5, 1)
foci at $(2 - \sqrt{10}, 1)$ and $(2 + \sqrt{10}, 1)$
asymptotes of $y - 1 = -\frac{1}{3}(x - 2)$ and $y - 1 = \frac{1}{3}(x - 2)$

B) center at (2, 1)
transverse axis is parallel to x–axis
vertices at (1, 1) and (3, 1)
foci at $(2 - \sqrt{10}, 1)$ and $(2 + \sqrt{10}, 1)$
asymptotes of $y - 1 = -3(x - 2)$ and $y - 1 = 3(x - 2)$

C) center at (2, 1)
transverse axis is parallel to y–axis
vertices at (2, –2) and (2, 4)
foci at $(2, 1 - \sqrt{10})$ and $(2, 1 + \sqrt{10})$
asymptotes of $y + 1 = -3(x + 2)$ and $y + 1 = 3(x + 2)$

D) center at (1, 2)
transverse axis is parallel to x–axis
vertices at (–2, 2) and (4, 2)
foci at $(1 - \sqrt{10}, 2)$ and $(1 + \sqrt{10}, 2)$
asymptotes of $y - 2 = -\frac{1}{3}(x - 1)$ and $y - 2 = \frac{1}{3}(x - 1)$

4 Find the Asymptotes of a Hyperbola

Find the asymptotes of the hyperbola.

1) $\frac{x^2}{25} - \frac{y^2}{9} = 1$

A) $y = \frac{5}{3}x$ and $y = -\frac{5}{3}x$

B) $y = \frac{9}{25}x$ and $y = -\frac{9}{25}x$

C) $y = \frac{25}{9}x$ and $y = -\frac{25}{9}x$

D) $y = \frac{3}{5}x$ and $y = -\frac{3}{5}x$

2) $y^2 - x^2 = 25$

A) $y = 5x$ and $y = -5x$

B) $y = x$ and $y = -x$

C) $y = \frac{1}{5}x$ and $y = -\frac{1}{5}x$

D) $y = \frac{1}{25}x$ and $y = -\frac{1}{25}x$

3) $\frac{(x-1)^2}{25} - \frac{(y+1)^2}{16} = 1$

A) $y = \frac{4}{5}(x-1)$ and $y = -\frac{4}{5}(x-1)$

B) $y + 1 = \frac{4}{5}(x-1)$ and $y + 1 = -\frac{4}{5}(x-1)$

C) $y - 1 = \frac{4}{5}(x+1)$ and $y - 1 = -\frac{4}{5}(x+1)$

D) $y + 1 = \frac{5}{4}(x-1)$ and $y + 1 = -\frac{5}{4}(x-1)$

4) $x^2 - y^2 + 6x - 2y - 8 = 0$

A) $y + 3 = (x+1)$ and $y + 3 = -(x+1)$

B) $y + 1 = \frac{1}{4}(x+3)$ and $y + 1 = -\frac{1}{4}(x+3)$

C) $y + 1 = (x+3)$ and $y + 1 = -(x+3)$

D) $y - 3 = (x-1)$ and $y - 3 = -(x-1)$

5 Work with Hyperbolas with Center at (h, k)

Find an equation for the hyperbola described.

1) Vertices $(\frac{1}{2}, -3)$ and $(-\frac{9}{2}, -3)$; asymptotes $y + 3 = \pm\frac{6}{5}(x+2)$

A) $\frac{4(x-2)^2}{25} - \frac{(y-3)^2}{9} = 1$

B) $\frac{(y+3)^2}{9} - \frac{4(x+2)^2}{25} = 1$

C) $\frac{(x+2)^2}{9} - \frac{4(y+3)^2}{25} = 1$

D) $\frac{4(x+2)^2}{25} - \frac{(y+3)^2}{9} = 1$

2) Foci at (-3, 4) and (7, 4); length of the transverse axis is 4

A) $\frac{(x-2)^2}{4} - \frac{(y-4)^2}{21} = 1$

B) $\frac{(y-4)^2}{29} - \frac{(x-2)^2}{4} = 1$

C) $\frac{(x-2)^2}{4} - \frac{(y-4)^2}{29} = 1$

D) $\frac{(y-4)^2}{21} - \frac{(x-2)^2}{4} = 1$

3) center at (6, 5); focus at (1, 5); vertex at (5, 5)

A) $(x - 5)^2 - \frac{(y - 6)^2}{24} = 1$

B) $\frac{(x - 5)^2}{24} - (y - 6)^2 = 1$

C) $(x - 6)^2 - \frac{(y - 5)^2}{24} = 1$

D) $\frac{(x - 6)^2}{24} - (y - 5)^2 = 1$

Find the center, transverse axis, vertices, foci, and asymptotes of the hyperbola.

4) $(x + 7)^2 - 49(y - 2)^2 = 49$

5) $36x^2 - y^2 - 576x - 4y + 1004 = 0$

6 Solve Applied Problems Involving Hyperbolas

Solve.

1) Two LORAN stations are positioned 274 miles apart along a straight shore. A ship records a time difference of 0.00108 seconds between the LORAN signals. (The radio signals travel at 186,000 miles per second.) Where will the ship reach shore if it were to follow the hyperbola corresponding to this time difference? If the ship wants to enter a harbor located between the two stations 22 miles from the master station, what time difference should it be looking for?

A) 37 miles from the master station, 0.0006183 seconds

B) 100 miles from the master station, 0.0012366 seconds

C) 100 miles from the master station, 0.0006183 seconds

D) 37 miles from the master station, 0.0012366 seconds

2) Two recording devices are set 4000 feet apart, with the device at point A to the west of the device at point B. At a point on a line between the devices, 300 feet from point B, a small amount of explosive is detonated. The recording devices record the time the sound reaches each one. How far directly north of site B should a second explosion be done so that the measured time difference recorded by the devices is the same as that for the first detonation?

A) 6944.3 feet

B) 652.94 feet

C) 1977.37 feet

D) 1626.76 feet

Ch. 6 Conics
Answer Key

6.1 Preliminaries
1 Know the Names of the Conics

1) C
2) B
3) A
4) C

6.2 The Parabola
1 Find the Equation of a Parabola

1) D
2) C
3) D
4) D
5) D
6) B
7) A
8) A
9) B
10) C
11) A
12) C

2 Graph Parabolas

1) C
2) A
3) C
4) D
5) D
6) B

3 Discuss the Equation of a Parabola

1) No Correct Answer Was Provided.

4 Work with Parabolas with Vertex at (h, k)

1) D
2) C
3) C

5 Solve Applied Problems Involving Parabolas

1) D
2) B
3) D

6.3 The Ellipse
1 Find the Equation of an Ellipse

1) A
2) A
3) B
4) D
5) A
6) B

2 Graph Ellipses

1) B
2) A
3) B
4) A
5) A
6) A

3 Discuss the Equation of an Ellipse

1) D
2) B
3) C
4) D
5) C
6) D
7) C
8) A

4 Work with Ellipses with Center at (h, k)

1) B
2) C
3) C
4) C
5) C
6) D

5 Solve Applied Problems Involving Ellipses

1) D
2) C
3) D
4) D

6.4 The Hyperbola

1 Find the Equation of a Hyperbola

1) D
2) C
3) D
4) D
5) C

2 Graph Hyperbolas

1) C
2) A
3) B
4) A
5) A
6) A

3 Discuss the Equation of a Hyperbola

1) A
2) D
3) D
4) C
5) A

4 Find the Asymptotes of a Hyperbola

1) D
2) B
3) B
4) C

5 Work with Hyperbolas with Center at (h, k)

1) D
2) A
3) C

4) $\frac{(x+7)^2}{49} - (y-2)^2 = 1$

center: (-7, 2)
transverse axis: parallel to x-axis
vertices: (0, 2), (-14, 2)
foci: (0.1, 2), (-14.1, 2)
asymptotes: $y - 2 = \pm\frac{1}{7}(x + 7)$

5) $\frac{(x-8)^2}{36} - \frac{(y+2)^2}{1296} = 1$

center: (8, -2)
transverse axis: parallel to x-axis
vertices: (14, -2), (2, -2)
foci: (44.5, -2), (-28.5, -2)
asymptotes: $y + 2 = \pm 6(x - 8)$

6 Solve Applied Problems Involving Hyperbolas

1) D
2) B

Ch. 7 Systems of Equations and Inequalities

7.1 Systems of Linear Equations: Substitution and Elimination

1 Solve Systems of Equations by Substitution

Verify that the values of the variables listed are solutions of the system of equations.

1)
$$\begin{cases} x + y = 2 \\ x - y = -6 \end{cases}$$
$x = -2, y = 4$

A) solution B) not a solution

2)
$$\begin{cases} x + y = 0 \\ x - y = -8 \end{cases}$$
$x = 4, y = 4$

A) solution B) not a solution

3)
$$\begin{cases} 3x + y = 7 \\ 4x + 3y = 16 \end{cases}$$
$x = 1, y = 4$

A) solution B) not a solution

4)
$$\begin{cases} 2x + y = -7 \\ 3x + 2y = -9 \end{cases}$$
$x = -5, y = -3$

A) solution B) not a solution

Solve the system of equations by substitution.

5)
$$\begin{cases} x + y = -4 \\ x - y = 11 \end{cases}$$

A) $x = 3.5, y = 7.5$ B) $x = 3.5, y = -7.5$ C) $x = 4, y = 3.5$ D) $x = 4, y = -7.5$

6)
$$\begin{cases} x + 7y = -2 \\ 3x + y = 34 \end{cases}$$

A) $x = 12, y = -2$ B) $x = 3, y = 7$ C) $x = -2, y = 3$ D) $x = 7, y = 12$

7)
$$\begin{cases} x + 3y = 3 \\ 9x - 8y = -8 \end{cases}$$

A) $x = 0, y = 0$ B) $x = 1, y = 1$ C) $x = 1, y = 0$ D) $x = 0, y = 1$

8)
$$\begin{cases} 5x - 2y = -1 \\ x + 4y = 35 \end{cases}$$

A) $x = 2, y = 9$ B) $x = 3, y = 9$ C) $x = 3, y = 8$ D) $x = 2, y = 8$

9)
$$\begin{cases} 2x + 3y = 6 \\ 4x = -12 \end{cases}$$
A) x = -4, y = 0　　B) x = -3, y = 4　　C) x = -3, y = -4　　D) x = 4, y = -3

10)
$$\begin{cases} 5x + 3y = 80 \\ 2x + y = 30 \end{cases}$$
A) x = 10, y = 0　　B) x = 0, y = 10　　C) x = 10, y = 10　　D) x = 0, y = 0

11)
$$\begin{cases} 8x + y = 0 \\ -8x + y = -16 \end{cases}$$
A) x = -1, y = -8　　B) x = 1, y = -8　　C) x = 1, y = 16　　D) x = -1, y = 8

12)
$$\begin{cases} x + y = 0 \\ 2x + 3y = -7 \end{cases}$$
A) x = -7, y = 7　　B) x = 6, y = -6　　C) x = -6, y = 6　　D) x = 7, y = -7

13)
$$\begin{cases} 3x + y = 13 \\ 2x - 7y = 24 \end{cases}$$
A) x = 5, y = -2　　B) x = 5, y = 2　　C) x = -5, y = -2　　D) x = -5, y = 2

14)
$$\begin{cases} \frac{x}{5} - y = -\frac{4}{5} \\ x + 8y = -4 \end{cases}$$
A) x = 0, y = -4　　B) x = -4, y = 0　　C) x = 4, y = 0　　D) x = 0, y = 4

15)
$$\begin{cases} \frac{x}{2} + \frac{2y}{3} = 32 \\ \frac{x}{4} - \frac{5y}{9} = 40 \end{cases}$$
A) x = 100, y = -27　　B) x = -100, y = -27　　C) x = 100, y = 27　　D) x = -100, y = 27

16)
$$\begin{cases} 2x + 9y = -8 \\ x + \frac{y}{3} = \frac{13}{3} \end{cases}$$
A) x = -5, y = -2　　B) x = 5, y = 2　　C) x = -5, y = 2　　D) x = 5, y = -2

Solve the system of equations by elimination.

1) $\begin{cases} 2x + 81y = 81 \\ 3x - 9y = -9 \end{cases}$

A) $x = 0, y = 0$ B) $x = 1, y = 1$ C) $x = 1, y = 0$ D) $x = 0, y = 1$

2) $\begin{cases} 5x - 2y = -1 \\ x + 4y = 35 \end{cases}$

A) $x = 3, y = 9$ B) $x = 2, y = 9$ C) $x = 2, y = 8$ D) $x = 3, y = 8$

3) $\begin{cases} x + y = -7 \\ x - y = 13 \end{cases}$

A) $x = 7, y = 3$ B) $x = 7, y = -10$ C) $x = 3, y = 10$ D) $x = 3, y = -10$

4) $\begin{cases} 3x + 8y = 7 \\ 3x + 2y = 31 \end{cases}$

A) $x = -4, y = 13$ B) $x = -13, y = 8$ C) $x = 13, y = -4$ D) $x = -13, y = 3$

5) $\begin{cases} 6x + 3y = 51 \\ 2x - 6y = 38 \end{cases}$

A) $x = -10, y = 3$ B) $x = 3, y = -10$ C) $x = 10, y = -3$ D) $x = -3, y = 10$

6) $\begin{cases} 3x - 5y = -12 \\ 6x + 8y = -24 \end{cases}$

A) $x = 0, y = 4$ B) $x = -4, y = 0$ C) $x = 0, y = -4$ D) $x = 4, y = 0$

7) $\begin{cases} 2x + 18y = -144 \\ 12x + 3y = 81 \end{cases}$

A) $x = 9, y = -9$ B) $x = -3, y = 9$ C) $x = -9, y = 9$ D) $x = 12, y = -12$

8) $\begin{cases} 5x + 3y = 80 \\ 2x + \ \ y = 30 \end{cases}$

A) $x = 0, y = 10$ B) $x = 10, y = 0$ C) $x = 10, y = 10$ D) $x = 0, y = 0$

9) $\begin{cases} \dfrac{7x}{12} - y = 10 \\ \dfrac{5x}{9} + 2y = 11 \end{cases}$

A) $x = 16, y = \frac{1}{2}$ B) $x = -18, y = -\frac{1}{2}$ C) $x = -16, y = -\frac{1}{2}$ D) $x = 18, y = \frac{1}{2}$

10) $\begin{cases} \frac{2}{5}x + \frac{7}{10}y = \frac{24}{5} \\ 4x + 2y = 68 \end{cases}$

A) $x = -19, y = 7$ B) $x = -19, y = 4$ C) $x = 19, y = -4$ D) $x = -4, y = 19$

3 Identify Inconsistent Systems of Equations Containing Two Variables

Solve the system of equations by substitution.

1) $\begin{cases} x + y = -9 \\ x + y = 1 \end{cases}$

A) $x = 0, y = -8$ B) $x = -9, y = 1$ C) $x = 0, y = 0$ D) inconsistent

2) $\begin{cases} 2x - 5y = 9 \\ 2x - 5y = -5 \end{cases}$

A) $x = \frac{9}{5}, y = \frac{-5}{2}$ B) $x = 9, y = -5$ C) $x = 2, y = 5$ D) inconsistent

Solve the system of equations by elimination.

3) $\begin{cases} 3x - 3y = 4 \\ 9x - 9y = 16 \end{cases}$

A) $x = \frac{5}{3}, y = -\frac{5}{3}$ B) $x = 3, y = 4$ C) $x = 4, y = 16$ D) inconsistent

4) $\begin{cases} 8x - 9y = 2 \\ -32x + 36y = -4 \end{cases}$

A) $x = 4, y = 2$ B) $x = 8, y = 2$ C) $x = 12, y = -\frac{27}{2}$ D) inconsistent

4 Express the Solution of a System of Dependent Equations Containing Two Variables

Solve the system of equations by substitution.

1) $\begin{cases} 6x + y = 7 \\ -24x - 4y = -28 \end{cases}$

A) $x = -6y + 7$, where y is any real number B) $y = -6x + 7$, where x is any real number

C) $y = 6x + 7$, where x is any real number D) inconsistent

Solve the system of equations by elimination.

2)

$$\begin{cases} \dfrac{x}{2} + \dfrac{y}{3} = 4 \\ \dfrac{x}{4} + \dfrac{y}{6} = 2 \end{cases}$$

A) $y = -\frac{3}{2}x + 12$, x is any real number

B) (0, 12)

C) $(-1, \frac{15}{2})$

D) inconsistent

5 Solve Systems of Three Equations Containing Three Variables

Verify that the values of the variables listed are solutions of the system of equations.

1)

$$\begin{cases} x + y + z = 4 \\ x - y + 3z = 2 \\ 2x + y + z = 1 \end{cases}$$

$x = 3, y = 4, z = -3$

A) solution

B) not a solution

2)

$$\begin{cases} x - y + 5z = 7 \\ 3x + z = 2 \\ x + 2y + z = 8 \end{cases}$$

$x = 0, y = 3, z = 2$

A) solution

B) not a solution

Solve the system of equations.

3)

$$\begin{cases} x + 4y + 4z = -6 \\ 2y + 5z = 7 \\ z = 3 \end{cases}$$

A) $x = -2, y = -4, z = 3$

B) $x = 3, y = -4, z = -2$

C) $x = -2, y = 3, z = -4$

D) inconsistent

4)

$$\begin{cases} x - y + 4z = 6 \\ 5x + z = 2 \\ x + 3y + z = 8 \end{cases}$$

A) $x = 0, y = 2, z = 2$

B) $x = 2, y = 2, z = 0$

C) $x = 2, y = 0, z = 2$

D) inconsistent

5)

$$\begin{cases} x - y + 5z = 1 \\ 5x + z = 0 \\ x + 3y + z = -3 \end{cases}$$

A) $x = 0, y = -1, z = 1$

B) $x = 0, y = 0, z = -1$

C) $x = 0, y = -1, z = 0$

D) inconsistent

6)

$$\begin{cases} 3x + 5y + z = 25 \\ 4x - 3y - z = -8 \\ 4x + y + 4z = -3 \end{cases}$$

A) $x = 1, y = -3, z = 5$ B) $x = 1, y = 5, z = -3$ C) $x = -3, y = 5, z = 1$ D) inconsistent

7)

$$\begin{cases} 7x + 7y + z = 1 \\ x + 8y + 8z = 8 \\ 9x + y + 9z = 9 \end{cases}$$

A) $x = 0, y = 1, z = 0$ B) $x = -1, y = 1, z = 1$ C) $x = 0, y = 0, z = 1$ D) $x = 1, y = -1, z = 1$

8)

$$\begin{cases} x + y + z = 0 \\ x - y + 3z = 2 \\ 2x + y + z = 3 \end{cases}$$

A) $x = -1, y = 3, z = -2$ B) $x = 3, y = -2, z = -1$ C) $x = -1, y = -2, z = 3$ D) inconsistent

9)

$$\begin{cases} x - y + z = -1 \\ x + y + z = 9 \\ x + y - z = 7 \end{cases}$$

A) $x = 5, y = 1, z = 3$ B) $x = 3, y = 1, z = 5$ C) $x = 3, y = 5, z = 1$ D) inconsistent

10)

$$\begin{cases} x + y + z = -5 \\ x - y + 5z = -19 \\ 2x + y + z = -2 \end{cases}$$

A) $x = -5, y = -3, z = 3$ B) $x = 3, y = -3, z = -5$ C) $x = -5, y = 3, z = -3$ D) inconsistent

11)

$$\begin{cases} 2x + 3y + z = 2 \\ 5x - 2y - z = -25 \\ 3x + y + 2z = -17 \end{cases}$$

A) $x = -5, y = 5, z = -4$ B) $x = -4, y = -5, z = 5$ C) $x = -4, y = 5, z = -5$ D) inconsistent

6 Identify Inconsistent Systems of Equations Containing Three Variables

Solve the system of equations.

1)

$$\begin{cases} x + y + z = 4 \\ x - y + 5z = 16 \\ 3x + 3y + 3z = -6 \end{cases}$$

A) $x = 5, y = -5, z = 4$ B) $x = -5, y = 4, z = 5$ C) $x = 5, y = 4, z = -5$ D) inconsistent

2)

$$\begin{cases} x - y + 5z = 4 \\ 2x \quad + z = 0 \\ -x + y - 5z = -16 \end{cases}$$

A) $x = 0, y = -4, z = 0$ B) $x = 0, y = 0, z = -4$ C) $x = 5, y = -4, z = 0$ D) inconsistent

7 Express the Solutions of a System of Dependent Equations Containing Three Variables

Solve the system of equations.

1)

$$\begin{cases} x + 4y - z = 3 \\ x + 5y - 2z = 5 \\ 3x + 12y - 3z = 9 \end{cases}$$

A) $x = -3z - 5$, and $y = z + 2$, where z is any real number

B) $x = 3z + 5$, and $y = z - 2$, where z is any real number

C) $x = z - 2$, and $y = -3z - 5$, where z is any real number

D) inconsistent

2)

$$\begin{cases} -x + y + 2z = 0 \\ x + 2y + z = 6 \\ -2x - y + z = -6 \end{cases}$$

A) $x = 2 - z$, and $y = z + 2$, where z is any real number

B) $x = z + 2$, and $y = z - 2$, where z is any real number

C) $x = z + 2$, and $y = 2 - z$, where z is any real number

D) inconsistent

3)

$$\begin{cases} 2x - y + 5z = -7 \\ x + y - 2z = -2 \\ x - y + 4z = -4 \end{cases}$$

A) $x = z + 3$, and $y = 3z + 1$, where z is any real number

B) $x = 3z + 1$, and $y = z - 3$, where z is any real number

C) $x = -3 - z$, and $y = 3z + 1$, where z is any real number

D) inconsistent

4)

$$\begin{cases} 2x + 4y - 2z = 0 \\ 3x + 5y \quad = 1 \end{cases}$$

A) $x = 1 - 5z$, and $y = 3z - \frac{1}{2}$, where z is any real number

B) $x = -2 - 7z$, and $y = -3z - 1$, where z is any real number

C) $x = 2 - 5z$, and $y = 3z - 1$, where z is any real number

D) $x = 2 + 7z$, and $y = -3z - 1$, where z is any real number

Use a graphing calculator to solve the system of equations.

1)
$$\begin{cases} y = \sqrt{3}x - 25\sqrt{5} \\ y = -0.3x + 27 \end{cases}$$

A) $x = 0.5$, $y = 14.76$ B) $x = 1.07$, $y = 31.27$ C) $x = 70.66$, $y = 14.76$ D) $x = 40.8$, $y = 14.76$

2)
$$\begin{cases} y = 1.93x - 40\sqrt{5} \\ y = -0.3x + 23 \end{cases}$$

A) $x = 62.89$, $y = 31.94$ B) $x = 93.52$, $y = 7.87$ C) $x = 97.32$, $y = 7.87$ D) $x = 50.42$, $y = 7.87$

3)
$$\begin{cases} 2.48x - y - 29.37 = 0 \\ -0.5x - y + 24 \quad = 0 \end{cases}$$

A) $x = 35.44$, $y = 15.05$ B) $x = 44.42$, $y = 15.05$ C) $x = 21.88$, $y = 24.9$ D) $x = 17.91$, $y = 15.05$

4)
$$\begin{cases} \sqrt{1.5}x - y - 25\sqrt{7} = 0 \\ -0.72x - y + 20 \quad = 0 \end{cases}$$

A) $x = 84.29$, $y = 37.08$ B) $x = 54.25$, $y = -11.89$ C) $x = 44.3$, $y = -11.89$ D) $x = 56.43$, $y = -11.89$

Solve the system of equations.

5)
$$\begin{cases} \dfrac{2}{x} + \dfrac{4}{y} = 7 \\ \dfrac{1}{x} - \dfrac{2}{y} = 4 \end{cases}$$

A) $x = \frac{15}{4}$, $y = -\frac{1}{8}$ B) $x = \frac{4}{15}$, $y = -8$ C) $x = -8$, $y = \frac{4}{15}$ D) $x = -\frac{1}{8}$, $y = \frac{15}{4}$

Solve the problem.

6) Find real numbers a, b, and c such that the graph of the function $y = ax^2 + bx + c$ contains the points (1, 1), (2, 4), and (-3, 29).

7) Find real numbers a, b, and c such that the graph of the function $y = ax^2 + bx + c$ contains the points (1, 2), (2, 11), and (-3, -14).

8) A retired couple has $200,000 to invest to obtain annual income. They want some of it invested in safe Certificates of Deposit yielding 5%. The rest they want to invest in AA bonds yielding 10% per year. How much should they invest in each to realize exactly $17,000 per year?

A) $150,000 at 10% and $50,000 at 5% B) $140,000 at 5% and $60,000 at 10%

C) $130,000 at 5% and $70,000 at 10% D) $140,000 at 10% and $60,000 at 5%

9) The Family Fine Arts Center charges $25 per adult and $10 per senior citizen for its performances. On a recent weekend evening when 507 people paid admission, the total receipts were $6855. How many who paid were senior citizens?

A) 298 senior citizens B) 209 senior citizens C) 119 senior citizens D) 388 senior citizens

10) A tour group split into two groups when waiting in line for food at a fast food counter. The first group bought 8 slices of pizza and 6 soft drinks for $29.16. The second group bought 7 slices of pizza and 5 soft drinks for $25.20. How much does one slice of pizza cost?

A) $1.26 per slice of pizza
B) $2.70 per slice of pizza
C) $1.76 per slice of pizza
D) $2.20 per slice of pizza

11) A flat rectangular piece of aluminum has a perimeter of 54 inches. The length is 11 inches longer than the width. Find the width.

A) 30 in.
B) 19 in.
C) 27 in.
D) 8 in.

12) A twin-engined aircraft can fly 1120 miles from city A to city B in 5 hours with the wind and make the return trip in 7 hours against the wind. What is the speed of the wind?

A) 48 mph
B) 16 mph
C) 32 mph
D) 64 mph

13) A bank teller has 50 $10 and $20 bills in her cash drawer. The value of the bills is $900. How many $10 bills are there?

A) 38 $10 bills
B) 10 $10 bills
C) 12 $10 bills
D) 40 $10 bills

14) The Paperback Trader is a book store that takes in used paperbacks for 20% of their cover price and sells them for 50% of their cover price. Pat brings in a stack of 14 paperback books to trade and gets $17.49 credit. Some of the books had a cover price of $7.96, the rest $4.96. She wants to get some Tom Clancy books having a cover price of $7.96. How many $7.96 books did she bring in and how many Clancy books can she get without paying any additional cash?

A) 8 $7.96 books, 2 Clancy books
B) 6 $7.96 books, 2 Clancy books
C) 6 $7.96 books, 5 Clancy books
D) 6 $7.96 books, 4 Clancy books

15) The perimeter of a parking lot is 500 yards. Find the dimensions of the lot if the length is 50 yards more than three times the width.

16) A store has a sale on workout gear. Mark bought three pairs of shorts and three T-shirts for $70.35 (before tax). Later, he went back and bought two more pairs of shorts and four more T-shirts for $63.90 (before tax). How much did the shorts and T-shirts cost?

17) A tea shop owner is mixing a blend of two teas, one of which costs $6.50 per pound, the other costing $4.00 per pound. The owner wants to have 20 pounds of a mixture that will sell for $5.50 per pound. How much of each type of tea should be used?

18) Jaya has $16,000 to invest. She invests part of it in an account paying 6% simple interest and the rest in an account paying 7% simple interest. If her annual income is $1080, how much does she have invested in each account?

19) A boat on a river goes 77 miles downstream in 2 hours and 45 minutes. The return trip takes 3 hours and 30 minutes. If the boat's speed is the same in each direction, find the speed of the boat and the speed of the current.

20) The Family Arts Center charges $24 for adults, $15 for senior citizens, and $6 for children under 12 for their live performances on Sunday afternoon. This past Sunday, the paid revenue was $10,794 for 746 tickets sold. There were 44 more children than adults. How many children attended?

A) 249 children
B) 395 children
C) 215 children
D) 272 children

21) Ron attends a cocktail party (with his graphing calculator in his pocket). He wants to limit his food intake to 100 g protein, 88 g fat, and 126 g carbohydrate. According to the health conscious hostess, the marinated mushroom caps have 3 g protein, 5 g fat, and 9 g carbohydrate; the spicy meatballs have 14 g protein, 7 g fat, and 15 g carbohydrate; and the deviled eggs have 13 g protein, 15 g fat, and 6 g carbohydrate. How many of each snack can he eat to obtain his goal?

A) 6 mushrooms, 4 meatballs, 2 eggs
B) 4 mushrooms, 2 meatballs, 6 eggs
C) 7 mushrooms, 5 meatballs, 3 eggs
D) 2 mushrooms, 6 meatballs, 4 eggs

22) A ceramics workshop makes wreaths, trees, and sleighs for sale at Christmas. A wreath takes 3 hours to prepare, 2 hours to paint, and 10 hours to fire. A tree takes 17 hours to prepare, 3 hours to paint, and 4 hours to fire. A sleigh takes 4 hours to prepare, 14 hours to paint, and 7 hours to fire. If the workshop has 121 hours for preparation time, 73 hours for painting, and 121 hours for firing, how many of each can be made?

A) 9 wreaths, 6 trees, 4 sleighs
B) 5 wreaths, 3 trees, 8 sleighs
C) 3 wreaths, 8 trees, 5 sleighs
D) 8 wreaths, 5 trees, 3 sleighs

23) Three shrimp boats supply the shrimp wholesalers on Hilton Head with fresh catch. The Annabelle takes 50% of its catch to Hudson's, 20% to Captain J's, and 30% to Mainstreet. The Curly Q takes 40% of its catch to Hudson's, 40% to Captain J's, and 20% to Mainstreet. The SloJoe takes 30% of its catch to Hudson's, 40% to Captain J's, and 30% to Mainstreet. One week Hudson's received 248.1 pounds of shrimp, Captain J's received 218.8 pounds, and Mainstreet received 161.1 pounds. How many pounds of shrimp did each boat catch?

A) Annabelle 193 lbs, Curly Q 273 lbs, SloJoe 162 lbs
B) Annabelle 193 lbs, Curly Q 162 lbs, SloJoe 273 lbs
C) Annabelle 273 lbs, Curly Q 193 lbs, SloJoe 162 lbs
D) Annabelle 162 lbs, Curly Q 273 lbs, SloJoe 193 lbs

24) To raise money, a group of three friends is each trying to sell 100 candles. Each friend has a package containing the same number of tapers, votives, and tealights. The tapers sell for $6.00 each, the votives sell for $1.50 each, and the tealights sell for $0.75 each. Maya sold all of the candles to raise $431.25, Lorrin sold all of the tapers and tealights to raise $378.75, while Sara sold all the votives, all the tealights, and half of the tapers to raise $251.25. How many of each type of candle did each package contain?

25) Meisha has $25,000 that she wants to invest. She invests it in accounts paying 12%, 7%, and 6% simple interest. The account paying 12% is a higher–risk account, so she wants the amount in that account to be half of the amount she has in the account paying 6% simple interest. If her annual interest is $1945, how much is invested at each rate?

26) Lexie wants to have an income of $9000 per year from investments. To that end she is going to invest $90,000 in three different accounts. These accounts pay 7%, 10%, and 14% simple interest. If she wants to have $10,000 more in the account paying 7% simple interest than she has in the account paying 14% simple interest, how much should go into each account?

27) Craig, Jenni, and Jade go to a store having a sale on workout gear. Craig buys three pairs of socks, two T–shirts, and one pair of shorts for $35.90. Jenni buys four pairs of socks, three T–shirts, and three pairs of shorts for $71.35. Jade buys one pair of socks, four T–shirts, and 2 pairs of shorts for $59.30. What is the price of each item?

28) A health shop owner made trail mix containing dried fruit, nuts, and carob chips. The dried fruit sells for $5.50 per pound, the nuts sell for $7.50 per pound, and the carob chips sell for $8.50 per pound. The shop owner mixed 50 pounds of the trail mix and sells it for $6.70 per pound. If the amount of nuts is five pounds more than the amount of carob ships, how much of each item was used for the trail mix?

7.2 Systems of Linear Equations: Matrices

1 Write the Augmented Matrix of a System of Linear Equations

Write the augmented matrix of the given system of equations.

1)
$$\begin{cases} 9x + 2y = 61 \\ 6x + 3y = 39 \end{cases}$$

A) $\left[\begin{array}{cc|c} 9 & 2 & 61 \\ 6 & 3 & 39 \end{array}\right]$
B) $\left[\begin{array}{cc|c} 61 & 2 & 9 \\ 39 & 6 & 3 \end{array}\right]$
C) $\left[\begin{array}{cc|c} 9 & 2 & 39 \\ 3 & 6 & 61 \end{array}\right]$
D) $\left[\begin{array}{cc|c} 9 & 6 & 61 \\ 2 & 3 & 39 \end{array}\right]$

2)
$$\begin{cases} 4x + 4y = 56 \\ 7y = 42 \end{cases}$$

A) $\left[\begin{array}{cc|c} 4 & 4 & 56 \\ 7 & 42 & 0 \end{array}\right]$
B) $\left[\begin{array}{cc|c} 7 & 0 & 42 \\ 4 & 4 & 4 \end{array}\right]$
C) $\left[\begin{array}{cc|c} 4 & 4 & 56 \\ 0 & 7 & 42 \end{array}\right]$
D) $\left[\begin{array}{cc|c} 56 & 4 & 4 \\ 42 & 0 & 7 \end{array}\right]$

3)
$$\begin{cases} 8x - y = 0 \\ -8x + y - 16 = 0 \end{cases}$$

A) $\left[\begin{array}{cc|c} 8 & -1 & 0 \\ -8 & 1 & -16 \end{array}\right]$
B) $\left[\begin{array}{cc|c} 8 & -1 & 0 \\ -8 & 1 & 16 \end{array}\right]$
C) $\left[\begin{array}{cc|c} 8 & -8 & 0 \\ -1 & 1 & 16 \end{array}\right]$
D) $\left[\begin{array}{cc|c} 8 & -8 & 16 \\ -1 & 1 & 0 \end{array}\right]$

4)
$$\begin{cases} 3.5x + 0.3y = 19 \\ 0.7x - 0.9y = -1 \end{cases}$$

A) $\left[\begin{array}{cc|c} 3.5 & 19 & 0.3 \\ 0.7 & -1 & -0.9 \end{array}\right]$
B) $\left[\begin{array}{cc|c} 19 & 0.3 & 3.5 \\ -1 & -0.9 & 0.7 \end{array}\right]$
C) $\left[\begin{array}{cc|c} 3.5 & 0.7 & 19 \\ -0.9 & 0.3 & -1 \end{array}\right]$
D) $\left[\begin{array}{cc|c} 3.5 & 0.3 & 19 \\ 0.7 & -0.9 & -1 \end{array}\right]$

5)
$$\begin{cases} 5x + 2z = 40 \\ -2y + 9z = 29 \\ 6x + 6y + 3z = 99 \end{cases}$$

A) $\left[\begin{array}{cc|c} 5 & 0 & 2 \\ 0 & -2 & 9 \\ 6 & 6 & 3 \end{array}\right]$
B) $\left[\begin{array}{ccc|c} 5 & 0 & 2 & 40 \\ 0 & -2 & 9 & 29 \\ 6 & 6 & 3 & 99 \end{array}\right]$
C) $\left[\begin{array}{ccc|c} 5 & 0 & 6 & 40 \\ 0 & -2 & 6 & 29 \\ 2 & 9 & 3 & 99 \end{array}\right]$
D) $\left[\begin{array}{ccc|c} 5 & 2 & 0 & 40 \\ -2 & 9 & 0 & 29 \\ 6 & 6 & 3 & 99 \end{array}\right]$

6)
$$\begin{cases} 4x + 7y + 8z = 63 \\ 3x + 4y + 9z = 64 \\ 2x + 5y + 9z = 63 \end{cases}$$

A) $\left[\begin{array}{ccc|c} 4 & 7 & 8 & 63 \\ 3 & 4 & 9 & 64 \\ 2 & 5 & 9 & 63 \end{array}\right]$
B) $\left[\begin{array}{ccc|c} 4 & 3 & 2 & 63 \\ 7 & 4 & 5 & 64 \\ 8 & 9 & 9 & 63 \end{array}\right]$
C) $\left[\begin{array}{cc|c} 4 & 7 & 8 \\ 3 & 4 & 9 \\ 2 & 5 & 9 \end{array}\right]$
D) $\left[\begin{array}{ccc|c} 63 & 8 & 7 & 4 \\ 64 & 9 & 4 & 3 \\ 63 & 9 & 5 & 2 \end{array}\right]$

7) $\begin{cases} 2x + 2y + 6z = 28 \\ 7x - 2y + 6z = 5 \\ 2x + 2y + 9z = 34 \\ 2x + 7y + 2z = 2 \end{cases}$

A) $\left[\begin{array}{ccc|c} 2 & 7 & 2 & 2 \\ 2 & -2 & 2 & 7 \\ 6 & 6 & 9 & 7 \\ 28 & 5 & 34 & 2 \end{array}\right]$

B) $\left[\begin{array}{ccc|c} 2 & 2 & 6 & 28 \\ 7 & -2 & 6 & 5 \\ 2 & 2 & 9 & 34 \\ 2 & 7 & 2 & 2 \end{array}\right]$

C) $\left[\begin{array}{ccc|c} 2 & 7 & 2 & 7 \\ 2 & -2 & 2 & 2 \\ 6 & 6 & 9 & 7 \\ 28 & 5 & 34 & 2 \end{array}\right]$

D) $\left[\begin{array}{ccc|c} 2 & 2 & 6 & 2 \\ 7 & -2 & 6 & 34 \\ 2 & 2 & 9 & 5 \\ 2 & 7 & 2 & 28 \end{array}\right]$

2 Write the System from the Augmented Matrix

Write the system of equations corresponding to the given augmented matrix.

1) $\left[\begin{array}{cc|c} 6 & 12 & 7 \\ 10 & 12 & 10 \end{array}\right]$

A) $\begin{cases} 12x + 6y = 7 \\ 10x + 12y = 10 \end{cases}$

B) $\begin{cases} 6x + 12y = 7 \\ 10x + 12y = 10 \end{cases}$

C) $\begin{cases} 6x + 12y = 7 \\ 12x + 10y = 10 \end{cases}$

D) $\begin{cases} 6x + 12y = -7 \\ 10x + 12y = -10 \end{cases}$

2) $\left[\begin{array}{cc|c} 2 & 4 & 10 \\ 2 & 2 & 22 \end{array}\right]$

A) $\begin{cases} 2x + 2y = 10 \\ 4x + 2y = 22 \end{cases}$

B) $\begin{cases} 2x + 4y = 10 \\ 2x + 2y = 22 \end{cases}$

C) $\begin{cases} 10x + 4y = -2 \\ 22x + 2y = -2 \end{cases}$

D) $\begin{cases} 2x + 22y = -4 \\ 4x + 10y = -2 \end{cases}$

3) $\left[\begin{array}{ccc|c} 1 & 0 & 0 & -6 \\ 0 & 1 & 0 & -2 \\ 0 & 0 & 1 & -5 \end{array}\right]$

A) $\begin{cases} x = 6 \\ y = 2 \\ z = 5 \end{cases}$

B) $\begin{cases} x = -1 \\ y = 3 \\ z = 0 \end{cases}$

C) $\begin{cases} x = -6 \\ y = -2 \\ z = -5 \end{cases}$

D) $\begin{cases} x = 0 \\ y = -8 \\ z = -11 \end{cases}$

4) $\left[\begin{array}{ccc|c} 8 & 9 & 2 & -2 \\ 2 & 0 & 5 & 4 \\ 3 & 5 & 0 & 2 \end{array}\right]$

A) $\begin{cases} 8x + 9y + 2z = -2 \\ 2x \qquad + 5z = 4 \\ 3x + 5y \qquad = 2 \end{cases}$

B) $\begin{cases} 8x - 9y + 2z = -2 \\ 2x \qquad + 5z = -4 \\ 3x + 5y \qquad = -2 \end{cases}$

C) $\begin{cases} 8x - 9y + 2z = -2 \\ 2x \qquad + 5z = -4 \\ 3x \qquad + 5z = -2 \end{cases}$

D) $\begin{cases} 8x + 9y + 2z = -2 \\ 2x \qquad + 5z = 4 \\ 3x \qquad + 5z = 2 \end{cases}$

5)

$$\left[\begin{array}{cccc|c} 1 & 0 & 0 & 0 & -1 \\ 0 & 1 & 0 & 0 & -9 \\ 0 & 0 & 1 & 0 & -1 \\ 0 & 0 & 0 & 1 & 0 \end{array}\right]$$

A) $\begin{cases} x_1 = 1 \\ x_2 = 9 \\ x_3 = 1 \\ x_4 = 0 \end{cases}$

B) $\begin{cases} x_1 = -1 \\ x_2 = -9 \\ x_3 = -1 \\ x_4 = 0 \end{cases}$

C) $\begin{cases} x_1 = 0 \\ x_2 = -10 \\ x_3 = -2 \\ x_4 = 1 \end{cases}$

D) $\begin{cases} x_1 = 0 \\ x_2 = -8 \\ x_3 = 0 \\ x_4 = -8 \end{cases}$

Determine whether the system corresponding to the given augmented matrix is consistent or inconsistent. If it is consistent, give the solution.

6)

$$\left[\begin{array}{ccc|c} 1 & 0 & 0 & -2 \\ 0 & 1 & 0 & 6 \\ 0 & 0 & 0 & -6 \end{array}\right]$$

A) consistent; $x = -2, y = 6$

B) consistent; $x = -2, y = 6, z = -6$

C) consistent; $x = 2, y = -6$

D) inconsistent

7)

$$\left[\begin{array}{ccc|c} 1 & 0 & 0 & 0 \\ 0 & 1 & 0 & 0 \\ 0 & 0 & 0 & 8 \end{array}\right]$$

A) consistent; $z = 8$

B) consistent; $x = 0, y = 8$

C) consistent; $x = 0, y = 0$

D) inconsistent

8)

$$\left[\begin{array}{ccc|c} 1 & 3 & 4 & 22 \\ & 3 & 3 & 21 \\ & & 1 & 2 \end{array}\right]$$

A) consistent; $x = 2, y = 5, z = -1$

B) consistent; $x = -1, y = 2, z = 5$

C) consistent; $x = -1, y = 5, z = 2$

D) inconsistent

3 Perform Row Operations on a Matrix

Perform the row operation(s) on the given augmented matrix.

1) $R_3 = 4r_1 + r_3$

$$\left[\begin{array}{ccc|c} -7 & -5 & -1 & -10 \\ 6 & -2 & 9 & 5 \\ 28 & -6 & 6 & 18 \end{array}\right]$$

2) (a) $R_2 = -5r_1 + r_2$
(b) $R_3 = -4r_1 + r_3$
(c) $R_3 = 5r_2 + r_3$

$$\left[\begin{array}{ccc|c} 1 & -3 & -5 & -2 \\ 5 & -5 & -4 & 5 \\ 4 & 5 & 4 & 6 \end{array}\right]$$

A) $\left[\begin{array}{ccc|c} 1 & -3 & -5 & -2 \\ 0 & 20 & 29 & 10 \\ 0 & 117 & 169 & 64 \end{array}\right]$

B) $\left[\begin{array}{ccc|c} 1 & -3 & -5 & -2 \\ 0 & 10 & 21 & 15 \\ 0 & 27 & 45 & 29 \end{array}\right]$

C) $\left[\begin{array}{ccc|c} 1 & -3 & -5 & -2 \\ 0 & 10 & 21 & 15 \\ 0 & 67 & 129 & 89 \end{array}\right]$

D) $\left[\begin{array}{ccc|c} 1 & -3 & -5 & -2 \\ 0 & -8 & -9 & 3 \\ 0 & -23 & -21 & -1 \end{array}\right]$

3) (a) $R_2 = 3r_1 + r_2$
(b) $R_3 = 4r_1 + r_3$
(c) $R_3 = 5r_2 + r_3$

$$\left[\begin{array}{ccc|c} 1 & -3 & -5 & 2 \\ -3 & -5 & 2 & 5 \\ -4 & -5 & 4 & 6 \end{array}\right]$$

A) $\left[\begin{array}{ccc|c} 1 & -3 & -5 & 2 \\ 0 & -14 & -13 & 11 \\ 0 & 3 & 69 & 69 \end{array}\right]$

B) $\left[\begin{array}{ccc|c} 1 & -3 & -5 & 2 \\ 0 & -14 & -13 & 11 \\ 0 & -63 & -41 & 69 \end{array}\right]$

C) $\left[\begin{array}{ccc|c} 1 & -3 & -5 & 2 \\ 0 & -14 & -13 & 11 \\ 0 & -87 & -81 & 69 \end{array}\right]$

D) $\left[\begin{array}{ccc|c} 1 & -3 & -5 & 2 \\ 0 & -4 & -13 & 11 \\ 0 & -21 & -29 & 25 \end{array}\right]$

4) (a) $R_2 = 3r_1 + r_2$
(b) $R_3 = -2r_1 + r_3$
(c) $R_3 = 4r_2 + r_3$

$$\left[\begin{array}{ccc|c} 1 & 2 & -3 & 4 \\ -3 & -5 & 8 & -10 \\ 2 & 0 & -1 & 4 \end{array}\right]$$

A) $\left[\begin{array}{ccc|c} 1 & 2 & -3 & -4 \\ 0 & 1 & -1 & 2 \\ 0 & 0 & 1 & 4 \end{array}\right]$

B) $\left[\begin{array}{ccc|c} 1 & 2 & -3 & 4 \\ 0 & 1 & -1 & 2 \\ 0 & 0 & 1 & 4 \end{array}\right]$

C) $\left[\begin{array}{ccc|c} 1 & 2 & -3 & 4 \\ 0 & 1 & 1 & -2 \\ 0 & 0 & 1 & 4 \end{array}\right]$

D) $\left[\begin{array}{ccc|c} 1 & 2 & -3 & -4 \\ 0 & 1 & -1 & 2 \\ 0 & 0 & 1 & -4 \end{array}\right]$

5) $R_2 = -2r_2 + r_1$

$$\left[\begin{array}{ccc|c} 2 & 4 & 6 & -8 \\ 1 & 2 & 3 & 6 \\ 4 & 6 & 7 & 1 \end{array}\right]$$

A) $\left[\begin{array}{ccc|c} 2 & 4 & 6 & -8 \\ 0 & 0 & 0 & 0 \\ 4 & 6 & 7 & 1 \end{array}\right]$

B) $\left[\begin{array}{ccc|c} 2 & 4 & 6 & -8 \\ 0 & 0 & 0 & -20 \\ 4 & 6 & 7 & 1 \end{array}\right]$

C) $\left[\begin{array}{ccc|c} 2 & 4 & 6 & -8 \\ -4 & -8 & -12 & 6 \\ 4 & 6 & 7 & 1 \end{array}\right]$

D) $\left[\begin{array}{ccc|c} 2 & 4 & 6 & -8 \\ -4 & -8 & -12 & -20 \\ 4 & 6 & 7 & 1 \end{array}\right]$

4 Solve Systems of Linear Equations Using Matrices

Solve the system of equations using matrices (row operations). If the system has no solution, say that it is inconsistent.

1)
$$\begin{cases} 4x + 3y = 7 \\ 6x + 4y = 8 \end{cases}$$

A) $x = 5, y = -2$ B) $x = -2, y = -5$ C) $x = -2, y = 5$ D) inconsistent

2)
$$\begin{cases} 5x + 2y = -2 \\ 4x \qquad = -8 \end{cases}$$

A) $x = -2, y = -4$ B) $x = -2, y = 4$ C) $x = 4, y = -2$ D) inconsistent

3)
$$\begin{cases} 3x + y = 14 \\ 5x + 3y = 18 \end{cases}$$

A) $x = 6, y = -4$ B) $x = -4, y = -6$ C) $x = -4, y = 6$ D) inconsistent

4)
$$\begin{cases} 5x - 8y = -7 \\ 25x - 40y = 2 \end{cases}$$

A) $x = -7, y = 2$ B) $x = 5, y = 5$ C) $x = 5, y = 2$ D) inconsistent

5)
$$\begin{cases} 8x + 5y - z = 96 \\ x + 6y - 2z = 44 \\ 9x + y + z = 82 \end{cases}$$

A) $x = 8, y = 3, z = 7$ B) $x = 8, y = 7, z = 3$ C) $x = -8, y = 7, z = 16$ D) inconsistent

6)
$$\begin{cases} -3x + 7y - z = -8 \\ x + 3y + 8z = 80 \\ -7x + y + z = -38 \end{cases}$$

A) $x = 7, y = 8, z = 3$ B) $x = 7, y = 3, z = 8$ C) $x = -7, y = 3, z = 14$ D) inconsistent

7)
$$\begin{cases} 6x - y + 2z = 36 \\ -8x - 8z = -88 \\ 4y + z = 15 \end{cases}$$

A) $x = 4, y = 7, z = 2$ B) $x = 4, y = 2, z = 7$ C) $x = -4, y = 2, z = 8$ D) inconsistent

8)
$$\begin{cases} -4x - 3y + 2z = -15 \\ 16x + 12y - 8z = 60 \\ -20x - 15y + 10z = -75 \end{cases}$$

A) $x = -\frac{3}{4}y + \frac{1}{2}z + \frac{15}{4}$, y is any real number, z is any real number

B) $x = -1, y = 3, z = -5$

C) $x = -20, y = 15, z = 10$

D) inconsistent

9)
$$\begin{cases} 3x - 2y + z = -7 \\ x + y - 2z = 12 \\ 3x + y - z = 10 \end{cases}$$

A) $x = 2, y = 3, z = -7$

B) $x = 1, y = -3, z = -16$

C) $x = 1, y = 3, z = -4$

D) $x = 5, y = \frac{13}{2}, z = -3$

10)
$$\begin{cases} -5x + 2z = 13 \\ x + 3y - 2z = -15 \\ 4x + 5y - z = -18 \end{cases}$$

11)
$$\begin{cases} 3x + y = 6 \\ 2x - y + z - w = 2 \\ z + w = 3 \end{cases}$$

A) $x = 3 - \frac{6}{5}w, y = 1 + \frac{2}{5}w, z = 3 - w$, where w is any real number

B) $x = 1 - \frac{2}{5}w, y = 3 + \frac{6}{5}w, z = 3 - w$, where w is any real number

C) $x = 1 + \frac{2}{5}w, y = 3 - \frac{6}{5}w, z = 3 - w$, where w is any real number

D) inconsistent

5 Additional Applications

Solve the problem.

1) Find real numbers a, b, and c such that the graph of the function $y = ax^2 + bx + c$ contains the points (-2, -4), (1, -1), and (3, -19).

2) Find the function $f(x) = ax^3 + bx^2 + cx + d$ for which $f(0) = -2$, $f(1) = 5$, $f(-1) = 3$, $f(2) = 4$.

A) $f(x) = -\frac{10}{3}x^3 + 6x^2 + \frac{13}{3}x - 2$

B) $f(x) = \frac{8}{3}x^3 + 4x^2 + \frac{5}{3}x - 2$

C) $f(x) = -8x^3 + 12x^2 + 5x - 6$

D) $f(x) = 10x^3 - 18x^2 - 13x + 6$

3) Melody has $45,000 to invest and wishes to receive an annual income of $4290 from this money. She has chosen investments that pay 5%, 8%, and 12% simple interest. Melody wants to have the amount invested at 12% to be double the amount invested at 8%. How much should she invest at each rate?

4) A company manufactures three types of wooden chairs: the Kitui, the Goa, and the Santa Fe. To make a Kitui chair requires 1 hour of cutting time, 1.5 hours of assembly time, and 1 hour of finishing time. A Goa chair requires 1.5 hours of cutting time, 2.5 hours of assembly time and 2 hours of finishing time. A Santa Fe chair requires 1.5 hours of cutting time, 3 hours of assembly time, and 3 hours of finishing time. If 41 hours of cutting time, 70 hours of assembly time, and 58 hours of finishing time were used one week, how many of each type of chair were produced?

5) Ron attends a cocktail party (with his graphing calculator in his pocket). He wants to limit his food intake to 120 g protein, 105 g fat, and 159 g carbohydrate. According to the health conscious hostess, the marinated mushroom caps have 3 g protein, 5 g fat, and 9 g carbohydrate; the spicy meatballs have 14 g protein, 7 g fat, and 15 g carbohydrate; and the deviled eggs have 13 g protein, 15 g fat, and 6 g carbohydrate. How many of each snack can he eat to obtain his goal?

A) 5 mushrooms, 2 meatballs, 8 eggs
B) 8 mushrooms, 5 meatballs, 2 eggs
C) 9 mushrooms, 6 meatballs, 3 eggs
D) 2 mushrooms, 8 meatballs, 5 eggs

6) The perimeter of a picture frame is 11 feet. If three times the height is equal to eight times the width, what are the dimensions of the frame?

7) Jenny receives $1270 per year from three different investments totaling $20,000. One of the investments pays 6%, the second one pays 8%, and the third one pays 5%. If the money invested at 8% is $1500 less than the amount invested at 5%, how much money has Jenny invested in the investment that pays 6%?

A) $1500
B) $10,000
C) $8500
D) $4500

7.3 Systems of Linear Equations: Determinants

1 Evaluate 2 by 2 Determinants

Find the value of the determinant.

1)
$$\begin{vmatrix} 3 & 8 \\ 5 & 5 \end{vmatrix}$$

A) 55
B) 25
C) -1
D) -25

2)
$$\begin{vmatrix} 6 & 1 \\ 8 & -4 \end{vmatrix}$$

A) -16
B) 38
C) 32
D) -32

3)
$$\begin{vmatrix} 1 & 3 \\ 3 & 2 \end{vmatrix}$$

A) 7
B) -7
C) -3
D) 11

4)
$$\begin{vmatrix} 12 & -7 \\ -4 & 3 \end{vmatrix}$$

A) -8
B) 8
C) 64
D) 4

Solve for x.

5)

$$\begin{vmatrix} 8 & x \\ 2 & 5 \end{vmatrix} = 32$$

6)

$$\begin{vmatrix} 5 & 9 \\ -2 & x \end{vmatrix} = 8$$

2 Use Cramer's Rule to Solve a System of Two Equations, Two Variables

Solve the system of equations using Cramer's Rule if it is applicable. If Cramer's Rule is not applicable, say so.

1)

$$\begin{cases} 4x + 5y = 8 \\ 2x + y = -2 \end{cases}$$

A) $x = -3, y = 4$ B) $x = 3, y = -4$ C) $x = 4, y = -3$ D) $x = -4, y = -3$

2)

$$\begin{cases} 2x + 4y = 52 \\ 4x + 2y = 50 \end{cases}$$

A) $x = 8, y = 9$ B) $x = -9, y = 8$ C) $x = -8, y = -9$ D) $x = 9, y = 8$

3)

$$\begin{cases} 3x + 3y = 24 \\ 3x - 2y = 14 \end{cases}$$

A) $x = -2, y = 6$ B) $x = 2, y = 6$ C) $x = 6, y = 2$ D) $x = -6, y = -2$

4)

$$\begin{cases} 4x - 7y = 5 \\ 2x + 5y = -3 \end{cases}$$

A) $x = \frac{2}{3}, y = \frac{1}{3}$ B) $x = \frac{2}{17}, y = -\frac{11}{17}$ C) $x = -\frac{2}{17}, y = \frac{11}{17}$ D) $x = \frac{23}{3}, y = -\frac{11}{3}$

5)

$$\begin{cases} 9x - 2y = 4 \\ 27x - 6y = 16 \end{cases}$$

A) $x = \frac{5}{9}, y = -\frac{5}{2}$ B) $x = 3, y = 4$ C) $x = 4, y = 16$ D) not applicable

3 Evaluate 3 by 3 Determinants

Find the value of the determinant.

1)

$$\begin{vmatrix} 1 & -1 & 3 \\ -5 & 3 & 2 \\ -3 & 4 & 2 \end{vmatrix}$$

A) -57 B) -39 C) -51 D) 39

2)

$$\begin{vmatrix} 5 & 3 & 3 \\ 1 & 1 & 6 \\ 3 & 2 & 6 \end{vmatrix}$$

A) -105 B) 3 C) -3 D) 177

3)

$$\begin{vmatrix} 4 & -1 & -3 \\ 3 & -2 & -4 \\ -1 & 5 & -2 \end{vmatrix}$$

A) 47 B) -113 C) -47 D) 55

4)

$$\begin{vmatrix} -2 & 5 & 4 \\ 3 & -2 & 1 \\ 1 & 6 & -3 \end{vmatrix}$$

A) 80 B) -12 C) -90 D) 130

Solve for x.

5)

$$\begin{vmatrix} x & -4 & -1 \\ -2 & 2 & 0 \\ -1 & -2 & 8 \end{vmatrix} = 10$$

6)

$$\begin{vmatrix} 5 & -3 & 1 \\ -2 & -2 & x \\ 8 & 2 & -1 \end{vmatrix} = 28$$

4 Use Cramer's Rule to Solve a System of Three Equations, Three Variables

Solve the system of equations using Cramer's Rule if it is applicable. If Cramer's Rule is not applicable, say so.

1)

$$\begin{cases} 2x - 4y - z = -18 \\ x - 4y + 3z = -5 \\ 6x + y + z = 27 \end{cases}$$

A) $x = 3, y = -5, z = -4$ B) $x = 5, y = 4, z = 5$ C) $x = 4, y = 3, z = 4$ D) $x = 3, y = 5, z = 4$

2)

$$\begin{cases} 6x - 3y - 5z = -9 \\ -7x + 7y - 2z = -19 \\ -4x - 3y - 4z = -63 \end{cases}$$

A) $x = 6, y = -5, z = -6$ B) $x = 5, y = 6, z = 5$ C) $x = 6, y = 5, z = 6$ D) $x = 7, y = 3, z = 6$

3)

$$\begin{cases} 3x - 3z = -15 \\ -3x + 7y + 2z = 13 \\ -8x - 3y = -35 \end{cases}$$

A) $x = 4, y = -1, z = -9$ B) $x = 4, y = 1, z = 9$ C) $x = 5, y = -1, z = 9$ D) $x = 1, y = 9, z = 1$

4)
$$\begin{cases} x - y + 3z = -2 \\ 2x \quad + z = 0 \\ -x + y - 3z = 6 \end{cases}$$

A) $x = 0, y = 0, z = 2$ B) $x = 0, y = 2, z = 0$ C) $x = 3, y = 2, z = 0$ D) not applicable

5 Know Properties of Determinants

Use properties of determinants to find the value of the second determinant, if the value of the first is known.

1)
$$\begin{vmatrix} x & y & z \\ u & v & w \\ 1 & -2 & 2 \end{vmatrix} = -68 \quad \begin{vmatrix} 1 & -2 & 2 \\ u & v & w \\ x & y & z \end{vmatrix} = ?$$

A) 0 B) -68

C) 68 D) cannot be determined

2)
$$\begin{vmatrix} x & y & z \\ u & v & w \\ 1 & -3 & 2 \end{vmatrix} = -19 \quad \begin{vmatrix} x & y & z \\ u & v & w \\ -3 & 9 & -6 \end{vmatrix} = ?$$

A) 57 B) 19 C) -19 D) -57

3)
$$\begin{vmatrix} x & y & z \\ u & v & w \\ 1 & -3 & 1 \end{vmatrix} = -39 \quad \begin{vmatrix} u & v & w \\ 3 & -9 & 3 \\ x & y & z \end{vmatrix} = ?$$

A) -117 B) 39 C) -39 D) 117

4)
$$\begin{vmatrix} x & y & z \\ u & v & w \\ 1 & 1 & -1 \end{vmatrix} = 11 \quad \begin{vmatrix} 1 & 1 & -1 \\ 2u & 2v & 2w \\ x-1 & y-1 & z+1 \end{vmatrix} = ?$$

A) 22 B) 11 C) -11 D) -22

5)
$$\begin{vmatrix} x & y & z \\ u & v & w \\ 1 & -2 & 2 \end{vmatrix} = -59 \quad \begin{vmatrix} x & y & z-x \\ u & v & w-u \\ 1 & -2 & 1 \end{vmatrix} = ?$$

A) 0 B) 59

C) -59 D) cannot be determined

6)
$$\begin{vmatrix} x & y & z \\ u & v & w \\ 1 & -2 & 1 \end{vmatrix} = 6 \quad \begin{vmatrix} x-2 & y+4 & z-2 \\ -2u-1 & -2v+2 & -2w-1 \\ 1 & -2 & 1 \end{vmatrix} = ?$$

A) 12 B) 6 C) -6 D) -12

7)

$$\begin{vmatrix} s & t & u \\ v & w & x \\ 4 & 2 & 8 \end{vmatrix} = 3 \qquad \begin{vmatrix} 32-s & 16-t & 64-u \\ v & w & x \\ 4 & 2 & 8 \end{vmatrix} = ?$$

A) -3 B) -24 C) 3 D) 24

8)

$$\begin{vmatrix} x & y & z \\ a & b & c \\ 2 & 4 & 5 \end{vmatrix} = 3 \qquad \begin{vmatrix} 2 & 4 & 5 \\ 3a & 3b & 3c \\ x-2 & y-4 & z-5 \end{vmatrix} = ?$$

A) 6 B) -9 C) 0 D) 9

7.4 Matrix Algebra

1 Find the Sum and Difference of Two Matrices

Perform the indicated operation, whenever possible.

1)

$$[1\ 4] + \begin{bmatrix} -3 \\ 9 \end{bmatrix}$$

A) $\begin{bmatrix} -2 \\ 13 \end{bmatrix}$

B) $[-2\ 13]$

C) $\begin{bmatrix} 1 & -3 \\ 4 & 9 \end{bmatrix}$

D) not defined

2)

$$\begin{bmatrix} 3 & 3 \\ 3 & -9 \\ -2 & 1 \end{bmatrix} + \begin{bmatrix} -7 & -2 \\ 2 & -1 \\ -5 & 3 \end{bmatrix}$$

A) $\begin{bmatrix} -4 & -9 \\ 5 & -10 \\ -7 & 4 \end{bmatrix}$

B) $\begin{bmatrix} -4 & 1 \\ -5 & -9 \\ -7 & -4 \end{bmatrix}$

C) $\begin{bmatrix} 10 & 5 \\ 1 & -8 \\ 3 & -2 \end{bmatrix}$

D) $\begin{bmatrix} -4 & 1 \\ 5 & -10 \\ -7 & 4 \end{bmatrix}$

3)

$$\begin{bmatrix} -1 & 0 \\ 3 & 2 \end{bmatrix} - \begin{bmatrix} -1 & 3 \\ 3 & 1 \end{bmatrix}$$

A) $[-2]$

B) $\begin{bmatrix} -2 & 3 \\ 6 & 3 \end{bmatrix}$

C) $\begin{bmatrix} 0 & -3 \\ 0 & 1 \end{bmatrix}$

D) $\begin{bmatrix} 0 & 3 \\ 0 & -1 \end{bmatrix}$

4)

$$\begin{bmatrix} -4 & -6 \\ -8 & 8 \end{bmatrix} + \begin{bmatrix} 1 & -8 \\ 9 & 3 \end{bmatrix} - \begin{bmatrix} -8 & -9 \\ 5 & -3 \end{bmatrix}$$

A) $\begin{bmatrix} 5 & -5 \\ -4 & 14 \end{bmatrix}$

B) $\begin{bmatrix} -11 & -23 \\ 6 & 8 \end{bmatrix}$

C) $\begin{bmatrix} -13 & -7 \\ -12 & 2 \end{bmatrix}$

D) $\begin{bmatrix} 3 & 11 \\ -22 & 8 \end{bmatrix}$

5)
$$\begin{bmatrix} 7 & -4 & 8 \\ -6 & 5 & -1 \\ 0 & 6 & -3 \end{bmatrix} - \begin{bmatrix} -2 & -6 & -1 \\ -7 & -4 & 3 \\ -3 & -9 & -5 \end{bmatrix}$$

A) $\begin{bmatrix} 5 & -10 & 7 \\ -13 & 1 & 2 \\ -3 & -3 & -5 \end{bmatrix}$

B) $\begin{bmatrix} 9 & 2 & 9 \\ 1 & 9 & 2 \\ 3 & 15 & -4 \end{bmatrix}$

C) $\begin{bmatrix} 9 & 2 & 9 \\ 1 & 9 & -4 \\ 3 & 15 & 2 \end{bmatrix}$

D) $\begin{bmatrix} 5 & -10 & 7 \\ -13 & 1 & -8 \\ -3 & -3 & 2 \end{bmatrix}$

6)
$$\begin{bmatrix} -4 & 6 & 7 \\ 3 & -5 & 12 \\ 7 & -11 & 14 \end{bmatrix} - \begin{bmatrix} 6 & 10 & -4 \\ -5 & 6 & -8 \\ 3 & 11 & 7 \end{bmatrix}$$

A) $\begin{bmatrix} 10 & 4 & -11 \\ -8 & 11 & -20 \\ -4 & 22 & -7 \end{bmatrix}$

B) $\begin{bmatrix} -10 & -4 & 11 \\ 8 & -11 & 20 \\ 4 & -22 & 7 \end{bmatrix}$

C) $\begin{bmatrix} -10 & -4 & 3 \\ -2 & -11 & 4 \\ 4 & -22 & 7 \end{bmatrix}$

D) $\begin{bmatrix} 2 & 16 & 3 \\ -2 & 1 & 4 \\ 10 & 0 & 21 \end{bmatrix}$

2 Find Scalar Multiples of a Matrix

Find the indicated expression.

1) Let $A = \begin{bmatrix} -3 & 5 \\ 0 & 2 \end{bmatrix}$. Find 4A.

A) $\begin{bmatrix} 1 & 9 \\ 4 & 6 \end{bmatrix}$

B) $\begin{bmatrix} -12 & 20 \\ 0 & 2 \end{bmatrix}$

C) $\begin{bmatrix} -12 & 20 \\ 0 & 8 \end{bmatrix}$

D) $\begin{bmatrix} -12 & 5 \\ 0 & 2 \end{bmatrix}$

2) Let $B = \begin{bmatrix} -1 & 6 & 7 & -3 \end{bmatrix}$. Find −4B.

A) $\begin{bmatrix} -4 & 24 & 28 & -12 \end{bmatrix}$

B) $\begin{bmatrix} -3 & 4 & 5 & -5 \end{bmatrix}$

C) $\begin{bmatrix} 4 & 6 & 7 & -3 \end{bmatrix}$

D) $\begin{bmatrix} 4 & -24 & -28 & 12 \end{bmatrix}$

3) Let $C = \begin{bmatrix} 2 \\ -2 \\ 8 \end{bmatrix}$. Find 1/2 C.

A) $\begin{bmatrix} 2 \\ -1 \\ 8 \end{bmatrix}$

B) $\begin{bmatrix} 1 \\ -2 \\ 8 \end{bmatrix}$

C) $\begin{bmatrix} 1 \\ -1 \\ 4 \end{bmatrix}$

D) $\begin{bmatrix} 4 \\ -4 \\ 16 \end{bmatrix}$

4) Let $A = \begin{bmatrix} 1 & 3 \\ 2 & 4 \end{bmatrix}$ and $B = \begin{bmatrix} 0 & 4 \\ -1 & 6 \end{bmatrix}$. Find 3A + B.

A) $\begin{bmatrix} 3 & 21 \\ 3 & 30 \end{bmatrix}$

B) $\begin{bmatrix} 3 & 13 \\ 5 & 18 \end{bmatrix}$

C) $\begin{bmatrix} 3 & 13 \\ 1 & 10 \end{bmatrix}$

D) $\begin{bmatrix} 3 & 7 \\ 5 & 10 \end{bmatrix}$

5) Let $C = \begin{bmatrix} 1 \\ -3 \\ 2 \end{bmatrix}$ and $D = \begin{bmatrix} -1 \\ 3 \\ -2 \end{bmatrix}$. Find C - 2D.

A) $\begin{bmatrix} 3 \\ -6 \\ 4 \end{bmatrix}$

B) $\begin{bmatrix} 3 \\ -9 \\ 6 \end{bmatrix}$

C) $\begin{bmatrix} -3 \\ 9 \\ -6 \end{bmatrix}$

D) $\begin{bmatrix} -1 \\ 3 \\ 4 \end{bmatrix}$

6) Let A = $\begin{bmatrix} -4 & 2 \end{bmatrix}$ and B = $\begin{bmatrix} 1 & 0 \end{bmatrix}$. Find 3A + 4B.

A) $\begin{bmatrix} -1 & 2 \end{bmatrix}$

B) $\begin{bmatrix} -12 & 4 \end{bmatrix}$

C) $\begin{bmatrix} -7 & 4 \end{bmatrix}$

D) $\begin{bmatrix} -8 & 6 \end{bmatrix}$

7) If A = $\begin{bmatrix} 2 & -1 \\ 7 & 9 \end{bmatrix}$ and B = $\begin{bmatrix} 5 & -3 \\ 4 & 7 \end{bmatrix}$, find -2A + 4B.

A) $\begin{bmatrix} -3 & -6 \\ 3 & 5 \end{bmatrix}$

B) $\begin{bmatrix} 7 & 4 \\ 11 & 13 \end{bmatrix}$

C) $\begin{bmatrix} -24 & -18 \\ -30 & -34 \end{bmatrix}$

D) $\begin{bmatrix} 16 & -10 \\ 2 & 10 \end{bmatrix}$

8) If A = $\begin{bmatrix} 7 & -5 \\ -8 & 1 \end{bmatrix}$ and B = $\begin{bmatrix} -9 & -3 \\ 2 & -8 \end{bmatrix}$, find -5A + 4B.

3 Find the Product of Two Matrices

Compute the product.

1)
$\begin{bmatrix} -2 & 3 \\ 2 & 2 \end{bmatrix}\begin{bmatrix} -2 & 0 \\ -1 & 2 \end{bmatrix}$

A) $\begin{bmatrix} 4 & -6 \\ -2 & 1 \end{bmatrix}$

B) $\begin{bmatrix} 4 & 0 \\ -2 & 4 \end{bmatrix}$

C) $\begin{bmatrix} 1 & 6 \\ -6 & 4 \end{bmatrix}$

D) $\begin{bmatrix} 6 & 1 \\ 4 & -6 \end{bmatrix}$

2)
$\begin{bmatrix} 1 & 3 & -1 \\ 4 & 0 & 3 \end{bmatrix}\begin{bmatrix} 3 & 0 \\ -1 & 1 \\ 0 & 3 \end{bmatrix}$

A) $\begin{bmatrix} 3 & -3 & 0 \\ 0 & 0 & 9 \end{bmatrix}$

B) $\begin{bmatrix} 0 & 0 \\ 12 & 9 \end{bmatrix}$

C) $\begin{bmatrix} 0 & 0 \\ 9 & 12 \end{bmatrix}$

D) not defined

3)
$\begin{bmatrix} 0 & -3 & 1 \\ 5 & -1 & 0 \end{bmatrix}\begin{bmatrix} 1 & 2 \\ 0 & 1 \\ 1 & -1 \end{bmatrix}$

A) $\begin{bmatrix} 0 & 10 \\ 0 & -1 \\ 0 & 0 \end{bmatrix}$

B) $\begin{bmatrix} 1 & 5 \\ -4 & 9 \end{bmatrix}$

C) $\begin{bmatrix} 10 & -5 & 1 \\ 5 & -1 & 0 \\ -5 & -2 & 0 \end{bmatrix}$

D) $\begin{bmatrix} 1 & -4 \\ 5 & 9 \end{bmatrix}$

4)
$\begin{bmatrix} -3 & -8 & 2 \\ -5 & 6 & 8 \end{bmatrix}\begin{bmatrix} -5 \\ 4 \\ 9 \end{bmatrix}$

A) $\begin{bmatrix} 1 & 121 \end{bmatrix}$

B) $\begin{bmatrix} -3 & -8 & 2 \\ -5 & 6 & 8 \\ -5 & 4 & 9 \end{bmatrix}$

C) $\begin{bmatrix} 1 \\ 121 \end{bmatrix}$

D) not defined

5)

$$\begin{bmatrix} -6 & 2 & 8 \end{bmatrix}\begin{bmatrix} 2 \\ 0 \\ -3 \end{bmatrix}$$

A) $\begin{bmatrix} -12 & 0 & -24 \end{bmatrix}$

B) $\begin{bmatrix} 90 \end{bmatrix}$

C) $\begin{bmatrix} -36 \end{bmatrix}$

D) $\begin{bmatrix} -12 \\ 0 \\ -24 \end{bmatrix}$

6)

$$\begin{bmatrix} -2 \\ 1 \\ -1 \end{bmatrix}\begin{bmatrix} 1 & -1 & 0 \end{bmatrix}$$

7)

$$\begin{bmatrix} 4 & 2 & -5 \\ -2 & 5 & -7 \end{bmatrix}\begin{bmatrix} -9 & 4 & -1 \\ -7 & -1 & 4 \\ -8 & 8 & 4 \end{bmatrix}$$

A) $\begin{bmatrix} -10 & 39 \\ -26 & -69 \\ -16 & -6 \end{bmatrix}$

B) $\begin{bmatrix} -36 & 8 & 5 \\ 14 & -5 & -28 \\ 16 & 40 & -28 \end{bmatrix}$

C) $\begin{bmatrix} -10 & -26 & -16 \\ 39 & -69 & -6 \end{bmatrix}$

D) $\begin{bmatrix} 4 & 2 & -5 \\ -2 & 5 & -7 \\ -9 & 5 & -7 \\ -7 & -1 & 4 \\ -8 & -8 & 4 \end{bmatrix}$

8)

$$\begin{bmatrix} -9 & -3 & -2 \\ 1 & -9 & -8 \\ -8 & -1 & 3 \end{bmatrix}\begin{bmatrix} 9 & 6 & -9 \\ 8 & -8 & -6 \\ -9 & 4 & 9 \end{bmatrix}$$

A) $\begin{bmatrix} -81 & -18 & 18 \\ 8 & 72 & 48 \\ 72 & -4 & 27 \end{bmatrix}$

B) $\begin{bmatrix} -9 & -3 & -2 \\ 1 & -9 & -8 \\ -8 & -1 & 3 \\ 9 & 6 & -9 \\ 8 & -8 & -6 \\ -9 & 4 & 9 \end{bmatrix}$

C) $\begin{bmatrix} -87 & 9 & -107 \\ -38 & 9 & -28 \\ 81 & -27 & 105 \end{bmatrix}$

D) $\begin{bmatrix} -87 & -38 & 81 \\ 9 & 46 & -27 \\ -107 & -28 & 105 \end{bmatrix}$

9)

$$\begin{bmatrix} 2 & -8 \\ 1 & 6 \\ 0 & 1 \end{bmatrix}\begin{bmatrix} -3 & 2 \\ 1 & -1 \\ 6 & 1 \end{bmatrix}$$

A) $\begin{bmatrix} -24 & 10 & 4 \\ 9 & -5 & 12 \\ 2 & -1 & 1 \end{bmatrix}$

B) $\begin{bmatrix} 9 & -1 & 4 \\ -24 & -5 & 1 \\ 2 & 10 & 12 \end{bmatrix}$

C) $\begin{bmatrix} 4 & 10 & -24 \\ 12 & -5 & 9 \\ 1 & -1 & 2 \end{bmatrix}$

D) not defined

10)

$$\begin{bmatrix} 3 & -2 & 1 \\ 0 & 4 & -3 \end{bmatrix}\begin{bmatrix} 5 & 0 \\ -2 & 1 \end{bmatrix}$$

A) $\begin{bmatrix} 15 & -10 & 5 \\ -6 & 8 & -5 \end{bmatrix}$

B) $\begin{bmatrix} 15 & 0 \\ 0 & 4 \end{bmatrix}$

C) $\begin{bmatrix} 15 & -6 \\ -10 & 8 \\ 5 & -5 \end{bmatrix}$

D) not defined

Solve the problem.

11) A company manufactures three types of wooden chairs at two different locations. In one week, the main location produces 18 Kitui chairs, 12 Goa chairs, and 6 Santa Fe chairs while the secondary location produces 10 Kitui chairs, 3 Goa chairs, and 5 Santa Fe chairs.
(a) Find a 2 by 3 matrix representing the above data.
(b) If each Kitui chair requires 25 board-feet of wood, a Goa chair requires 29 board-feet of wood, and a Santa Fe requires 30 board-feet of wood, find a 3 by 1 matrix representing the amount of material.
(c) Multiply the 2 by 3 matrix found in part a and the 3 by 1 matrix found in part b to get a 2 by 1 matrix showing the week's usage of material at both locations.

12) The matrix $\begin{bmatrix} 922 \\ 435 \end{bmatrix}$ represents one week's usage of wood (in board-feet) at two locations of a company that manufactures chairs. Due to differences in transportation costs, the first location pays \$0.89 per board-foot and the second location pays \$0.94 per board-foot.
(a) Find a 1 by 2 matrix representing the cost of the wood.
(b) Multiply the given matrix and the matrix found in part a to determine the total cost of wood for that week.

4 Find the Inverse of a Matrix

Each matrix is non singular. Find the inverse of the matrix. Be sure to check your answer.

1)

$$\begin{bmatrix} 2 & 4 \\ 1 & -5 \end{bmatrix}$$

A) $\begin{bmatrix} \frac{5}{14} & \frac{2}{7} \\ \frac{1}{14} & -\frac{1}{7} \end{bmatrix}$

B) $\begin{bmatrix} \frac{1}{7} & \frac{2}{7} \\ \frac{1}{14} & -\frac{5}{14} \end{bmatrix}$

C) $\begin{bmatrix} \frac{5}{14} & -\frac{2}{7} \\ \frac{1}{14} & \frac{1}{7} \end{bmatrix}$

D) $\begin{bmatrix} -\frac{5}{14} & -\frac{2}{7} \\ -\frac{1}{14} & \frac{1}{7} \end{bmatrix}$

2)

$$\begin{bmatrix} -4 & 2 \\ -4 & -5 \end{bmatrix}$$

A) $\begin{bmatrix} -\frac{5}{28} & \frac{1}{14} \\ -\frac{1}{7} & -\frac{1}{7} \end{bmatrix}$

B) $\begin{bmatrix} \frac{1}{7} & -\frac{1}{7} \\ -\frac{5}{28} & -\frac{1}{14} \end{bmatrix}$

C) $\begin{bmatrix} -\frac{1}{7} & -\frac{1}{14} \\ \frac{1}{7} & -\frac{5}{28} \end{bmatrix}$

D) $\begin{bmatrix} -\frac{5}{28} & -\frac{1}{14} \\ \frac{1}{7} & -\frac{1}{7} \end{bmatrix}$

3)
$$\begin{bmatrix} 0 & -1 \\ 6 & 6 \end{bmatrix}$$

A) $\begin{bmatrix} 0 & \frac{1}{6} \\ -1 & 1 \end{bmatrix}$ B) $\begin{bmatrix} 1 & \frac{1}{6} \\ -1 & 0 \end{bmatrix}$ C) $\begin{bmatrix} -1 & 0 \\ 1 & \frac{1}{6} \end{bmatrix}$ D) $\begin{bmatrix} 1 & -\frac{1}{6} \\ 1 & 0 \end{bmatrix}$

4)
$$\begin{bmatrix} 5 & -6 \\ 0 & 2 \end{bmatrix}$$

A) $\begin{bmatrix} 0 & \frac{1}{2} \\ \frac{1}{5} & \frac{3}{5} \end{bmatrix}$ B) $\begin{bmatrix} \frac{1}{2} & \frac{3}{5} \\ 0 & \frac{1}{5} \end{bmatrix}$ C) $\begin{bmatrix} \frac{1}{5} & -\frac{3}{5} \\ 0 & \frac{1}{2} \end{bmatrix}$ D) $\begin{bmatrix} \frac{1}{5} & \frac{3}{5} \\ 0 & \frac{1}{2} \end{bmatrix}$

5)
$$\begin{bmatrix} 3 & 1 \\ 2 & 0 \end{bmatrix}$$

A) $\begin{bmatrix} -\frac{3}{2} & \frac{1}{2} \\ 1 & 0 \end{bmatrix}$ B) $\begin{bmatrix} 1 & -\frac{3}{2} \\ 0 & \frac{1}{2} \end{bmatrix}$ C) $\begin{bmatrix} 0 & \frac{1}{2} \\ 1 & -\frac{3}{2} \end{bmatrix}$ D) $\begin{bmatrix} 0 & -\frac{1}{2} \\ -1 & -\frac{3}{2} \end{bmatrix}$

6)
$$\begin{bmatrix} 3 & -3 & 1 \\ -2 & 2 & -1 \\ -4 & 5 & -2 \end{bmatrix}$$

A) $\begin{bmatrix} 1 & -1 & 0 \\ -2 & -3 & 0 \\ 0 & -2 & -1 \end{bmatrix}$ B) $\begin{bmatrix} 1 & -1 & 1 \\ -2 & -3 & 0 \\ 0 & -2 & 1 \end{bmatrix}$ C) $\begin{bmatrix} 1 & -1 & 1 \\ 0 & -2 & 1 \\ -2 & -3 & 0 \end{bmatrix}$ D) $\begin{bmatrix} 1 & -3 & 1 \\ 0 & -2 & 1 \\ -2 & -3 & 0 \end{bmatrix}$

7)
$$\begin{bmatrix} 1 & 0 & 0 \\ 4 & 1 & 0 \\ 0 & -2 & 1 \end{bmatrix}$$

A) $\begin{bmatrix} 1 & 0 & 0 \\ 4 & 1 & 0 \\ 0 & 0 & -2 \end{bmatrix}$ B) $\begin{bmatrix} 1 & 0 & 0 \\ -4 & 1 & 0 \\ -8 & 2 & 1 \end{bmatrix}$ C) $\begin{bmatrix} 1 & 0 & 0 \\ -2 & -1 & 0 \\ 8 & 4 & 1 \end{bmatrix}$ D) $\begin{bmatrix} 1 & -2 & 8 \\ 0 & 1 & 1 \\ 0 & 0 & 1 \end{bmatrix}$

Show that the matrix has no inverse.

8)
$$\begin{bmatrix} 4 & -14 \\ -2 & 7 \end{bmatrix}$$

Use a graphing utility to find the inverse, if it exists, of the matrix. Round answers to two decimal places.

9)

$$\begin{bmatrix} -16 & 3 & 28 \\ 5 & -14 & 15 \\ 34 & 25 & 2 \end{bmatrix}$$

A) $\begin{bmatrix} -0.02 & 0.03 & 0.02 \\ 0.02 & -0.04 & 0.02 \\ 0.02 & 0.02 & 0.01 \end{bmatrix}$

B) $\begin{bmatrix} -0.01 & 0.03 & 0.02 \\ 0.02 & -0.03 & 0.01 \\ 0.02 & 0.02 & 0.01 \end{bmatrix}$

C) $\begin{bmatrix} -0.02 & 0.03 & 0.02 \\ 0.02 & -0.04 & 0.01 \\ 0.02 & 0.02 & 0.00 \end{bmatrix}$

D) $\begin{bmatrix} -0.01 & 0.02 & 0.01 \\ 0.02 & -0.03 & 0.01 \\ 0.02 & 0.02 & 0.01 \end{bmatrix}$

5 Solve Systems of Equations Using Inverse Matrices

Solve the system using the inverse matrix method.

1)

$$\begin{cases} 2x + 6y = 2 \\ 2x - y = -5 \end{cases}$$

A) $x = -1, y = 2$ B) $x = -2, y = 1$ C) $x = 1, y = -2$ D) $x = 2, y = -1$

2)

$$\begin{cases} x + 3y = -8 \\ 14x + 4y = 2 \end{cases}$$

A) $x = -3, y = 1$ B) $x = 3, y = -1$ C) $x = -1, y = 3$ D) $x = 1, y = -3$

3)

$$\begin{cases} -3x + 9y = 9 \\ 3x + 2y = 13 \end{cases}$$

A) $x = 2, y = 3$ B) $x = -3, y = -2$ C) $x = -2, y = -3$ D) $x = 3, y = 2$

4)

$$\begin{cases} -5x + 3y = 8 \\ 3x - 6y = -30 \end{cases}$$

A) $x = 2, y = 6$ B) $x = -6, y = -2$ C) $x = -2, y = -6$ D) $x = 6, y = 2$

5)

$$\begin{cases} bx + 4y = 6 \\ bx + 3y = 4 \end{cases} \quad b \neq 0$$

A) $x = \frac{2}{b}, y = 2$ B) $x = \frac{2}{b}, y = -2$ C) $x = 2, y = \frac{-2}{b}$ D) $x = \frac{-2}{b}, y = 2$

6)

$$\begin{cases} x + 2y + 3z = 9 \\ x + y + z = 12 \\ 2x + 2y + z = -11 \end{cases}$$

The inverse of $\begin{bmatrix} 1 & 2 & 3 \\ 1 & 1 & 1 \\ 2 & 2 & 1 \end{bmatrix}$ is $\begin{bmatrix} -1 & 4 & -1 \\ 1 & -5 & 2 \\ 0 & 2 & -1 \end{bmatrix}$.

A) $x = 36, y = -24, z = -11$
B) $x = 46, y = 47, z = 13$
C) $x = 50, y = -73, z = 35$
D) $x = 3, y = -46, z = 44$

7)

$$\begin{cases} x + 2y + 3z = -3 \\ x + y + z = 9 \\ x - 2z = 2 \end{cases}$$

The inverse of $\begin{bmatrix} 1 & 2 & 3 \\ 1 & 1 & 1 \\ 1 & 0 & -2 \end{bmatrix}$ is $\begin{bmatrix} -2 & 4 & -1 \\ 3 & -5 & 2 \\ -1 & 2 & -1 \end{bmatrix}$.

A) $x = 31, y = -53, z = 19$
B) $x = 40, y = -50, z = 19$
C) $x = -3, y = 0, z = 0$
D) $x = 32, y = 40, z = 17$

8)

$$\begin{cases} x + 2y + 3z = -3 \\ x + y + z = 10 \\ -x + y + 2z = 5 \end{cases}$$

The inverse of $\begin{bmatrix} 1 & 2 & 3 \\ 1 & 1 & 1 \\ -1 & 1 & 2 \end{bmatrix}$ is $\begin{bmatrix} 1 & -1 & -1 \\ -3 & 5 & 2 \\ 2 & -3 & -1 \end{bmatrix}$.

A) $x = -23, y = 38, z = 18$
B) $x = -18, y = 69, z = -41$
C) $x = 12, y = 51, z = 29$
D) $x = -3, y = 40, z = -10$

9)

$$\begin{cases} 4x + 2z = -20 \\ x - y - 5z = 5 \\ -3x - 2y - z = 12 \end{cases}$$

10)

$$\begin{cases} 2x + 4y - 5z = -8 \\ x + 5y + 2z = -1 \\ 3x + 3y + 3z = 15 \end{cases}$$

A) $x = 5, y = 2, z = -2$
B) $x = 5, y = -2, z = 2$
C) $x = -5, y = -2, z = -2$
D) $x = 2, y = 5, z = 2$

Use a graphing utility to solve the system of equations. Round answers to two decimal places.

11)
$$\begin{cases} 25x + 61y - 12z = 15 \\ 18x - 12y + 7z = -3 \\ 3x + 4y - z = 12 \end{cases}$$

A) $x = -22.55, y = -6.06, z = 4.56$
B) $x = 4.56, y = -6.06, z = -22.55$
C) $x = -22.55, y = 4.56, z = -6.06$
D) $x = 4.56, y = -22.55, z = -6.06$

7.5 Partial Fraction Decomposition

1 Decompose $\frac{P}{Q}$, Where Q Has Only Nonrepeated Linear Factors

Tell whether the given rational expression is proper or improper. If improper, rewrite it as the sum of a polynomial and a proper rational expression.

1) $\frac{x}{4 - x}$

A) improper; $1 - \frac{4}{4 - x}$
B) improper; $1 + \frac{4}{4 - x}$
C) improper; $-1 + \frac{4}{4 - x}$
D) proper

2) $\frac{x - 5}{x^2 + 8}$

A) improper; $\frac{5}{x + 2\sqrt{2}} - \frac{5}{x - 2\sqrt{2}}$
B) improper; $\frac{5}{x + 2\sqrt{2}} + \frac{5}{x - 2\sqrt{2}}$
C) improper; $\frac{x - 5}{(x + 2\sqrt{2})(x - 2\sqrt{2})}$
D) proper

Write the partial fraction decomposition of the rational expression.

3) $\frac{x - 6}{(x - 3)(x - 4)}$

A) $\frac{3}{x - 3} + \frac{2}{x - 4}$
B) $\frac{-2}{x - 3} + \frac{3}{x - 4}$
C) $\frac{3}{x - 3} + \frac{-2}{x - 4}$
D) $\frac{2}{x - 3} + \frac{-3}{x - 4}$

4) $\frac{x}{(x - 3)(x - 4)}$

A) $\frac{3}{x - 3} + \frac{-4}{x - 4}$
B) $\frac{-3}{x - 3} + \frac{-4}{x - 4}$
C) $\frac{-4}{x - 3} + \frac{3}{x - 4}$
D) $\frac{-3}{x - 3} + \frac{4}{x - 4}$

5) $\frac{x}{x^2 + 5x + 6}$

A) $\frac{3}{x + 3} + \frac{2}{x + 2}$
B) $\frac{2}{x + 3} + \frac{-3}{x + 2}$
C) $\frac{-3}{x + 3} + \frac{2}{x + 2}$
D) $\frac{3}{x + 3} + \frac{-2}{x + 2}$

6) $\frac{11x^2 - x - 20}{x(x + 1)(x - 1)}$

A) $\frac{20}{x} + \frac{-4}{x + 1} + \frac{-5}{x - 1}$
B) $\frac{20}{x} + \frac{-4}{x + 1} + \frac{5}{x - 1}$
C) $\frac{20}{x} + \frac{4}{x + 1} + \frac{-5}{x - 1}$
D) $\frac{20}{x} + \frac{-5}{x + 1} + \frac{4}{x - 1}$

7) $\dfrac{12x^2 + 162x + 384}{(x + 8)(x + 2)(x + 11)}$

A) $\dfrac{8}{x+8} + \dfrac{2}{x+2} + \dfrac{2}{x+11}$

B) $\dfrac{-8}{x+8} + \dfrac{2}{x+2} + \dfrac{2}{x+11}$

C) $\dfrac{-8}{x+8} - \dfrac{2}{x+2} - \dfrac{2}{x+11}$

D) $\dfrac{8}{x+8} + \dfrac{2}{x+2} - \dfrac{2}{x+11}$

8) $\dfrac{2x - 5}{x^2 - 5x - 6}$

A) $\dfrac{9}{x+2} + \dfrac{1}{x-3}$

B) $\dfrac{17}{x+6} - \dfrac{3}{x-1}$

C) $\dfrac{1}{x-3} + \dfrac{1}{x-2}$

D) $\dfrac{1}{x-6} + \dfrac{1}{x+1}$

2 Decompose $\dfrac{P}{Q}$, Where Q Has Repeated Linear Factors

Write the partial fraction decomposition of the rational expression.

1) $\dfrac{-10x^2 - 22x - 9}{(x + 2)(x + 1)^2}$

A) $\dfrac{3}{x+2} + \dfrac{-5}{x+1} + \dfrac{5}{(x+1)^2}$

B) $\dfrac{5}{x+2} + \dfrac{-5}{x+1} + \dfrac{-3}{(x+1)^2}$

C) $\dfrac{-5}{x+2} + \dfrac{5}{x+1} + \dfrac{3}{(x+1)^2}$

D) $\dfrac{-5}{x+2} + \dfrac{-5}{x+1} + \dfrac{3}{(x+1)^2}$

2) $\dfrac{100 - 15x}{x^3 - 10x^2 + 25x}$

A) $\dfrac{4}{x} + \dfrac{5}{x-5} + \dfrac{-4}{(x-5)^2}$

B) $\dfrac{-4}{x} + \dfrac{4}{x-5} + \dfrac{5}{(x-5)^2}$

C) $\dfrac{4}{x} + \dfrac{-4}{x-5} + \dfrac{5}{(x-5)^2}$

D) $\dfrac{4}{x} + \dfrac{-4}{x-5} + \dfrac{10}{(x-5)^2}$

3) $\dfrac{7x^3 - 2}{x^2(x + 1)^3}$

4) $\dfrac{x + 4}{x^3 - 2x^2 + x}$

A) $\dfrac{4}{x} + \dfrac{-4}{x-1} + \dfrac{9}{(x-1)^2}$

B) $\dfrac{-4}{x} + \dfrac{4}{x-1} + \dfrac{5}{(x-1)^2}$

C) $\dfrac{4}{x} + \dfrac{5}{x-1} + \dfrac{-4}{(x-1)^2}$

D) $\dfrac{4}{x} + \dfrac{-4}{x-1} + \dfrac{5}{(x-1)^2}$

5) $\dfrac{x + 1}{(x - 2)^2(x + 4)}$

A) $\dfrac{\frac{1}{2}}{(x - 2)^2} + \dfrac{-\frac{1}{12}}{x + 4}$

B) $\dfrac{\frac{1}{12}}{x - 2} + \dfrac{\frac{1}{2}}{(x - 2)^2} + \dfrac{-\frac{1}{12}}{x + 4}$

C) $\dfrac{12}{x - 2} + \dfrac{2}{(x - 2)^2} + \dfrac{-12}{x + 4}$

D) $\dfrac{-1}{x - 2} + \dfrac{\frac{1}{4}x}{(x - 2)^2} + \dfrac{-\frac{1}{4}}{x + 4}$

3 Decompose $\frac{P}{Q}$, Where Q Has Only Nonrepeated Irreducible Quadratic Factors

Write the partial fraction decomposition of the rational expression.

1) $\dfrac{10x + 2}{(x - 1)(x^2 + x + 1)}$

A) $\dfrac{4}{x - 1} + \dfrac{-4x + 2}{x^2 + x + 1}$

B) $\dfrac{4}{x - 1} + \dfrac{-4}{x + 1} + \dfrac{2}{x - 1}$

C) $\dfrac{4}{x - 1} + \dfrac{2x - 4}{x^2 + x + 1}$

D) $\dfrac{-4}{x - 1} + \dfrac{4x + 2}{x^2 + x + 1}$

2) $\dfrac{x^2 - 111}{x^4 - x^2 - 72}$

A) $\dfrac{1}{x + 3} + \dfrac{1}{x - 3} - \dfrac{7}{x^2 + 8}$

B) $\dfrac{1}{x + 3} - \dfrac{1}{x - 3} + \dfrac{7}{x^2 + 8}$

C) $\dfrac{1}{x + 3} - \dfrac{1}{x - 3} - \dfrac{7}{x^2 + 8}$

D) $\dfrac{1}{x + 3} + \dfrac{1}{x - 3} + \dfrac{7}{x^2 + 8}$

3) $\dfrac{x^2 - 56}{x^4 + 5x^2 - 36}$

4) $\dfrac{3x - 2}{x^3 - 1}$

A) $\dfrac{3}{x - 1} + \dfrac{-3(x - 7)}{x^2 + x + 1}$

B) $\dfrac{3}{(x - 1)^2} + \dfrac{1}{(x - 1)^3}$

C) $\dfrac{\frac{1}{2}}{x - 1} + \dfrac{\frac{5}{2}}{x + 1}$

D) $\dfrac{\frac{1}{3}}{x - 1} + \dfrac{-\frac{1}{3}x + \frac{7}{3}}{(x^2 + x + 1)}$

4 Decompose $\frac{P}{Q}$, Where Q Has Repeated Irreducible Quadratic Factors

Write the partial fraction decomposition of the rational expression.

1) $\frac{2x^3 + 2x^2}{(x^2 + 5)^2}$

A) $\frac{2x + 2}{x^2 + 5} + \frac{10x - 10}{(x^2 + 5)^2}$ B) $\frac{2x + 2}{x^2 + 5} + \frac{-10x - 10}{(x^2 + 5)^2}$ C) $\frac{2x + 2}{x^2 + 5} + \frac{10x + 10}{(x^2 + 5)^2}$ D) $\frac{2x - 2}{x^2 + 5} + \frac{-10x + 10}{(x^2 + 5)^2}$

2) $\frac{x^2 + 2x + 1}{(x^2 + 4)^2}$

A) $\frac{-1}{x^2 + 4} + \frac{2x - 3}{(x^2 + 4)^2}$ B) $\frac{1}{x^2 + 4} + \frac{-x - 3}{(x^2 + 4)^2}$ C) $\frac{x + 1}{x^2 + 4} + \frac{2x - 3}{(x^2 + 4)^2}$ D) $\frac{1}{x^2 + 4} + \frac{2x - 3}{(x^2 + 4)^2}$

3) $\frac{3x^3 + 17x - 6}{(x^2 + 5)^2}$

A) $\frac{3x}{x^2 + 5} + \frac{2x + 6}{(x^2 + 5)^2}$ B) $\frac{3x + 1}{x^2 + 5} + \frac{2x - 6}{(x^2 + 5)^2}$ C) $\frac{3x}{x^2 + 5} + \frac{-2x - 6}{(x^2 + 5)^2}$ D) $\frac{3x}{x^2 + 5} + \frac{2x - 6}{(x^2 + 5)^2}$

4) $\frac{4x^3 - 3x^2 + 5x - 12}{(x^2 + 2)^3}$

A) $\frac{x}{x^2 + 2} + \frac{4x - 3}{(x^2 + 2)^2} + \frac{-3x - 6}{(x^2 + 2)^3}$ B) $\frac{4x + 3}{(x^2 + 2)^2} + \frac{-3x + 6}{(x^2 + 2)^3}$

C) $\frac{4x - 3}{(x^2 + 2)^2} + \frac{-3x - 6}{(x^2 + 2)^3}$ D) $\frac{x + 1}{x^2 + 2} + \frac{4x - 3}{(x^2 + 2)^2} + \frac{-3x - 6}{(x^2 + 2)^3}$

7.6 Systems of Nonlinear Equations

1 Solve a System of Nonlinear Equations Using Substitution

Solve using substitution.

1)
$$\begin{cases} x^2 + y^2 = 61 \\ x + y = -11 \end{cases}$$

A) x = -5, y = -6; x = -6, y = -5 B) x = 5, y = 6; x = 6, y = 5

C) x = 5, y = -6; x = 6, y = -5 D) x = -5, y = 6; x = -6, y = 5

2)
$$\begin{cases} xy = 72 \\ x + y = -17 \end{cases}$$

A) x = 8, y = -9; x = 9, y = -8 B) x = -8, y = -9; x = -9, y = -8

C) x = -8, y = 9; x = -9, y = 8 D) x = 8, y = 9; x = 9, y = 8

3)

$$\begin{cases} x^2 + y^2 = 61 \\ x - y = 1 \end{cases}$$

A) x = -6, y = -5; x = -5, y = -6

B) x = 6, y = -5; x = 5, y = -6

C) x = 6, y = 5; x = -5, y = -6

D) x = -6, y = 5; x = -5, y = 6

4)

$$\begin{cases} y = x^2 - 8x + 16 \\ x + y = 6 \end{cases}$$

A) x = 2, y = 4; x = 5, y = 1

B) x = 4, y = 2

C) x = 2, y = 8; x = 5, y = 1

D) x = -2, y = 8; x = -5, y = 11

5)

$$\begin{cases} y = -x^2 + 13 \\ x^2 + y^2 = 25 \end{cases}$$

A) x = 9, y = 94; x = 16, y = 269

B) x = 4, y = -3; x = 3, y = 4; x = -3, y = 4; x = -4, y = -3

C) x = 4, y = -3; x = -4, y = -3

D) x = 3, y = 4; x = -3, y = 4

6)

$$\begin{cases} x^2 + y^2 = 4 \\ x + y = 2 \end{cases}$$

A) x = 0, y = 2; x = 2, y = 0

B) x = 2, y = -2; x = -2, y = -2

C) x = 0, y = 0; x = 2, y = -2

D) x = 0, y = -2; x = -2, y = 0

7)

$$\begin{cases} xy = 20 \\ x + y = 9 \end{cases}$$

A) x = 20, y = 1; x = 1, y = 20

B) x = 5, y = 4; x = 4, y = 5

C) x = 10, y = 2; x = 2, y =10

D) x = 6, y = 3; x = 3, y = 6

8)

$$\begin{cases} x^2 + y^2 = 169 \\ x + y = 17 \end{cases}$$

A) x =-12, y = -5; x = -5, y = -12

B) x = 12, y = 5; x = 5, y = 12

C) x = 12, y = -5; x = 5, y = -12

D) x = -12, y = 5; x = -5, y = 12

9)

$$\begin{cases} xy - x^2 = -20 \\ x - 2y = 3 \end{cases}$$

A) x = 5, y = 1; x = -8, $y = -\frac{11}{2}$

B) x = 5, y = 1; $x = -\frac{11}{2}$, y = -8

C) x = -5, y = -1; x = 8, $y = \frac{11}{2}$

D) x = -5, y = -1; $x = \frac{11}{2}$, y = 8

10)

$$\begin{cases} x^2 - y^2 = 39 \\ x - y = 3 \end{cases}$$

A) $x = -8, y = 5$ B) $x = -8, y = -5$ C) $x = 8, y = -5$ D) $x = 8, y = 5$

11)

$$\begin{cases} y = 6x^2 - 5x \\ y = 2x + 3 \end{cases}$$

A) $x = \frac{1}{6}, y = \frac{10}{3}$; $x = 1, y = 5$ B) $x = \frac{3}{2}, y = 6$; $x = -\frac{1}{3}, y = -\frac{7}{3}$

C) $x = \frac{1}{3}, y = \frac{11}{3}$; $x = -\frac{3}{2}, y = 0$ D) $x = -\frac{1}{2}, y = 2$; $x = 1, y = 5$

2 Solve a System of Nonlinear Equations Using Elimination

Solve using elimination.

1)

$$\begin{cases} x^2 + y^2 = 52 \\ x^2 - y^2 = 20 \end{cases}$$

A) $x = 6, y = 4$; $x = -6, y = 4$; $x = 6, y = -4$; $x = -6, y = -4$

B) $x = 6, y = -4$; $x = 6, y = 4$

C) $x = 6, y = 4$; $x = 4, y = 6$; $x = -6, y = -4$; $x = -4, y = -6$

D) $x = -6, y = -4$; $x = -4, y = -6$

2)

$$\begin{cases} 2x^2 - 5y^2 = -37 \\ 4x^2 + 3y^2 = 43 \end{cases}$$

A) $x = -2, y = -3$; $x = -3, y = -2$

B) $x = 2, y = 3$; $x = 3, y = 2$; $x = -2, y = -3$; $x = -3, y = -2$

C) $x = 2, y = 3$; $x = -2, y = 3$; $x = 2, y = -3$; $x = -2, y = -3$

D) $x = 2, y = -3$; $x = 2, y = 3$

3)

$$\begin{cases} x^2 + y^2 = 100 \\ x^2 - y^2 = 100 \end{cases}$$

A) $x = -10, y = 0$; $x = -10, y = 10$ B) $x = 10, y = 0$; $x = 10, y = 10$

C) $x = 10, y = 0$; $x = -10, y = 0$ D) $x = 10, y = 10$; $x = -10, y = 10$

4)

$$\begin{cases} x^2 + y^2 = 64 \\ \frac{x^2}{64} + \frac{y^2}{9} = 1 \end{cases}$$

A) $x = 0, y = -3$; $x = 0, y = 3$ B) $x = -8, y = 0$; $x = 8, y = 0$

C) $x = 0, y = -8$; $x = 0, y = 8$ D) inconsistent

5)

$$\begin{cases} 3x^2 + 2y^2 = 89 \\ x^2 - 2y^2 = -21 \end{cases}$$

6)

$$\begin{cases} 2x^2 + y^2 = 17 \\ 3x^2 - 2y^2 = -6 \end{cases}$$

A) $x = 2, y = 3;\ x = 2, y = -3;\ x = -2, y = 3;\ x = -2, y = -3$

B) $x = 1, y = 3;\ x = -1, y = -3$

C) $x = 2, y = -3;\ x = -2, y = 3$

D) $x = 1, y = 3;\ x = 1, y = -3;\ x = -1, y = 3;\ x = -1, y = -3$

7)

$$\begin{cases} 2x^2 + xy - y^2 = 3 \\ x^2 + 2xy + y^2 = 3 \end{cases}$$

A) $x = \frac{-2}{3}, y = \frac{1}{3};\ x = \frac{2\sqrt{3}}{3}, y = -\frac{1}{3}$

B) $x = \frac{2\sqrt{3}}{3}, y = -\frac{\sqrt{3}}{3};\ x = -\frac{2\sqrt{3}}{3}, y = \frac{\sqrt{3}}{3}$

C) $x = \frac{2\sqrt{3}}{3}, y = \frac{\sqrt{3}}{3};\ x = -\frac{2\sqrt{3}}{3}, y = -\frac{\sqrt{3}}{3}$

D) $x = \frac{-2}{3}, y = -\frac{1}{3};\ x = \frac{2\sqrt{3}}{3}, y = \frac{1}{3}$

3 Additional Applications

Graph the equations of the system. Then solve the system to find the points of intersection.

1)

$$\begin{cases} y = x^2 - 12x + 36 \\ y = -x + 6 \end{cases}$$

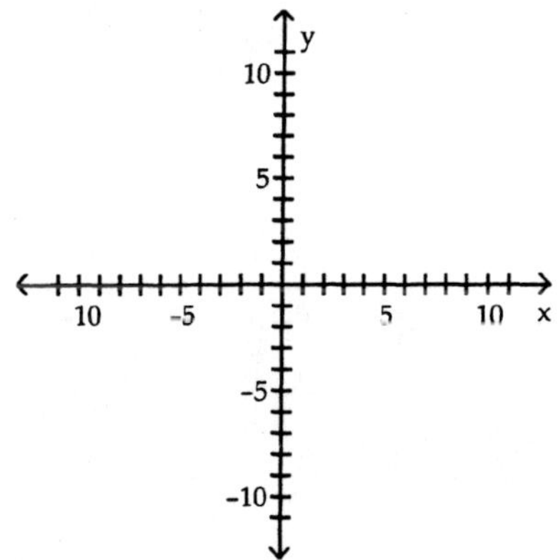

2)

$$\begin{cases} x^2 + y^2 = 9 \\ x^2 - y^2 = 9 \end{cases}$$

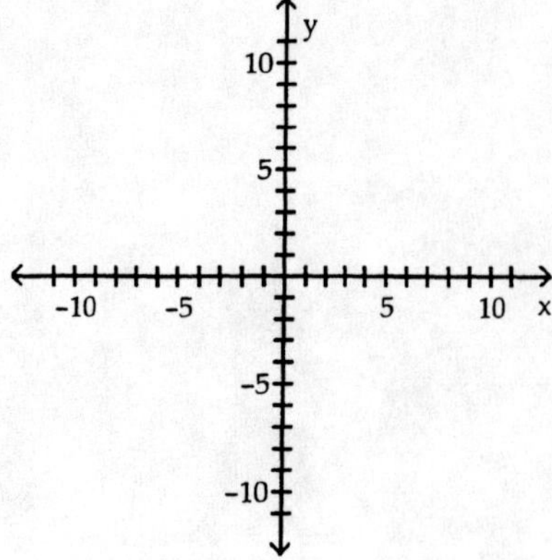

3)

$$\begin{cases} x^2 + y^2 = 36 \\ y = x^2 - 6 \end{cases}$$

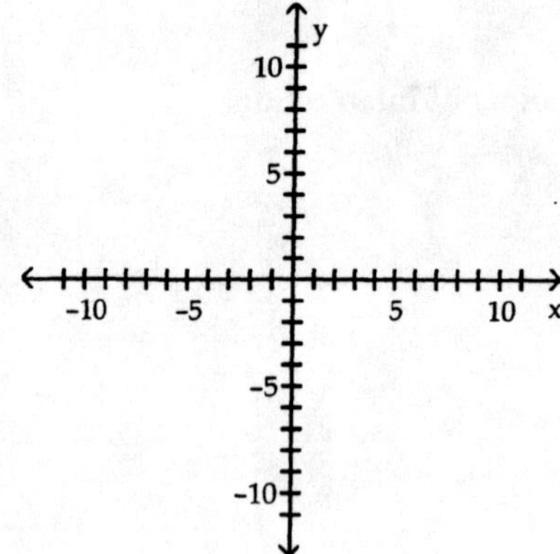

4)

$$\begin{cases} y = \sqrt{x} \\ y = 6 - x \end{cases}$$

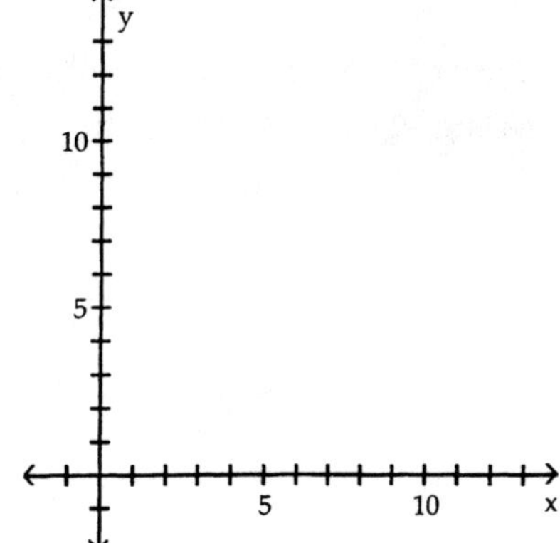

5)

$$\begin{cases} (x + 3)^2 + (y - 2)^2 = 4 \\ y^2 - 4y - x - 3 = 0 \end{cases}$$

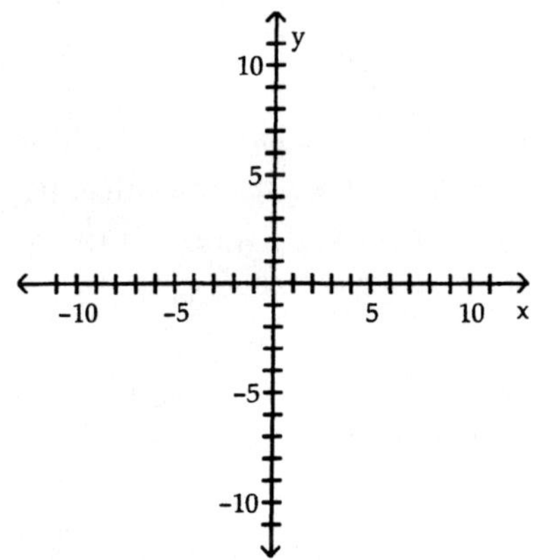

Solve the system of equations.

6)

$$\begin{cases} \ln x = 3 \ln y \\ 3^x = 27^y \end{cases}$$

A) $x = 3\sqrt{3}, y = \sqrt{3}$ B) $x = \sqrt{3}, y = 9$ C) $x = 9, y = \sqrt{3}$ D) $x = \sqrt{3}, y = 3\sqrt{3}$

Use a graphing utility to solve the system of equations. Round answers to two decimal places.

7)

$$\begin{cases} x^3 + y^2 = 2 \\ x^2 y = 4 \end{cases}$$

A) $x = 2.14, y = -1.37$ B) $x = -1.37, y = 2.14$ C) $x = 2.14, y = 1.37$ D) $x = 1.37, y = 2.14$

Solve the problem.

8) The sum of the squares of two numbers is 145. The sum of the two numbers is –1. Find the two numbers.

A) -9 and 8
B) 8 and 9; or –9 and –8
C) -9 and 8; or –8 and 9
D) –8 and 9

9) The sum of the squares of two numbers is 68. The difference of the two numbers is 6. Find the two numbers.

A) 2 and 8
B) –8 and –2
C) –8 and 2; or –2 and 8
D) –8 and –2; or 2 and 8

10) The difference of two numbers is 5 and the difference of their squares is 55. Find the numbers.

11) A right triangle has an area of 55 square inches. The square of the hypotenuse is 221. Find the lengths of the legs of the triangle. Round your answer to the nearest inch.

A) 5 in. and 22 in.
B) 10 in. and 11 in.
C) 20 in. and 5.5 in.
D) 100 in. and 121 in.

12) The perimeter of a rectangle is 38 inches and its area is 78 square inches. What are its dimensions?

A) 5 in. by 12 in.
B) 7 in. by 12 in.
C) 5 in. by 14 in.
D) 6 in. by 13 in.

13) A rectangular piece of tin has an area of 792 square inches. A square of 4 inches is cut from each corner, and an open box is made by turning up the ends and sides. If the volume of the box is 1568 cubic inches, what were the original dimensions of the piece of tin?

A) 26 in. by 40 in.
B) 22 in. by 36 in.
C) 14 in. by 24 in.
D) 18 in. by 32 in.

14) The diagonal of the floor of a rectangular office cubicle is 2 ft longer than the length of the cubicle and 5 ft longer than twice the width. Find the dimensions of the cubicle. Round to the nearest tenth, if necessary.

A) width = 2 ft, length = 9 ft
B) width = 9.7 ft, length = 22.4 ft
C) width = 3.9 ft, length = 9.7 ft
D) width = 4 ft, length = 11 ft

15) In a 1–mile race, the winner crosses the finish line 13 feet ahead of the second–place runner and 30 feet ahead of the third–place runner. Assuming that each runner maintains a constant speed throughout the race, by how many feet does the second–place runner beat the third–place runner? (5280 feet in 1 mile.)

A) 4.01 ft
B) –13.07 ft
C) 17.04 ft
D) –17.1 ft

16) A person at the top of a 600 foot tall building drops a yellow ball. The height of the yellow ball is given by the equation $h = -16t^2 + 600$ where h is measured in feet and t is the number of seconds since the yellow ball was dropped. A second person, in the same building but on a lower floor that is 312 feet from the ground, drops a white ball 3 seconds after the yellow ball was dropped. The height of the white ball is given by the equation $h = -16(t - 3)^2 + 312$ where h is measured in feet and t is the number of seconds since the yellow ball was dropped. Find the time that the balls are the same distance above the ground and find this distance.

A) 5 sec; 200 ft
B) 3.5 sec; 404 ft
C) 4.5 sec; 276 ft
D) 4 sec; 344 ft

7.7 Systems of Inequalities

1 Graph an Inequality

Graph the inequality.

1) $x > -9$

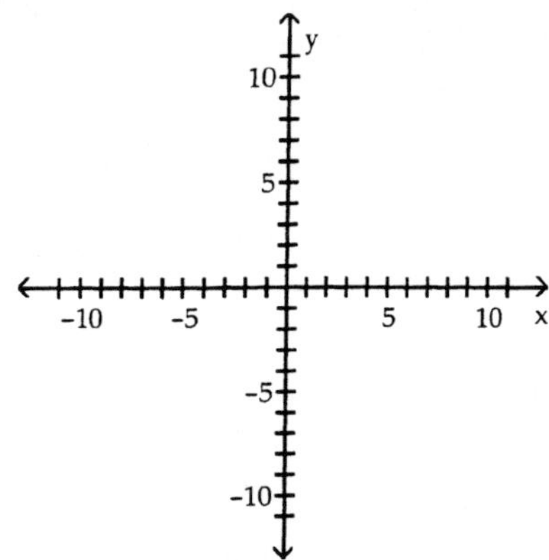

A)

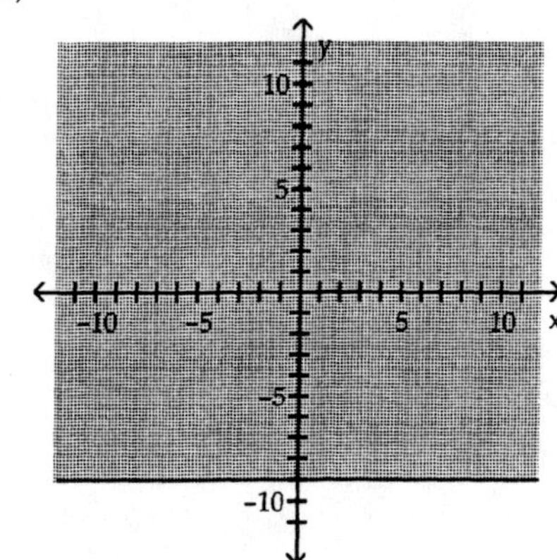

B)

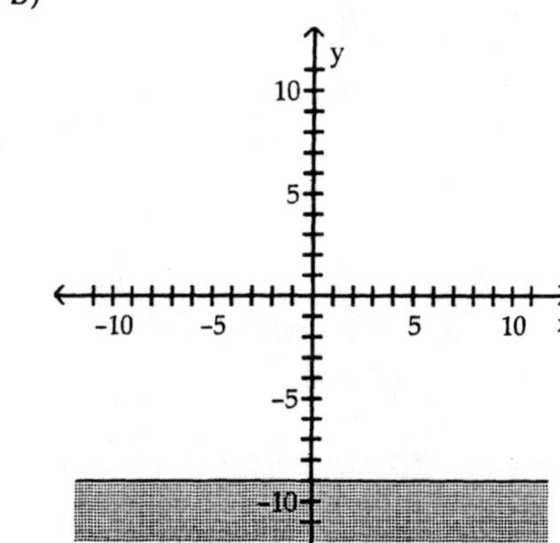

C)

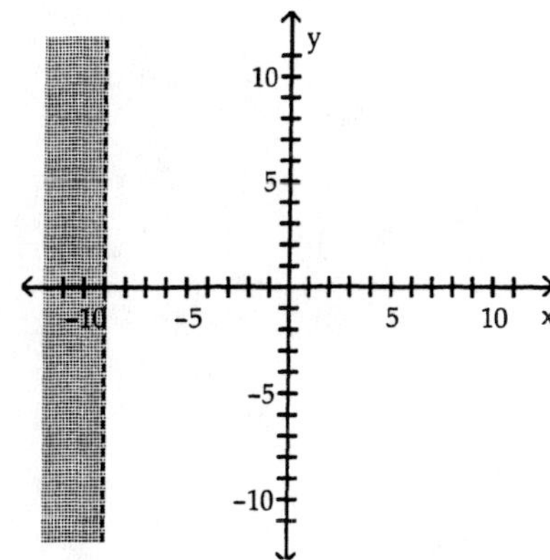

D)

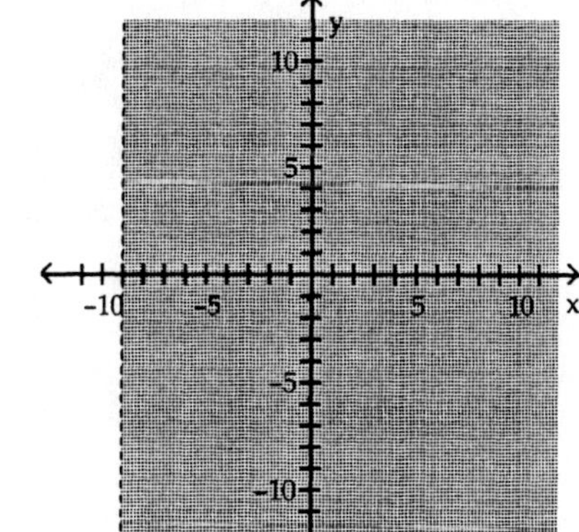

2) $y \le -7$

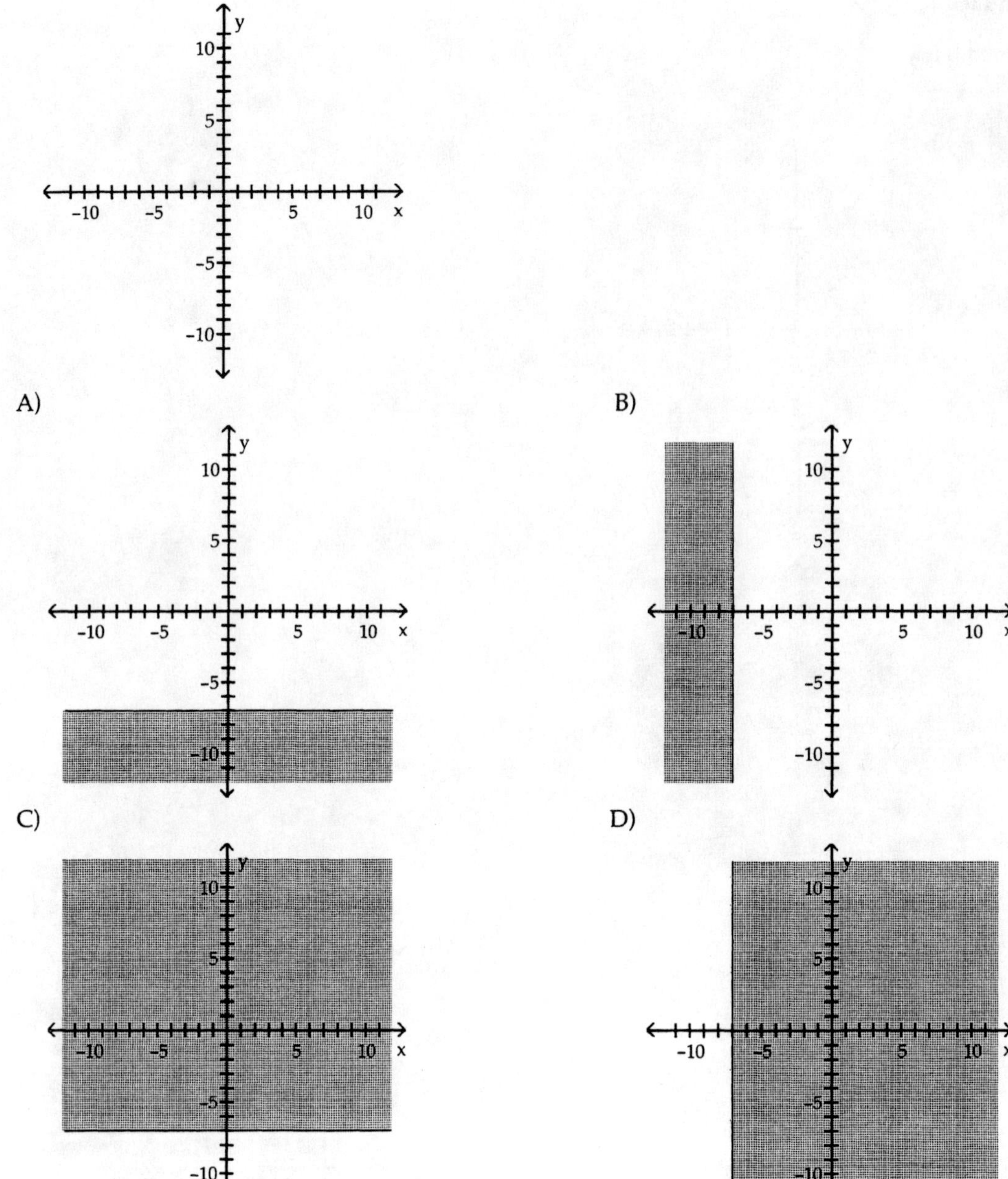

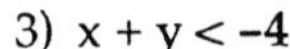

3) $x + y < -4$

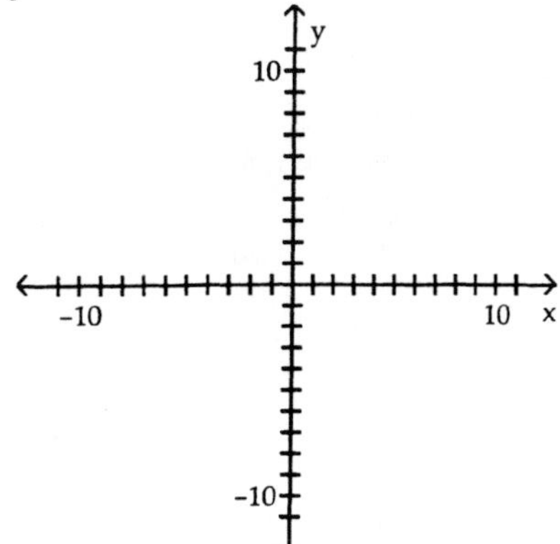

y
10
-10
10
x
-10

y
10
-10
10
x
-10

y
10
-10
10
x
-10

y
10
-10
10
x
-10

y
10
-10
10
x
-10

A)

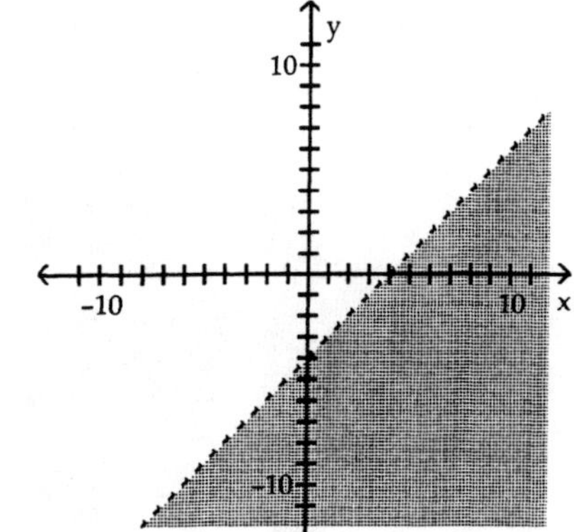

B)

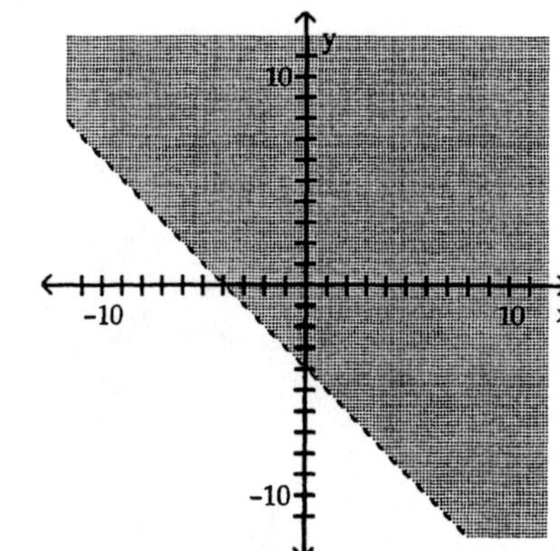

C)

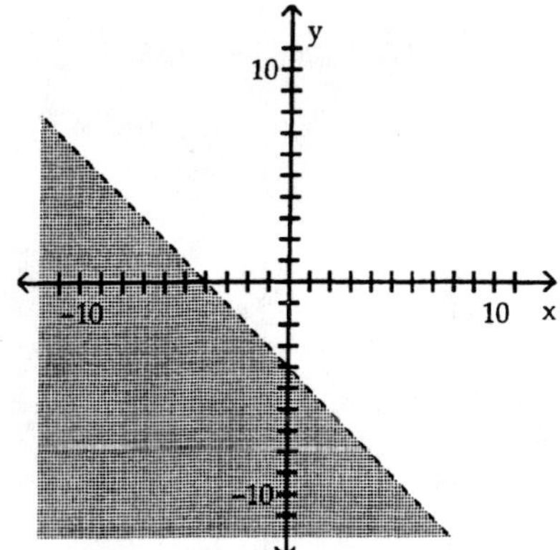

D)

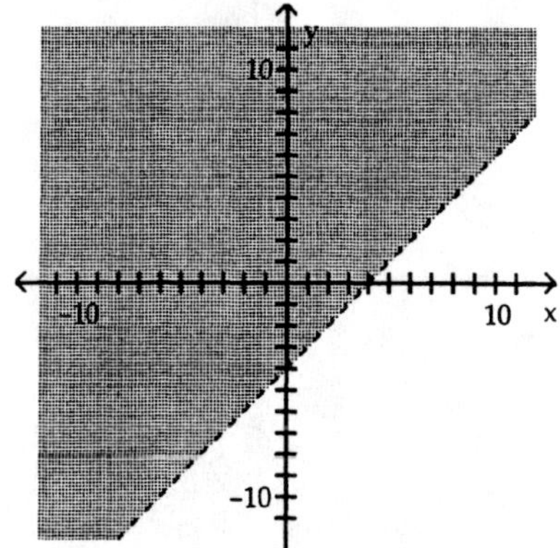

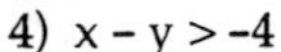
4) $x - y > -4$

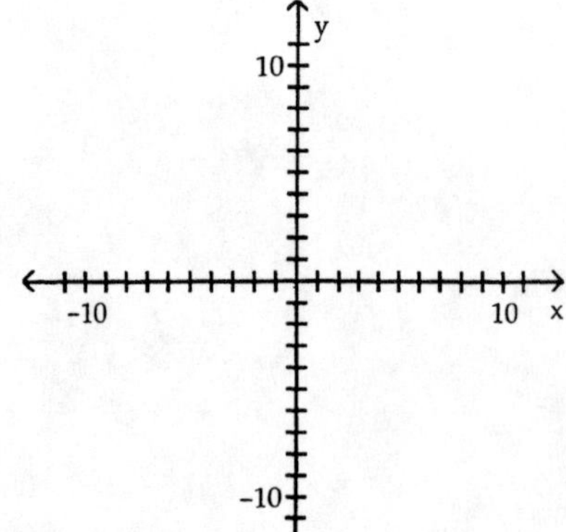

y
10
-10
10
x
-10

y
10
-10
10
x
-10

y
10
-10
10
x
-10

y
10
-10
10
x
-10

y
10
-10
10
x
-10

A)

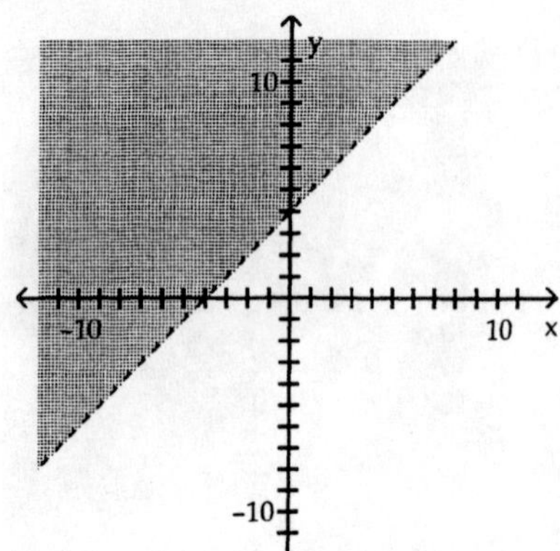

B)

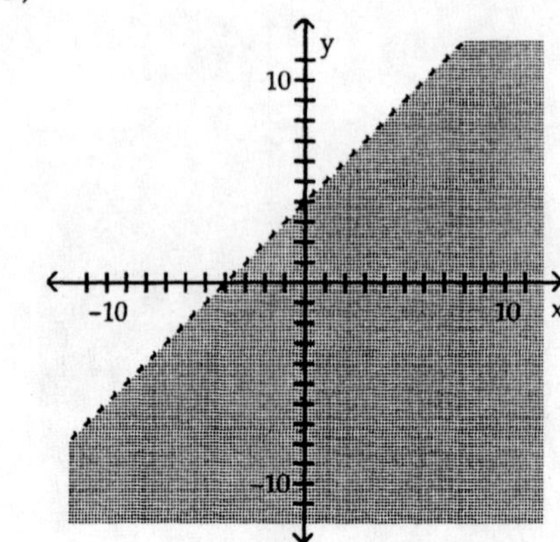

C)

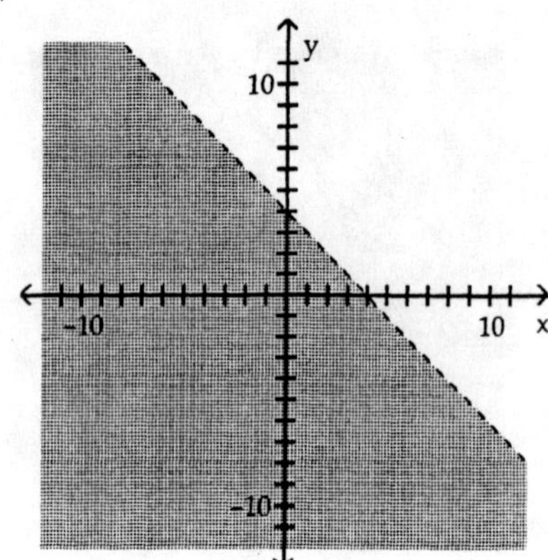

D)

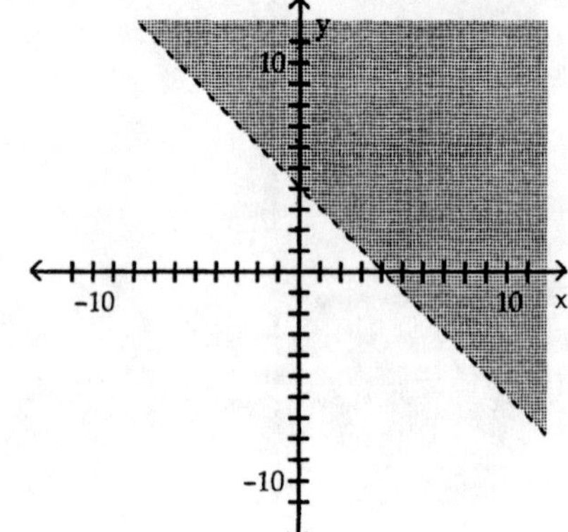

5) $x - y < -2$

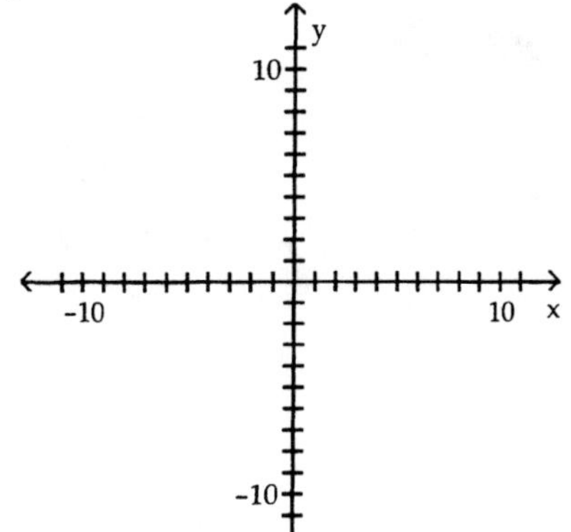

A)

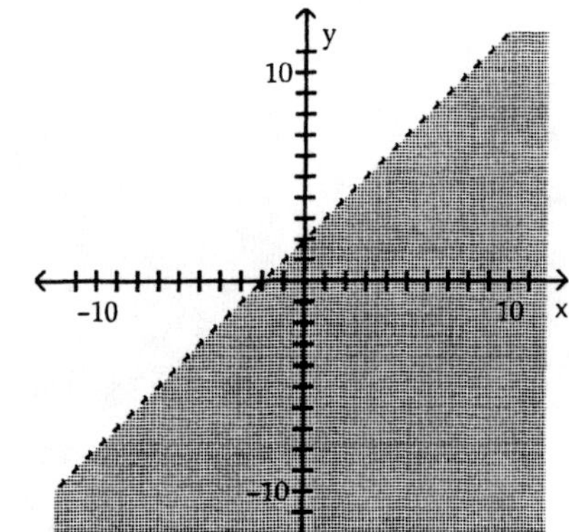

B)

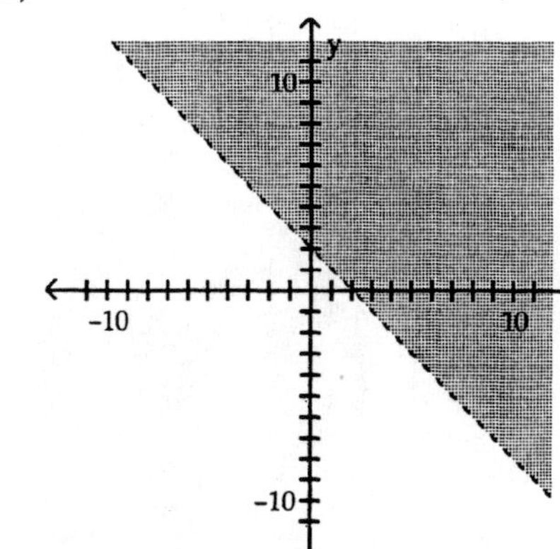

C)

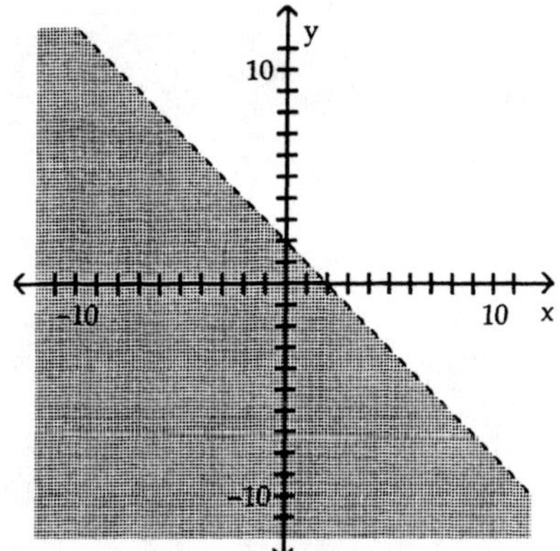

D)

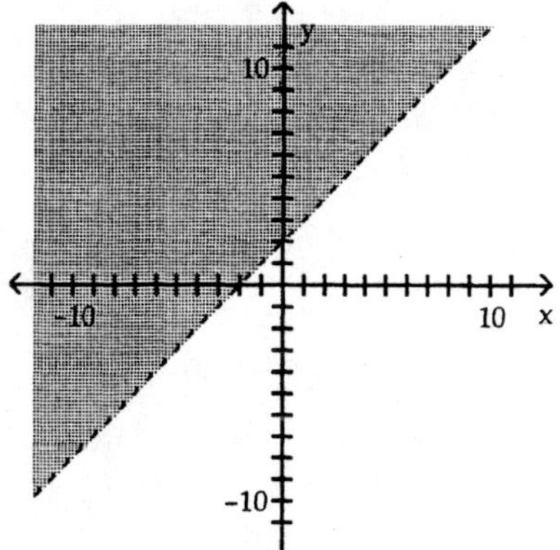

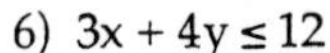
6) $3x + 4y \le 12$

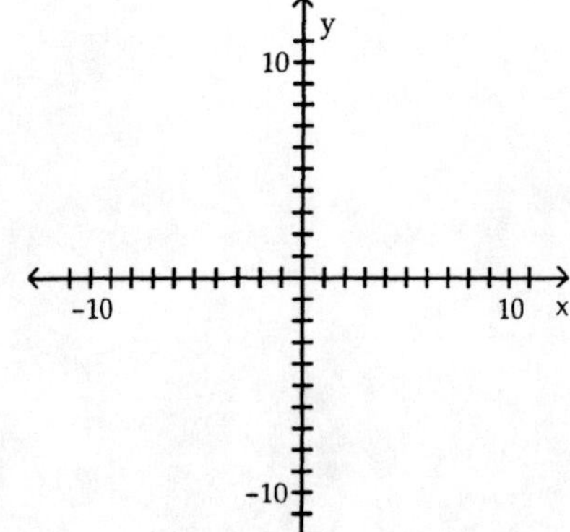

A)

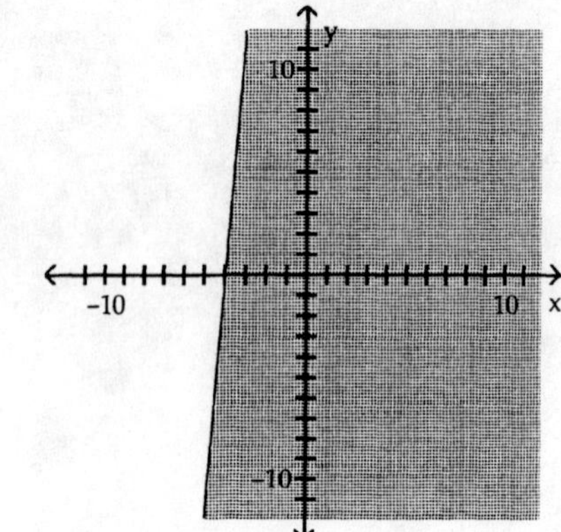

B)

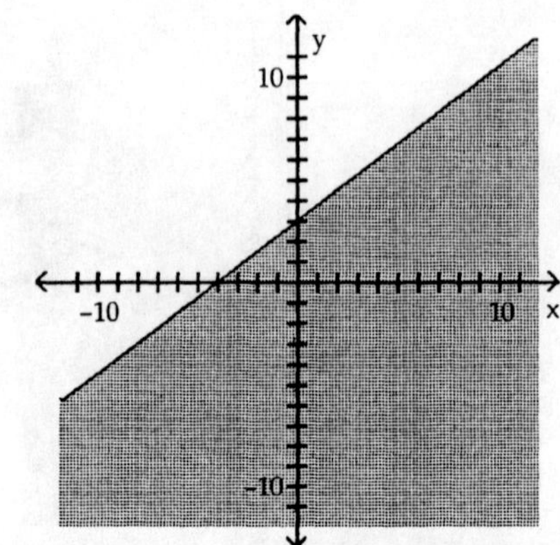

C)

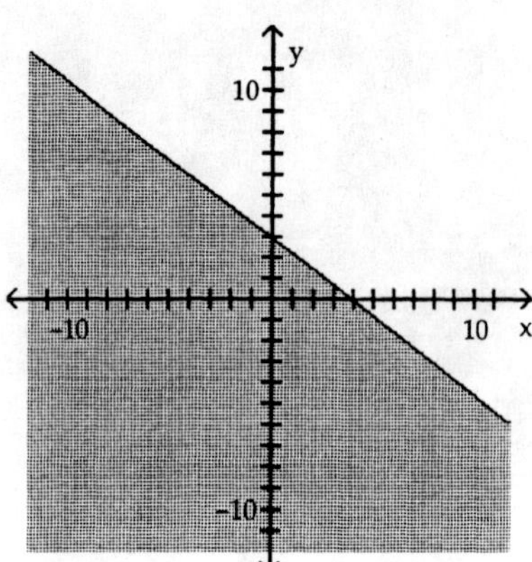

D)

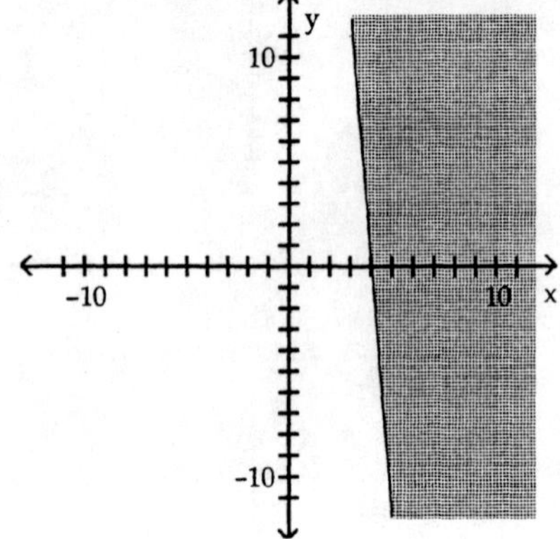

7) $-2x - 3y \leq -6$

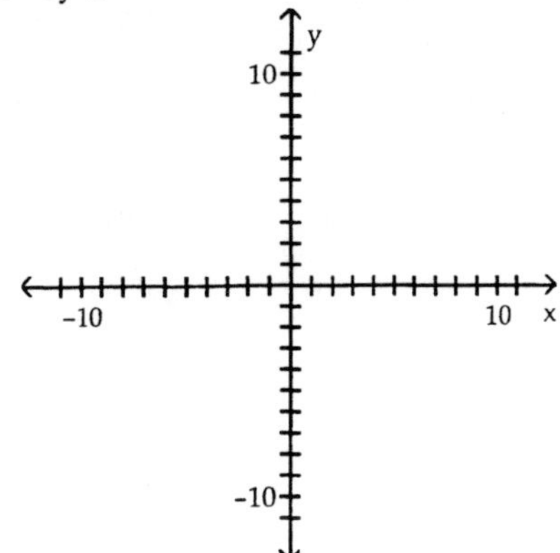

A)

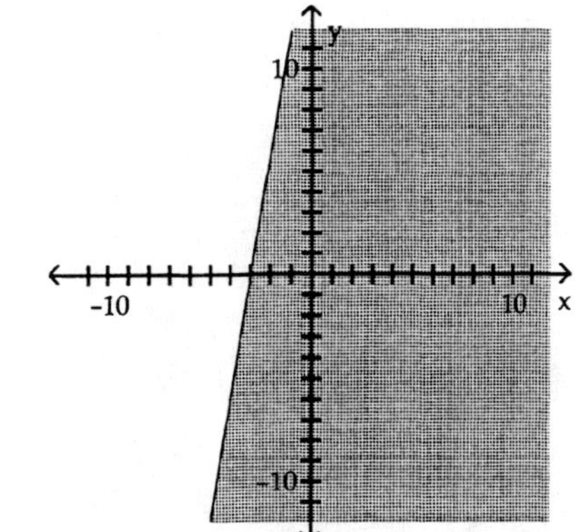

B)

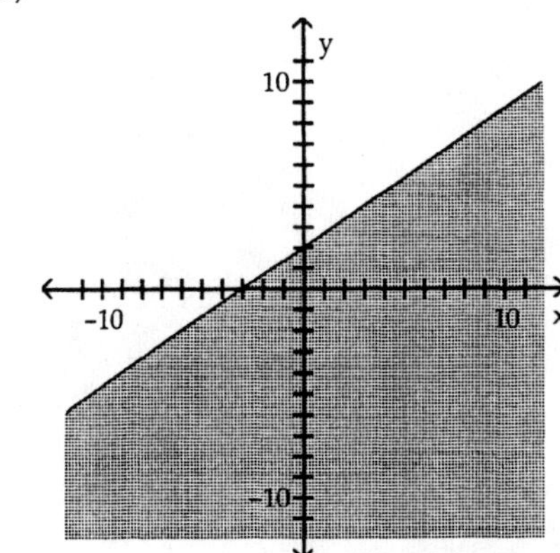

C)

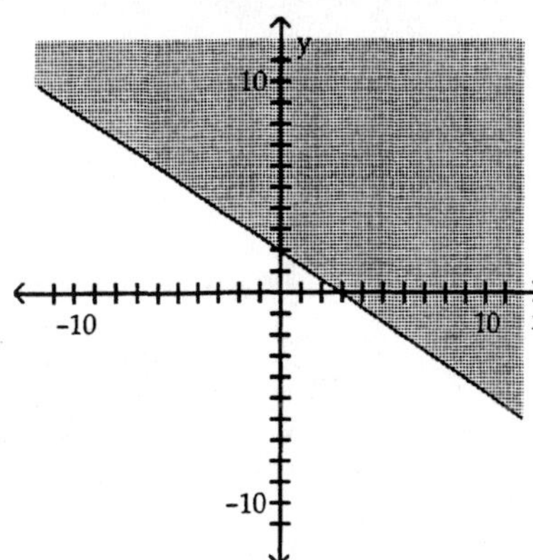

D)

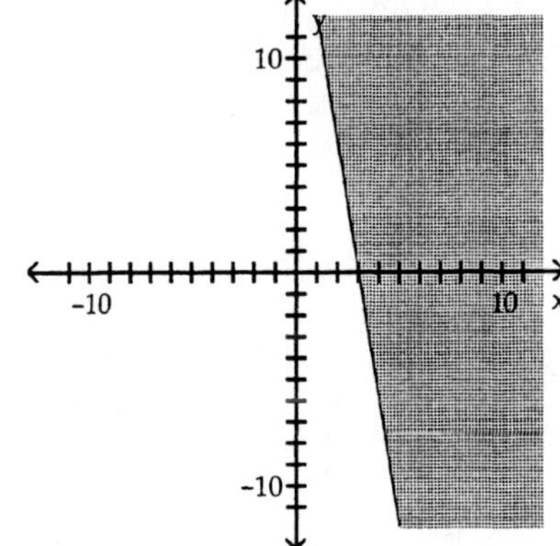

8) $x^2 + y^2 \leq 1$

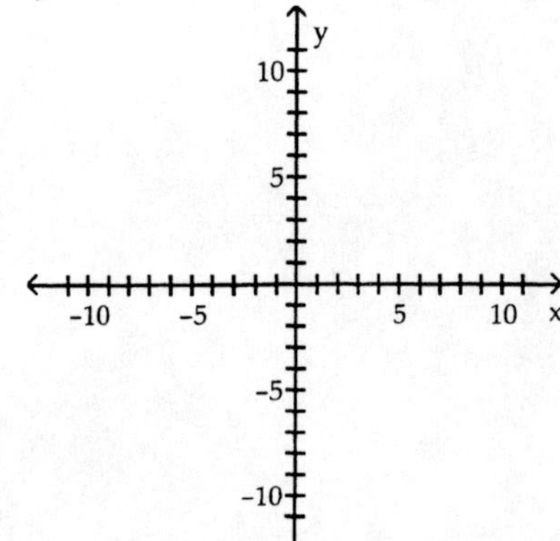

A)

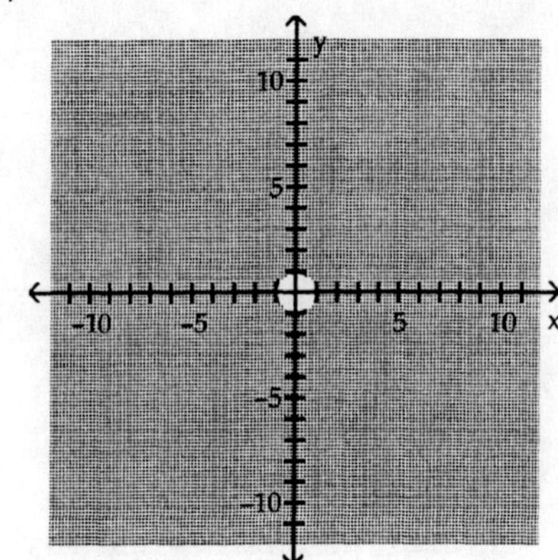

B)

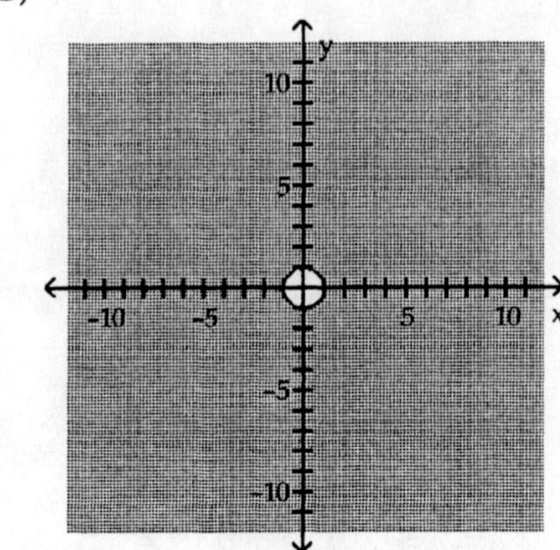

C)

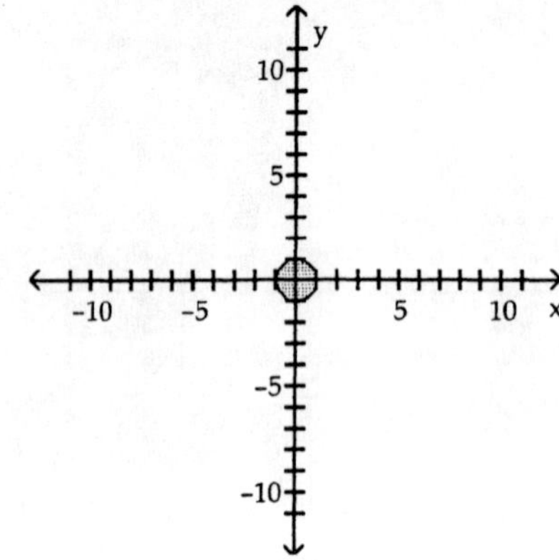

D)

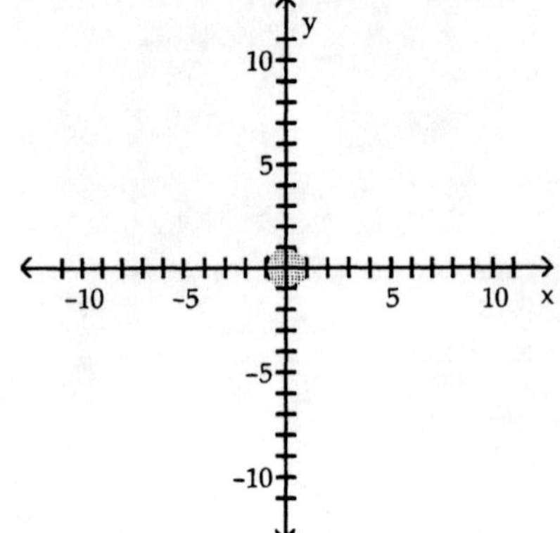

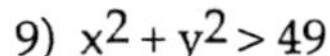

9) $x^2 + y^2 > 49$

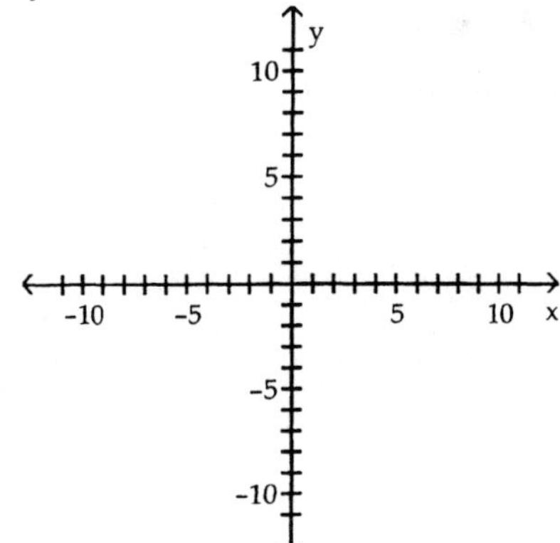

A)

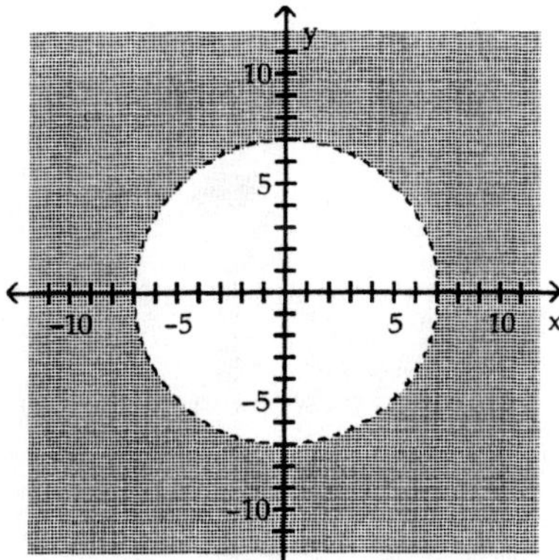

B)

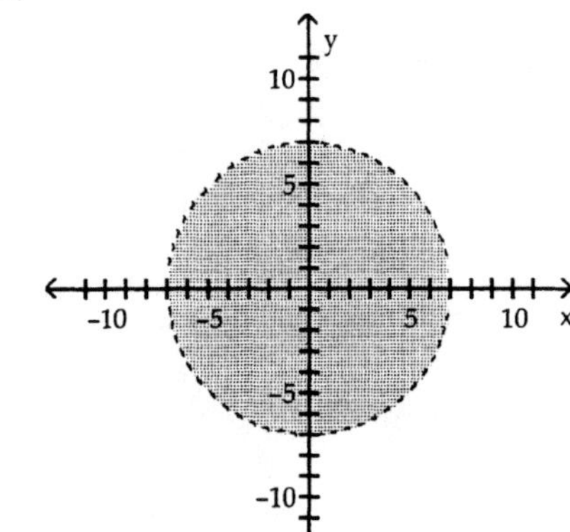

C)

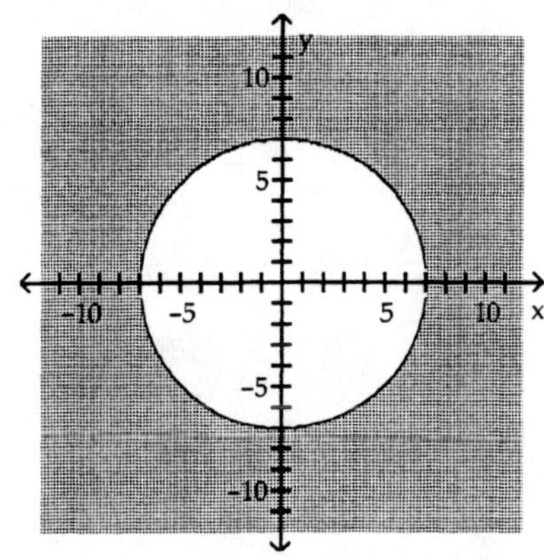

D)

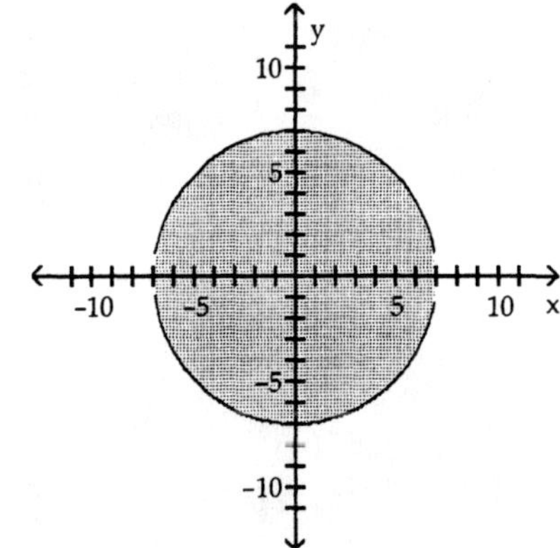

10) $y > x^2 - 1$

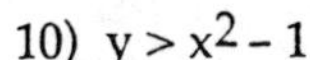

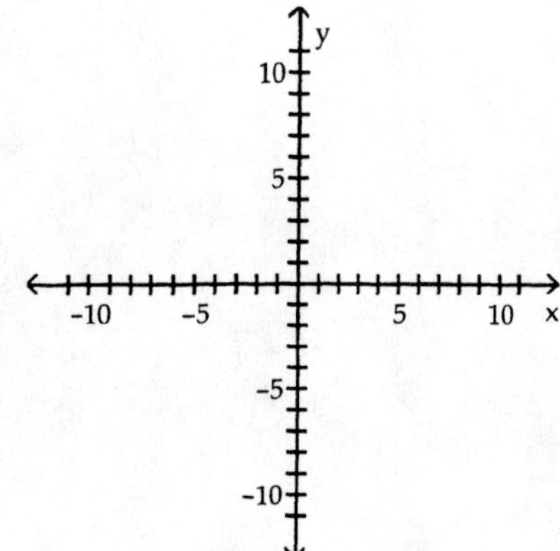

A)

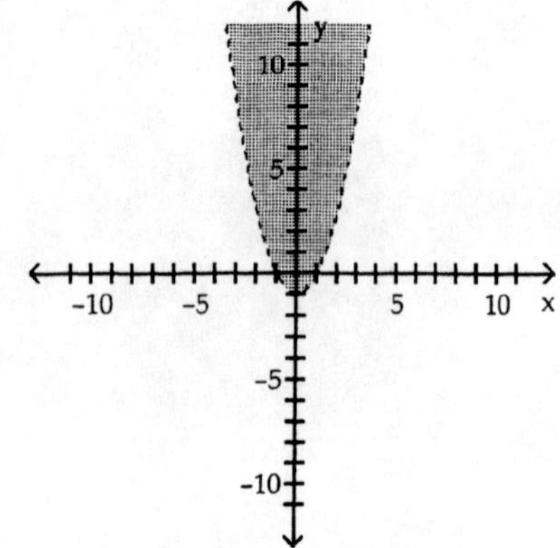

B)

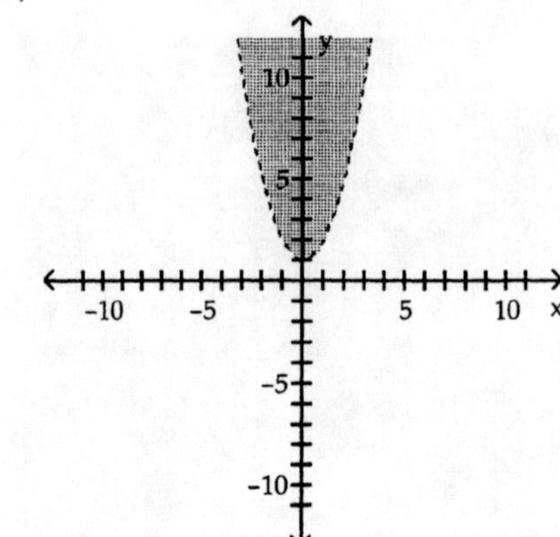

C)

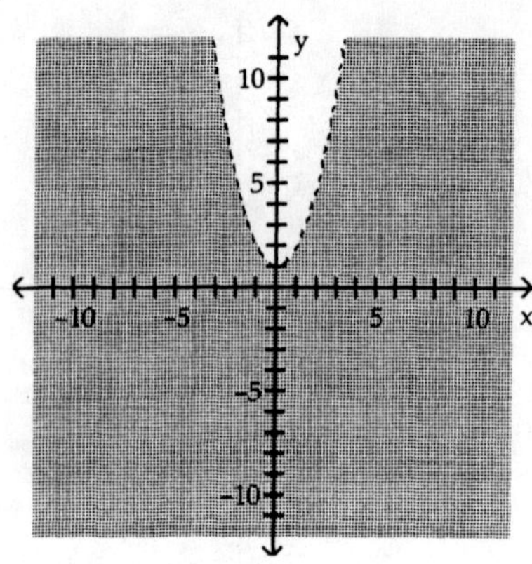

D)

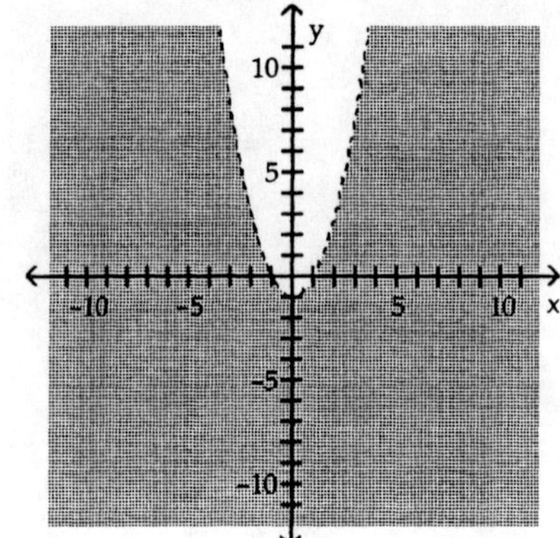

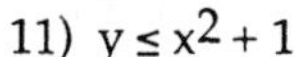

11) $y \le x^2 + 1$

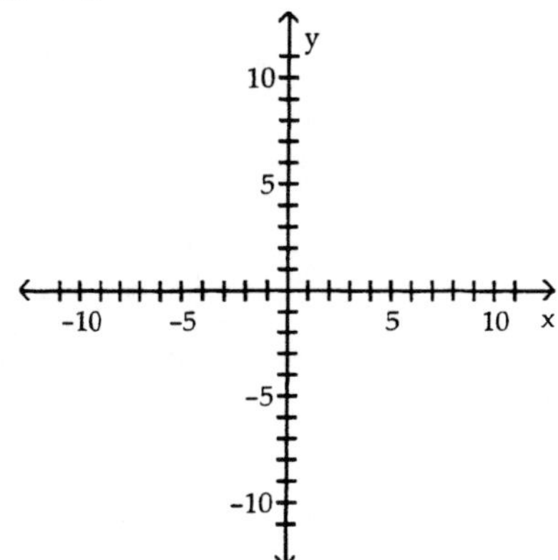

A)

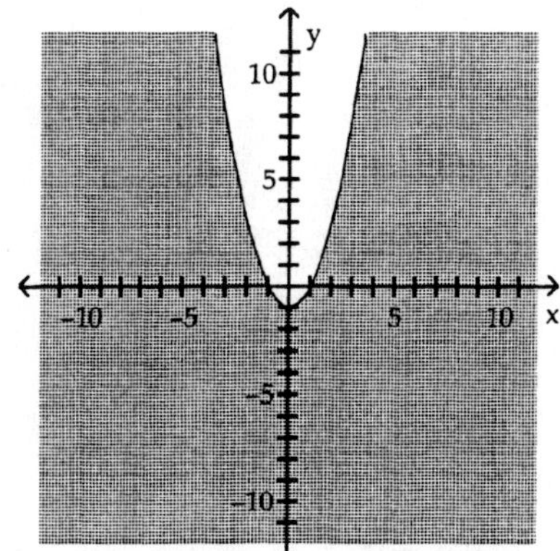

B)

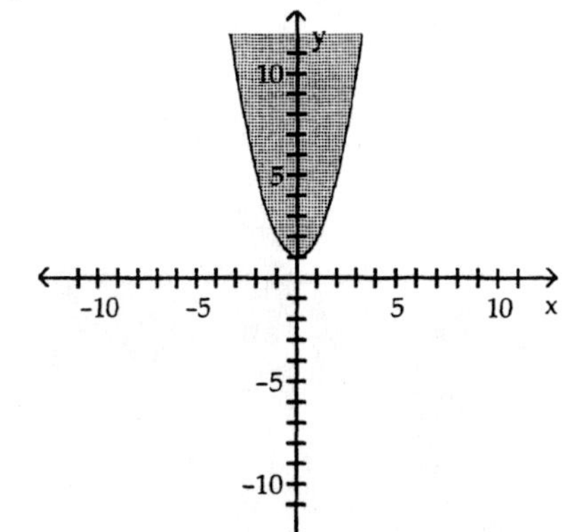

C)

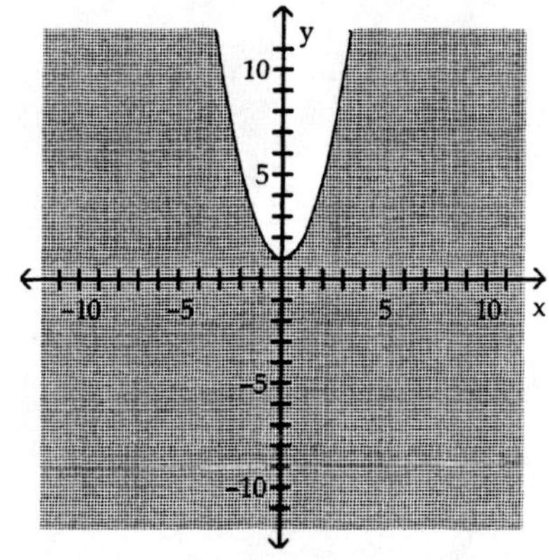

D)

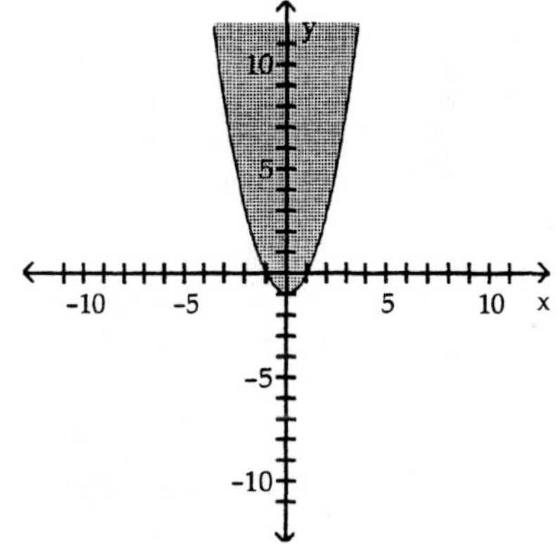

12) $xy \leq 5$

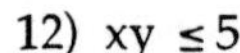

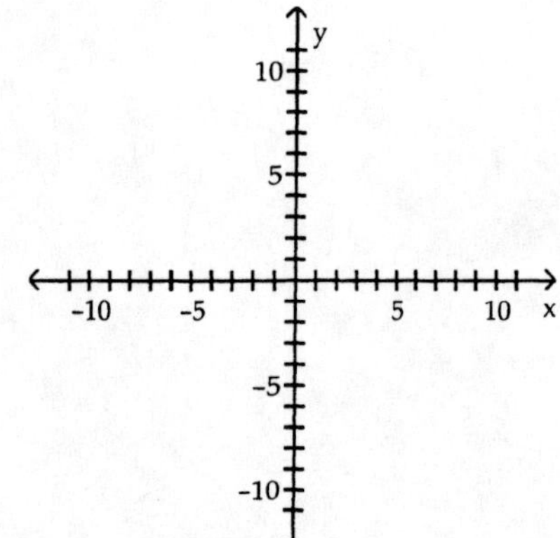

A)

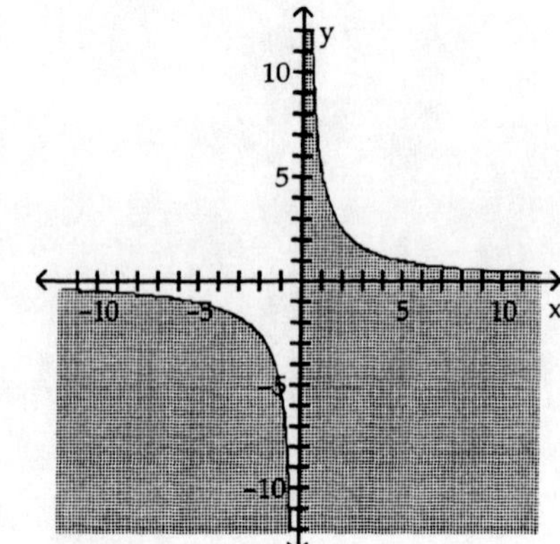

B)

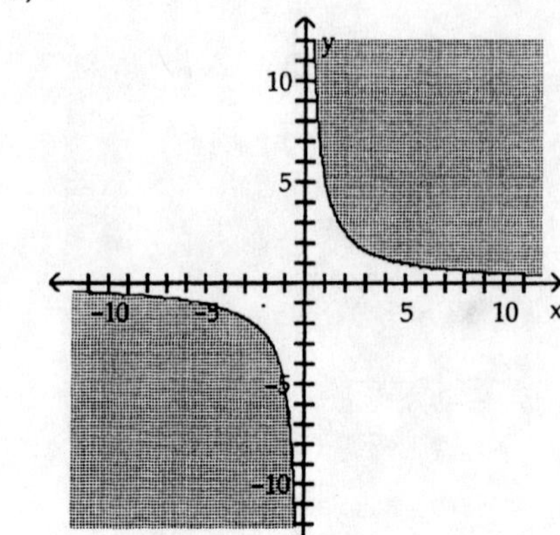

C)

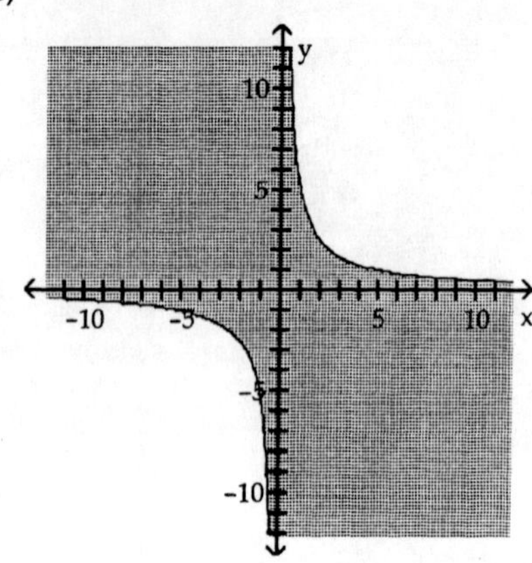

D)

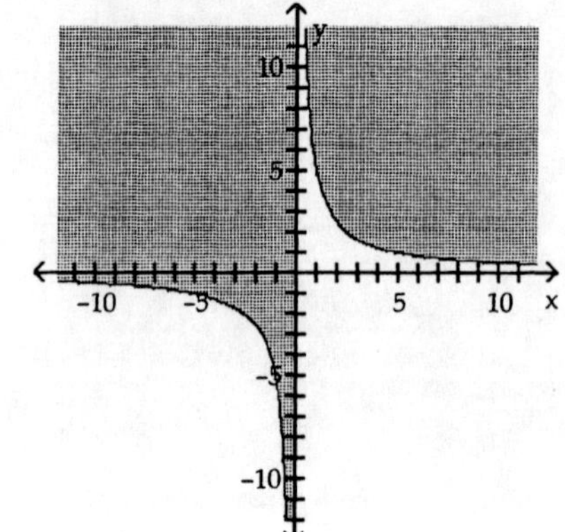

2 Graph a System of Inequalities

Graph the system of inequalities.

1)

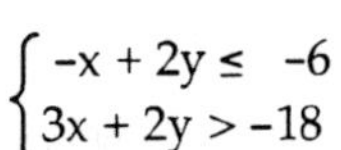

$$\begin{cases} -x + 2y \le -6 \\ 3x + 2y > -18 \end{cases}$$

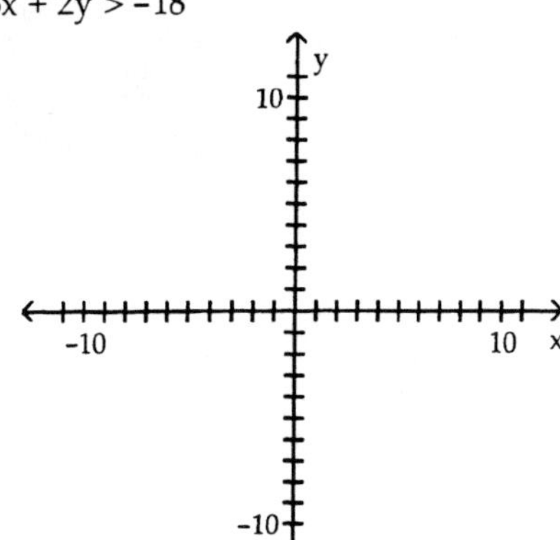

A)

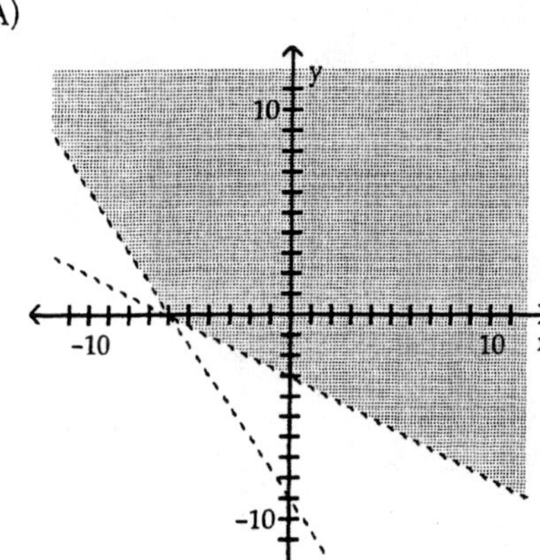

B)

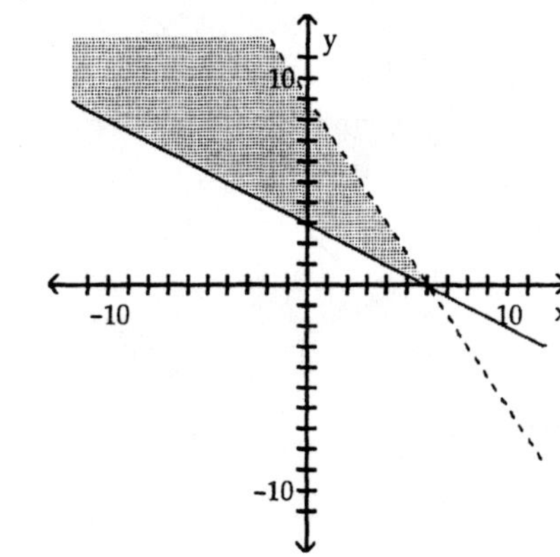

C)

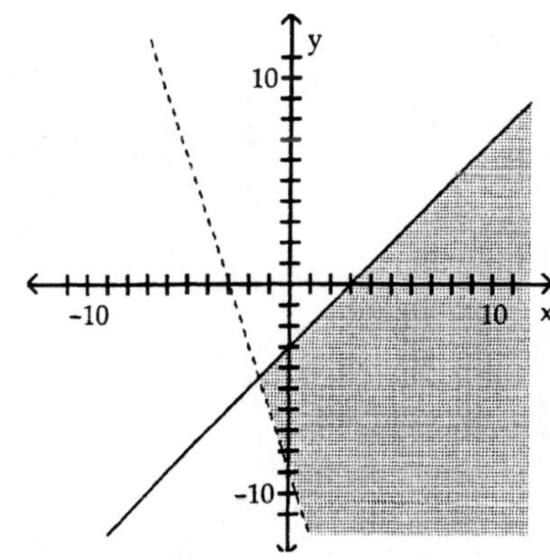

D)

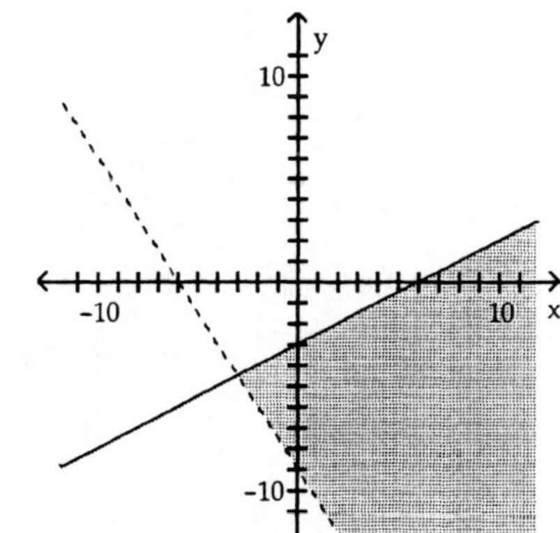

2)

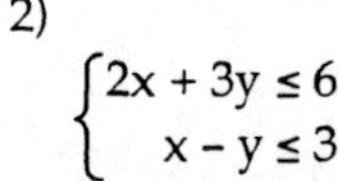

$$\begin{cases} 2x + 3y \le 6 \\ x - y \le 3 \end{cases}$$

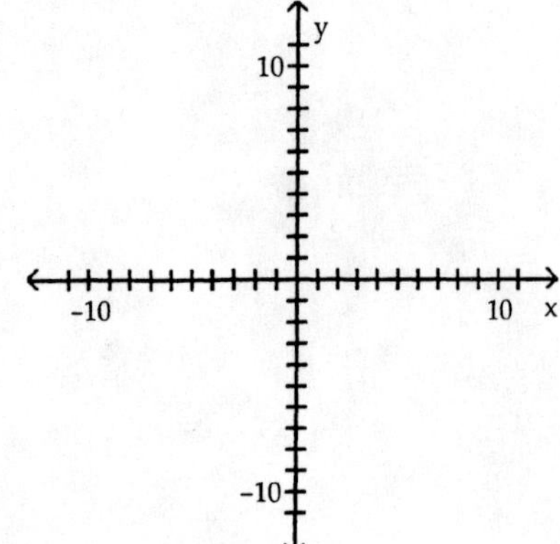

A)

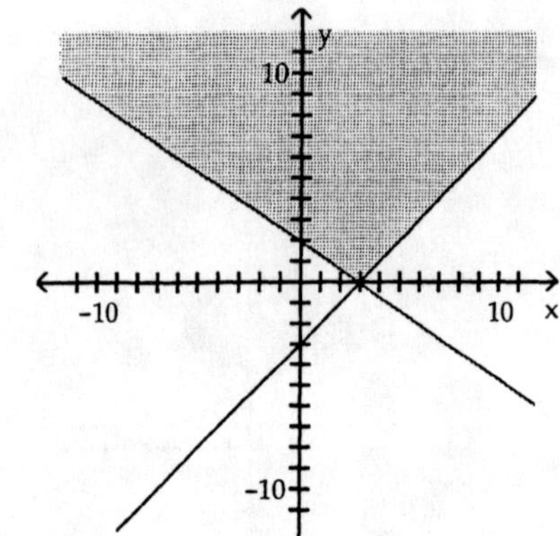

B)

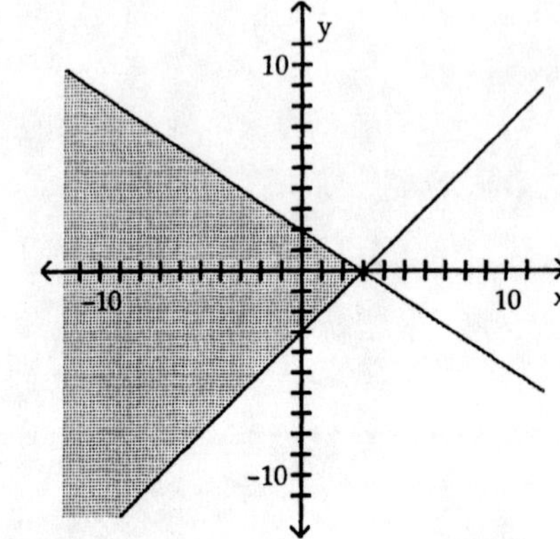

C)

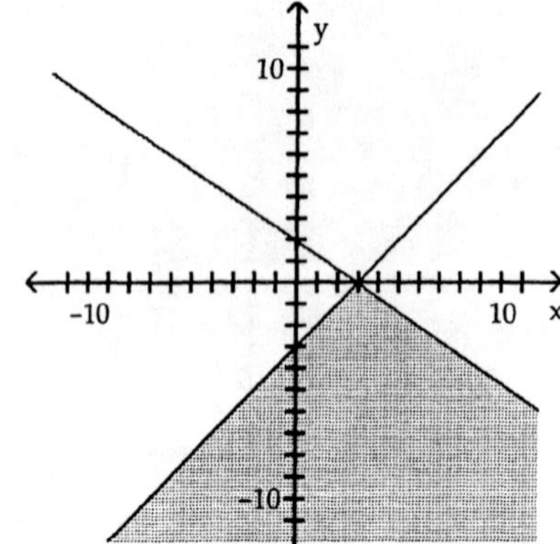

D)

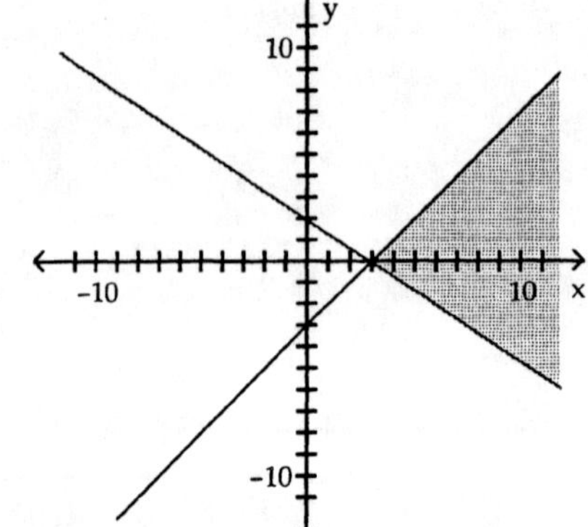

3)

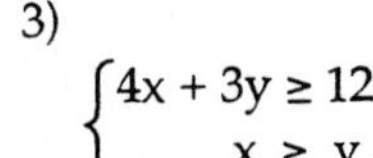

$$\begin{cases} 4x + 3y \geq 12 \\ x \geq y \end{cases}$$

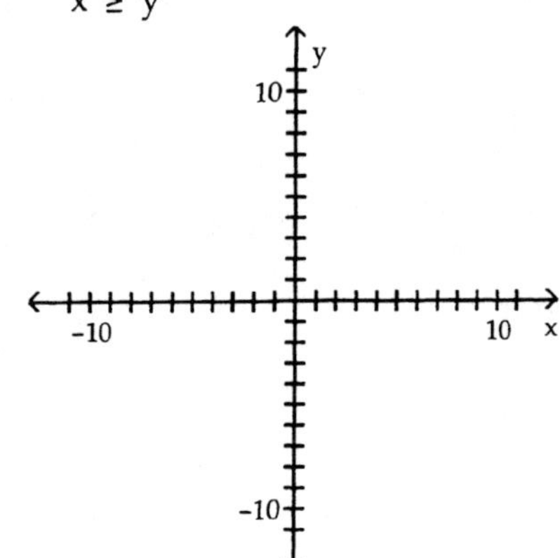

A)

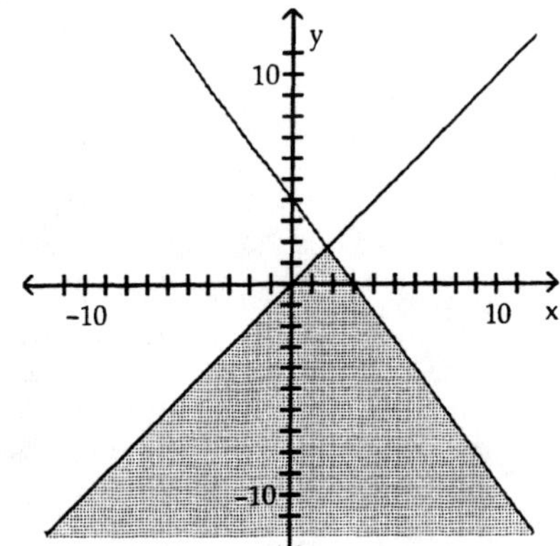

B)

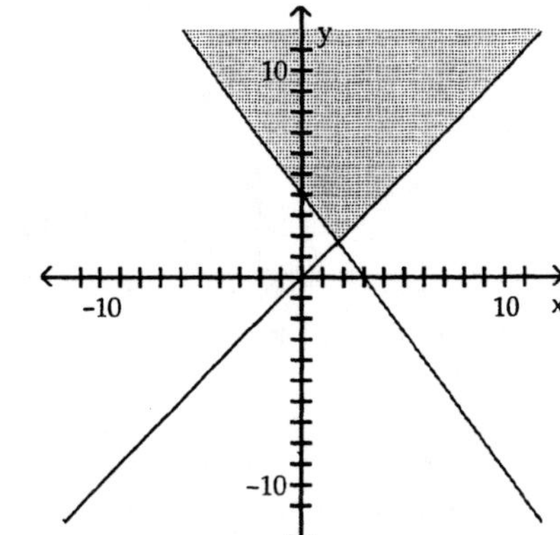

C)

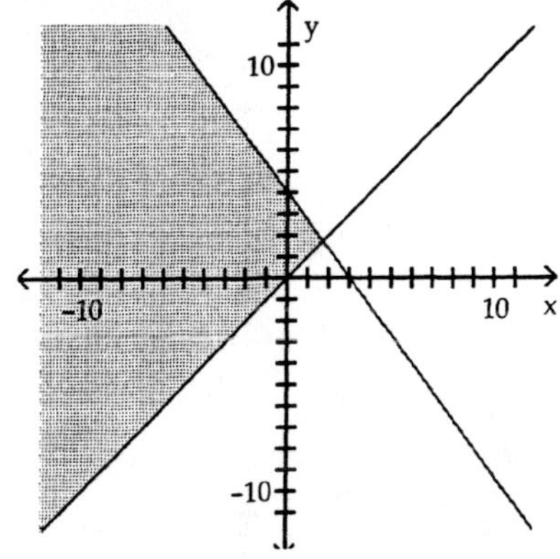

D)

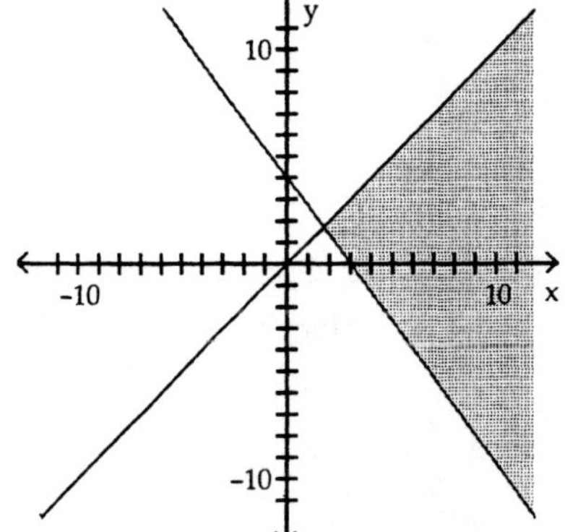

4)

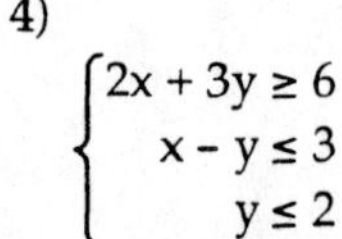

$$\begin{cases} 2x + 3y \geq 6 \\ x - y \leq 3 \\ y \leq 2 \end{cases}$$

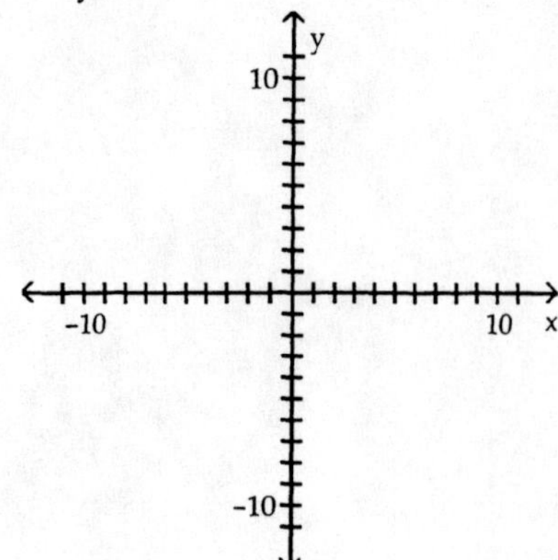

A)

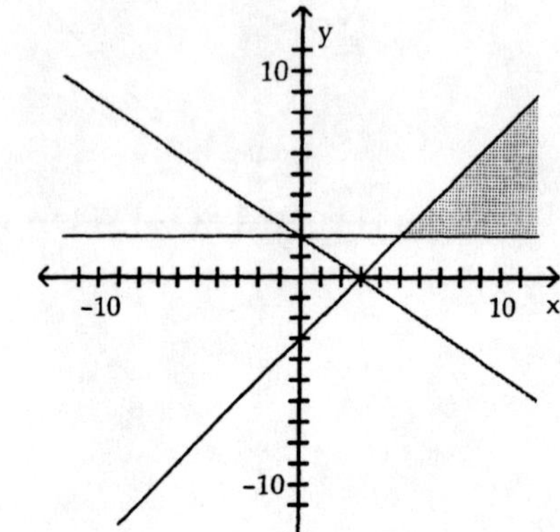

B)

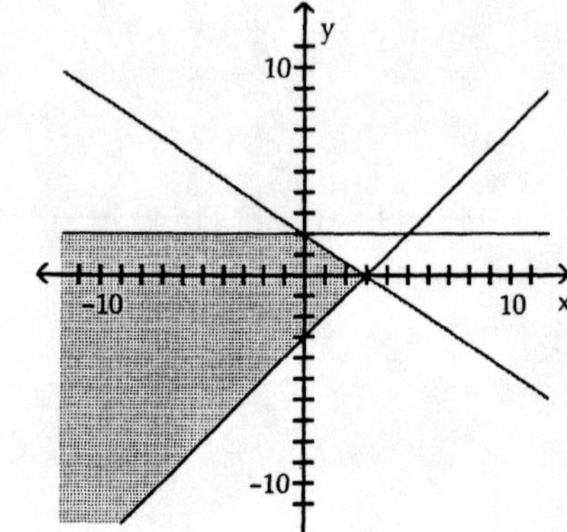

C)

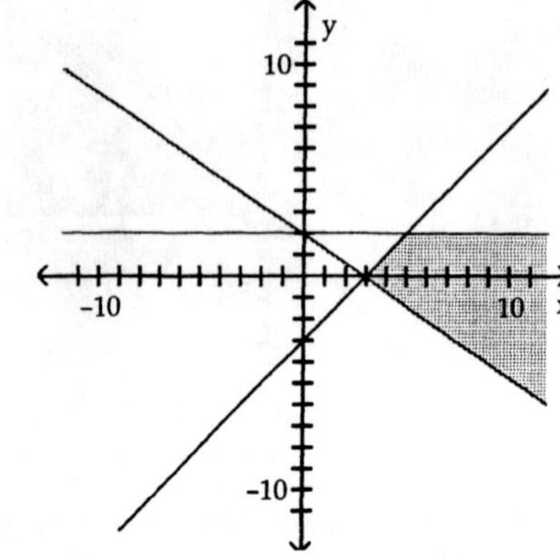

D)

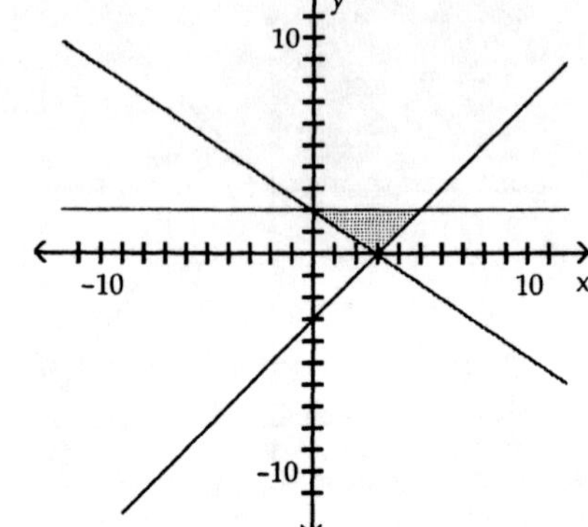

5)

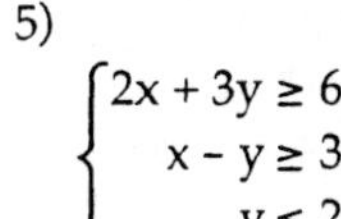

$$\begin{cases} 2x + 3y \geq 6 \\ x - y \geq 3 \\ y \leq 2 \end{cases}$$

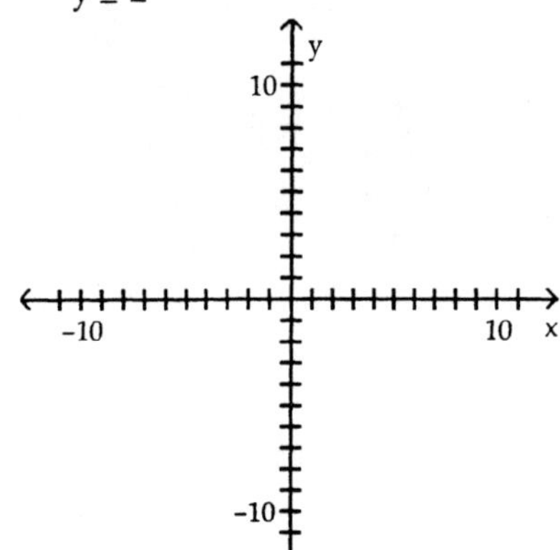

A)

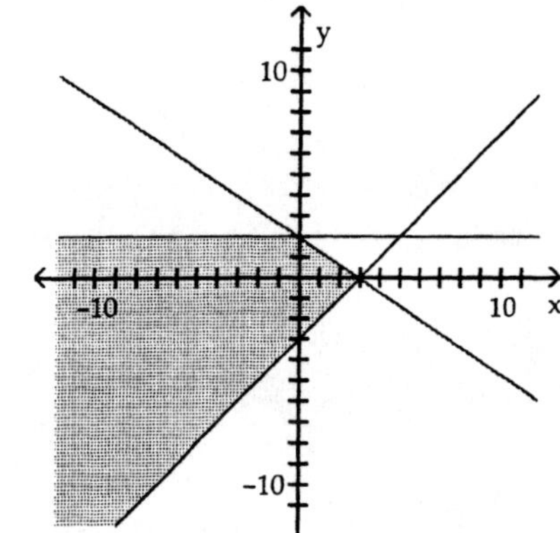

B)

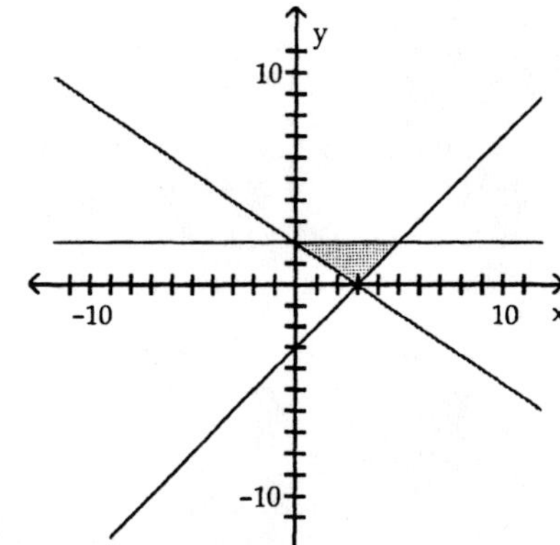

C)

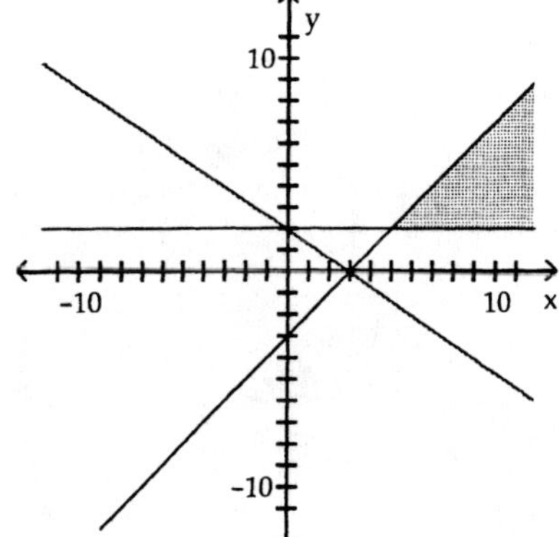

D)

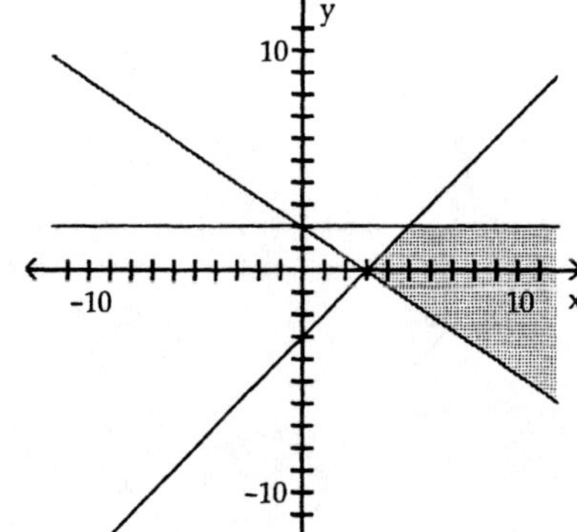

6)

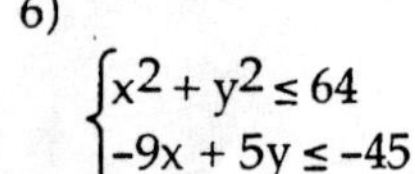

$$\begin{cases} x^2 + y^2 \le 64 \\ -9x + 5y \le -45 \end{cases}$$

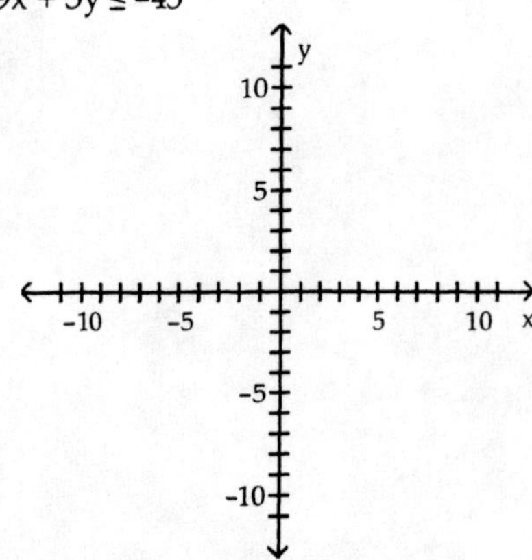

A)

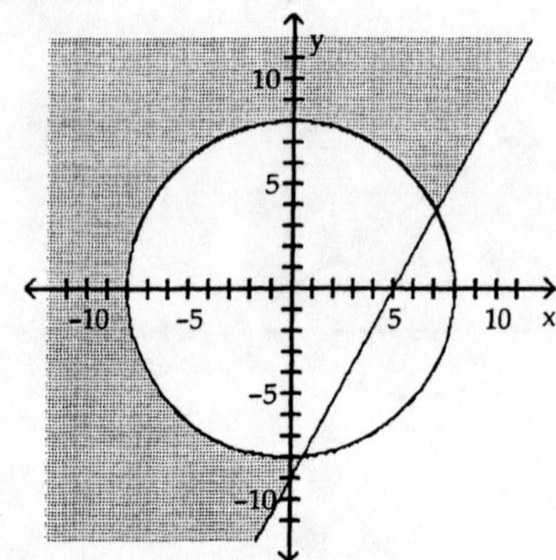

B)

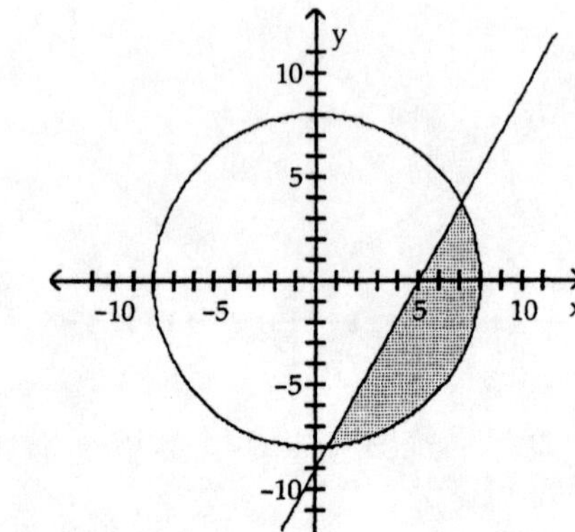

C)

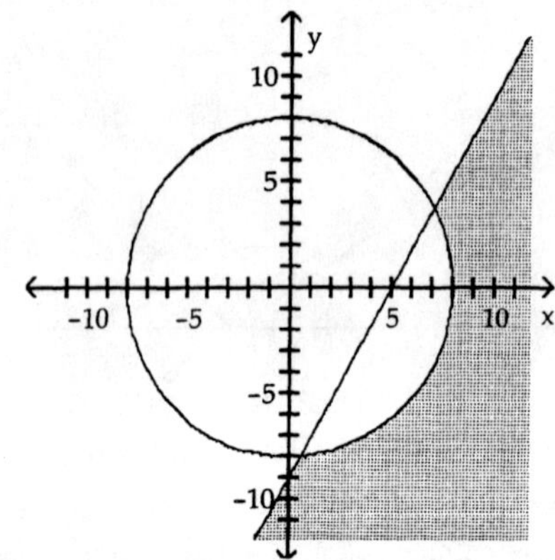

D)

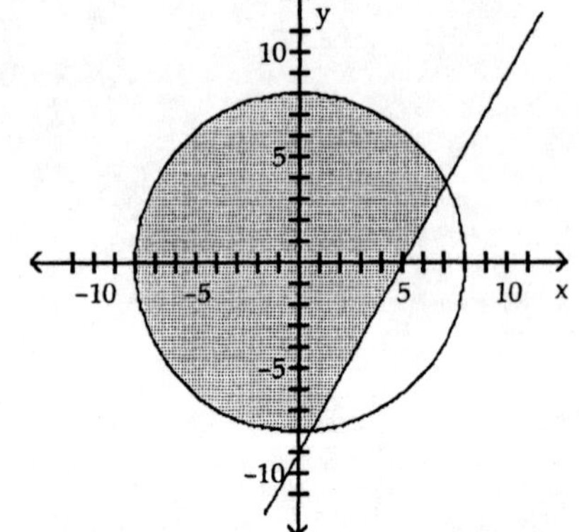

7)

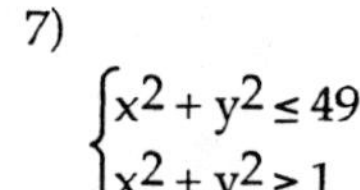

$$\begin{cases} x^2 + y^2 \le 49 \\ x^2 + y^2 \ge 1 \end{cases}$$

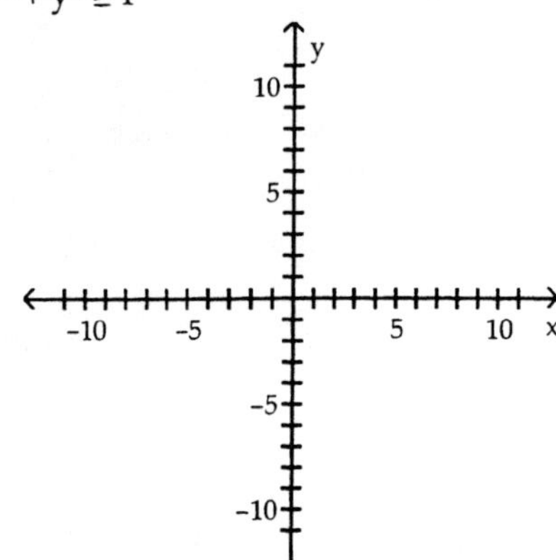

A)

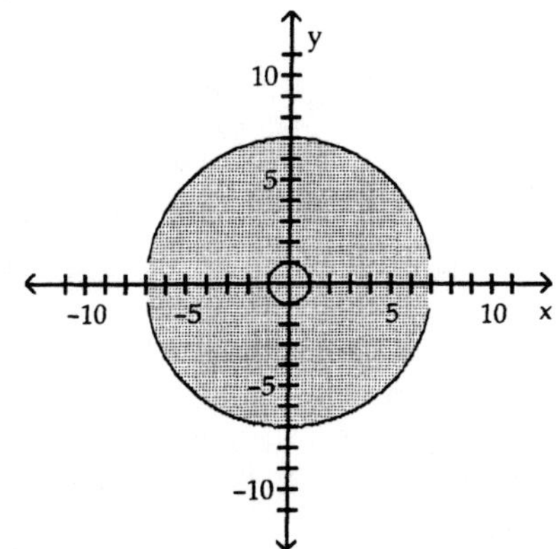

B)

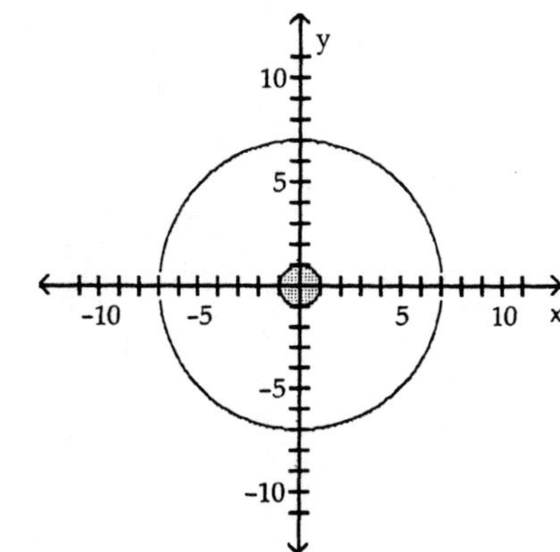

C)

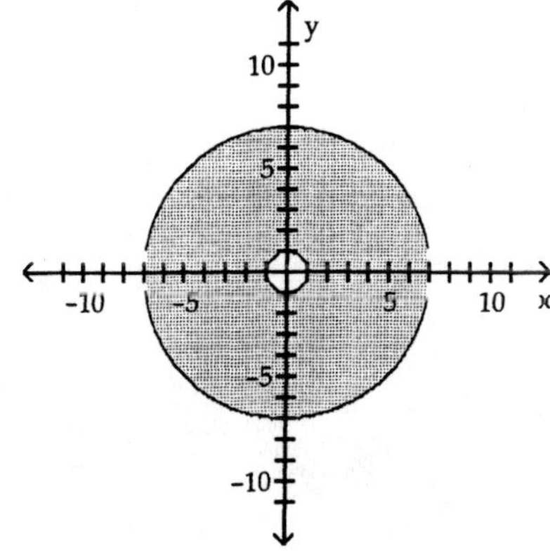

D) no solution

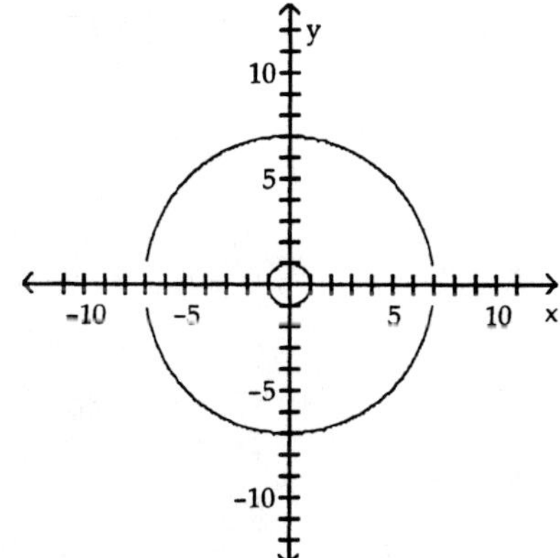

8)

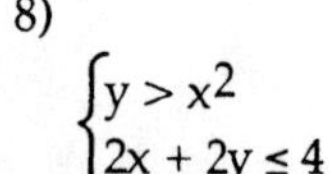

$$\begin{cases} y > x^2 \\ 2x + 2y \le 4 \end{cases}$$

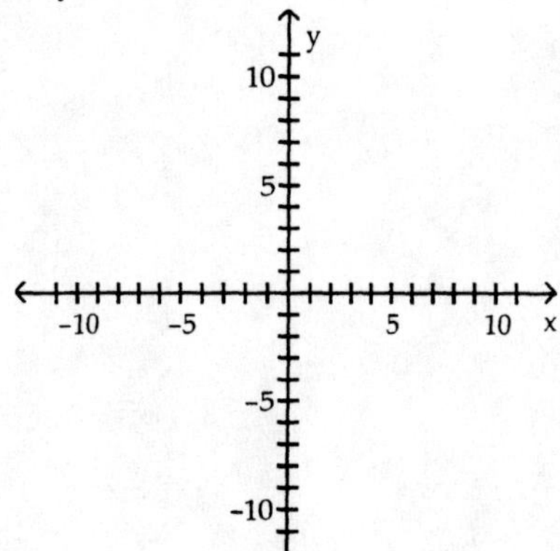

A)

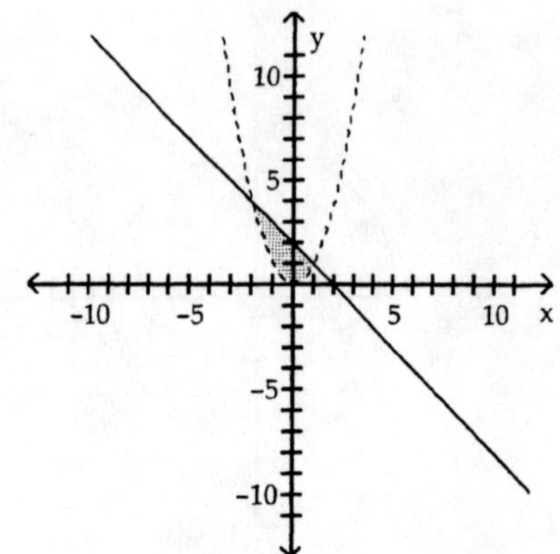

B)

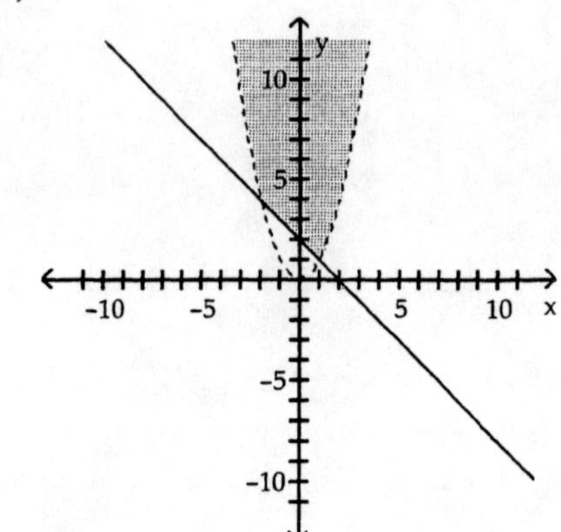

C)

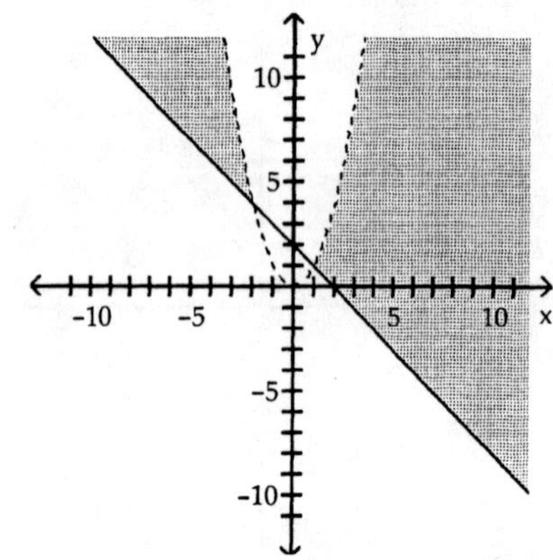

D)

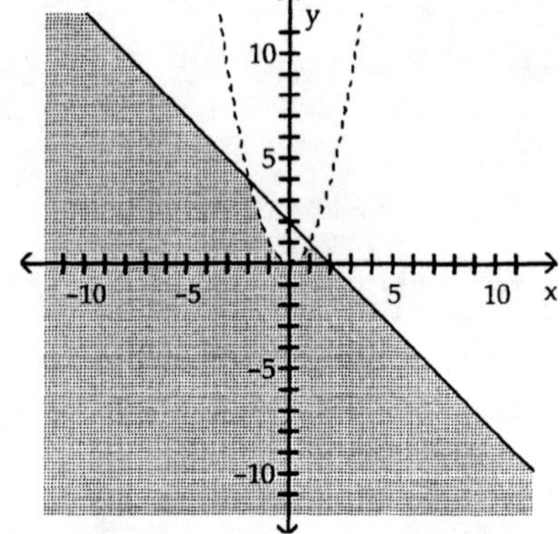

9)

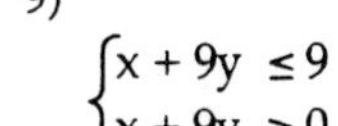

$$\begin{cases} x + 9y \leq 9 \\ x + 9y \geq 0 \end{cases}$$

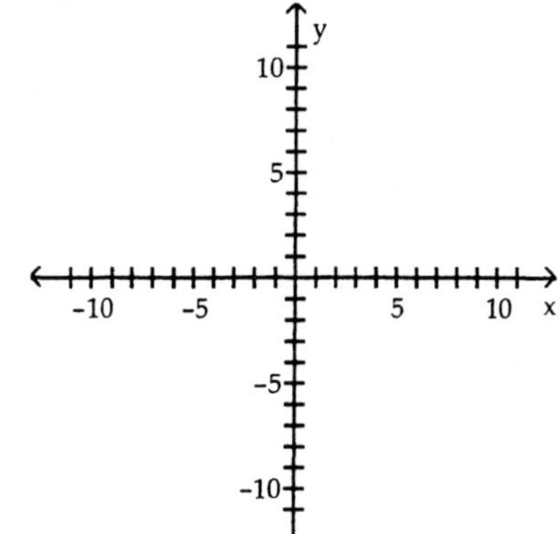

A)

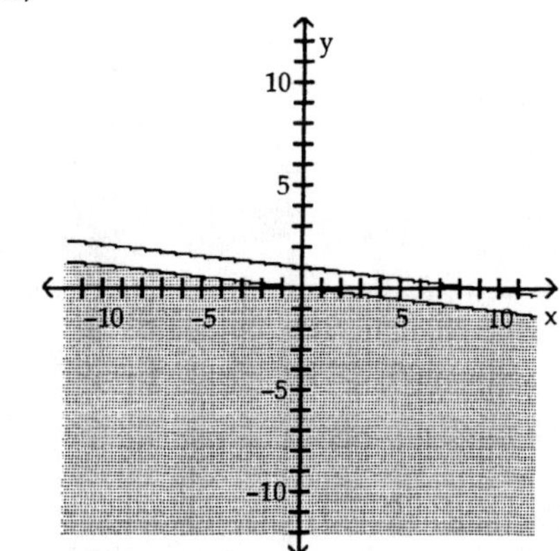

B)

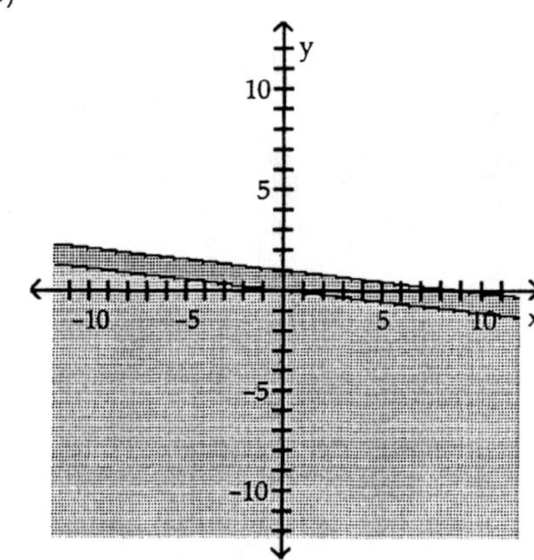

C)

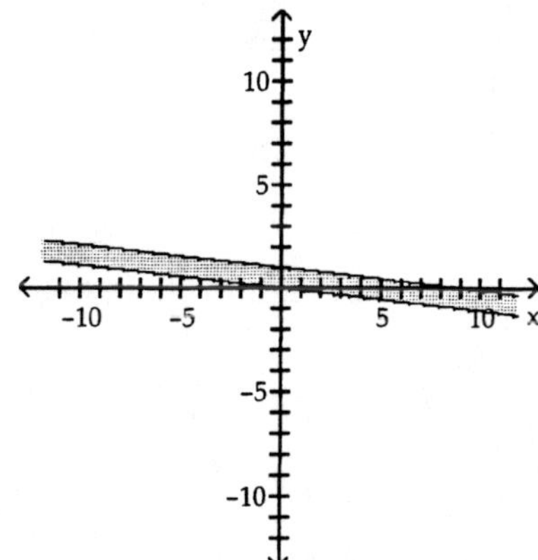

D) no solution

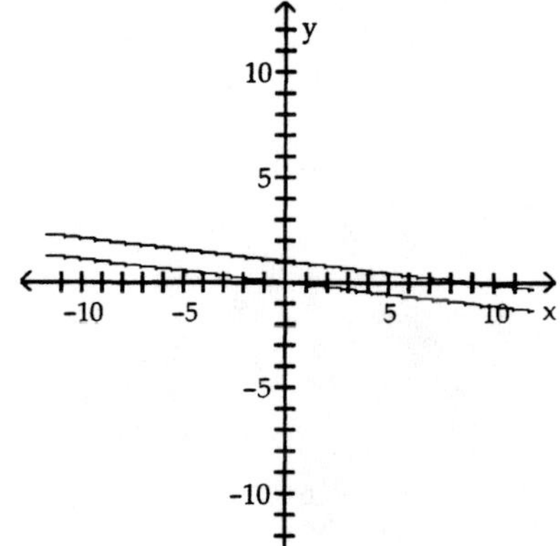

10)

$$\begin{cases} 9x + y \geq 9 \\ 9x + y \geq 0 \end{cases}$$

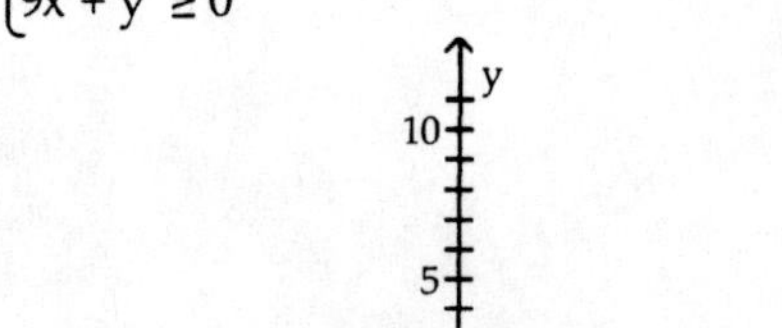

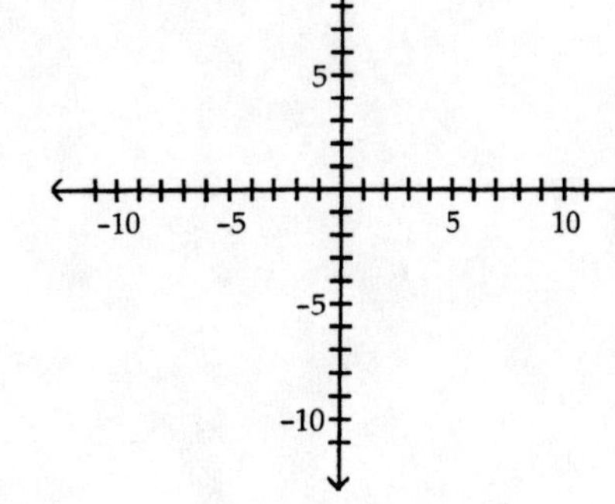

A)

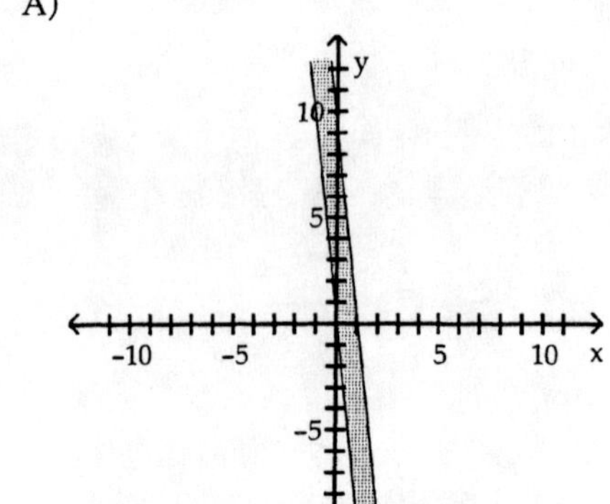

B)

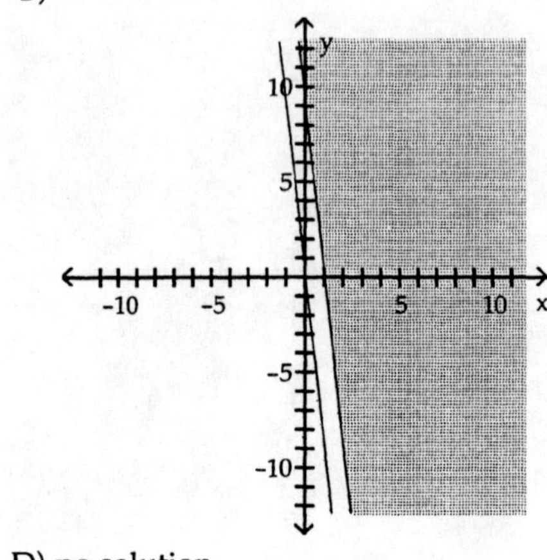

C)

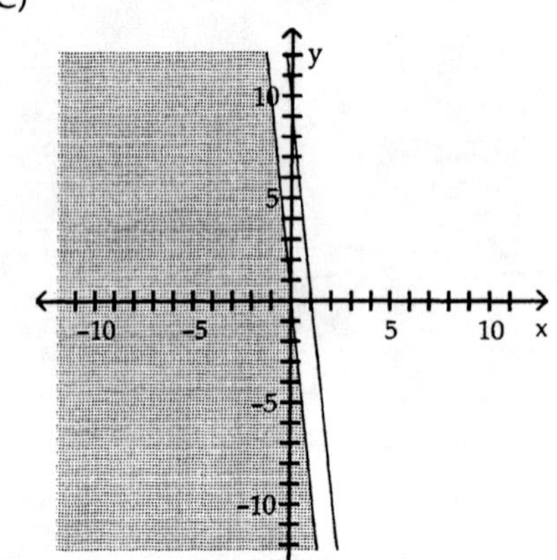

D) no solution

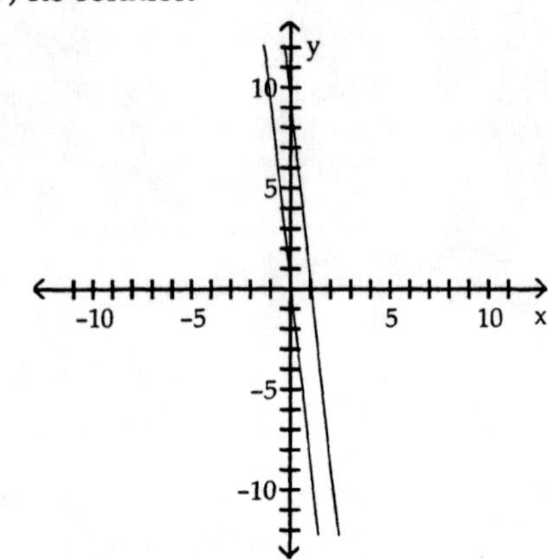

Graph the system of linear inequalities. Tell whether the graph is bounded or unbounded, and label the corner points.

11) $\begin{cases} x \geq 0 \\ y \geq 0 \\ x + y \leq 6 \\ x + y \geq 5 \end{cases}$

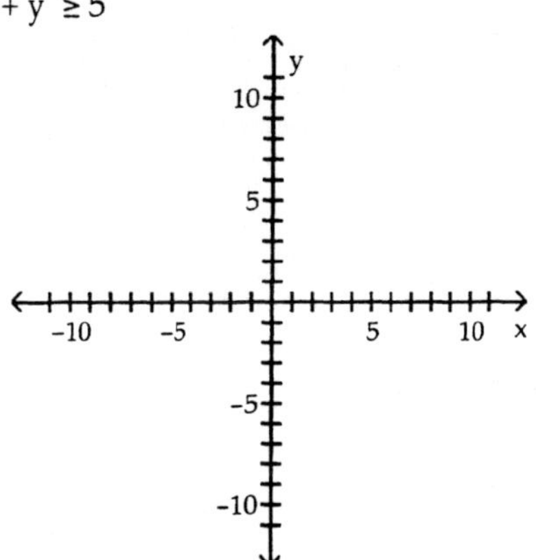

A) unbounded; corner points (6, 0), (0, 6)

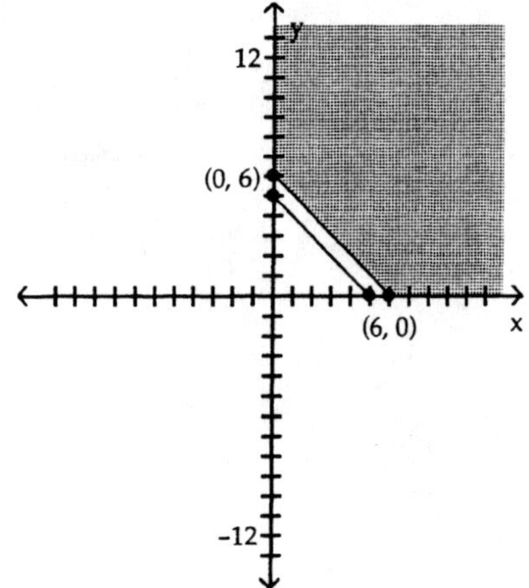

B) unbounded; corner points (0, 0), (0, 5), (5, 0)

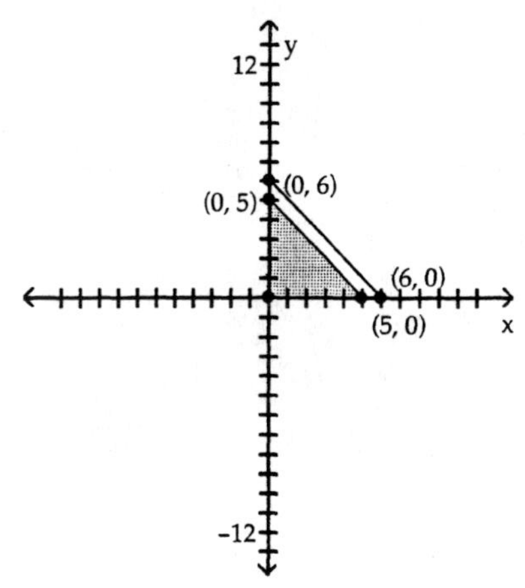

C) bounded; corner points (6, 0), (0, 6), (0, 5), (5, 0)

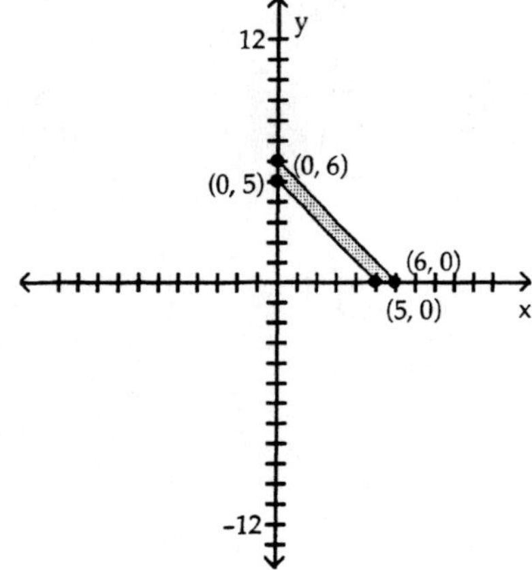

D) no solution

Graph the region indicated by graphing the system of inequalities. Label all points of intersection.

12)

$$\begin{cases} y \geq x^2 - 8 \\ y \leq -x^2 \end{cases}$$

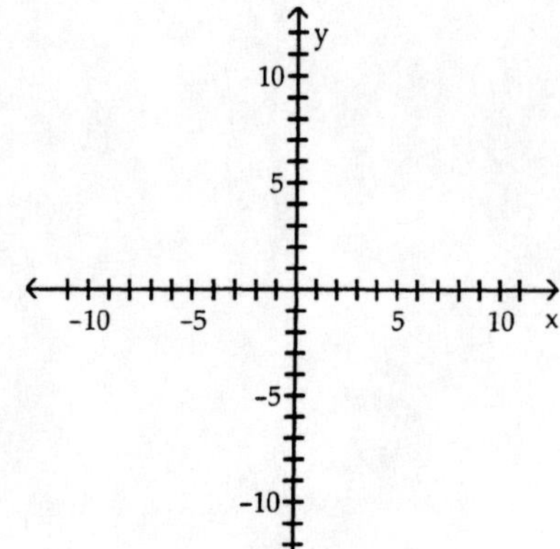

A)

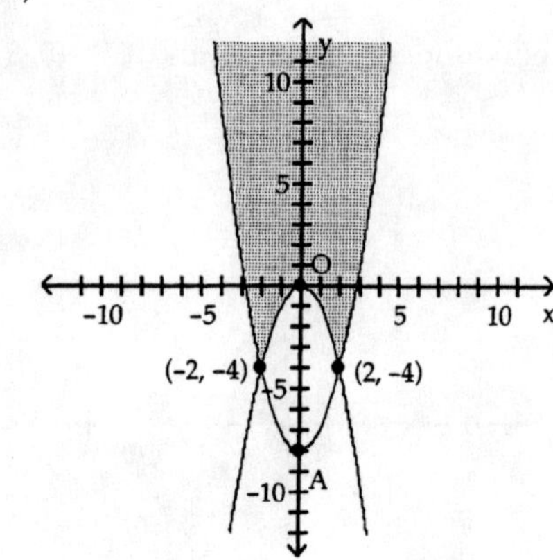

O(0, 0), A(0, -8)

B)

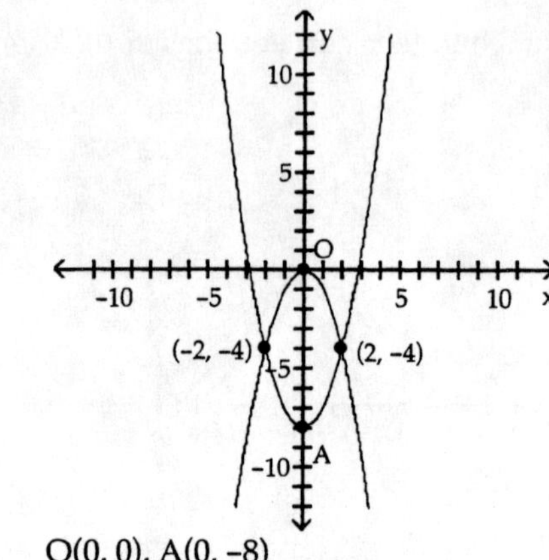

O(0, 0), A(0, -8)

C)

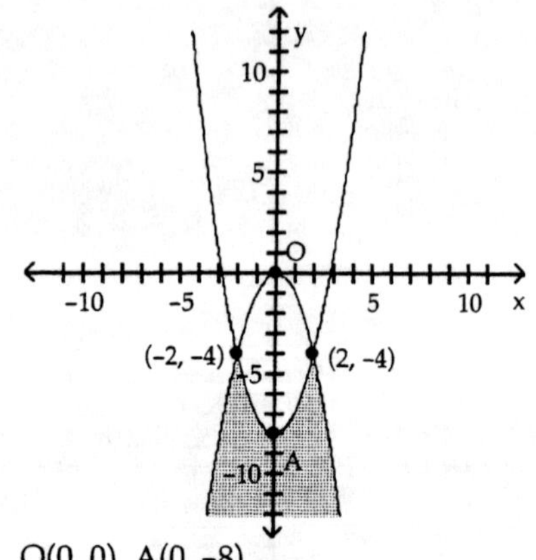

O(0, 0), A(0, -8)

D)

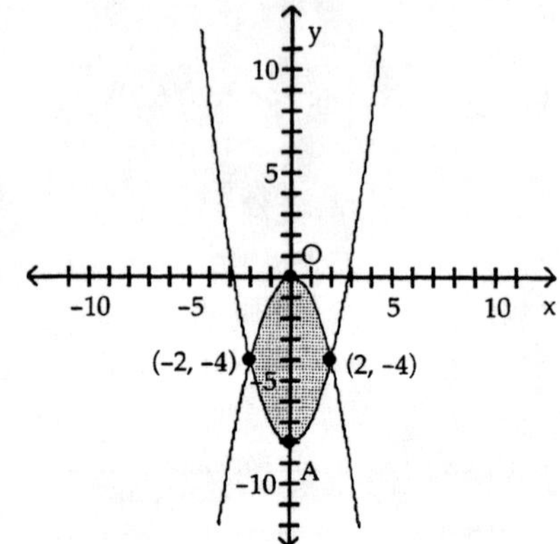

O(0, 0), A(0, -8)

Write a system of linear inequalities that has the given graph.

1)

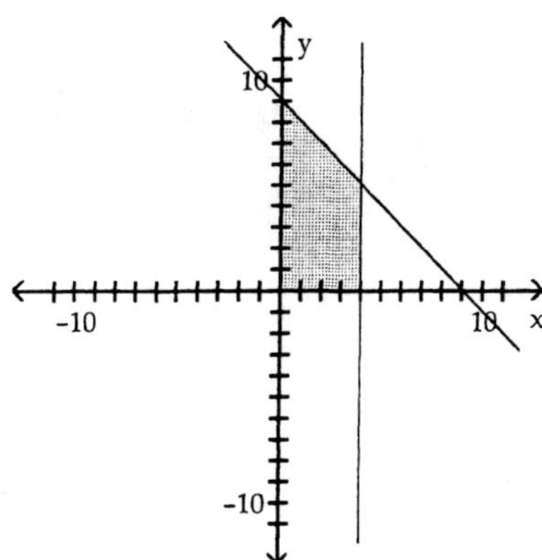

A) $\begin{cases} y \geq 0 \\ x \geq 0 \\ x \leq 4 \\ y + x \geq 9 \end{cases}$

B) $\begin{cases} y \geq 0 \\ x \geq 0 \\ x \leq 4 \\ y + x \leq 9 \end{cases}$

C) $\begin{cases} x \leq 4 \\ y + x \leq 9 \end{cases}$

D) $\begin{cases} y \geq 0 \\ x \geq 0 \\ x \leq 9 \\ y + x \leq 4 \end{cases}$

2)

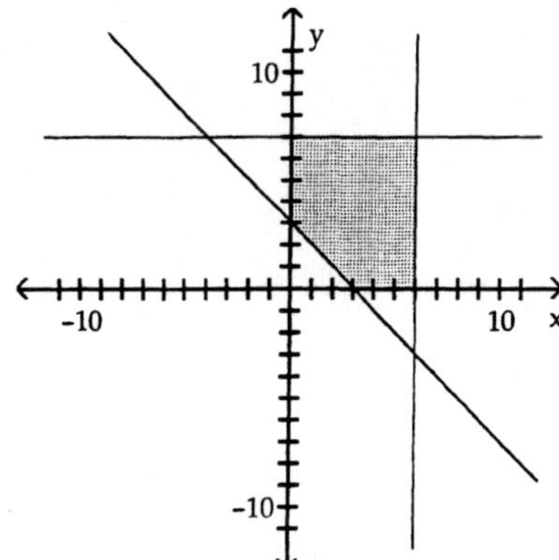

A) $\begin{cases} x \leq 6 \\ y \leq 7 \\ x + y \geq 3 \end{cases}$

B) $\begin{cases} y \geq 0 \\ x \geq 0 \\ x \leq 6 \\ y \leq 7 \\ x + y \geq 3 \end{cases}$

C) $\begin{cases} y \geq 0 \\ x \geq 0 \\ y \leq 7 \\ x + y \geq 3 \end{cases}$

D) $\begin{cases} y \geq 0 \\ x \geq 0 \\ x \leq 7 \\ x + y \geq 3 \end{cases}$

3)

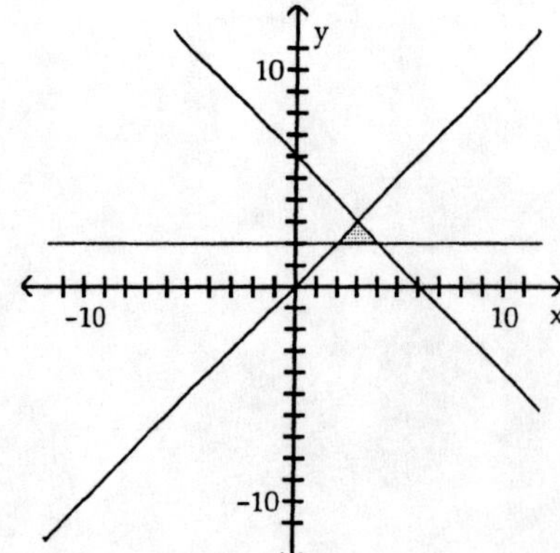

A) $\begin{cases} y \geq 0 \\ x \geq 0 \\ y \leq 2 \\ y \leq x \\ y + x \leq 6 \end{cases}$

B) $\begin{cases} y \geq 2 \\ y \geq x \\ y + x \leq 6 \end{cases}$

C) $\begin{cases} x \geq 0 \\ y \geq 2 \\ y \leq x \\ y + x \leq 6 \end{cases}$

D) $\begin{cases} y \geq 0 \\ x \geq 0 \\ y \leq 2 \\ y + x \leq 6 \end{cases}$

4)

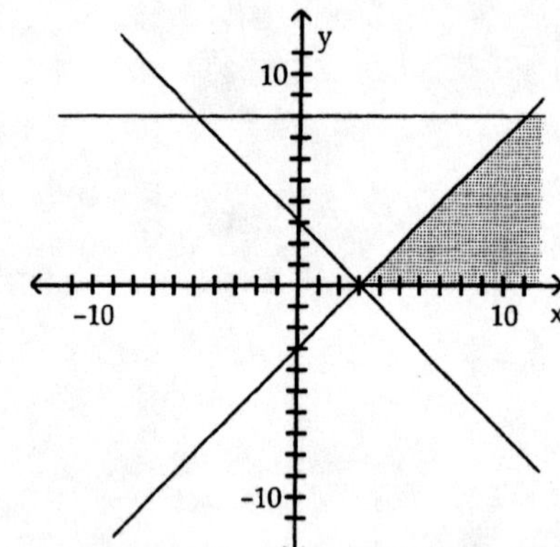

A) $\begin{cases} y \geq 0 \\ y \leq 8 \\ y \leq x - 3 \\ y \geq 3 - x \end{cases}$

B) $\begin{cases} x \geq 0 \\ y \geq 0 \\ y \leq 8 \\ y \leq x - 3 \\ y \leq 3 - x \end{cases}$

C) $\begin{cases} y \geq 0 \\ x \leq 8 \\ y \leq x - 3 \\ y \geq 3 - x \end{cases}$

D) $\begin{cases} x \geq 0 \\ y \leq 8 \\ y \leq x - 3 \\ y \geq 3 - x \end{cases}$

Solve the problem.

5) A coffee store has available 75 pounds of A grade coffee and 120 pounds of B grade coffee. These will be blended into 1 pound packages as follow: an economy blend that contains 4 ounces of A grade coffee and 12 ounces of B grade coffee and a superior blend that contains 8 ounces of A grade coffee and 8 ounces of B grade coffee. Using x to denote the number of packages of the economy blend and y to denote the number of packages of the superior blend, write a system of linear inequalities that describes the possible number of packages of each blend. Graph the system and label the corner points.

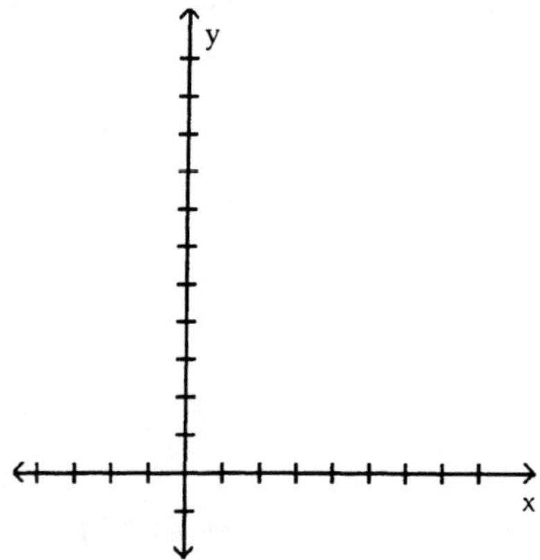

7.8 Linear Programming

1 Set up a Linear Programming Problem

Set up the linear programming problem.

1) The Jillson's have up to $75,000 to invest. They decide that they want to have at least $40,000 invested in stable bonds yielding 6% and that no more than $20,000 should be invested in more volatile bonds yielding 12%.
(a) Using x to denote the amount of money invested in the stable bonds and y the amount invested in the more volatile bonds, write a system of linear inequalities that describes the possible amounts of each investment.
(b) Graph the system and label the corner points.

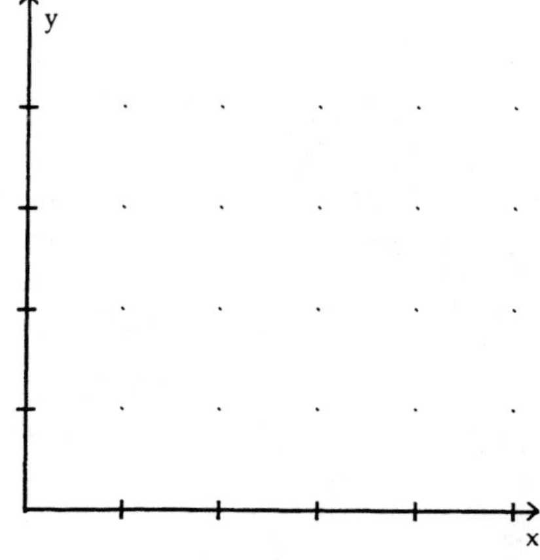

2) Charlene baby–sits for \$4 per hour. She also works as a tutor for \$7 per hour. Because of school, her parents only allow her to work 13 hours per week. How many hours can Charlene tutor and baby–sit and still make at least \$50 per week? (a) Write a system of inequalities for this situation.
(b) Graph the solution set.

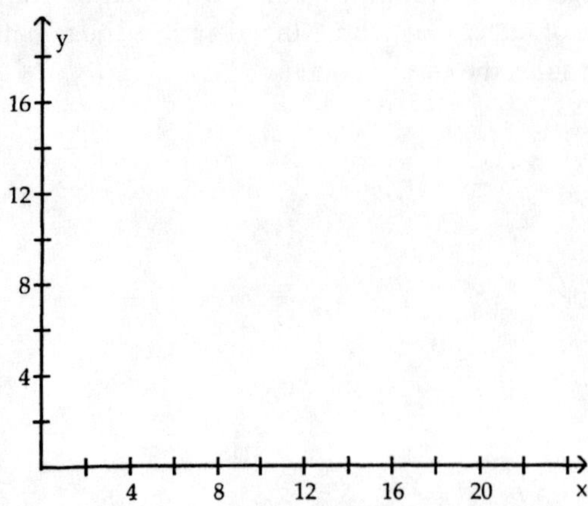

3) Eric's Carpentry manufactures two types of bookshelves that are 4 feet tall and 3 feet wide, a basic model and a deluxe model. Each basic bookshelf requires 1.5 hours for assembly and 1 hour for finishing; each deluxe model requires 2.5 hours for assembly and 1 hour for finishing. Two assemblers and one finisher are employed by the company, and each works 40 hours per week.
(a) Using x to denote the number of basic bookcases and y to denote the number of deluxe bookcases, write a system of linear inequalities that describes the possible number of each model of bookcase that can be manufactured in a week.
(b) Graph the system and label the corner points.

4) The liquid portion of a diet is to provide at least 300 calories, 36 units of vitamin A, and 90 units of vitamin C daily. A cup of dietary drink X provides 60 calories, 12 units of vitamin A, and 10 units of vitamin C. A cup of dietary drink Y provides 60 calories, 6 units of vitamin A, and 30 units of vitamin C. Set up a system of linear inequalities that describes the minimum daily requirements for calories and vitamins. Let x = number of cups of dietary drink X, and y = number of cups of dietary drink Y. Write all the constraints as a system of linear inequalities.

A)
$$\begin{cases} 60x + 60y \geq 300 \\ 12x + 6y \geq 36 \\ 10x + 30y \geq 90 \\ x \geq 0 \\ y \geq 0 \end{cases}$$

B)
$$\begin{cases} 60x + 60y \geq 300 \\ 12x + 6y > 36 \\ 10x + 30y \geq 90 \end{cases}$$

C)
$$\begin{cases} 60x + 60y > 300 \\ 12x + 6y > 36 \\ 10x + 30y > 90 \\ x > 0 \\ y > 0 \end{cases}$$

D)
$$\begin{cases} 60x + 60y \leq 300 \\ 12x + 6y \leq 36 \\ 10x + 30y \leq 90 \end{cases}$$

2 Solve a Linear Programming Problem

Find the maximum or minimum value of the given objective function of a linear programming problem. The figure illustrates the graph of the feasible points.

1) $z = 3x + 8y$. Find maximum and minimum.

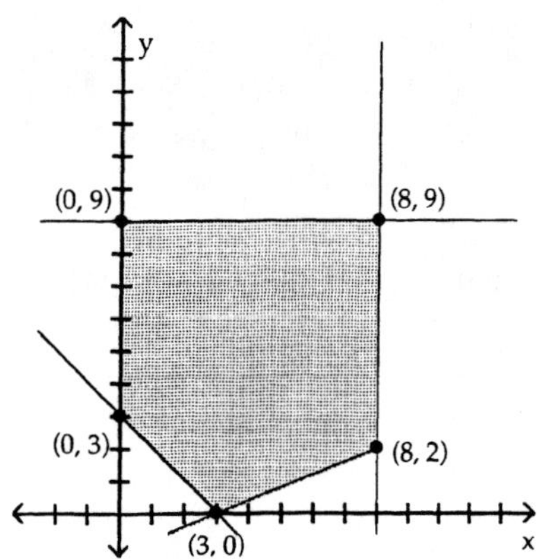

A) maximum value: 96; minimum value: 24
B) maximum value: 96; minimum value: 9
C) maximum value: 40; minimum value: 24
D) maximum value: 40; minimum value: 9

Find the maximum or minimum value of the given objective function of a linear programming problem. The figure illustrates the graph of feasible points.

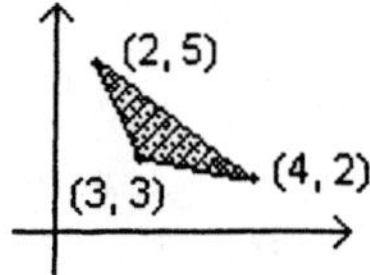

2) $z = x + 9y$. Find maximum.

A) maximum: 30
B) maximum: 47
C) maximum: 22
D) maximum: 38

3) $z = -x - 9y$. Find maximum.

A) maximum: −38
B) maximum: −22
C) maximum: −47
D) maximum: −30

4) $z = 8x + 9y$. Find minimum.

A) minimum: 51
B) minimum: 61
C) minimum: 50
D) no minimum

5) $z = x + 9y + 5$. Find minimum.

A) minimum: 52
B) minimum: 27
C) minimum: 36
D) no minimum

6) $z = -6x - y$. Find maximum.

A) maximum: −26
B) maximum: −17
C) maximum: −21
D) no maximum

Solve the linear programming problem.

7) Maximize and minimize $z = 18x + 10y$ subject to:
$x \geq 0, \quad y \geq 0, \quad 2x + 3y \geq 6, \quad x \leq 10, \quad y \leq 5.$

A) maximum: 230; minimum: 20
B) maximum: 50; minimum: 30
C) maximum: 180; minimum: 30
D) maximum: 230; minimum: 180

8) Minimize $z = 6x + 5y + 17$ subject to:
$x \geq 0, \quad y \geq 0, \quad x + y \geq 1.$

A) minimum: 17 B) minimum: 23 C) minimum: 22 D) minimum: 28

9) Maximize and minimize $z = 5x - 15y$ subject to:
$x \geq 0, \quad y \geq 0, \quad 4x + 5y \leq 30, \quad 4x + 3y \leq 20, \quad x \leq 5, \quad y \leq 8.$

A) maximum: 25; minimum: 0 B) maximum: 25; minimum: -90

C) maximum: -68.75; minimum: -90 D) maximum: -90; minimum: 0

Solve the problem.

10) Mrs. White wants to crochet beach hats and baby afghans for a church fund-raising bazaar. She needs 7 hours to make a hat and 3 hours to make an afghan and she has 59 hours available. She wants to make no more than 13 items and no more than 10 afghans. The bazaar will sell the hats for $14 each and the afghans for $7 each. How many of each should she make to maximize the income for the bazaar? What is the maximum income?

A) 8 hats, 5 afghans, $147 B) 7 hats, 6 afghans, $140

C) 5 hats, 8 afghans, $126 D) 10 hats, 3 afghans, $161

11) An office manager needs to buy new filing cabinets. Cabinet A costs $4, takes up 3 square feet of floor space, and holds 11 cubic feet of files. Cabinet B costs $5, takes up 1 square foot, and holds 9 cubic feet. He has only $35 to spend and the office has room for no more than 18 square feet of cabinets. How many of each can he buy to maximize storage capacity? What is the maximum storage capacity?

A) 6 cabinets A, 0 cabinet B, 66 cubic feet B) 5 cabinets A, 3 cabinets B, 82 cubic feet

C) 0 cabinet A, 7 cabinets B, 63 cubic feet D) 0 cabinet A, 6 cabinets B, 54 cubic feet

12) A candy company has 125 pounds of cashews and 160 pounds of peanuts which they combine into two different mixes. The deluxe mix has half cashews and half peanuts and sells for $6 per pound. The economy mix has one third cashews and two thirds peanuts and sells for $4.40 per pound. How many pounds of each mix should be prepared for maximum revenue?

A) 90 deluxe, 35 economy B) 125 deluxe, 0 economy

C) 270 deluxe, 70 economy D) 180 deluxe, 105 economy

13) A doctor has told a sick patient to take vitamin pills. The patient needs at least 6 units of vitamin A and at least 6 units of vitamin B. The red vitamin pills cost 10¢ each and contain 1 unit of A and 2 units of B. The blue vitamin pills cost 25¢ each and contain 2 units of A and 1 unit of B. To avoid indigestion, the patient should take no more than 4 red pills and no more than 7 blue pills. How many pills should the patient take each day to minimize costs?

A) 4 red and 7 blue B) 4 red and 1 blue C) 2 red and 2 blue D) 0 red and 6 blue

14) A doctor has told a patient to take vitamin pills. The patient needs at least 12 units of vitamin A and at least 8 units of vitamin D. The red vitamin pills cost 20¢ each and contain 3 units of A and 1 unit of D. The blue vitamin pills cost 35¢ each and contain 2 units of A and 2 units of D. To avoid indigestion, the patient should take no more than 6 red pills and no more than 9 blue pills. How many pills should the patient take each day to minimize costs?

A) 6 red and 1 blue B) 2 red and 3 blue C) 6 red and 9 blue D) 0 red and 9 blue

15) Your computer supply store sells two types of laser printers. The first type, A, has a cost of $86 and you make a $45 profit on each one. The second type, B, has a cost of $130 and you make a $35 profit on each one. You expect to sell at least 100 laser printers this month and you need to make at least $3850 profit on them. How many of what type of printer should you order if you want to minimize your cost?

16) The Jillson's have up to $75,000 to invest. They decide that they want to have at least $25,000 invested in stable bonds yielding 6% and that no more than $45,000 should be invested in more volatile bonds yielding 12%. How much should they invest in each type of bond to maximize income if the amount in the more volatile bond should not exceed the amount in the more stable bond? What is the maximum income?

17) The Fiedler family has up to $130,000 to invest. They decide that they want to have at least $40,000 invested in stable bonds yielding 5.5% and that no more than $60,000 should be invested in more volatile bonds yielding 11%. How much should they invest in each type of bond to maximize income if the amount in the stable bond should not exceed the amount in the more volatile bond? What is the maximum income?

18) Eric's Carpentry manufactures two types of bookshelves that are 4 feet tall and 3 feet wide, a basic model and a deluxe model. Each basic bookshelf requires 1.5 hours for assembly and 1 hour for finishing; each deluxe model requires 2.5 hours for assembly and 1 hour for finishing. Two assemblers and one finisher are employed by the company, and each works 40 hours per week. Eric can sell more basic models than deluxe models, so he wants the number of basic models produced to be 50% more than the number of deluxe models produced. If he makes $50 profit on the basic models and $65 profit on the deluxe models, how many should he make to maximize the profit? What is the maximum profit?

19) Joely's Tea Shop, a store that specializes in tea blends, has available 45 pounds of A grade tea and 70 pounds of B grade tea. These will be blended into 1 pound packages as follows: A breakfast blend that contains one third of a pound of A grade tea and two thirds of a pound of B grade tea and an afternoon tea that contains one half pound of A grade tea and one half pound of B grade tea. If Joely makes a profit of $1.50 on each pound of the breakfast blend and $2.00 profit on each pound of the afternoon blend, how many pounds of each blend should she make to maximize profits? What is the maximum profit?

20) An artist is creating a mosaic that cannot be larger than the space allotted which is 4 feet tall and 6 feet wide. The mosaic must be at least 3 feet tall and 5 feet wide. The tiles in the mosaic have words written on them and the artist wants the words to all be horizontal in the final mosaic. The word tiles come in two sizes: The smaller tiles are 4 inches tall and 4 inches wide, while the large tiles are 6 inches tall and 12 inches wide. If the small tiles cost $3.50 each and the larger tiles cost $4.50 each, how many of each should be used to minimize the cost? What is the minimum cost?

Ch. 7 Systems of Equations and Inequalities
Answer Key

7.1 Systems of Linear Equations: Substitution and Elimination

1 Solve Systems of Equations by Substitution

1) A
2) B
3) A
4) B
5) B
6) A
7) D
8) C
9) B
10) C
11) B
12) D
13) A
14) B
15) A
16) D

2 Solve Systems of Equations by Elimination

1) D
2) D
3) D
4) C
5) C
6) B
7) A
8) C
9) D
10) C

3 Identify Inconsistent Systems of Equations Containing Two Variables

1) D
2) D
3) D
4) D

4 Express the Solution of a System of Dependent Equations Containing Two Variables

1) B
2) A

5 Solve Systems of Three Equations Containing Three Variables

1) B
2) A
3) A
4) A
5) C
6) B
7) C
8) B
9) C
10) B
11) C

6 Identify Inconsistent Systems of Equations Containing Three Variables

1) D
2) D

7 Express the Solutions of a System of Dependent Equations Containing Three Variables

1) A
2) C
3) C
4) C

8 Additional Applications

1) D
2) D
3) D
4) C
5) B
6) a = 2, b = -3, c = 2
7) a = 1, b = 6, c = -5
8) D
9) D
10) B
11) D
12) C
13) B
14) D
15) 200 yd by 50 yd
16) shorts: $14.95; T-shirts: $8.50
17) 12 lbs of the $6.50/lb tea; 8 lbs of the $4.00/lb tea
18) $4000 at 6%; $12,000 at 7%
19) boat: 25 mph; current: 3 mph
20) B
21) A
22) D
23) D
24) 60 tapers, 35 votives, 25 tealights
25) $6500 at 12%, $5500 at 7%, $13,000 at 6%
26) $40,000 at 7%, $20,000 at 10%, $30,000 at 14%
27) socks: $2.50 per pair; T-shirts: $7.95 each; shorts: $12.50 per pair
28) fruit: 25 lbs; nuts: 15 lbs; carob chips: 10 lbs

7.2 Systems of Linear Equations: Matrices

1 Write the Augmented Matrix of a System of Linear Equations

1) A
2) C
3) B
4) D
5) B
6) A
7) B

2 Write the System from the Augmented Matrix

1) B
2) B
3) C
4) A
5) B
6) D
7) D
8) C

3 **Perform Row Operations on a Matrix**

1)

$$\left[\begin{array}{ccc|c} -7 & -5 & -1 & -10 \\ 6 & -2 & 9 & 5 \\ 0 & -26 & 2 & -22 \end{array}\right]$$

2) C
3) C
4) B
5) B

4 **Solve Systems of Linear Equations Using Matrices**

1) C
2) B
3) A
4) D
5) B
6) B
7) B
8) A
9) C
10) $x = -1, y = -2, z = 4$
11) C

5 **Additional Applications**

1) $a = -2, b = -1, c = 2$
2) A
3) \$9000 at 5%, \$12,000 at 8%, and \$24,000 at 12%
4) 14 Kitui chairs, 10 Goa chairs, 8 Santa Fe chairs
5) B
6) 1.5 ft by 4 ft
7) A

7.3 Systems of Linear Equations: Determinants

1 **Evaluate 2 by 2 Determinants**

1) D
2) D
3) B
4) B
5) 4
6) -2

2 **Use Cramer's Rule to Solve a System of Two Equations, Two Variables**

1) A
2) A
3) C
4) B
5) D

3 **Evaluate 3 by 3 Determinants**

1) B
2) B
3) A
4) D
5) 5
6) 0

4 **Use Cramer's Rule to Solve a System of Three Equations, Three Variables**

1) D
2) C
3) B
4) D

5 Know Properties of Determinants

1) C
2) A
3) A
4) D
5) C
6) D
7) A
8) B

7.4 Matrix Algebra

1 Find the Sum and Difference of Two Matrices

1) D
2) D
3) C
4) A
5) C
6) B

2 Find Scalar Multiples of a Matrix

1) C
2) D
3) C
4) B
5) B
6) D
7) D
8)
$$\begin{bmatrix} -71 & 13 \\ 48 & -37 \end{bmatrix}$$

3 Find the Product of Two Matrices

1) C
2) B
3) D
4) C
5) C
6)
$$\begin{bmatrix} -2 & 2 & 0 \\ 1 & -1 & 0 \\ -1 & 1 & 0 \end{bmatrix}$$
7) C
8) D
9) D
10) D
11) (a) $\begin{bmatrix} 18 & 12 & 6 \\ 10 & 3 & 5 \end{bmatrix}$ (b) $\begin{bmatrix} 25 \\ 29 \\ 30 \end{bmatrix}$ (c) $\begin{bmatrix} 978 \\ 487 \end{bmatrix}$
12) (a) $\begin{bmatrix} 0.89 & 0.94 \end{bmatrix}$ (b) $1229.48

4 Find the Inverse of a Matrix

1) A
2) D
3) B
4) D
5) C
6) C
7) B

8) $\left[\begin{array}{cc|cc} 4 & -14 & 1 & 0 \\ -2 & 7 & 0 & 1 \end{array}\right] \rightarrow \left[\begin{array}{cc|cc} 1 & -\frac{7}{2} & \frac{1}{4} & 0 \\ -2 & 7 & 0 & 1 \end{array}\right] \rightarrow \left[\begin{array}{cc|cc} 1 & -\frac{7}{2} & \frac{1}{4} & 0 \\ 0 & 0 & \frac{1}{2} & 1 \end{array}\right]$

9) A

5 Solve Systems of Equations Using Inverse Matrices

1) B
2) D
3) D
4) A
5) D
6) C
7) B
8) B
9) $x = -4, y = 1, z = -2$
10) B
11) B

7.5 Partial Fraction Decomposition

1 Decompose $\frac{P}{Q}$, Where Q Has Only Nonrepeated Linear Factors

1) C
2) D
3) C
4) D
5) D
6) A
7) A
8) D

2 Decompose $\frac{P}{Q}$, Where Q Has Repeated Linear Factors

1) D
2) C
3) $\frac{6}{x} - \frac{2}{x^2} - \frac{6}{x+1} + \frac{3}{(x+1)^2} - \frac{9}{(x+1)^3}$
4) D
5) B

3 Decompose $\frac{P}{Q}$, Where Q Has Only Nonrepeated Irreducible Quadratic Factors

1) A
2) B
3) $\frac{1}{x+2} - \frac{1}{x-2} + \frac{5}{x^2+9}$
4) D

4 Decompose $\frac{P}{Q}$, Where Q Has Repeated Irreducible Quadratic Factors

1) B
2) D
3) D
4) C

7.6 Systems of Nonlinear Equations

1 Solve a System of Nonlinear Equations Using Substitution

1) A
2) B

3) C
4) A
5) B
6) A
7) B
8) B
9) A
10) D
11) B

2 **Solve a System of Nonlinear Equations Using Elimination**

1) A
2) C
3) C
4) B
5) $x = \sqrt{17}, y = \sqrt{19}$; $x = -\sqrt{17}, y = \sqrt{19}$; $x = \sqrt{17}, y = -\sqrt{19}$; $x = -\sqrt{17}, -y = \sqrt{19}$
6) A
7) C

3 **Additional Applications**

1)

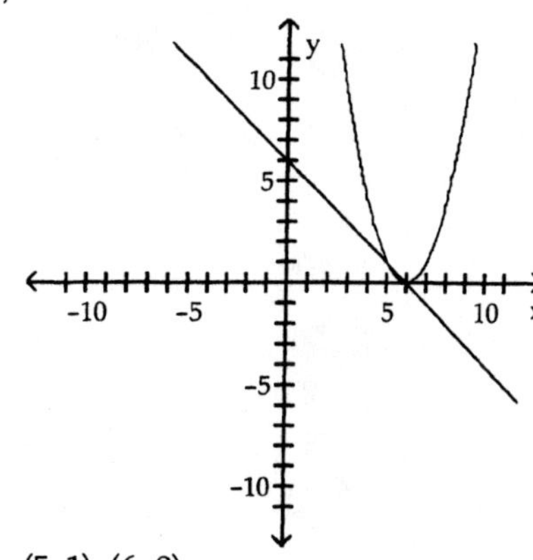

(5, 1), (6, 0)

2)

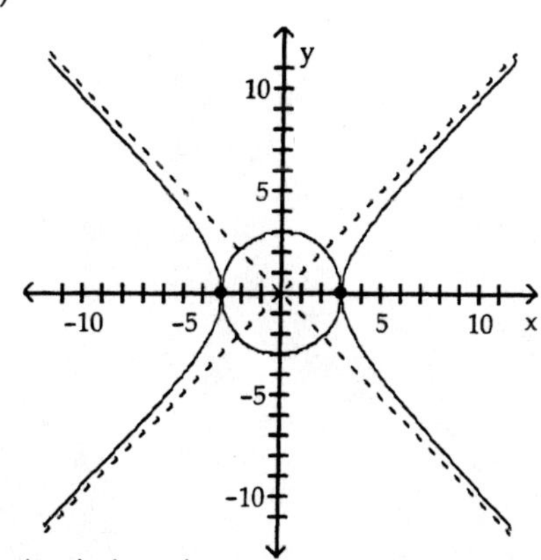

(3, 0), (–3, 0)

3)

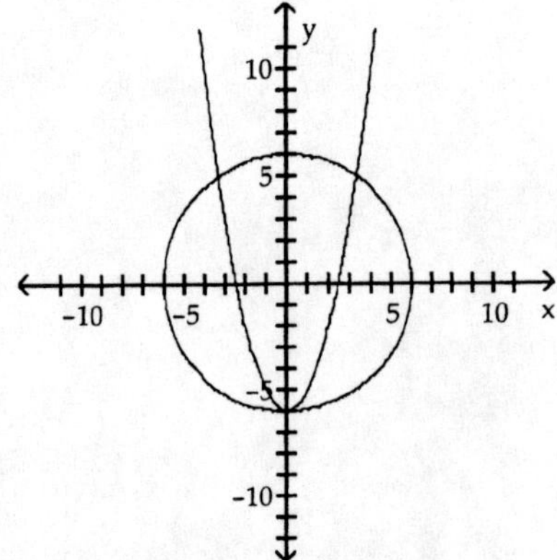

(0, −6), ($\sqrt{11}$, 5), ($-\sqrt{11}$, 5)

4)

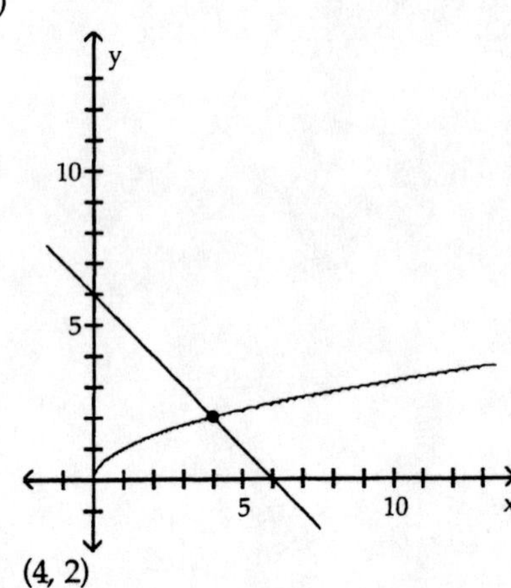

(4, 2)

5)

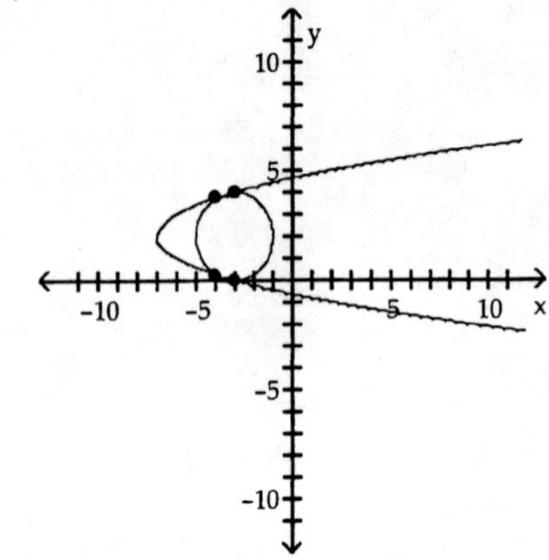

(−3, 4), (−3, 0), (−4, $2 + \sqrt{3}$), (−4, $2 - \sqrt{3}$)

6) A
7) B
8) A
9) D
10) 8 and 3
11) B
12) D
13) B
14) B
15) C
16) C

7.7 Systems of Inequalities

1 Graph an Inequality

1) D
2) A
3) C
4) B
5) D
6) C
7) C
8) C
9) A
10) A
11) C
12) C

2 Graph a System of Inequalities

1) D
2) B
3) B
4) D
5) D
6) B
7) C
8) A
9) C
10) B
11) C
12) D

3 Additional Applications

1) B
2) B
3) C
4) A
5)

$$\begin{cases} x \geq 0 \\ y \geq 0 \\ 3x + 2y \leq 480 \\ x + 2y \leq 300 \end{cases}$$

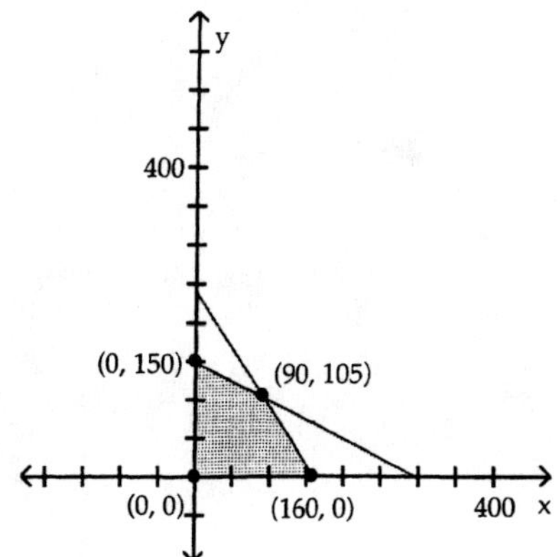

7.8 Linear Programming

1 Set up a Linear Programming Problem

1) (a)

$$x + y \leq 75{,}000$$
$$x \geq 40{,}000$$
$$y \leq 20{,}000$$
$$x \geq 0$$
$$y \geq 0$$

(b)

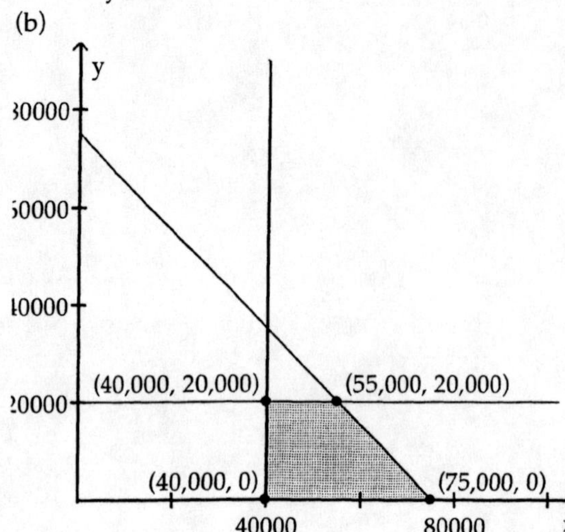

2) (a)

$$\begin{cases} x \geq 0 \\ y \geq 0 \\ 4x + 7y \geq 50 \\ x + y \leq 13 \end{cases}$$

(b)

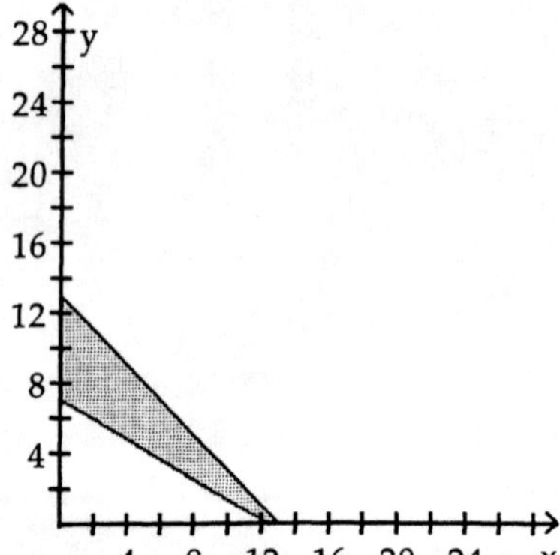

3) (a)

$$\begin{cases} 1.5x + 2.5y \le 80 \\ x + y \le 40 \\ x \ge 0 \\ y \ge 0 \end{cases}$$

(b)

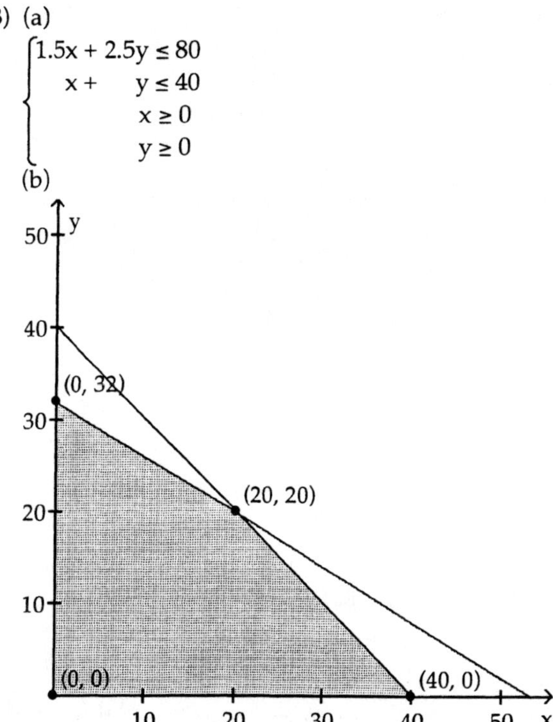

4) A

2 **Solve a Linear Programming Problem**

1) B
2) B
3) B
4) C
5) B
6) B
7) A
8) C
9) B
10) C
11) B
12) D
13) B
14) B
15) order 100 type A printers
16) \$37,500 in the stable bonds and \$37,500 in the volatile bonds; maximum income \$6750
17) \$60,000 in the stable bonds and \$60,000 in the volatile bonds; maximum income \$9900
18) 24 basic models, 16 deluxe models; maximum profit \$2240
19) 75 packages of the breakfast blend, 40 packages of the afternoon blend; maximum profit \$192.50
20) 0 small tile, 6 large tiles; minimum cost \$27

Ch. 8 Sequences; Induction; the Binomial Theorem

8.1 Sequences

1 Write the First Several Terms of a Sequence

Write out the first five terms of the sequence.

1) $\{n - 1\}$

A) −4, −3, −2, −1, 0 B) 0, 1, 2, 3, 4 C) 1, 2, 3, 4, 5 D) −1, 0, 1, 2, 3

2) $\{4n - 2\}$

A) 2, 3, 4, 5, 6 B) 2, 6, 10, 14, 18 C) 6, 10, 14, 18, 22 D) −2, −6, −10, −14, −18

3) $\{2(4n - 1)\}$

A) 6, 12, 18, 24, 30 B) 3, 7, 11, 15, 19 C) 6, 14, 22, 30, 38 D) −2, 6, 14, 22, 30

4) $\{n^2 - n\}$

A) 1, 4, 9, 16, 25 B) 2, 6, 12, 20, 30 C) 0, 2, 6, 12, 20 D) 0, 3, 8, 15, 24

5) $\{2^n\}$

A) 4, 8, 16, 32, 64 B) 2, 4, 8, 16, 32 C) 1, 4, 9, 16, 25 D) 1, 2, 4, 8, 16

6) $\left\{(-1)^{n-1}\left(\dfrac{n+3}{2n-1}\right)\right\}$

A) $4, -\frac{5}{3}, \frac{6}{5}, -1, \frac{8}{9}$ B) $-4, \frac{5}{3}, -\frac{6}{5}, 1, -\frac{8}{9}$ C) $-4, \frac{5}{3}, \frac{6}{5}, -1, \frac{8}{9}$ D) $4, \frac{5}{3}, \frac{6}{5}, 1, \frac{8}{9}$

7) $\left\{\dfrac{n}{n^2+2}\right\}$

A) $\frac{1}{3}, \frac{1}{3}, \frac{3}{8}, \frac{2}{5}, \frac{5}{12}$ B) $\frac{1}{3}, \frac{1}{3}, \frac{3}{11}, \frac{2}{9}, \frac{5}{27}$ C) $\frac{1}{2}, \frac{1}{3}, \frac{3}{8}, \frac{2}{5}, \frac{5}{12}$ D) $\frac{1}{4}, \frac{1}{3}, \frac{3}{8}, \frac{2}{5}, \frac{5}{12}$

8) $\left\{\dfrac{(2n-1)!}{(n)!}\right\}$

A) 2, 6, 40, 1260, 72,576 B) 1, 3, 20, 210, 3024 C) 2, 3, 20, 210, 3024 D) 1, 6, 40, 1260, 72,576

The given pattern continues. Write down the nth term of the sequence suggested by the pattern.

9) 4, 20, 36, 52, 68, ...

A) $a_n = 4(16)^{n-1}$ B) $a_n = 16n - 3$ C) $a_n = 4(4n - 3)$ D) $a_n = 12n - 16$

10) 1, 5, 9, 13, 17, ...

A) $a_n = 3n - 4$ B) $a_n = 4n - 3$ C) $a_n = 1(4)^{n-1}$ D) $a_n = n + 4$

11) 0, 2, 6, 12, 20, ...

A) $a_n = 2n - 2$ B) $a_n = n^2 - n$ C) $a_n = 4n - 6$ D) $a_n = 2^{n-1} - 1$

12) 4, 16, 64, 256, 1024, ...

A) $a_n = 12n$ B) $a_n = 4 + 12(n - 1)$ C) $a_n = 4^{n-1} + 3$ D) $a_n = 4^n$

13) $\frac{1}{1}, \frac{1}{4}, \frac{1}{9}, \frac{1}{16}, \frac{1}{25}, \ldots$

A) $a_n = \left(\frac{1}{2}\right)^{n-1}$ B) $a_n = \frac{1}{n^2}$ C) $a_n = \frac{1}{3n-2}$ D) $a_n = \frac{1}{n^{n-1}}$

Find the indicated term for the sequence.

14) $a_n = 4(4n - 1)$; a_5

A) 85 B) 64 C) 80 D) 76

15) $a_n = 2n - 2$; a_7

A) 0 B) 12 C) 16 D) 14

16) $a_n = n^2 - n$; a_{15}

A) 210 B) −210 C) 240 D) 15

17) $a_n = 3^n$; a_4

A) 81 B) 64 C) 27 D) 12

18) $a_n = \frac{1}{n^2}$; a_8

A) 64 B) $\frac{1}{16}$ C) $\frac{1}{64}$ D) $\frac{1}{4}$

2 Write the Terms of a Sequence Defined by a Recursive Formula

The sequence is defined recursively. Write the first four terms.

1) $a_1 = 2$ and $a_n = a_{n-1} - 4$ for $n \geq 2$

A) 2, 0, −4, −8 B) 2, −2, −6, −10 C) −4, −8, −12, −16 D) 2, 6, 10, 14

2) $a_1 = 4$ and $a_n = 2a_{n-1}$ for $n \geq 2$

A) 6, 12, 24, 48 B) 4, 8, 16, 32 C) 4, 7, 6, 5 D) 4, 10, 18, 34

3) $a_1 = 3$ and $a_n = 2a_{n-1} + 1$ for $n \geq 2$

A) 3, 6, 12, 24 B) 3, 7, 13, 25 C) 3, 7, 15, 31 D) 3, 5, 9, 17

4) $a_1 = 2$, $a_2 = 5$ and $a_n = a_{n-2} - 3a_{n-1}$ for $n \geq 3$

A) 2, 5, 1, 2 B) 2, 5, −13, 44 C) 2, 5, −1, −16 D) 2, 5, 17, −46

5) $a_1 = 135$ and $a_{n+1} = \frac{1}{3}(a_n)$ for $n \geq 2$

Solve the problem.

6) Given that $a_1 = -4$, $a_2 = -4$ and $a_{n+2} = a_{n+1} - 4a_n$, what is the fifth term of this recursively defined sequence?

A) −20 B) 764 C) 28 D) −308

7) Karen has a balance of \$1000 on a department store credit card that charges 1.5% interest per month on any unpaid balance. She can afford to pay \$150 toward the balance each month. Her balance each month, after making a \$150 payment, is given by the recursively defined sequence

$B_0 = \$1000 \qquad B_n = 1.015B_{n-1} - 150$

Determine Karen's balance after making the first payment. That is, determine B_1.

8) Jake bought a truck by taking out a loan for \$26,500 at 0.25% interest per month. Jake's regular monthly payment is \$567, but he decides to pay an extra \$75 toward the balance each month. His balance each month, after making his payment, is given by the recursively defined sequence

$B_0 = \$26{,}500 \qquad B_n = 1.0025B_{n-1} - 642$

Determine Jake's balance after making the first payment. That is, determine B_1.

9) A wildlife refuge currently has 100 deer in it. A local wildlife society decides to add an additional 2 deer each month. It is already known that the deer population is growing 12% per year. The size of the population is given by the recursively defined sequence

$p_0 = 100 \qquad p_n = 1.01p_{n-1} + 2$

How many deer are in the wildlife refuge at the end of the second month? That is, what is p_2?

10) Lexington Reservoir has 300 million gallons of water. About 0.07% of the water is lost to evaporation every week. About 150,000 gallons of water enter the reservoir every week. The amount of water in the reservoir at the end of each week is given by the recursively defined sequence

$w_0 = 300 \qquad w_n = (0.9993)w_{n-1} + 0.15$

Determine the amount of water in the reservoir at the beginning of the second week. That is, determine w_2.

11) Maria deposited \$1500 in an independent retirement account at her bank. She earns 0.5% interest per month on the balance. Each month, she deposits \$100 in the account. Her balance each month after making a \$100 deposit, is given by the recursively defined sequence

$B_0 = \$1500 \qquad B_n = 1.005B_{n-1} + 100$

If she made the initial deposit on September 30, and makes each monthly deposit on the last day of the month, how much money will be in the account at the end of the year for Maria to count as a deduction for that year's federal income taxes? That is, determine B_3.

3 Use Summation Notation

Write out the sum. Do not evaluate.

1)

$$\sum_{k=1}^{4} (k + 3)$$

A) 7 + 8 + 9 + 10 B) 3 + 4 + 5 + 6 C) 4 + 5 + 6 + 7 D) 4 + 5 + 6 + 7 + ...

2)

$$\sum_{k=1}^{5} \frac{k+1}{k+2}$$

A) $\frac{1}{2}+\frac{2}{3}+\frac{3}{4}+\frac{4}{5}+\frac{5}{6}$ B) $\frac{1}{3}+\frac{2}{4}+\frac{3}{5}+\frac{4}{6}$ C) $\frac{2}{3}+\frac{3}{4}+\frac{4}{5}+\frac{5}{6}+\frac{6}{7}$ D) $\frac{6}{7}$

3)

$$\sum_{k=1}^{6} (6k - 3)$$

A) $3 + 9 + 15 + 21 + 27 + 33$ B) 75

C) $1 + 3 + 9 + 15 + 21 + 27$ D) $3 + 9 + 15 + 21 + 27 + 33 + \ldots$

4)

$$\sum_{k=1}^{4} (k^2 - 4k - 3)$$

A) $-6 - 7 - 6 - 3$ B) $-6 - 7 - 6 - 3 + \ldots$ C) $1 - 6 - 7 - 6$ D) $-6 + 7 - 6 + 3$

Express the sum using summation notation.

5) $3^2 + 4^2 + 5^2 + \cdots + 8^2$

A) $\sum_{k=3}^{n} k^2$ B) $\sum_{k=4}^{8} (k-1)^2$ C) $\sum_{k=3}^{8} k^2$ D) $\sum_{k=1}^{8} k^2$

6) $2 + 4 + 6 + \cdots + 14$

A) $\sum_{k=1}^{7} k^2$ B) $\sum_{k=0}^{7} 2k$ C) $\sum_{k=1}^{7} 2k$ D) $\sum_{k=1}^{7} 2k^2$

7) $\frac{1}{3}+\frac{1}{2}+\frac{3}{5}+\cdots+\frac{15}{17}$

A) $\sum_{k=0}^{15} \frac{k}{k+2}$ B) $\sum_{k=2}^{15} \frac{k}{k+1}$ C) $\sum_{k=1}^{n} \frac{k}{k+2}$ D) $\sum_{k=1}^{15} \frac{k}{k+2}$

8) $\frac{3}{4} - \frac{9}{16} + \frac{27}{64} - \cdots + (-1)^{(7+1)}\left(\frac{3}{4}\right)^7$

A) $\sum_{k=1}^{7} (-1)^{(k+1)}\left(\frac{3}{4}\right)^k$ B) $\sum_{k=1}^{7} \left(\frac{3}{4}\right)^k$

C) $\sum_{k=1}^{8} (-1)^{(k+1)}\left(\frac{3}{4}\right)^k$ D) $\sum_{k=1}^{n} (-1)^{(k+1)}\left(\frac{3}{4}\right)^k$

4 Find the Sum of a Sequence

Find the sum of the sequence.

1) $\sum_{k=1}^{11} 3$

A) 8 B) 3 C) 11 D) 33

2) $\sum_{k=1}^{5} k$

A) 5 B) 20 C) 4 D) 15

3) $\sum_{k=3}^{6} (4k - 3)$

A) 63 B) 33 C) 51 D) 60

4) $\sum_{k=1}^{5} (k - 12)$

A) -18 B) -7 C) -45 D) -38

5) $\sum_{k=4}^{7} 3k$

A) 21 B) 45 C) 33 D) 66

6) $\sum_{k=1}^{4} 2^k$

A) 18 B) 30 C) 20 D) 14

7) $\sum_{k=3}^{5} (k^2 - 4)$

A) 12 B) 0 C) 35 D) 38

8) $\sum_{k=2}^{4} k(k - 8)$

A) -28 B) -50 C) -43 D) -6

9)

$$\sum_{k=1}^{4} \left(-\frac{1}{4}\right)^k$$

A) $-\frac{47}{256}$ B) $\frac{85}{256}$ C) $-\frac{51}{256}$ D) $\frac{51}{256}$

10)

$$\sum_{k=1}^{4} (-1)^k(k + 18)$$

A) -82 B) 82 C) 2 D) 74

11)

$$\sum_{k=2}^{5} (-1)^{k+1}(k + 8)^2$$

A) -534 B) 46 C) 351,788 D) 534

12)

$$\sum_{k=1}^{4} (-1)^k \cdot 9k$$

A) 6561 B) 90 C) -90 D) 18

13)

$$\sum_{k=1}^{16} (2k + 7)$$

A) 53 B) 384 C) 272 D) 400

8.2 Arithmetic Sequences

1 Determine If a Sequence Is Arithmetic

Determine whether the sequence is arithmetic.

1) 4, 12, 36, 108, 972, ...

A) arithmetic B) not arithmetic

2) 2, -3, -8, -13, -18, ...

A) arithmetic B) not arithmetic

3) 2, 4, 6, 10, 12, ...

A) arithmetic B) not arithmetic

4) 2, -1, -4, -7, -10, ...

A) arithmetic B) not arithmetic

5) 5, -15, 45, -135, 405, ...

A) arithmetic B) not arithmetic

6) 3, 7, 11, 15, 19, ...

A) arithmetic
B) not arithmetic

An arithmetic sequence is given. Find the common difference and write out the first four terms.

7) {10 - 4n}

A) d = 4; 6, 10, 14, 18
B) d = -4; 6, 4, 0, -4
C) d = -4; -4, -8, -12, -16
D) d = -4; 6, 2, -2, -6

2 Find a Formula for an Arithmetic Sequence

Find the nth term and the indicated term of the arithmetic sequence whose initial term, a, and common difference, d, are given.

1) a = 7; d = 3
a_n = ?; a_{13} = ?

A) $a_n = 4 + 3n$; $a_{13} = 34$
B) $a_n = 7 + 3n$; $a_{13} = 43$
C) $a_n = 4 - 3n$; $a_{13} = 43$
D) $a_n = 4 + 3n$; $a_{13} = 43$

2) a = 82; d = -2
a_n = ?; a_7 = ?

A) $a_n = 84 - 2n$; $a_7 = 52$
B) $a_n = 84 - 2n$; $a_7 = 70$
C) $a_n = 82 - 2n$; $a_7 = 52$
D) $a_n = 82 - 2n$; $a_7 = 70$

3) a = 5; d = -3
a_n = ?; a_{16} = ?

A) $a_n = 8 - 3n$; $a_{16} = -22$
B) $a_n = 5 - 3n$; $a_{16} = -40$
C) $a_n = 8 + 3n$; $a_{16} = -40$
D) $a_n = 8 - 3n$; $a_{16} = -40$

4) a = -9; d = 6
a_n = ?; a_7 = ?

A) $a_n = -15 + 6n$; $a_7 = 75$
B) $a_n = -15 + 6n$; $a_7 = 27$
C) $a_n = -15 - 6n$; $a_7 = 27$
D) $a_n = -9 + 6n$; $a_7 = 27$

Find the indicated term of the sequence.

5) The twentieth term of the arithmetic sequence -13, -8, -3, ...

A) -108
B) 82
C) -113
D) 87

6) The twenty-third term of the arithmetic sequence 0, 3, 6, ...

A) 72
B) 69
C) 66
D) 44

7) The twelfth term of the arithmetic sequence 28, 22, 16, ...

A) -44
B) 94
C) -38
D) -66

8) The twenty-second term of the arithmetic sequence $-2\sqrt{3}$, $-7\sqrt{3}$, $-12\sqrt{3}$, ...

A) $103\sqrt{3}$
B) $108\sqrt{3}$
C) $-107\sqrt{3}$
D) $-112\sqrt{3}$

Find the first term, the common difference, and give a recursive formula for the arithmetic sequence.

9) 8th term is 75; 15th term is 61

A) $a_1 = 91,\ d = -2,\ a_n = a_{n-1} - 2$
B) $a_1 = 89,\ d = -2,\ a_n = a_{n-1} - 2$
C) $a_1 = 89,\ d = 2,\ a_n = a_{n-1} + 2$
D) $a_1 = 91,\ d = 2,\ a_n = a_{n-1} + 2$

10) 10th term is -69; 15th term is -109

A) $a_1 = 3,\ d = 8,\ a_n = a_{n-1} + 8$
B) $a_1 = 11,\ d = 8,\ a_n = a_{n-1} + 8$
C) $a_1 = 3,\ d = -8,\ a_n = a_{n-1} - 8$
D) $a_1 = 11,\ d = -8,\ a_n = a_{n-1} - 8$

11) 8th term is 68; 13th term is 118

A) $a_1 = -12,\ d = 10,\ a_n = a_{n-1} + 10$
B) $a_1 = -2,\ d = 10,\ a_n = a_{n-1} + 10$
C) $a_1 = -12,\ d = 10,\ a_n = a_{n-1} - 10$
D) $a_1 = -2,\ d = 10,\ a_n = a_{n-1} - 10$

12) 6th term is -10; 15th term is -46

A) $a_1 = -10,\ d = -4,\ a_n = a_{n-1} - 4$
B) $a_1 = -30,\ d = 4,\ a_n = a_{n-1} + 4$
C) $a_1 = 10,\ d = 4,\ a_n = a_{n-1} + 4$
D) $a_1 = 10,\ d = -4,\ a_n = a_{n-1} - 4$

Find the indicated term using the given information.

13) $a = 6, d = 5;\ a_8$

A) 41 B) 46 C) -34 D) -29

14) $a = 7, d = -1;\ a_{44}$

A) -36 B) 50 C) -37 D) 51

15) $a = 8, d = \frac{7}{3};\ a_{16}$

A) 43 B) - 27 C) $\frac{136}{3}$ D) $-\frac{88}{3}$

16) 13, 10, 7, ... ; a_{27}

A) 91 B) -65 C) -68 D) 94

17) $a_{11} = 17, a_{19} = 41;\ a_1$

A) -7 B) -10 C) 3 D) -13

18) $a_{17} = -46, a_{10} = -25;\ a_5$

A) -3 B) -13 C) -10 D) 2

19) $a_{21} = \frac{107}{4}, a_{13} = \frac{67}{4};\ a_4$

A) $\frac{11}{2}$ B) $\frac{27}{4}$ C) $\frac{5}{4}$ D) 7

Solve the problem.

20) The population of a town is increasing by 500 inhabitants each year. If its population at the beginning of 1990 was 25,180, what was its population at the beginning of 1999?

A) 452,880 inhabitants B) 29,180 inhabitants C) 226,440 inhabitants D) 29,680 inhabitants

3 Find the Sum of an Arithmetic Sequence

Find the sum of the arithmetic sequence.

1) $-3 + 1 + 5 + 9 + 13 + ... + (4n - 7)$

A) $n(2n + 5)$ B) $n(2n - 5)$ C) $n(4n + 7)$ D) $n(4n - 7)$

2) $(-6) + (-1) + 4 + 9 + ... + 39$

A) 165 B) 160 C) 170 D) 330

3) $(-9) + (-1) + 7 + 15 + ... + 71$

Solve the problem.

4) A theater has 22 rows with 20 seats in the first row, 23 in the second row, 26 in the third row, and so forth. How many seats are in the theater?

A) 1166 seats B) 1133 seats C) 2332 seats D) 2266 seats

5) A brick staircase has a total of 18 steps The bottom step requires 121 bricks. Each successive step requires 4 less bricks than the prior one. How many bricks are required to build the staircase?

A) 1530 bricks B) 1566 bricks C) 3132 bricks D) 2790 bricks

6) Suppose you just received a job offer with a starting salary of $37,000 per year and a guaranteed raise of $1500 per year. How many years will it be before you've made a total (or aggregate) salary of $1,025,000?

7) A local civic theater has 22 seats in the first row and 21 rows in all. Each successive row contains 3 additional seats. How many seats are in the civic theater?

A) 1092 seats B) 1070 seats C) 790 seats D) 1010 seats

8.3 Geometric Sequences; Geometric Series

1 Determine If a Sequence Is Geometric

Determine whether the sequence is geometric.

1) 4, 12, 36, 108, 324, ...

A) geometric B) not geometric

2) 3, 1, −1, −3, −5, ...

A) geometric B) not geometric

3) 3, 5, 7, 11, 13, ...

A) geometric B) not geometric

4) 4, −12, 36, −108, 324, ...

A) geometric B) not geometric

5) 7, 4, 1, -2, -5, ...

A) geometric

B) not geometric

6) 2, 6, 18, 54, 162, ...

A) geometric

B) not geometric

7) $\frac{1}{5}, \frac{1}{8}, \frac{1}{11}, \frac{1}{14}, \ldots$

A) geometric

B) not geometric

If the sequence is geometric, find the common ratio. If the sequence is not geometric, say so.

8) 3, 12, 48, 192, 768

A) 4 B) 12 C) 1/4 D) not geometric

9) 3, -9, 27, -81, 243

A) -3 B) -12 C) 3 D) not geometric

10) $\frac{4}{3}, \frac{16}{3}, \frac{64}{3}, \frac{256}{3}, \frac{1024}{3}$

A) 4 B) 10 C) 12 D) not geometric

11) 47, 28.2, 16.92, 10.152, 6.09

A) -0.6 B) 1.67 C) 6.09 D) 0.6

12) $\frac{3}{4}, \frac{3}{8}, \frac{3}{16}, \frac{3}{32}, \frac{3}{64}$

A) 2 B) 12 C) $\frac{1}{2}$ D) $\frac{1}{12}$

13) $\frac{1}{10}, \frac{1}{15}, \frac{1}{20}, \frac{1}{25}$

A) $\frac{1}{5}$ B) $5\frac{1}{5}$ C) 5 D) not geometric

2 Find a Formula for a Geometric Sequence

A geometric sequence is given. Find the common ratio and write out the first four terms.

1) $\left\{2\left(\frac{1}{2}\right)^{n-1}\right\}$

A) $a_n = 2 \cdot (2)^{n-1}$

$r = 2;\ 2, 4, 8, 16, \ldots$

B) $a_n = \frac{1}{2}(2)^{n-1}$

$r = \frac{1}{2};\ 2, 4, 8, 16, \ldots$

C) $a_n = 2\left(\frac{1}{2}\right)^{n-1}$

$r = \frac{1}{2};\ 2, 1, \frac{1}{2}, \frac{1}{4}, \ldots$

D) $a_n = 2\left(\frac{1}{4}\right)^{n-1}$

$r = \frac{1}{4};\ 2, \frac{1}{2}, \frac{1}{8}, \frac{1}{32}, \ldots$

Determine whether the given sequence is arithmetic, geometric, or neither. If arithmetic, find the common difference. If geometric, find the common ratio.

2) $\{3n - 4\}$

A) geometric; $r = 3$ B) arithmetic; $d = 3$ C) arithmetic; $d = -4$ D) neither

3) $\{3n^2\}$

A) geometric; $r = \frac{3}{2}$ B) geometric; $r = 3$ C) arithmetic; $d = 3$ D) neither

4) $\left\{\left(\frac{4}{3}\right)^n\right\}$

A) arithmetic; $d = \frac{4}{3}$ B) geometric; $r = \frac{3}{4}$ C) geometric; $r = \frac{4}{3}$ D) neither

Find the fifth term and the nth term of the geometric sequence whose initial term, a, and common ratio, r, are given.

5) $a = 8$; $r = 4$

A) $a_5 = 2048$; $a_n = 8 \cdot (4)^n$

B) $a_5 = 2048$; $a_n = 8 \cdot (4)^{n-1}$

C) $a_5 = 8192$; $a_n = 8 \cdot (4)^n$

D) $a_5 = 8192$; $a_n = 8 \cdot (4)^{n-1}$

6) $a = 2$; $r = \frac{1}{4}$

A) $a_5 = \frac{1}{128}$; $a_n = 2 \cdot \left(\frac{1}{4}\right)^{n-1}$

B) $a_5 = \frac{1}{512}$; $a_n = 2 \cdot \left(\frac{1}{4}\right)^{n}$

C) $a_5 = \frac{1}{128}$; $a_n = 2 \cdot \left(\frac{1}{4}\right)^{n}$

D) $a_5 = \frac{1}{512}$; $a_n = 2 \cdot \left(\frac{1}{4}\right)^{n-1}$

7) $a = 7$; $r = -2$

A) $a_5 = 112$; $a_n = 7 \cdot (-2)^n$

B) $a_5 = -56$; $a_n = 7 \cdot (-2)^{n-1}$

C) $a_5 = 112$; $a_n = 7 \cdot (-2)^{n-1}$

D) $a_5 = -56$; $a_n = 7 \cdot (-2)^n$

8) $a = -3$; $r = -5$

A) $a_5 = 375$; $a_n = -3 \cdot (-5)^{n-1}$

B) $a_5 = -1875$; $a_n = -3 \cdot (-5)^{n-1}$

C) $a_5 = -1875$; $a_n = -3 \cdot (-5)^n$

D) $a_5 = 375$; $a_n = -3 \cdot (-5)^n$

Solve the problem.

9) Find the 10th term of the geometric sequence $\frac{1}{3}$, 1, 3, ...

A) 6561 B) 177, 147 C) 19683 D) 2187

10) For the geometric sequence 2, 1, $\frac{1}{2}$, $\frac{1}{4}$, ... , find a_n.

A) $a_n = 2^{n-2}$ B) $a_n = \left(\frac{1}{2}\right)^{n-2}$ C) $a_n = \left(\frac{1}{2}\right)^{1-n}$ D) $a_n = 2^{1-n}$

11) For the geometric sequence 64, 16, 4, 1, ... , find a_n.

12) A new piece of equipment cost a company \$47,000. Each year, for tax purposes, the company depreciates the value by 25%. What value should the company give the equipment after 7 years?

A) \$11 B) \$6274 C) \$8365 D) \$3

13) A particular substance decays in such a way that it loses half its weight each day. How much of the substance is left after 7 days if it starts out at 32 grams?

A) $\frac{1}{4}$ gram B) $\frac{1}{2}$ gram C) 2 grams D) 4 grams

14) A bicycle wheel rotates 500 times in a minute as long as the rider is pedalling. If the rider stops pedalling, the wheel starts to slow down. Each minute it will rotate only 2/3 as many times as in the preceding minute. How many times will the wheel rotate in the 5th minute after the rider's feet leave the pedals? Round your answer to the nearest unit.

A) 6 times B) 66 times C) 99 times D) 0 times

15) A pendulum bob swings through an arc 70 inches long on its first swing. Each swing thereafter, it swings only 63% as far as on the previous swing. What is the length of the arc after 10 swings? Round your answer to two decimal places, if necessary.

A) 0.69 inches B) 1.09 inches C) 0.43 inches D) 396.9 inches

3 Find the Sum of a Geometric Sequence

Find the sum of the sequence.

1) $\frac{1}{7}+\frac{3}{7}+\frac{3^2}{7}+\frac{3^3}{7}+\cdots+\frac{3^{n-1}}{7}$

A) $-\frac{2}{7}(1-3^n)$ B) $-\frac{1}{7}(1-3^n)$ C) $-\frac{1}{14}(1-3^n)$ D) $\frac{1}{14}(1-3^n)$

2)

$\sum_{k=1}^{n} 5\cdot 7^{k-1}$

A) $-\frac{5}{6}(1-7^{n-1})$ B) $-\frac{5}{6}(1-7^n)$ C) $-30(1-7^n)$ D) $-5(1-7^n)$

Use a graphing utility to find the sum of the geometric sequence. Round answer to two decimal places, if necessary.

3) $4+12+36+108+324+\cdots+4\cdot 3^8$

A) 39,344 B) 39,366 C) 39,364 D) 39,401

4) $-6-12-24-48-96-\cdots-6\cdot 2^{12}$

A) −49,144 B) −49,146 C) −49,109 D) −49,166

5) $7-21+63-189+567-\cdots+7\cdot(-3)^{13}$

A) −8,370,194 B) −8,370,200 C) −8,370,187 D) −8,370,196

6) $\frac{1}{5}+\frac{2}{5}+\frac{2^2}{5}+\frac{2^3}{5}+\cdots+\frac{2^{13}}{5}$

A) 6553.4 B) 3276.4 C) 6553.2 D) 3276.6

7)

$$\sum_{k=1}^{5} 2(3)^k$$

A) 1230 B) 24 C) 87 D) 726

8)

$$\sum_{k=1}^{5} 2(-3)^k$$

A) -1230 B) -366 C) 12 D) 75

9)

$$\sum_{k=1}^{8} \frac{1}{4} \cdot (-2)^{k-1}$$

A) -22.75 B) -19.5 C) -21.25 D) -21.75

Solve the problem.

10) Initially, a pendulum swings through an arc of 3 feet. On each successive swing, the length of the arc is 0.8 of the previous length. After 10 swings, what total length will the pendulum have swung (to the nearest tenth of a foot)?

4 Find the Sum of a Geometric Series

Find the sum of the infinite geometric series.

1) $5 + \frac{5}{4} + \frac{5}{16} + \cdots$

A) $\frac{5}{4}$ B) 5 C) $\frac{25}{4}$ D) $\frac{20}{3}$

2) $1 - \frac{1}{4} + \frac{1}{16} - \cdots$

A) $\frac{3}{4}$ B) $-\frac{1}{4}$ C) $\frac{4}{5}$ D) 1

3) $18 + 6 + 2 + \cdots$

A) -9 B) 18 C) 27 D) 26

4) $-30 - 5 - \frac{5}{6} - \cdots$

A) -36 B) $-\frac{215}{6}$ C) 6 D) -30

5)

$$\sum_{k=1}^{\infty} 4\left(\frac{2}{3}\right)^{k-1}$$

A) 16 B) 12 C) $\frac{8}{3}$ D) 4

6) $(0.996) + (0.996)^2 + (0.996)^3 + \ldots$

Solve the problem.

7) A pendulum bob swings through an arc 80 inches long on its first swing. Each swing thereafter, it swings only 70% as far as on the previous swing. How far will it swing altogether before coming to a complete stop?

A) 133 inches B) 229 inches C) 267 inches D) 114 inches

8) A ball is dropped from a height of 25 feet. Each time it strikes the ground, it bounces up to 0.7 of the previous height. The total distance the ball has travelled before the second bounce is $25 + 2(25 \cdot 0.7)$ feet, and the total distance the ball has travelled before bounce $n + 1$ is

$$25 + \sum_{k=1}^{n} 50\left(0.7^k\right) \text{ feet.}$$

Use facts about infinite geometric series to calculate the total distance the ball has travelled by the time it has stopped bouncing.

9) A ping-pong ball is dropped from a height of 9 ft and always rebounds $\frac{1}{3}$ of the distance fallen. Find the total sum of the rebound heights of the ball.

A) 6 ft B) 4.5 ft C) 13.5 ft D) 3 ft

10) Write 0.272727... as a fraction. (Find the sum of the repeating decimal.)

A) $\frac{6}{11}$ B) $\frac{3}{11}$ C) $\frac{1}{37}$ D) $\frac{3}{10}$

5 Solve Annuity Problems

Solve the problem.

1) Jack decided to put \$600 into an IRA account every 3 months at a rate of 6% compounded quarterly. Find a recursive formula that represents his balance at the end of each quarter. How long will it be before the value of the account is \$100,000? What will be the balance in 30 years when Jack retires?

A) $B_0 = 600 \quad B_n = (1 + \frac{0.06}{3})B_{(n-1)} + 600$; 18 years; \$151,860

B) $B_0 = 600 \quad B_n = (1 + \frac{0.6}{4})B_{(n-1)} + 600$; 7 years; \$4 million

C) $B_0 = 600 \quad B_n = (1 + \frac{0.06}{12})B_{(n-1)} + 600$; 10 years; \$6.8 million

D) $B_0 = 600 \quad B_n = (1 + \frac{0.06}{4})B_{(n-1)} + 600$; 21 years; \$202,355

2) After working for 25 years you would like to have \$800,000 in an annuity for early retirement. If the annual interest rate is 7.5%, compounded monthly, what will your monthly deposit need to be?

A) \$3088.48 B) \$2666.67 C) \$911.93 D) \$1146.98

3) You deposit \$150 each 6 months into an annuity with an annual interest rate of 13.5%, compounded semiannually.

a) What is the balance after 20 years?

b) What is the balance after 40 years?

4) Christine contributes $100 each month to her 401(k). To the nearest dollar, what will be the value of Christine's 401(k) in 20 years if the per annum rate of return is assumed to be 10% compounded monthly.

8.4 Mathematical Induction

1 Prove Statements Using Mathematical Induction

Use the Principle of Mathematical Induction to show that the statement is true for all natural numbers n.

1) $3 + 8 + 13 + ... + (5n - 2) = \frac{n}{2}(5n + 1)$

2) $11 + 22 + 33 + ... + 11n = \frac{11n(n + 1)}{2}$

3) $1 + 6 + 6^2 + ... + 6^{n-1} = \frac{6^n - 1}{5}$

4) $n^2 - n + 2$ is divisible by 2

5) $\frac{1}{2} + \frac{1}{4} + \frac{1}{8} + \frac{1}{16} + ... + \frac{1}{2^n} = 1 - \frac{1}{2^n}$

6) Use the Principle of Mathematical Induction to show that the statement "5 is a factor of $7^n - 2^n$" is true for all natural numbers. (Hint: $7^{k+1} - 2^{k+1} = 7(7^k - 2^k) + 5 \cdot 2^k$)

7) Show that the formula

$$2 + 4 + 6 + 8 + \cdots + 2n = n^2 + n + 3$$

obeys Condition II of the Principle of Mathematical Induction. That is, show that if the formula is true for some natural number k, it is also true for the next natural number k + 1. Then show that the formula is false for n = 1.

8) Show that the statement "$n^2 - n + 3$ is a prime number" is true for n = 1, but is not true for n = 3.

Use mathematical induction to prove the statement is true for all positive integers n.

9) $6 + 12 + 18 + ... + 6n = 3n(n + 1)$

10) $2 + 2 \bullet \frac{1}{5} + 2 \bullet (\frac{1}{5})^2 + ... + 2 \bullet (\frac{1}{5})^{n-1} = \frac{2(1 - (\frac{1}{5})^n)}{1 - \frac{1}{5}}$

11) $1^2 + 4^2 + 7^2 + ... + (3n - 2)^2 = \frac{n(6n^2 - 3n - 1)}{2}$

12) $1 \bullet 2 + 2 \bullet 3 + 3 \bullet 4 + ... + n(n + 1) = \frac{n(n + 1)(n + 2)}{3}$

13) $\left(1 - \frac{1}{2}\right)\left(1 - \frac{1}{3}\right)...\left(1 - \frac{1}{n + 1}\right) = \frac{1}{n + 1}$

14) $3 + 2 \bullet 3 + 3 \bullet 3 + \ldots + 3n = \dfrac{3n(n+1)}{2}$

15) $(7^2)^n = 7^{2n}$

Solve the problem.

16) Let P(n) represent the statement:

$-3 + 3 + 9 + \ldots + (6n - 9) = 3n^2 - 6n$

In the proof that P(n) is true for all integers n, $n \geq 1$, what term must be added to both sides of P(k) to show P(k +1) follows from P(k)?

A) $6k - 3$ B) $6k + 3$ C) $P(k + 1)$ D) $6k - 9$

17) Let P(n) represent the statement:

$-2 + 4 + 10 + \ldots + (6n - 8) = 3n^2 - 5n$

In the proof that P(n) is true for all integers n, $n \geq 1$, what term must be added to both sides of P(k) to show P(k +1) follows from P(k)?

8.5 The Binomial Theorem

1 Evaluate a Binomial Coefficient

Evaluate the expression.

1) $\binom{10}{8}$

A) 22 B) 90 C) 1 D) 45

2) $\binom{12}{12}$

A) 479,001,600 B) 1 C) 0 D) 2

3) $\binom{151}{2}$

A) 11,325 B) 149 C) 22,650 D) $\dfrac{151!}{149!}$

4) $\binom{7}{0}$

A) 5040 B) 1 C) 0 D) 2

Find the indicated coefficient or term.

5) The coefficient of x in the expansion of $(4x + 5)^3$

A) 300 B) 25 C) 240 D) 600

6) The coefficient of x in the expansion of $(2x + 3)^5$

A) 1215 B) 540 C) 270 D) 810

7) The coefficient of $\dfrac{1}{x}$ in the expansion of $\left(2x + \dfrac{1}{x}\right)^3$

8) The coefficient of x^8 in the expansion of $(x^2 - 3)^7$

A) −2835 B) 2835 C) 945 D) −945

9) The 3rd term in the expansion of $(4x + 7)^3$

A) $336x^2$ B) $1176x$ C) 49 D) $588x$

10) The 5th term in the expansion of $(2x + 5)^5$

A) $2500x^2$ B) $1250x$ C) $6250x$ D) 15,625

2 Expand a Binomial

Expand the expression using the Binomial Theorem.

1) $(x - 1)^6$

A) $x^6 - 6x^5 - 15x^4 - 20x^3 - 15x^2 - 6x - 6$
B) $x^6 - 6x^5 + 30x^4 - 120x^3 + 30x^2 - 6x + 1$
C) $x^6 - 6x^5 + 15x^4 - 20x^3 + 15x^2 - 6x + 1$
D) $x^6 - 6x^5 - 30x^4 - 120x^3 - 360x^2 - 720x + 720$

2) $(x - 2)^5$

A) $x^5 - 10x^4 + 40x^3 - 80x^2 + 80x - 32$
B) $x^5 - 10x^4 + 80x^3 - 160x^2 + 80x - 2$
C) $x^5 - 10x^4 + 80x^3 - 160x^2 + 80x - 32$
D) $x^5 - 10x^4 + 40x^3 - 80x^2 + 80x - 2$

3) $(4x + 3)^3$

A) $64x^3 + 144x^2 + 108x + 27$
B) $64x^3 + 144x^2 + 144x + 27$
C) $16x^2 + 24x + 9$
D) $16x^6 + 12x^3 + 729$

4) $(5x + 4)^4$

A) $625x^3 + 2000x^2 + 2400x + 1280$
B) $625x^4 + 2000x^3 + 2400x^2 + 1280x + 256$
C) $2500x^4 + 8000 x^3 + 2400x^2 + 5120x + 256$
D) $(25x^2 + 20x + 16)^4$

5) $(4x + 1)^5$

A) $1024x^5 + 20x^4 + 160x^3 + 160x^2 + 20x + 1$
B) $1024x^5 + 256x^4 + 64x^3 + 16x^2 + 4x + 1$
C) $1024x^5 + 1280x^4 + 640x^3 + 160x^2 + 20x + 1$
D) $(16x^2 + 8x + 1)^5$

6) $(2x^2 - 1)^3$

A) $16x^8 + 8x^6 - 12x^4 + 6x^2 - 1$
B) $8x^3 - 12x^2 + 6x - 1$
C) $(4x^4 - 4x^2 + 1)^3$
D) $8x^6 - 12x^4 + 6x^2 - 1$

7) $(4x - 5y)^3$

A) $64x^3 - 80x^2y + 100xy^2 - 125y^3$
B) $64x^3 - 240x^2y + 300xy^2 - 125y^3$
C) $16x^3y - 40x^2y^2 + 25xy^3$
D) $16x^3y - 20x^2y^2 + 25xy^3$

8) $(2x + y)^6$

A) $64x^6 + 192x^5y + 240x^4y^2 + 160x^3y^3 + 240x^2y^4 + 192xy^5 + 64y^6$
B) $2x^6 + 12x^5y + 30x^4 y^2 + 40x^3y^3 + 12xy^5 + y^6$
C) $64x^6 + 192x^5y + 240x^4 y^2 + 160x^3 y^3 + 60x^2y^4 + 12xy^5 + y^6$
D) $64x^6 + 192x^5y + 480x^4y^2 + 960x^3y^3 + 1440x^2y^4 + 12xy^5 + y^6$

9) $(w - s)^6$

A) $w^6 - s^6$

B) $w^6 - 6w^5s + 15w^4s^2 - 20w^3s^3 + 15w^2s^4 - 6ws^5 + s^6$

C) $w^6 - 6w^5s - 30w^4s^2 + 120w^3s^3 + 360w^2s^4 - 720ws^5 - 720s^6$

D) $w^6 - 8w^5s + 17w^4s^2 - 22w^3s^3 + 17w^2s^4 - 8ws^5 + s^6$

10) $(g - 2h)^3$

11) $\left(x - \frac{3}{\sqrt{x}}\right)^4$.

Solve the problem.

12) Use the Binomial Theorem to approximate $(1.01)^5 = (1 + 10^{-2})^5$ to 7 decimal places.

13) Use the Binomial Theorem to approximate $(0.98)^6 = (1 - 2(10^{-2}))^6$ to 5 decimal places.

14) A cube with side length s units has a volume equal to s^3. Use the Binomial Theorem to calculate the difference in volume if the side length is increased by h units. That is, expand and simplify the expression $(s + h)^3 - s^3$. How much (to three decimal places) does the volume of a cube of side length 2 inches change if this side is increased by 0.03 of an inch?

Ch. 8 Sequences; Induction; the Binomial Theorem
Answer Key

8.1 Sequences

1 Write the First Several Terms of a Sequence

1) B
2) B
3) C
4) C
5) B
6) A
7) B
8) B
9) C
10) B
11) B
12) D
13) B
14) D
15) B
16) A
17) A
18) C

2 Write the Terms of a Sequence Defined by a Recursive Formula

1) B
2) B
3) C
4) B
5) 135, 45, 15, 5
6) A
7) $865
8) $25,924.25
9) 106 deer
10) 299,880 gallons
11) $1824.12

3 Use Summation Notation

1) C
2) C
3) A
4) A
5) C
6) C
7) D
8) A

4 Find the Sum of a Sequence

1) D
2) D
3) D
4) C
5) D
6) B
7) D
8) C
9) C
10) C

11) B
12) D
13) B

8.2 Arithmetic Sequences

1 Determine If a Sequence Is Arithmetic

1) B
2) A
3) B
4) A
5) B
6) A
7) D

2 Find a Formula for an Arithmetic Sequence

1) D
2) B
3) D
4) B
5) B
6) C
7) C
8) C
9) B
10) C
11) B
12) D
13) A
14) A
15) A
16) B
17) D
18) C
19) A
20) D

3 Find the Sum of an Arithmetic Sequence

1) B
2) A
3) 341
4) B
5) B
6) 20 years
7) A

8.3 Geometric Sequences; Geometric Series

1 Determine If a Sequence Is Geometric

1) A
2) B
3) B
4) A
5) B
6) A
7) B
8) A
9) A
10) A
11) D

12) C
13) D

2 Find a Formula for a Geometric Sequence

1) C
2) B
3) D
4) C
5) B
6) A
7) C
8) B
9) A
10) B
11) $a_n = \left(\frac{1}{4}\right)^{n-4}$
12) B
13) A
14) B
15) B

3 Find the Sum of a Geometric Sequence

1) C
2) B
3) C
4) B
5) A
6) D
7) D
8) B
9) C
10) approximately 13.4 feet

4 Find the Sum of a Geometric Series

1) D
2) C
3) C
4) A
5) B
6) 249
7) C
8) $141\frac{2}{3}$ feet
9) B
10) B

5 Solve Annuity Problems

1) D
2) C
3) a) $28,081.98
b) $411,032.84
4) $75,937

8.4 Mathematical Induction

1 Prove Statements Using Mathematical Induction

1) First we show that the statement iss true when n = 1.

For n = 1, we get $3 = \frac{(1)}{2}(5(1) + 1) = 3.$

This is a true statement and Condition I is satisfied.

Next, we assume the statement holds for some k. That is,

$3 + 8 + 13 + ... + (5k - 2) = \frac{k}{2}(5k + 1)$ is true for some positive integer k.

We need to show that the statement holds for k + 1. That is, we need to show that

$$3 + 8 + 13 + ... + (5(k + 1) - 2) = \frac{k+1}{2}(5(k + 1) + 1).$$

So we assume that $3 + 8 + 13 + ... + (5k - 2) = \frac{k}{2}(5k + 1)$ is true and add the next term, 5(k + 1) - 2, to both sides of the equation.

$$\begin{aligned} 3 + 8 + 13 + ... + (5k - 2) + 5(k + 1) - 2 &= \frac{k}{2}(5k + 1) + 5(k + 1) - 2 \\ &= \frac{1}{2}[k(5(k + 1) - 4) + 10(k + 1) - 4] \\ &= \frac{1}{2}[5k(k + 1) - 4k + 10(k + 1) - 4] \\ &= \frac{1}{2}[(k + 1)(5k + 10) - 4(k + 1)] \\ &= \frac{k+1}{2}(5k + 6) \\ &= \frac{k+1}{2}(5(k + 1) + 1). \end{aligned}$$

Conditions II is satisfied. As a result, the statement is true for all natural numbers n.

2) First, we show the statement is true when n = 1.

For n = 1, we get $11 = \frac{11(1)((1)+1)}{2} = 11$.

This is a true statement and Condition I is satisfied.

Next, we assume the statement holds for some k. That is,

$11 + 22 + 33 + \ldots + 11k = \frac{11k(k+1)}{2}$ is true for some positive integer k.

We need to show that the statement holds for k + 1. That is, we need to show that

$$11 + 22 + 33 + \ldots + 11(k+1) = \frac{11(k+1)(k+2)}{2}.$$

So we assume that $11 + 22 + 33 + \ldots + 11k = \frac{11k(k+1)}{2}$ is true and add the next term, 11(k + 1), to both sides of the equation.

$$\begin{aligned}
11 + 22 + 33 + \ldots + 11k + 11(k+1) &= \frac{11k(k+1)}{2} + 11(k+1) \\
&= 11\left(\frac{k(k+1)}{2} + k + 1\right) \\
&= 11\left(\frac{k(k+1)}{2} + \frac{2(k+1)}{2}\right) \\
&= 11 \cdot \frac{k(k+1) + 2(k+1)}{2} \\
&= 11 \cdot \frac{(k+1)(k+2)}{2} \\
&= \frac{11(k+1)(k+2)}{2}.
\end{aligned}$$

Conditions II is satisfied. As a result, the statement is true for all natural numbers n.

3) First, we show that the statement is true when n = 1.

For n = 1, we get 1 (or $6^{[(1) - 1]}$) $= \frac{6^{(1)} - 1}{5} = \frac{5}{5} = 1$.

This is a true statement and Condition I is satisfied.

Next, we assume the statement holds for some k. That is,

$1 + 6 + 6^2 + ... + 6^{k-1} = \frac{6^k - 1}{5}$ is true for some positive integer k.

We need to show that the statement holds for k + 1. That is, we need to show that

$1 + 6 + 6^2 + ... + 6^k = \frac{6^{k+1} - 1}{5}$.

So we assume that $1 + 6 + 6^2 + ... + 6^{k-1} = \frac{6^k - 1}{5}$ is true and add the next term, 6^k, to both sides of the equation.

$$\begin{aligned} 1 + 6 + 6^2 + ... + 6^{k-1} + 6^k &= \frac{6^k - 1}{5} + 6^k \\ &= \frac{6^k - 1 + 5 \cdot 6^k}{5} \\ &= \frac{6 \cdot 6^k - 1}{5} \\ &= \frac{6^{k+1} - 1}{5}. \end{aligned}$$

Conditions II is satisfied. As a result, the statement is true for all natural numbers n.

4) First, we show that the statement is true when n = 1.

For n = 1, $n^2 - n + 2 = (1)^2 - (1) + 2 = 2$.

This is a true statement and Condition I is satisfied.

Next, we assume the statement holds for some k. That is,

$k^2 - k + 2$ is divisible by 2 is true for some positive integer k.

We need to show that the statement holds for k + 1. That is, we need to show that

$(k + 1)^2 - (k + 1) + 2$ is divisible by 2.

So we assume $k^2 - k + 2$ is divisible by 2 and look at the expression for n = k + 1.

$$\begin{aligned} (k + 1)^2 - (k + 1) + 2 &= k^2 + 2k + 1 - k - 1 + 2 \\ &= (k^2 - k + 2) + 2k \end{aligned}$$

Since $k^2 - k + 2$ is divisible by 2, then $k^2 - k + 2 = 2m$ for some integer m. Hence,

$$\begin{aligned} (k + 1)^2 - (k + 1) + 2 &= (k^2 - k + 2) + 2k \\ &= 2m + 2k \\ &= 2(m + k). \end{aligned}$$

Conditions II is satisfied. As a result, the statement is true for all natural numbers n.

5) When n = 1, the left side of the statement is $\frac{1}{2^n} = \frac{1}{2^1} = \frac{1}{2}$, and the right side of the statement is $1 - \frac{1}{2^n} = 1 - \frac{1}{2^1} = 1 - \frac{1}{2} = \frac{1}{2}$, so the statement is true when n = 1.

Assume the statement is true for some natural number k. Then,

$$\frac{1}{2} + \frac{1}{4} + \frac{1}{8} + \frac{1}{16} + \dots + \frac{1}{2^k} + \frac{1}{2^{k+1}} = \left(1 - \frac{1}{2^k}\right) + \frac{1}{2^{k+1}} = 1 - \frac{1}{2^k}\left(1 - \frac{1}{2}\right) = 1 - \frac{1}{2^{k+1}}.$$

So the statement is true for k + 1. Conditions I and II are satisfied; by the Principle of Mathematical Induction, the statement is true for all natural numbers.

6) When n = 1, $7^n - 2^n = 7^1 - 2^1 = 5$, so the statement is true when n = 1. Assume the statement is true for some natural number k. That is, $7^k - 2^k = 5m$ for some integer m. Then,

$$7^{k+1} - 2^{k+1} = 7(7^k - 2^k) + 5 \cdot 2^k = 7(5m) + 5 \cdot 2^k = 5(7m + 2^k).$$

So the statement is true for k + 1. Conditions I and II are satisfied; by the Principle of Mathematical Induction, the statement is true for all natural numbers.

7) Assume the statement is true for some natural number k. Then

$$\begin{aligned} 2 + 4 + 6 + 8 + \cdots + 2k + 2(k + 1) &= (k^2 + k + 3) + 2(k + 1) \\ &= (k^2 + 3k + 2) + 3 \\ &= (k + 2)(k + 1) + 3 \\ &= ((k + 1) + 1)(k + 1) + 3 \\ &= (k + 1)^2 + (k + 1) + 3 \end{aligned}$$

So the statement is true for k + 1.

However, when n = 1, the left side of the statement is 2n = 2(1) = 2, and the right side of the statement is $n^2 + n + 3 = 1^2 + 1 + 3 = 5$, so the formula is false for n = 1.

8) When n = 1, $n^2 - n + 3 = 1^2 - 1 + 3 = 3$, which is a prime number, so the statement is true when n = 1. When n = 3, $n^2 - n + 3 = 3^2 - 3 + 3 = 9$, which is not a prime number, so the statement is not true for n = 3.

9) Answers will vary.
10) Answers will vary.
11) Answers will vary.
12) Answers will vary.
13) Answers will vary.
14) Answers will vary.
15) Answers will vary.

16) A
17) 6k - 2

8.5 The Binomial Theorem

1 Evaluate a Binomial Coefficient

1) D
2) B
3) A
4) B
5) A
6) D
7) 6
8) D
9) D
10) C

2 Expand a Binomial

1) C
2) A
3) A
4) B

5) C
6) D
7) B
8) C
9) B
10) $g^3 - 6g^2h + 12gh^2 - 8h^3$
11) $x^4 - 12x^{5/2} + 54x - \frac{108}{\sqrt{x}} + \frac{81}{x^2}$
12) 1.0510101
13) 0.88584
14) $3s^2h + 3sh^2 + h^3$; 0.365 cubic inches

Ch. 9 Counting and Probability

9.1 Sets and Counting

1 Find All the Subsets of a Set

Write down all the subsets of the given set.

1) {1, 3, 9, 10}

A) {1}, {3}, {9}, {10}, {1, 3}, {1, 9}, {1, 10}, {3, 9}, {3, 10}, {9, 10}, {1, 3, 9}, {1, 3, 10}, {1, 9, 10}, {3, 9, 10}, ∅

B) {1}, {3}, {9}, {10}, {1, 3}, {1, 9}, {1, 10}, {3, 9}, {3, 10}, {9, 10}, {1, 3, 9}, {1, 3, 10}, {1, 9, 10}, {3, 9, 10}, {1, 3, 9, 10}, ∅

C) {1}, {3}, {9}, {10}, {1, 3}, {1, 9}, {1, 10}, {3, 9}, {3, 10}, {1, 3, 9}, {1, 3, 10}, {1, 9, 10}, {3, 9, 10}, {1, 3, 9, 10}, ∅

D) {1}, {3}, {9}, {10}, {1, 3}, {1, 9}, {1, 10}, {3, 9}, {3, 10}, {9, 10}, {1, 3, 9}, {1, 3, 10}, {1, 9, 10}, {3, 9, 10}, {1, 3, 9, 10}

2) {1, α, 6, π}

A) {1}, {α}, {6}, {π}, {1, α}, {1, 6}, {1, π}, {α, 6}, {α, π}, {6, π}, {1, α, π}, {1, 6, π}, {α, 6, π}, {1, α, 6, π}, ∅

B) {1}, {α}, {6}, {π}, {1, α}, {1, 6}, {1, π}, {α, 6}, {α, π}, {6, π}, {1, α, 6}, {1, α, π}, {1, 6, π}, {α, 6, π}, {1, α, 6, π}

C) {1}, {α}, {6}, {π}, {1, α}, {1, 6}, {1, π}, {α, 6}, {α, π}, {6, π}, {1, α, 6}, {1, α, π}, {1, 6, π}, {α, 6, π}, {1, α, 6, π}, ∅

D) {1}, {α}, {6}, {π}, {1, α}, {1, 6}, {1, π}, {α, 6}, {α, π}, {6, π}, {1, α, 6}, {1, α, π}, {1, 6, π}, {α, 6, π}, ∅

3) {a}

A) a B) {a, b} C) {a} D) ∅, {a}

4) {a, b}

2 Find the Intersection and Union of Sets

Let A = {q, s, u, w, y}, B = {q, s, y, z}, and C = {v, w, x, y, z}. Find the indicated set.

1) A ∪ B

A) {q, s, y, z} B) {q, s, u, w, y, z} C) {q, s, y} D) {q, s, u, y, z}

2) A ∪ (B ∩ C)

A) {q, y, z} B) {q, s, u, w, y, z} C) {q, w, y} D) {q, r, w, y, z}

3) B ∩ (A ∪ C)

A) {q, w, y} B) {q, r, w, y, z} C) {q, s, u, w, y, z} D) {q, s, y, z}

Find the indicated set.

4) Let A = {2, 5, 8, 17, 24} and B = { 1, 5, 18, 21, 24}.
Find A ∩ B.

A) {1, 2, 5, 8, 17, 18, 21, 24} B) ∅
C) {5, 24} D) none of these

5) Let A = {2, 7, 9, 12, 15}, B = {3, 9, 11, 15, 20}, and C = {4, 8, 9, 10, 12, 14}.
Find $(A \cap B) \cap C$.

A) {9, 12, 15}
B) {9}
C) {2, 3, 4, 7, 8, 9, 10, 11, 12, 14, 15, 20}
D) {9, 12}

6) Let A = {0, 3, 5, 8}, B = {1, 4, 6, 9}, and C = {0, 4, 5, 6}.
Find $(A \cup C) \cap B$.

A) {0, 1, 4, 5, 6, 9}
B) {4, 6}
C) ∅
D) {0, 1, 3, 4, 5, 6, 8, 9}

Solve the problem.

7) A group of friends were discussing vegetables they liked. Andy liked only broccoli, beets, spinach, and mushrooms. Brad only liked corn, potatoes, and mushrooms. Carl only liked broccoli, spinach, eggplant, and mushrooms. David only liked beets, spinach, corn, and mushrooms. Which vegetable(s) do all of them like? Which vegetable(s) do both Carl and David like?

A) {mushrooms}, {corn, spinach, mushrooms}
B) {mushrooms}, {spinach, mushrooms}
C) {mushrooms}, {beets, corn, broccoli, eggplant, spinach, mushrooms}
D) {mushrooms}, {beets, corn, broccoli, eggplant}

3 Find the Complement of a Set

Let U = {q, r, s, t, u, v, w, x, y, z}, A = {q, s, u, w, y}, B = {q, s, y, z}, and C = {v, w, x, y, z}. Find the indicated set.

1) $\overline{(A \cup B)}$

A) {r, s, t, u, v, w, x, z}
B) {r, t, v, x}
C) {s, u, w}
D) {t, v, x}

2) $\overline{(A \cap B)}$

A) {t, v, x}
B) {s, u, w}
C) {q, s, t, u, v, w, x, y}
D) {r, t, u, v, w, x, z}

3) $\overline{C} \cup \overline{A}$

A) {s, t}
B) {q, s, u, v, w, x, y, z}
C) {q, r, s, t, u, v, x, z}
D) {w, y}

4) $\overline{C} \cap \overline{A}$

A) {q, r, s, t, u, v, x, z}
B) {w, y}
C) {r, t}
D) {q, s, u, v, w, x, y, z}

5) $A \cap \overline{B}$

A) {q, s, t, u, v, w, x, y}
B) {r, s, t, u, v, w, x, z}
C) {t, v, x}
D) {u, w}

6) $\overline{A} \cup B$

A) {q, s, t, u, v, w, x, y}
B) {q, r, s, t, v, x, y, z}
C) {s, u, w}
D) {r, s, t, u, v, w, x, z}

Find the indicated set.

7) Let U = Universal set = {0, 1, 2, 3, 4, 5, 6, 7, 8, 9}, A = {2, 3, 4, 5, 8}, B = {1, 4, 6, 7, 9} and C = {2, 3, 4, 6}.
Find $\overline{A \cap B}$.

A) {1, 2, 3, 4, 5, 6, 7, 8, 9}
B) {0}
C) {0, 1, 2, 3, 5, 6, 7, 8, 9}
D) {4}

8) Let U = Universal set = {0, 1, 2, 3, 4, 5, 6, 7, 8, 9}, A = {2, 4, 5, 6, 7}, B = {1, 3, 5, 8, 9} and C = {2, 4, 5, 6}. Find $\overline{A \cap B}$.

A) {2, 4, 5, 6}
B) {0, 1, 2, 3, 4, 6, 7, 8, 9}
C) ∅
D) {1, 2, 3, 4, 6, 7, 8, 9}

4 Count the Number of Elements in a Set

Solve the problem.

1) If n(A) = 4, n(B) = 9, and n(A ∩ B) = 2, find n(A ∪ B).

A) 11 B) 13 C) 12 D) 10

2) If n(A) = 32, n(B) = 93, and n(A ∪ B) = 109, find n(A ∩ B).

A) 16 B) 8 C) 48 D) 18

3) If n(B) = 36, n(A ∩ B) = 7, and n(A ∪ B) = 63, find n(A).

A) 36 B) 32 C) 34 D) 27

4) If n(A) = 15, n(A ∪ B) = 43, and n(A ∩ B) = 11, find n(B).

A) 28 B) 39 C) 40 D) 38

5) If n(A ∪ B) = 82, n(A) = 56, n(B) = 65, find n(A ∩ B).

A) 9 B) 121 C) 43 D) 39

6) If n(A) = 16, n(A ∩ B ∩ C) = 4, n(A ∩ C) = 10, n($A \cap \overline{B}$) = 11, n(B ∩ C) = 6, n($\overline{A} \cap \overline{B} \cap \overline{C}$) = 20, n($C \cap \overline{B}$) = 13, and n(B ∪ C) = 23, find n(U).

A) 57 B) 52 C) 48 D) 45

Use the information given in the figure.

7)

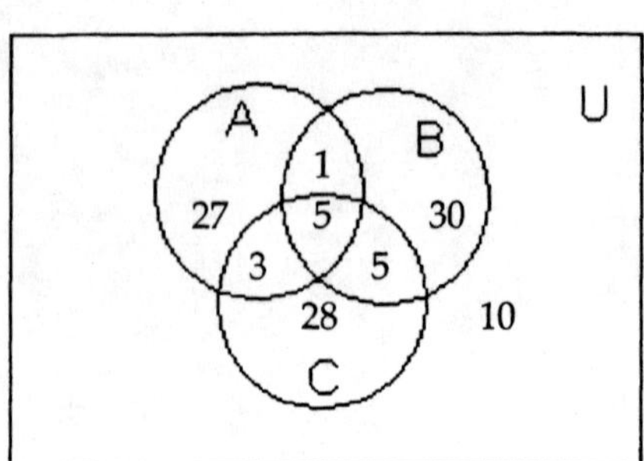

How many are in set A?

A) 27 B) 46 C) 36 D) 31

8) 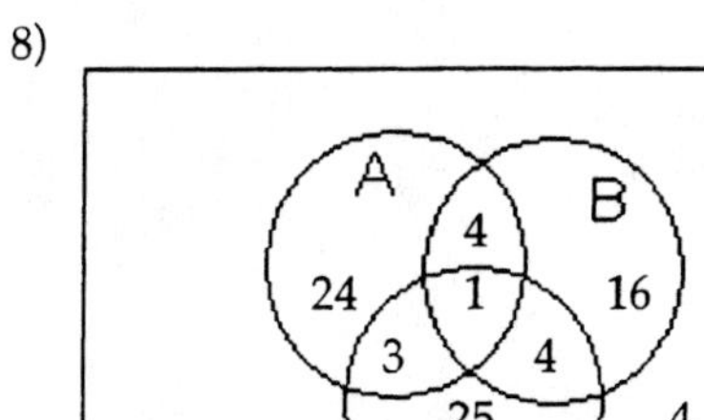

How many are in B or C?

A) 52   B) 45   C) 5   D) 53

9)

How many are in B but not in A?

A) 31   B) 36   C) 28   D) 37

10)

How many are not in C?

A) 51   B) 58   C) 53   D) 60

11)

How many are in A and B and C?

A) 84 B) 71 C) 3 D) 13

12)

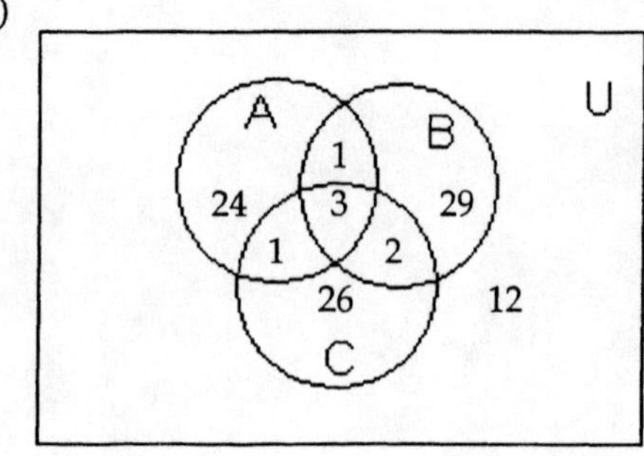

How many are in A or B or C?

A) 3 B) 86 C) 7 D) 98

Solve the problem.

13) Among a group of 72 investors, 20 owned shares of Stock A, 30 owned shares of Stock B, 36 owned shares of Stock C, 9 owned shares of both Stock A and Stock B, 7 owned shares of Stock A and Stock C, 14 owned shares of Stock B and Stock C, and 5 owned shares of all three. How many investors did not have shares in any of the three? How many owned shares of either Stock A or Stock C but not Stock B?

A) 11; 34 B) 11; 29 C) 16; 29 D) 11; 31

14) In a survey of 59 hospital patients, 23 said they were satisfied with the nursing care, 22 said they were satisfied with the medical treatment, and 6 said they were satisfied with both. How many patients were satisfied with neither? How many were satisfied with only the medical treatment?

A) 20; 22 B) 17; 16 C) 26; 22 D) 20; 16

15) In a survey of 359 computer buyers, 176 put price as a main consideration, 183 put performance as a main consideration, and 54 listed both price and performance. How many computer buyers listed other considerations? How many looked only for performance?

A) 54; 183 B) 122; 129 C) 108; 183 D) 54; 129

16) In a survey of 184 vacationers in a popular beach resort town, 74 indicated they would consider buying a home there, 79 would consider buying a beach villa, 68 would consider buying a lot, 23 would consider both a home and a villa, 33 would consider both a home and a lot, 25 would consider both a villa and a lot, and 13 would consider all three. How many vacationers would not consider any of the three? How many would consider only a home?

A) 31; 31 B) 31; 23 C) 44; 41 D) 31; 44

17) A survey of 2111 credit card users indicated that 917 had bought books online, 1026 had bought music online, 445 had bought pet supplies online, 107 had bought both books and music, 205 had bought both books and pet supplies, 164 had bought both music and pet supplies, and 77 had bought all three. How many credit card users did not buy any of the three? How many bought either books or pet supplies but not music?

A) 122; 835 B) 122; 912 C) 122; 963 D) 199; 835

18) In survey of 50 households, 25 responded that they have an HDTV television, 35 responded that they had a multimedia personal computer and 15 responded they had both. How many households had neither an HDTV television nor a multimedia personal computer?

A) 5 B) 15 C) 25 D) 35

19) In a telephone survey conducted by a marketing firm for a rock star, 50 indicated they would purchase a CD at a concert, 25 indicated they would purchase a T-shirt, 15 indicated they would purchase both, and 50 indicated they would purchase neither. How many people participated in the survey?

A) 65 B) 110 C) 125 D) 140

9.2 Permutations and Combinations

1 Solve Counting Problems Using the Multiplication Principle

Solve the problem.

1) A man has 9 shirts and 10 ties. How many different shirt and tie arrangements can he wear?

A) 90 B) 180 C) 100 D) 81

2) A restaurant offers a choice of 4 salads, 5 main courses, and 4 desserts. How many possible 3-course meals are there?

A) 13 possible meals B) 80 possible meals C) 20 possible meals D) 160 possible meals

3) Lisa has 4 skirts, 7 blouses, and 3 jackets. How many 3-piece outfits can she put together assuming any piece goes with any other?

A) 28 possible outfits B) 14 possible outfits C) 168 possible outfits D) 84 possible outfits

4) How many 6-symbol codes can be formed using 3 different symbols? Repeated symbols are allowed.

A) 20 B) 3 C) 729 D) 120

5) A certain mathematics test consists of 20 questions. Goldie decides to answer the questions without reading them. In how many ways can Goldie fill in the answer sheet if the possible answers are true and false?

A) 190 B) 1,048,576 C) 400 D) 40

6) A student must choose 1 of 4 mathematics electives, 1 of 5 science electives, and 1 of 7 programming electives. How many possible course selections are there?

A) 280 course selections B) 16 course selections
C) 20 course selections D) 140 course selections

2 Solve Counting Problems Using Permutations

Find the value of the permutation.

1) P(5, 3)

A) 30 B) 20 C) 4 D) 60

2) P(4, 0)

A) 1 B) 48 C) 2 D) 24

3) P(5, 1)

A) 1 B) 120 C) 24 D) 5

4) P(10, 10)

A) 1,814,400 B) 3,628,800 C) 2 D) 1

Solve the problem.

5) In how many ways can 6 people be lined up?

A) 30 ways B) 1440 ways C) 360 ways D) 720 ways

6) In how many ways can 6 volunteers be assigned to 6 booths for a charity bazaar?

A) 1440 ways B) 720 ways C) 30 ways D) 360 ways

7) A night watchman visits 12 offices every night. To prevent others from knowing when he will be at a particular office, he varies the order of his visits. In how many ways can this be done?

A) 4096 B) 239,500,800 C) 479,001,600 D) 792

8) How many 3-letter codes can be formed using the letters A, B, C, D, and E? No letter can be used more than once.

A) 60 B) 20 C) 40 D) 10

9) How many 2-letter codes can be formed using the letters A, B, C, D, E, and F? No letter can be used more than once.

A) 360 B) 720 C) 30 D) 15

10) How many 3-digit numbers can be formed using the digits 1, 2, 3, 4, 5, 6, 7, 8, 9, and 0? No digit can be used more than once.

A) 1,209,600 B) 604,800 C) 120 D) 720

11) A civil light opera production of Carousel is holding auditions for the roles of Billy Bigelow, Jigger Craigin, and Enoch Snow. If 35 men try out for these roles, how many ways are there to cast the roles?

A) 39,304 B) 6545 C) 42,875 D) 39,270

12) In how many ways can Linda arrange 4 of her 6 new CDs into 4 available slots in her CD holder?

A) 360 B) 30 C) 60 D) 15

13) A combination lock has 25 numbers on it. How many different 3-digit lock combinations are possible if no digit can be repeated?

A) 2300 B) 13,800 C) 4600 D) 600

14) The town hall has 6 bells in its bell tower. Before sunset 4 bells are rung in sequence. No bell is rung more than once. How many sequences are possible?

A) 60 B) 30 C) 15 D) 360

15) A club elects a president, vice-president, and treasurer. How many sets of officers are possible if there are 10 members and any member can be elected to each position? No person can hold more than one office.

A) 360 B) 240 C) 720 D) 5040

16) How many different license plates can be made using 2 letters followed by 4 digits selected from the digits 0 through 9, if letters and digits may be repeated?

A) 36 B) 6,760,000 C) 260 D) 8

17) How many 2-letter words (real or imaginary)can be formed from the letters of the word HYPERBOLAS if no letter is to be used more than once?

A) 45 B) 90 C) 1,814,400 D) 3,628,800

18) How many 6-symbol computer passwords can be formed if the first two symbols are letters (either upper or lower case) and the rest of the symbols are digits? (Letters and digits may be repeated.)

A) 3, 276,000 B) 27,040,000 C) 6,760,000 D) 13,366,080

3 Solve Counting Problems Using Combinations

Find the value of the combination.

1) C(8, 3)

A) 6720 B) 240 C) 168 D) 56

2) C(7, 7)

A) 5040 B) 0.5 C) 1260 D) 1

Solve the problem.

3) From 8 names on a ballot, a committee of 3 will be elected to attend a political national convention. How many different committees are possible?

A) 336 B) 56 C) 6720 D) 168

4) A hot dog stand sells hot dogs with cheese, relish, chili, tomato, onion, mustard, or ketchup. How many different hot dogs can be concocted using any 5 of the extras?

A) 1260 B) 42 C) 2520 D) 21

5) A stack of 9 different cards are shuffled and spread out face down. If 4 cards are turned face up, how many different 4-card combinations are possible?

A) 126 B) 3024 C) 1512 D) 15,120

6) In how many ways can a committee of three men and four women be formed from a group of 11 men and 11 women?

A) 554,400 B) 110 C) 54,450 D) 7,840,800

7) An exam consists of 9 multiple-choice questions and 6 essay questions. If the student must answer 6 of the multiple-choice questions and 4 of the essay questions, in how many ways can the questions be chosen?

A) 1260 B) 261,273,600 C) 21,772,800 D) 1296

8) Mary finds 9 fish at a pet store that she would like to buy, but she can afford only 5 of them. In how many ways can she make her selection? How many ways can she make her selection if he decides that one of the fish is a must?

A) 15,120; 1680 B) 7560; 840 C) 3024; 1680 D) 126; 70

9) In a student government election, 7 seniors, 2 juniors, and 3 sophomores are running for election. Students elect four at-large senators. In how many ways can this be done?

A) 42 B) 11,880 C) 495 D) 19,958,400

10) How many 5-card poker hands consisting of three 9's and two cards that are not 9's are possible in a 52-card deck?

A) 2652 B) 5304 C) 2256 D) 4512

4 Solve Counting Problems Using Permutations Involving n Nondistinct Objects

Solve the problem.

1) How many different ways can 8 balls be pulled out of an urn if 3 are white, 3 are blue, and 2 are red?

A) 56 B) 21 C) 560 D) 18

2) How many different 10-letter words (real or imaginary) can be formed from the letters in the word ACCOUNTING?

A) 90,720 B) 907,200 C) 3,628,800 D) 1,814,400

3) How many different 11-letter words (real or imaginary) can be formed from the letters of the word MISSISSIPPI? Leave your answer in factorial form.

9.3 Probability

1 Construct Probability Models

Solve the problem.

1) In a probability model, which of the following numbers could be the probability of an outcome:

$0,\ 0.2,\ -0.01,\ -\frac{1}{3},\ \frac{1}{2},\ \frac{5}{4},\ 1,\ 1.5$

A) $0,\ 0.2,\ -0.01,\ -\frac{1}{3},\ \frac{1}{2},\ 1$ B) $0,\ 0.2,\ \frac{1}{2},\ 1$

C) $0.2,\ \frac{1}{2},\ 1$ D) $0,\ 0.2,\ -0.01,\ 1,\ 1.5$

Determine whether the following is a probability model.

2)

Outcome	Probability
Red	0.17
Blue	0.21
Green	0.31
White	0.31

A) Yes B) No

3)

Outcome	Probability
Red	0.16
Blue	0.27
Green	0.34
White	0.50

A) Yes B) No

4)

Outcome	Probability
Red	0.12
Blue	0.15
Green	0.30
White	0.28

A) Yes B) No

5)

Outcome	Probability
Red	−0.22
Blue	0.29
Green	0.35
White	0.14

A) Yes B) No

6)

Outcome	Probability
Jim	0
Tom	0
Bill	1
Carl	0

A) Yes B) No

7)

Outcome	Probability
Golfing	0.14
Skiing	0.12
Swimming	0.13
Biking	0.30
Hiking	0.31

A) Yes B) No

Construct a probability model for the experiment.

8) Tossing two fair coins once

9) Tossing one fair coin three times

10) Spinner I has 4 sections of equal area, numbered from 1 to 4, and Spinner II has 4 sections of equal area, labeled Red, Yellow, Green, Blue. Spin Spinner I and then spin Spinner II.

11) Rolling a 6-sided fair die once

Solve the problem.

12) A twelve-sided die is weighted so that only the numbers 1 through 4 will appear and they will occur with the same probability. What probability should be assigned to each face?

13) A coin is weighted so that heads is 9 times as likely as tails to occur. What probability should be assigned to heads? to tails?

2 Compute Probabilities of Equally Likely Outcomes

Solve the problem.

1) A bag contains 6 red marbles, 9 blue marbles, and 2 green marbles. What is the probability of choosing a blue marble when one marble is drawn?

A) $\frac{9}{17}$ B) $\frac{3}{5}$ C) $\frac{2}{17}$ D) $\frac{6}{17}$

2) A 6-sided die is rolled. What is the probability of rolling a number less than 3?

A) $\frac{1}{3}$ B) $\frac{2}{3}$ C) $\frac{1}{6}$ D) $\frac{1}{2}$

3) Two 6-sided dice are rolled. What is the probability the sum of the two numbers on the dice will be 3?

A) $\frac{1}{18}$ B) $\frac{17}{18}$ C) 2 D) $\frac{1}{2}$

4) A bag contains 13 balls numbered 1 through 13. What is the probability of selecting a ball that has an even number when one ball is drawn from the bag?

A) 6 B) $\frac{2}{13}$ C) $\frac{6}{13}$ D) $\frac{13}{6}$

5) A bag contains 8 red marbles, 7 white marbles, and 6 blue marbles. Find the probability of obtaining a white marble in a single draw.

A) $\frac{7}{20}$ B) $\frac{1}{7}$ C) $\frac{1}{4}$ D) $\frac{1}{3}$

6) What is the probability that the arrow will land on an odd number?

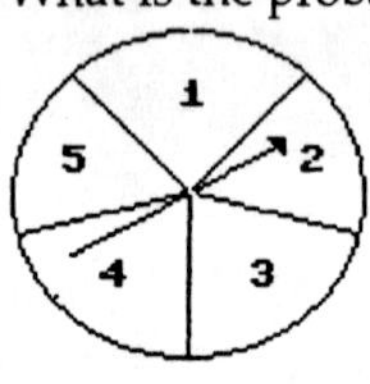

A) 1 B) $\frac{3}{5}$ C) $\frac{2}{5}$ D) 0

3 Use the Addition Rule to Find Probabilities

Solve the problem.

1) Give the probability that the roll of a die will show 1 or 6.

A) 1 B) $\frac{1}{6}$ C) $\frac{1}{3}$ D) 2

2) Give the probability that the roll of a die will show a number less than 4.

A) $\frac{1}{2}$ B) $\frac{6}{7}$ C) $\frac{2}{3}$ D) $\frac{1}{3}$

3) Each of ten tickets is marked with a different number from 1 to 10 and put in a box. If you draw a ticket from the box, what is the probability that you will draw 1, 6, or 3?

A) 1 B) $\frac{1}{10}$ C) $\frac{1}{6}$ D) $\frac{3}{10}$

4)

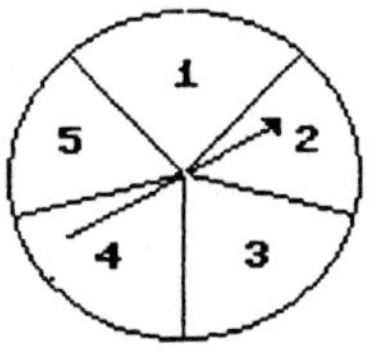

What is the probability that the arrow will land on 4 or 2?

A) $\frac{2}{3}$ B) $\frac{2}{5}$ C) 4 D) $\frac{3}{4}$

5) A lottery game has balls numbered 1 through 21. What is the probability of selecting an even numbered ball or a 13?

A) $\frac{10}{11}$ B) $\frac{3}{7}$ C) $\frac{10}{21}$ D) $\frac{11}{21}$

6) One digit from the number 8,494,224 is written on each of seven cards. What is the probability of drawing a card that shows 8 or 2?

A) $\frac{8}{7}$ B) $\frac{2}{7}$ C) $\frac{4}{7}$ D) $\frac{3}{7}$

7) The psychology lab at a college is staffed by 7 male doctoral students, 6 female doctoral students, 13 male undergraduates, and 12 female undergraduates. If a person is selected at random from the group, find the probability that the selected person is an undergraduate or a female.

A) $\frac{9}{19}$ B) $\frac{25}{38}$ C) $\frac{31}{38}$ D) $\frac{1}{2}$

8) A card is drawn from a well-shuffled deck of 52 cards. What is the probability of drawing an ace or a 6?

A) $\frac{2}{13}$ B) $\frac{13}{2}$ C) 7 D) $\frac{7}{26}$

9) A spinner has regions numbered 1 through 18. What is the probability that the spinner will stop on an even number or a multiple of 3?

A) 15 B) $\frac{2}{3}$ C) $\frac{1}{3}$ D) 1

10) Of 100 students, 24 are taking Calculus, 30 are taking French, and 13 are taking both Calculus and French. If a student is picked at random, what is the probability that the student is taking Calculus or French?

A) $\frac{41}{100}$ B) $\frac{3}{50}$ C) $\frac{54}{100}$ D) $\frac{67}{100}$

11) The table below represents the number of deaths per 100 cases for an illness having a median mortality of four years and a right–skewed distribution over time. What is the probability of living more than 12 years after diagnosis of the disease?

Years after Diagnosis	Number deaths
1–2	15
3–4	35
5–6	16
7–8	9
9–10	6
11–12	4
13–14	2
15+	13

A) 0.19 B) 0.85 C) 0.13 D) 0.15

4 Use the Complement Rule to Find Probabilities

Solve the problem.

1) A bag contains 6 red marbles, 2 blue marbles, and 1 green marble. What is the probability of choosing a marble that is not blue when one marble is drawn from the bag?

A) $\frac{9}{7}$ B) $\frac{7}{9}$ C) $\frac{2}{9}$ D) 7

2) What is the probability that in a group of 15 people at least 2 people have the same birthday? Assume that there are 365 days in a year. If necessary, round to three decimal places.

A) 0.205 B) 0.005 C) 0.133 D) 0.253

3) During July in Jacksonville, Florida, it is not uncommon to have afternoon thunderstorms. On average, 9.8 days have afternoon thunderstorms. What is the probability that a randomly selected day in July will not have a thunderstorm? Round to two decimal places, if necessary.

A) 0.67 B) 0.9 C) 0.68 D) 0.32

4) Determine the probability of having no boys in a three–child family.

A) $\frac{1}{3}$ B) $\frac{1}{4}$ C) $\frac{1}{2}$ D) $\frac{1}{8}$

Ch. 9 Counting and Probability
Answer Key

9.1 Sets and Counting
1 Find All the Subsets of a Set

1) B
2) C
3) D
4) $\emptyset$, {a}, {b}, {a, b}

2 Find the Intersection and Union of Sets

1) B
2) B
3) D
4) C
5) B
6) B
7) B

3 Find the Complement of a Set

1) B
2) D
3) C
4) C
5) D
6) B
7) C
8) B

4 Count the Number of Elements in a Set

1) A
2) A
3) C
4) B
5) D
6) C
7) C
8) D
9) A
10) B
11) C
12) B
13) D
14) D
15) D
16) A
17) C
18) A
19) B

9.2 Permutations and Combinations
1 Solve Counting Problems Using the Multiplication Principle

1) A
2) B
3) D
4) C
5) B
6) D

2 Solve Counting Problems Using Permutations

1) D
2) A
3) D
4) B
5) D
6) B
7) C
8) A
9) C
10) D
11) D
12) A
13) B
14) D
15) C
16) B
17) B
18) B

3 Solve Counting Problems Using Combinations

1) D
2) D
3) B
4) D
5) A
6) C
7) A
8) D
9) C
10) D

4 Solve Counting Problems Using Permutations Involving n Nondistinct Objects

1) C
2) B
3) $\frac{11!}{4!4!2!}$

9.3 Probability

1 Construct Probability Models

1) B
2) A
3) B
4) B
5) B
6) A
7) A
8) S = {HH, HT, TH, TT}; each outcome has the probability of $\frac{1}{4}$
9) S = {HHH, HTH, HHT, HTT, THH, TTH, THT, TTT}; each outcome has the probability of $\frac{1}{8}$
10) S = {1 Red, 1 Yellow, 1 Green, 1 Blue, 2 Red, 2 Yellow, 2 Green, 2 Blue, 3 Red, 3 Yellow, 3 Green, 3 Blue, 4 Red, 4 Yellow, 4 Green, 4 Blue}; each outcome has the probability of $\frac{1}{16}$
11) S = {1, 2, 3, 4, 5, 6}; each outcome has the probability of $\frac{1}{6}$

12) The faces numbered 1 through 4 will each have the probability $\frac{1}{4}$; faces numbered higher than 4 will each have probability 0.

13) $P(H) = \frac{9}{10}$; $P(T) = \frac{1}{10}$

2 Compute Probabilities of Equally Likely Outcomes

1) A
2) A
3) A
4) C
5) D
6) B

3 Use the Addition Rule to Find Probabilities

1) C
2) A
3) D
4) B
5) D
6) D
7) C
8) A
9) B
10) A
11) D

4 Use the Complement Rule to Find Probabilities

1) B
2) D
3) C
4) D

Ch. 10 (Appendix) Graphing Utilities

10.1 The Viewing Rectangle

1 The Viewing Rectangle

Determine the coordinates of the points shown. Tell in which quadrant the point lies. Assume the coordinates are integers.

1)

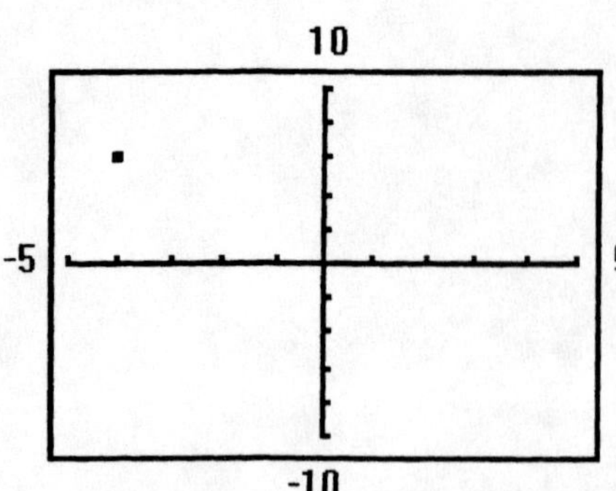

A) (-4, 6); quadrant II B) (-4, 3); quadrant I C) (-4, 6); quadrant I D) (-4, 3); quadrant II

2)

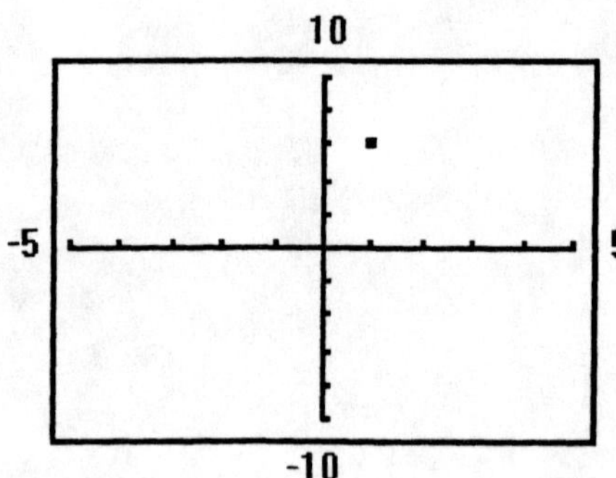

A) (1, 6); quadrant II B) (1, 3); quadrant I C) (1, 6); quadrant I D) (1, 3); quadrant II

3)

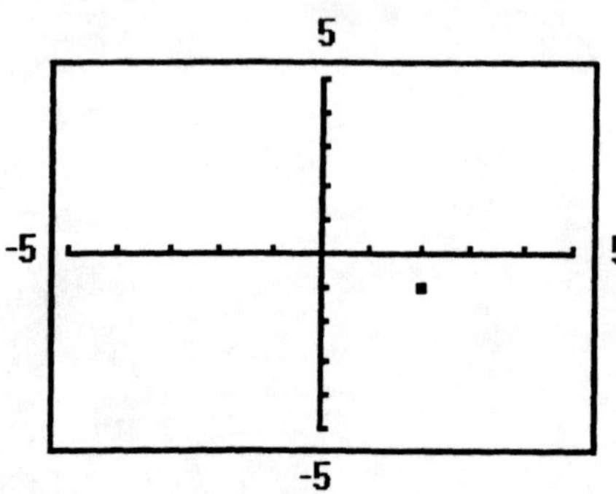

A) (2, -2); quadrant IV B) (2, -1); quadrant III
C) (2, -2); quadrant III D) (2, -1); quadrant IV

4)

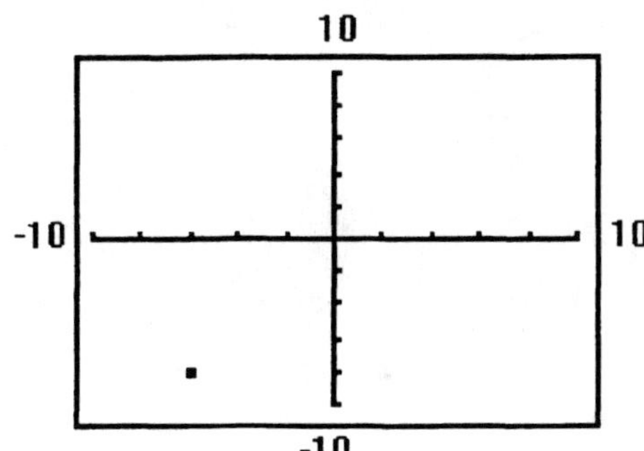

A) (-6, -4); quadrant IV

B) (-6, -8); quadrant III

C) (-6, -8); quadrant IV

D) (-6, -4); quadrant III

Determine the viewing window used.

5)

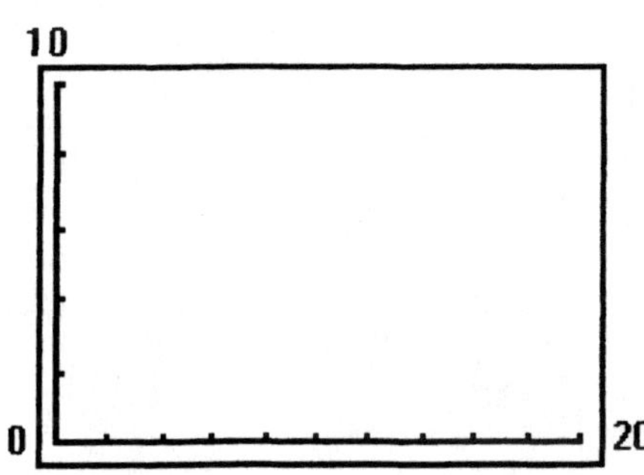

A)

WINDOW
Xmin=0
Xmax=10
Xscl=2
Ymin=0
Ymax=20
Yscl=2
Xres=1

B)

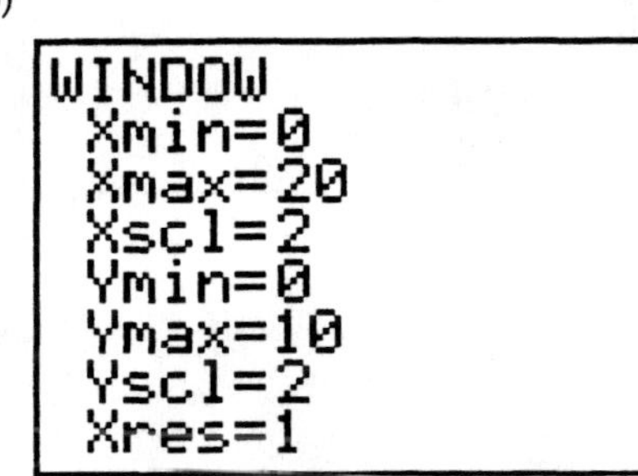

C)

WINDOW
Xmin=0
Xmax=20
Xscl=1
Ymin=0
Ymax=10
Yscl=1
Xres=1

D)

WINDOW
Xmin=0
Xmax=10
Xscl=1
Ymin=0
Ymax=20
Yscl=1
Xres=1

6)

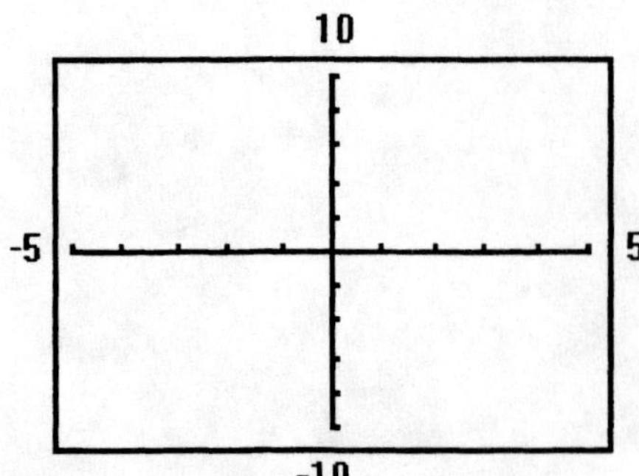

A)

```
WINDOW
 Xmin= -5
 Xmax=5
 Xscl=1
 Ymin= -10
 Ymax=10
 Yscl=2
 Xres=1
```

B)

```
WINDOW
 Xmin= -5
 Xmax=5
 Xscl=1
 Ymin= -10
 Ymax=10
 Yscl=1
 Xres=1
```

C)

```
WINDOW
 Xmin= -5
 Xmax=5
 Xscl=5
 Ymin= -10
 Ymax=10
 Yscl=5
 Xres=1
```

D)

```
WINDOW
 Xmin= -5
 Xmax=5
 Xscl=5
 Ymin= -10
 Ymax=10
 Yscl=10
 Xres=1
```

7)

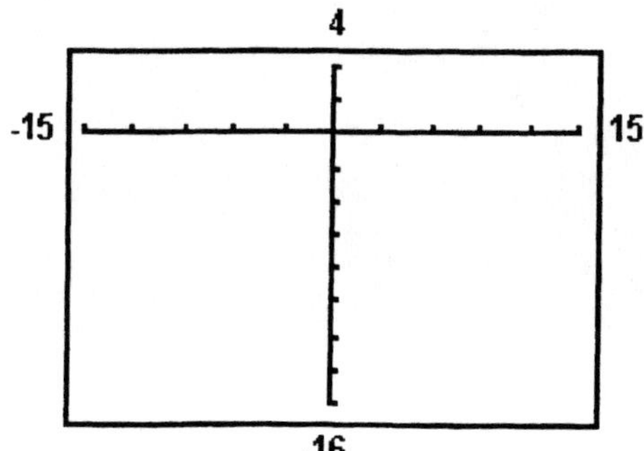

A)

```
WINDOW
 Xmin=-15
 Xmax=15
 Xscl=5
 Ymin=-16
 Ymax=4
 Yscl=2
 Xres=1
```

B)

```
WINDOW
 Xmin=-16
 Xmax=4
 Xscl=2
 Ymin=-15
 Ymax=15
 Yscl=5
 Xres=1
```

C)

```
WINDOW
 Xmin=-16
 Xmax=4
 Xscl=2
 Ymin=-15
 Ymax=15
 Yscl=3
 Xres=1
```

D)

```
WINDOW
 Xmin=-15
 Xmax=15
 Xscl=3
 Ymin=-16
 Ymax=4
 Yscl=2
 Xres=1
```

8)

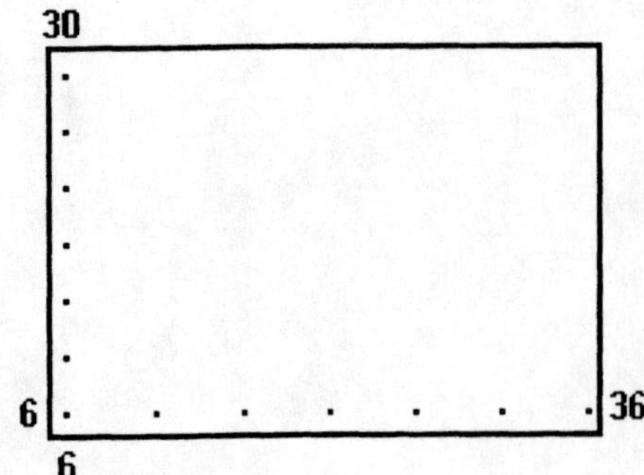

A)

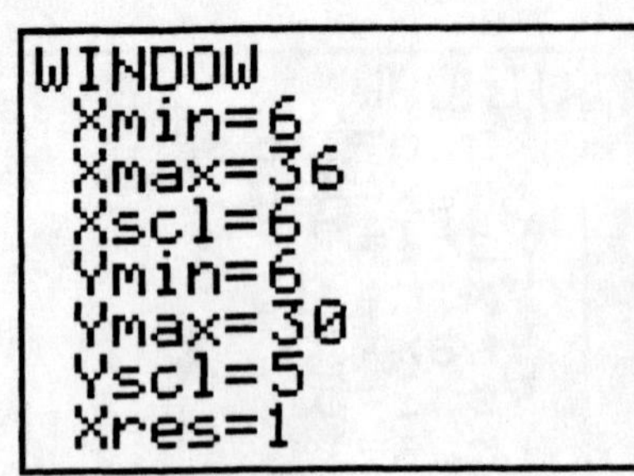
WINDOW
Xmin=6
Xmax=36
Xscl=6
Ymin=6
Ymax=30
Yscl=5
Xres=1

B)

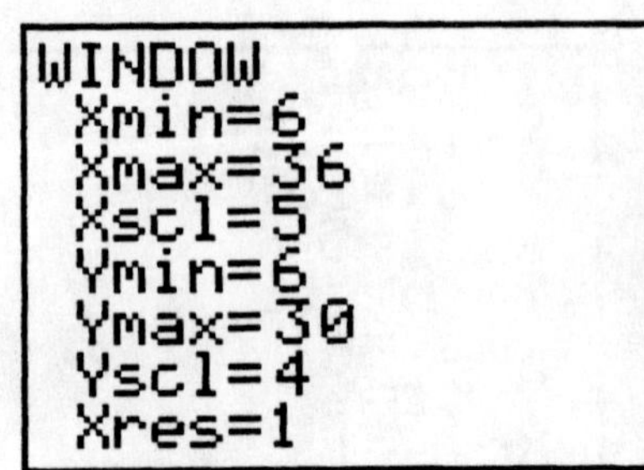
WINDOW
Xmin=6
Xmax=36
Xscl=5
Ymin=6
Ymax=30
Yscl=4
Xres=1

C)

WINDOW
Xmin=6
Xmax=30
Xscl=5
Ymin=6
Ymax=36
Yscl=6
Xres=1

D)

WINDOW
Xmin=6
Xmax=30
Xscl=4
Ymin=6
Ymax=36
Yscl=5
Xres=1

Select a setting so that the given points will lie within the viewing rectangle.

9) (-70, 2), (-60, 9), (130, 0)

A) X min = -80
X max = 140
X scl = 10
Y min = 1
Y max = 10
Y scl = 1

B) X min = -70
X max = 140
X scl = 10
Y min = -1
Y max = 10
Y scl = 1

C) X min = -80
X max = 140
X scl = 10
Y min = -1
Y max = 10
Y scl = 1

D) X min = -1
X max = 10
X scl = 1
Y min = -80
Y max = 140
Y scl = 10

Find the length of the line segment. Assume that the endpoints of the segment have integer coordinates.

10)

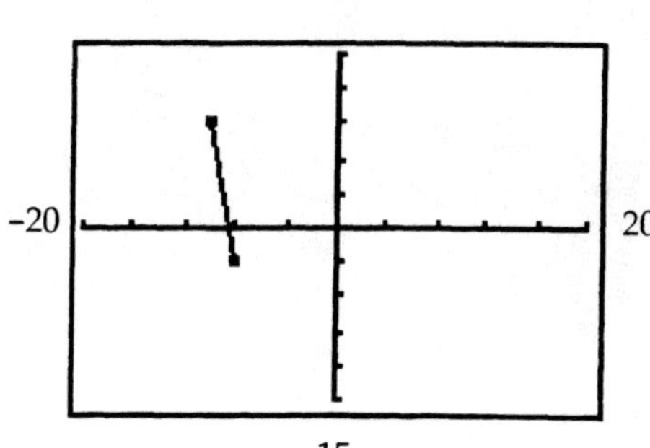

A) $2\sqrt{37}$ B) $4\sqrt{10}$ C) $2\sqrt{10}$ D) $6\sqrt{10}$

11)

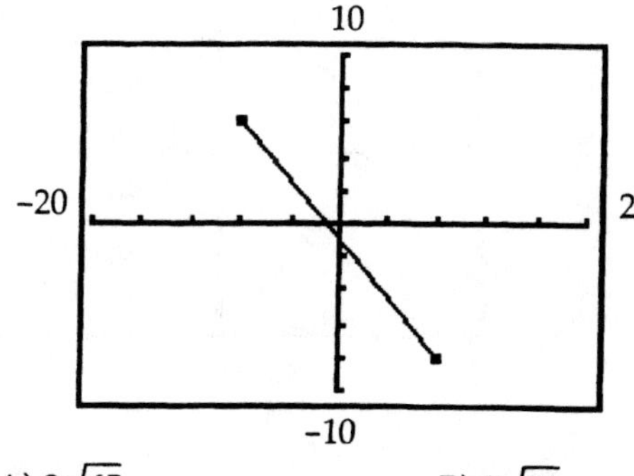

A) $2\sqrt{65}$ B) $2\sqrt{53}$ C) $2\sqrt{113}$ D) $2\sqrt{5}$

10.2 Using a Graphing Utility to Graph Equations

1 Using a Graphing Utility to Graph Equations

Graph the equation using the indicated viewing window.

1) $y = x + 4$;
X min = -5
X max = 5
X scl = 1
Y min = -4
Y max = 4
Y scl = 1

A)

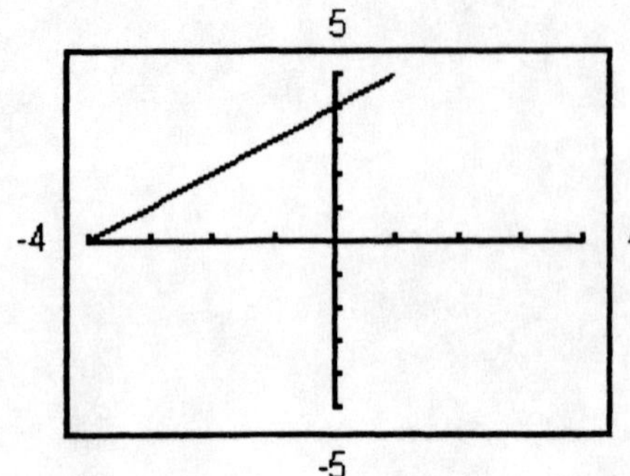

B)

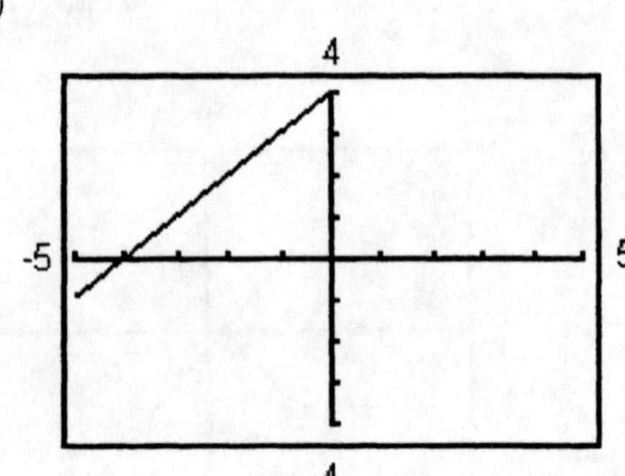

C)

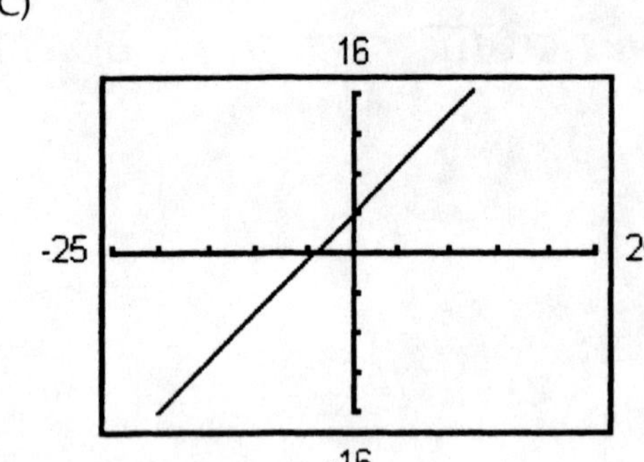

D)

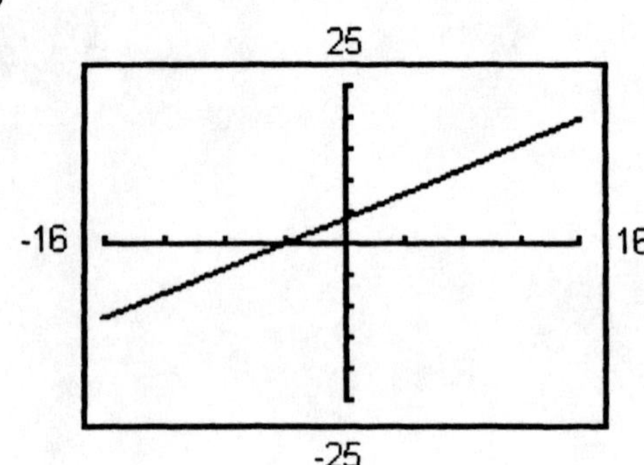

2) $2x + 3y = 6$;
X min = −10
X max = 10
X scl = 2
Y min = −8
Y max = 8
Y scl = 2

A) 

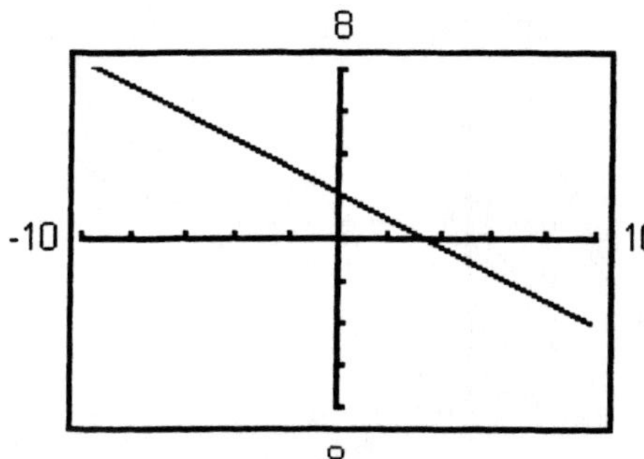

B) 

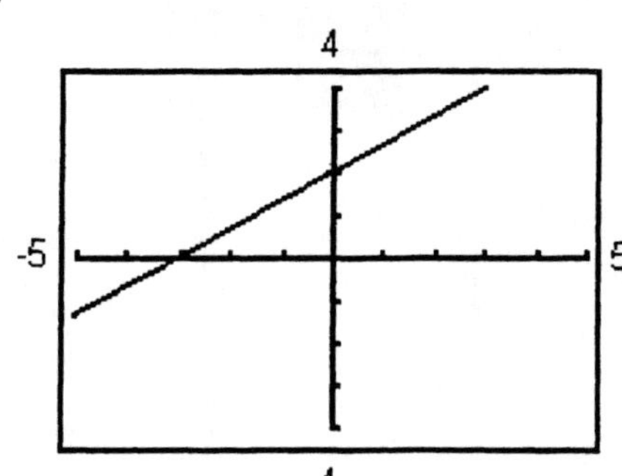

C)

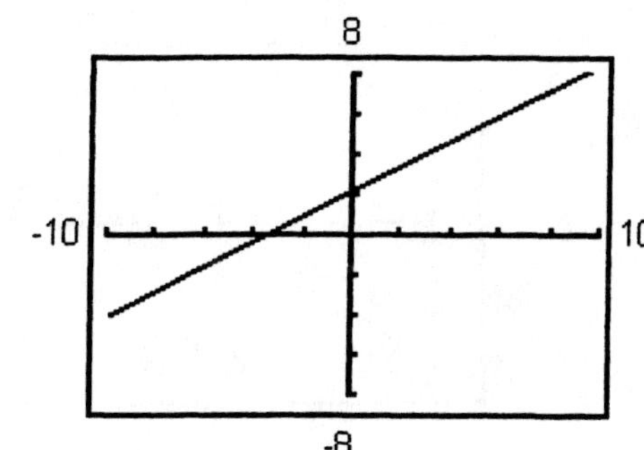

D)

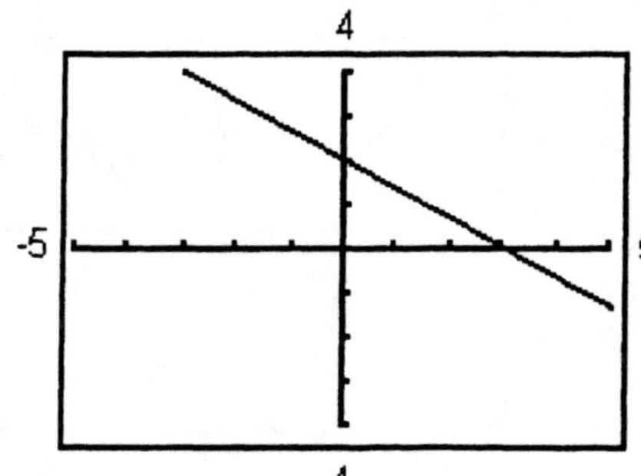

3) $y = x^2 + 2$;
X min = -5
X max = 5
X scl = 1
Y min = -20
Y max = 20
Y scl = 5

A)

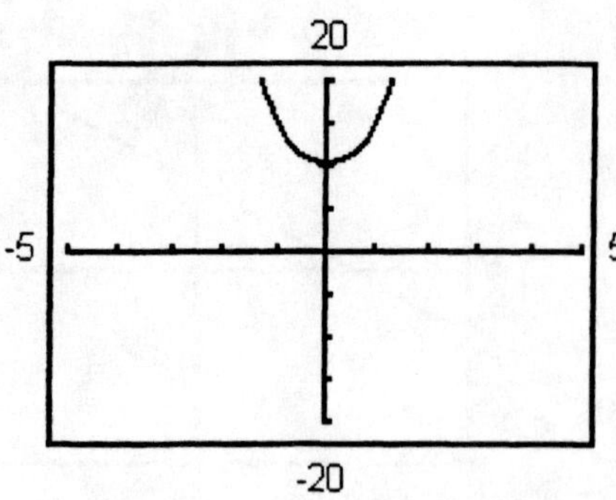

B)

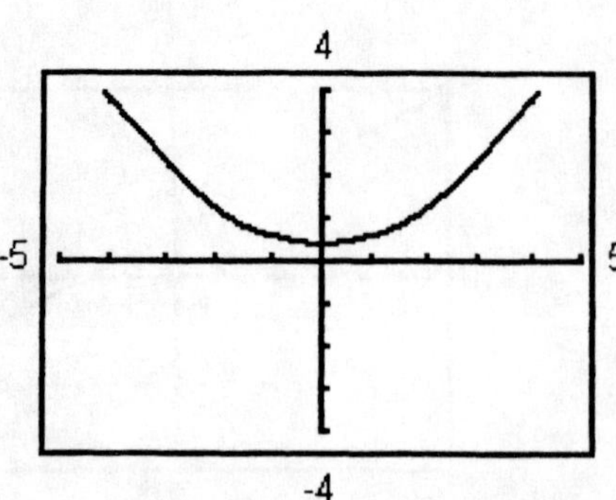

C)

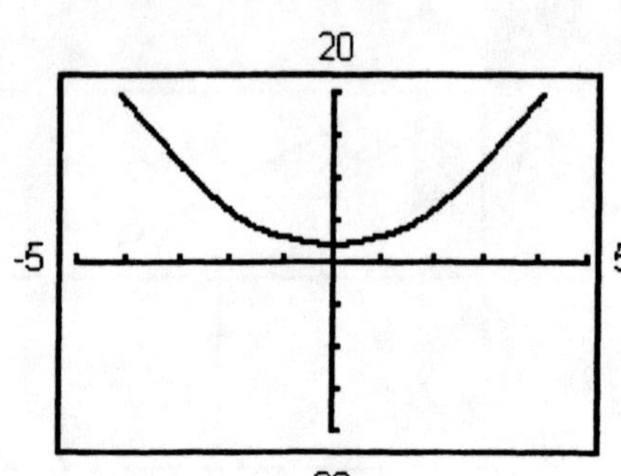

D)

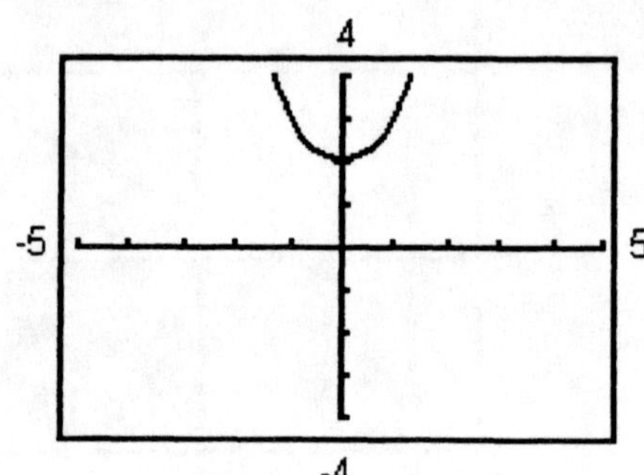

10.3 Using a Graphing Utility to Locate Intercepts and Check for Symmetry

1 Using a Graphing Utility to Locate Intercepts and Check for Symmetry

Use ZERO (or ROOT) to approximate the smaller of the two x-intercepts of the equation. Express the answer rounded to two decimal places.

1) $y = x^2 + 5x - 2$

A) -5.37 B) 5.37 C) 0.37 D) -0.37

Use ZERO (or ROOT) to approximate the positive x-intercepts of the equation. Round to two decimal places.

2) $y = x^3 + 3x^2 - 5x - 7$

A) 1.83 B) 1.91 C) 1.76 D) 3.83

3) $y = x^3 + 3.3x^2 - 5.2x - 6.3$

A) 4.18 B) 1.75 C) 1.57 D) 0.86

4) $y = x^4 + 1.5x^3 - 8.31x^2 - 3.27x + 8.39$

A) 1.19 and 3.44 B) 0.87 and 2.21 C) 0.94 and 2.18 D) 8.39

10.4 Using a Graphing Utility to Solve Equations

1 Using a Graphing Utility to Solve Equations

Use ZERO (or ROOT) to find the solutions to the equation. Round to two decimal places.

1) $x^2 + 6x - 10 = 0$

A) x = 1.18, -7.54 B) x = 1.18, -7.19 C) x = 1.53, -7.54 D) x = 1.36, -7.36

Use INTERSECT to find the solution to the equation. Round to two decimal places, if necessary.

2) $2(x - 3) = 5(x - 2)$

A) x = 4.67 B) x = -1.33 C) x = -4.67 D) x = 1.33

10.5 Square Screens

1 Square Screens

Determine whether the given viewing rectangle will result in a square screen.

1)

X min = 0
X max = 12
X scl = 1
Y min = -2
Y max = 6
Y scl = 2

A) Yes B) No

10.6 Using a Graphing Utility to Graph Inequalities

1 Using a Graphing Utility to Graph Inequalities

1) There are no exercises for this objective.

10.7 Using a Graphing Utility to Solve Systems of Linear Equations

1 Using a Graphing Utility to Solve Systems of Linear Equations

1) There are no exercises for this objective.

Ch. 10 (Appendix) Graphing Utilities
Answer Key

10.1 The Viewing Rectangle
1 The Viewing Rectangle

1) A
2) C
3) D
4) B
5) B
6) A
7) D
8) B
9) C
10) A
11) C

10.2 Using a Graphing Utility to Graph Equations
1 Using a Graphing Utility to Graph Equations

1) B
2) A
3) C

10.3 Using a Graphing Utility to Locate Intercepts and Check for Symmetry
1 Using a Graphing Utility to Locate Intercepts and Check for Symmetry

1) A
2) A
3) B
4) C

10.4 Using a Graphing Utility to Solve Equations
1 Using a Graphing Utility to Solve Equations

1) D
2) D

10.5 Square Screens
1 Square Screens

1) A

10.6 Using a Graphing Utility to Graph Inequalities
1 Using a Graphing Utility to Graph Inequalities

1) No Correct Answer Was Provided.

10.7 Using a Graphing Utility to Solve Systems of Linear Equations
1 Using a Graphing Utility to Solve Systems of Linear Equations

1) No Correct Answer Was Provided.